CIRCUITS, SIGNALS, AND SYSTEMS FOR BIOENGINEERS

THIRD EDITION

This is a volume in the
ACADEMIC PRESS SERIES IN BIOMEDICAL ENGINEERING

JOSEPH BRONZINO, SERIES EDITOR
TRINITY COLLEGE—HARTFORD, CONNECTICUT

CIRCUITS, SIGNALS, AND SYSTEMS FOR BIOENGINEERS

A MATLAB®-BASED INTRODUCTION

THIRD EDITION

JOHN SEMMLOW

ACADEMIC PRESS

An imprint of Elsevier

Academic Press is an imprint of Elsevier
125 London Wall, London EC2Y 5AS, United Kingdom
525 B Street, Suite 1800, San Diego, CA 92101-4495, United States
50 Hampshire Street, 5th Floor, Cambridge, MA 02139, United States
The Boulevard, Langford Lane, Kidlington, Oxford OX5 1GB, United Kingdom

Notices
Knowledge and best practice in this field are constantly changing. As new research and experience broaden our understanding, changes in research methods, professional practices, or medical treatment may become necessary.

Practitioners and researchers must always rely on their own experience and knowledge in evaluating and using any information, methods, compounds, or experiments described herein. In using such information or methods they should be mindful of their own safety and the safety of others, including parties for whom they have a professional responsibility.

To the fullest extent of the law, neither the Publisher nor the authors, contributors, or editors, assume any liability for any injury and/or damage to persons or property as a matter of products liability, negligence or otherwise, or from any use or operation of any methods, products, instructions, or ideas contained in the material herein.

MATLAB® is a trademark of The MathWorks, Inc. and is used with permission. The MathWorks does not warrant the accuracy of the text or exercises in this book. This book's use or discussion of MATLAB® software or related products does not constitute endorsement or sponsorship by The MathWorks of a particular pedagogical approach or particular use of the MATLAB® software.

Library of Congress Cataloging-in-Publication Data
A catalog record for this book is available from the Library of Congress

British Library Cataloguing-in-Publication Data
A catalogue record for this book is available from the British Library

ISBN: 978-0-12-809395-5

For information on all Academic Press publications visit our website at https://www.elsevier.com/books-and-journals

Working together
to grow libraries in
developing countries

www.elsevier.com • www.bookaid.org

Publisher: Katey Birtcher
Acquisition Editor: Steve Merken
Sr. Content Development Specialist: Nate McFadden
Production Project Manager: Sujatha Thirugnana Sambandam
Cover Designer: Matthew Limbert

Typeset by TNQ Books and Journals

Dedicated to
Susanne Oldham
Who has shown me, and shared with me, so much of life.
Let the adventure continue...

Contents

APPENDICES

Preface to the Third Edition

This edition was motivated in large part by the continuously changing educational environment in biomedical engineering. The wide-ranging expertise expected of a competent biomedical engineer continues to grow, putting additional stress on curricula and courses. This text provides a comprehensive introductory coverage of linear systems, including circuit analysis. Not all of the material can be covered in a single 3-h course; rather the book is designed to allow the instructor to choose those topics deemed most essential and to be able to find them in a single, coherent text.

NEW TO THIS EDITION

The third edition presents a number of modifications over the previous text primarily in the form of new material. Each chapter now begins with a section describing and listing the goals of that chapter and ends with a summary of topics covered. In addition, several new chapters have been added.

Noise is ubiquitous in biomedical measurements and nonlinearity is inherent in most biomedical systems. These important measurement features required a new chapter on stochastic, nonstationary, and nonlinear systems and signals. A chapter on basic image analysis has been added along with introductory material. This chapter extends some one-dimensional signal processing tools to two dimensions. Although the only discrete systems a biomedical engineer is likely to encounter are digital filters, their importance in signal processing justifies introducing the z-transform before describing digital filters. Finally, a more traditional approach is followed in this edition in which time-domain concepts such as convolution are presented before introducing frequency-domain analyses.

Many biomedical engineering programs cover circuits in a separate course (usually in electrical engineering). Yet there may come a time when programs include this material, so a section on circuits remains at the end of the book. Chapter 15 on electronics is not usually included in a linear systems course, but since the students will have the necessary background to analyze useful circuits, I thought it worth including. Not all instructors will choose to include these new topics, but they are relevant and I feel they should be available as options. The chapters on circuits also include sections on lumped-parameter mechanical systems that should be considered optional.

Retained from earlier editions is the strong reliance on MATLAB®. This software is an essential adjunct to understanding signals and systems, and this book is not meant to be used without this powerful pedagogical tool. Another concept from earlier editions is the development of some of the deeper concepts, such as the Fourier transform and the transfer function, using an intuitive approach. For example, the Fourier transform is presented in the context of correlation between an unknown signal and a family of sinusoidal probing functions.

The general philosophy of this text is to introduce and develop concepts through computational methods that allow students to explore operations such as correlations, convolution, the Fourier transform, and the transfer function. I also include a few examples and problems that follow more traditional methods of mathematical rigor. A few more intriguing problems have been sprinkled into the problem sets for added interest and challenge, and many more problems are now based on biological examples.

ANCILLARIES

The text comes with a number of educational support materials. For purchasers of the text, a website contains downloadable data and MATLAB® functions needed to solve the problems, a few helpful MATLAB® routines, and all of the MATLAB® examples given in the book. Since many of the problems are extensions or modifications of examples, these files can be helpful in reducing the amount of typing required to carry out an assignment. Please visit www.elsevierdirect.com, search on "Semmlow," select the link to "Circuits, Signals, and Systems for Bioengineers - 3rd Edition," and click on the "companion site" link on the book's web page to download this material.

For instructors, an educational support package is available that contains a set of PowerPoint files that include all of the figures and most of the equations along with supporting text for each chapter. This package also contains the solutions to the problem sets and some sample examinations. The package is available for download from the publisher at: www.textbooks.elsevier.com.

John L. Semmlow, PhD
New Brunswick, NJ 2017

Acknowledgments

This text benefited from comments and ideas from several biomedical engineering educators from across the globe. I am particularly grateful to Rebeca Goya Esteban in the Department of Signal Theory and Communication of Rey Juan Carlos University, Madrid, who carefully reviewed a number of chapters and offered advice, corrections, and suggestions. I would also like to recognize contributions and suggestions from Hananeh Esmailbeigi, Clinical Assistant Professor, Department of Bioengineering at the University of Illinois at Chicago (one of my former hangouts); Yong K. Kim, Chancellor Professor of Bioengineering at UMass Dartmouth; Hatice Ozturk, Teaching Associate Professor of Electrical, Computer and Biomedical Engineering at North Carolina State University; and Ana Paula Rocha, Assistant Professor of Mathematics at the University of Porto, Portugal. Three graduate students in the Department of Biomedical Engineering at Rutgers University assisted with the editing of the manuscript: Cosmas Mwikirize, Hwan June Kang, and Matthew Richtmyer. Despite their help, I suspect some errors still slipped through for which I am exclusively to be blamed. Finally, I want to express my thanks to Susanne Oldham (to whom this book is dedicated) for her patient editing and support and to Peggy Christ who demonstrated great understanding during the far too long preparation of this book.

John L. Semmlow, PhD
New Brunswick, NJ, 2017

PART I

SIGNALS

1

The Big Picture: Bioengineering Signals and Systems

1.1 WHY BIOMEDICAL ENGINEERS NEED TO ANALYZE SIGNALS AND SYSTEMS

With only slight exaggeration, just about everything we encounter can be called a system. Part of this is due to the broad definition of a system: a collection of processes or components that interact with a common purpose. In this book, we are interested in signals as well as systems, so for our purposes a system is defined as a collection of processes or components that operate on, or originate, one or more signals. The human body offers many examples of well-defined systems devoted to a common purpose. The cardiovascular system delivers oxygen-carrying blood to the peripheral tissues. The pulmonary system is devoted to exchanging gases (primarily O_2 and CO_2) between the blood and air. The mission of the renal system is to regulate water and ion balance and adjust the concentration of ions and molecules. Mass communication is the objective of the endocrine system with the distribution of signaling molecules through the blood stream, whereas the nervous system performs tightly controlled communication using neurons and axons to process and transmit information coded as electrical impulses.

Many medical specialties, particularly areas of internal medicine, emphasize a specific physiological system. Cardiologists specialize in the cardiovascular system, neurologists in the nervous system, ophthalmologists in the visual system, nephrologists in the kidneys, pulmonologists in the respiratory system, gastroenterologists in the digestive system, and endocrinologists in the endocrine system. Most other medical specialties are based on common tools or approaches such as surgery, radiology, and anesthesiology, whereas one specialty, pediatrics, is based on the type of patient.

Given this systems-based approach to physiology and medicine, it is not surprising that early bioengineers applied their engineering tools to some of these systems. Early applications of systems analysis included the quantification and explanation of the oscillatory respiratory patterns known as Cheyne–Stokes respiration, understanding the neural control factors underlying the speed and accuracy of the eye movements, and explaining the mechanism of pupil oscillations called hippus. The nervous system, with its apparent similarity to early computers, was another favorite target of bioengineers, as was the cardiovascular system with its obvious links to hydraulics and fluid dynamics. Some of these early efforts are described in the sections on system and analog models. As bioengineering has expanded into areas of molecular biology, systems on the cellular, and even subcellular, levels have utilized the tool of system analysis.

Irrespective of the type of biological system, it needs to interact with other systems and we bioengineers need a way of interacting with these systems. Signals, or more specifically "biosignals," carry information and mediate this intrasystem communication. By definition, signals carry information that we can use; if a "signal" contains no useful information, it is "noise."[1] (Quotes are used to highlight terms that are particularly important and should be an integral part of a bioengineer's vocabulary.) Physicians and biomedical researchers (who can be viewed as large, complex systems) use the information contained in biosignals to determine or diagnose the state of a physiological system. Such biosignals arise as changes in various biological or physiological variables. Common signals measured in diagnostic medicine include: electrical activity of the heart, muscles, and brain; blood pressure; heart rate; blood gas concentrations and concentrations of other blood components; and sounds generated by the heart and its valves.

Often, it is desirable to send signals into a biological system for purposes of experimentation or therapy. We often use the term "stimulus" for signals directed at a specific physiological system; if an output signal is evoked by these inputs, we term it a "response." In this scenario, the biological system acts like an input–output system, a classic model used in systems analysis and illustrated in Figure 1.1.

Examples of well-defined input–output physiological systems include the knee-jerk reflex, where the input is a mechanical force and the output is mechanical motion, and the pupillary light reflex, where the input is light and the output is a mechanical change in the iris muscles. Drug treatments can be included in this input–output description, where the input is the molecular configuration of the drug and the output is the therapeutic benefit (if any). Such system-analytic representations are further explored in the sections on systems and analog modeling.

Systems that produce an output without the need for an input stimulus, like the electrical activity of the heart, are considered biosignal "sources." (Although the electrical activity of the heart can be moderated by stimuli, such as exercise, the basic signal does not require a specific stimulus.) Input-only systems, like write-only memory, are not very useful since the purpose of an input signal is usually to produce some sort of response. An exception

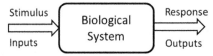

FIGURE 1.1   A classic systems view of a physiological system that receives an external stimulus or input signal that evokes a response or output.

[1]We sometimes use the term "random signal" which is a signal that does not carry information, so it really should be called "random noise." Put this down to sloppy terminology and just substitute "noise" for "signal" when you see it.

is the placebo, which is designed to produce no physiological response. (Nonetheless, it sometimes produces substantive results probably by interacting through complex, poorly understood neurological processes.)

Since all of our interactions with physiological systems are through biosignals, the characteristics of these signals are of great importance. Gleaning more information from these signals with signal processing tools is also very important in bioengineering. Indeed, much of modern medical technology is devoted to either extracting new physiological signals from the body or gaining more information from existing biosignals.

1.1.1 Goals of This Book

The tools of MATLAB make it possible for bioengineers to better analyze and understand signals and systems, the bread and butter of our professional lives. This book helps you to make the most of this powerful computer language in its applications to signal processing and systems analysis. The objective of this book is to give the reader a fundamental understanding of the field traditionally known as "linear systems analysis," but with concepts and applications in bioengineering. The basic ideas of linear systems analysis are well established and well understood. They can be divided into areas of signal analysis, system analysis, and the application of systems to signals known as signal processing.

The goal of signal analysis is to extract information from a signal by identifying features that are of particular interest. This includes basic and refined descriptions of a signal's time representation and a breakdown of its frequency content. This book covers the how and why of examining the correlations that occur between different portions of a signal. We find that correlation is a useful tool in signal analysis: we use it to determine similarities between two signals. When we use correlation to compare a signal and the (almost mystical) sinusoidal waveform, we discover that a bunch of sinusoids can give an alternative, yet complete representation of almost any signal. Moreover, applying a little algebra to this alternative "sinusoidal" representation can give us the "frequency spectrum" (or just "spectrum") of the signal.

In systems analysis, the basic goal is to be able to describe, quantitatively, the response of a system to a wide variety of stimuli. We will first learn to define system behavior analytically using differential equations. However, as is typical of engineers, we will find an easy way solve these equations using algebra. When digital systems are involved, they are described using difference equation rather than differential equations, but again we solve them algebraically. Surpassing these traditional analysis methods in ease and power is continuous system simulation implemented here with MATLAB's powerful system simulation tool, Simulink. This tool allows us to describe the behavior of very complex systems, even those that include nonlinear elements.

Mixing signal and system analysis, we use some special systems to alter signals. Applying these systems to a signal can produce an altered signal that is more useful to us. Often this increased utility is achieved by removing distracting components from the signal (i.e., noise). For example, we might remove frequencies in the signal that do not contribute to the signal's information content. Using a system to remove unwanted frequencies is known as "filtering" and the special system that does this is called, logically, a filter.

1.2 BIOSIGNAL: SIGNALS PRODUCED BY LIVING SYSTEMS

Biosignals are of utmost importance to us: much of our work, be it clinical or in research, involves the measurement, processing, analysis, and display of these signals. This section discusses some of the common physiological sources of biosignals and how these signals are measured.

1.2.1 Biological Signal Sources

Different physiological variations are supported by different energy forms. Table 1.1 summarizes the different energy forms that carry biological information, including the energy, the physiological variable that is changing, and the names of the biosignals associated with those variables.

Again, signals involve change. Within the body, communication between physiological systems uses signals encoded as changes in electrical or chemical energy. [2] Chemical energy encodes information by changing the concentration of the chemical within a "physiological compartment," for example, the concentration of a hormone in the blood. When speed is important, "bioelectric" signals are used. Since the body does not have many free electrons, it relies on ions, notably Na^+, K^+, and Cl^-, as the primary charge carriers.

TABLE 1.1   Energy Forms and Associated Information-Carrying Variables (i.e., Signals)

Energy	Variable Measured	Biosignal
Chemical	Chemical activity and/or concentration	Blood ions, O_2, CO_2, pH, hormonal concentrations, and other chemistry
Mechanical	Position	Muscle movement
	Force, torque, or pressure	Cardiovascular pressures, muscle contractility
		Valve and other cardiac sounds
Electrical	Voltage (potential energy of charge carriers)	EEG, ECG, EMG, EOG, ERG, EGG, GSR
	Current (energy in charge carrier flow)	
Thermal[a]	Temperature	Body temperature, thermography

ECG, electrocardiogram; EEG, electroencephalogram; EGG, electrogastrogram; EMG, electromyogram; EOG, electrooculogram; ERG, electroretinogram; GSR, galvanic skin response.
[a]Thermal energy is the average kinetic energy of the molecules involved so it is actually mechanical energy.

[2]Outside the body, information is commonly transmitted and processed as variations in electrical energy, although mechanical energy was used in the 18th and 19th centuries to send messages. The semaphore telegraph used the position of one or more large arms placed on a tower or high point to encode letters of the alphabet. These arm positions could be observed at some distance (on a clear day), and relayed onward if necessary. Information processing can also be accomplished mechanically, as in the early numerical processors constructed by Babbage in the early and mid-19th century. In the mid to late 20th century, mechanically based digital components were tried using variations in fluid flow.

1.2.2 Biotransducers and Common Physiological Measurements

Outside the body, electrically based signals are so useful that signals carried by other energy forms are usually converted to electrical energy when significant transmission or processing tasks are required. So even if we identify a physiological signal carrying information that we want, we usually need to convert that signal to one carried by electrical energy: an important, and often the first step in gathering information for clinical or research use. The energy conversion task is done by a device generally termed a "transducer," or "biotransducer."

A transducer is a device that converts energy from one form to another. By this definition, a light bulb or a motor is a transducer. In signal processing applications, the purpose of energy conversion is to transfer information, not to transform energy as with a light bulb or a motor. In physiological measurement systems, all transducers are so-called input transducers: they convert nonelectrical energy into an electronic signal. An exception to this is the electrode, a transducer that converts electrical energy from ionic to electronic form. The output of a biotransducer is a voltage (or current) whose amplitude is proportional to the measured energy. Figure 1.2 shows a transducer that converts acoustic sounds from the heart to electric signals. This "cardiac microphone" uses piezoelectric elements to convert the mechanical sound energy to electrical energy.

The energy that is converted by the input transducer may be generated by the physiological process itself as with heart sounds, or it may be energy that is indirectly related to the physiological process, although sometimes the energy being produced is only indirectly related to the system of interest, Figure 1.3A. For example, cardiac internal pressures are usually measured using a pressure transducer placed on the tip of a catheter, which is introduced into the appropriate chamber of the heart. The measurement of electrical activity in the heart, muscles, or brain provides other examples of direct measurement of physiological energy. For

FIGURE 1.2 A cardiac microphone that converts the mechanical sound energy produced by the beating heart into an electrical signal is shown. The device uses a piezoelectric disk that produces a voltage when it is deformed by movement of the patient's chest. The white foam pad is specially designed to improve the coupling of energy between the chest and the piezoelectric disk.

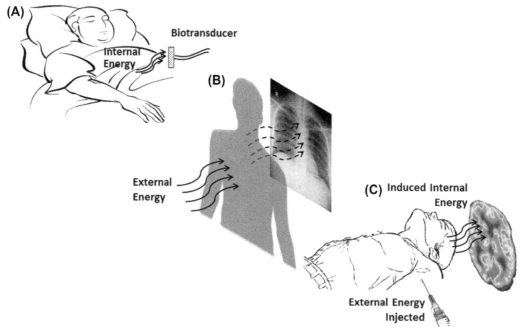

FIGURE 1.3 Three strategies for obtaining measurements from the body (or other physiological systems). (A) The body generates the energy, such as in heart sounds or electrically excitable tissue. (B) External energy is passed through the body, which is differentially blocked. For example, x-rays are differentially absorbed by body tissue and bone. In another example, external pressure applied to the arm induces sound generation used in the standard measurement of blood pressure. (C) The body is converted to a signal source by an external substance or energy source. For example, in positron emission tomography (PET), a radioisotope is introduced into the body. In magnetic resonance imaging, electromagnetic radiation and a strong magnetic field convert tissue protons into miniature radio transmitters.

these measurements, the bioelectric energy is already electrical and only needs to be converted from ionic to electronic current using an "electrode."[3] An early device for monitoring the electrocardiography (ECG) is shown in Figure 1.4. The interface between the body and the electrical monitoring equipment, the electrode, is buckets filled with saline ("E" in Figure 1.4).

In another approach, energy is injected from an external source and a difference in tissue absorption or reflection is measured by external transducers, Figure 1.3B. Examples include x-rays, computer-aided tomography, and ultrasound scans.

In the third approach, external energy is introduced in the physiological system, which then becomes an energy source that can be monitored externally, Figure 1.3C. Such energy can be in the form of short-lived radioisotopes introduced into the body that then accumulate

[3]Measurements of bioelectric energy are usually named using the rubric ExG, where the "x" represents the physiological process that produces the electrical energy: ECG, electrocardiogram; EEG, electroencephalogram; EMG, electromyogram; EOG, electrooculogram; ERG, electroretinogram; and EGG, electrogastrogram. An exception to this terminology is the galvanic skin response, GSR, the electrical activity generated by the skin.

FIGURE 1.4 An early electrocardiography machine.

in a target system to be detected by radiation detectors (PET). In the best known example of this approach, body protons are induced to become radio transmitters by placing them in strong magnetic fields and exciting them with external radio frequency (rf) radiation, an approach known as magnetic resonance imaging (MRI).

The biotransducer is often the most critical element in the system, since it defines the accuracy or resolution of the measurement and acts as an interface between the life process and the rest of the system. The transducer establishes the risk, or "invasiveness," of the overall system. For example, an imaging system based on differential absorption of x-rays, such as a computed tomography (CT) scanner, is considered more invasive than an imaging system based on ultrasonic reflection, since CT uses ionizing radiation that may have an associated risk. (The actual risk of ionizing radiation is still an open question, and imaging systems based on x-ray absorption are considered minimally invasive.) Both ultrasound and x-ray imaging would be considered less invasive than, for example, monitoring internal cardiac pressures through cardiac catheterization in which a small catheter is threaded into the heart chamber. Indeed, fame and fortune await the bioengineer who solves any of the many outstanding problems in biomedical measurement; for example, the noninvasive measurement of internal cardiac pressures or intracranial pressure.

1.2.3 Signal Representation: Continuous (Analog) Signals Versus Discrete (Digital) Signals

Signals fall into two broad categories: one-dimensional signals, where the dimension is usually time, and two-dimensional signals, where the dimensions are usually spatial. Here we are primarily concerned with one-dimensional signals, but many of the operations we apply to one-dimensional signals are easily extended to two-dimensional signals. Both types of signals can be further classified based on the way they are represented. These signal

TABLE 1.2   Signal Representations

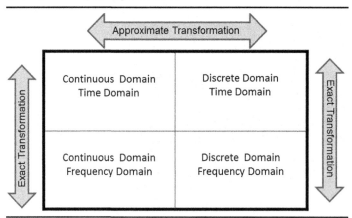

representations are unique: a signal is represented in either the continuous or discrete domain, and in either the time or frequency domain. This results in the possible combinations shown in Table 1.2. Methods for transferring between time and frequency domains are described for continuous signals in Chapter 3 and for discrete signals in Chapter 4. These transformations are exact: there is no loss of signal information through the transfer. Transformation between continuous and discrete signals is applied only in the time domain and is described later. This transformation is only approximate and, depending on the way it is done, there could be a loss of signal information.

Signals also have some important properties that we need to understand before we analyze them. Important signal properties include deterministic (or chaotic) versus stochastic; linear versus nonlinear; and stationary or time-invariant versus nonstationary. These properties are not necessarily unique: a signal can be basically linear but still have some nonlinearity properties and the same could be said of the other properties. Usually, signal properties are due to the system that created the signal, so we discuss these properties in Section 1.4 along with system properties.

As the name implies, continuous signals vary in time and amplitude in a continuous manner, whereas discrete signals have discrete levels of amplitude and exist at discrete moments in time. Continuous signals are often called "analog signals" and are commonly found in the real world. Discrete signals, often called "digital signals," are found inside computers or in discrete logic circuits.[4] Analog or digital signals can be defined by their mathematical domain. In the continuous domain, information is represented by a signal's amplitude (i.e., the value), which can change continuously over time.[5] In the digital domain, a signal is

---

[4]The central nervous system uses both. Action potentials transmitted through the axons are discrete, whereas the nerve body maintains a continuously variable, analog voltage.

[5]Analog signals may also be encoded as a continuous change in frequency, so-called frequency modulation (FM), or change in phase, termed phase modulation, but such encoding schemes are not common in nature and are not found in biosignals.

represented by numbers and different encoding schemes can be used. Signals are usually represented as a series of discrete numbers sampled at discrete points in time, so they are doubly discrete: in time and in value.

1.2.3.1 Analog Signals

An analog signal can be mathematically represented by the equation:

$$x(t) = f(t) \tag{1.1}$$

where $f(t)$ is some function of time and can be quite complex. For an electronic signal, $x(t)$ is the value of the voltage (occasionally current) at a given time. Some signals are so complicated that it is impossible to find a mathematical expression for $f(t)$ and the signal must be presented graphically. Again we emphasize that all signals are by nature "time-varying," since a time-invariant value contains no information: it is just some meaningless number.[6]

Often, an analog signal encodes the information as a linear change in signal amplitude. For example, a temperature transducer might encode room temperature into voltage as shown in Table 1.3 below:

This encoding could be defined by the equation:

$$\text{temperature} = 2 \times \text{voltage amplitude} - 10 \tag{1.2}$$

Analog signals and linear encoding are common in consumer electronics, such as within hi-fi amplifiers, although many applications that traditionally used analog encoding, such as television and sound and video recording, now primarily use digital signals. An interesting exception is the resurgence of music recorded as an analog signal on vinyl (i.e., records), including analog playback, as it is thought by some to have better sound characteristics. In addition, analog encoding remains important to the biomedical engineer because many physiological systems use analog encoding, and most biotransducers generate

TABLE 1.3 Relationship Between Temperature and Voltage of a Hypothetical Temperature Transducer

Temperature (°C)	Voltage (volts)
−10	0.0
0.0	5.0
+10	10.0
+20	15.0

[6]Modern information theory makes explicit the difference between information and meaning. The latter depends on the receiver, that is, the device or person for which the information is intended. Many students have attended lectures with a considerable amount of information that, for them, had little meaning. This text strives valiantly for both information and meaning.

analog signals. In living systems, not all analog signals are linearly encoded, although we often make that assumption as an approximation to allow us to use advanced signal analysis methods. In this book, the assumption is that all analog signals are linearly encoded unless otherwise stated.

1.2.3.2 Digital Signals

Signals may originate in the analog domain, but they usually need to be represented, processed, and stored in a digital computer. To transform a signal from the analog to digital domain requires that the continuous analog signal be converted to a series of numbers through a process known as "sampling." Since digital numbers can only represent discrete or specific amplitudes, the analog signal must also be sliced up in amplitude, a process called "quantization." Hence digitizing or sampling an analog signal requires slicing the signal two ways: in time and in amplitude. The details of the sampling process and its limitations are discussed in the next two sections, but essentially the continuous analog signal, $x(t)$, is sliced up into a sequence of discrete numbers, usually at equal time intervals, T_s. Such time intervals are called the "sampling interval" as shown in Figure 1.5.

1.2.3.2.1 TIME SAMPLING

A time-sampled, quantized signal, also referred to as a "digitized signal" or simply "digital signal," can be easily stored in a digital computer. A digital signal, $x[k]$, is just a series of discrete numbers: $x[k] = x_1, x_2, x_3, \ldots x_N$. These sequential numbers approximate, after

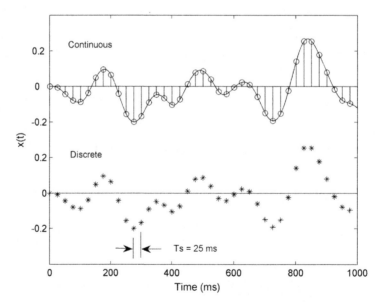

FIGURE 1.5 Digitizing a continuous signal, upper curve, requires slicing the signal in time and amplitude. The result is a series of discrete numbers (* points) that approximate the original signal, but can be stored in a computer. Time slicing is known as sampling and amplitude slicing (or rounding) as quantization. Here the interval between time slices, the sample interval, is 25 ms.

rounding, the value of the analog signal at a discrete point in time determined by the sample interval, T_s, or by the sampling frequency, f_s:

$$t = nT_s = n/f_s \tag{1.3}$$

where n is the position of the number in the sequence, t is the signal time represented by the number, and f_s is the inverse of the sample interval:

$$f_s = 1/T_s \tag{1.4}$$

Usually this series of numbers would be stored in sequential memory locations with $x[1]$ followed by $x[2]$, then $x[3]$, etc. As is common, we use brackets to identify a discrete or digital variable, $x[k]$, whereas parentheses are used for continuous variables, $x(t)$. [7]

In some advanced signal processing techniques, it is useful to think of the sequence of n numbers representing a signal as a single vector pointing to the combination of n numbers in an n-dimensional space. Fortunately, this challenging concept (imagine a 256-dimensional space, or even a space with only 5 dimensions) is not used here except in passing, but it gives rise to use of the term "vector" when describing a series of numbers such as $x[k]$. This terminology is also used by MATLAB, so in this text the term "vector" translates to "sequence or array of numbers" often representing a time signal.

Time slicing samples the continuous waveform, $x(t)$, at discrete points in time, nT_s, where T_s is the sample interval and $n = 1,2,3,...$. The consequences of time slicing depend on the signal characteristics and the sampling time, and are discussed in Chapter 4.

1.2.3.2.2 QUANTIZATION

Slicing the signal amplitude in discrete levels, quantization, is shown in Figure 1.6. The equivalent number can only approximate the level of the analog signal, and the degree of approximation depends on the range of numbers used and the amplitude of the analog signal. For example, if the signal is converted into an 8-bit binary number, the range of numbers is 2^8 or 256 discrete values. If the analog signal amplitude ranges between 0.0 and 5.0 V, then the quantization interval is 5/256 or 0.0195 V. If the analog signal is continuous in value, as is usually the case, it can only be approximated by a series of binary numbers representing the approximate analog signal level at discrete points in time as seen in Figure 1.6. The errors associated with quantization are described in Chapter 3.

---

[7]Note that the MATLAB programming language used throughout this text uses parentheses to index a variable (i.e., x(1), x(2) x(3)...) even though MATLAB variables are clearly digital. MATLAB reserves brackets for concatenation of numbers or variables. In addition, MATLAB uses the integer 1, not zero, to indicate the first number in a variable sequence. This can be a source of confusion when programming some equations that use zero as the index of the first number in a series. Hence x[0] in an equation translates to x(1) in a MATLAB program. (All MATLAB variables and statements used in the book are typeset in a sans serif typeface for improved clarity.)

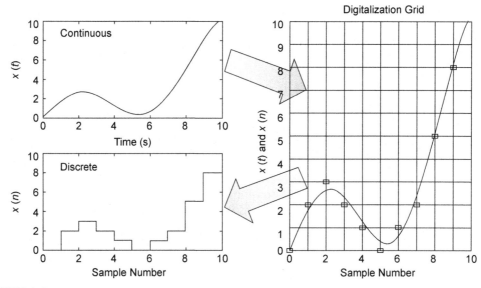

FIGURE 1.6 Digitizing a continuous signal, upper left, requires slicing the signal in time and amplitude, right. The result, lower left, is a series of numbers that approximate the original signal as a series of discrete levels at discrete time values. This digitizing operation is also known as analog-to-digital conversion.

EXAMPLE 1.1

A 12-bit analog-to-digital converter (ADC) advertises an accuracy of $\pm$ the least significant bit (LSB). If the input range of the ADC is 0 to 10 V, what is the accuracy of the ADC in analog volts?

Solution: If the input range is 10 V, then the analog voltage represented by the LSB is:

$$V_{LSB} = \frac{V_{max}}{2^{Nu\ bits} - 1} = \frac{10}{2^{12} - 1} = \frac{10}{4095} = .0024 \ V$$

Hence the accuracy would be $\pm.0024$ V.

It is relatively easy, and common, to convert between the analog and digital domains using electronic circuits specially designed for this purpose. Many medical devices acquire the physiological information as an analog signal, then convert it to digital format using an "analog-to-digital converter" ("ADC") for subsequent computer processing. For example, the electrical activity produced by the heart can be detected using properly placed electrodes, and the resulting signal, the electrocardiogram (ECG), is an analog encoded signal. This signal might undergo some "preprocessing" or "conditioning" using analog electronics, but would eventually be converted to a digital signal using an ADC for more complex, computer-based processing and storage. In fact, conversion to digital format is usually done even when the data are only stored for later use.

Transformation from the digital to the analog domain is possible using a "digital-to-analog" converter ("DAC"). Most PCs include both ADCs and DACs as part of a sound card. This circuitry is specifically designed for the conversion of audio signals, but can be used for other analog signals. Data transformation cards and USB-driven devices designed as general-purpose ADCs and DACs are readily available and offer greater flexibility in sampling rates and conversion gains. These devices generally provide multichannel ADCs (usually 8–16 channels) and several DAC channels.

In this text, the basic concepts that involve signals are often introduced or discussed in terms of analog signals, but most of these concepts apply equally well to the digital domain, assuming that the digital representation of the original analog signal is accurate. The equivalent digital domain equation is presented alongside the analog equation to emphasize the equivalence. Many of the problems and examples use a computer, so they are necessarily implemented in the digital domain even if they are presented as analog domain problems.

Is a signal that has been transformed from the continuous to the discrete domain the same? Clearly not; just compare the two different signals in Figure 1.5. Yet in signal analysis we often operate on discrete signals converted from an analog signal with the expectation (or assumption) that the discrete version is essentially the same as the original continuous signal. If they are not the same, is there at least some meaningful relationship between the two? The definitive answer is, maybe. The conditions necessary for the existence of a meaningful relationship between a continuous signal and its discrete version are described in Chapter 4. For now we will assume that all computer-based signals used in examples and problems are accurate representations of their associated continuous signals. In Chapter 4 we look at the consequences of the analog-to-digital conversion process in more detail and establish rules for when a digitized signal can be taken as a truthful representation of the original analog signal.

1.2.3.3 *Time and Frequency Domains*

We are already familiar with time domain signals from our discussion of the transformation of time domain signals between analog and digital domains. Frequency domain signals can also exist, or at least be analyzed, in either domain. Signals are represented in the frequency domain as two functions of frequency: a magnitude and a phase.

Figure 1.7 shows a section of an electroencephalography (EEG) signal in the time domain (left) and frequency domain (right). To completely represent the signal in the frequency domain, two functions are required, a magnitude (upper right) and a phase (lower right). The time domain representation is not easy to interpret; it looks like a disorganized bunch of squiggles. You could say the same for the phase; it is usually difficult to interpret. Because it is needed to fully represent the time domain signal, it is determined, even though it is not informative. On the other hand, the magnitude does show some interesting features. Specifically, there are peaks around 1–3 Hz and 16–18 Hz, and a large peak around 7–8 Hz. These peaks in frequency are well known to neurologists: the 1–3 Hz peaks are called theta waves, the 16–18 Hz peaks are called beta waves, and the 7–8 Hz peaks are called alpha waves, which have been associated with states of meditation. The important point is that in this case, these features could not be easily detected in the time domain representation (although they are there), but are very easy to spot in the magnitude of the frequency spectrum.

Because the frequency domain signals are functions of frequency, there are called spectra, the magnitude spectrum and phase spectrum. Although the time and frequency domain representation hold the same information, they do not store it in the same way so that one or the other may be more diagnostically useful in a given situation. It is fairly easy to transform from one domain to the other using techniques presented in Chapters 3 and 4.

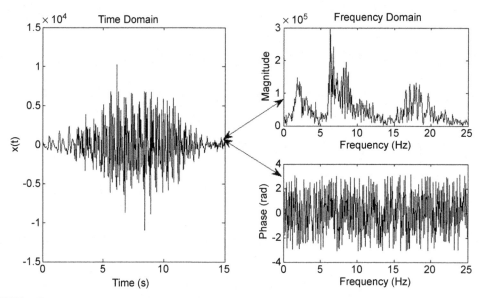

FIGURE 1.7 Time and (left) frequency domain representations of a 15-s segment of an EEG signal. The two representations are equivalent in terms of the information they contain, but they do present this information differently. Features seen in the frequency domain magnitude plot are not obvious in the time domain. These features, distinct peaks, have clinical relevance.

1.2.4 Two-Dimensional Signals ~ Images

The signals we have discussed thus far are one-dimensional functions of time that are represented in MATLAB as arrays or vectors. It is possible to have multiple signals that have some common origin and they should be treated together. A common biomedical example is the multilead EEG signal, where the electrical activity of the brain is recorded from 16, 32, or even 64 different locations on the head. Some analyses described in the next chapter treat these signals in combination, looking at features common to these signals as well as analyzing them individually. Multiple signals are often represented as a matrix where each row is a different signal.

Images in the digital domain, images that have been sampled, can be considered two-dimensional signals and can be represented as matrices. Each number in the matrix represents the intensity of a small region of space. These regions are usually square and are called pixels, so each number in the matrix represents a pixel. The distance between pixels is the spatial interval, the two-dimensional equivalent of the sample interval of a digitized signal.

MATLAB has a special toolbox for image processing that provides a number of useful routines for manipulating images, but image processing can be done using standard MATLAB routines. As an example, we generate a simple image: one consisting of vertical bars that vary sinusoidally in the horizontal direction. When considered and displayed as an image, a matrix consisting of identical rows of sinusoids looks like a sinusoidally varying pattern of vertical bars. Such an image is called a "sinusoidal grating" and is used in experiments on the human visual system.

EXAMPLE 1.2

Construct the image of sine wave bars having four cycles across the image. Make the image 100 by 100 pixels.

Solution: The strategy is straightforward; construct a single sine wave in a 100-point array, then make the vertical bars by copying that sine wave into sequential rows of a matrix. If you are not a MATLAB guru, constructing the sine wave is a nontrivial achievement, but very worthwhile as we will be doing similar operations throughout this book. If your MATLAB is a little rusty, jump ahead and check out Example 1.6, where the only goal it to construct and display a sine wave.

To generate the 100-point, four-cycle sine wave, we take advantage of MATLAB's ability to operate on a string of numbers (i.e., arrays) with a single statement. To construct a one-cycle sine wave, all we need is an array of numbers that go from 0 to 2π. If that array is labeled n, then the command, x = sin(n); would give us a one-cycle sine wave in array x where x is the same length as n.[8] Since we want a 4-cycle sine wave, array n should range between 0 and 8π. Since we want that 4-cycle sine wave to be 100 points long, we need n to be 100 points long. We can generate a 100-point array whose numbers range between 0 and 8π with the command: n = (0:99)*8*pi/100;. To be more general, we could define the array for a variable length, N.

```
N = 100;              % Number of points in the array, 100 points
n = (0:N-1)8*pi/N ;   % N-point array with numbers between 0 and 0.99
```

Then:

```
x = sin(n);     % Construct array with 4-cycle sine wave
```

More commonly, we would put the 8*pi in the sin argument because as we see later, that makes the frequency of the sine wave explicit. So the instructions to construct the sine wave become:

```
N = 100;              % Number of points in the array, 100 points
n = (0:N-1)/N ;       % N-point array with numbers between 100-point array
x = sin(8*pi*n);      % Construct array with 4-cycle sine wave
```

Duplicating that sine wave over 100 rows of a matrix is straightforward. If we call the matrix I (for image):

```
for k = 1:N
  I(k,:) = y; % Duplicate 100 times
end
```

When displayed as an image, a matrix consisting of identical rows of sinusoids would look like a sinusoidally varying pattern of vertical bars. Such an image is called a "sinusoidal grating" and is used in experiments on the human visual system. There are a number of ways to display a matrix as an image in MATLAB. In this example, we use MATLAB's `pcolor` and smooth the resulting display through interpolation using the MATLAB routine `shading interp`. This will generate a smooth image with color variation reflecting the sine wave's value. Images in this book are limited to black and white or levels of gray so we convert the color scale to a "grayscale" image using the command `colormap (bone)`.[9] The complete program becomes:

```
% Example 1.2 Construct the image of a sine wave grating having 4 cycles
%
N = 100;              % Number of pixels per line and number of lines
x = (0:N-1)/N;         % Spatial vector
y = sin(8*pi*x);      % Four cycle sine wave
for k = 1:N
  I(k,:) = y;         % Duplicate 100 times
end
pcolor(I);            % Display image
shading interp;       % Use interpolation
colormap(bone);       % Use a grayscale color map
```

Result: The image produced by this program is shown in Figure 1.8. The sinusoidal variation in gray levels is compromised somewhat in the printing process.

FIGURE 1.8 A sinusoidal grating generated in Example 1.3.

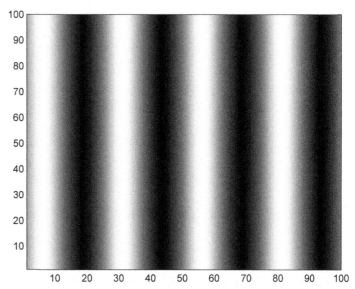

[8]As mentioned in the Preface, MATLAB code is distinguished in this book by a different `typeface`.
[9]Other MATLAB colormaps can be used in problem solutions to get some fun, near psychedelic effects! Try `colormap(jet)`, for example.

The grating in Figure 1.8 has a sinusoidal variation in intensity across the image. Since this is an image and not a time plot, the horizontal axis is position, not time, and the variation is a function of spatial position. The frequency of a sine wave that varies with position is called "spatial frequency" and is in units of cycles/distance. If, in this book, Figure 1.8 is printed to have a width of 4 inches, then the horizontal frequency of the sine wave would be 1.0 cycle/in. There is no variation vertically so the vertical frequency of this image is 0.0 cycles/in. One of the problems generates an image that has a sinusoidal variation in both dimensions.

Just about any MATLAB operation can be applied to an image, at least when it is represented as a standard MATLAB variable; conceptually the image is just another matrix to MATLAB. The next example shows some basic mathematical and threshold operations applied to an magnetic resonance (MR) image of the brain.

EXAMPLE 1.3

This example implements some standard MATLAB operations to a matrix that is actually an MR image of the brain. This image is stored as variable I in file `brain.mat`. We lighten the image by adding a constant to the entire array and enhance the contrast by multiplying the matrix by a constant. Finally, we threshold the image and make the darker regions white and all others as black (inverted with respect to the normal MATLAB coding of dark and light).

Solution: The file is loaded in the usual way using `load brain.mat`. The image is scaled to be between 0 (black) and 1.0 (white), a convention used by MATLAB for scaling image data. Since `pcolor` normalizes the data each time it is called, we will fix the display range to be between 0 and 1 using the `caxis` routine: `caxis([0 1])`. After displaying the original image as in the preceding example, we will lighten the image by increasing all the pixel values in the image by a value of 0.3 (remember 1.0 is white) and display. We then multiply the original image by 1.75 to increase its contrast and display. To apply the threshold operation, we examine every pixel and generate a new image that contains a 1.0 (i.e., white) in the corresponding pixel if the original pixel is below 0.25 and a 0 (i.e., black) otherwise. This image is then displayed.

```
% Example 1.3  Example to apply some mathematical and threshold operations
%  to an MR image % of the brain.
%
load brain;           % Load image
subplot(2,2,1);       % Display the images 2 by 2
pcolor(I);            % Display original image
shading interp;       % Use interpolation
colormap(bone);       % Grayscale color map
caxis([0 1]);         % Fix pcolor scale
title('Orignal Image');
%
subplot(2,2,2);
I1 = I + .3;          % Brighten the image by adding 0.3
  .......displayed and titled as above.......
%
subplot(2,2,3);
I1 = 1.75*I;          % Increase image contrast by multiplying by 1.75
  .......displayed and titled as above.......
%
```

```
subplot(2,2,4);
[r, c] = size(I);        % Get image dimensions
thresh = 0.25;           % Set threshold
for k = 1:r
  for j = 1:c
    if I(k,j) < thresh   % Test each pixel separately
      I1(k,j) = 1;        % If low make corresponding pixel white (1)
    else
      I1(k,j) = 0;        % Otherwise make it black (0)
    end
  end
end
.......displayed and titled as above.......
```

Results: The four images produced by this program are shown in Figure 1.9. The changes produced by each of these operations are clearly seen. Many other operations could be performed on this, or any other image, using standard MATLAB routines. A few are explored in the problems and others in Chapter 11.

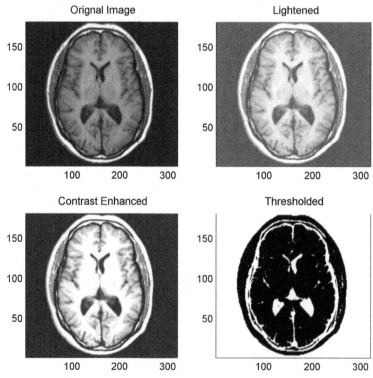

FIGURE 1.9   Images of the brain produced in Example 1.3. The MR image (upper left) has been brightened (upper right), contrast enhanced (lower left), and thresholded (lower right). In the thresholded image, the darker regions are shown as white and all other regions as black.

1.3 NOISE

All signals carry information as variations in energy (see Table 1.1). However, the reverse is not necessarily true: variations in energy do not always carry information, or at least not the information we want. We call these useless variations noise. When we make a real-world measurement, we usually get a combination of signal and noise. As bioengineers, we put a lot of effort into separating these components.

Where there is signal there is noise. Noise can originate in a variety of sources, including the biological system being measured, the measurement process, or the measurement environment. Random processes are known as "stochastic processes" and the resulting signal is said to be "stochastic" or can show stochastic behavior. When referring to signals, the terms "noise" or "stochastic behavior" are pretty much synonymous. The term "drift" also describes a noise, but the word implies that the stochastic behavior occurs at lower frequencies.

The extent of noise in a signal can vary widely and in the next chapter, we quantify the amount of noise in a signal as a ratio of the signal over noise in equivalent units. Occasionally the randomness will be at such a low level that it is of little concern, but more often it is this randomness that limits the amount of information we can get out of the signal. This is especially true for physiological measurements since they are subject to many potential sources of noise. Noise often limits the diagnostic value of a medical instrument and many devices employ advanced signal conditioning circuitry and signal processing algorithms in efforts (not always successful) to compensate.

1.3.1 Biological Noise Sources

Noise or stochastic behavior in biomedical measurements has four possible origins: (1) physiological indeterminacy; (2) environmental noise or interference; (3) measurement or transducer artifact; and (4) electronic noise.

Dealing with physiological indeterminacy is tough because the information you desire is based on measurements that are also influenced by other biological processes. For example, an assessment of respiratory function based on the measurement of blood pO_2 could be confused by other physiological mechanisms that alter blood pO_2, such as heart rate. Trying to determine pressure inside the heart from measurements on peripheral vessels will be affected in an unpredictable way by characteristics of the blood vessels and tissues. Physiological indeterminacy can be a very difficult problem to solve, sometimes requiring a total redesign of the approach.

Environmental noise can come from sources external or internal to the body. A classic example is the measurement of the fetal ECG signal where the desired signal is corrupted by the mother's ECG. Since it is not possible to describe the specific characteristics of environmental noise, typical noise reduction approaches such as filtering (described in Chapter 9) are not usually successful. Sometimes environmental noise can be reduced using "adaptive filters" or "noise cancellation" techniques that adjust their filtering properties based on the current situation.

Measurement artifact is produced when the biotransducer responds to energy modalities other than those desired. For example, the ECG recordings that reflect cardiac electrical activity are made using electrodes placed on the skin. These electrodes are also sensitive to movement, so-called motion artifact, where the electrodes respond to mechanical movement as well as to the electrical activity of the heart. This artifact is not usually a problem when the ECG is recorded in the doctor's office, but it can be if the recording is made during a patient's normal daily living, as in a 24-hour "Holter" recording.[10] Measurement artifacts can sometimes be successfully addressed by modifications in transducer design. For example, aerospace research has led to the development of electrodes that are quite insensitive to motion artifact.

Electronic noise is due to stochastic processes within the transducer elements or associated electronics. Unlike other sources of variability, the characteristics of these processes are pretty well known. Appropriate transducer and electronic design can often reduce this noise. Since electronic noise is well defined and can be addressed through appropriate design decisions, the next section examines these noise sources in more detail.

The various sources of noise or variability along with their causes and possible remedies are presented in Table 1.4. Note that in three of four instances, appropriate transducer design may aid in the reduction of the variability or noise, another example of the important role the transducer plays in the overall performance of a medical device.

1.3.2 Noise Properties: Additive Gaussian Noise

Noise is a common enemy and since it is useful to know your enemy, we should try to understand its properties. Like signals, noise can have different properties, but one set of properties, additive Gaussian noise, is of particular interest to biomedical engineers. The properties of Gaussian noise are well known: it adds a random variable to your signal and that variable has a Gaussian or normal distribution. In Chapter 2, we develop a simple signal processing technique that can reduce additive Gaussian noise in some situations.

TABLE 1.4 Sources of Variability

Source	Cause	Potential Remedy
Physiological indeterminacy	Measurement indirectly related to variable of interest	Modify overall approach
Environmental (internal or external)	Other sources of similar energy	Apply noise cancellation
		Alter transducer design
Artifact	Transducer responds to other energy sources	Alter transducer design
Electronic	Thermal or shot noise	Alter transducer or electronic design

[10]A Holter monitor is a battery-operated portable device that continuously measures and records the ECG for 24–48 h. The patient maintains his or her daily routine and keeps a diary of any cardiac abnormalities. Patient-reported symptoms are then compared with the cardiac electrical activity at the time of the event.

Since noise is random, a time function or time plot is not particularly useful. It is more common to discuss other properties of noise such as its probability distribution, its variance, or its frequency characteristics as described in Chapter 2. Although noise can take on a variety of different probability distributions, the "Central Limit Theorem" implies that most noise will have a Gaussian or normal distribution.[11] The Central Limit Theorem states that when noise is generated by a large number of independent sources, it will have a Gaussian probability distribution regardless of the probability distribution of the individual sources.

Figure 1.10 provides a dramatic demonstration of the Central Limit Theorem at work. In Figure 1.10A, the distribution of 20,000 uniformly distributed random numbers between

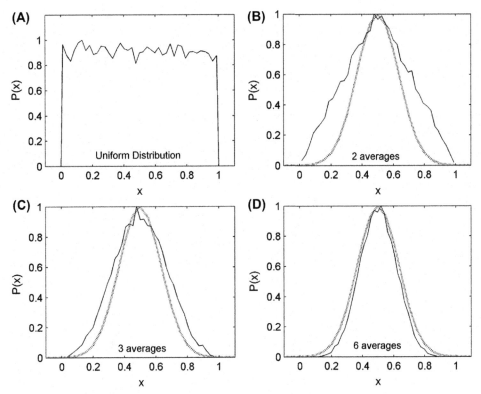

FIGURE 1.10 (A) The distribution of 20,000 uniformly distributed random numbers. The distribution function is approximately flat between 0 and 1. (B) The distribution of the same size data set where each number is the average of just two of the uniformly distributed numbers. The dashed line is the Gaussian distribution for comparison. (C) Each number in the data set is the average of three uniformly distributed numbers. (D) When only six uniformly distributed numbers are averaged to produce a number in the data set, the distribution becomes very close to Gaussian. (Note: this averaging trick is used by some computer algorithms to produce a Gaussian distribution from more easily generated uniformly distributed pseudo-random numbers.)

[11]Both terms are commonly used. We use the term "Gaussian" to avoid the value judgment implied by the word "normal"!

0 and +1 is shown (generated using MATLAB's rand function). A uniformly distributed variable has an equal probability of any value between 0 and 1, so its distribution is approximately flat between these limits. But now if we produce a new data set that is the average of just two uniformly distributed random numbers, we get a distribution that begins to look Gaussian, Figure 1.10B (the dashed line is a Gaussian distribution for comparison). When the data set is produced from the average of only six uniformly distributed random numbers, the resulting distribution is very nearly Gaussian, Figure 1.10D.

The probability distribution of a Gaussianly distributed variable, x, with zero mean is specified in the well-known equation:

$$p(x) = \frac{1}{\sigma\sqrt{2\pi}}e^{-x^2/2\sigma^2} \tag{1.5}$$

This is the distribution produced by the MATLAB function randn. (Note: The MATLAB routine rand produces uniformly distributed data with a mean of 0.5, whereas randn produces Gaussianly distributed data with a mean of 0.0. The Gaussian distributions shown in Figure 1.10 do not have zero mean because they were generated by averaging uniform data with a mean of 0.5.)

EXAMPLE 1.4

Use a large data set generated by randn to determine if these data have a Gaussian probability distribution. Also estimate the probability distribution of the data produced by rand.

Solution: A straightforward way to estimate the distribution function of an existing data set is to construct a "histogram" of the data set. A histogram is done by dividing a data set into a number of ranges, or "bins." For example, if the values of a data set ranged between 0 and 99, you could divide this range into five bins, each having a range of values: 0—19, 20—39, etc., Figure 1.11. Next, you

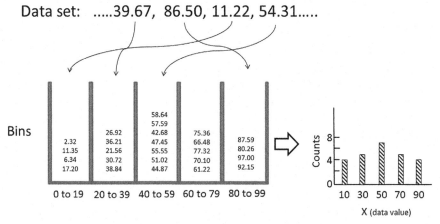

FIGURE 1.11   Construction of a histogram. Data are binned into five set ranges (left). After all the data in the set are binned, the number of data points in each bin is counted and plotted against the range value(s) (right).

count the number of specific values in your data set that fall within each of five ranges, Figure 1.11 (left). Then plot the number of counts in each range against the range, or the middle value of the range, Figure 1.11 (right), to produce the histogram. In other words, the histogram is a plot of the number of data points that fall within a given range (i.e., fall within a given bin) against the range, or mean value, of that range. You usually decide how many bins to use. More bins means smaller ranges for each bin, which produces higher resolution, but the bins will also have fewer counts, and if the data set is small, the histogram will be noisier. Usually the mean value of a range is used and is plotted on the horizontal axis with counts on the vertical axis.[12] Bar-type plots are commonly used for plotting histograms.

Construction of histograms is such an important operation that MATLAB has a routine to calculate and plot them. As is typical of MATLAB routines, hist has a number of options. The most useful calling structure for this example is:

```
[ht,xout] = hist(x,nu_bins);    % Calculate the histogram of data in x
```

where x is the data set and nu_bins is the number of bins desired. The outputs are the histogram, ht, and an array, xout, useful in plotting the histogram. This array gives the range of each bin as the mean of its range. So the statement plot(nu_bins, ht) plots the histogram and correctly scales the horizontal axis plotting (although as mentioned, it is more common to use a bar type plot as in the example below). Arrays that make plotting and scaling the axes easier (usually the horizontal axis) are often found as outputs in MATLAB routines and you benefit from using them.

This example first constructs a large (20,000-point) data set of Gaussianly distributed random numbers using randn, then uses hist to calculate the histogram and plot the results. This procedure is then repeated (but not shown) using rand to produce a uniformly distributed data set.

```
% Example 1.4 Evaluation of the distribution of data produced by MATLAB's
% rand and randn functions.
%
N = 20000;                      % Number of data points
nu_bins = 40;                   % Number of bins
y = randn(1,N);                 % Generate random Gaussian noise
[ht,xout] = hist(y,nu_bins);    % Calculate histogram
ht = ht/max(ht);                % Normalize histogram to 1.0
bar(xout, ht, 'c');             % Plot as bar graph (use color)
   ....... Label axes and title .......
   .......Repeat for rand ........
```

Results: The bar graphs produced by this example are shown in Figure 1.12. Note the approximately Gaussian distribution for the randn function, Figure 1.12A, and close to flat for the rand function, Figure 1.12B. One of the problems explores the distributions obtained using different data lengths.

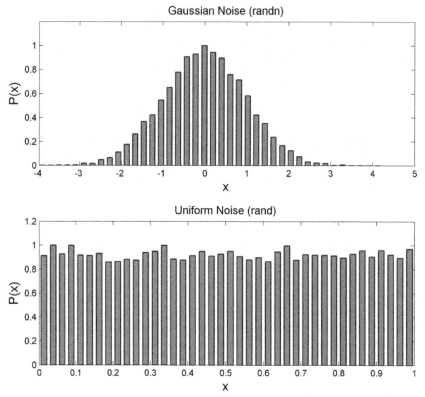

FIGURE 1.12   (A) The distribution of a 20,000-point data set produced by the MATLAB random number routine randn. As is seen here, the distribution is quite close to the theoretical Gaussian distribution. (B) The distribution of a 20,000-point data set produced by rand.

---

[12]The vertical axis of a histogram may also be labeled "Number" (or N), "Occurrences," "Frequency of occurrence," or shortened to "Frequency." The latter term should not be confused with "frequency" when describing cycles per second as is the most common use of the word.

1.3.2.1 Electronic Noise

Any biomedical engineer involved with devices will eventually encounter the additive Gaussian noise that is inherent in electronics. This noise falls into two, source-dependent, classes: "thermal" or "Johnson" noise, and "shot" noise. The former is produced primarily in resistors or resistance materials, whereas the latter is related to voltage barriers associated with semiconductors.

The good thing about electronic noise is we know a lot about it: not only where it comes from, but also how much noise to expect from a given component. We also know its frequency characteristics: its value at different frequencies. Both Johnson and shot noise contain energy over a broad range of frequencies, often extending from DC (i.e., 0 Hz) to 10^{12}–10^{13} Hz. White light contains energy at all frequencies, or at least at all frequencies

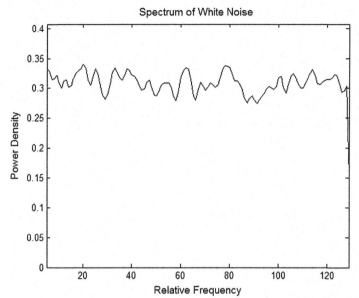

FIGURE 1.13 A plot of the energy in white noise as a function of frequency. The noise has a fairly flat spectral characteristic: it has nearly the same energy over the frequencies plotted. This equal-energy characteristic gives rise to the term "white noise." Techniques for producing a general signal's spectral plot are covered in Chapter 3.

we can see, so such broad-spectrum noise is also referred to as "white noise" since it contains energy at all frequencies, or all the frequencies we ever worry about. Figure 1.13 shows a plot of the energy in a simulated white noise waveform (actually, an array of random numbers) plotted against frequency. Again, this is a plot of energy against frequency, so it is called a spectrum.

1.3.2.1.1 JOHNSON NOISE

Johnson or thermal noise is produced by resistance sources and the amount of noise generated is related to the resistance and to the temperature:

$$V_J = \sqrt{4kT\,R\,BW}\ \text{volts} \tag{1.6}$$

where R is the resistance in ohms, T is the temperature in degrees Kelvin, and k is Boltzman's constant ($k = 1.38 \times 10^{-23}$ J/K). A temperature of 310 K is often used as room temperature, in which case $4kT = 1.7 \times 10^{-20}$ J. Here BW is the range of frequencies that is included in the signal. This range of frequencies is termed "bandwidth" and is better defined in Chapter 4. This frequency range is usually determined by the characteristics in the measurement system, often the filters used in the system. Since noise is spread over all frequencies, the greater the signal bandwidth, the greater the noise in any given situation.

If noise current is of interest, the equation for Johnson noise current can be obtained from Equation 1.6 in conjunction with Ohm's law:

$$I_J = \sqrt{4kT\,BW/R}\ \text{amps} \tag{1.7}$$

In practice, there will be limits imposed on the frequencies present within any waveform (including noise waveforms), and these limits are used to determine bandwidth. In the problem set at the end of the chapter, bandwidth, BW, is explicitly stated. It is common to specify BW as a frequency range in units of hertz, which are actually units of inverse seconds: i.e., $Hz = 1/s$.[13]

Since bandwidth is not always known in advance, it is common to describe a relative noise; specifically, the noise that would occur if the bandwidth were 1.0 Hz. Such relative noise specification can be identified by the unusual units required: $volts/\sqrt{Hz}$ or $amps/\sqrt{Hz}$.

1.3.2.1.2 SHOT NOISE

Shot noise is defined as current noise and is proportional to the baseline current through a semiconductor junction:

$$I_s = \sqrt{2qI_d\,BW}\ \text{amps} \tag{1.8}$$

where q is the charge on an electron (1.662×10^{-19} coul.), and I_d is the baseline semiconductor current. In photodetectors, the baseline current that generates shot noise is termed the "dark current," and this was the motivation for the letter "d" in the current symbol, I_d, in Equation 1.8. As with Johnson noise, the noise is spread across all frequencies, so the bandwidth, BW, must be specified to obtain a specific value or else a relative noise can be specified in $amps/\sqrt{Hz}$.

When multiple noise sources are present, as is often the case, their voltage or current contributions to the total noise add as the square root of the sum of the squares, assuming that the individual noise sources are independent. For voltages:

$$V_T = \sqrt{V_1^2 + V_2^2 + V_3^2 + \dots V_N^2} \tag{1.9}$$

A similar equation applies to current noise.

EXAMPLE 1.5

A 20 mA current flows through both a diode (i.e., a semiconductor) and a 200-Ω resistor, Figure 1.14. What is the net current noise, i_n? Assume a bandwidth of 1 MHz (i.e., 1×10^6 Hz).

Solution. Find the noise contributed by the diode using Equation 1.7 and the noise contributed by the resistor using Equation 1.6 and then combine them using Equation 1.9.

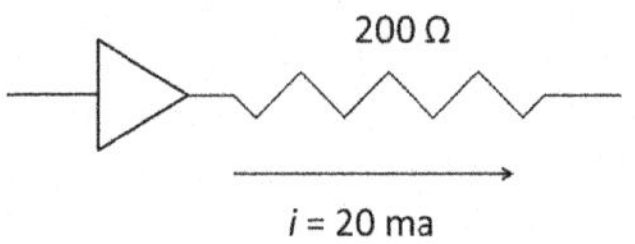

FIGURE 1.14   Simple electric circuit consisting of a resistor and diode used in Example 1.5.

---

[13]In fact, frequencies and bandwidths were given in units of "cycles per second" until the 1970s when it was decided that an inverse definition, a "per" definition, was not appropriate for such an important unit.

$$i_{nd} = \sqrt{2q\overline{I_d}BW} = \sqrt{2(1.66 \times 10^{-19})(20 \times 10^{-3})10^6} = 8.15 \times 10^{-8} \text{ amps}$$

$$i_{nR} = \sqrt{4kTBW/R} = \sqrt{1.7 \times 10^{-20}(10^6/200)} = 9.22 \times 10^{-9} \text{ amps}$$

$$i_{nT} = \sqrt{i_{nd}^2 + i_{nR}^2} = \sqrt{6.64 \times 10^{-15} + 8.46 \times 10^{-17}} = 8.20 \times 10^{-8} \text{ amps}$$

Note that most of the current noise is coming from the diode, so the addition of the resistor's current noise does not contribute much to the noise current. Also the arithmetic in this example could be simplified by calculating the square of the noise current (i.e., not taking the square roots) and using those values to get the total noise.

1.4 BIOLOGICAL SYSTEMS

"Consider the source" is a good rule for signal processors, since signal features often reflect their source. Biological systems, biosystems, produce or modify biosignals, and the properties of these systems transfer to their signals. Many important system (and signal) behaviors are dichotomous; that is, they are mutually exclusive: linear versus nonlinear, time invariant versus nonstationary, and deterministic or stochastic. But a system can exhibit both behaviors in different mixes. A major conundrum in biosignal analysis is that, although most biological systems are to some degree nonlinear, nonstationary, and stochastic, our most powerful analytical tools apply only to systems that are the opposite: deterministic, linear, and time invariant. For this reason, we often analyze biosignals, and the underlying biosystems, as if they were deterministic, linear, time-invariant systems. Such systems are given the short-hand term "LTI" systems. There is some exciting ongoing work in the analysis of nonlinear signals; at least we have some methods to determine if a signal contains nonlinearity, but unfortunately, these nonlinear tests are not always definitive and are often fooled by noise.

1.4.1 Deterministic Versus Stochastic Signals and Systems

Deterministic systems and signals can be described by a system of differential equations, at least in principle. These equations may be very complicated and they may not always be known. A deterministic signal has no randomness in its behavior, and it is possible to predict the future behavior of the signal from past values. The noise-free sine wave given by Equation 1.14 is an example of a deterministic signal. We can predict with perfect accuracy the value of a sine wave at any time knowing only the frequency, amplitude, and phase of the sine wave.

Stochastic behavior (a.k.a., noise) is inherently random: it cannot be described by a set of equations. In the next chapter we describe some general properties of a stochastic signal such as the mean, variance, and distribution function. Again, whether or not a system or signal is to be considered deterministic or stochastic depends on how much noise is in the signal. In practice, even signals produced by deterministic systems are corrupted by some noise so their future values may not be exactly predictable. Many signal processing techniques have been specifically designed to mitigate the influence of stochastic behavior.

1.4.1.1 Chaotic Signals and Systems

The signal from a chaotic system appears to combine properties from both random and deterministic signals but in fact chaotic processes are deterministic, not stochastic. There are three criteria used to define a chaotic system: (1) steady-state values are not fixed and do not repeat: they are not periodic; (2) system behavior is very sensitive to initial conditions; and (3) the system behavior is not simply a response to a random stimulus. There are methods to test if an apparently random signal is actually chaotic, many based on the first two criteria. Unfortunately, these tests are often less than definitive, especially if noise is intermixed with the chaotic signal.

A popular example of a chaotic system comes from the study of the variability in species populations. An equation that predicts animal populations fairly well is a simple quadratic equation called the *logistic difference* equation or *logistic map*, since it maps population values at generation n to values at the next generation, $n + 1$. In other words, the equation relates the population $x[n + 1]$ to the previous generation's population $x[n]$. The logistic equation is given as:

$$x[n] = rx[n](1 - x[n]) \qquad (1.10)$$

where $x[n]$ is the population at generation n (normalized to 1.0), $x[n + 1]$ is the population of the next generation, and r is the so-called driving parameter, which is related to growth. For small values of r, less than 3.0, Equation 1.14 produces smooth monotonic changes from one generation to the next, which eventually converge to a stable value. Above 3.0, the population alternates between two levels and, as r increases further, it alternates between 4, then 8, then 16 different population values: a behavior known as period doubling. Above a value of 3.57, the generation-to-generation fluctuation becomes apparently random and is shown in Figure 1.15. Although this fluctuation appears to be random, it is not since it is still completely determined by Equation 1.10. This behavior is explored in a problem at the end of the chapter.

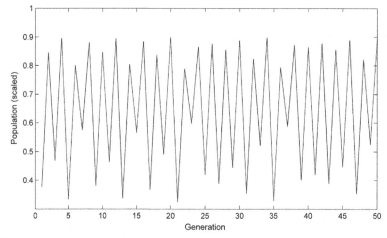

FIGURE 1.15 Value of the logistic equation, Equation 1.10 as a function of n where $r = 3.6$. The chaotic signal appears to be random, but is deterministic as it is generated by an equation, a rather simple one.

Chaos analysis is being applied in a number of different areas. In addition to ecology and epidemiology, chaos analysis has been applied to epileptic seizures, cardiac rhythms, patterns of nervous activity, fluctuation in leukocytes of patients with leukemia, and basic neural function. For example, it has been suggested that the basic operations of the nervous system employ chaos. In the cardiovascular system, chaotic heart rhythms are considered positive and some disease processes have been shown to reduce this chaos. Finally, studies are finding that some of the noise associated with biological systems may actually be chaotic behavior.

1.4.1.2 *Fractal Signals and Images*

Fractal signals are deterministic signals that look the same, or at least similar, at every magnification level, or "scale." Such signals are said to be "self-similar," "scale-invariant," or "scale free" because the pattern is much the same no matter what scale is used to view the pattern. One classic example is the outline of a coast. From high above the earth, coastlines have random-looking patterns, but as you get closer, and closer, they still have a similar random-looking pattern. Although fractal signals are only occasionally found in biology, there are many examples of fractal images. Figure 1.16 shows a fractal branching pattern. For each descending generation, a line branches into two shorter, diverging lines, and this progression continues through 16 generations. Two levels of increasing magnification show similar patterns, the hallmark of a fractal image. The bronchi of the lungs and coronary vascularization exhibit similar fractal patterns. Fractals, like chaos, demonstrate how complex structures can emerge from a simple strategy.

In fractal signals, small segments are similar to larger segments scaled-up in amplitude. The relationship between the change in timescale and change in amplitude scale often follows a power relationship. Finding the value for this scaling can be useful in research, and changes

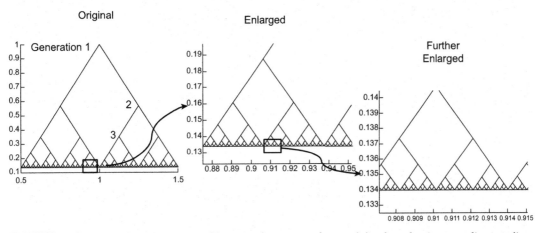

FIGURE 1.16 A fractal pattern generated by a simple strategy where each line branches into two divergent lines as it descends. The original pattern and two magnifications are shown. The similarity at each scale is emblematic of fractals.

in this relationship may indicate the onset or prognosis of disease. This promise motivates some major ongoing research to increase our understanding of these intriguing signal properties. Methods that search for fractal properties in signals are described in Chapter 10.

1.4.2 Deterministic Signal Properties: Periodic, Aperiodic, and Transient

Many of the signals discussed in this text can be divided into three general classes: "periodic," "aperiodic," and step-like or "transient."[14] Periodic signals repeat exactly after a fixed period of time known as the "period," T. The formal mathematical definition of a periodic signal is:

$$x(t) = x(t + T) \tag{1.11}$$

Frequency can be expressed in either radians or hertz (the units formerly known as "cycles per second") and the two are related by 2π:

$$\omega_p = 2\pi f_p \tag{1.12}$$

Both forms of frequency are used in the text and you should be comfortable with both, although frequency in hertz is most common in engineering. Frequency is the inverse of the period, T:

$$f_p = \frac{1}{T} \tag{1.13}$$

The sine wave is an example of a periodic signal, one that repeats in an identical manner following some time period T, or frequency ω_p, or frequency f_p:

$$x(t) = A \sin(\omega_p t) = A \sin(2\pi f_p t) = A \sin\left(\frac{2\pi t}{T}\right) \tag{1.14}$$

The sine wave is rather boring: if you have seen one cycle you have seen them all. Worse, the boredom goes on forever since the sine wave of Equation 1.14 is infinite in duration. Since the signal is completely defined by A (amplitude) and f_p (frequency), if neither changes over time, this signal does not convey any information. These limitations notwithstanding, sine waves (and cosine waves) are the foundation of many signal analysis techniques. Some reasons why sine waves are so important in signal processing are given in Chapter 3, but part of their importance stems from their simplicity.

Other common periodic signals include the square wave, pulse train, and sawtooth as shown in Figure 1.17. Since all periods of a periodic signal are identical, only one period is

---

[14]All time-varying signals could be called "transient" in the sense that they are always changing. However, in describing signals, the term transient is often reserved for signals that change from one level to another and never return to their initial value.

required to completely define a periodic signal, and any operations on such a signal need only be performed over one cycle. For example, to find the average value of a periodic signal, it is only necessary to determine the average over one period.

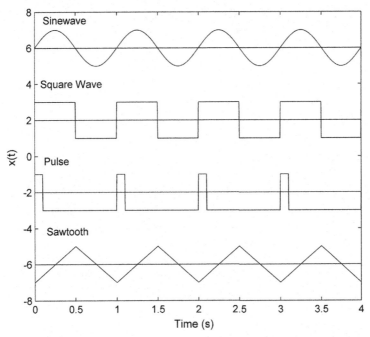

FIGURE 1.17 Four different periodic signals having the same, 1 s, period: sine wave, square wave, pulse train, and sawtooth.

EXAMPLE 1.6

Use MATLAB to construct a sinusoidal signal having an amplitude of 5.5 and a frequency of 10 Hz. Construct the sine wave in a 1000-point array. Plot two cycles of this signal as a function of time. Assume a sampling frequency of $f_s = 1000$ Hz.

Solution: This example is a simplification of Example 1.4, but it uses a formalism better adapted to signals as opposed to images. Throughout this text, we encounter the need to construct waveforms to illustrate various operations and principles; waveforms having specific properties such as shape or frequency. Given this ongoing need, let us establish a general approach that we can fall back on to make the coding of constructed signals easier.

Since we are working in the digital domain, we use a MATLAB array to hold the sine wave. Usually, the first step is to determine an appropriate length for the array: N, the number of data points in the array. In this problem, the array length is assigned as 1000 points, a value we often use as a sort of default. Before constructing the sine wave, we often construct a time vector[15] of the same length. To construct this time vector, we use a variation of Equation 1.3:

$$t = n/f_s$$

where n is a MATLAB variable going from 1 to N (or 0 to N − 1). This gives the MATLAB code lines:

```
fs = 1000;        % Define the sample frequency
N = 1000;         % Define the total number of points (given)
t = (0-N-1)/fs;   % Construct the time vector16
```

Next we construct the sine wave in a MATLAB vector, let us call it x, using Equation 1.8 (the middle version on the right hand side):

```
A = 5.5;               % Sine wave amplitude (given)
fp = 10;               % Sine wave frequency (given)
x = A*sin(2*pi*f*t);   % Construct the sine wave
```

Since we are requested to plot just two cycles of the sine wave, we need to limit the number of points plotted. Note that from rearranging Equation 1.7 one cycle is: $T = 1/f_p = 1/10 = 0.1$ s; so two cycles is $2T = 0.2$ s. The easiest way to do this conceptually is to plot the entire vector then as MATLAB's xlim to limit the time axis to the first 2 s:

```
plot(t,x);        % Plot sine wave as a function of time
xlim([0 0.2]);    % Limit the plot to the first 0.2 sec
```

We also add labels to the two axes. Given how easy this is to do in MATLAB, it should be a part of any graphical output, as it adds significant clarity to graphs. Putting this all together gives the solution code:

```
% Example 1.6  Plot two cycles of a 10 Hz sine wave
%
fs = 1000;                % Define the sample frequency
N = 1000;                 % Define the total number of points (given)
t = (0-N-1)/fs;           % Construct the time vector
A = 5.5;                  % Sine wave amplitude (given)
fp = 10;                  % Sine wave frequency (given)
x = A*sin(2*pi*fp*t);     % Define the sine wave
plot(t,x,'k');            % Plot sine wave
xlim([0 0.2]);            % Limit plot to first 0.2 sec
xlabel('Time (sec)');     % Label time axis
ylabel('Amplitude');      % Label y axis
```

Results/comments: The output of this program is shown in Figure 1.18 to be a 10 Hz sine wave having an amplitude of 5.5. Since we are not told what the sine wave represents (e.g., voltage, current, other?), the y-axis units are arbitrary. Note that once we take the trouble of setting up the

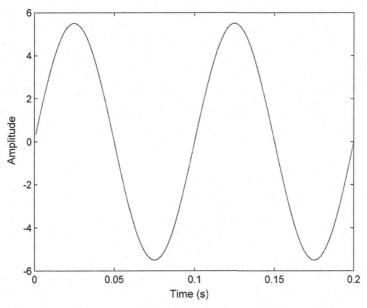

FIGURE 1.18   The 10 Hz sine wave produced by the MATLAB code in Example 1.2.

appropriate time vector, t, the code that produces the sine wave is the same as the equation for a sine wave given by Equation 1.14. We frequently find this to be the case when implementing waveform equations in MATLAB.

[15]As mentioned, we often refer to arrays as "vectors." The motivation for using this term is given in the next chapter, but for now assume the two words are synonymous.
[16]In this book, we use the typeface shown here for MATLAB variables and code.

Aperiodic signals exist over some finite time frame, but are zero before and after that time frame. Figure 1.19 shows a few examples of aperiodic signals. The assumption is that these signals are zero outside the period shown. A mathematically convenient way to view aperiodic signals is as periodic signals where the period goes to infinity. Operations on aperiodic signals need only be applied to their finite, nonzero segments.

The third class of signals, transient signals, includes signals that change, often dramatically, but never return to an initial level. A few examples are given in Figure 1.20, including the commonly used step signal. These signals do not end at the plot, but continue to infinity. These signals are the hardest to treat mathematically. They stretch to infinity, but are not periodic, so any mathematical operation must be carried out from zero to infinity. For example, it is impossible to calculate an average value for any of the signals in Figure 1.20 since these signals are nonzero to infinity, so their average values must also infinite. Moreover, it is not possible to represent such signals in a computer since they would require an infinite memory. However, methods for dealing with these signals exist and are discussed in Chapter 7.

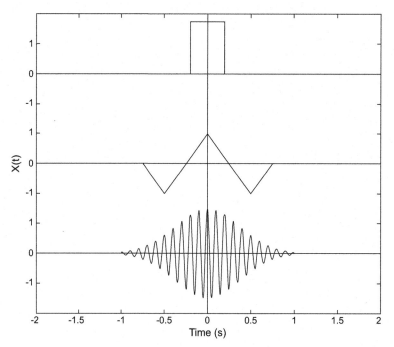

FIGURE 1.19   Three examples of aperiodic signals. It is common to show these signals centered on $t = 0$, especially when they are symmetric.

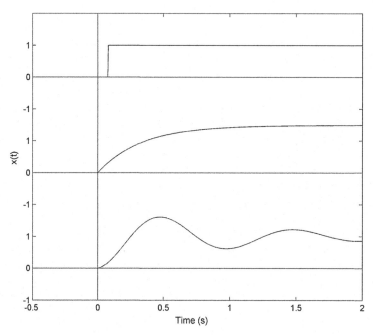

FIGURE 1.20   Three examples of transient or step-like signals. These signals do not end at the end of the plot but continue to infinity. The step signal is common in signal processing.

I. SIGNALS

1.4.2.1 *Time-Invariant Versus Nonstationary Signals and Systems*

The name defines it, time invariant: signals and systems that do no change over time. Time-invariant systems have fixed equations that are not functions of time. Time-invariant systems produce time-invariant signals. Time-invariant signals can change with time (most do), but they are called time invariant because they can be described by equations that do not change over time. In other words, a time-invariant signal can be "time varying" (i.e., a function of time) as long as the equation defining that signal does not change. For example, the sine wave defined by Equation 1.14 is time varying, but since Equation 1.14 is fixed, it is a time-invariant signal.

The mathematical definition of a time-invariant function, f, is given as:

$$\begin{aligned} y &= f(x) \quad \text{where } f \text{ is a linear function, then}: \\ y(t-T) &= f(x(t-T)) \end{aligned} \tag{1.15}$$

Nonstationarity is associated with systems or signals that change their basic statistical properties with time; if these signals have defining equations, those change over time. If signals or systems are nonstationary, then their statistical properties may be the same or they may change over time. Statistical properties of a signal that vary over time could include the mean or average value, the range of variability (variance), or the correlations between the signal and different delayed versions of the signal.

It would be nice if biological systems did not change over time, yet many do. As the systems change so do the signals they produce. The EEG signal, the electrical activity of the brain, is a classic example of a nonstationary signal. Its statistical properties are dependent on internal stated of the brain. Figure 1.21 shows a segment of an EEG signal where

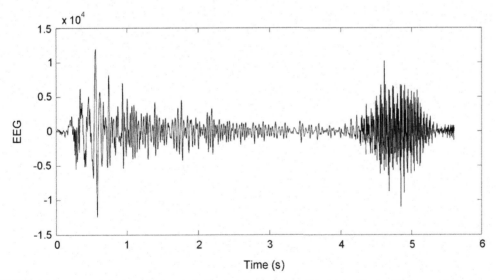

FIGURE 1.21 A segment of an EEG signal, the electrical activity of the brain. The basic characteristics appear to change over the course of the signal, an indication of nonstationarity.

nonstationary behavior can be directly observed: the basic nature of the signal appears to be different over different segments of the signal. In Chapter 10, techniques for testing signals for nonstationarity are described and applied to signals such as the EEG signal in Figure 1.21.

Nonstationarities present a moving target to any analysis approach. If dealt with at all, it is usually on an ad hoc basis, depending strongly on the type of nonstationarity. Most methods to deal with nonstationary signals are quite advanced and depend on the specific nonstationarity, but some approaches are described in Chapter 10. For example, if the mean or average value of the signal is varying, it may be possible to compensate for, or eliminate, this variation. Taking the derivative, which removes the mean along with any variation in mean, is one approach. Another is to estimate the variation in signal mean and subtract it out. Such an approach is called "detrending" and an example is presented in Chapter 10. Another popular approach that can be applied to a wide range of nonstationary behavior is simply to limit the signal analyses to time segments of the signal where it can be considered stationary.

1.4.3 Causal Versus Noncausal Signals and Systems

Causality is a fundamental Newtonian concept that responses are caused by stimuli. In systems, output signals are caused by input signals. In other words, responses cannot come before the thing they are responding to. Stated mathematically, the output of a system at $t = t_0$ can only be due to an input signal, $x(t)$, at $t \leq t_0$. This seems rather self-evident and unimportant, as all physical systems must be causal. But if the data are stored in a computer, it is possible to perform operations that include data points from both the past and future. Consider an operation on a MATLAB array; assume the following statement is in a `for` loop: `y(k) = x(k-4) - x(k+4)`; A new array, `y`, is being generated from points before and after the equivalent point (`k`) in the new array. If the data were a time series, then this operation would be equivalent to the time series equation:

$$y(t) = x(t-4) - x(t+4) \tag{1.16}$$

The output signal is based, in part on a signal from the future, $x(t+4)$. So this is a "noncausal" system. Again, noncausal systems can only be implemented using signals that have already occurred, for example, those stored in a computer. Physical systems must be causal and computer-based systems that produce real-time outputs must also be causal.

Causal signals are presumed to begin at $t = 0$, so the mathematical definition for a casual signal is:

$$x(t) = 0 \quad \text{for } t < 0 \tag{1.17}$$

Noncausal signals, those that exist for $t < 0$, are frequently shown here. This is just a graphical convention; all real-world signals are causal.

1.4.4 Linear Versus Nonlinear Signals and Systems

The concept of linearity has a rigorous definition, but the basic idea is one of "proportionality of response." If you double the stimulus into a linear system, you get twice the response. One way of stating this proportionality property mathematically is:

$$y = f(x) \quad \text{where } f \text{ is a linear function, then :}$$
$$ky = f(kx) \quad \text{where } k \text{ is a constant}$$
(1.18)

In other words, the output scales in proportion to the input.
Also, if f is a linear function:

$$f(x_1(t)) + f(x_2(t)) = f(x_1(t) + x_2(t))$$
(1.19)

In addition, if: $y = f(x)$ and $z = \frac{df(x)}{dx}$, then $\frac{df(kx)}{dx} = k\left(\frac{df(x)}{dx}\right) = kz$.

Similarly, if: $y = f(x)$ and $z = \int f(x)dx$, then $\int f(kx)dx = k\int f(x)dx = kz$.

Derivation and integration are linear operations. When they are applied to linear functions, these operations preserve linearity. Recall that systems that are both linear and time invariant are referred to as linear time invariant, or LTI, systems.

In a nonlinear system, the concept of proportionality (Equation 1.18) does not hold, nor does the summation of two signals expressed on Equation 1.19. Consider the simple example of a nonlinear system described by:

$$y(t) = x^2(t)$$
(1.20)

If the input to this system is value A, then the output is A^2, whereas the output for an input of value B is B^2. If this system were linear, the output to the combined input of $A + B$ would be $A^2 + B^2$, but in this system it is: $(A + B)^2 = A^2 + 2AB + B^2$.

If you have a biological system that you can isolate, drive with a controlled stimulus, and measure the response, then you might be able to test for nonlinearity using Equation 1.18. For example, you could input signals having different amplitudes (k in Equation 1.18) and determine if the output amplitude follows proportionally. This is the approach used in the next example.

EXAMPLE 1.7

The MATLAB routine `unknown1.m` takes a single input argument and produces a single output; i.e., `out = unknown1(in)`. Consider this routine as representing an input—output system and design a test to see if it is linear.

Solution: It is not possible to test if a system is linear over all possible inputs. Indeed most real systems are nonlinear if the amplitude range is large enough. Moreover, some systems may be linear for some frequencies, but nonlinear for others. Here we test to see if the system is linear for a 10 Hz sine wave over a range of amplitudes of 1–100 (arbitrary units).

First, generate a 10 Hz sine wave using the same approach as in Example 1.6. To reuse some of the code in that example, we will use an array length of 1000 points and a sampling frequency of $f_s = 1000$ Hz. Again these numbers were chosen arbitrarily, but they are appropriate for representing a 10 Hz signal. We use this sine wave as the input to unknown1.m and then plot the peak values of the output as a function of the input. If the sine wave is linear, the output scales in proportion with the input and the plot should be a straight line as per Equation 1.18.

```
% Example 1.7 to test nonlinearity of 'system' unknown.m
% System will be evaluated using a 10 Hz sine wave over an
%  amplitude range of 1-100.
% Use a 1000 point array to store the signal and assume a sampling
%  frequency of 1000 Hz
%
fs = 1000;                      % Sampling frequency in Hz
N = 1000;                       % Array for sine wave
t = (1:N)/fs;                   % Generate a time array using Equation 1.3
fp = 10;                        % Sine wave frequency in Hz
sine wave = sin(2*pi*fp*t);     % Generate 10 Hz sine wave having
                                % an amplitude of 1 using Equation 1.8
for k = 1:100                   % Test 100 different amplitudes
  in = sine wave * k;           % Adjust sine wave amplitude
  out = unknown1(in);           % Input sine wave to unknown system
  input_ampl(k) = max(in);      % Save input amplitude
  out_ampl(k) = max(out);       % Get output/input peak values
end
plot(input_ampl,out_ampl);      % Plot results
         ..........labels.............
```

Discussion and Results: After defining the sample frequency, fs, data array length, N, and sine wave frequency, fp, we generate a time array, t, using Equation 1.3. This is our approach in all MATLAB problems that involve a time function. We then use Equation 1.8 to produce the sine wave array. Using a for loop, we multiply the sine wave by numbers 1 through 100 and take the result as input to the routine unknown1.m. For each input amplitude, we save the maximum input and output signal amplitude.

The input and output signals are shown in Figure 1.22A to be sinusoidal, but the phase of the input (solid line) and the output (dotted line) are different. A plot of the output maximum amplitude as a function of the input maximum value, Figure 1.22B, is a straight line. This indicates that the output scales in proportion with the input as described by Equation 1.18. So although the unknown system is still unknown, we can say that it is linear, at least over the range of inputs tested.

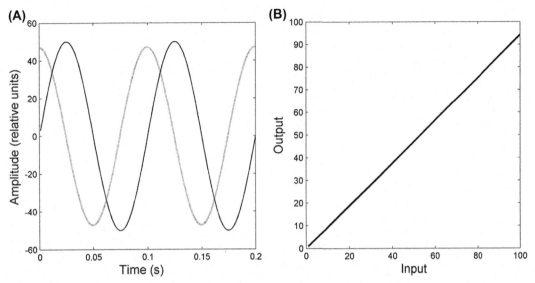

FIGURE 1.22 (A) A time plot showing the input signal to an unknown system (*solid line*) and the resulting output signal (*dashed line*). The input and output signals are 1 s long (1000 samples × 1/1000 samples per sec = 1 s), but only the first 0.2 s are shown for clarity. (B) A plot of the output signal's maximum amplitude versus the input signal's maximum amplitude. The resulting straight line indicates that the unknown process is linear as described by Equation 1.18.

If you have a biosignal, but do not have access to the system that generated that signal, testing for nonlinearity is more difficult. Signals from a nonlinear system may contain irregularities that appear more complicated than signals from linear systems and there are a range of tests that evaluate this complexity. Some of these tests are described in Chapter 10. Unfortunately, not all complex biosignals are from nonlinear systems and the development of definitive tests for nonlinearity is an active field of biomedical signal processing research.

1.4.5 Biosystems Modeling

Models are simplified representations of a system, simplified to better understand the function and/or operation of that system. Some might call a qualitative description of a biological system, such as found in basic physiology textbooks, a model, but as engineers, our models should be quantitative; that is, models that are supported by mathematical equations. It is usually possible to find solutions to these equations (sometimes with the help of a computer) and then we can check model predictions against the behavior of the real system.

Sometimes the model is a collection of equations, most often differential equations, but in a complex system these equations can be difficult to relate to actual biological system and its underlying components. Here we give examples of two modeling approaches that use elements as stand-ins for the differential equations: the "analog model" and the "systems model." These approaches differ in the type of elements used, but both can be represented graphically. In analog models, electrical or mechanical elements are used to represent a biological feature; in system models, an element represents the stimulus–response behavior of a related biological component.[17] In both models, the elements are alternative representations of differential equations, but they add an identity and a visual component that augments, and we hope clarifies, the relationship between the model and the real-world system.

1.4.5.1 Analog Models

In analog models, individual electrical or mechanical elements represent a biological function. For example, an analog muscle model might use springs and friction elements to represent a muscle's elastic and viscous elements. The mechanical elements have the same defining differential equations as the biological mechanical properties they represent. When mechanical elements are used to represent biological mechanics, model forces and velocities represent biological forces and velocities.

A model of skeletal muscle that features three mechanical elements is shown in Figure 1.23. The model is based on pioneering work on isolated muscle preparations by A. V. Hill in the 1930s. The spring labeled PE is the parallel elastic element of muscle and the spring labeled SE is the series elasticity. The element labeled CE is the contractile element, which is responsible for generating the muscle force. In Chapter 12, we find that elastic elements functionally relate force to the integral of velocity. In the model here, both elastic elements are nonlinear, so the relationship between the force they produce and the velocity at which they are stretched is nonlinear. The contractile element is also nonlinear and generates a force that is dependent on both the velocity and initial length of the element. Simulations of this model have shown that it can represent many important contractile features of muscle.

Analog models can also use electrical elements as long as the analogous element has the same defining equations as the real-world element it purports to represent. When electrical elements are used, voltage is a stand-in for force (or pressure) and current represents velocity.[18] In Chapter 7, we note that a resistor imposes a proportional relationship between voltage and current where the constant of proportionality is the resistance in ohms (i.e.,

FIGURE 1.23 A three-element model of muscle based on early experimental work of A.V. Hill. Simulations of this model have shown that it can represent the important features of muscle contraction.

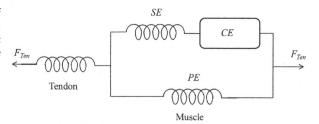

---

[17]"Analog" in this use describes the fact that the model elements are *analogous* to the biological function they represent. An analog model is conceptually continuous (like an analog signal), but so is a systems model. When simulated using a computer, both models are transported to the digital domain, but they are still conceptualized as continuous.

[18]This analogy is enhanced by the similarity between the major variables in the electrical and cardiovascular system. Voltage is a pressure that drives electrons and current is the flow of electrons. This dualism between mechanical and electrical variables is discussed in Chapter 12.

Ohms law: $V = RI$). When used to represent features of the cardiovascular system, a resistor describes relationships between cardiac pressures (or force) and blood flow.

Blood vessels present a resistance to flow and, if they were rigid, the relationship between pressure and flow could be completely described by an analogous resistance. However, blood vessels can expand, particularly larger vessels; they have what is known as "compliance" in addition to resistance. This feature can be represented by a capacitor since the pressure versus flow equation for mechanical compliance is similar to voltage versus current equation for a capacitor.

Combining resistance with compliance leads to the two-element Windkessel model shown in Figure 1.24A. This model was originally proposed by the German physiologist Otto Frank in 1899 to represent the load presented to the heart by the aorta and blood vessels. In modified form, it is still in use today. The purpose of this model is to quantify the mechanical load on the heart and from this representation predict the relationship between cardiac pressure and blood flow. The original model represented this load as two passive elements: a parallel resistor and a capacitor driven by pressure $P(t)$, the pressure generated by the heart, Figure 1.24A. The resistor, R, represents the combined resistance of all the blood vessels (in units of mm Hg s/cm^3), although the primary contribution comes from the smaller peripheral vessels. The capacitor, C, represents the net compliance of the blood vessels (in units of cm^3/mm Hg), which is mainly due to the aorta. Using a circuits simulation program, this model can be solved for any $P(t)$ to give blood flow from the heart $F(t)$ (in units of cm^3/s.).

The two-element Windkessel model was successful at modeling flow during diastole, but not systole, motivating the addition of other elements. Figure 1.24B shows a four-element Windkessel model. The series impedance, Z, represents what is known as the "characteristic impedance" of the aorta (impedance is discussed in Chapters 12–14). The parallel inductor, L, represents the inertia of the blood in the aorta. The four-element model represents a good approximation to the real system.

The Windkessel model is an example of a "lumped-parameter" model because it lumps features that are spread throughout the vascular system into single element. Frank's original motivation for this model was to predict cardiac output (i.e., flow) given the pressure waveform, and this it does quite well. It cannot tell you about the distribution of pressures and

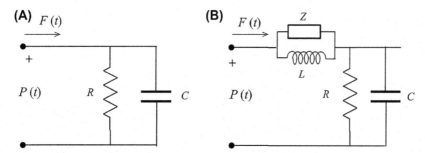

FIGURE 1.24 (A) A two-element Windkessel model of the load imposed on the heart by the blood vessels. This model can be used to predict aortic blood flow, $F(t)$, given the cardiac pressure, $P(t)$. (B) A more elaborate four-element Windkessel model that includes an additional resistor to model the peripheral aorta and an inductor to represent the inertia of the blood.

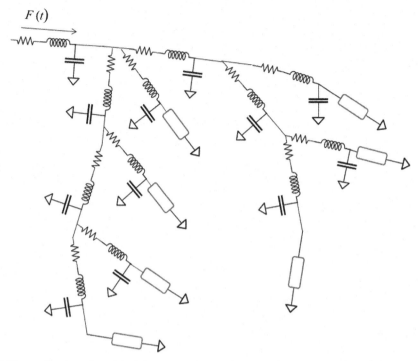

FIGURE 1.25 A distributed model of the coronary arterial system that can provide information on the distribution of pressure and flows in different arteries. This model was developed Wang, his mentor Welkowitz, and colleagues at Rutgers University. *Adapted from Wang et al., 1989. Med. Biol. Engr. Comp. 27:416.*

flows in the cardiovascular system. For that, a "distributed model" would be necessary. An example of a distributed model of the primary coronary arteries is shown in Figure 1.25. This model was used to predict coronary blood flow in the study of the sounds that might be produced by partially blocked arteries, a highly prevalent disorder known as coronary artery disease.

1.4.5.2 Systems Models (Transfer Function Models)

In a systems model, a biological process is represented by an element that mimics the stimulus–response behavior of that process. The stimulus is the input to a model element, whereas the response is the element's output. The elements are composed, internally, of differential equations and may include nonlinear terms. These equations are constructed to parody the stimulus–response behavior of the biological process. Because the elements describe how an input is transferred to the output, the elements are also known as "transfer functions." (The transfer function is introduced in Chapter 6.) Systems models are really good at representing the flow of information through a complex biological system, but they are less clear on how biological processes mediate a specific stimulus–response operation. The elements of a systems model describe what the process does, but not how.

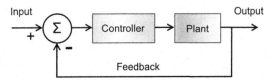

FIGURE 1.26 A common system model structure that includes a controller that governs an operation known as a plant (as in a chemical or manufacturing plant). A feedback pathway lets the controller know what the plant is doing. The controller then modifies its control signals to make the plant achieve the desired end as indicted by the input signal. This configuration is known as a "feedback control system."

Systems models date from the 1950s when control engineers turned their attention to biological systems. The classic structure of a systems model is shown in Figure 1.26. These models are divided into two major subdivisions: components that control an operation and components that implement that operation. The latter are referred to as the "plant," a term held over from the application of systems modeling to large manufacturing facilities such as a chemical plant. Logically, the collection of control components is called the "controller." A "feedback" pathway connects the output of the plant back to the controller. This feedback signal lets the controller know what the plant is doing so that it can adjust its control to achieve some desired end, or output. This desired end is indicated by the input signal.

A classic example of a feedback control system is a climate control system such as a home thermostat. The plant, the effector mechanism, is an air conditioner and/or furnace and the controller is a thermostat. The desired output is room temperature, say 70°. The feedback pathway sends information about the plant output, which is the room temperature, to the controller. The controller compares the plant output (room temperature) with a desired setting, to determine if it should activate the air conditioner or furnace and to what extent.

Systems models have been applied to a wide range of physiological mechanisms, including muscle-based control, fluid compartment models (see Chapter 10), drug uptake dynamics, and some endocrine systems. The first system models were of muscle-based control systems, including physiological motor control systems, the respiratory system, and the cardiovascular system. These systems fit the controller–plant paradigm quite well: nervous system components serve as the controller and muscle mechanisms are the plant. For motor control systems, the desired output is a movement; for respiratory and cardiovascular systems, the outputs are blood gas levels.

The control of human eye movements requires high precision and high speed. Eye movement positional errors are less than a degree and a typical movement takes less than 400 ms. (Small movements approach velocities of 600 deg/s!) This extraordinary performance attracted some of the earliest research in muscle control modeling. Figure 1.27 shows a model of the control of "vergence" eye movements where the eyes move in opposition to view targets at different depths. The plant in this model includes the dynamics of the extraocular muscles and mechanics of the eyeball. The input to the plant is a neural signal delivered by the oculomotor neurons in the brainstem and the output is the angle at which the two eyes converge. The plant component is actually a differential equation that represents this

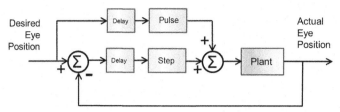

FIGURE 1.27   A systems model of the vergence eye movement control system; the ocular responses that drive the eye in opposition to fixate targets at different depths. A rapid response pulse control signal gets the eyes close to the desired position. It does not rely on feedback and is not subject to the longer delays involved with the feedback signal. Feedback is used to drive the step signal and is used to achieve high levels of positional accuracy, but more slowly. *Adapted from Lee et al., 2012. J. Eye Movement Res., 5:1.*

input—output relationship; that is, the relationship between nerve signal amplitude and the angular position of the eyes.

Although all system elements have a defining equation, they are not always indicated in the element graphic. Whether shown or not, the defining equation describes the element's dynamic properties: how its output varies over time to any given input. In the case of the eye movement plant, these properties have been determined with good accuracy for a range of eye movements, through neurophysiological and behavioral studies. The plant element, like the overall model, provides a tool for unifying experimental findings and for connecting theory and experiment.

The controller features two separate pathways that reflect the underlying neurophysiology. The upper pathway develops a pulse-like signal that originates from so-called burst cells in the brainstem. This neural signal starts the eyes moving rapidly, bringing them close to their final destination. The lower pathway produces a step-like signal originating from "tonic cells" found in the same region of the brainstem as the burst cells. This signal holds the eyes in a position at the final value, and by using the feedback signal, provides fine tuning that results in a very accurate final position.

The feedback signal is explicitly shown to project from the output position back to the input (through appropriate sensory mechanisms), and when compared with the desired output position, provides an error signal that implements the fine adjustment needed to achieve the high position accuracy of these movements. Since this error signal is obtained by comparing the desired position with the negative of the feedback signal, this configuration is sometimes referred to as a "negative feedback control system."

In theory, the vergence eye controller could rely only on the feedback signal to drive the eyes, but because of delays in neural processing, the resultant system would be either very slow or unstable. To solve this problem, the neural controller in the brain adds a pulse signal to increase the speed of the controlled movement. The model configuration of Figure 1.27 shows that the pulse signal is not influenced by the feedback signal. This rapid response pulse signal brings the eyes quickly, but only approximately, to the desired position. This component is not affected by the neural delays associated with feedback. The step signal does use feedback to achieve a high level of accuracy, but at a much slower pace. Finally, the plant consists of differential equations.

Again, each of the lines in Figure 1.27 represents a signal pathway with the direction of the signal indicated by the arrow. Each box represents a set of differential equations that describe how the input signal is transformed into an output signal. The behavior of this model can be evaluated through simulation: programming the operation of the model on a computer. As described in Chapter 9, MATLAB has an effective, easy-to-use simulation tool called Simulink.

In some systems, the desired response is a fixed constant so that the only signal to the controller is the feedback signal. An example is temperature regulation in the human body. The desired response is a fixed constant temperature (98.6°F). Although body temperature is maintained by a complex plant that involves a number of different mechanisms (peripheral blood flow, muscle activity), the controller resides in the brain and relies on core body temperature to develop the necessary neural control signals.

1.5 SUMMARY

It is a great time to be a biomedical engineer, but you need some specific tools to carry out your tasks. The primary purpose of this book is to develop two important biomedical skill sets: signal analysis and systems analysis. In addition, circuit analysis skills are presented in the final sections of the book. I cannot present all the many tools in these skill sets, but we do cover the most important and the most fundamental.

From traditional reductionist viewpoint, living things are described in terms of component systems. Traditional physiological systems, such as the cardiovascular, endocrine, and nervous systems, are large, complex, and composed of numerous subsystems. Biosignals provide communication between systems and subsystems and are our primary source of information on the behavior of these systems. Interpretation and transformation of signals is a major focus of this text.

Biosignals, like all signals, must be carried by some form of energy. Common biological energy sources include chemical, mechanical, and electrical energy (in the body, electrical signals are carried by ions, not electrons). Sometimes these signals can be obtained directly from their biological source as in cardiac auscultation (i.e., listening to the heart with a stethoscope). Other approaches look at how the physiological system interacts with an external energy source such as in ultrasound and CT scanning. Finally, as in MRI, external energy can be introduced into the body and this originally external energy creates the biological signal source.

Measuring a signal usually entails converting it to an electric signal to be compatible with computer processing. Signal conversion is achieved by a device known as an input transducer, and when the energy is from a living system, it is termed a biotransducer. Creation of a digital (or discrete-time) signal requires slicing a continuous signal into discrete levels of both amplitude and time and is accomplished with an electronic device known, logically, as an analog-to-digital converter. Amplitude slicing adds noise to the signal and the noise level is dependent on the size of the slice, but typical slices are so small that this noise is usually ignored. Time slicing produces a more complicated transformation that is fully explored in Chapter 4.

Signals have a number of important properties that we need to understand in order to work with them effectively. Many of the properties are shared by, and often caused by the systems that produce (or modify) those signals. Some signals and systems are stationary or time invariant: they follow the same basic patterns for all time; their properties, and sometimes the complete signal, can be described by equations. Signals that change their basic nature over time are called nonstationary signals. Sometimes we can adjust for these changes, but often we just try to pick out sections of the signal that are stationary. In Chapter 10, we present a few techniques for analyzing nonstationary signals.

Signals and systems can also be causal or noncausal. Noncausal systems produce signal outputs that respond to parts of the input signal that have not yet occurred; at least part of the system's response is driven by a signal from the future. Although such systems would be great for analyzing the stock market, they do not exist in the real world. You can construct a noncausal system in a computer if it is acting on prestored data.

Systems, and their resultant signals, can be linear or nonlinear. Linear systems respond proportionally: double the input and you get double the output. In Chapter 5, we discover that linearity allows us to use a powerful analysis tool called "superposition." Nonlinear systems are more difficult to analyze, although we find in Chapter 9 that simulation methods allow us to predict the response of systems that have well-defined nonlinearities. It can be very difficult just finding out if a signal has nonlinear properties; however, such properties can be of great diagnostic value. In Chapter 10, we examine a few tools recently developed to identify signal nonlinearity.

Signals can be deterministic or stochastic. Deterministic signals are totally predictable and can be represented by an equation. These are the signals we generally rely on throughout this book. Deterministic signals are necessarily time invariant (their statistical properties do not change with time) and can be divided into three classes: periodic signals that regularly repeat, aperiodic signals that occur only once for a finite time, and transient signals that make a step-like change and never return to their baseline or starting level. Stochastic signals are unpredictable, but if they are time invariant, some of their properties such as mean and variance can be uniquely defined.

By definition, a signal is what you want and noise is everything else; all signals contain some noise. Since we are generally not interested in a random signal, stochastic signals are usually regarded as noise. But a deterministic signals can also be noise if it is unwanted. Noise has many sources and only electronic noise can be evaluated based on known characteristics of the instrumentation. Since electronic noise exists at all frequencies, reducing the frequency ranges of a signal is bound to reduce noise. That provides the motivation for filtering a signal as detailed in Chapter 9.

Biosystems modeling is a powerful analytical tool for investigating living systems. Two very different models have been used to represent various physiological systems: analog models and system models. In analog models, analogous electrical or mechanical elements are used to represent specific biological processes (analog as used here refers to the analogy made between the living process and the element and has a different meaning than when used in analog signal). In systems models, each element represents a relationship between the input and the output signals of a process. This relationship is often called a transfer function since it effectively transfers the input to the output (see Chapter 6).

The strength of analog models is that they have a more intuitive relationship to the physiological process. The elasticity of muscle is represented by a spring, its mass by a mass element; there is a close relationship between the physical activity of the process and that of the element. In system models, no such relationship exists: a muscle is represented by its transfer function: a simple input—output relationship where nerve activation is the input and force, or movement, the output. The strength of systems models is the simplification provided by the transfer function and that information flow is shown explicitly. For this reason, systems models are favored when large biosystems are involved and where the influence of one process on another is of particular concern. Moreover, systems models can be easily simulated (as described in Chapter 10), which has led to a growing number of biological applications.

The goal of this book is to present the most important and fundamental of the many powerful signal and systems analysis tools available to biomedical engineers. It is a mixed blessing, you have a lot of options with which to attack life science problems, but you need to know a lot to use these tools effectively. Still, if you wanted it easy, you would have chosen something other than engineering!

PROBLEMS

1. The file on the CD `quantized.mat` contains a signal, x, and an amplitude sliced version of that signal, y, and time vector (for plotting) in seconds, t. This signal has been sliced into 16 different levels. In Chapter 3, we find that amplitude slicing is like adding noise to a signal. Plot the two signals superimposed. Find the effective "noise signal" by subtracting y from x and plot on the same graph. Now find the RMS value of this noise and compare it to the RMS value of the original signal. As with all graphs, label the axes. Also use the time vector to correctly scale the horizontal axis. (Hint: RMS means "root mean squared" the equation, and MATLAB code just follows the name (root mean squared); take the square root of the mean of the signal squared: RMS_x = sqrt(mean(x.2));)

2. Image generation. Using the approach shown in Example 1.2, construct a sinusoidal grating having a spatial frequency of five cycles per horizontal distance. Transpose this image and display. To transpose, use the MATALAB® operator. You should have a sinusoidal grating with horizontal strips. Then multiply each row in the transposed image by the original sine wave and display. This should generate a checkerboard pattern.

3. Load the image of the brain found in `brain.mat`, display the original and apply several mathematical operations. (1) Invert the image: make black white and white black. (2) Apply a nonlinear transformation: make a new image that is the square root of the pixel value of the original. (Note: this requires only one line of MATLAB code.) (3) Create a doubly thresholded image. Set all values below 0.25 in the original image to zero (black), all values above 0.5 to 1.0 (white), and anything in between to 0.5 (gray). (In this exercise, it is easier not to use `caxis` to set the grayscale range. Just let the `pcolor` routine automatically adjust to the range of your transformed images.) These images should be plotted using the `bone` colormap to reproduce the grayscale image

with accuracy, but after plotting out the figure you could apply other colormaps such as jet, hot, or hsv for some interesting pseudocolor effects. Just type colormap(hot), etc.

4. Repeat Example 1.4 using Gaussian arrays that are 100, 500, 1000, and 5000 points long. Plot the distribution functions generated by the MATLAB hist routine using a bin width of 40. Use subplot to combine the plots. (Hint: A for loop can significantly reduce the amount of code needed.)

5. Repeat Problem 4, with a single Gaussian array of 200 points. Construct four different histograms using 10, 20, 30, and 40 bins. Plot as in Problem 4.

6. Follow Example 1.6 to construct plots of a 2.5 Hz sine wave and a 1.5 Hz cosine wave. Make the peak amplitude of both 20. Use a 500-point array ($N = 500$) and make the sampling frequency 250 Hz. Plot the two waveforms in different colors superimposed and label both axes. Also plot a zero centerline.

7. Generate the sine and cosine wave used in Problem 6, following the procedure used in Example 1.6, but make the frequency of both 2.5 Hz. Use an array of 5000 points and an fs of 2500. Find the distribution of the sine wave (or cosine wave, they are the same) by taking the histogram. Use 40 bins. The result should be shaped like a "U" since the sine (or cosine) spends most of its time, so to speak, at the two extremes. Now combine the sine and cosine by simply adding them together and construct the histogram. Note that the distribution of the combined waveform is very different, approaching a Gaussian distribution. Another example of the Central Limit Theorem in action.

8. A resistor produces 10 μV noise (i.e., 10×10^{-6} volts noise) when the room temperature is 310 K and the bandwidth is 1 kHz (i.e., 1000 Hz). What current noise would be produced by this resistor?

9. The noise voltage out of a 1-MΩ (i.e.,10^6-ohm) resistor is measured using a digital voltmeter as 1.5 μV at a room temperature of 310 K. What is the effective bandwidth of the voltmeter?

10. A 3 mA current flows through both a diode (i.e., a semiconductor) and a 20,000-Ω (i.e. 20-kΩ) resistor. What is the net current noise, i_n? Assume a bandwidth of 1 kHz (i.e. 1×10^3 Hz). Which of the two components is responsible for producing the most noise?

11. Use MATLAB to evaluate the logistic equation (Equation 1.10) for four different values of r: 1.25, 2.25, 3.2, and 3.6. Evaluate the first 50 generations (use a for loop to increment n from 1 to 50) and start with an initial value for x of 0.02. Plot the population x as a function of generation, n. Use subplot to put the four plots together. Label the plots appropriately.

12. Construct the aperiodic function similar to that shown in the upper trace in Figure 1.19. The waveform should be 1.0 from −0.5 to +0.5 s and zero elsewhere. Use a 400-point array and set the sampling frequency so the time scale is from −1 to +1 s. Label and scale the time axis correctly. (Hint: Use Equation 1.3 to get f_s where $n = N = 400$ and $t = $ total time $= 2.0$ s. You can construct the time vector for plotting as in Example 1.2, but will need to subtract to get it to begin at −1.0 s.)

13. Follow the procedure in Example 1.7 to determine if the unknown process "unknown2.m" is linear or over what range of inputs it could be considered approximately linear.

Signal Analysis in the Time Domain

2.1 GOALS OF THIS CHAPTER

Signals are the foundation of information processing, transmission, and storage. Signals also are the basis for communication between biological processes, Figure 2.1, and for our interaction with those processes. If we are going to work with them, we need to understand them.

To recap some basic concepts from Chapter 1, signals differ in the way they are represented and they also exhibit different properties. Signal representations are unique; a signal

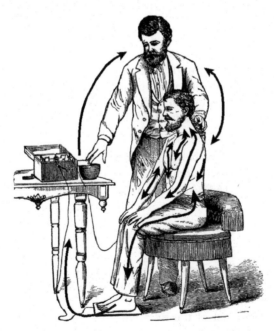

FIGURE 2.1   Signals course relentlessly through the body. As with all signals, they must be carried by some form of energy. Biosignals are carried by electrical energy in the nervous system or by chemical energy as molecular signatures in the endocrine system and other biosystems. Measurement and analysis of biosignals is fundamental to diagnostic medicine and biomedical research.

Circuits, Signals, and Systems for Bioengineers
http://dx.doi.org/10.1016/B978-0-12-809395-5.00002-3

TABLE 2.1 Major Signal Properties

Signal Properties	
Linear	Nonlinear
Deterministic	Stochastic (random)
Stationary (time invariant)	Nonstationary

is either analog or digital, time domain or frequency domain. Signal representations are summarized in Table 1.2. In this chapter, we work only with the time domain representations of signals.

Signal properties are less definitive. A signal can contain a mixture of properties; for example, a signal can be considered deterministic yet still contain some random behavior.[1] The basic signal properties introduced in Chapter 1 are summarized in Table 2.1.

In this chapter we will:

- Master some very basic, and not so basic, measurements applied to the time domain representations of signal and noise.
- Define measurements that quantify the strength of the signal and that quantify the ratio of signal to noise in a waveform.
- Define measurements that define a signal's stochastic properties (i.e., its variability or randomness).
- Describe a powerful method for reducing the influence of noise under certain circumstances.
- Define and work with the most important waveform of all, the sinusoid.
- Introduce some more involved time domain analytic techniques based on the concept of correlation that quantifies the similarity between two signals or functions.
- Apply these correlations techniques to shifted versions of two signals, or between different segments of the same signal.

In Chapter 3, we use correlations between signals and some carefully selected sinusoids to transform signals to the frequency domain. Signal comparisons using correlation are very useful and, since we do it in MATLAB, not all that difficult.

2.2 TIME DOMAIN MEASUREMENTS

Sometimes all we need to do is look at a signal and we can get the information we need. For example, a radiologist can often make a diagnosis by just looking at an image. Similarly a cardiologist can sometimes make diagnoses after a quick, qualitative examination of an electrocardiography signal. But most of the time we must probe deeper. We may need to get

---

[1]In a strict sense, any signal that contains randomness is a stochastic signal, but if the randomness is small enough that it can be ignored, we could consider it deterministic.

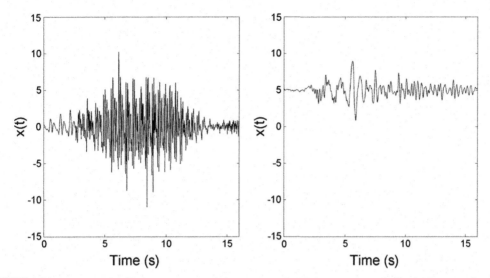

FIGURE 2.2 Two clearly different electroencephalogram signals. The one on the right has higher overall values because it has a higher mean level, but the signal on the left appears more energetic because it has more fluctuation.

quantitative measurements from a signal, some numbers that summarize its medically useful features. We begin with a discussion of the two most basic measurements you can make on a signal: its average value and its effective strength.

2.2.1 Mean and Root Mean Squared Signal Measurements

Figure 2.2 shows two electroencephalogram (EEG) signals that are obviously different. The signal on the right has higher overall values, but because it has less fluctuation it appears less energetic than the one on the left. The two most basic signal measurements, the mean and root mean squared (rms) value, quantify these two differences. The equations we use to calculate these measurements depend on whether the domain used to represent the signal is discrete or continuous. Although the same measurement requires a different equation for discrete versus continuous signals, the two equations are related.

For a discrete signal that is represented by a series of numbers, determination of mean value is intuitive; it is the average of all the numbers in the series. To determine the average of a series of numbers, simply add the numbers together and divide by the length of the series (i.e., the number of numbers in the series). For a series of N numbers:

$$x_{avg} = \bar{x} = \frac{1}{N} \sum_{n=1}^{N} x_n \qquad (2.1)$$

where n is an integer indicating a specific member in the series and ranging from 1 to N.[2] Some equations show summation, or similar operations, as ranging between 0 and $N - 1$.

---

[2]In this text, the letters n, k, and i are used as an index to identify a specific member in a series of numbers.

However, in MATLAB the index, n, must be nonzero, so for compatibility we usually use a range of 1 to N. The bar over the x in Equation 2.1 stands for "the average of" whatever variable it is over, in this case the average of x.

For signals represented in the continuous domain, summation becomes integration. The discrete set of numbers, x_n, becomes a continuous time series, $x(t)$, and the length of the series, N, becomes a fixed time period, T. This leads to the equation:

$$\bar{x} = \frac{1}{T} \int_0^T x(t)dt \tag{2.2}$$

In general, although different equations are required for the same operation in the discrete and continuous domains, these equations are related. Specifically, summation in the discrete domain becomes integration in the continuous domain, discrete variables become continuous variables, and the number of points used to represent a discrete signal becomes a fixed time period. Moreover, as defined in Chapter 1, the variable index is related to time, $n = t f_s$, and the total number of points is related to the total time, $N = T f_s$, where f_s is the sampling frequency. These equivalences are shown in Table 2.2 and provide a simple algorithm for converting from discrete domain equations to time domain equations and vice versa. Whether signals in the two domains contain the same information is another matter that is covered in Chapter 4.

The other basic difference between the two signals in Figure 2.2 is their range of variability. The signal on the left clearly shows greater and more prevalent fluctuations. This signal property is quantified by the rms value. The equation for the rms of a signal follows this measurement's name: first square the signal, then take its mean, then take the square root of this mean:

$$x_{rms} = \left[\frac{1}{N} \sum_{n=1}^N x_n^2 \right]^{1/2} \tag{2.3}$$

TABLE 2.2   Relationship Between Discrete and Continuous Operations

Operation	Discrete	Continuous
Summation/integration	$\sum_{n=1}^N x_n$	$\int_{t=0}^T x(t)$
Variable name	x_n	$x(t)$
	$n = t f_s$	$t = n/f_s$
Signal length	N	T
	$N = T f_s$	$T = N/f_s$

The continuous version of the equation is obtained by following the simple rules in Table 2.2.

$$\bar{x}_{rms} = \left[\frac{1}{T}\int_0^T x(t)^2 dt\right]^{1/2} \tag{2.4}$$

Note that the discrete equation (Equation 2.3) is easy to implement on a computer (even easier in MATLAB as shown later), but the continuous equation (Equation 2.4) needs to be solved "analytically," or by hand with paper and pencil. The next example finds the rms value of a simple signal, a sine wave, in both the continuous and discrete domains.

EXAMPLE 2.1

Find the rms value in both the continuous and discrete domains of the sinusoidal signal:

$x(t) = A\sin\left(\frac{2\pi t}{T}\right) = A\sin(2\pi ft)$.

Analytical Solution: Since this signal is periodic, with each period being the same as the previous one, it is sufficient to apply the rms equation over a single period. (This is true for most operations on sinusoids.) Applying Equation 2.4 along with a little calculus including a good table of integrals:

$$\bar{x}(t)_{rms} = \left[\frac{1}{T}\int_0^T x(t)^2 dt\right]^{1/2} = \left[\frac{1}{T}\int_0^T \left(A\sin\left(\frac{2\pi t}{T}\right)\right)^2 dt\right]^{1/2}$$

$$= \left[\frac{1}{T}\frac{A^2 T}{2\pi}\left(-\cos\left(\frac{2\pi t}{T}\right)\sin\left(\frac{2\pi t}{T}\right) + \frac{\pi t}{T}\right)\Big|_0^T\right]^{1/2}$$

$$= \left[\frac{A^2}{2\pi}(-\cos(2\pi)\sin(2\pi) + \pi + \cos 0 \sin 0)\right]^{1/2}$$

$$= \left[\frac{A^2\pi}{2\pi}\right]^{1/2} = \left[\frac{A^2}{2}\right]^{1/2} = \frac{A}{\sqrt{2}} = 0.707A$$

Hence, there is a proportional relationship between the "peak-to-peak" amplitude of a sinusoid (2A in this example) and its rms value: the rms value is $1/\sqrt{2}$ (rounded here to 0.707) times the amplitude, A. This is only true for sinusoids. For other waveforms, you need to apply the defining equation (Equation 2.4).

Discrete Solution: Here we take advantage of a useful MATLAB routine that calculates the mean (i.e., average) of an array.

```
xm = mean(x);     % Evaluate mean of x
```

where x is an array that in our case will contain the sine wave signal. (Note that if x is a matrix, the output is a row vector that is the mean of each column of the matrix. We use this powerful feature of this routine in a later example.)

To solve the problem on a computer, we first need to generate the sine wave and we need definitive values for f (or T) and A. One of the few downsides of computer solutions is that we cannot solve a problem with general variables. Arbitrarily we assign $f = 4$ Hz and $A = 1.0$. Also arbitrarily, we choose an N of 500 and a sampling frequency of 500 Hz. We then follow the approach used in Example 1.6, generate a time array, and use that to produce the sine wave

```
N = 500;                  % Number of points (arbitrary)
fs = 500;                 % Sampling frequency in Hz (arbitrary)
f = 4;                    % Sine wave frequency in Hz (arbitrary)
t = (0:N-1)/fs;           % Generate time array N points long (or: t = (0:N)/fs;)
x = sin(2*pi*f*t);        % Generate sine wave
```

Next, perform the rms operation using the sqrt and mean routines.

```
rms = sqrt(mean(x.^2));   % Calculate rms
disp(rms)                 % and display
```

Note that the .^ operator must be used for squaring x since we want each point squared separately. The program then becomes:

```
% Example 2.1 find the rms value of a discrete sine wave.
%
fs = 500;                 % Sample interval
N = 500;                  % Number of points
f = 4;                    % Frequency of sine wave
t = (1:N)/fs;             % Generate the time vector
x = sin(2*pi*f*t);        % Generate sine wave
rms = sqrt(mean(x.^2));   % Calculate rms
disp (rms)                % and display
```

Result: The calculated rms value determined by this program was 0.7071, which is the same as that found analytically if the amplitude, A, of the sine wave equals 1.0. Although the MATLAB approach is easier for most of us, it does not provide the general solution of the analytical approach.

2.2.1.1 Decibels

It is common to compare the intensity of two signals using ratios (i.e., V_{Sig1}/V_{Sig2}), particularly if the two are the signal and noise components of a waveform. Moreover, these signal ratios are commonly represented in units of "decibels" or "dB." When decibels are applied to ratios, they are dimensionless since whatever units are involved cancel. So decibels are not really units, but a logarithmic scaling of dimensionless ratios. The decibel has several features: (1) it provides a measurement of the effective power, or rather power ratio; (2) the log operation compresses the range of values (for example, a range of 1–1000 becomes a range of 1–3 in log units); (3) when numbers or ratios are to be multiplied, they are simply added if they are in log units; and (4) the logarithmic characteristic is similar to human perception. It is this latter feature that motivated Alexander Graham Bell to develop the logarithmic unit called the "Bel." Audio power increments in logarithmic Bels were perceived as equal increments by the human ear. The Bel turned out to be inconveniently large, so it has been replaced by the decibel: $dB = 1/10$ Bel.

Although the dB unit was originally defined only in terms of a ratio, the convenience of dB units has motivated their use in applications that do not involve ratios. In such cases, the dB actually has a dimension, the dimension of the signal (dB volts, dB dynes, etc), but these units are often not specifically stated. So when you see dB, you need to look at the context to see if it has dimensions.

When applied to a power measurement, the decibel is defined as 10 times the log of the power ratio:

$$P_{dB} = 10 \log\left(\frac{P_2}{P_1}\right) dB \tag{2.5}$$

As described in the next paragraph, the power of a signal is proportional to the amplitude (in rms) squared. So when it is applied to the voltage ratio, the decibel is defined as 20 times the log of the voltage ratio since $10 \log x^2 = 20 \log x$. The same is true when it is applied to just a voltage (i.e., not a ratio). So the definition of dB when applied to voltage (or other nonpower units) is:

$$v_{dB} = 10 \log\left(\frac{v_2^2}{v_1^2}\right) = 20 \log\left(\frac{v_2}{v_1}\right) \text{ or}$$

$$v_{dB} = 10 \log\left(v_{rms}^2\right) = 20 \log(v_{rms}) \tag{2.6}$$

To demonstrate that power is proportional to voltage squared, consider the power that can be produced by voltage v. To produce energy, it is necessary to feed the voltage into a resistor, or a resistor-like element, that consumes energy. (Recall from basic physics that resistors convert electrical energy into thermal energy, i.e., heat.) The power (energy per unit time) induced in the resister from the voltage is given by:

$$P = \frac{v_{rms}^2}{R} \tag{2.7}$$

where R is the resistance. Equation 2.7 shows that the power transferred to a resistor by a given voltage depends, in part, on the value of the resistor. Assuming a nominal resistor value of 1 Ω, the power will be equal to the voltage squared; however, for any resistor value, the power transferred will still be proportional to the voltage squared. When dB units are used to describe a ratio of voltages, the value of the resistor is irrelevant, since the resistor values will cancel out, assuming that the two values of R represent the same resistor:

$$v_{dB} = 10 \log\left(\frac{v_2^2/R}{v_1^2/R}\right) = 10 \log\left(\frac{v_2^2}{v_1^2}\right) = 20 \log\left(\frac{v_2}{v_1}\right) \tag{2.8}$$

If dB units are used to express the intensity of a single signal, then the units will be proportional to the log power in the signal.

To convert a voltage from dB to rms, use the inverse of the defining equation (Equation 2.8):

$$v_{rms} = 10^{v_{dB}/20} \qquad (2.9)$$

Putting units in dB is particularly useful when comparing ratios of signal and noise to determine the signal-to-noise ratio as shown in the next section.

EXAMPLE 2.2

A sinusoidal signal is fed into an *attenuator* that reduces the intensity of the signal. The input signal has a peak amplitude of 2.8 V and the output signal is measured at 2 V peak amplitude. Find the ratio of output to input voltage in dB. Compare the power-generating capabilities of the two signals in linear units.

Solution: Convert each peak voltage to rms, then apply Equation 2.8 to the given ratio. Also calculate the ratio without taking the log.

$$V_{rms\,dB} = 20 \log\left(\frac{V_{out\,rms}}{V_{in\,rms}}\right) = 20 \log\left(\frac{2.0 \times 0.707}{2.8 \times 0.707}\right)$$

$$V_{rms\,dB} = -3 \text{ dB}$$

The power ratio is:

$$\text{Power ratio} = \frac{V^2_{out\,rms}}{V^2_{in\,rms}} = \frac{(2 \times 0.707)^2}{(2.8 \times 0.707)^2} = 0.5$$

Analysis: The ratio of the amplitude of a signal coming out of a process to that going into the process is known as the *gain*, and is often expressed in dB. When the gain is <1, it means there is actually a loss, or reduction, in signal amplitude. In this case, the signal loss is 3 dB, so the gain of the attenuator is actually −3 dB. To confuse yourself further, you can reverse the logic and say that the attenuator has an attenuation (i.e., loss) of +3 dB. In this example, the power ratio is 0.5, meaning that the signal coming out of the attenuator has half the power-generating capabilities of the signal that goes in. An attenuation of 3 dB is equivalent to a loss of half the signal's energy. Of course, it is not really necessary to convert the peak voltages to rms since a ratio of these voltages is taken and the conversion factor (0.707) cancels out.

2.2.1.2 Signal-to-Noise Ratio

Most waveforms consist of signal plus noise mixed together. As noted previously, signal and noise are relative terms, relative to the task at hand: the signal is what you want from the waveform, whereas the noise is everything else. Often the goal of signal processing is to separate a signal from noise, or to identify the presence of a signal buried in noise, or to detect features of a signal buried in noise.

The "signal-to-noise ratio" or "SNR" quantifies the relative amount of signal and noise present in a waveform. As the name implies, this is simply the ratio of signal to noise, both measured in rms amplitude, and is often expressed in dB:

$$SNR = 20 \log \left(\frac{\text{signal}}{\text{noise}} \right) dB \qquad (2.10)$$

To convert from dB scale to a linear scale:

$$SNR_{\text{Linear}} = 10^{dB/20} \qquad (2.11)$$

When expressed in dB, a positive number means the signal is greater than the noise and a negative number means the noise is greater than the signal. Some typical dB ratios and their corresponding linear values are given in Table 2.3.

To give a feel of what some SNR dB values actually mean, Figure 2.3 shows a sinusoidal signal with various amounts of white noise. When the SNR is +10 dB, the sine wave can be readily identified. This is also true when the SNR is +3 dB. When the SNR is −3 dB you can still spot the signal, at least if you know what to look for. When the SNR decreases to −10 dB, it is impossible to determine if a sine wave is present visually.

2.2.2 Variance and Standard Deviation

Signal properties such as the mean and rms apply to both deterministic and stochastic signals, but some signal properties are dedicated to describing a signal's random behavior. The most common description of signal randomness is the "sample variance," s^2 (also commonly written as σ^2). The variance is a measure of signal fluctuation or variability and is insensitive to the signal's average value. The calculation of variance for both continuous and discrete signals is given as:

$$s^2 = \frac{1}{T} \int_0^T (x(t) - \bar{x})^2 dt \qquad (2.12)$$

TABLE 2.3 Equivalence Between dB and Linear Signal-to-Noise Ratios

SNR in dB	SNR Linear
+20	$v_{\text{signal}} = 10 \, v_{\text{noise}}$
+3	$v_{\text{signal}} = 1.414 \, v_{\text{noise}}$
0	$v_{\text{signal}} = v_{\text{noise}}$
−3	$v_{\text{signal}} = 0.707 \, v_{\text{noise}}$
−20	$v_{\text{signal}} = 0.1 \, v_{\text{noise}}$

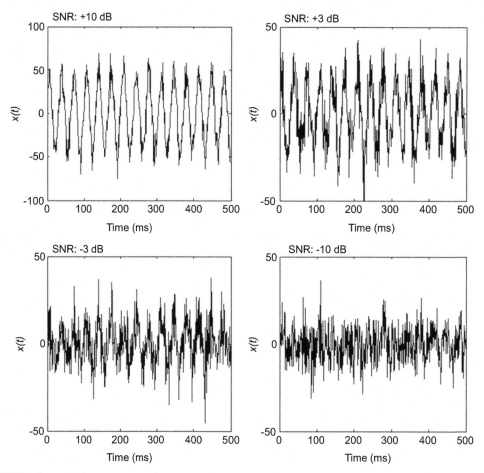

FIGURE 2.3 A 30 Hz sine wave with varying amounts of added noise. The sine wave is barely discernable when the SNR is −3 dB and is not visible when the SNR is −10 dB.

$$s^2 = \frac{1}{N-1} \sum_{n=1}^{N} (x(t) - \bar{x})^2 \tag{2.13}$$

where $\bar{x}$ is the signal mean (Equation 2.1). Comparing Equation 2.13[3] with Equation 2.3, we see that if the signal mean is zero (i.e., $\bar{x} = 0$), then the variance is almost the same as the rms value. The difference is that in rms you divide the summation by the signal length $\left(\frac{1}{N}\right)$, whereas in the variance calculation you divide the summation by the signal length minus 1, i.e., $\left(\frac{1}{N-1}\right)$. The reason for this subtle difference, which becomes ever smaller as N increases, has to do with the definition of variance. Variance is defined in terms of the probability

---

[3]Some statisticians prefer to use the symbol s_{N-1}^2 to indicate that the summation is normalized by $\frac{1}{N-1}$.

distribution of a random variable, which is generally unknown in real-world situations. So the variance measurements produced by Equations 2.12 and 2.13 are estimations of the true variance that are obtained from a limited sample. They also include estimations of the sample mean, $\bar{x}$, in those equations. Dividing by $N - 1$ reduces the bias produced by approximating the mean and results in what is known as the "unbiased sample variance." This is the equation most commonly used to estimate the variance from real data samples. Despite the similarities between the calculations of rms (for zero mean signals) and variance, they come from very different traditions (statistics versus measurement) and are used to describe conceptually different aspects of a signal: variance describes signal variability and rms describes signal magnitude.

The "standard deviation" is another measure of a signal's variability and is simply the square root of the variance:

$$ s = \left[\frac{1}{T} \int_0^T (x(t) - \bar{x})^2 dt \right]^{\frac{1}{2}} \tag{2.14} $$

$$ s = \left[\frac{1}{N-1} \sum_{n=1}^N (x_n - \bar{x})^2 \right]^{\frac{1}{2}} \tag{2.15} $$

Calculating the variance or standard deviation of a signal using MATLAB is as easy as calculating the mean.

```
xv = var(x);      % Evaluate variance of x

xsd = std(x);     % Evaluate standard deviation of x
```

A MATLAB example that shows the estimation of standard deviation is given next.

EXAMPLE 2.3

Generate a 5000-point array of Gaussianly distributed (i.e., normal) numbers and calculate the standard deviation of those numbers. Plot the histogram of that data set along with a vertical line indicating plus and minus one standard deviation. Repeat with uniformly distributed data, but remove the mean before calculating the standard deviation and plotting.

Solution: Use randn to generate the 5000-point array and std to calculate the standard deviation. Use hist to construct the histogram (see Example 1.4). Plot the histogram and the standard deviation lines. Use rand to generate the uniformly distributed data, subtract the mean, and repeat the standard deviation determination and plotting.

```
% Example 2.3 Program to calculate the standard deviation of a set of Gaussian
% and uniform random numbers, plot the histogram and standard deviations.
%
```

```
N = 5000;                          % Number of data points
nu_bins = 15;                      % Number of bins for histogram
x = randn([1,N]);                  % Generate Gaussian random numbers
[H,bin] = hist(x,nu_bins);         % Generate histogram
xsd = std(x);                      % Calculate standard deviation
subplot(2,1,1);                    % Select plot
plot(bin,H,'--k'); hold on;        % and plot
plot([xsd xsd],[0 max(H)],'k'); % Plot straight lines
plot([-xsd -xsd],[0 max(H)],'k');
   .........label and title........
x = rand([1,N]);                   % Generate uniform random numbers
x = x-mean(x);                     % Remove mean.
   ........repeat standard deviation and plotting.......
```

Results: The plots generated by this example are shown in Figure 2.4. The standard deviation of the Gaussian random numbers is 1.0, which is expected since this is the distribution created by randn. The uniformly distributed number random numbers range between ±0.5 (after the mean is removed) and the standard deviation of these numbers is approximately 0.3.

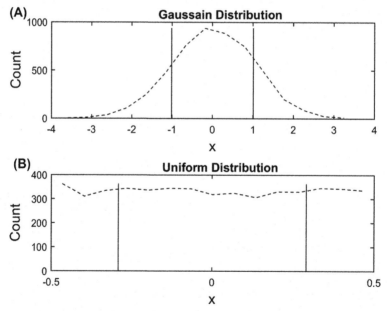

FIGURE 2.4   (A) Histogram of Gaussian random members with vertical lines indicating plus and minus one standard deviation. (B) Histogram of random numbers uniformly distributed between ± 0.5 with standard deviations.

2.2.3 Averaging for Noise Reduction

When multiple measurements are made, multiple values or signals are obtained. If these measurements are combined or added together, the means add so that the combined value or signal has a mean that is the average of the individual means. The same is true for the variance: the variances add and the average variance of the combined measurement is the mean of the individual variances.

$$\overline{s^2} = \frac{1}{N} \sum_{n=1}^{N} s_n^2 \qquad (2.16)$$

When a number of signals are added or averaged together, it can be shown that the standard deviations of the noise components are reduced by a factor equal to $1/N$, where N is the number of measurements that are averaged.

$$\overline{s_{Noise}} = \frac{s}{\sqrt{N}} \qquad (2.17)$$

In other words, averaging multiple measurements from the same source, will reduce the standard deviation of the measurement's variability or noise by the square root of the number of averages. For this reason, it is common to make multiple measurements whenever possible and average the results. The ability of averaging to reduce noise is demonstrated in the next example of a design simulation.

EXAMPLE 2.4

Use of averaging to reduce noise or measurement variability. You are working for a company that produces a device that measures body temperature using infrared reflection. The basic measurement is known to produce Gaussianly distributed errors with an expected standard deviation of 2.0 degree. In other words, the measurement has Gaussian added noise with a standard deviation of 2.0. You want to demonstrate, using a simulation approach, that by averaging measurements you can effectively reduce the measurement error.

Solution: To accomplish this demonstration, use MATLAB to simulate the measurement with added Gaussian noise having a standard deviation of 2.0 degree. Since the `randn` routine produces a Gaussian distribution with a standard deviation of 1.0, simulate the measurement using:

```
actual_temperature = 98.6;                    % Actual body temperature
x_measured = actual_temperature + 2*randn;    % Simulated measurement
```

To simulate averaging you could use a `for` loop to generate multiple measurements and average these together. Assume you would like to assess averages of 4, 8, 16, and 32.

```
actual_temperature = 98.6   % Actual temperature
N_avg = [4 8 16 32];        % Number of averages
for k1 = 1:4                % Repeat for each average
```

```
for k = 1:N_avg(k1)
   x_measured(k) = actual_temperature + 2*randn; % Individual observations
 end
 x_avg = mean(x);        % Measurement average
 disp(['Nu_avg: ',num2str(N_avg(k1)),' Avg: ',num2str(x_avg)])  % Display results
end
```

A typical run might output:

```
Nu_avg: 4 Avg: 98.3401
Nu_avg: 8 Avg: 98.575
Nu_avg: 16 Avg: 98.4914
Nu_avg: 32 Avg: 98.6797
```

Since the measurement is random, the averages are random, although we see that the result of 32 averages is closer to the actual temperature than the average of only four individual measurements. However, you want to assess the noise associated with each measurement, that is, the standard deviation of each average. Since this is a simulation, you can just repeat the average measurements a large number of times and calculate the standard deviation of all those averages. This means adding an internal loop leading to the solution:

```
actual_temperature = 98.6;        % Actual value of measurement.
N_avg = [4,8,16,32];              % Number of averages
for k1 = 1:4
  for k2 = 1:100
    for k = 1:N_avg(k1)
      x_measurement(k) = temperature + 2*randn;     % Individual observations
    end
    x_avg(k2) = mean(x_measurement);                % Measurement average
  end
  sd = std(x_avg);        % Average measurement standard deviation
  disp(['Nu_avg: ',num2str(N_avg(k1)),' Std: ',num2str(sd)])    % Display result
end
```

Result: This simulation produces:

```
Nu_avg: 4  Std: 1.1685 deg
Nu_avg: 8  Std: 0.74813 deg
Nu_avg: 16  Std: 0.50366 deg
Nu_avg: 32  Std: 0.33155 deg
```

Since the original noise (i.e., measurement error) had a standard deviation of 2.0 degree, if this noise were reduced by the $\sqrt{N}$, then the resultant noise should be: $2/\sqrt{4} = 1.0$, $2/\sqrt{8} = 0.707$, $2/\sqrt{16} = 0.5$, and $2/\sqrt{32} = 0.354$. Although there is some randomness in the result, this simulation clearly demonstrates that averaging reduces the noise in the final measurement and the reduction is very close to that predicted by Equation 2.17. Since you thought of the idea of using averaging to reduce the noise and then demonstrated to management that it actually works, you can expect a promotion in the near future.

2.2.4 Ensemble Averaging

The use of averaging to reduce noise can be applied to entire signals, a technique known as "ensemble averaging." Ensemble averaging is a simple yet powerful signal processing technique for reducing noise, but you need to be able to make multiple observations on essentially the same system. These multiple observations could come from multiple sensors, but in many biomedical applications they come from repeated responses to the same stimulus. This only works if each signal can be taken as an exact repetition (not counting the added noise) of the other signals. Many biological responses change with repeated stimuli, even if the stimuli are identical (i.e., many biosystems are nonstationary), but a few systems produce highly repetitive responses to even a large number of repeating stimuli.

To construct an ensemble average you must have a time reference, some point that represents a corresponding time in all ensemble signals. For example, the onset time of a repeating stimulus could be used as a timing reference. Figure 2.5 shows three discrete signals

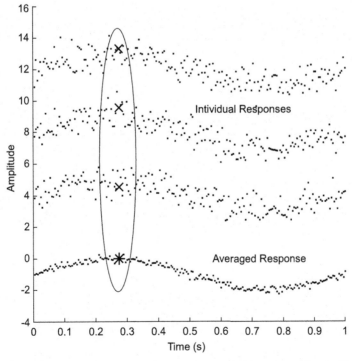

FIGURE 2.5 Three individual signals are used to construct an ensemble average. The signals are shown already aligned to the same time frame. To construct the average response, corresponding points (those that have the same timing with respect to $t = 0$) in the individual responses (x-points) are averaged to make a single point in the averaged response (*-point). This process is repeated for all samples in the individual responses. For these signals, $N = 200$.

consisting of sinusoids and Gaussian noise. (These signals are plotted using separate points to emphasize that they are discrete.) The signals have been aligned so they all share a common time frame. When the individual signals are aligned in time, an ensemble average signal is constructed by averaging over corresponding points from the individual signals. In Figure 2.5, three corresponding points (x-points) are averaged to construct one point in the averaged response (*-point). To construct the entire averaged response, this averaging is repeated for each sample in the time frame. (In the case of Figure 2.5, there are 200 samples in the time frame so the averaging process must be repeated 200 times.) The implementation of ensemble averaging in MATLAB is very easy as illustrated by the example.

A classic biomedical engineering example of the application of ensemble averaging is the visual evoked response (VER) in which a visual stimulus produces a small neural signal embedded in the EEG. Usually this signal cannot be detected in the EEG signal, but by averaging hundreds of EEG signals, time referenced to a visual stimulus, you can determine the visually evoked signal.

To produce an ensemble average, you need multiple response signals and some sort of timing reference that is closely linked to these responses. The timing reference shows how to align the individual responses before averaging. This approach is illustrated in the following example.

EXAMPLE 2.5

Find the average response of a number of individual VERs. The basic signal is an EEG signal recorded near the visual cortex. The stimulus was a repetitive light flash. Individual recordings have too much noise[4] to see the neural response produced by the stimulus. The MATLAB file ver.mat contains 100 EEG signals in a matrix variable ver. These signals were recorded immediately after the flash and digitized using a sample interval of 0.005 s. (i.e, 5.0 ms). Construct and plot the ensemble average and also plot one of the individual responses to show what the actual EEG signal looked like.

Solution: Use the MATLAB averaging routine mean. If this routine is given a matrix variable, it averages each column. If the various signals are arranged as rows in the matrix, the mean routine will produce the ensemble average (assuming that they are properly aligned). To determine the orientation of the data, we check the size of the data matrix. Normally the number of signal data points is greater than the number of signals.[5] Since we want the signals to be in rows, if the number of rows is greater than the number of columns, we transpose the data using the MATLAB transposition operator (i.e., the ' symbol).

```
% Example 2.5  Example of ensemble averaging
load ver;                           % Get visual evoked response data;
fs = 1/.005;                        % Sample frequency from sample interval
[nu,N] = size(ver);                 % Get data matrix size
if nu > N
        ver = ver';                 % Transpose if necessary
end
t = (1:length(ver))/fs;             % Generate time vector
%
```

```
subplot(1,2,1);
plot(t,ver(3,:));              % Plot   individual   record   (select   record   3
arbitrarily)
        ......label axes........
% Construct and plot the ensemble average
avg = mean(ver);               % Calculate ensemble average
subplot(1,2,2)
plot(t,avg);                   % Plot ensemble average
        ......label axes........
```

Results: Calculation of the ensemble average could not be easier: only one MATLAB command. But in this example, the hard work, aligning the signals, had already been done. In a practical situation, you might trigger the data acquisition process using the same signal that triggers the light stimulus.

This program produces the results shown in Figure 2.6. Figure 2.6A shows a single response, and the evoked response is not even remotely discernable. In Figure 2.6B the VER from the ensemble average of 100 individual responses is clearly visible.

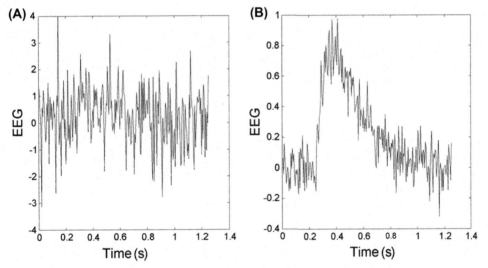

FIGURE 2.6   (A) The raw EEG signal showing a single response to an evoked stimulus. (B) The ensemble average of 100 individual signals showing the VER.

[4]Actually this noise is not really noise but other brain activity, but if it is not of interest, it is noise.
[5]This trick is not fool-proof, but if you guess wrong on the orientation of the signal, it will be obvious in the graphs. You can then try the other orientation.

You might ask if ensemble averaging reduces noise as much as single variable averaging illustrated in Example 2.4. In other words, does the reduction in standard deviation of the averaged signal follow the prediction of Equation 2.17: a reduction proportional to the square root of the number of signals averaged? If so, we would expect that the noise in Figure 2.6B is 10 times less than the raw EEG data in Figure 2.6A. A problem at the end of this chapter explores this question.

2.3 THE BASIC WAVEFORM: SINUSOIDS

In the book *The Martian*, Mark Watney tells us that duct tape does indeed work in a vacuum and that "duct tape is like magic and should be worshiped." As we discover in the next chapter, the unprepossessing sinusoid approaches the same mythic importance in signal analysis. Now we take a closer look at some of the definitions and mathematical operations associated with this important waveform.

2.3.1 Sinusoids as Real-Valued Signals

A sinusoidal signal is any signal that follows a basic sine wave pattern at a single frequency. It can be either real valued or complex (i.e., using complex number representation). Complex sinusoidal signals are described in the next section. Of course all real-world signals are real valued. Describing a sinusoidal signal in the complex domain is mathematically a little more complicated, but surprisingly, complex representation makes sinusoidal math easier.

The most basic representation of a sinusoidal signal was given in Eq. 1.14 in Chapter 1 and is repeated here:

$$x(t) = A \sin(\omega_p t) = A \sin(2\pi f_p t) = A \sin\left(\frac{2\pi t}{T}\right) \tag{2.18}$$

where A is the signal amplitude. Three different methods for describing the sine wave's frequency are shown: ω_p is the frequency in radians per second; f_p is the frequency in hertz; and T is the period in seconds. Frequency in hertz is most common in engineering, but we should be comfortable with any of the representations in Equation 2.18.

To move to a more general sinusoidal representation, we note that sine wave–like signals can also be represented by cosines, and the two are related.

$$A \cos(\omega t) = A \sin\left(\omega t + \frac{\pi}{2}\right) = A \sin(\omega t + 90°)$$

$$A \sin(\omega t) = A \cos\left(\omega t - \frac{\pi}{2}\right) = A \cos(\omega t - 90°) \tag{2.19}$$

The right-hand representations (i.e., $A \sin(\omega t + 90°)$ and $A \cos(\omega t - 90°)$) have confused units. The first part of the sine argument, ωt, is in radians, whereas the second part is in degrees. Nonetheless, this is fairly common usage and we use it a lot; however, in MATLAB the trig function arguments must be in radians. To convert between degrees and radians. note that 2π radians $= 360$ degrees, so 1 radian $= 360/2\pi$ degrees, and 1 degree $= 2\pi/360$ radians.

A general real-valued sinusoidal signal is just a sine or cosine with a nonzero phase term:

$$x(t) = A\sin(2\pi ft + \theta) = A\sin(\omega t + \theta) = A\sin\left(\frac{2\pi t}{T} + \theta\right) \text{ or}$$

$$x(t) = A\cos(2\pi ft + \theta) = A\cos(\omega t + \theta) = A\cos\left(\frac{2\pi t}{T} + \theta\right)$$

(2.20)

where again the phase θ is usually expressed in degrees, even though the frequency descriptor (ω, or $2\pi ft$, or less common $2\pi t/T$) is expressed in hertz or radians. Many of the sinusoidal signals we use are versions of Equation 2.20. Note that to completely describe a sinusoid, we need only three numbers: amplitude A, phase θ, and frequency f (or ω). Example 2.6 uses MATLAB to generate two sinusoids that differ by 60 degree.

EXAMPLE 2.6

Use MATLAB to generate and plot two sinusoids that are 60 degrees out of phase. Calculate, analytically, the difference in time between the two sinusoids assuming they have a frequency of 2.0 Hz.

Solution, MATLAB: Since no details are given except the sinusoidal frequency, this example is mainly an exercise in specifying signal parameters. For sampling frequency and data length we arbitrarily select an f_s of 1.0 kHz and N of 500. The 500 data points are certainly enough to generate a smooth plot and with an f_s of 1.0 kHz we get a signal length of 0.5 s. This will produce 1 cycle of our 2 Hz sinusoids, which should make for a nice plot.

We can generate one sinusoid using the techniques in Examples 1.6 and 2.1 with zero phase angle (i.e., $\theta = 0$ in Equation 2.20). The second sinusoid is similarly generated, except that we make the phase 60 degree. (i.e., $\theta = 60$ in Equation 2.20). We do need to convert the phase from deg to rad for MATLAB. Since 1 degree $= 2\pi/360$ radians, we simply let MATLAB multiply our 60 degree by 2*pi/360.

```
% Example 2.6 Plot two sinusoids 60 deg out of phase
%
fs = 1000;                  % Assumed sampling frequency
N = 500;                    % Number of points
t = (0:N-1)/fs;             % Generate time vector
f = 2;                      % Sinusoid frequency
phase = 60*(2*pi/360);      % 60 deg phase; convert to radians
x1 = sin(2*pi*f*t);         % Construct sinusoids
x2 = sin(2*pi*f*t + phase); % One with a 60 phase shift
hold on;
plot(t,x1);                 % Plot the unshifted sinusoid
plot(t,x2,'--');            % Plot using a dashed line
plot([0 .5], [0 0]);        % Plot horizontal line
    .......label axes.......
```

Results, MATLAB: The graph produced by this simple example program is shown in Figure 2.7.

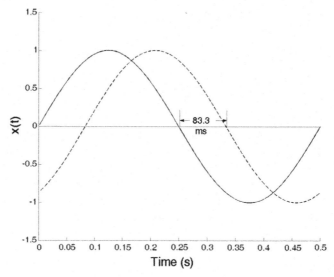

FIGURE 2.7 Two sinusoids with a phase difference of 60 degree. For 2 Hz sinusoids that translates to a time difference of 83 ms.

Solution, Analytical: To convert the difference in phase angle to a difference in time, note that the phase angle varies through 360 degree during the course of one period, and a period is T sec long. Hence to calculate the time difference between the two sinusoids given the phase angle θ:

$$t_d = \frac{\theta}{360} T = \frac{\theta}{360f} \tag{2.21}$$

For 2 Hz sinusoids, $T = 0.5$ s, so substituting T into Equation 2.21:

$$t_d = \frac{\theta}{360} T = \frac{60}{360} 0.5 = 0.0833 \text{ sec.} = 83.3 \ m \text{ sec.}$$

EXAMPLE 2.7

Find the time difference between two sinusoids:

$$x_1(t) = \cos(4t + 30), \quad \text{and} \quad x_2(t) = 2\sin(4t)$$

Solution: This has the added complication that one waveform is defined in terms of the cosine function and the other in terms of the sine function. So we need to convert them both to either a sine or cosine. Here we convert to cosines:

$$x_2(t) = 2\cos(4t - 90)$$

So the total angle between the two sinusoids is $30 - (-90) = 120$ degree and since the period is:

$$T = \frac{1}{f} = \frac{2\pi}{\omega} = \frac{2\pi}{4} = 1.57 \text{ sec.}$$

and the time delay is:

$$t_d = \frac{\theta}{360}T = \frac{120}{360}1.57 = 0.523 \text{ sec.}$$

Equation 2.20 describes a fairly intuitive way of thinking about a sinusoid as a sine wave with a phase shift. Alternatively, a cosine could just as well be used instead of the sine as also shown in Equation 2.20. Here we will use both.

Sometimes it is mathematically convenient to represent a sinusoid as a combination of a pure sine and a pure cosine rather than a single sinusoid with a phase angle. This representation can be achieved using the well-known trigonometric identity for the difference of two arguments of a cosine function:

$$\cos(x - y) = \cos(x)\cos(y) + \sin(x)\sin(y) \tag{2.22}$$

Based on this identity, the equation for a sinusoid can be written as:

$$C\cos(2\pi f t - \theta) = C\cos(\theta)\cos(2\pi f t) + C\sin(\theta)\sin(2\pi f t)$$
$$= a\cos(2\pi f t) + b\sin(2\pi f t) \tag{2.23}$$

where:

$$a = C\cos(\theta) \quad \text{and} \quad b = C\sin(\theta) \tag{2.24}$$

Note that θ is defined as negative in these equations. To convert the other way, from a sine and cosine to a single sinusoid with amplitude C and angle θ, start with Equation 2.24. $a = C\cos(\theta)$ and $b = C\sin(\theta)$: to determine C, square both equations and sum:

$$a^2 + b^2 = C^2\left(\cos^2\theta + \sin^2\theta\right) = C^2$$
$$C = \sqrt{a^2 + b^2} \tag{2.25}$$

The calculation for θ given a and b is:

$$\frac{b}{a} = \frac{C\sin(\theta)}{C\cos(\theta)} = \tan(\theta) \quad \text{and...}$$
$$\theta = \tan^{-1}\left(\frac{b}{a}\right) \tag{2.26}$$

Again, Equation 2.26 finds $-\theta$ as defined in Equation 2.23.

Care must be taken in evaluating Equation 2.26 to ensure that θ is determined to be in the correct quadrant based on the signs of a and b. Many calculators and MATLAB's `atan` function will evaluate θ as in the first quadrant (0–90 degree) if the ratio of b/a is positive, and in the fourth quadrant (270–360 degree) if the ratio is negative. If both a and b are positive, then

θ is between 0 and 90 degree, but if b is positive and a is negative, then θ is in the second quadrant between 90 and 180 degree, and if both a and b are negative, θ must be in the third quadrant between 180 and 270 degree. Finally, if b is negative and a is positive, then θ must belong in the fourth quadrant between 270 and 360 degree. You can avoid all this worry by using MATLAB's atan2(b,a) routine, which correctly outputs the arctangent of the ratio of b over a (in radians) in the appropriate quadrant.

To add sine waves simply add their amplitudes. The same applies to cosine waves:

$$a_1 \cos(\omega t) + a_2 \cos(\omega t) = (a_1 + a_2)\cos(\omega t)$$
$$a_1 \sin(\omega t) + a_2 \sin(\omega t) = (a_1 + a_2)\sin(\omega t)$$

(2.27)

To add two sinusoids (i.e., $C \sin(\omega t + \theta)$ or $C \cos(\omega t - \theta)$), convert them to sines and cosines using Equation 2.23, add sines to sines and cosines to cosines (Equation 2.27), and convert back to a single sinusoid if desired.

EXAMPLE 2.8

Convert the sum of a sine and cosine wave, $x(t) = -5 \cos(10t) - 3 \sin(10t)$ into a single sinusoid.
Solution: Apply Equations 2.25 and 2.26

$$a = -5 \quad \text{and} \quad b = -3$$

$$C = \sqrt{a^2 + b^2} = \sqrt{(-5)^2 + (-3)^2} = 5.83$$

$$\theta = \tan^{-1}\left(\frac{b}{a}\right) = \tan^{-1}\left(\frac{-3}{-5}\right) = 31 \text{ deg},$$

But θ must be in the third quadrant since both a and b are negative:

$$\theta = 31 + 180 = 211 \text{ deg}.$$

Therefore, the single sinusoid representation would be:

$$x(t) = C \cos(\omega t - \theta) = 5.83 \cos(10t - 211°)$$

Using the above-mentioned equations, any number of sines, cosines, or sinusoids can be combined into a single sinusoid provided they are all at the same frequency. This is demonstrated in Example 2.9.

EXAMPLE 2.9

Combine $x(t) = 4 \cos(2t + 30°) + 3 \sin(2t + 60°)$ into a single sinusoid.
Solution: First expand each sinusoid into a sum of cosine and sine, then algebraically add the cosines and sines, then recombine them into a single sinusoid. Be sure to convert the sine into a cosine (recall Equation 2.19: $\sin(\omega t) = \cos(\omega t - 90°)$) before expanding this term.

$4 \cos(2t + 30) = a \cos(2t) + b \sin(2t)$
where : $a = C \cos(\theta) = 4 \cos(-30) = 3.5$ and $b = C \sin(\theta) = 4 \sin(-30) = -2$
$4 \cos(2t + 30) = 3.5 \cos(2t) - 2 \sin(2t)$

(Note that θ is actually negative in the above equations.)

Convert the sine to a cosine then decompose the sine into a cosine plus a sine:

$$3\sin(2t + 60) = 3\cos(2t - 30) = 2.6\cos(2t) + 1.5\sin(2t)$$

Combine cosine and sine terms algebraically:

$$4\cos(2t + 30) + 3\sin(2t + 60) = (3.5 + 2.6)\cos(2t) + (-2 + 1.5)\sin(2t)$$

$$= 6.1\cos(2t) - 0.5\sin(2t)$$

$$= C\cos(2t + \theta) \quad \text{where:} \quad C = \sqrt{6.1^2 + (-0.5)^2} \quad \text{and} \quad \theta = \tan^{-1}\left(\frac{-.5}{6.1}\right)$$

$$C = 6.1; \quad \theta = 4.7 \ \left(\text{Since } b \text{ is negative, } \theta \text{ is in 4th quadrant}\right) \text{ so } \theta = -4.7 \text{ deg}$$

$$x(t) = 6.1\cos(2t - 4.7°)$$

This approach can be extended to any number of sinusoids as shown in the problems.

2.3.2 Complex Representation of Sinusoids

Why use complex numbers? If you are willing to deal with complex numbers and variables (and you are), sinusoidal representation becomes a lot simpler. The beauty of complex numbers and complex variables is that the real and imaginary parts are orthogonal. Orthogonality is discussed later, but the importance for complex numbers is that the real and imaginary parts do not interact; they are independent as we explain in the next paragraph.

Recall that a complex number combines a real number and an imaginary number. Real numbers are what we commonly use, whereas imaginary numbers are the result of square roots of negative numbers and are represented as real numbers multiplied by $\sqrt{-1}$. In mathematics, $\sqrt{-1}$ is represented by the letter i, but engineers use the letter j, reserving the letter i for current. So, although 5 is a real number, $j5$ (i.e., $5\sqrt{-1}$) is an imaginary number. A complex number is just a combination of a real and imaginary number such as $5 + j5$. A complex variable simply combines a real and an imaginary variable: $z = x + jy$. The arithmetic of complex numbers and some of their important properties are reviewed in Appendix E. We use complex variables and complex arithmetic extensively in later chapters, so it is worthwhile to review these operations.

Complex numbers are illustrated graphically as inhabiting orthogonal horizontal and vertical axes as shown in Figure 2.8. Clearly you can change one component of a complex number without changing the other. This means that a complex number is really two separate numbers rolled into one and a complex variable is two separate variables.

Getting two numbers for the price of one would not be such a big deal except that for a given frequency, f, we only need two numbers to completely describe a sinusoid: A and θ in Equation 2.20 or a and b in Equation 2.23. This suggests that it might be possible to describe a sinusoid with only a single complex number. Thanks to the Swiss mathematician

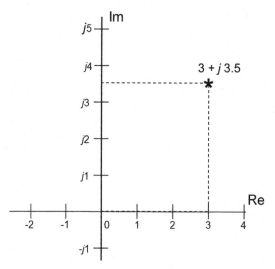

FIGURE 2.8 A graphical representation of real and imaginary numbers that illustrates their orthogonality, or independence. A change in the value of an imaginary number does not affect the real number and vice versa.

Leonhard Euler (pronounced "oiler"),[6] we know how do to just that. Euler's identity relates a complex exponential to a real cosine and imaginary sine.[7]

$$e^{\pm jx} = \cos x \pm j \sin x \qquad (2.28)$$

Note that this equation looks like the representation for a sinusoid given by Equation 2.23 except that the sine term is imaginary. Although this difference necessitates some extra mathematical features, it is well worth the result: the ability to represent a sinusoid by a single number or variable.

MATLAB variables can be either real or complex and MATLAB deals with them appropriately. A typical MATLAB number is identified by its real and imaginary part: x = 2 + 3i or x = 2 + 3j. MATLAB assumes that both i and j stand for the complex operator unless they are defined to mean something else in your program. Another way to define a complex number is using the complex routine: z = complex(x,y) where x and y are two arrays. The first argument, x, is taken as the real part of z and the second, y, is taken as the imaginary part of z. To find the real and imaginary components of a complex number or array, use real(z) to get the real component and imag(z) to get the imaginary component. To find the polar components of a complex variable use abs(z) to get the magnitude component and angle(z) to get the angle component (in radians).

[6]The use of the symbol *e* for the base of the natural logarithmic system is a tribute to Euler's extraordinary mathematical contributions.

[7]The derivation for this equation is given in Appendix A.

EXAMPLE 2.10

Demonstration of the Euler equation in MATLAB. Generate one cycle of the complex sinusoid given in Equation 2.28 using a 1.0-s time vector. Assume $f_s = 500$ Hz. Plot the real and imaginary parts of this function superimposed to produce a cosine and sine wave.

Solution: Construct the time vector given the sample frequency and total desired time. Although you are not given the number of points the time vector is easily determined by constructing a vector between 0 and 1.0 with steps of $1/f_s$: t = (0:1/fs:1.0); . Then generate e^{-jx} as in Equation. 2.28 where $x = 2\pi t$ so that the complex sinusoid is one cycle (x will range from 0 to 2π as t goes from 0 to 1.0). Plot the real and imaginary part using MATLAB's real and imag routines.

```
% Example 2.10 Demonstration of the Euler equation in MATLAB
%
fs = 500;                        % Sampling frequency
t = (0:1/fs:1);                  % Time vector
z = exp(-j*2*pi*t);              % Complex sinusoid (Equation 2.28)
plot(t,real(z),'k',t,imag(z),':k');    % Plot result
    .......labels........
```

Result: The sine and cosine wave produced by this program are seen in Figure 2.9.

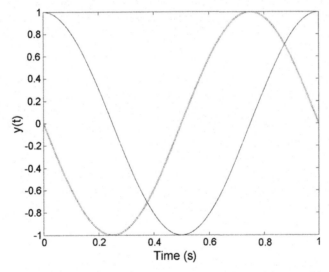

FIGURE 2.9 A sine wave and cosine wave produced from a single complex variable using Euler's identity (Equation 2.28).

The convenience of representing a sinusoid by a single complex exponential will find important applications in two sinusoidally based analysis techniques: Fourier analysis described in the next chapter and phasor analysis described in Chapter 6.

2.4 TIME DOMAIN ANALYSIS

The basic measurements of mean, variance, standard deviation, and rms generally do not capture the important features of a signal. For example, if we have digital versions of the EEG signals shown in Figure 2.2, we can easily compute their mean, variance, standard deviation, and rms value, but these would not tell us much about the signals or the neural processes that created them. More insight might be gained by comparing these signals with some reference signal(s).

Comparing a signal with a reference signal, or perhaps a group or "family" of reference signals, is an oft used tool in signal analysis. Such reference signals, or signal families, tend to be much less complicated than the signal such as a sine wave or family of sine waves.[8] A quantitative comparison can tell you how much your complicated signal is like a simpler, easier to understand reference signal or family. If enough comparisons are made with a well-chosen reference family, these comparisons can actually provide an alternative representation of the signal. Sometimes this new representation of the signal is more informative or enlightening than the original.

2.4.1 Comparison Through Correlation

We are constantly making subjective comparisons in shopping, job hunting, dating, and many other aspects of daily life, but in signal analysis we need objective, quantitative comparisons. We need to get a number that describes how well one signal compares with another. One of the most common ways of quantitatively comparing two functions is through mathematical correlation. Applied to signals, correlation seeks to quantify how much one signal is like another. This sounds like just what we need, and mathematical correlation does a pretty good job of describing similarity, but once in a while it breaks down. For example, a sine wave and cosine wave are pretty much alike in that they both exhibit similar oscillatory behavior, yet in Figure 2.10 we show their mathematical correlation is zero.

Correlation between different signals is illustrated in Figure 2.10, which shows various pairs of waveforms and the correlation between them. The lack of correlation between two sinusoids shows that correlation does not always measure general similarity. Mathematically they are as un-alike as possible, even though they have similar behavioral patterns.

The linear correlation between two digital functions or signals can be obtained using the Pearson correlation coefficient defined as:

$$r_{xy} = \frac{1}{(N-1)s_x s_y} \sum_{n=1}^{N} (x_n - \bar{x})(y_n - \bar{y}) \tag{2.29}$$

where r_{xy} is a common symbol in signal analysis for the correlation between x and y. (For the Pearson correlation coefficient, the symbol ρ_{xy} is also used.) The variables x and y could represent any two waveforms. Again $\bar{x}$ and $\bar{y}$ are the means of x and y (Equations 2.3 and 2.4),

---

[8]Sine waves are pretty simple. We know they can be completely described by only three numbers. In the next chapter, we show that sine waves are simple in other respects; they contain energy at only one frequency.

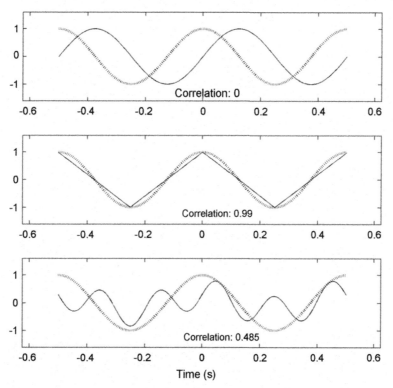

FIGURE 2.10 Three pairs of signals and the correlation between them as given by the Pearson correlation coefficient defined in Equation 2.29. The high correlation between the sine wave and triangular wave (center) correctly expresses the similarity between them, but the zero correlation between the sinusoids in the upper plot does not reflect their general similarity.

and s_x and s_y are the standard deviations of x and y (Equations 2.14 and 2.15). This equation scales the correlation coefficient to be between ± 1, but if we are not concerned with the scale, then it is not necessary to normalize by the standard deviations and $N - 1$. If, in addition, the means of the signal and reference are zero, correlations can be determined using a simpler equation:

$$r_{xy} = \frac{1}{N} \sum_{n=1}^{N} x[n]y[n]$$

$$r_{xy} = \frac{1}{N} \left(x[1]y[1] + x[2]y[2] + x[3]y[3] \ldots + x[N]y[N] \right) \tag{2.30}$$

where r_{xy} is the unscaled correlation between x and y. Since this is a relative correlation, the summation can be used directly without scaling by $1/N$. This is the basic normalized correlation. If continuous functions are involved, the summation becomes an integral and the discrete functions, $x[n]$ and $y[n]$, become continuous functions, $x(t)$ and $y(t)$:

$$r_{xy} = \frac{1}{T} \int_0^T x(t)y(t)dt \qquad (2.31)$$

Both Equations 2.30 and 2.31 use integration (or summation) and scaling (dividing by T or N) to get the average of the product between signals over some range. The correlation values produced by Equations 2.30 and 2.31 will not range between ± 1, but they do give relative values that are proportional to the linear correlation of x and y. The correlation of r_{xy} will have the largest possible positive value when the two functions are identical and the largest negative value when the two functions are exact opposites of one another (i.e., one function is the negative of the other). The average product produced in these equations is zero when the two functions are mathematically completely dissimilar.

The only difference between the Pearson correlation coefficient given in Equation 2.29 and the correlation given in Equation 2.30 is the normalization that makes the correlation values range between ± 1.0. To convert an unnormalized correlation to a Pearson normalization:

$$r_{xy\ Pearson} = \frac{r_{xy\ un-norm}}{(N-1)s_x s_y} = \frac{r_{xy-un-norm}}{(N-1)\sqrt{s_x^2 s_y^2}} \qquad (2.32)$$

where the standard deviations, s, are defined in Equations 2.14 and 2.15. Note that we can also use the square root of the variances s^2, as defined in Equations 2.12 and 2.13. The term "correlation coefficient" implies this Pearson normalization, whereas the term "correlation" is used more loosely and could mean normalized or unnormalized correlation.

Some deeper concepts are associated with correlation and vectors. Recall that in Chapter 1, it is mentioned that a signal sequence made up of n numbers could be thought of as a vector in n-dimensional space. Running with this concept, correlation becomes the projection of one signal vector upon the other: it is the mathematical equation for projecting an n-dimension x vector upon an n-dimension y vector. When two functions are uncorrelated, they are also said to be orthogonal and their n-dimensional vectors are perpendicular. Such vectors have a projection of zero on one another. In fact, a good way to test if two functions are orthogonal is to evaluate their correlation. When this concept of vector projection is used to describe correlation, it is common to refer to a family of reference signals or functions as a *basis* or *basis functions*. Hence "projecting a signal on a basis" is just a cool way of saying "correlating a signal with a reference family." Moreover, even if you like thinking of signals as n-dimensional vectors, you still use Equations 2.29 or 2.30 to compute their correlation.

Covariance computes the variance that is shared between two (or more) signals. Covariance is usually defined in discrete notation as:

$$Cov = \frac{1}{N-1} \sum_{n=1}^N (x_n - \bar{x})(y_n - \bar{y}) \qquad (2.33)$$

The equation for covariance is similar to the discrete form of the Pearson correlation except that it is not normalized by $\frac{1}{s_x s_y}$.

As seen from the above-mentioned equations, correlation operations have both continuous and discrete versions so that both analytical and computer solutions are possible. The next two examples solve for the correlation between two waveforms using both the continuous, analytical approach and a computer algorithm.

EXAMPLE 2.11

Use Equation 2.31 (continuous form) to find the correlation (unnormalized) between the sine wave and the square wave shown in Figure 2.11A. Also find the correlation between a cosine wave and a square wave shown in Figure 2.11B. All waveforms have amplitudes of 1.0 V (peak-to-peak) and periods of 4.0 s.

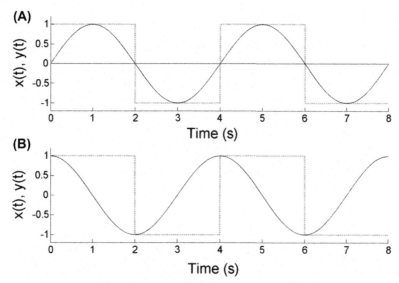

FIGURE 2.11 (A) Sine wave and square wave. (B) Cosine wave and square wave. The unbiased correlation between these waveforms is found in Example 2.11.

Solution. Sine and square wave. Apply Equation. 2.31, but use symmetry to make it easier. Since the waveforms are periodic, we only need evaluate over one period. Moreover, the correlation in the second half of the 1-s period equals the correlation in the first half, so it is only necessary to calculate the correlation period in the first half.

$$Corr = \frac{1}{T}\int_0^T x(t)y(t)dt = \frac{2}{T}\int_0^{T/2}(1)\sin\left(\frac{2\pi t}{T}\right)dt = \frac{2}{T}\frac{T}{2\pi}\left(-\cos\left(\frac{2\pi t}{T}\right)\right)\Big|_0^{T/2}$$

$$Corr = \frac{1}{\pi}(-\cos(\pi) - -\cos(0)) = \frac{2}{\pi}$$

Solution. Cosine and square wave. A good look at two waveforms shows that the correlation in the first half is equal and opposite to that in the second half, so we can guess the total correlation is zero. But to have more fun with basic calculus, we work it all out, calculating the correlation over both half cycles.

$$Corr = \frac{1}{T}\int_0^T x(t)y(t)dt = \frac{1}{4}\left(\int_0^2 (1)\cos\left(\frac{2\pi t}{4}\right)dt + \int_2^4 (-1)\cos\left(\frac{2\pi t}{4}\right)dt\right)$$

$$Corr = \frac{1}{4}\left(\frac{4}{2\pi}\left(\sin\left(\frac{2\pi t}{4}\right)\right)\Big|_0^2 - \frac{4}{2\pi}\left(\sin\left(\frac{2\pi t}{4}\right)\right)\Big|_2^4\right)$$

$$Corr = \frac{1}{2\pi}(\sin(\pi) - \sin(0) - \sin(2\pi) + \sin(\pi)) = \frac{1}{2\pi}(0 - 0 - 0 + 0) = 0$$

Analysis: We find correlation between a sine and square wave but none between a cosine and square wave. This is another example of how correlation does not always represent similarity. We find out how to get around this problem using a technique called "cross-correlation" (see Section 2.4.4).

The next example uses MATLAB to confirm the analytical results found in Example 2.11.

EXAMPLE 2.12

Use Equation 2.31 and MATLAB to confirm the result given in Example 2.11.

Solution: Construct a time vector 500 points long ranging from 0.0 to 4 s (construct only one period of the waveforms.). Use the time vector to construct a 0.25 Hz sine wave and a cosine wave. Then evaluate the correlation between the two by direct application of Equation 2.31.

```
% Example 2.12 Confirm the results of Example 2.11.
N = 500;                          % Number of points
Tt = 4.0                          % Desired total time
f = 0.25;                         % Wave frequency in Hz
fs = N/Tt;                        % Calculate sampling frequency
t = (0:N-1)/fs;                   % Time vector from 0 (approx.) to 4 sec
x = sin(2*pi*f*t);                % 0.25 Hz sine wave
y = cos(2*pi*f*t);                % 0.25 Hz cosine wave
z = [ones(1,N/2) -ones(1,N/2)];   % 0.25 Hz square wave
rxz = mean(x.*z);                 % Correlation (Equation 2.30) x and z
rxy = mean(x.*y);                 % Correlation x and y
disp([rxz rxy])                   % Output correlations
```

Analysis: In Example 1.3 we learned how to construct a time vector given the number of points and sampling frequency (N and f_s) or the total time and sampling frequency (T_t and f_s). Here we are presented with a third possibility, the number of points and total time (N and T_t). To calculate this time vector, the appropriate sampling frequency was determined from the total points and total time: $f_s = N/T_t$. Note that Equation 2.30 takes a single line of MATLAB code.

Result: The MATLAB program produces:

```
rxz = 0.63659    rxy = -3.1627e-017
```

The correlation between the sine and square wave, rxz, is very close to that found analytically: $2/\pi = 0.6366$. The correlation of the square and cosine wave, rxy is close to zero. This is an example where a hand calculation is more accurate than the computer since we know the true correlation should be 0.0. Computer round-off errors result in a small, meaningless, negative correlation.

2.4.2 Orthogonal Signals and Orthogonality

Orthogonal signals and functions can be very useful signal processing tools. In common usage, "orthogonal" means perpendicular: if two lines are orthogonal, they are perpendicular. In the graphical representation of complex numbers shown in Figure 2.8, the real and imaginary components are perpendicular to one another; hence, they are orthogonal. But what makes two signals orthogonal? The formal definition for orthogonal signals is that their inner product (also called the dot product) is zero:

$$\int_{-\infty}^{\infty} x(t)y(t)dt = 0 \tag{2.34}$$

For discrete signals, this becomes:

$$\sum_{n=1}^{N} x[n]y[n] = 0 \tag{2.35}$$

These equations[9] are the same as the correlation equations (Equations 2.30 and 2.31); the missing normalization by $1/N$ could be multiplied out since one side is zero. So signals that are orthogonal are uncorrelated and vice versa, and any correlation method, the Pearson correlation or the unnormalized versions, could test if two signals are orthogonal.

An important characteristic of signals that are orthogonal (i.e., uncorrelated) is that when they are combined or added together they do not interact with one another. This noninterference comes in handy when we represent a complicated signal with a collection of more basic signals. If the basic signals are orthogonal, you can determine each one independently without worrying about the other signals in the collection. Orthogonality simplifies many calculations where multiple signals are involved. Some analyses could not be done, at least not practically, using nonorthogonal signals. Orthogonality is not limited to just two signals. Whole families exist where each signal is orthogonal to all other members in the family. Such families of orthogonal signals are called "orthogonal sets."

2.4.3 Correlations Between Signals

MATLAB has functions for determining the correlation or covariance that are particularly useful when several signals are compared. A matrix of correlations among multiple signals can be calculated using `corrcoef`. Similarly, for covariances use `cov`. The covariance is the correlation normalized by N given by Equation 2.30, whereas the matrix of correlations is the Pearson correlation normalized as in Equation 2.29. For the latter case, correlation values

---

[9]In Chapter 1 and in the discussion of correlation, we mentioned that a digital signal composed of N samples can be considered a single vector in N-dimensional space. In this representation, two orthogonal signals would have vectors that actually are perpendicular. So the concept of orthogonal holds for the vector representation of signals as well.

run between ± 1, with 1.0 indicating identical signals and −1.0 indicating identical signals, but with one inverted with respect to the other. The calls are similar for both functions:

```
Rxx = corrcoef(X);        % Signal correlations
S = cov(X);               % Signal covariances
```

where X is a matrix that contains the various signals to be compared in columns. Some options are available as explained in the associated MATLAB help file. The output, Rxx, of corrcoef is an n-by-n matrix where n is the number of signals (i.e., columns). The diagonals of this matrix represent the Pearson correlation of the signals with themselves and therefore must be 1.0. The off-diagonals represent the Pearson correlation coefficients of the various combinations. For example, r_{12} is the correlation between signals 1 and 2. Since the correlation of signal 1 with signal 2 is the same as that of signal 2 with signal 1, $r_{12} = r_{21}$, and in general $r_{m,n} = r_{n,m}$, so the matrix will be symmetrical about the diagonals:

$$
r_{x,y} = \begin{bmatrix} 1 & r_{1,2} & \cdots & r_{1,N} \\ r_{2,1} & 1 & \cdots & r_{2,N} \\ \vdots & \vdots & \ddots & \vdots \\ r_{N,1} & r_{N,2} & \cdots & 1 \end{bmatrix} \tag{2.36}
$$

where N is the number of signals being compared.

The cov routine produces a similar output, except the diagonals are the variances of the various signals and the off-diagonals are the covariances, the same values given by Equation 2.33.

$$
S = \begin{bmatrix} cov_{1,1} & cov_{1,2} & \cdots & cov_{1,N} \\ cov_{2,1} & cov_{2,1} & \cdots & cov_{2,N} \\ \vdots & \vdots & \ddots & \vdots \\ cov_{N,1} & cov_{n,2} & \cdots & cov_{N,N} \end{bmatrix} \tag{2.37}
$$

Example 2.13 uses covariance and correlation analysis to determine if sines and cosines of different frequencies are orthogonal. Again, two orthogonal signals have zero correlation. Either covariance or correlation could be used to determine if signals are orthogonal. Example 2.13 uses both.

EXAMPLE 2.13

Determine if 1.0 Hz sine and cosine waves are orthogonal and if a 2.0 Hz cosine wave is orthogonal to a 1.0 Hz cosine wave. Of course, we already know the answer, but this example shows how well the cov and corrcoef routines perform. The 1 and 2 Hz cosine waves are called "harmonically related," as they have frequencies that are multiples. Also determine if a 1.0 Hz sawtooth is orthogonal to the sinusoidal waveforms. Make the peak-to-peak amplitude of the four signals ±1.0.

Solution: Generate a 1000-point, 1.0-s time vector using the approach in Example 2.12. Use this time vector to generate a data matrix with four columns representing the 1.0 Hz cosine and sine waves, the 2.0 Hz sine wave, and the 1.0 Hz sawtooth. To generate a sawtooth, use the MATLAB routine sawtooth(2*pi*t), to generate a 1.0 Hz sawtooth wave. Apply the covariance and correlation MATLAB routines (i.e., cov and corrcoef) and display results.

```
% Example 2.13
% Application of the covariance matrices to sinusoids that
% are orthogonal and a sawtooth
%
N = 1000;                   % Number of points
Tt = 1;                     % desired total time
fs = N/Tt;                  % Calculate sampling frequency
t = (0:N-1)/fs;             % Time vector from 0 (approx.) to 1 sec
X(:,1) = cos(2*pi*t)';      % Generate a 1 Hz cosine
X(:,2) = sin(2*pi*t)';      % Generate a 1 Hz sine
X(:,3) = cos(4*pi*t)';      % Generate a 2 Hz cosine
X(:,4) = saw(2*pi*t)';      % Generate a 1 Hz sawtooth
%
S = cov(X)                  % Print covariance matrix
Rxx = corrcoef(X)           % and correlation matrix
```

Analysis: The program defines a time vector in the standard manner. The program then generates the three sinusoids using this time vector in conjunction with sin and cos functions, arranging the signals as columns of X. The program then uses the sawtooth routine to generate a 1.0 Hz sawtooth having the same number of samples as the sinusoids. The four signals are placed in a single matrix variable X. The program then determines the covariance and correlation matrices of X.

Results: The output from this program is a covariance and correlation matrix. The covariance matrix is:

```
S =
    0.5005  -0.0000  -0.0000  -0.0010
   -0.0000   0.5005   0.0000   0.3186
   -0.0000   0.0000   0.5005   0.0000
   -0.0010   0.3186   0.0000   0.3337
```

and

```
Rxx =
    1.0000  -0.0000  -0.0000  -0.0024
   -0.0000   1.0000   0.0000   0.7797
   -0.0000   0.0000   1.0000  -0.0024
   -0.0024   0.7797  -0.0024   1.0000
```

The diagonals of the covariance matrix give the variance of the four signals. These are consistent for the sinusoids and slightly less for the sawtooth. The correlation matrix shows similar results except that the diagonals are now 1.0 since these reflect the correlation of the signals with themselves.

The covariance and correlation between the various signals are given by the off-diagonals. The off-diagonals show that the sinusoids have no correlation with one another, a reflection of their orthogonal relationship. The 1.0 Hz sine wave is correlated with the sawtooth, but there is negligible correlation between the sawtooth and the cosine wave or the 2.0 Hz cosine wave.

2.4.4 Shifted Correlations: Cross-correlation

Many of the real-world signals we are called upon to analyze are quite complex. Check out the EEG signal shown in Figure 2.2 (left side). One way to deal with these complex signals is to see how much they are like less complicated signals by using our new-found correlation tools and comparing the EEG signal with some less complicated signals such as sinusoids. Comparing with sinusoids has the advantage as it is easy to interpret the results; sinusoids represent oscillatory behavior so, when we compare a signal with a sinusoid at a given frequency, we are actually searching for oscillatory behavior at that frequency. But the lack of correlation between sine and cosine at the same frequency becomes troubling. Figure 2.12 illustrates the problem finding the correlation between a cosine and a sine with a phase shift (i.e., a sinusoid). The correlation depends on the phase shift of the sine wave. In Figure 2.12A the sine is not shifted and there is zero correlation, but when the sine is shifted by 45 degrees (Figure 2.12B) the correlation coefficient is 0.71, and at a 90-degree shift, the correlation coefficient is 1.0 (Figure 2.12C). Figure 2.12D shows that the correlation as a function of sinusoidal shift is itself a sine wave ranging between ± 1.

This complicates a search for oscillatory behavior. To test for oscillatory (i.e., sinusoidal) behavior at, say, 14 Hz, we might compare the EEG with a 14 Hz sine wave. Maybe the EEG is highly oscillatory at 14 Hz, but it is more like a cosine wave than a sine wave. The correlation with a sine wave could be very low and we would mistakenly assume there was no oscillatory behavior around 14 Hz. Okay, we could correlate the signal with both a cosine and sine, but what if the sinusoidal behavior had a phase shift that placed it halfway between a sine and cosine wave (i.e., 45 degree)?[10]

To restate the problem, when using a reference signal to search for a particular behavior we could miss it depending on the time position (or phase) of the reference signal. Figure 2.12D shows correlations as a function of time shift of the sine wave reference signal and shows that at some shift (0.125 s in this particular situation) it is an exact match. Whenever we search a target signal for a particular behavior using a reference signal, we could try correlation at a bunch of different time shifts and take the maximum correlation as representing the true similarity between reference and signal. If we are to apply this shifting/ correlation approach, we need to decide how much to shift the reference signal between correlations. We do not want to shift the reference by so much that we pass over the shift

---

[10]Actually this could work because Equations 2.23 and 2.24 show that a general sinusoid, $C \cos(2\pi f t - \theta)$, can be represented by a combined sine and cosine function. But if we were using some other waveform as our basis of comparison, it might not.

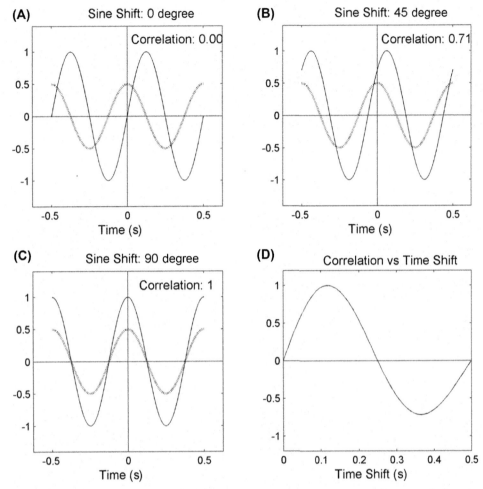

FIGURE 2.12 (A) The correlation between a 2 Hz cosine reference (dashed) and an unshifted 2.0 Hz sine wave is zero. (B) When the sine wave is time shifted by the equivalent of 45 degree, the Pearson correlation is 0.71. (C) When the sine wave is time shifted equal to 90 degree the sine wave becomes a cosine wave and the correlation is 1.0. (D) Plotting the correlation as a function of the time shift shows a sinusoidal variation. The peak value, 1.0, comes at 0.125 s, which happens to be a phase shift of 90 degrees. At a time shift of 0.25 s the sine wave is a sine wave again (but inverted) so the correlation is back to zero. At a time shift of 0.375 s the sine wave (now shifted by 270 degree) becomes an inverted cosine wave so it has a correlation of −1.0 with a cosine wave.

that gives maximum correlation between the reference and signal. If we are dealing with digital data the minimum shift would be one data point. If we made the shift just one data point we would be sure to hit the shift for maximum correlation. Of course, this means doing a lot of correlations, but it is the computer, not us, doing the work. We try this strategy out in the next example by searching for oscillatory behavior in the EEG signal at two frequencies: 6.5 and 14 Hz.

EXAMPLE 2.14

Find the correlation between the EEG signal given in Figure 2.2 (left side) and sinusoids at 6.5 and 14 Hz. The EEG data is stored in file EEG_data1.mat and the signal was sampled at 100 Hz. Is the oscillatory behavior greater at 14 Hz sinusoid or 6.5 Hz? Also generate two plots of the EEG signal superimposed on each of the two sinusoids shifted for best correlation. Scale the plots to display all the signals nicely.

Solution: We know how to perform correlation using Equation 2.29 or the unnormalized version (Equation 2.30). We are only asked to assess the relative correlation at two different frequencies, so we only need to do an unnormalized correlation.

Time shifting should be easy, just digital bookkeeping, but we need to find out what to do at the end points. This is a recurring problem in digital signal processing and we describe some common strategies in Section 2.4.4.1. Most commonly, when a signal is shifted, zeros are added to the ends as needed. We discuss this in more detail later, but here, since the reference waveform is periodic, we warp the points around: tack the end points onto the beginning. An example of this is shown in Figure 2.13 where a 2.0 Hz cosine wave (solid line) is shifted left (dashed line). The end points are wrapped around so it still looks like a sinusoid, just phase shifted. Such a shift is easy to do in MATLAB as shown in the code below.

```
% Example 2.14 Find the correlation between the EEG signal given in Figure 1.7 and
%  sinusoids at 6.5 and 14 Hz. Is the oscillatory behavior greater at 6.5 or 14 Hz?
%
load eeg_data1;          % Get the EEG data (in variable 'eeg')
N = length(eeg);         % Number of EEG data points
fs = 100;                % Sampling frequency of EEG data (given)
f = [6.5 14];            % Frequencies of reference signals
t = (0:N-1)/fs;          % Time vector
```

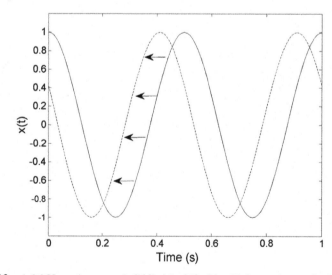

FIGURE 2.13   A 2.0 Hz cosine wave (*solid line*) is shifted by 10 data points to the left (*dashed line*).

```
for k1 = 1: 2
  x = cos(2*pi*f(k1)*t);      % Generate reference signal at desired frequency
  for k = 2:N                 % Perform N-1 correlations
    y = [x(k:end), x(1:k-1)]; % Shift reference circularly
    rxy(k) = mean(eeg.*y);    % Correlation  as a function of shift k
  end
  [corr(k1),shift(k1)] = max(rxy);   % Find maximum correlation
end
%
% Plotting section
for k = 1:2
  subplot(2,1,k);                 % Plot the two sine waves separately
  x = cos(2*pi*f(k)*t);           % Recreate the reference signal for plotting
  y = [x(shift(k):end), x(1:shift(k)-1)];   % Shift reference
  y = y * (max(eeg)/max(y))/4;    % Scale the sinusoid for good viewing
  plot(t,eeg); hold on;           % Plot EEG
  plot(t,y);                      % Plot shifted reference
  xlim([2 2.6]);                  % Scale time axis for better viewing
  ....labels and text.........
end
```

Analysis: The analysis section uses a for loop to perform the operations at the two frequencies. In each loop, a reference signal having the desired frequency is constructed to be the same length as the data. An inner `for` loop performs $N-1$ correlations between the shifted data and the EEG signal, storing the results in an array `rxy`. The wraparound shift is pretty easy to implement: a reference signal is constructed by removing the first $k-1$ points and tacking them on the end. The correlation is done as in Example 2.12. An array is constructed that contains the $N-1$ correlations where the array index indicates the shift. The maximum correlation and corresponding index are found using the `max` operator and saved.

The plotting section also uses a for loop. The reference signal is reconstructed and shifted to the position that gives the best correlation. Since the reference signal is much smaller than the EEG signal, it is scaled to be $^1/_4$ the maximum height of the EEG signal for good viewing. Both the scaled reference signal and the EEG signal are then plotted using the time vector. For better viewing the time axis is expanded to cover a period of 2.0–2.6 s. Labels and text are then added.

Results: The results produced by this program are shown in Figure 2.14. The 14 Hz sinusoidal reference has a correlation that is three times that of the 6.5 Hz reference. Figure 2.14 also shows that while the EEG signal has a lot going on, the two scaled, shifted reference signals (dashed curves) are making every effort to match the EEG signal. Perhaps sinusoids at different frequencies would better capture the oscillatory behavior of the EEG signal, but clearly no single sinusoid could represent this signal. You might think that some collection of sinusoidal references, taken together, might provide a pretty accurate representation of the EEG signal; you would be right and that is the basis of the next chapter.

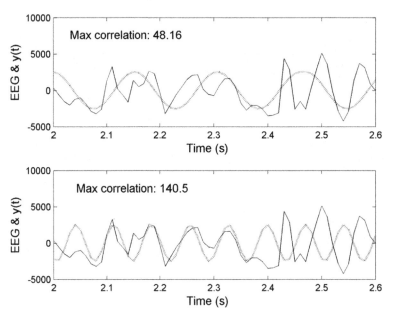

FIGURE 2.14   Plots generated by the code in Example 2.14. Both plots show a 0.6-s segment of the EEG signal shown in Figure 1.7 (*solid curves*). In the upper graph, a 6.5 Hz sinusoidal reference signal is shown shifted to best match the EEG signal (*dashed line*) and in the lower graph the shifted 14 Hz signal is shown. The shifted reference signals provide the best match possible given that they are very simple signals. These correlations show that the 14 Hz reference is a better match to the EEG signal as suggested by the signal segments shown here.

Repeated shifting and correlation of a reference signal across the entire target signal seems to work well in as much as we seem to get the best match possible. We call this approach "cross-correlation." Note that it works even if the reference signal is shorter than the target, as is sometimes the case. An equation for cross-correlation can be derived from the basic correlation equation (Equation 2.30). We just need to introduce another variable that accounts for the shift. It does not matter which function is shifted, the results would be the same. Just as in Example 2.14, the correlation operation (Equation 2.30) is performed repeatedly for different time shifts, k, and the result is a function $r_{xy}[k]$, a series of correlations for different values of k:

$$r_{xy}[k] = \frac{1}{N} \sum_{n=1}^{N} y[n]\, x[n+k] \tag{2.38}$$

where k is the shift variable[11]. To get all possible combinations between the target and reference signals, the shift variable, k, usually ranges over positive and negative integers to some max value K: $k = 0, \pm1, \pm2,\ldots\pm K$. This shifts the reference signal in both directions

---

[11]The true correlation sequence uses the estimation, or the expected value operator, applied to the product of the two signals: $y[n]$ and $x[n-k]$. However, the estimation operator requires the probability distribution functions of the two signals, which are generally not available, so applications of cross-correlation to real-world signals use Equation 2.38.

with respect to the target signal. The variable k is often called the "lags" and for the output sequence, $r_{xy}[k]$, it specifies the shift for a given correlation. If the cross-correlated signals are originally time functions, lags can be converted to time in seconds. Note the continued use of r_{xy} to indicate cross-correlation, pretty much a universal symbol for any type of correlation.

2.4.4.1 Zero Padding

The value of K is often equal to N (or $N - 1$), but can be less. Now we return to the end point problem. If the two signals are the same length, Figure 2.15A, as soon as we shift either signal just one position, we run out of matching points, Figure 2.15B. We need to extend the signals to create matching points. In Example 2.14 we got around this problem by using a wraparound technique: it was as if we were extending the periodic sinusoid with additional periods. With nonperiodic signals, this approach does not make as much sense. The more general approach is to extend the signals by adding zeros at the ends, Figure 2.15C. Extending a signal (or any data set) with zeros to permit calculation beyond its nominal end is called *zero padding*. This may seem like cheating, but it usually works as discussed in the next chapter.

The correlation calculation requires the two arrays be the same length, so the signals must completely overlap. To evaluate all possible alignment of the two signals, we have to extend the unshifted signal by $N - 1$ points at the beginning of the signal and another $N - 1$ points at the end of the signal. (It is $N - 1$ because we start and end with a one-point overlap.) We achieve this signal extension by zero padding as shown in Figure 2.16.

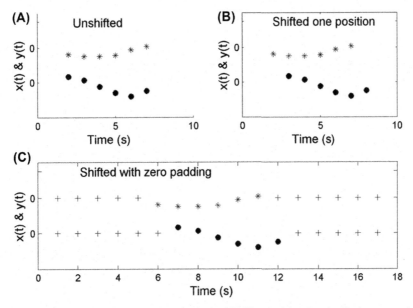

FIGURE 2.15 (A) Two signals are aligned for correlation. (B) If one of the signals (the lower one in this case) is shifted even one position as in cross-correlation, matching points are missing at both ends. (C) A common solution to this problem is to add zeros on to each end of the data, a technique known as zero padding.

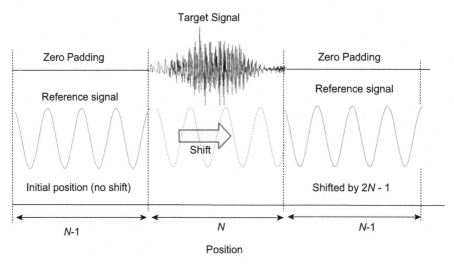

FIGURE 2.16 An example of shifting and zero padding used in cross-correlation. To cover all possible relative positions between the two signals, one signal (the sinusoid in this case) has to be shifted by $2N - 1$ positions. To have the complete overlap required by the correlation algorithm, the unshifted signal needs to be extended by $N - 1$ points at both the beginning and the end. Initially there is a one-point overlap between the reference and original, unextended signal at the inner left dashed line. After the last shift, there is a one-point overlap between the original signal and the final position of the reference signal at the right inner dashed line.

In the Signal Processing Toolbox, MATLAB features a routine that performs cross-correlation called xcorr. The calling structure is:

```
[rxy lags] = xcorr(x,y,maxlags);    % Perform crosscorrelation
```

where x and y are the target and reference signals (makes no difference which is which) and maxlags is an optional argument indicating the maximum number of shifts (the default is $2N - 1$ where N is the length of the larger input signal). The routine uses zero padding to generate the additional points. The cross-correlation found in rxy and lags is a vector the same length as rxy and contains the corresponding lags. This is useful in finding the lag that corresponds to a given correlation (such as the maximum correlation) or for plotting.

The xcorr routine has a lot of bells and whistles, including options such as various ways to bias the correlation, but we can produce a similar routine by revising the code in Example 2.14. So in the next example we modify the code in Example 2.14 and make a routine similar to xcorr. We use that routine to find the correlation between sinusoidal reference signals and the EEG signal, but rather than just compare at two sinusoids, let us make the comparison over a range of sinusoids from 1 to 25 Hz in 0.5-Hz increments. We also compare our routine to MATLAB's xcorr for these same reference signals.

EXAMPLE 2.15

Develop a routine to apply cross-correlation to a pair of signals. Perform the correlation over all possible configurations of the two signals, i.e., $2N - 1$ correlations shifting one signal incrementally. Use this routine to cross-correlation the EEG signal with a collection of sinusoidal reference signals ranging from 1 to 25 Hz in 0.5-Hz increments. Plot the maximum correlation at each reference signal frequency as a function of frequency.

Solution: Correlation routine: The only modification to the code in Example 2.14 is to extend one of the signals with zeros. As a first effort, this crosscorr routine assumes that both signals are the same length and that the lags include all possible combinations (i.e., maxlags $= 2N - 1$). In this routine, we arbitrarily selected signal x for padding, adding $N - 1$ zeros at both the beginning and the end.

```
function [rxy lags] = crosscorr(x,y)
% Function to perform crosscorrelation similar to MATLAB's xcorr
% ......help comments describing input and output arguments .......
%
ly = length(y);             % Length of signals (both same length)
maxlags = 2*ly - 1;         % Compute maxlags from data length
x = [zeros(1,ly-1) x zeros(1,ly-1)];     % Zero pad signal x (could have been y)
for k = 1:maxlags
  x1 = x(k:k+ly-1);         % Constructed shifted signal
  rxy(k) = mean(x1.*y);     % Correlation (Equation 2.30)
  lags(k) = k - ly;         % Compute lags (useful for plotting)
end
```

Solution: Main code: Again we adopt the code from Example 2.14, replacing the cross-correlation with a call to our new routine and adjusting the frequency vector, f, to range from 1 to 25 in 0.5 increments. We also add a call to MATLAB's xcorr routine using the "biased" option, which uses the same scaling as our cross-correlation algorithm.

```
% Example 2.15 Find the correlation between the EEG signal and sinusoids ranging
%   from 1 to 25 Hz in 0.5-Hz increments.
%
load eeg_data1;      % This and next 3 statements identical to Example 2.13
N = length(eeg);     % Number of points
fs = 100;            % Sampling frequency of data
t = (0:N-1)/fs;      % Time vector
f = (1:0.5:25);      % Frequencies of reference signals
%
for k = 1:length(f)
  x = cos(2*pi*f(k)*t);            % Generate reference signal
```

```
rxy = crosscorr(x,eeg);              % Compute crosscorrelation
max_corr(k)= max(rxy);               % Find maximum correlation
rxy_M = xcorr(x,eeg,'biased');       % Find MATLAB crosscorrelation
max_corr_M(k) = max(rxy_M);          % MATLAB's max correlation
end
plot(f,max_corr,'k'); hold on;       % Plot crosscorr results
plot(f,max_corr_M,'*k');             % Plot MATLAB results as *-points
   ......labels.......
```

Results. Figure 2.17 shows maximum cross-correlation between the EEG signal and a sinusoidal reference signal as a function of the frequency of that reference signal. The "*" points represent the results obtained from MATLAB's xcorr routine and are identical to the results with crosscorr. The plot represents the relative amount of oscillatory behavior in the EEG signal as a function of frequency. Note a particularly active region around 8 Hz. This oscillatory activity is well known to neurophysiologists and is called the alpha wave.

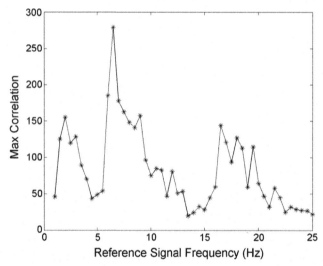

FIGURE 2.17   The plot generated by the code in Example 2.15, which shows the maximum cross-correlation between a reference sinusoid and the EEG signal of Figure 2.2 (left side). The solid line connects values obtained by the crosscorr routine and the "*"-points values are obtained from MATLAB's xcorr routine. The plot represents the amount of oscillation in the EEG signal as a function of frequency. The peak around 8 Hz is known as the alpha wave.

For continuous signals, the time shifting is continuous and the correlation becomes a continuous function of the time shift. This leads to an equation for cross-correlation that is an extension of Equation 2.31 adding a time shift variable, τ:

$$r_{xy}(\tau) = \frac{1}{T} \int_0^T y(t)x(t + \tau)dt \tag{2.39}$$

where variable τ is a continuous variable of time that specifies the time shift of $x(t)$ with respect to $y(t)$. The variable τ is analogous to the lags variable k in Equation 2.38. It is a variable of time with respect to the cross-correlation function r_{xy}, but not the time variable used to define the signals, which is denoted by the symbol "t". The τ time variable is sometimes referred to as a "dummy time variable." There is nothing particularly dumb about it, it is just a secondary time variable we use when we need two different time variables. Note that like the digital version, the continuous cross-correlation function requires multiple integrations, one integration for every value of τ. Since τ is continuous that would be an infinite number of integrations, but it is possible to evaluate Equation 2.39 analytically, at least for some very simple signals. An example of evaluating an equation similar to Equation 2.39 analytically is given in Chapter 5.

All real-world applications of cross-correlation are done in the digital domain where software such as MATLAB does all the work. It may be of some value to do at least one evaluation manually because it provides insight into the algorithm used to implement the digital cross-correlation equation, Equation 2.38.

EXAMPLE 2.16

Evaluate the cross-correlation of the two short digital signals, $x[n]$ and $y[n]$ shown in Figure 2.18 without using a computer.

Solution: To allow for all possible shifts, we need to zero pad one of the signals. Since $y[n]$ contains four samples and $x[n]$ contains three samples, we can add two zeros to each side of $y[n]$ and shift the shorter signal. The discrete cross-correlation equation (Equation 2.38) begins with one sample of $x[n]$ overlapping one sample of $y[n]$, Figure 2.19 (upper left). The calculation continues as $x[n]$ marches rightward in subsequent graphs. In this example, we use no normalization.

Results: From Figure 2.19, we can see that the unnormalized cross-correlation function is:

$$r_{xy} = [0.5, 1.10, 1.13, 0.665, 0.35, 0.125]$$

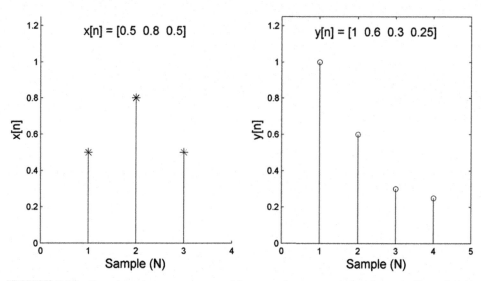

FIGURE 2.18 Two short data sequences used for manual cross-correlation shown in Example 2.16.

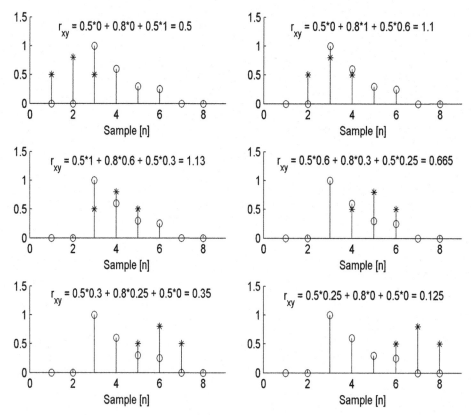

FIGURE 2.19 Results from Example 2.16, illustrating manual calculation of the cross-correlation function of the two short signals shown in Figure 2.18. The signal $x[n]$ ("*" points) is shifted across signal $y[n]$ ("o" points). The first position of $x[n]$ is at the padded zeros, two points to the left of the first valid $y[n]$ point. The highest correlation is found for a shift of three in the middle-left plot, which seems reasonable.

As in Example 2.16, cross-correlation is often used to determine the similarity between a signal and a reference waveform when the relative position for best match is unknown. Simply take the maximum value of the cross-correlation function. If needed, the shift corresponding to the maximum correlation can be determined from the lags variable. Another useful biomedical application for cross-correlation is shown in the next example: finding the time delay between two signals.

EXAMPLE 2.17

File `neural_data.mat` contains two waveforms, `x` and `y`, that were recorded from two different neurons in the brain with a sampling interval of 0.2 ms (i.e., $f_s = 5$ kHz). They are believed to be involved in the same neural operation, but are separated by one or more neuronal junctions that impart delay to one of the signals. Plot the original data, determine if they are related and, if so, the time delay between them.

Solution: Take the cross-correlation between the two signals. Find the maximum correlation and the time shift at which that maximum occurs. We use the maximum correlation to tell us if the two signals are related and the shift to determine the time delay. We can use the routine `crosscorr` developed in Example 2.15 to compute the cross-correlation, but we should scale the correlation to be between ± 1 using the square root of the variances, Equation 2.32. This will better enable us to tell if the signals are related.

```
% Example 2.17 Cross-correlation of two neural signals.
%
load neural_data.mat;                  % Load data
fs = 1/0.0002;                         % Sample frequency
t = (1:length(x))/fs;                  % Calc. time vector
subplot(2,1,1);
plot(t,y,'k',t,x,':k');                % Plot data
     ........label and title.......
[rxy,lags] = crosscorr(x',y');         % Compute crosscorrelation

rxy = rxy/sqrt(var(x)*var(y));         % Scale by Pearson's; Equation 2.32

[max_corr, max_shift] = max(rxy);      % Get max correlation
time_delay = lags(max_shift)/fs;       % Delay in sec.
subplot (2,1,2);
plot(lags/fs,rxy,'k');                 % Plot crosscorrelation
     .......Labels and title.......
```

Analysis: In the call to `crosscorr` we had to transpose the two vectors since our routine assumes they are row vectors, but in fact they were column vectors. Many MATLAB routines check input vectors for orientation and adjust as needed and we modify crosscorr to do this in the next example. After cross-correlation, we find the maximum correlation using MATLAB's max operator. The lag corresponding to the max shift gives us the shift between the two signals in data points. To convert data points to time, as always, we multiply by T_s or divide by f_s as done here.

Result: The two signals are shown in Figure 2.20A. The dashed signal, x, follows the solid line signal, y. The cross-correlation function is seen in Figure 2.20B. Although the time at which the peak occurs is difficult to determine visually because of the large scale of the horizontal axis, the peak is found to be 0.013 s from the lag at max correlation. The max correlation is 0.45, which indicates that they are likely involved in some common neural operation.

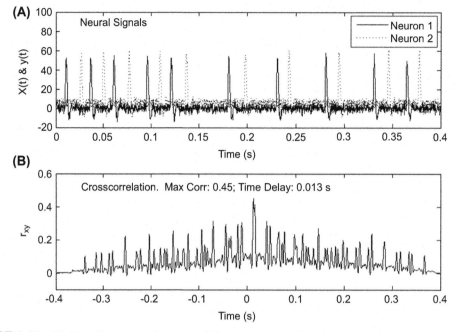

FIGURE 2.20 (A) Recordings made from two different neurons believed to be involved in the same neural operation, but delayed by an intervening synapse(s). Background noise common to such recordings is seen in both signals. (B) Cross-correlation between the two neural signals shows a peak just to the right of the center that corresponds to a shift of 0.013 s. The Pearson correlation at this point is 0.45.

2.4.4.2 Cross-correlating Signals Having Different Lengths

Unlike correlation, cross-correlation can be done using signals that have different lengths and is often used for probing waveforms with short reference signals. For example, given the signals seen in Figure 2.21, we might ask: "does any portion of the larger signal contain

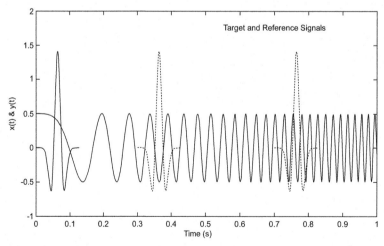

FIGURE 2.21 In Example 2.18, the short signal on the left is used as a reference signal to probe the longer target signal using cross-correlation.

something similar to the short reference signal shown on the left, and if so, how similar and where?" We answer this question using cross-correlation in our next example.

EXAMPLE 2.18

Evaluate the long signal shown in Figure 2.21 by probing it with the short signal on the left side. Determine if this target signal contains a segment that is similar to the reference, and if so, when and how much. Both signals are found as x (target) and y (reference) in file chirp_signal.mat and have a sampling interval of 1.0 ms.

Solution: Main Program: Follow the same procedure as in Example 2.18 except we need to modify our cross-correlation routine to allow for different signal lengths. While we are at it, we modify it to permit input signals as either row or column vectors. First we present the main code followed by the modified cross-correlation routine.

```
% Example 2.18 Cross-correlation of two signals of unequal length
%
fs = 1/.001;                            % Sampling frequency (1/Ts)
load chirp_signal;                      % Load data
[rxy,lags] = crosscorr1(x,y);           % Crosscorrelate (modified)
plot(lags/fs,rxy,'k');                  % Plot data
[max_corr max_shift] = max(rxy);        % Find max values
time_delay = lags(max_shift)/fs;        % Delay in sec.
title(['B) Max Corr: ',num2str(max_corr,2),' Time delay: ',...
   num2str(time_delay,2),' sec']);      % Put results on plot
.......label and axis.......
```

Solution: Revised Cross-correlation Routine: The revised cross-correlation routine incorporates three improvements: it can accept signal vectors as either row or column vectors, it can deal with signals of different length, and it can do a Pearson normalization if requested. The input arguments now include an optional third term: if this variable is set to "p," a Pearson normalization is done; if set to "s," a standard normalization is done (i.e., Equation 2.38.).

After checking the number of arguments and setting the default normalization if need be, the routine tests the orientation of the two input signals and transposes them if they are not already row vectors. The next section ensures that signal y is less than or equal to signal x. This makes the end point problem easier; we can always move y across an extended version of x.[12] Signal x is then zero padded at each end with the number of points in y − 1, the same as done in the previous routine. The final modification in crosscorr1 is to apply the Pearson normalization using Equation 2.32. We now have a smarter, more flexible, easier to use cross-correlation routine.

```
function [rxy lags] = crosscorr1(x,y, normalization)
% Function to perform crosscorrelation similar to MATLAB's xcorr
% This version does not assume x and y are the same length or are row vectors
%
if nargin < 3
```

```
  normalization = 's';              % Standard normalization (1/N). Default
end
%
% Insure input signals are row vectors
[N,lx] = size(x);                   % Rearrange both vectors as row vectors if needed
if N > lx                           % Rearrange as row vector
  x = x';
  lx = N;
end
[N,ly] = size(y);
if N > ly                           % Rearrange as row vector
  y = y';
  ly = N;
end
ly = length(y);                     % Get new vector lengths
lx = length(x);
%
% If the input signals have unequal lengths, make y the shorter signal
if lx < ly
  temp = x;                         % Make y the shorter signal so the padded
  x = y;                            % signal will always be x
  y = temp;                         % Swap vectors x and y
end
Nx= length(x);                      % Re-establish vector lengths
Ny = length(y);
maxlags = Nx + Ny - 1;
x = [zeros(1,Ny-1) x zeros(1,Ny-1)];      % Zero pad signal x
var_y = var(y);                     % Get variance of x for possible Pearson norm.
for k = 1:maxlags
  x1 = x (k:k+ly-1);                % Shift signal x

  rxy(k) = mean(x1.*y);            % Correlation (Equation 2.30)

  lags(k) = k - ly;                % Compute lags
  if normalization == 'p'          % If requested, use Pearson normalization

    rxy(k) = rxy(k)/sqrt(var(x)*var_y);   % Equation 2.32

  end
end
```

Result: The cross-correlation of the original chirp signal and the shorter reference shows a cross-correlation function that oscillates at the same frequency as the sinusoid. Note that the cross-correlation function decreases toward zero at both ends, an inevitable consequence of zero padding. (The correlations at very large and small shifts involve more zeros, Figure 2.15.) A maximum correlation of 0.66 occurs at 0.325 s, Figure 2.22 (vertical line). On comparison of the signal and reference, the time and value of max correlation looks about right.

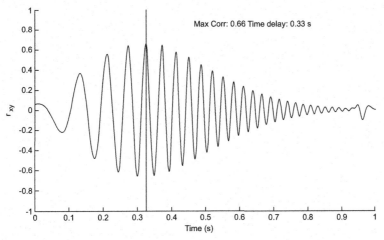

FIGURE 2.22 The cross-correlation function between the target and reference signals in Figure 2.21. The maximum Pearson correlation is moderately high at 0.66 and occurs at 0.33 s as indicated by the vertical line.

[12]The MATLAB xcorr routine simply zero pads the shorter signal to make the two equal in length. That may seem less efficient, but xcorr uses a trick to gain speed. It actually implements cross-correlation in the frequency domain. We discuss this approach in Section 5.3 on convolution because MATLAB also uses this trick to calculate convolution.

We are now quite skilled in cross-correlation, but what if we have only one signal? That is the featured topic of the next section.

2.4.5 Autocorrelation

If we have only one signal, why not try cross-correlating it with itself? Easy to do, although not so obvious what it would mean. To correlate a signal with itself, just make a copy of the single signal and apply cross-correlation (i.e., `crosscorr1(x,x)`). What does it mean? Self cross-correlation is called "autocorrelation" and basically it describes how well a signal correlates with shifted versions of itself. This could be useful in searching for repeating segments in a signal.

Autocorrelation can also be used to determine how signal data points correlate with their neighbors. As the lag (i.e., shift) increases, signal points are compared with more distant neighbors. Determining how neighboring segments relate to one another provides some insight into how the signal was generated. For example, suppose a signal remains highly correlated with itself over an extended period of time. That signal must have been produced, or modified, by a process that uses previous signal values to determine future values (at least in part). Such a process can be described as having memory, since it must remember past values of the signal to shape the signal's current values. The longer the memory, the more the signal will remain partially correlated with shifted versions of itself. Conversely, if the correlation decays after just a few lags, the system producing or acting on a signal has a short memory. If the correlation goes to zero after a shift of only one data point (i.e., lag ≥ 1), the system has no memory and that signal is actually white noise.

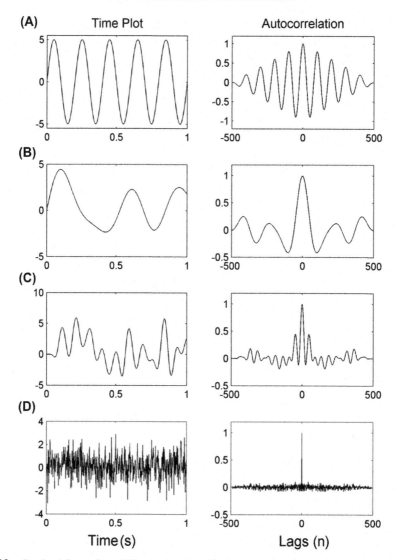

FIGURE 2.23 On the left are four different signals with their autocorrelation functions on the right. (A) A truncated sinusoid. The autocorrelation function is a cosine wave that decays due to the finite length of the signal. The autocorrelation function of an infinite sine wave would be a nondiminishing cosine wave. (B) A slowly varying signal has a slowly decaying autocorrelation function indicating that the process that acted on that signal has a relatively long memory. (C) A rapidly varying process has a rapidly decaying autocorrelation function pointing to a process with a shorter memory. (D) A random Gaussian signal has an autocorrelation function that drops to near zero for all nonzero lags.

To derive the autocorrelation equation, simply substitute the same variable for x and y in Equation 2.38 or Equation 2.39:

$$r_{xx}[k] = \frac{1}{N} \sum_{n=1}^{N} x[n]x[n+k] \tag{2.40}$$

$$r_{xx}(\tau) = \frac{1}{T} \int_0^T x(t)x(t+\tau)dt \qquad (2.41)$$

where r_{xx} is the autocorrelation function and k and τ are the lag or shift variables.

Figure 2.23 shows the autocorrelation of several different waveforms. It is common to normalize the autocorrelation function to 1.0 at lag 0 (no shift) when the signal is being correlated with itself. This is the case for the autocorrelation functions in Figure 2.23. The autocorrelation of a sine wave is another cosine since the correlation varies sinusoidally with the lag. If this sine wave were infinity long the cosine would not decay, but with finite (i.e., real-world) signals, autocorrelation, like cross-correlation, requires zero padding at the signal's end points, causing the cosine function to decay at larger lags. (Again, the larger lags necessarily include more zeros in the correlation computation.)

Although all zero padded signals have an autocorrelation that decays with increasing lag,[13] the rate of that decay depends on how rapidly the signal decorrelates with itself. Decorrelation depends on how rapidly the signal fluctuates in the time domain. A rapidly varying signal, Figure 2.23C, decorrelates quickly: the average correlation between neighbors falls off rapidly with distance. You could say this signal has a poor memory of its past values and is probably the product of a process with a short memory. For slowly varying signals, the autocorrelation falls slowly, as in Figure 2.23B. For a Gaussian random signal, the correlation falls to zero instantly for all nonzero lags, both positive and negative, Figure 2.23D. This tells us that for these random signals, every data point is uncorrelated with its neighbors. Such a random signal has no memory of its past and was not operated on by a process with memory.

Since shifting the waveform with respect to itself produces the same results no matter which way the function is shifted, the autocorrelation function will be symmetrical about lag zero. Mathematically, the autocorrelation function is an even function:

$$r_{xx}(-\tau) = r_{xx}(\tau) \qquad (2.42)$$

The maximum value of r_{xx} clearly occurs at zero lag, where the waveform is fully correlated with itself. If the autocorrelation is normalized by the variance that is common, the value will be one at zero lag. (Since in autocorrelation the same function is involved twice, the normalization equation given in Equation 2.32 reduces to $1/s^2$.)

When autocorrelation is implemented on a computer, it is usually considered a special case of cross-correlation. That is the case here where crosscorr1 is used with two identical inputs.

```
[rxx, lags] = crosscorr2(x,x,'a');     % Autocorrelation
```

The crosscor2 routine is a minor modification of crosscorr1 constructed in Example 2.18. This modification gives the option of the common autocorrelation normalization: an 'a' as

---

[13]Instead of zero padding, it is theoretically possible to extend a signal with a wraparound approach, such as that used in Example 2.14, and this might make sense if the signal was periodic. However, this would be a special case and would require modifying the standard cross-correlation routines.

the third argument normalizes the rxx to be 1.0 at zero shift (i.e., when lags = 0). A simple application of autocorrelation is shown in the next example.

EXAMPLE 2.19

Evaluate and plot the autocorrelation function of the respiratory signal resp in file resp.mat ($f_s = 125$ Hz). To better view the decrease in correlation at small shifts, plot only shifts between ±20 s.

Solution: Load the respiratory signal and use crosscorr2 to generate the autocorrelation function. Plot the result against lags in seconds (i.e., divide the lags variable by f_s). Since the respiratory signal is expected to be highly oscillatory, we anticipate the autocorrelation function to be like that of a sinewave (Figure 2.23A) and to decay slowly. We scale the x axis to show only the first ±20-s lags, which should show the correlation between 6 and 12 neighbors (assuming a respiratory period of approximately 1.5–3.5 s).

```
% Example 2.19 Program to plot autocorrelation function of respiratory data
%
load resp;                          % Get data (resp)
fs = 125;                           % Sample frequency 100 Hz
[rxx,lags] =crosscorr2(resp,resp,'a');  % Autocorrelation
plot(lags/fs, rxx); hold on;        % Plot autocorrelation function
plot([lags(1) lags(end)], [0 0]);   % Plot zero line
axis([-20 20 -0.2 1.2]);            % Scale x-axis to be between ± 20 sec
        .......title and labels.......
```

Result: The autocorrelation function of the respiratory signal is shown in Figure 2.24. As expected, the signal decorrelates slowly and shows an oscillatory pattern. The period of oscillation is

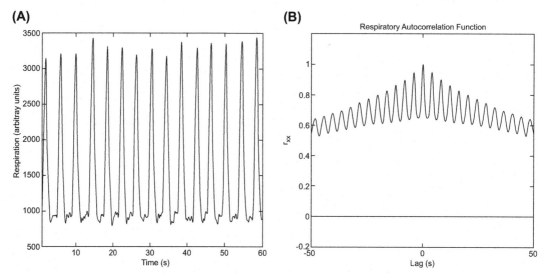

FIGURE 2.24 (A) Respiratory signal used in Example 2.19. (One minute shown.) (B) Autocorrelation function of the respiratory signal. The autocorrelation decorrelates slowly and shows oscillation. The oscillatory period of approximately 3 s reflects the subject's respiratory rate and the slow decay in correlation shows the subject had a regular breathing pattern when these data were taken.

approximately 3 s and reflects the respiratory period. (A respiratory period of 3 s corresponds to 20 breaths/min.) This example shows how autocorrelation can be used to search for oscillatory behavior in a signal. The slow decorrelation indicates that this subject had even breathing that changed very little cycle-to-cycle.

2.4.6 Autocovariance and Cross-covariance

Two operations closely related to autocorrelation and cross-correlation are autocovariance and cross-covariance. It is the same relationship we saw between correlation and covariance: in covariance operations the means are subtracted from the input signals. For cross-covariance, the discrete and continuous equations are:

$$C_{xy}[k] = \frac{1}{N} \sum_{n=1}^{N} \left(x[n] - \overline{x[n]} \right) \left(y[n+k] - \overline{y[n]} \right) \tag{2.43}$$

$$C_{xy}(\tau) = \frac{1}{T} \int_{0}^{T} \left(x(t) - \overline{x(t)} \right) \left(y(t+\tau) - \overline{y(t)} \right) dt \tag{2.44}$$

Again k ranges from 0 to $\pm K$.

For autocovariance the discrete and continuous equations become:

$$C_{xx}[k] = \frac{1}{N} \sum_{n=1}^{N} \left(x[n] - \overline{x[n]} \right) \left(x[n+k] - \overline{x[n]} \right) \tag{2.45}$$

$$C_{xx}(\tau) = \frac{1}{T} \int_{0}^{T} \left(x(t) - \overline{x(t)} \right) \left(x(t+\tau) - \overline{x(t)} \right) dt \tag{2.46}$$

The autocovariance function can be thought of as measuring the memory or self-similarity of the deviation of a signal about its mean level. Similarly, the cross-covariance function is a measure of the similarity of the deviation of two signals about their respective means. If signal means are zero, the correlation and covariance operations are identical. Only occasionally do we stumble across biosignals that have a nonzero mean. One such signal is heart rate, and analysis of heart rate variability has generated considerable interest because so many physiological processes influence the heart rate. An example of the application of autocovariance to the analysis of heart rate variability is given next.

We now understand that autocorrelation and autocovariance describe how one segment of data is correlated, on average, with adjacent segments. As mentioned previously, such correlations could be due to memory-like properties in the process that generated the data. Many physiological processes are repetitive, such as respiration and heart rate, yet vary somewhat on a cycle-to-cycle basis. Autocorrelation and cross-correlation can be

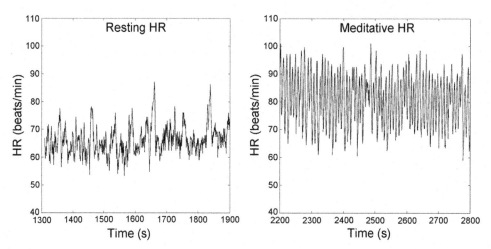

FIGURE 2.25 Ten minutes of beat-by-beat heart rate data taken from a normal resting subject and one who is meditating. Differences are substantial, with the meditative subject showing a higher overall heart rate and greater heat-to-beat fluctuations. In Example 2.20, we use autocovariance to analyze heart rate variability in the resting condition.

used to explore this variation. In the analysis of heart rate variability, autocovariance can be used to tell us if these variations are completely random or if there is some correlation between beats, or over several beats. For a variability analysis, we want to use autocovariance, not autocorrelation, since we are interested in heart rate variability, not heart rate per se. (Remember that autocovariance subtracts the mean value of the heart rate from the data and analyzes only the variation.)

Figure 2.25 shows the time plots of instantaneous heart rate in beats per minute taken under normal and meditative conditions. These data are found as HR_pre.mat (preliminary) and HR_med.mat (meditative) and are from the PhysioNet database. (Goldberger et al., 2000). Clearly the meditators have a higher average heart rate with more variability. These differences are explored in later examples, but here we look at the rate of variation and its correlation over successive beats.

EXAMPLE 2.20

Determine correlations in the heart rate variability of the resting subject whose heart rate is shown in Figure 2.25 (left side). The variability in the meditative subject is more interesting, so it is investigated in one of the problems at the end of the chapter.

Solution: Load the heart rate data taken during the two conditions. The file Hr_pre.mat contains the variable hr_pre, the instantaneous (beat-by-beat) heart rate. Since the heart rate is determined each time a heartbeat occurs, it is not evenly time-sampled and a second variable t_pre contains the time at which each beat is sampled. For this problem, we determine the autocovariance as a function of heart beat and we do not need the time variable. We can find the autocovariance function using crosscorr2 as in autocorrelation as long as we first subtract the mean heart rate from the signals.

We then plot the autocovariance function and limit the x axis to ± 30 successive beats to better observe the decrease in covariance with successive beats.

```
% Example 2.20 Use autocovariance to determine the correlation
% of heart rate variation between heart beats
%
load Hr_pre;                              % Load normal HR data
[cov_pre,lags_pre] = crosscorr2(hr_pre - mean(hr_pre),...
  hr_pre-mean(hr_pre),'a');               % Autocovariance
plot(lags_pre,cov_pre,'k'); hold on;      % Plot resting autocovariance
plot([lags_pre(1) lags_pre(end)], [0 0],'k'); % Plot a zero line
axis([-30 30 -0.2 1.2]);                  % Limit x-axis to ± 30 beats
```

Results: The autocovariance function Figure 2.26 shows that there is some average correlation between adjacent heartbeats out to five to eight beats.

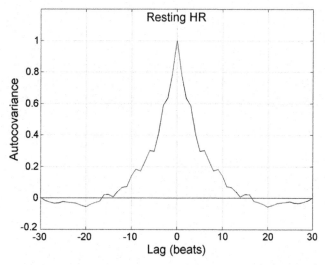

FIGURE 2.26   Autocovariance function of the heart rate variability of a normal resting subject. Some beat-to-beat correlation is seen between neighboring beats up to about five to eight beats away.

2.5 SUMMARY

The sinusoidal waveform is arguably the single most important waveform in signal processing. Some of the reasons for this importance are provided in the next chapter. Because of their importance, it is essential to know the mathematics associated with sines, cosines, and general sinusoids. Since we work with both real-valued and complex sinusoids, it is important to understand both representations and the associated math.

Some basic measurements that can be made of signals include mean values, standard deviations, variances, and rms values, all easily implemented in MATLAB. Averaging is a very powerful tool for noise reduction. If multiple observations of a physiological response can be obtained, entire responses can be averaged in an approach known as ensemble averaging. Isolating some brain activity such as its electrical response to a visual stimulus, the VER, requires averaging hundreds of responses to a repeating stimulus. Recovering the very weak evoked response that is heavily buried in the EEG is not possible without ensemble averaging.

Although the basic measurements describe some fundamental signal features, they do not provide much information on signal content or meaning. A common approach to obtain more information is to probe a signal by correlating it with one or more reference waveforms. One of the most popular probing signals is the sinusoid, and sinusoidal correlation is covered in detail in the next chapter. Sometimes a signal will be correlated with another signal in its entirety, a process known as correlation (or the closely related covariance). Zero correlation between signal and reference does not necessarily mean they have nothing in common, only that the signals are mathematically orthogonal. Signals and families of signals that are orthogonal are particularly useful in signal processing because, when they are used in combination, each orthogonal signal can be treated separately: it does not interact with the other signals.

Sines and cosines have much in common; both exhibit oscillatory behavior, but they are orthogonal and their mutual correlation is zero. This presents a problem if you are probing a target signal using sinusoids: an oscillatory pattern could be missed if the target signal's oscillation was out of phase with the sinusoidal reference. A solution that works for sinusoids as well as other reference signals is to shift the reference probe so that all possible phases are correlated with the target signal. Correlation with shifting is called cross-correlation and should be used whenever we want to establish the general similarity between a reference and target signal. Cross-correlation not only quantifies the similarity between the reference and target, but also shows where the match is the greatest. Hence, cross-correlation can also be used to measure the time delay between two similar signals.

A signal can be correlated with shifted versions of itself, a process known as autocorrelation. The autocorrelation function describes the time period for which a signal remains partially correlated with itself, and this relates to the structure of the signal. A signal consisting of random noise decorrelates immediately, whereas a slowly varying signal will remain correlated over a longer period. Correlation, cross-correlation, and autocorrelation have related operations called covariances where the signals' baseline is removed before correlation. All correlations and covariances are easy to implement in MATLAB and a single routine handles these operations.

PROBLEMS

1. Use Equation 2.4 to analytically determine the rms value of a "square wave" with amplitude of 1.0 V and a period 0.2 s.

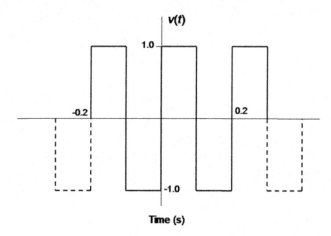

2. Generate one cycle of the square wave similar to the one shown above in a 500-point MATLAB array. Determine the RMS value of this waveform. When you take the square of the data array be sure to use a period before the up arrow so that MATLAB does the squaring point by point (i.e., x.^2).

3. Use Equation 2.4 to analytically determine the rms value of the waveform shown below with amplitude of 1.0 V and a period 0.5 s. (Hint: use the symmetry of the waveform to reduce the calculation required.)

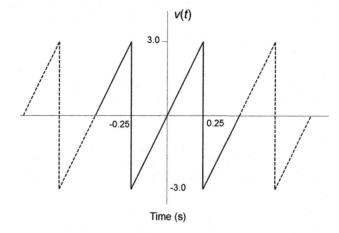

4. Generate the waveform shown for Problem 3 above in MATLAB. Use 1000 points to produce one period. Take care to determine the appropriate time vector. (Constructing the function in MATLAB is more difficult than the square wave of Problem 2, but can still be done in one line of code.) Calculate the rms value of this waveform as in Problem 2. Plot this function to ensure you have constructed it correctly.

5. Fill a MATLAB vector array with 4000 Gaussianly distributed numbers (i.e., `randn`) and another with 1000 uniformly distributed numbers (i.e., `rand`). Find the mean and standard deviation of both sets of numbers. Modify the array of uniformly distributed numbers to have a mean value of zero. Confirm it has a mean of zero and recalculate the standard deviation.

6. The website file `amplitude_slice.mat` contains a signal, `x`, and an amplitude sliced version of that signal `y` ($f_s = 1$ kHz). This signal has been sliced into 16 levels. In Chapter 4, we find that amplitude slicing is like adding noise to a signal. Plot the two signals superimposed and find the rms value of the noise added to the signal by the quantization process. (Hint: subtract the sliced signal from the original and take the rms value of this difference.)

7. If a signal is measured as 2.5 V and the noise is 28 mV (28×10^{-3} V), what is the SNR in dB?

8. A single sinusoidal signal is found in a large amount of noise. (If the noise is larger than the signal, the signal is sometimes said to be "buried in noise.") If the rms value of the noise is 0.5 V and the SNR is 10 dB, what is the rms amplitude of the sinusoid?

9. The file `signal_noise.mat` contains a variable `x` that consists of a 1.0-volt peak sinusoidal signal buried in noise. What is the SNR for this signal and noise? Assume that the noise rms is much much greater than the signal rms.

10. Load the data in `ensemble_data.mat`, which contains a data matrix. The data matrix contains 100 responses of a signal in noise. In this matrix, each row is a separate response. Plot several randomly selected samples of these responses. Is it possible to identify the signal from any single record? Construct and plot the ensemble average for these data. Also construct and plot the ensemble standard deviation.

11. In this problem, we evaluate the noise reduction produced by ensemble averaging. Load the VER data variable `ver`, along with the actual, noise-free VER in variable `actual_ver`. Both these variables can be found in file `Prob2_11_data.mat`. The visual response data set consists of 1000 responses, each 100 points long. The sample interval is 5 ms. Construct an ensemble average of 25, 100, and 1000 responses. The two variables are in the correct orientation and do not have to be transposed. Subtract the noise-free variable (`actual_ver`) from an individual evoked response and the three ensemble averages to get an estimate of the noise in the three waveforms. Compute the standard deviations of the unaveraged waveform with the three averaged waveforms. Output the unaveraged standard deviation to a table of the three averaged standard deviations and the theoretical standard deviation predicted by Equation 2.17. How does this compare with the reduction predicted theoretically by Equation 2.17? Note there are practical limits to the noise reduction that you can obtain by ensemble averaging. Can you explain what might limit the continued reduction in noise as the number of responses in the average gets very large?

12. Two 10 Hz sine waves have a relative phase shift of 30 degree. What is the time difference between them? If the frequency of these sine waves doubles, but the time difference stays the same, what is the phase difference between them?

13. Convert $x(t) = -5 \cos(5t) + 6 \sin(5t)$ into a single sinusoid, i.e., $A \sin(5t + \theta)$.

14. Convert $x(t) = 30 \sin(2t + 50)$ into sine and cosine components. (Angles should always be in degrees unless otherwise specified.)

15. Convert $x(t) = 5 \cos(10t + 30) + 2 \sin(10t - 20) + 6 \cos(10t + 80)$ into a single sinusoid as in Problem 13.

16. Find the delay between $x_1(t) = \cos(10t + 20)$ and $x_2(t) = \sin(10t - 10)$.

17. Equations 2.23 and 2.24 were developed to convert a sinusoid such as $\cos(\omega t - \theta)$ into a sine and cosine wave. Derive the equations to convert a sinusoid based on the sine function, $\sin(\omega t + \theta)$, into a sine and cosine wave. (Hint: Use the appropriate identity from Appendix C.)

18. (A) Given the complex number $C = -5 - j3$, find the angle using the MATLAB `atan` function. Remember MATLAB trig. functions use radians for both input and output. Note the quadrant-related error in the result. (B) Now use the `atan2` function to find the angle. Note that the arguments for this function are `atan2(b,a)`. (C) Finally, find the angle of C using the MATLAB `angle` function (i.e., `angle(C)`). Note that the angle is the same as that found by the `atan2` function. Also evaluate the magnitude of C using the MATLAB `abs` function (i.e., `abs(C)`).

19. Modify the complex exponential in Example 2.10 to generate and plot a cosine wave of amplitude 5.0 that is shifted by 45 degree. (Hint: Since the cosine is desired, you will only need to plot the real part and modify the magnitude and phase of the exponential. Remember MATLAB uses radians.)

20. Use Equation 2.31 to show that the correlation between $\sin(2\pi t)$ and $\cos(2\pi t)$ is zero. Do this both analytically and using MATLAB.

21. Use Equation 2.31 to find the correlation between the two waveforms shown below analytically.

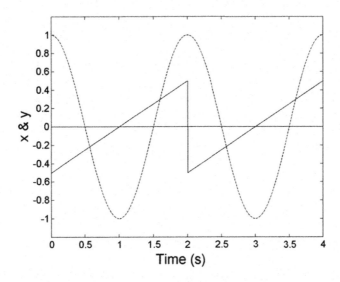

22. Use cross-correlation to find the delay between the 10 Hz sine waves described in Problem 12. (All cross-correlation problems use MATLAB.) Use a sample frequency of

2 kHz and total time of 0.5 s. Remember MATLAB trig. functions use radians. (Note that the second part of Problem 12, finding the phase when the frequency doubled, would be difficult to do in MATLAB. There are occasions where analytical solutions are preferred.)

23. Use cross-correlation to find the time delay in seconds of the two sinusoids of Problem 16. (You choose the sampling frequency and number of data points.)

24. Use cross-correlation to find the phase shift between $x(t)$ in Problem 15 and a sine wave of the same frequency. Plot $x(t)$ and the sine wave and the cross-correlation function and find the lag at which the maximum (or minimum) correlation occurs. You choose the sample frequency and number of points. (Hint: Define $x(t)$ in MATLAB by simply writing the equation found in Problem 15 and define the second signal as sin(10t)). After finding the time shift, T_d, in seconds, convert to a phase shift. (Hint: Note that the period of $x(t)$ is $T_p = 1/(10/2\pi)$ s and the phase is the ratio of the time delay to the period, times 360, i.e., $\theta = 360\frac{T_d}{T_p}$).

25. The file two_var.mat contains two variables x and y. Is either of these variables random? Are they orthogonal to each other? (Use any valid method to determine orthogonality.)

26. The file prob2_26_data.mat contains a signal x ($f_s = 500$ Hz). Determine if this signal contains a 50 Hz sine wave and if so at what time(s). (Hint. The max operator will determine only one peak. If the plot of r_{xy} suggests multiple peaks are possible, you may want to apply the max operator to selected segments of r_{xy}.)

27. Is the cross-correlation function of two random Gaussian variables ($N = 500$) itself a Gaussian random variable? Show. (Make your evidence definitive!)

28. The file prob2_28_data.mat contains a variable x that is primarily noise but may contain a periodic function. Plot x with the correct time axis ($f_s = 1$ kHz). Can you detect any structure in this signal? Apply autocorrelation and see if you can detect a periodic process in r_{xx}. If so, what is the frequency of this periodic process? It will help to expand the x axis to see detail. The next chapter presents a more definitive approach for detecting periodic processes and their frequencies.

29. Develop a program along the lines of Example 2.19 to determine the correlation in heart rate variability during meditation. Load file Hr_med.mat, which contains the heart rate in variable hr_med and the time vector in variable t_med. Calculate and plot the autocovariance. The result will show that the heart rate under meditative conditions contains some periodic elements. Can you determine the frequency of these periodic elements?

30. We know that the autocorrelation function of Gaussain random noise goes to zero for all nonzero lags, Figure 2.23D. What would you expect from a uniformly distributed random numbers? Compare the autocorrelation functions of Gaussianly and uniformly distributed random numbers. If the autocorrelation functions are different, explain.

31. This is an example of memory. Construct an array, x, of Gaussian random numbers ($N = 2000$). Construct a new signal from this array where each data point is a five-point running average of x (i.e., y(1) = mean(x(1:5)); y(2) = mean(x(2:6)); ... y(N-5) = mean(x(N-5:N));

Plot the autocorrelation functions of both the Gaussian and averaged signals. Expand the lag axis to ±12 lags to observe the effect of memory on the autocorrelation function.

Signal Analysis in the Frequency Domain: The Fourier Series and the Fourier Transformation

3.1 GOALS OF THIS CHAPTER

In this chapter we determine how to find the sinusoidal components of a general signal. But this is something we have already done in Example 2.15 when we used correlation between an electroencephalography (EEG) signal and sinusoids to search for oscillatory behavior.[1] Here we develop a computationally more efficient approach to do the same thing. More importantly, we dig deeper into the greater significance of those sinusoidal components. We also address some issues we glossed over in Example 2.15. For example, in our search for oscillatory behavior, we compared the EEG signal with a group of sinusoids ranging in frequencies between 1 and 25 Hz in 0.5-Hz intervals. Could we have missed some oscillatory behavior with this arbitrary range of frequencies and frequency increments? Should we have used a broader range of frequencies and/or finer intervals, or perhaps used fewer sinusoids and larger intervals? Here we offer a definitive answer to these questions.

We also find that correlating a signal with a series of sinusoids gives us more than just a measure of the signal's oscillatory behavior. If we choose the right combination of sinusoids, and we use enough of them, the correlation coefficients become an equivalent, alternative representation of the signal. In other words, the correlation coefficients give a complete representation of the signal and it is possible to reconstruct the original signal from just these coefficients. This works only if we use enough sinusoids at the right frequencies, but the technique for finding the right correlation coefficients is straightforward.

The correlation coefficient signal representation can be very useful both in providing a new description of the signal and in certain operations applied to the signal. The correlation

---

[1]We actually used cross-correlation between a cosine and an EEG signal. But in cross-correlation, the waveforms are shifted with respect to one another, so it is as if we were performing simple correlations between the EEG and sinusoids that have a range of phase shifts. We then took the maximum cross-correlation as the best match between the EEG and a shifted sinewave.

coefficient representation of a signal is called the "frequency domain" representation since it is based on the correlations with sinusoids having a range of frequencies. Converting a signal from its time domain representation to its frequency domain representation is an example of a class of operations known as "signal transformations" or just "transformations." Again, the two representations are completely interchangeable. You can go from the time to the frequency domain and back again with no constraints; you just have to follow the mathematical rules.

In Example 2.15 we used digital signals in the digital domain, but in this chapter we also determine sinusoidal correlations analytically from continuous signals in the continuous domain. In fact, time—frequency domain transformations were first developed in the continuous domain and laboriously worked out by hand before the advent of the digital computer. We do a few easy examples on continuous domain operations because as engineers we should know the inner workings of this important transformation. Fortunately, all real-world time—frequency transformations are done in the digital domain on a computer.

To summarize, in this chapter we will:

- Show how to decompose any periodic waveform into a series of sinusoidal components and how to do the opposite, recombine the sinusoidal components into a waveform.
- Demonstrate how sinusoidal decomposition leads to the frequency characteristics of the spectrum of a waveform.
- Describe the Fourier transform using complex notation.
- Use the Fourier transform to find the frequency domain representation of a periodic waveform.
- Show one method for tracking the changes in waveform's spectrum over time.

3.2 TIME—FREQUENCY DOMAINS: GENERAL CONCEPTS

We use the Fourier series theorem, commonly known as the Fourier transform, to move between the time and frequency domains. The Fourier series theorem is one of the most important and far-reaching concepts presented in this book. Its several versions are explored using both analytical solutions in the continuous domain and computer algorithms in the digital domain. For a digital signal, both time and frequency representations consist of a sequence of discrete numbers. In the analog domain, the time domain representation is a continuous function of time, but it can also be described using a time plot. The frequency domain representation of an analog signal depends on the type of signal. If it is periodic, the frequency domain representation is a discrete series corresponding to specific frequencies, but if it is aperiodic, the frequency domain representation becomes a continuous function of frequency. Either way, the frequency plot describes the signal's spectrum, and the terms "spectrum" or "spectral" are inherently frequency domain terms. Bandwidth is also a frequency domain term, as it describes the characteristics of a signal's spectrum.

There are several motivations for transforming a signal into the frequency domain. Sometimes the spectral representation of a signal is more understandable than the time domain representation. The EEG time domain signal, such as shown in Figure 1.7, is quite complicated. The time domain signal may not be random, but whatever structure it contains is not apparent in the time response. For example, the frequency domain representation determined in Example 2.15[2] revealed a structure in which certain frequencies contained more oscillatory behavior (i.e., more energy) than other frequencies. Electroneurophysiologists often use the spectral representation to study EEG waves and have identified some especially important frequencies. For example, an oscillation found between 8 and 12 Hz is called an "alpha wave" and is present when a subject is at rest with eyes closed, but not tired or asleep. Other oscillatory features revealed in the EEG frequency spectrum include the beta wave between 12 and 30 Hz associated with the muscle contractions or sensory feedback in static motor control, and the "delta wave" between 0 and 4 Hz, usually associated with deep sleep.

Systems can also be represented in the time and/or frequency domain. The spectral representation of a system provides a good description of how that system operates on signals that pass through it. For systems, the frequency domain is the more intuitive: it shows how the system alters each frequency component in the signal. Specifically, the system's spectrum shows if the signal energy at a given frequency is increased, decreased, or left the same and how it might be shifted in time. The same mathematical tools used to convert signals between the time and frequency domains are used to flip between the two system representations and, again, there are no constraints.

3.3 TIME–FREQUENCY TRANSFORMATION OF CONTINUOUS SIGNALS

Sinusoids are the most basic of waveforms and that makes them of particular value in signal analysis. Because of their simplicity, it is easy to represent them in either the time or frequency domain and we use them as a gateway between the two domains. Here we list four properties of sinusoids that we use for time–frequency transformations and two other properties that will come in handy in system analysis.

3.3.1 Sinusoidal Properties in the Time and Frequency Domains

1. Sinusoids have energy at only one frequency: the frequency of the sinusoid. This property is unique to sinusoids. Because of this property, sinusoids are sometimes referred to as "pure": the sound a sinusoidal wave produces is perceived as pure or basic. (Of the common orchestral instruments, the flute produces the most sinusoidal-like tone.)
2. Sinusoids are completely described by three parameters: magnitude (amplitude), phase, and frequency. If frequency is fixed, then only two variables are required, amplitude and phase. Moreover, if you use complex variables, these two parameters (amplitude and phase) can be rolled into a single complex variable.

We already know how to describe sinusoids in the time domain (e.g., Equation 1.14). Combining Property 1 with Property 2 gives us the frequency domain representation of a

---

[2]As we discover later in this chapter, Example 2.15 found only half of the frequency domain representation of the EEG signal. However, the half that is missing is not very informative and is often not displayed. But it is needed if we want to reconstruct the EEG signal from its frequency domain representation.

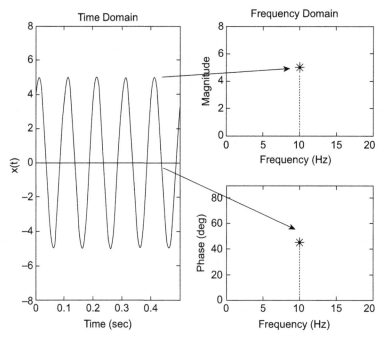

FIGURE 3.1   A sinusoid is completely represented by its magnitude and phase at a given frequency and its spectrum can be represented as single points on two plots: magnitude and phase against frequency.

sinusoid; i.e., its spectrum. It is uniquely defined by two points: one point on a plot of magnitude versus frequency and another on a plot of phase versus frequency, Figure 3.1. Both points are plotted against the frequency of the sinusoid. Moving sinusoids between the time and frequency domain is trivial. To go from the time to frequency domain: take the amplitude and phase of the sinusoid and plot them against its frequency in the magnitude and phase plots. To go from the frequency to time domain, find the sinusoidal amplitude from the magnitude plot, its phase from the phase plot, and the frequency from either plot and enter these into an equation such as Equation 1.14.

3.3.2 Sinusoidal Decomposition: The Fourier Series

If we could transform a general signal into sinusoids, we could use those sinusoidal components to determine the frequency domain representation of that signal. This brings us to the third important property of sinusoids: the "Fourier series."

3. All periodic signals[3] can be broken down into a series of sinusoids, plus a constant to account for any offset or DC component.

$$x(t)_{\text{periodic}} = C_0 + \text{sinusoid}_1 + \text{sinusiod}_2 + \text{sinusiod}_3 + \dots \text{sinusoid}_M \qquad (3.1)$$

---

[3]Recall that a periodic signal repeats itself exactly after some period of time, T: $x(t) = x(t + T)$.

where C_0 is the constant representing any offset or bias in the signal. The only constraint on the signal is that it must be periodic, or assumed to be periodic. Moreover, all the sinusoids in this series will be at the same or multiples of the frequency of the periodic signal. Hence, each sinusoidal component in the series will take the form:

$$C_m \cos(2\pi m f_1 t + \theta_m) \tag{3.2}$$

where f_1 is a constant equal to the frequency of the periodic signal ($f_1 \equiv f_p = 1/T$) and m is the reference number of the sinusoidal component and is an integer, $m = 1, 2, 3, \ldots$. The two variables are those that define the sinusoid (along with frequency): C_m, the amplitude of the mth sinusoid in the series, and θ_m, the phase angle of the mth sinusoid. A waveform at a multiple frequency of another waveform is termed a "harmonic" and a series containing sinusoids at multiple frequencies is called a harmonic series.[4] The sinusoidal series number m is also referred to as the harmonic number.

Substituting the format for a general sinusoid in Equation 3.2 into the series of Equation 3.1 leads to the formal Fourier series equation:

$$x(t) = \frac{C_0}{2} + \sum_{m=1}^{M} C_m \cos(2\pi m f_1 t + \theta_m) \tag{3.3}$$

Equation 3.3[5] works in both directions. You can represent any periodic signal by a sinusoidal series and you can construct any periodic signal by simply adding together harmonically related sinusoids. Of course, any sinusoidal component of the series must have a specific magnitude and phase (i.e., C_m and θ_m) to represent a given signal and you might need a large number of such sinusoids to accurately reproduce the signal, but it can always be done.[6] You can readily flip back and forth, convert the signal to a sinusoidal series, then sum the series to reconstruct the signal. This flipping between time and sinusoidal representation of a signal is illustrated in Figure 3.2. In this example the reconstructed signal (Figure 3.2 right side) is not perfect, but only three sinusoids were used to represent the signal.

3.3.3 Fourier Series Analysis and the Fourier Transform

Combining the sinusoidal decomposition properties of Property 3 with the frequency representation of sinusoids in Property 2 gives us a powerful and useful transformation: the ability to convert any periodic signal from the time domain into the frequency domain (and vice versa). Just decompose the periodic time domain signal into equivalent series of sinusoids,

---

[4]In music, an octave is the doubling in frequency of a musical tone. So in a harmonic series, each component is an octave apart from its higher and lower frequency neighbors.

[5]The first term of this equation is the mean value of $x(t)$. Different normalizations of the Fourier series can be found in the literature, but for the normalization used here C_0 is defined as twice the mean value (see Equation 3.10), hence it is divided by 2 in this equation.

[6]In a few cases, such as the square wave, you need an infinite number of sinusoidal waves. As shown in Figure 3.6, a finite series will exhibit an oscillatory artifact.

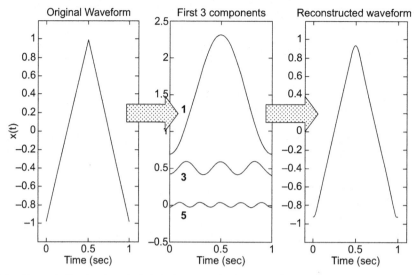

FIGURE 3.2 The periodic signal on the left (only one cycle shown) is decomposed into three harmonic sinusoids in the middle figure. (Recall that Figure 3.1 shows that these three signals can be represented by three points on a magnitude and phase frequency plots.) When the three sinusoids are added together, they produce a fair reconstruction of the original time domain signal (right side plot). More components would give a more accurate representation of the time single and a better reconstruction. (In Figure 3.5 we try reconstructing a periodic square wave using different numbers of sinusoids.)

and represent each sinusoid in the frequency domain as a magnitude and phase plot, Figure 3.3.

The only question is, how do we get the appropriate sinusoidal components? We already know the frequencies of the sinusoidal components; they are harmonics of the signal's frequency based on its period: $f_1 = 1/T$. (The base frequency, f_1, is also known as the "fundamental frequency.") The frequency associated with a given component is the fundamental frequency, f_1, times the harmonic number m:

$$f = \frac{m}{T} = mf_1 \tag{3.4}$$

For example, each C_m in Equation 3.3 would appear as a single point on the magnitude (Figure 3.3, upper right graph) plot, whereas each θ_m shows as a single point on the phase curve (Figure 3.3, lower right graph). These frequency plots provide the frequency domain representation, just as the original plot of $x(t)$ is the time domain representation.

We still need to find the magnitude and phases, the appropriate C_m's and θ_m's in Equation 3.3, but we already have an idea on how to find components in a time signal: use cross-correlation.[7] To summarize, sinusoids found through correlation provide a gateway between

[7]In practice we use alternative methods that are easier or faster. In the analog domain, we apply standard correlation (not cross-correlation) to a sine/cosine representation of Equation 3.3 and in the digital domain we apply correlation to a complex representation of the sinusoid (recall the Euler identity).

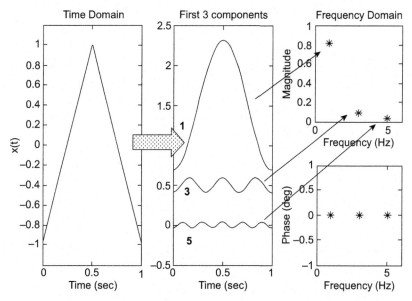

FIGURE 3.3   By combining the simplicity of a sinusoid's frequency representation (Property 2, a magnitude and a phase point on the frequency plot) with harmonic decomposition (Property 3) we can convert any periodic signal from a time domain plot to a frequency domain plot. If enough sinusoidal components are used, the frequency domain representation is equivalent to the time domain representation.

the time and frequency representations for any periodic function. We work out the details of calculating the correlation coefficients in the next section, but first we list some other useful sinusoidal properties.

4. As found in Chapter 2, harmonically related sinusoids are orthogonal (see Example 2.13). For Fourier series analysis this means that the values we calculate for a given Fourier series component (i.e., the C_m's and θ_m's) will be independent from the values of all other components. Practically, this means that if we decide to decompose a waveform into 10 harmonically related sinusoids, but later decide to decompose it into 12 sinusoids to attain more accuracy, the addition of more components will not change the C_m and θ_m values of the 10 components we already have.

The remaining two properties are not used in this chapter, but are useful in later discussions of systems analysis.

5. The calculus operations of differentiation and integration change only the magnitude and phase of a sinusoid. The result of either operation is still a sinusoid at the same frequency. For example, the derivative of a sine is a cosine, which is just a sine with a 90-degree phase shift; the integral of a sine is a negative cosine, the same as a sine with a -90-degree phase shift. In general, the derivative or integral of any sinusoid, $A \sin(2\pi ft + \theta)$, is unchanged except that A is scaled and θ is shifted by ± 90 degrees.

6. If the input to any linear system is a sinusoid, the output is another sinusoid at the same frequency irrespective of the complexity of the system. (Linear systems are defined

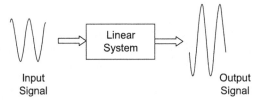

FIGURE 3.4 The output of any linear system driven by a sinusoid is a sinusoid at the same frequency. Only the magnitude and phase of the output differs from the input.

and studied in Chapters 5 and 6.) The only difference between the input and output is the magnitude, A, and phase, θ, of the sinusoid, Figure 3.4. This is because linear systems are composed of elements that can be completely defined by scaling, derivative, or integral operations and, as noted in Chapter 6, these operations modify only the magnitude and phase of a sinusoid.

3.3.4 Finding the Fourier Coefficients

The Fourier series equation, Equation 3.3, can give us the frequency domain representation of any signal, $x(t)$. For each sinusoidal term (i.e., each value of harmonic number m), the coefficient C_m contributes one magnitude frequency point, and the coefficient θ_m provides one magnitude phase point.[8] So all we have to do is find those coefficients. In principle we could use cross-correlation, but that is difficult to do analytically (and not efficient when we use the computer). Instead we split the sinusoidal term in Equation 3.3 into sine and cosine terms using Equations 2.23 and 2.24. The Fourier series equation of Equation 3.3 then becomes:

$$x(t) = \frac{a_0}{2} + \sum_{m=1}^{\infty} a_m \cos(2\pi m f_1 t) + \sum_{m=1}^{\infty} b_m \sin(2\pi m f_1 t) \tag{3.5}$$

where $a_0 \equiv C_0$ in Equation 3.3, and from Equation 2.24, noting that θ is defined as negative in those equations: $a_m = C_m \cos(-\theta) = C_m \cos(\theta)$ and $b_m = C_m \sin(-\theta) = -C_m \sin(\theta)$. In Equation 3.5 the a and b coefficients are the amplitudes of the cosine and sine components of $x(t)$. The Fourier series equation using a and b coefficients is referred to as the "rectangular representation," whereas the equation using C and θ is called the "polar representation." Rectangular to polar conversion is done using Equations 2.25 and 2.26 and are repeated here slightly modified:

$$C_m = \sqrt{a_m^2 + b_m^2} \tag{3.6}$$

---

[8]Note that the frequency domain representation of a continuous periodic signal is discrete: it is a series of discrete coefficients, C_m and θ_m. Later in this chapter we find that the frequency representation of a continuous aperiodic signal is continuous.

$$\theta_m = \tan^{-1}\left(\frac{-b_m}{a_m}\right) \tag{3.7}$$

Again, Equation 2.26 solves for negative θ, but in the Fourier series equations, Equations 3.2 and 3.3, θ is positive. By making b_m negative in Equation 3.7, the resulting θ becomes positive.

Using the rectangular representation of the Fourier series (Equation 3.5), we can find the a_m and b_m coefficients by simple correlation. We apply Equation 2.31, the basic correlation equation in the continuous domain, where $y(t)$ is the sine or cosine term and $x(t)$ is the signal (or vice versa):

$$a_m = \frac{2}{T}\int_0^T x(t)\cos(2\pi m f_1 t)dt \quad m = 1,2,3,\dots \tag{3.8}$$

$$b_m = \frac{2}{T}\int_0^T x(t)\sin(2\pi m f_1 t)dt \quad m = 1,2,3,\dots \tag{3.9}$$

These correlations are calculated for each harmonic number, m, to obtain the Fourier series coefficients, the a_m's and b_m's, representing the cosine and sine amplitudes at the associated frequencies, $2\pi m f_1$. (A formal derivation of Equations 3.8 and 3.9 are given in Appendix A-2.) Equations that calculate the Fourier series coefficients from $x(t)$ are termed "analysis equations." Equation 3.5 works in the other direction, generating $x(t)$ from the a's and b's, and is known as the "synthesis equation."

The factor of 2 is used because in the Fourier series, the coefficients, a_m and b_m, are defined in Equation 3.3 as amplitudes, not correlations. However, there is really no agreement on how to scale the Fourier equations, so you might find these equations with other scalings. MATLAB's approach is avoidance: its routine for determining these coefficients uses no scaling.[9] When using MATLAB to find a_m and b_m, it is up to you to scale the output as you wish.

Our usual strategy for transforming a continuous signal into the frequency domain is to first calculate the a and b coefficients, then convert them to C and θ coefficients using Equations 3.6 and 3.7. We use this approach because it is easier to work out the correlations using sines and cosines (i.e., Equations 3.8 and 3.9) than cross-correlating against a sinusoidal term. That said, for complicated signals, it can still be quite difficult to solve for the Fourier coefficients equations analytically. Fortunately, for all real-world signals, these equations are implemented on a computer.

The constant term in Equation 3.5, $a_0/2$, is the same as $C_0/2$ in Equation 3.3 and is also known as the "DC term." It accounts for the offset or bias in the signal. If the signal has zero mean, as is often the case, then $a_0 = C_0 = 0$. Otherwise, the value of the DC term is just twice the mean:

$$a_0 = \frac{2}{T}\int_0^T x(t)dt \tag{3.10}$$

[9]Although MATLAB determines the Fourier series coefficients using an algorithm that involves the complex representation of a sinusoid, the output of this algorithm is essentially the a and b coefficients which are the real and imaginary parts of the output variable.

The reason a_0 and C_0 are calculated as twice the average value is for them to be compatible with the Fourier analysis equations of Equations 3.8 and 3.9, which also involve a factor of 2. To offset this doubling, a_0 and/or C_0 are divided by 2 in the synthesis equations, Equations 3.3 and 3.5 (logical, if a bit confusing).

Sometimes, $2\pi m f_1$ is stated in terms of radians, where $2\pi m f_1 = m\omega_1$. Using frequency in radians makes the equations look cleaner, but in engineering practice frequency is measured in hertz. Both are used here. Another way of writing $2\pi m f_1$ is to combine the mf_1 into a single term, f_m, so the sinusoid is written as $C_m \cos(2\pi f_m t + \theta_m)$ or in terms of radians as $C_m \cos(\omega_m t + \theta_m)$. So an equivalent representation of Equation 3.3 is:

$$x(t) = \frac{C_0}{2} + \sum_{m=1}^{\infty} C_m \cos(\omega_m t + \theta_m) \tag{3.11}$$

Equations 3.8 and 3.9 are also sometimes written in terms of a period normalized to 2π. This can be useful when working with a generalized $x(t)$ without the requirement for defining a specific period.

$$a_m = \frac{2}{\pi} \int_{-\pi}^{\pi} x(t)\cos(mt)dt \quad m = 1, 2, 3, \ldots \tag{3.12}$$

$$b_m = \frac{2}{\pi} \int_{-\pi}^{\pi} x(t)\sin(mt)dt \quad m = 1, 2, 3, \ldots \tag{3.13}$$

Here we always work with time functions that have a specific period to make our analyses correspond more closely to real-world applications. So Equations 3.12 and 3.13 are for reference only.

To implement the integration and correlation in Equations 3.8 and 3.9, there are a few constraints on $x(t)$. First, $x(t)$ must be capable of being integrated over its period; specifically:

$$\int_0^T |x(t)|dt < \infty \tag{3.14}$$

Unfortunately this constraint rules out a large class of interesting signals: transient signals as described in Section 1.4.2 and shown in Figure 1.20. Recall, these are signals that change, rapidly or slowly, but do not repeat and do not return to a baseline in a finite amount of time. Because the change lasts forever, the integral in Equation 3.14 must be taken between 0 to ∞ and is not finite. In some cases, you might be able to recast a transient signal into a periodic signal and this is explored in one of the problems.

The second constraint is that, although $x(t)$ can have discontinuities, those discontinuities must be finite in number and have finite amplitudes. Finally, the number of maxima and minima must also be finite. These three criteria are sometimes referred to as the "Dirichlet conditions" and are met by many real-world signals.

A brief note about terminology: The analysis (Equations 3.8 and 3.9) and their discrete equivalents should be called "Fourier series analysis," but they are often called the "Fourier transform," especially when implemented in the discrete domain on a computer. Technically, "Fourier transform" should be reserved for the analysis of continuous aperiodic signals. Likewise the synthesis equation, Equation 3.5, and its discrete equivalent are usually called the "inverse Fourier transform." This usage is so common that it is pointless to make distinctions between the Fourier transform and Fourier series analysis.

The more sinusoids included in the summations of Equations 3.3 or 3.5, the better the representation of the signal, $x(t)$. For an exact representation of a continuous signal, the summation should be infinite, but in practice the number of sine and cosine components that have meaningful amplitudes is limited. Often only a few sinusoids are required for a decent representation of the signal. Figure 3.5 shows the reconstruction of a square wave by a different number of sinusoids: 3, 9, 18, and 36. The square wave is one of the most difficult waveforms to represent using a sinusoidal series because of the sharp transitions. Figure 3.5 shows that the reconstruction is fairly accurate even when the summation contains only nine sinusoids.

The square wave reconstructions shown in Figure 3.5 become sharper as more sinusoids are added, but still contain oscillations. These oscillations, termed Gibbs artifacts, occur whenever a finite sinusoidal series is used to represent a signal with discontinuity. They

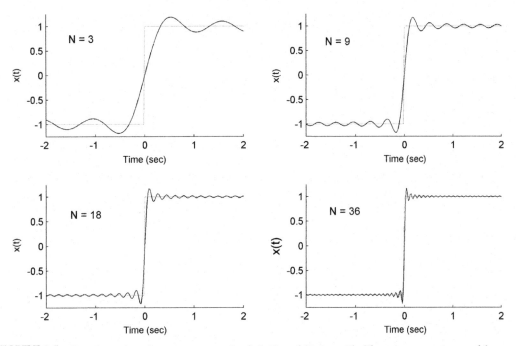

FIGURE 3.5   Reconstruction of a square wave using 3, 9, 18, and 36 sinusoids. The square wave is one of the most difficult signals to represent with a sinusoidal series. The oscillations seen in the sinusoidal approximations are known as "Gibbs artifacts." They increase in frequency, but do not diminish in amplitude, as more sinusoids are added to the summation.

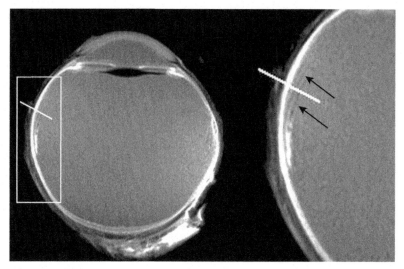

FIGURE 3.6   MR image of the eye showing Gibbs artifacts near the left boundary of the eye. This section is enlarged on the right side. *Image courtesy of Susan and Lawrence Strenk of MRI Research, Inc.*

increase in frequency when more sinusoidal terms are added so the largest overshoot moves closer to the discontinuity. Gibbs artifacts occur in a number of circumstances that involve truncation. They can be found in MR images when there is a sharp transition in the image and the resonance signal is truncated during data acquisition. They are seen as subtle dark ridges adjacent to high contrast boundaries as shown in the MR image of Figure 3.6. We encounter Gibbs artifacts again due to truncation of digital filter coefficients in Chapter 9.

3.3.5 Symmetry

Some waveforms are symmetrical or antisymmetrical about $t = 0$ so that one or the other of the Fourier series coefficients, either the a_m's or b_m's, will be zero. If the waveform has mirror symmetry about $t = 0$, that is, $x(t) = x(-t)$, Figure 3.7 (upper plot), then multiplications with all sine functions will be zero, so the b_m terms will be zero. Such mirror symmetry functions are termed "even functions." Functions with antisymmetry, $x(t) = -x(t)$, Figure 3.7 (middle plot), are "odd functions," and all multiplications with cosines will be zero so the a_m coefficients will be zero. Finally, functions that have half-wave symmetry will have no even coefficients, so both a_m and b_m will be zero for $m = $ even. These are functions where the second half of the period looks like the first half, but inverted, i.e., $x(t - T/2) = -x(t)$, Figure 3.7 (lower plot). Functions having half-wave symmetry are also even functions. These symmetries are useful not only for simplifying the task of solving for the coefficients manually, but also for checking solutions done on a computer. Table 3.1 and Figure 3.7 summarize these properties.

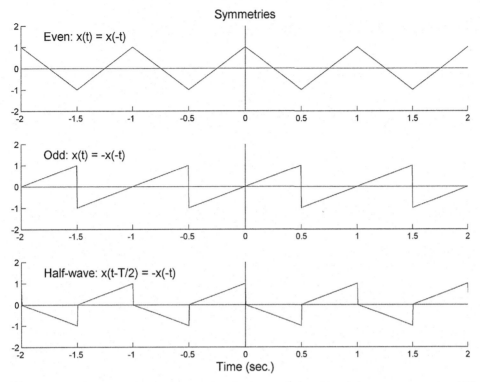

FIGURE 3.7 Waveform symmetries. Each waveform has a period of 1 s. Upper plot: even symmetry; middle plot: odd symmetry; lower plot: half-wave symmetry. Note that the lower waveform also has even symmetry.

TABLE 3.1 Function Symmetries

Function Name	Symmetry	Coefficient Values
Even	$x(t) = x(-t)$	$b_m = 0$
Odd	$x(t) = -x(-t)$	$a_m = 0$
Half-wave	$x(t) = x(t - t/2)$	$a_m = b_m = 0$; for m even

EXAMPLE 3.1

Find the Fourier series of the triangle waveform shown in Figure 3.8. The equation for this waveform is:

$$x(t) = \begin{cases} t & 0 < t \leq 0.5 \\ 0 & 0.5 < t \leq 1.0 \end{cases}$$

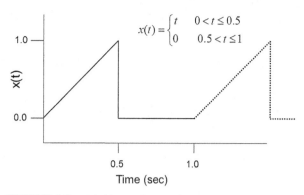

FIGURE 3.8 A half-triangle waveform used in Example 3.1.

Find the first four magnitude and phase components (i.e., $m = 1, 2, 3, 4$). Construct frequency plots of the magnitude components.

Solution: Use Equations 3.8 and 3.9 to find the a_m and b_m coefficients. Then convert to magnitude, C_m, and phase, θ_m, using Equations 3.6 and 3.7. Plot C_m and θ_m against the associated frequency, $f = mf_1$.

Start by evaluating either the sine or cosine coefficients; we begin with the sine coefficients using Equation 3.9:

$$b_m = \frac{2}{T} \int_0^T x(t)\sin(2\pi m f_1 t)dt = \frac{2}{1} \int_0^{0.5} t\sin(2\pi m t)dt$$

$$= \frac{2}{4\pi^2 m^2}[\sin(2\pi m t) - 2\pi m t\cos(2\pi m t)]\big|_0^{0.5}$$

$$= \frac{2}{4\pi^2 m^2}[\sin(\pi m) - \pi m\cos(\pi m)] = \frac{-1}{2\pi^2 m^2}[\cos(\pi m)]$$

$$b_m = \frac{1}{2\pi}; \frac{-1}{4\pi}; \frac{1}{6\pi}; \frac{-1}{8\pi}; \dots = 0.159; -0.080; 0.053; -0.040; \dots$$

To find the cosine coefficients, use Equation 3.8:

$$a_m = \frac{2}{T} \int_0^T x(t)\cos(2\pi m f_1 t)dt = \frac{2}{1} \int_0^{0.5} t\cos(2\pi m t)dt$$

$$= \frac{2}{4\pi^2 m^2}[\cos(2\pi m t) - 2\pi m t\sin(2\pi m t)]\big|_0^{0.5}$$

$$= \frac{2}{4\pi^2 m^2}[\cos(\pi m) - \pi m\sin(\pi m) - 1] = \frac{-1}{2\pi^2 m^2}[\cos(\pi m) - 1]$$

$$a_m = \frac{-1}{\pi^2}; \ 0; \ \frac{-1}{9\pi^2}; \ 0; \dots = -0.101; 0; -0.0118; 0; \dots$$

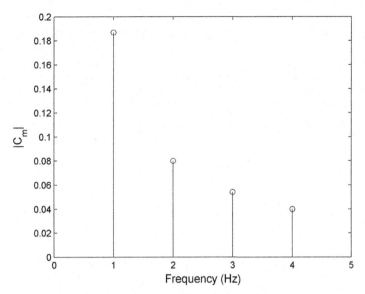

FIGURE 3.9 Magnitude plot (*circles*) of the frequency characteristics of the waveform given in Example 3.1. Only the first four components are shown because these are the only ones calculated in the example. These components were determined by taking the square root of the sum of squares of the first four cosine and sine terms (*a*'s and *b*'s).

To express the Fourier coefficients in terms of magnitude, C_m, and phase, θ_m, use Equations 3.6 and 3.7. Care must be taken in computing the phase angle to ensure that it represents the proper quadrant[10]:

$$C_m = \sqrt{a_m^2 + b_m^2} = 0.187, 0.080, 0.054, 0.040$$

$$\theta_m = \tan^{-1}\left(\frac{-b_m}{a_m}\right) = -122.42, 90.00, -101.72, 90.00 \text{ degrees}$$

For a complete description of $x(t)$ we need more components and also the a_0 term.

The a_0 term is twice the average value of $x(t)$, which can be obtained using Equation 3.10:

$$C_0 = a_0 = \frac{2}{T}\int_0^T x(t)dt = \frac{2}{1}\int_0^{0.5} t\,dt = \frac{2t^2}{2}\bigg|_0^{0.5} = 0.25$$

$$\frac{a_0}{2} = \bar{x}(t) = \frac{1}{T}\int_0^T x(t)dt = \frac{1}{1}\int_0^{.5} t\,dt = \frac{t^2}{2}\bigg|_0^{.5} = .125$$

To plot the spectrum of this signal, or rather a partial spectrum, we plot the sinusoid coefficients, C_m and θ_m, against frequency. Since the period of the signal is 1.0 second, the fundamental frequency, f_1, is 1 Hz. Therefore the first four values of m represent 1, 2, 3, and 4 Hz. The magnitude plot is shown in Figure 3.9.

---

[10]The arctangent function on some calculators does not take into account the signs of b_m an a_m in which case it is up to you to figure out the proper quadrant. This is also true of MATLAB's atan function. However, MATLAB does have a function, atan2(b,a), that takes the signs of b and a into account and produces an angle (in radians) in the proper quadrant.

The four components computed in Example 3.1 are far short of that necessary for a reasonable frequency representation, but when we implement Fourier series analysis on a computer we solve for hundreds, or even thousands, of components. We can check on how good the representation is by reconstructing the signal using only these four components. We do that in the next example with the help of MATLAB.

EXAMPLE 3.2

Reconstruct the waveform of Example 3.1 using the four components found in that example. Use the polar representation (i.e., magnitude and phase) of the Fourier series equation, Equation 3.3, to reconstruct the signal and plot the time domain reconstruction.

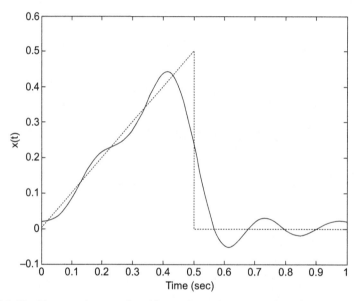

FIGURE 3.10   The waveform produced by applying the magnitude and phase version of the Fourier series equation (Equation 3.3) to the four magnitude and phase components found analytically in Example 3.1. The Fourier summation (*solid line*) produces a rough approximation of the original waveform (*dotted line*), even though only four components are used.

Solution: Apply Equation 3.3 directly using the four magnitude and phase components found in the last example. Remember to add the DC components, a_0. Plot the result of the summation.

```
% Example 3.1 Reconstruct the waveform of Example 3.1
fs = 500;                          % Assumed sample frequency
N = 500;                           % Number of points for 1 sec.
t = (1:N)/N;
C = [0.187 0.08 0.054 0.04];       % Component magnitudes
theta = [-122 90 -101 90]*2*pi/360; % Component phase (in radians)
```

```
%
x = zeros(1,N);
for f = 1:4                          % Add the 4 terms of Equation 3.3
   x = x + C(f)*cos(2*pi*f*t + theta(f));   % using appropriate A and theta
end
x = x + 0.125;                       % Add the DC term
plot(t,x,'k');                       % Plot the result
     .......labels.......
```

The result produced by this program is shown in Figure 3.10, which shows it to be a rough approximation of the original signal.

Fourier series analysis is not the only way to transform a time signal into the frequency domain, but it is the most general approach and the most often used. Other approaches require you to make some assumptions about the signal. The digital version of Equations 3.8 and 3.9 can be calculated with great speed using an algorithm known as the "fast Fourier transform," abbreviated "FFT." There is also an "inverse fast Fourier transform," abbreviated "IFFT," which implements the synthesis equation (Equation 3.3 or 3.5).

The resolution of a spectrum can be loosely defined as the difference in frequencies that a spectrum can resolve: that is, how close two signal frequencies can get and still be identified as two frequencies in the frequency domain. This resolution clearly depends on the frequency spacing between harmonic numbers, which in turn is equal to $1/T$ (Equation 3.4). Hence the longer the signal period, the better the spectral resolution. Later we find this holds for digital data as well.

3.3.6 Complex Representation

It is possible to rewrite the Fourier analysis and synthesis equations using the complex representation of the sinusoid as given by Euler's identity (Equation 2.28). You might wonder why we would do this given that we have a perfectly good set of equations for moving between the time and frequency domains. There are actually four reasons: (1) the complex equations are more succinct and there is only one analysis equation; (2) in a few cases the complex equations are easier to solve analytically; (3) computer algorithms use the complex equations; and (4) this is the way you will see the Fourier transform equations written in research papers.

The complex representation of the Fourier series analysis can be derived directly from Equation 3.5 using only algebra and Euler's identity (Equation 2.28). We start with the exponential definitions of the sine and cosine functions which come directly from Euler's identity:

$$\cos(2\pi m f_1 t) = \frac{1}{2}\left(e^{+j2\pi m f_1 t} + e^{-j2\pi m f_1 t}\right) \text{ and}$$

$$\sin(2\pi m f_1 t) = \frac{1}{j2}\left(e^{+j2\pi m f_1 t} - e^{-j2\pi m f_1 t}\right)$$

(3.15)

Recall the Fourier series synthesis equation (Equation 3.5) repeated here:

$$x(t) = \frac{a_0}{2} + \sum_{m=1}^{\infty} a_m \cos(2\pi m f_1 t) + \sum_{m=1}^{\infty} b_m \sin(2\pi m f_1 t)$$

Substituting the complex definitions for the sine and cosines into this equation and expanding:

$$x(t) = \frac{a_0}{2} + \sum_{n=1}^{\infty} \left(\frac{a_m}{2} e^{j2\pi m f_1 t} + \frac{a_m}{2} e^{-j2\pi m f_1 t} + \frac{b_m}{2j} e^{j2\pi m f_1 t} - \frac{b_m}{2j} e^{-j2\pi m f_1 t} \right)$$

Moving the j term to the numerator using the fact that $1/j = -j$:

$$x(t) = \frac{a_0}{2} + \sum_{n=1}^{\infty} \left(\frac{a_m}{2} e^{j2\pi m f_1 t} + \frac{a_m}{2} e^{-j2\pi m f_1 t} - \frac{jb_m}{2} e^{j2\pi m f_1 t} + \frac{jb_m}{2} e^{-j2\pi m f_1 t} \right)$$

Rearranging:

$$x(t) = \frac{a_0}{2} + \sum_{n=1}^{\infty} \left(\left(\frac{a_m}{2} - \frac{jb_m}{2} \right) e^{j2\pi m f_1 t} + \left(\frac{a_m}{2} + \frac{jb_m}{2} \right) e^{-j2\pi m f_1 t} \right)$$

Collecting terms into positive and negative summations of m:

$$x(t) = \frac{a_0}{2} + \sum_{m=1}^{\infty} \left(\frac{a_m - jb_m}{2} \right) e^{j2\pi m f_1 t} + \sum_{m=-\infty}^{-1} \left(\frac{a_{-m} + jb_{-m}}{2} \right) e^{j2\pi m f_1} \tag{3.16}$$

This can be combined into a single equation going from negative to positive infinity:

$$x(t) = \sum_{m=-\infty}^{m=\infty} X_m e^{j2\pi m f_1 t} \tag{3.17}$$

where the new coefficient, X_m is defined as:

$$X_{+m} = \frac{a_m - jb_m}{2}; \quad X_{-m} = \frac{a_m + jb_m}{2}; \quad X_0 = a_0 \tag{3.18}$$

Equation 3.17 is the complex Fourier series synthesis equation. Note the DC term, a_0, is included in the complex number series X_m, which also incorporates both the a_m and b_m coefficients.

To find the Fourier series analysis equation, substitute the original correlation equations for a_m and b_m (Equations 3.8 and 3.9) into Equation 3.18:

$$X_m = \frac{2}{2T} \int_0^T x(t) \cos(2\pi m f_1 t) dt - \frac{j2}{2T} \int_0^T x(t) \sin(2\pi m f_1 t) dt$$

Combining:

$$X_m = \frac{1}{T} \int_0^T x(t)[\cos(2\pi m f_1 t) - j\sin(2\pi m f_1 t)]dt \tag{3.19}$$

The term in the brackets in Equation 3.19 has the form $\cos(x) - j\sin(x)$. This is equal to e^{-jx} as given by Euler's identity (Equation 2.28). Hence the term in the brackets can be replaced by a single exponential, which leads to the complex form of the Fourier series:

$$X_m = \frac{1}{T} \int_0^T x(t)e^{-j2\pi m f_1 t}\, dt \quad m = 0, \pm 1, \pm 2, \pm 3, \ldots \tag{3.20}$$

Alternatively, since $f_1 = 1/T$

$$X_m = \frac{1}{T} \int_0^T x(t)e^{\frac{-j2\pi m t}{T}}dt \quad m = 0, \pm 1, \pm 2, \pm 3, \ldots \tag{3.21}$$

This is the most common representation of Fourier series analysis commonly called the Fourier transform. It is also the equation you will see whenever this transform is used and referenced in a research paper. In the complex representation, the harmonic number m must range from minus to plus infinity as in Equation 3.20, essentially to account for the negative signs introduced by the complex definition of sines and cosines (Equation 3.15). However, the $a_0/2$ term (the average or DC value) does not require a separate equation, since when $m = 0$ the exponential becomes: $e^{-j0} = 1$, and the integral computes the average value. Finally, only one analysis equation is needed, as the a_m and b_m coefficients are embedded in X_m as detailed in the following discussion.

The negative frequencies implied by the negative m terms are often dismissed as mathematical contrivances since negative frequency has no meaning in the real world. However, when a signal is digitized these negative frequencies do have consequences as shown later. In fact, C_m is symmetrical about zero frequency so the negative frequency components are just mirror images of the positive components. Since they are the same components just in reverse order, they do not provide additional information, but they do double the summations, so the normalization is now $1/T$, not the $2/T$ used for the noncomplex analysis equations. (Recall that there are different normalization strategies and, although MATLAB uses the complex form to compute the Fourier transform, it does not normalize the summation. You are expected to apply whatever normalization you wish.)

The complex variable C_m combines the cosine and sine coefficients. To get the a_m's and b_m's from C_m just double the real and imaginary parts:

$$a_m = 2\text{Re}(X_m); \quad b_m = 2\text{Im}(X_m) \tag{3.22}$$

The magnitude C_m and phase θ_m representation of the Fourier series can be found from C_m using complex algebra:

$$|X_m| = \sqrt{\text{Re}(X_m)^2 + \text{Im}(X_m)^2} = \sqrt{\frac{a_m^2 + b_m^2}{2^2}} = \frac{1}{2}\sqrt{a_m^2 + b_m^2} = 0.5C_m \tag{3.23}$$

$$\text{Angle}(C_m) = \tan^{-1}\left(\frac{\text{Im}(X_m)}{\text{Re}(X_m)}\right) = \tan^{-1}\left(\frac{\dfrac{-b_m}{2}}{\dfrac{a_m}{2}}\right) = \tan^{-1}\left(\frac{-b_m}{a_m}\right) = \theta \qquad (3.24)$$

The magnitude of X_m is equal to $1/2\,C_m$, the magnitude of the sinusoidal components, and the angle of C_m is equal to the phase, θ_m, of the sinusoidal components. Not only does X_m contain both sine and cosine coefficients, but these components can also readily be obtained in either the polar (X's and θ's, most useful for plotting) or the rectangular form (a's and b's). (Remember, the factors of 2 in Equation 3.22 and 0.5 in Equation 3.23 are based on the normalization strategy presented here.) Finally, X_0 is just the DC term (i.e., C_0 or a_0) so it must be real (the b_0 equivalent is zero).

EXAMPLE 3.3

(A) Find the Fourier series of the pulse waveform shown in Figure 3.11 with a period T, and amplitude V_p, and a pulse width of W. Use Equation 3.20, the complex equation. (B) Use MATLAB to plot the solution for $T = 1$ s; $V_p = 1.0$ and two values of pulse width: $W = 0.005$ and $W = 0.005$ s. Plot the first ± 100 harmonics.

(A) Solution: Apply Equation 3.20 directly, except since the signal is an even function ($x(t) = x(-t)$), it is easier to integrate from $t = -T/2$ to $+T/2$.

$$X_m = \frac{1}{T}\int_{-T/2}^{T/2} x(t)e^{-j2\pi mf_1 t}\,dt = \frac{1}{T}\int_{-W/2}^{W/2} V_p e^{-j2\pi mf_1 t}dt$$

$$= \frac{V_p}{T(-j2\pi mf_1)}\left[e^{-j2\pi mf_1 W/2} - e^{j2\pi mf_1 W/2}\right] = \left[\frac{V_p}{\pi mf_1 T}\right]\frac{\left[e^{j2\pi mf_1 W/2} - e^{-j2\pi mf_1 W/2}\right]}{2j}$$

where the second term in brackets is $\sin(2\pi mf_1 W/2) = \sin(\pi mf_1 W)$. So the expression for C_m becomes:

$$X_m = \frac{V_p}{\pi mf_1 T}\sin(\pi mf_1 W) \quad m = 0, \pm1, \pm2, \pm3, \ldots$$

It is common to rearrange this equation in the form $\frac{\sin(x)}{x}$.

$$X_m = \left(\frac{V_p W}{T}\right)\frac{\sin(\pi mf_1 W)}{\pi mf_1 W} \quad m = 0, \pm1, \pm2, \pm3, \ldots$$

(B) Solution MATLAB: Although this particular function of X_m is real and does not have an imaginary component, it still represents both magnitude and phase components. Plotting these components in MATLAB is straightforward. We enter the values given for T, V_p, and W. We set the harmonic number, m, to range between ±100. Three MATLAB functions are then used to simplify the code: the sinc function to evaluate $\sin(\pi x)/\pi x^{11}$; the abs (absolute value) function to get the

---

[11] The function $\sin(x)/x$ is also known as sinc(x). The MATLAB function includes π in the numerator and dominator, i.e., in MATLAB sinc(x) $= \sin(\pi x)/\pi x$.

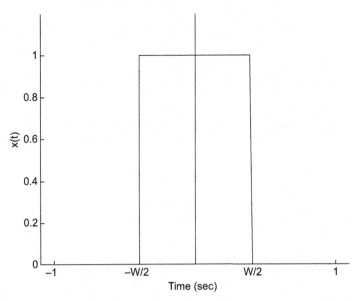

FIGURE 3.11   Pulse wave used in Example 3.3.

magnitude of C_m; and the `angle` function to get the phase of X_m. MATLAB always gives angles in radians so we convert to degrees by multiplying the angle by $360/2\pi$.

```
%  Example 3.3 Plot the magnitude spectrum of two pulse waveforms
%  found from the complex form of the Fourier series analysis in Example 3.3
%
T = 1.0;                   % Period (sec)
f1 = 1/T;                  % Pulse wave fundamental frequency
Vp = 1.0;                  % Pulse amplitude, Vp
W = [0.01 0.05];           % Pulse widths
m = -100:100;              % Harmonic numbers
%
for k = 1:length(W);
  Xm = ((Vp*W(k))/T)*sinc(m*f1*W(k));       % Solution eq.
  Mag = abs(Xm);                   % Get magnitude spectrum
  Phase = angle(Xm)*360/(2*pi);    % Get phase spectrum (in deg)
  subplot(2,2,2*k-1); hold on
  plot(m,Mag,'.k');                % Plot magnitude spectrum
  .....labels, title, and zero frequency line.......
  subplot(2,2,2*k); hold on;
  plot(m,Phase,'.k');              % Plot the phase spectrum
  .....labels, title, and zero frequency line.......
end
```

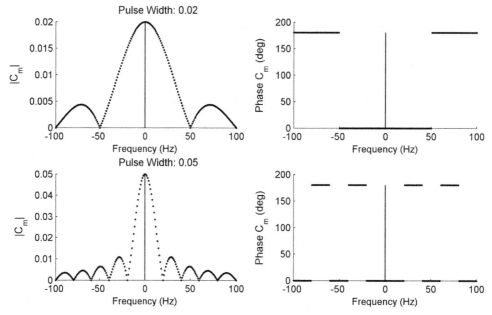

FIGURE 3.12   The frequency spectra of two pulse functions having different pulse widths found using the complex Fourier series analysis. The spectra consist of individual points spaced $f_1 = 1/T = 1.0$ Hz apart. The longer pulse produces a spectrum with a shorter peak (bottom figures). Both spectra have the shape of $|\sin(x)/x| \equiv |\text{sinc}(x)|$. When $\sin(x)$ switches between positive and negative, the phase spectra alternate between 0 and 180 degrees. Note the inverse relationship between pulse width and the width of the spectrum: the narrower the pulse the broader the spectrum and vice versa.

Result: The magnitude and phase spectra of the two pulses are shown Figure 3.12. These spectra are discrete consisting of individual points spaced $1/T$ or 1.0 Hz apart. Both curves take the shape of the magnitude of function $\sin(x)/x$, also known as $\text{sinc}(x)$. Note the inverse relationship between pulse width and spectrum width: a wider time domain pulse produces a narrower frequency domain spectrum. This inverse relationship is typical of time–frequency transformations. Later we will see that in the limit, as the pulse width $\rightarrow 0$, the spectrum becomes infinitely broad, i.e., a flat horizontal line.

EXAMPLE 3.4

Find the Fourier series of the waveform shown in Figure 3.13 using the complex form of the equation. The waveform is periodic and can be described over one period as: $x(t) = e^{-2.4t} - 1$.

Solution: Apply Equation 3.20 directly, noting that the equation for the waveform is:

$$x(t) = e^{2.4t} - 1 \quad \text{for } 0 < t \leq 1 \text{ and } f_1 = 1/T = 1.$$

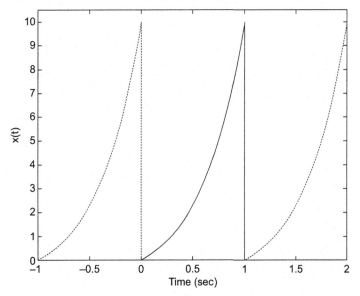

FIGURE 3.13 Periodic waveform used in Example 3.4.

$$X_m = \frac{1}{T} \int_0^T x(t)e^{-j2\pi mf_1 t}dt = \frac{1}{T} \int_0^T (e^{2.4t} - 1)e^{-j2\pi mf_1 t}dt$$

$$X_m = \int_0^1 (e^{2.4t}e^{-j2\pi mt} - e^{-j2\pi mt})dt = \int_0^1 e^{t(2.4-j2\pi m)}dt - \int_0^1 e^{-j2\pi mt}dt$$

$$X_m = \left.\frac{e^{t(2.4-j2\pi m)}}{2.4 - j2\pi m}\right|_0^1 - \left.\frac{e^{j2\pi m}}{-j2\pi m}\right|_0^1$$

$$X_m = \frac{e^{2.4-j2\pi m} - 1}{2.4 - j2\pi m} - \frac{e^{-j2\pi m} - 1}{-j2\pi m}$$

If we use Euler's identity to substitute in cosine and sine terms for $e^{-j2\pi m}$ ($e^{-j2\pi m} = \cos(2\pi m) + j \sin(2\pi m)$), we see that the sine term is always zero and the cosine term always 1 if m is an integer or zero. Hence, $e^{-j2\pi m} = 1$ for $m = 0, \pm1, \pm2, \pm3,\dots$. This gives:

$$X_m = \frac{e^{2.4}e^{-j2\pi m}}{2.4 - j2\pi m} - \frac{e^{-j2\pi m} - 1}{-j2\pi m} = \frac{e^{2.4} - 1}{2.4 - j2\pi m} - \frac{1-1}{-j2\pi m} = \frac{e^{2.4} - 1}{2.4 - j2\pi m} \quad m = 0, \pm1, \pm2, \pm3, \dots$$

The problem set contains additional applications of the complex Fourier series equation.

3.3.7 The Continuous Fourier Transform

The Fourier series analysis is a good approach to determining the frequency or spectral characteristics of a periodic waveform, but what if the signal is not periodic? Most real signals are not periodic, and for many physiological signals, such as the EEG signal introduced in the first

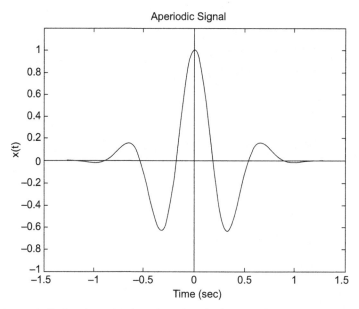

FIGURE 3.14 An aperiodic function exists for a finite length of time and is zero everywhere else. Unlike a periodic sine wave, you are seeing the complete signal here; this is all there is!

chapter, only a part of the signal is available. The segment of EEG signal shown previously is just a segment of the actual recording, but no matter how long the record, the EEG signal existed before the recording and will continue after the recording session ends (unless the EEG recording session was so traumatic as to cause untimely death!). Dealing with very long signals generally entails estimations or approximations, but if the signal is continuous and aperiodic, an extension of the Fourier series analysis can be used. An aperiodic signal is one that exists for a finite period of time and is zero at all other times, Figure 3.14.

To extend the Fourier series analysis to aperiodic signals, these signals are treated as periodic, but with a period that goes to infinity, (i.e., $T \rightarrow \infty$). If the period becomes infinite, then $f_1 = 1/T \rightarrow 0$; however, mf_1 does not go to zero since m goes to infinity. This is a case of limits: as T gets longer and longer, f_1 becomes smaller and smaller as does the increment between harmonics, Figure 3.15. In the limit, the frequency increment mf_1 becomes a continuous variable f. All the various Fourier series equations described above remain pretty much the same for aperiodic functions, only the $2\pi mf_1$ goes to $2\pi f$. If radians are used instead of Hz, then $m\omega_1$ goes to ω in these equations. The equation for the continuous Fourier transform in complex form is:

$$\lim_{\substack{T \rightarrow \infty \\ f \rightarrow 0}} X_m = \int_0^T x(t)e^{-j2\pi mf_1 t}dt = \int_{-\infty}^{\infty} x(t)e^{-2\pi ft}dt$$

$$X(f) = \int_{-\infty}^{\infty} x(t)e^{-2\pi ft}dt \text{ or } X(\omega) = \int_{-\infty}^{\infty} x(t)e^{-\omega t}dt \qquad (3.25)$$

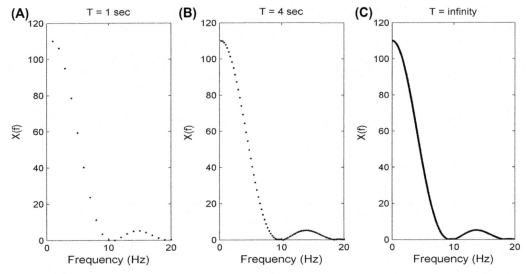

FIGURE 3.15   The effect of increasing the period, T, on the frequency spectrum of a signal. (A) When the period is relatively short (1 s), the spectral points are widely spaced at 1 Hz. (B) When the period becomes longer (4 s), the points are spaced closer together at intervals of 0.25 Hz. (C) When the period becomes infinite, the points become infinitely close together and what was a discrete set of points becomes a continuous curve.

Or for the noncomplex equations in terms of the sine and cosine:

$$a(f) = \int_{-\infty}^{\infty} x(t)\cos(2\pi ft)dt \ \text{ or } \ a(\omega) = \int_{-\infty}^{\infty} x(t)\cos(\omega t)dt$$

$$b(f) = \int_{-\infty}^{\infty} x(t)\sin(2\pi ft)dt \ \text{ or } \ b(\omega) = \int_{-\infty}^{\infty} x(t)\sin(\omega t)dt \tag{3.26}$$

These transforms produce a continuous function as an output and it is common to denote these terms with capital letters. Also the transform equation is no longer normalized by the period since $1/T \to 0$. Although the transform equation is theoretically integrated over times between $\pm\infty$, the actual limits will only be over the nonzero values of $x(t)$.

Although it is rarely used in analytical computations, the inverse continuous Fourier Transform is given in the following complex format. This version is given in radians, ω, instead of $2\pi f$ (just for variety):

$$x(t) = \frac{1}{2\pi} \int_{-\infty}^{\infty} X(\omega)e^{j\omega t}d\omega \tag{3.27}$$

where ω is frequency in radians.

EXAMPLE 3.5

Find the Fourier transform of the pulse in Example 3.3 assuming the period, T, goes to infinity.

Solution: We could apply Equation 3.20, but since $x(t)$ is an even function we could also use only the cosine equation of the Fourier transform in Equation 3.26 knowing that all of the sine terms would be zero for the reasons given in Section 3.3.5 on symmetry.

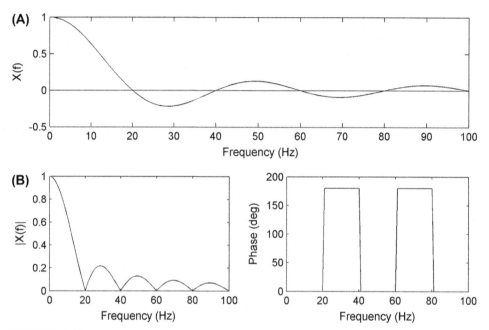

FIGURE 3.16 (A) The complex spectrum of an aperiodic pulse as determined in Example 3.5. $W = 0.05$ s. The complex function $X(f)$ is shown in the upper plot. (B) The magnitude and phase are shown in the lower plots. The vertical scale is not the same as in Figure 3.12 because the spectrum cannot be normalized by $1/T$ since $T \rightarrow \infty$.

$$X(f) = \int_{-\infty}^{\infty} x(t)\cos(2\pi ft)dt = \int_{-W/2}^{W/2} V_p \cos(2\pi ft)dt = \frac{V_p}{2\pi f}\sin(2\pi f)|_{t=-W/2}^{t=W/2}$$

$$X(f) = \frac{V_p}{2\pi f}(\sin(2\pi f\ W/2) - \sin(-2\pi f\ W/2)) = \frac{V_p}{2\pi f}(\sin(\pi f\ W) + \sin(\pi f\ W)) \qquad (3.28)$$

$$X(f) = \frac{V_p}{2\pi f}(2\sin(\pi f\ W)) = \frac{V_p \sin(\pi f\ W)}{\pi f}$$

Result: A plot of $X(f)$ is shown in Figure 3.16A. Note that the solution is similar to the solution found for a periodic pulse wave in Example 3.3. Again this solution is real, but it still represents both magnitude and phase components. The magnitude is the absolute value of $X(f)$, Figure 3.16B, whereas the phase alternates between 0 degree when the function is positive and 180 degrees when the function is negative.

A special type of pulse plays an important role in systems analysis: a pulse that is very short but maintains a pulse area of 1.0. For this special pulse, termed an "impulse function," the pulse width, W, theoretically goes to 0 but the amplitude of the pulse becomes infinite in such a way that the area of the pulse stays at 1.0. The spectrum of an impulse can be determined from Equation 3.28. For small angles $\sin(x) \approx x$, so in the solution above, as W gets small:

$$\sin(\pi f\, W) \to \pi f\, W$$

The frequency spectrum of this special pulse becomes:

$$X(f) = \frac{V_p}{\pi f}\pi f\, W = V_p W = 1 \tag{3.29}$$

since $V_p W$ is the area under the pulse and by definition the area of an impulse function equals 1.0. So the frequency spectrum of this special pulse, the impulse function, is a constant value of 1.0 for all frequencies. You might imagine that a function with this spectrum would be particularly useful. You might also speculate that a pulse with these special properties, infinitely short, infinitely high, yet still with an area of 1.0, could not be realized, and you would be right. However, it is possible to generate a real-world pulse that is short enough and high enough to function as an impulse at least for all practical purposes. The approach for generating a real-world impulse function in particular situations is provided in Chapter 7.

3.4 TIME–FREQUENCY TRANSFORMATION IN THE DISCRETE DOMAIN

3.4.1 The Discrete Fourier Transform

Most Fourier analysis is done using a digital computer and is applied to discrete data. Although there are differences between the continuous and discrete Fourier transforms (DFTs), the basic concepts are the same: they both use correlation with sinusoids and their unique frequency properties to transfer signals from the discrete time to discrete frequency domain. In fact, if you applied the continuous analysis and synthesis blindly (replacing summation with integration of course), you would get similar results. However, there are a few important differences: the Fourier series is always finite, and there are some tricks to improve the efficiency and speed.

Discrete signals differ from continuous signals in two fundamental ways: they are time and amplitude sampled as discussed in Chapter 1, and they are always finite. The discrete version of the Fourier analysis equation is termed the "discrete time Fourier series" or, more commonly, the "DFT." The discrete Fourier synthesis equation is known as the "inverse discrete Fourier series" or the "inverse discrete Fourier transform" (IDFT).

The DFT analysis equation can be derived directly from Equation 3.21 noting that integration becomes summation and the time variables t and T become:

$$t = nf_s \text{ and } T = Nf_s$$

If we ignore the normalization by $1/T$, this gives the un-normalized analysis equation:

$$X[m] = \sum_{n=0}^{N-1} x[n] e^{\frac{-j2\pi mnf_s}{Nf_s}} = \sum_{n=0}^{N-1} x[n] e^{\frac{-j2\pi mn}{N}} \quad m = 0, \pm 1, \pm 2, \pm 3 \dots \pm M \tag{3.30}$$

where n is the index of the signal array, N is the array length, and m is the harmonic number. The number of harmonics evaluated, M, must be less than or equal to $N - 1$. In the most common implementation of this equation, $M = N - 1$. Although no normalization is used in Equation 3.30, it is common to normalize the summation by $1/N$, the equivalent of $1/T$. (Occasionally when we compare results between the complex (Equation 3.30) and noncomplex equations (Equations 3.8 and 3.9), we normalize by $2/N$ to match the noncomplex normalization strategy.) The equation for the IDFT is quite similar and is given in Table 3.3, Equation 3.37.

As with the continuous version of the Fourier series, Equation 3.30 produces a series of complex numbers that describe the amplitude and phase of a harmonic series of sinusoids. To relate the sinusoidal frequencies to the original analog signal, note that if a periodic signal of period T is digitized at a sample frequency of f_s into N samples, then T is equivalent to N/f_s. The fundamental frequency becomes:

$$f_1 = \frac{1}{T} = \frac{f_s}{N} \tag{3.31}$$

So the component frequencies when related to the original signal are:

$$f = mf_1 = \frac{mf_s}{N} = \frac{m}{T} \tag{3.32}$$

In the next example, we apply Equation 3.30 to a signal to obtain the magnitude spectra. Since we are using the computer to calculate the coefficients we can make the signal as complicated as we want, the computer does not care. Nevertheless, we fall back on the pulse wave used in Example 3.3. Since the phase spectrum is not very interesting we only plot the magnitude spectrum, again for ± 100 components. In addition, we normalize the coefficients by $1/N$ to match that used in Equation 3.20.

EXAMPLE 3.6

Construct the pulse wave signal similar to that used in Example 3.3 with $T = 1$ s, $V_p = 1.0$, and $W = 0.05$. Assume $f_s = 1$ kHz and use $N = 1000$ points so the period of the waveform is 1.0 s. Since the transform is symmetrical about $f = 0$, calculate only positive coefficients using Equation 3.30 ($m = 0, 1, 2, 3 \dots 100$). Plot only the first 100 components of the magnitude spectrum for comparison with the magnitude spectrum in Figure 3.12. Scale the coefficients by $1/N$ to match the analytical coefficients. Finally, ensure that the horizontal axis has the correct frequencies.

Solution: Generate the waveform by constructing an array of zeros ($N = 1000$ for a 1.0 s period) then make the first $W*f_s$ points equal to 1.0 to produce a pulse of width W. Although this pulse waveform is shifted with respect to the one in Figure 3.11, as shown later in this chapter, such a time shift does not affect the magnitude spectrum, only the phase spectrum (and we are not asked to plot the phase spectrum). We normalize the coefficient by $1/N$ to match the scaling used in the analytical equation (Equation 3.20). To get the correct frequencies, we generate a frequency vector $f = mf_1 = mf_s/M$.

```
% Example 3.6 Find and plot the magnitude spectrum of a pulse waveform
% using the complex form of the Fourier series analysis (Equation 3.30)
%
fs = 1000;              % Sample frequency
N = 1000;               % Data array length for a 1.0 sec period.
W = 0.05;               % Pulse width
M = 100;                % Number of coefficients to calculate
x = zeros(1,N);         % Generate pulse waveform
PW = round(W*fs);       % Number of points in pulse
x(1:PW) = 1;            % Set pulse
%
% Apply complex Fourier series analysis
for m = 0:M-1
% Add 1 to m when used as an index
  Xf = sum(x.*exp(-j*2*pi*m(1:N)/N));  % Find complex coefficients, Equation 3.30
  Mag(m+1) = abs(Xf)/N;                % Find magnitude, Equation 3.23, and normalize
  f(m+1) = m*fs/N;                     % Generate frequency vector, Equation 3.32
end
plot(f,Mag,'k.');                      % Plot results
```

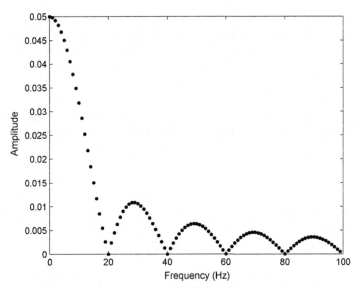

FIGURE 3.17 The spectrum generated by applying the discrete Fourier transform to a pulse waveform similar to that used in Example 3.3. The magnitude spectrum is identical to that found analytically and plotted in Figure 3.12.

Results: The magnitude spectrum generated by the program is shown in Figure 3.17 to match the one found analytically and shown in Figure 3.12. Note that our frequency vector correctly scales the frequency axis.

Example 3.6 shows how easy it is to determine a signal's spectrum using the discrete Fourier series equation, Equation 3.30. The same code could be used to find the magnitude spectrum of the most complex signal. However, in practice the DFT is normally implemented using the FFT.

Again, the underlying assumption of the discrete Fourier series is that the digitized signal represents one period of a periodic function. This is rarely the case in real situations and the assumption does produce some artifacts, particularly if the data segment is short. These artifacts and their potential remediation are discussed in the next chapter.

Application of Equation 3.30 produces a series of frequency components spaced $1/T$ apart, Equation 3.32. After the Fourier series is calculated, we sometimes pretend that the data set in the computer is actually an aperiodic function, that the original (undigitized) signal was zero for all time except for the segment captured in the computer. Under this aperiodic assumption, we can legitimately connect the points of the Fourier series together when plotting, since the difference between points would go to zero if the period was really infinite. This produces an apparently continuous curve and motivates the term digital Fourier transform (DFT) instead of digital Fourier series. Although the aperiodic assumption and associated term DFT are commonly used, we should understand that whenever we do a Fourier transform on a computer, we are implicitly assuming a periodic signal and what we get is a series of numbers not a continuous function.[12]

The number of different Fourier transforms has led to some confusion regarding terminology and there is a tendency to call all these operations "Fourier transforms" or, even more vaguely, "FFTs." Using appropriate terminology reduces confusion about what you are actually doing, and may even impress less informed bioengineers. To aid in being linguistically correct, Table 3.2 summarizes the terms and abbreviations used to describe the various aspects of Fourier analyses equations, and Table 3.3 does the same for the Fourier synthesis equation.

Tables 3.2 and 3.3 show a potentially confusing number of options for converting between the time and frequency domains. Of these options, there are only two real-world choices: the analysis and synthesis discrete Fourier transformations. In the interest of simplicity, the eight options are itemized in Table 3.4 (*sans* equations) with an emphasis on when they are applied.

Only two of these equations are used in real-world problems: the discrete Fourier series, usually just called the Fourier transform or DFT, and its inverse. These are the only equations that can be implemented on a computer. However, you need to know about all the versions, because they are sometimes referenced in engineering articles. If you do not understand these variations, you do not really understand the Fourier transform.

---

[12]In some theoretical analyses, infinite data are assumed, in which case the n/N in Equation 3.30 does become a continuous variable f as $N \to \infty$. In these theoretical analyses the data string, $x[n]$, really is aperiodic and the modified equation that results is the "discrete time Fourier transform" (the DTFT as opposed to just the DFT for periodic data). Since no computer can hold infinite data, the discrete time Fourier transform is of theoretical value only and is not used in calculating the spectrum.

TABLE 3.2 Analysis Equations (Forward Transform)[a]

$x(t)$ periodic and continuous: Fourier Series:	$x[n]$ periodic and discrete: Discrete Fourier Transform (DFT) or Discrete Time Fourier Series[b,c]:
$$X(m) = \frac{1}{T}\int_0^T x(t)e^{-j2\pi m f_1 t}dt$$ (3.33 and 3.21) $f_1 = 1/T$ $m = 0, \pm 1, \pm 2, \pm 3,\dots$	$$X[m] = \sum_{n=0}^{N-1} x[n]\,e^{-j2\pi mn/N}$$ (3.34 and 3.30) $m = 0, \pm 1, \pm 2, \pm 3,\dots \pm M\ (M \leq N)$
$x(t)$ aperiodic and continuous: Fourier Transform or Continuous Time Fourier Transform:	$x[n]$ aperiodic and discrete: Discrete Time Fourier Transform (DTFT)[d]:
$$X(f) = \int_{-\infty}^{\infty} x(t)e^{-j2\pi ft}dt$$ (3.35 and 3.25)	$$X[f] = \sum_{n=-\infty}^{\infty} x[n]\,e^{-j2\pi nf}$$ (3.36)

[a]Often ω is substituted for 2πf to make the equation shorter. In such cases the Fourier series integration is carried out between 0 and 2π or from −π to +π.
[b]Alternatively, $2\pi mnT_s/N$ is sometimes used instead of $2\pi mn/N$ where T_s is the sampling interval.
[c]Sometimes this equation is normalized by 1/N and then no normalization term is needed in the inverse DFT.
[d]The DTFT cannot be calculated with a computer since the summations are infinite. It is used only in theoretical problems as an alternative to the DFT.

TABLE 3.3 Synthesis Equations (Reverse Transform)[a]

Inverse Fourier Series:	Inverse Discrete Fourier Transform (DFT) or Inverse Discrete Time Fourier Series[b,c]:
$$x(t) = \sum_{n=-\infty}^{\infty} X(m)e^{j2\pi f_1 t} \quad m = 0, \pm 1, \pm 2, \pm 3 \dots$$ (3.37)	$$x[n] = \frac{1}{N}\sum_{n=0}^{N-1} X[m]e^{j2\pi mn/N}$$ (3.38)
Inverse Fourier Transform or Inverse Continuous Time Fourier Transform:	Inverse Discrete Time Fourier Transform[d,e]:
$$x(t) = \int_{-\infty}^{\infty} X(f)e^{j2\pi ft}df$$ (3.39 similar to 3.27)	$$x[n] = \frac{1}{2\pi}\int_0^{2\pi} X[f]e^{j2\pi nf}df$$ (3.40)

[a]Often ω is substituted for 2πf to make the equation shorter. In such cases the Fourier series integration is carried out between 0 and 2π or from −π to +π.
[b]Alternatively, $2\pi mnT_s/N$ is sometimes used instead of $2\pi mn/N$ where T_s is the sampling interval.
[c]Sometimes this equation is normalized by 1/N and then no normalization term is needed in the inverse DFT.
[d]The DTFT cannot be calculated with a computer since the summations are infinite. It is used only in theoretical problems as an alternative to the DFT.
[e]Evaluation of the integral requires an integral table because the summations that produce X[f] are infinite in number.

TABLE 3.4   Real-World Application of Fourier Transform Equations

Analysis Type [Equation]	Application
Continuous Fourier series [Equation 3.31 and 3.33]	Analytical (i.e., textbook) problems involving periodic continuous signals
Discrete Fourier Series [Equation 3.30 and 3.34]	*All real-world problems.* (Implemented on a computer using the FFT algorithm)
Continuous Fourier Transform [Equation 3.25 and 3.35]	Analytical (i.e., textbook) problems involving aperiodic continuous signals
Discrete Time Fourier Transform [Equation 3.36]	None. Theoretical only
Inverse Fourier Series [Equation 3.37]	Analytical (i.e., textbook) problems to reconstruct a signal from the Fourier series. Not common even in textbooks. (Used in Example 3.3, but computer implemented.)
Inverse Discrete Fourier Series [Equation 3.38]	*All real-world problems.* (Implemented on a computer using the inverse FFT algorithm)
Inverse Fourier Transform [Equation 3.27 and 3.39]	Analytical (i.e., textbook) problems to reconstruct a signal from the continuous Fourier Transform. Not common even in textbooks. (No applications in this book)
Inverse Discrete Time Fourier Transform [Equation 3.40]	None. Theoretical only

Although the DFT is normally implemented on a computer, it is occasionally implemented manually as a textbook exercise. For some students, it is helpful to see a step-by-step implementation of the DFT, so the next example is dedicated to them.

EXAMPLE 3.7

Find the DFT of the discrete periodic signal shown in Figure 3.18 manually.

Solution: The signal has a period of seven points beginning at $n = -3$ and running to $n = +3$:

$$x[-3] = 0, x[-2] = 1, x[-1] = 0.5, x[0] = 0, x[1] = -0.5, x[2] = -1, x[3] = 0$$

The Fourier transform equation is given in Equation 3.30 and restated here with $N = 7$ so n ranging between ± 3. The coefficients will range between ± 6 (recall the maximum coefficient $M \leq N - 1$):

$$X[m] = \sum_{n=1}^{N} x[n]e^{\frac{-j2\pi mn}{N}} = \sum_{n=-3}^{3} x[n]e^{\frac{-j2\pi mn}{7}} \quad m = 0, \pm 1\pm, 2, \pm 3, \ldots 6$$

Recall expanding the summation and substituting in the seven values of $x[n]$:

$$X[m] = \sum_{n=1}^{N} x[n]e^{\frac{j2\pi mn}{N}} = x[-3]e^{j2\pi m3/7} + x[-2]e^{j2\pi m2/7} + x[-1]e^{j2\pi m1/7} + x[0]e^{j2\pi m0}$$

$$+x[1]e^{-j2\pi m1/7} + x[2]e^{-j2\pi m2/7} + x[3]e^{-j2\pi m3/7}$$

$$= 0e^{j\pi m6/7} + 1e^{j\pi mn4/7} + 0.5e^{j\pi m2/7} + 0e^{j\pi m0} - 0.5e^{-j\pi m2/7} - 1e^{-j\pi m4/7} + 0e^{-j\pi m6/7}$$

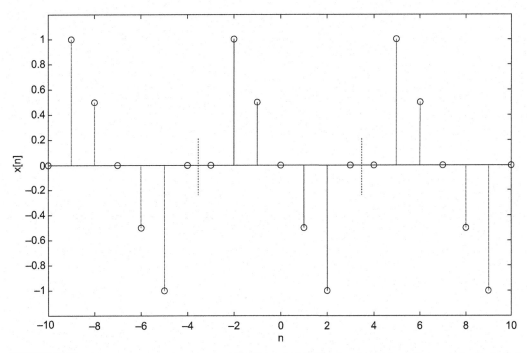

FIGURE 3.18   Discrete periodic waveform used in Example 3.7. The signal consists of discrete points indicated by the *circles*. The *lines* are simply to improve visibility. Discrete data are frequently drawn in this manner.

Collecting terms and rearranging:

$$X[m] = e^{j\pi m 4/7} - e^{-j\pi m 4/7} + 0.5\left(e^{j\pi m 2/7} - e^{-j\pi m 2/7}\right)$$

Noting that: $\sin x = \frac{e^{jx} - e^{-jx}}{2j}$

$$X[m] = 2j\sin(4/7\pi m) + j\sin(2/7\pi m) \quad m = 0, \pm 1, \pm 2, \ldots \pm 6$$

The problem set offers a couple of additional opportunities to solve for the DFT manually.

3.4.2 MATLAB Implementation of the Discrete Fourier Transform

MATLAB has a routine that uses the FFT algorithm to implement Equation 3.30 very quickly:

```
Xf = fft(x,n)      % Calculate the Fourier Transform
```

where x is the input waveform and Xf is a complex vector providing the sinusoidal coefficients. (Recall, it is common to use capital letters for the Fourier transform variable.) The first term of Xf is real and is the un-normalized DC component; you need to divide by the signal length, N, to get the actual DC component. The second term in Xf is the complex representation of the fundamental sinusoidal component; the third term represents the second harmonic; and so on. The argument n is optional and is used to modify the length of data analyzed: if n is less than the length of x, then the analysis is performed over the first n points. If n is greater than the length of x, then the signal is padded with trailing zeros to equal n.

There is one downside to the fft routine. The FFT algorithm requires the data length to be a power of 2. Although the MATLAB routine will interpolate if need be, calculation time will go up and how much depends on data length. The algorithm is fastest if the data length is a power of 2, or if the length has many prime factors. For example, on a slow machine, a 4096-point FFT takes 2.1 s, but requires 7 s if the sequence is 4095 points long and 58 s if the sequence is 4097 points long. If at all possible, it is best to stick with data lengths that are powers of 2.

The magnitude and phase spectra are calculated from the complex output Xf using abs(Xf) and angle(Xf), respectively (see Example 3.3). Again, the angle routine gives phase in radians so as to convert to the more commonly used degrees scale by $360/(2\pi)$.

EXAMPLE 3.8

Construct the waveform used in Example 3.2 (repeated below) and determine the Fourier transform using both the MATLAB fft routine and a direct implementation of the defining equations (Equations 3.8 and 3.9).

$$x(t) = \begin{cases} t & 0 < t \le 0.5 \\ 0 & 0.5 < t \le 1.0 \end{cases}$$

Solution: We need a total time of 1 s; let us assume a sample frequency of 256 Hz. This leads to a value of N that equals 256, which is a power of 2 as preferred by the fft routine. The MATLAB fft routine does no scaling. To compare the output of fft with the analytical results obtained from Equations 3.8 and 3.9, we must normalize by $2/N$.

The DC, or zero, frequency component is automatically calculated by MATLAB's fft routine, but it needs to be calculated separately when using Equation 3.10.

```
% Example 3.8 Find the Fourier transform of half triangle waveform
% used in Example 3.2. Use both the MATLAB fft and a direct
% implementation of Equations 3.8 and 3.9
%
T = 1;                          % Total time
F1 = 1/T;                       % Fundamental frequency
fs = 256;                       % Assumed sample frequency
N = fs*T;                       % Calculate number of points
M = 21;                         % Number of coefficients
t = (1:N)*T/N;                  % Generate time vector
f = (0:M)*fs/N;                 %  and frequency vector for plotting
x = [t(1:N/2) zeros(1,N/2)];    % Generate signal, ramp followed by zeros
%
Xf = fft(x);                    % Take Fourier transform, scale
Mag = abs(Xf)*2/N;              % Normalize
Phase = -angle(Xf)*360/(2*pi);
%
plot(f(1:M),Mag(1:M),'.'); hold on;  % Plot only first 20 coefficients plus DC
......labels....
```

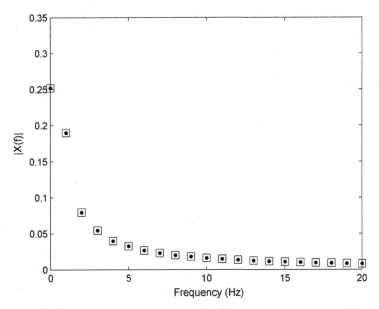

FIGURE 3.19 Magnitude frequency spectra produced by the MATLAB `fft` routine (*dots*) and a direct implementation of the Fourier transform equations, Equations 3.8 and 3.9 (*squares*).

```
%
% Calculate discrete Fourier Transform using basic equations
C(1) = (2/N)*sum(x);                    % DC Component. Equation 3.10
for m = 1:21
  a(m) = (2/N)*sum(x.*(cos(2*pi*m*f1*t)));   % Equation 3.8
  b(m) = (2/N)*sum(x.*(sin(2*pi*m*f1*t)));   % Equation 3.9
  C(m+1) = sqrt(a(m).^2 + b(m).^2);          % Equation 3.6
  theta(m) = atan2(b(m),a(m))* 360/(2*pi))   % Equation 3.7
end
disp([a(1:4)' b(1:4)' C(2:5)' Mag(2:5)' theta(2:5)' Phase(1:4)'])
plot(f(1:20),C(1:20),'sr');              % Plot superimposed
```

Results: The spectrum produced by the two methods is identical as seen by the perfect overlap of points and squares in Figure 3.19.

The numerical values produced by this program are given in Table 3.5.

TABLE 3.5 Numerical Results From Example 3.8

m	a_m (Analytical)[a]	b_m (Analytical)	C_m (Analytical)	Mag(fft)	Phase (Analytical)	Phase (fft)
1	−0.1033 (−0.101)	0.1591 (0.159)	0.1897 (0.157)	0.1897	−122.98 (−122.42)	−121.57
2	0.0020 (0.0)	−0.0796 (−0.080)	0.0796 (0.080)	0.0796	88.59 (90.0)	91.40
3	−0.0132 (−0.011)	0.0530 (0.053)	0.0546 (0.054)	0.0546	−103.99 (−101.72)	−99.77
4	0.002 (0.0)	−0.0398 (−0.040)	0.0398 (0.040)	0.0398	87.18 (90.0)	92.81

[a]*Manual values are from Example 3.1.*

Analysis: Both methods produce identical magnitude spectra and similar phase spectra (compare C_m and *Phase* from the noncomplex analysis with the FFT results, Mag(fft) and Phase(fft)). Note that both the atan2 and angle routines give the angle in radians, so in the program they are multiplied by $360/2\pi$ to convert to degrees.

The magnitudes and phases found by both methods closely match the values determined analytically (manually) in Example 3.1. However, the computer-determined phase angles for components 2 and 4 are not exactly zero as found analytically. This difference is due to small computational errors caused by rounding. This shows that analytical solutions done by hand can be more accurate than computer results!

How fast is the FFT? The next example compares the speed of direct implementation for the complex and noncomplex analysis equations against the FFT.

EXAMPLE 3.9

You are on a desert island with your laptop and your survival depends on finding the frequency spectra of a signal that is in your laptop. Unfortunately, you find your laptop does not have the FFT algorithm. "No problem," you say, "I will simply code the DFT algorithm given in Equation 3.30 and used in Example 3.6." (Of course you have copy of this textbook.) Then you find that for some mysterious reason your laptop cannot handle complex numbers. "No problem," you say, "I will simply code the discrete version of the real-valued Fourier series equations Equations 3.8 and 3.9." Naturally you expect the program will be considerably slower than the FFT algorithm, but you hope to find the solution before your battery dies.

To evaluate what you would be up against, compare the run times of the three possible approaches to evaluating the DFT. Your signal, x in file desert_island.mat, consists of approximately 8000 points sampled at 2 kHz. For the two non-fft approaches, calculate the magnitude and phase of the first 4000 components. (MATLAB's fft algorithm evaluates all N possible components.) Plot 4000 points of the magnitude spectra produced by the three methods superimposed. We use MATLAB's tic and toc to determine the run times.

Solution: Load the signal file. For the noncomplex DFT, reuse the code in Example 3.8 and for the complex FT that in Example 3.6. Apply the three methods in succession and time the executions. Plot the resulting magnitude spectra using different symbols. Again, since we want to compare results with Equations 3.8 and 3.9, we need to normalize by $2/N$.

```
% Example 3.9  Comparison of three different approaches to the DFT.
%
load desert_island.mat;  % Get signal
N = length(x);           % Number of samples
fs = 2000;               % Sample frequency (known)
f1 = fs/N;               % Fundamental frequency, Equation 3.31
M = 4000;                % Number of coefficients to evaluate
f = (0:M-1)*fs/N;        % Frequency vector for plotting
%
```

```
% Apply non-complex FT analysis
tstart = tic;                % Start timer
t = (1:N)/fs;                % Time vector
C(1) = mean(x);              % DC Component, Equation 2.10
for m = 1:M-1
  a(m+1) = sum(x.*(cos(2*pi*m*f1*t)));   % Equation 3.8
  b(m+1) = sum(x.*(sin(2*pi*m*f1*t)));   % Equation 3.9
end
C = sqrt(a.^2 + b.^2)/N;                  % Find magnitude Equation 3.6
theta = (360/(2*pi)) * atan2(b,a);        % Find phase, Equation 3.7
disp(['Non-complex FT duration: ',num2str(toc(tstart))]);    % Display timing
plot(f,C,'ok'); hold on;     % Plot non-complex magnitude spectrum
clear C theta a b;           % To insure clean slate for next eval.
%
% Apply complex FT
tstart = tic;                % Start timer
for m = 1:M
  Xf(m) = sum(x.*exp(-j*2*pi*m*(1:N)/N));   % Complex coefficients, Equation 3.30
end
Mag = abs(Xf)/N;             % Magnitude, Equation 3.23
Theta = angle(Xf)*360/(2*pi);  % Phase
disp(['Complex FT duration: ',num2str(toc(tstart))]);  % Display timing
plot(f,Mag,'+k');            % Plot complex magnitude spectrum

clear Xf Mag Theta;          % To insure clean slate for next eval.
%
% FFT
tstart = tic;                % Start timer
Xf = fft(x);
Mag = abs(Xf)/N;             % Find Magnitude, Equation 3.23
Theta = angle(Xf)*360/(2*pi);  % Find phase
disp(['FFT duration: ',num2str(toc(tstart))]);   % Display timing
plot(f,Mag(1:M),'sqk');      % Plot FFT magnitude spectrum
```

Results: The magnitude spectra produced by the three methods are shown in Figure 3.20 to be identical (the three symbols overlap). The spectra look strange, with two plateaus, but the three methods agree so they must be correct. No doubt the spectra represent some sort of code that tells you something important to your survival. Good luck.

Although the results are the same, the computation times shown Table 3.6 are quite different. The complex analysis is a little faster than the noncomplex analysis, but the FFT is approximately 1000 times faster than the other approaches. The signal length, N, was actually 8192 points, which is a power of 2 (2^{13}), ideal for the FFT algorithm. Extending the data by only 1 point slowed the FFT algorithm to between 0.003 and 0.005 s and made the timing highly variable. Nonetheless, the FFT is

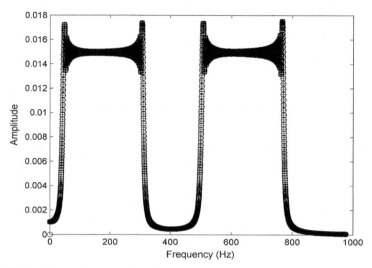

FIGURE 3.20   Magnitude spectra found for three algorithms that evaluate the Fourier transform analysis equations. The signal had $N = 8192$ and the first 4000 coefficients are shown (obviously not all 4000 points are actually plotted.) The symbols used for the different approaches are: "o" for noncomplex analysis; "+" for complex analysis, and a square for the FFT; however, the points overlap and it is difficult to see the individual symbols.

TABLE 3.6   Run-Time of Three Algorithms for the Fourier Transform Analysis Equations

Approach	Time for Execution (s)
Noncomplex analysis	3.55
Complex analysis	2.99
Fast Fourier transform	0.0027

really fast and the clear method of choice for converting from the time to frequency domain. Details of the FFT spectra are discussed in the next section.

3.4.3 Details of the DFT Spectrum

We have established that the three algorithms produce identical results, but the great speed of the FFT makes it the clear choice (moreover it is easier to code, just one line). We will take advantage of the FFT to examine some important properties of the DFT and the Fourier transform in general. To explore the properties of the Fourier transform, we sometimes use a test signal consisting of sinusoids and white noise. Here we present a MATLAB routine that generates multiple sinusoids accompanied by white noise. The routine, sig_noise, is found on the accompanying materials, and has a calling structure:

```
[x,t] = sig_noise([f],[SNR],N);    % Generate sinusoidal signals in noise
```

where f specifies the frequency of the sinusoid(s) in hertz, SNR specifies the desired noise associated with the sinusoid(s) in decibels, and N is the number of points in the signal. If f is a vector, then a number of sinusoids are generated, each with a signal to noise ratio (SNR) specified by SNR assuming it is a vector. If SNR is scalar, its value is used for the SNR of all the frequencies generated. The output waveform is in x and t is a time vector useful in plotting. The routine assumes a sample frequency of $f_s = 1$ kHz.

3.4.3.1 *DFT Spectral Redundancy*

The DFT produces twice as many points as contained in the input signal since both the magnitude and phase have the same number of points as the signal. Since the information content of the signal and its Fourier transform are the same, some on these points must be redundant. We have yet to look at a plot of the full spectrum generated by the FFT; that is, one with the same number of components as the signal (i.e., $M = N - 1$, in Equation 3.30). In the next example, we demonstrate that the upper half of the magnitude and phase spectra are mirror images of the lower half and therefore redundant.

EXAMPLE 3.10

Use fft to find the magnitude spectrum of a signal consisting of a single 250 Hz sine wave and white noise with an SNR of −14 dB. Plot the full magnitude spectrum (i.e., $M = N - 1$).

Solution: Use sig_noise to generate the waveform and use fft to obtain the complex Fourier transform. Use abs to find the magnitude spectrum and plot the full spectrum. Since we are not comparing our results with other methods, we use no normalization (the MATLAB default).

```
% Example 3.10   Determine the magnitude spectrum of a noisy waveform.
% N = 1024;                    % Number of data points
fs = 1000;                     % 1 kHz fs assumed by sig_noise.
f = (0:N-1)*fs/N;              % Frequency vector for plotting
% Generate signal using 'sig_noise'
%    250 Hz sin plus white noise; N data points; SNR = -14 dB
[x,t] = sig_noise (250,-14,N); % Generate signal and noise
%
Xf = fft(x);                   % Calculate FFT
Mf = abs(Xf);                  % Calculate the magnitude
plot(f,Mf);                    % Plot the magnitude spectrum
     .......label and title.......
```

Analysis: The program is straightforward. After constructing the signal using the routine sig_noise, the program takes the Fourier transform with fft and then plots the magnitude (using abs) versus frequency using a frequency vector to correctly scale the frequency axis. The frequency vector ranges from 0 to N-1 to include the DC term (which happens to be zero in this example.)

Results: The spectrum is shown in Figure 3.21 and the peak related to the 250 Hz sine wave is clearly seen. The spectrum above $f_s/2$ (i.e., 500 Hz, dashed line) is a mirror image of the lower half of the spectrum (as explained in Section 4.2.1). Theoretically the background spectrum should be a constant value since it is due to white noise, which has equal energy at all frequencies. As is typical

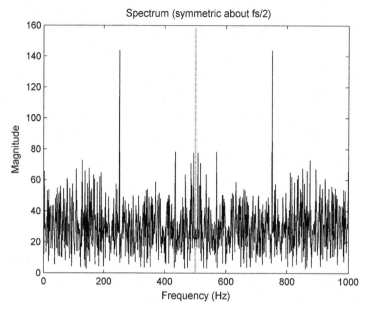

FIGURE 3.21   Plot produced by the previous MATLAB program. The peak at 250 Hz is apparent. The sampling frequency of these data is 1 kHz and the spectrum is symmetric about $f_s/2$ (*dashed line* at 500 Hz) for reasons described later. Normally only the first half of this spectrum would be plotted. (Signal characteristics: $f_{sine} = 250$ Hz; SNR $= -14$ dB; $N = 1024$.)

of noise spectra, the background is highly variable with occasional peaks that could be mistaken for signals. A better way to determine the spectrum of white noise is to use an averaging strategy described in the next chapter.

3.4.3.2 *Phase Wrapping*

The `fft` routine produces a complex spectrum that includes both the magnitude and phase information. Calculating and plotting both the magnitude and phase spectra seems straightforward, but the latter can be problematic. The output of the `angle` routine is limited to $\pm 2\pi$; larger values "wrap around" to fall within that limit. Thus a phase shift of $3\pi/2$ would be rendered as $\pi/2$ by the `angle` routine, as would a phase shift of $5\pi/2$. So when a signal's spectrum results in large phase angles, this wraparound characteristic produces jumps on the phase curve. An example of this is seen in Figure 3.22, which shows the magnitude and phase spectra of a pulse signal. The phase plot, this time plotted in radians, shows a sharp transition of approximately 2π whenever the phase reaches $-\pi$ (lower *dashed line*, Figure 3.22B).

The correct phase spectrum could be recovered by subtracting 2π from the curve every time there is a sharp transition. This is sometimes referred to as unwrapping the phase spectrum. It would not be difficult to write a routine to check for large jumps in the phase spectrum and subtract 2π from the rest of the curve whenever they are found. But whenever there is a common problem, you can bet that MATLAB has a routine to address it. The MATLAB

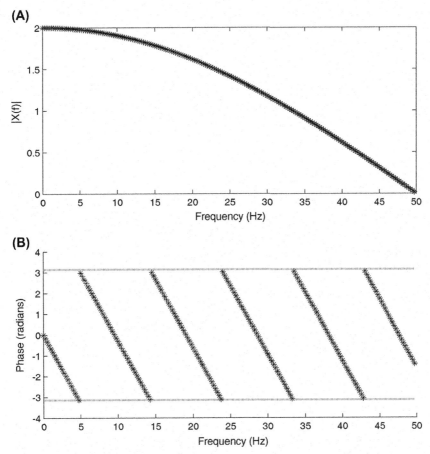

FIGURE 3.22   (A) Magnitude spectrum of a pulse waveform. (B) Phase spectrum in radians of the pulse waveform showing upward jumps of approximately 2π whenever the phase approaches $-\pi$. The *dashed horizontal lines* represent $\pm\pi$.

routine unwrap checks for jumps greater than π and does the subtraction whenever they occur. (The threshold for detecting a discontinuity can be modified as described in the unwrap help file.) The next example illustrates the use of unwrap for unwrapping the phase spectrum.

3.4.3.3 The Effect of Time Shifts on the Fourier Transform

Shifting a signal in time has a well-defined effect on the Fourier transform. This can be described by comparing the Fourier transform of two signals: an unshifted signal $x(t)$ and signal time shifted by an amount τ, $x(t - \tau)$. Using the Fourier transform for a continuous aperiodic function, Equations 3.25 and 3.35 (again we use ω instead of $2\pi f$ for variety):

$$X(f) = \int_{-\infty}^{\infty} x(t)e^{-j\omega t}dt$$

For signal time shifted τ sec:

$$X(f) = \int_{-\infty}^{\infty} x(t - \tau)e^{-j\omega(t)}dt$$

Substituting $u = t - \tau$ and $t = u + \tau$:

$$X(f) = \int_{-\infty}^{\infty} x(u)e^{-j\omega(u+\tau)}du = e^{-j\omega\tau}\int_{-\infty}^{\infty} x(u)e^{-j\omega(u)}du = X(f)_{unshifted}e^{-j\omega\tau} \qquad (3.41)$$

So the effect of a time shift, τ, in the frequency domain is to multiply the Fourier transform of the unshifted signal by $e^{-j\omega\tau}$. The magnitude of this term is 1.0 and the phase is $-\omega\tau$ or $-2\pi f\tau$. So shifting a signal in time has no effect on the magnitude spectrum, but does decrease the phase spectrum: the phase decreases linearly with frequency and in proportion to the time shift. The time-shift influence on the DFT is illustrated empirically in the next example.

EXAMPLE 3.11

Evaluate and plot the magnitude and phase of the waveform shown in Figure 3.14, but for two different time shifts. One signal should have the waveform symmetrical about $t = 0$ and the other centered in the signal array. Plot the two signals and their spectra. Plot the phase shift in degrees with and without unwrapping. Use the MATLAB unwrap routine.

Solution: The hardest part of this example is constructing the two signals. The waveform in Figure 3.14 is called a "wavelet" and is symmetrical. The equation for the right-hand side is:

$$x(t) = e^{-t^2} \cos\left(2\pi\sqrt{\frac{2}{\ln 2}}t\right) \qquad (3.42)$$

There are several ways to construct the two waveforms. For a signal that is symmetrical about $t = 0$, we recall that all DFT signals are periodic. To make the signal symmetrical about $t = 0$, we can use Equation 3.42 to build the positive half of the wavelet waveform at the beginning of the array, then construct its reversed version at the end of the array, with some zero points in between. We can use MATLAB's fliplr routine to reverse the waveform. This can be implemented using the MATLAB code:

```
morlet = (exp(-t1.^2).* cos(wo1*t1));      % Generate Morlet wavelet
x1 = [morlet, zeros(1,2*N1), fliplr(morlet)];   % Construct using t=0 symmetry
```

where N1 is 512 points and is the same length as the wavelet, morlet. When plotted this may not look like a waveform that is symmetrical, but it is if you consider that it represents a periodic signal. For the centered waveform, we put the flipped and unflipped version of Equation 3.42 in sequence with zeros placed symmetrically on either side.

```
x1 = [zeros(1, N1), fliplr(morlet), morlet, zeros(1,N1)];   % Construct centered
```

Once we have constructed the two signals, the rest is straightforward following methods used in previous examples. We find the length of the signals, N, construct frequency, and time vectors for plotting, find the complex spectra, and extract the magnitude and phase spectra. We find and plot both the wrapped and unwrapped phase spectra. As with most signals, only the lower frequencies are of interest, so we plot only the first 30 frequency components (found by trial and error to be significantly greater than zero). We arbitrarily assume a sample frequency, f_s, of 150 Hz.

```
% Example 3.11 Evaluate and plot the magnitude and phase of signal based on
% the waveform shown in Figure 3.14. Construct two signals with two different time
shifts.
% One waveform should be symmetrical about t = 0 and the other centered in the
signal array.
%
fs =150;                          % Assumed sample frequency
%
% Generate waveforms based on Morlet wavelet
N1 = 512;                         % Wavelet number of points
t1 = ((0:(N1/2)-1)/fs)*2;         % Show =/- 40 sec of the wavelet
wo1 = pi*sqrt(2/log2(2));         % Wavelet constant
mor = exp(-t1.^2).* cos(wo1*t1);      % Generate Morlet wavelet
x1 = [mor, zeros(1,2*N1), fliplr(mor)];    % Symmetrical about t = 0
x2 = [zeros(1,N1), fliplr(mor), mor, zeros(1,N1)]; % Centered
%
N = length(x1);                   % Signal number of points
t = (1:N)/fs;                     % Time vector for plotting
f = (0:N-1)*fs/(N-1);             % Frequency vector for plotting
%
subplot(1,2,1);                   % Plot the two signals
plot(t,x1)                        % Signal 1 time domain
   .......labels .........
subplot(1,2,2);
plot(t,x2)                        % Signal 2 time domain
   .......labels and new figure.........
%
Xf = fft(x1);                     % Signal 1 complex spectrum
Mag = abs(Xf);                    % Get magnitude
Phase = angle(Xf)*360/(2*pi);     % Get phase in deg not unwrapped
Phase_unwrap = unwrap(angle(Xf))*360/(2*pi); % Unwrapped phase in deg.

subplot(2,2,1);                   % Plot Signal 1 spectrum
plot(f(1:30),Mag(1:30),'.');      % Magnitude spectrum. Only first 30 components
   .......labels and title........
subplot(2,2,2); hold on;
  plot(f(1:30),Phase(1:30),'+'); % Plot wrapped phase. Use + points.
  plot(f(1:30),Phase_unwrap(1:30),'.'); % Plot unwrapped phase
   .......labels .........
```

```
%
Xf1 = fft(x2);                      % Signal 2complex spectrum
Mag = abs(Xf1);                     % Get magnitude
Phase = angle(Xf1)*360/(2*pi);      % Get phase in deg not unwrapped
Phase_unwrap = unwrap(angle(Xf1))*360/(2*pi);     % Unwrapped phase in deg.
subplot(2,2,3);               % Plot Signal 2 spectrum
plot(f(1:30),Mag(1:30),'.');      % Magnitude spectrum
  .......labels .........
subplot(2,2,4); hold on;
  plot(f(1:30),Phase(1:30),'+'); % Plot wrapped phase. Use + points.
  plot(f(1:30),Phase_unwrap(1:30),'.'); % Plot unwrapped phase
  .......labels .........
```

Results: The time domain representation of the two waveforms is Figure 3.23. The waveforms have the expected shape and position.

The spectra of the two signals are shown in Figure 3.24. The magnitude spectra of the two signals are identical (Figure 3.24, left-hand plots). Shifting the position of the waveform does not affect the magnitude spectrum: it is not affected by time shifts of the waveform. However, the two phase spectra are quite different. The phase of the signal that is symmetrical about $t = 0$ is constant and essentially zero (Figure 3.24 upper right) for all frequencies so the wrapped and unwrapped phase points are the same. The signal with the centered waveform produces a phase shift that increases linearly with frequency (Figure 3.24 lower right). Because the time shift is so large, the decrease in phase shift is enormous (note the scale of the vertical axis in Figure 3.24 lower right). If the phase is not unwrapped (+ points), it warps around and looks like a set of alternating points. The

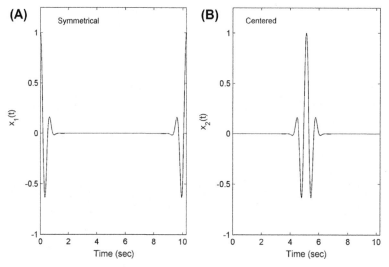

FIGURE 3.23   The two signals used in Example 3.11. Both are based on the Morlet wavelet. (A) The waveform is symmetrical about $t = 0$ (recall these are periodic signals). (B) The waveform is centered in the signal array.

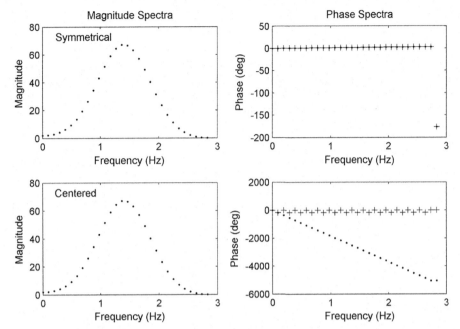

FIGURE 3.24 The magnitude and phase spectra of the signals shown in Figure 3.23. Left-hand plots: The magnitude spectra of the two signals are identical. The magnitude spectrum is not influenced by time shifts. Right-hand plots: The phase spectra of the signal symmetrically placed about $t = 0$ is constant at zero for all frequencies (upper right plot) and hence the same for wrapped and unwrapped data. The centered signal, when it is unwrapped, shows a dramatic linear decrease in phase with increasing frequency (dots, lower right), but this behavior is not apparent in the unwrapped data (+ points, lower right).

unwrapped version (dots) shows the dramatic decrease in phase with increasing frequency (Figure 3.24, lower right). The take-home message is always unwrap the phase data unless you are absolutely sure it will never be greater or less than 2π.

3.4.3.4 Respiratory Example

The next example applies the DFT to a signal related to respiration and uses the spectrum to find the breathing rate.

EXAMPLE 3.12

Find the magnitude and phase spectrum of the respiratory signal found in variable `resp` of file `Resp.mat` (data from PhysioNet, Golberger et al., 2000). The data have been sampled at 125 Hz. Plot only frequencies between the fundamental and 2 Hz. The magnitude spectrum should have a maximum energy peak corresponding to the frequency of the average respiration rate. Find this peak and use it to estimate the breaths per minute of this subject.

Solution: Loading the data and taking the DFT are straightforward, as is finding the spectral peak. The trick in this example is to find the series number, m, that corresponds to 2 Hz. From Equation 3.32 , $f = mf_s/N$. So 2 Hz corresponds to series number $m_{2Hz} = f\,N/f_s = 2N/f_s$.

To find the peak frequency, use the second output of MATLAB's max routine to find the series number m_{max} that corresponds to the maximum spectral energy. The second output argument of this routine gives the index of the maximum value, in this case the index of m_{max}. The frequency corresponding index m_{max} is $f_{max} = m_{max}f_s/N$ (Equation 3.32, but you can also get this from your frequency vector). The average interval between breaths is $t_{breath} = 1/f_{max}$. This represents the average time between each breath so this time divided into 60 gives breaths per min: $BPM = 60/t_{max}$.

In all of this remember that in the fft output, X(m), the first frequency component starts at X(2) as X(1) holds the DC component.

```
% Example 3.12   Example applying the FT to a respiration signal.
%
load Resp                   % Get respiratory signal
fs = 125;                   % Sampling frequency in Hz
max_freq = 2;               % Max desired plotting frequency in Hz
N = length(resp);           % Length of respiratory signal
f = (1:N)*fs/N;             % Construct frequency vector
t = (1:N)/fs               % Construct the time vector
plot(t,resp)                % Plot the time signal
    .......labels.......
Xf = fft(resp);             % Calculate the spectrum
m_2Hz = round(2N/fs);       % Find m for 2 Hz
subplot(2,1,1);
plot(f(1:m_2Hz-1),abs(Xf(2:m_2Hz)));      % Plot magnitude spectrum
    .......labels.......
phase = unwrap(angle(Xf))*360/(2*pi);     % Calculate phase spectrum in deg
%
subplot(2,1,2);
plot(f(1:m_2Hz-1),phase(2:m_2Hz));        % Plot phase spectrum
    .......labels.......
[peak m_max] = max(abs(Xf(2:m_Hz)));      % Find m at max magnitude peak
f_max = f(m_max)                          % Calculate and display frequency
time_max= 1/f_max                         % Calculate and display max time
breath_min = 60/ t_max                    % Calculate and display breath/min
```

Results: The time data are shown in Figure 3.25 to be a series of peaks corresponding to maximum inspiration. The spectrum plot produced by the program is shown in Figure 3.26. A peak is shown around 0.3 Hz, with a second peak around twice that frequency, a harmonic of the

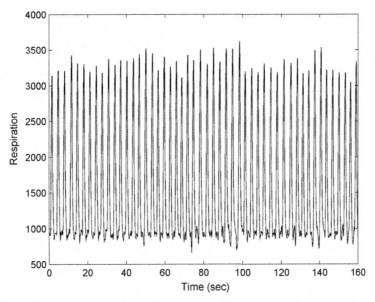

FIGURE 3.25 Respiratory signal used in Example 3.12. The peaks correspond to maximum inspiration.

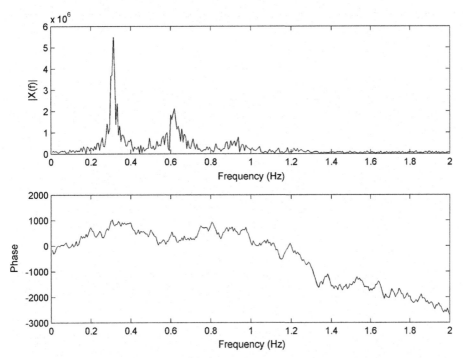

FIGURE 3.26 The magnitude and phase spectrum of the respiratory signal seen in Figure 3.25. The peak around 0.3 Hz corresponds to the interval between breaths. This is equivalent to 18.75 breaths per min.

first peak. The frequency of the peak was found to be 0.3125 Hz corresponding to an interbreath interval of 3.2 s giving rise to a respiratory rate of 18.75 breaths/min. Of course the respiratory rate could have been found from the time domain data by looking at intervals between the peaks in Figure 3.26, and then taking averages, but this would have been more difficult to code.

3.4.4 The Inverse Discrete Fourier Transform

The inverse Fourier transform performs the reverse Fourier transform and constructs a waveform from its Fourier coefficients. It implements the discrete version of the Fourier synthesis equation, Equation 3.38. It would not be difficult to code Equation 3.38 using a slightly modified version of Example 3.6. But, as usual, it is easier to use MATLAB's inverse Fourier transform routine, ifft. An opportunity to code a direct implementation of Equation 3.38 and show you are as good as MATLAB is provided in one of the problems.

The format of MATLAB's ifft routine is:

```
x = ifft(Xf,N);     % Inverse Fourier transform
```

where Xf is the Fourier coefficients and N is an optional argument limiting the number of coefficients to use. A few other less common options are described in the associated help file.

EXAMPLE 3.13

The final example uses the Morlet waveform used in Example 3.11, with the waveform initially on the left side of the signal array. We take the Fourier transform, and then modify the resulting spectrum before reconstructing a time domain signal using the inverse Fourier transform. Specifically, we multiply the phase spectrum by 4. In Example 3.11, we saw that a time shift alters only the phase; specifically, a shift that increases time increases the downward slope of the phase spectrum. Since the phase curve in Example 3.11 is linear, multiplying the phase curve by 4 increases the downward slope of the phase curve by 2 and should shift the Morlet waveform to the right, but not change its shape.

Solution: The first part of the program is identical to Example 3.11 with a slight change in the construction of the signal: the Morlet waveform is placed at the beginning of the signal array. This is done just for variety and the visual effect when the waveform is time shifted. The DFT is taken and the magnitude and phase are determined. The phase is left in radians and not unwrapped since we want to convert back using the inverse DFT after modification. The phase spectrum is multiplied by 4 and the complex spectrum is constructed from the combination of modified phase spectrum and unmodified magnitude spectrum.

MATLAB's inverse Fourier transform routine, iff(Y), expects the input, Y, to be in the complex form:

$$X_m = a_m + jb_m;$$

where: $a_m = C_m \cos(\theta_m)$ and $b_m = -C_m \sin(\theta_m)$ where C_m is the magnitude and θ_m is the modified phase (see Equation E-1, Appendix E).

```
%  Example 3.13 Take the Fourier transform of the Morlet waveform used in Example 3.11.
%  Modify the phase spectrum then reconstruct the time signal using the inverse
%  Fourier transform.
%
fs =150;                    % Assumed sample frequency. Same as Example 3.11.
N1 = 512;                   % Next 4 lines, same as Example 3.11
wo1 = pi*sqrt(2/log2(2));             % Const. for wavelet time scale
t1 = ((0:(N1/2)-1)/fs)*2;             % Time vector for Morlet wavelet
mor = (exp(-t1.^2).* cos(wo1*t1));    % Generate Morlet wavelet
x = [fliplr(mor), mor, zeros(1,2*N1)];   % Construct signal
%
N = length(x);              % Signal number of points
t = (1:N)/fs;              % Time vector for plotting
f = (0:N-1)*fs/N;           % Frequency vector for plotting
%
subplot(1,2,1);
plot(t,x);                 % Plot original signal
   ........labels and axis.......
%
X = fft(x);                % Signal complex spectrum
x1 = ifft(X);              % Unmodified inverse Fourier transform
Mag = abs(X);              % Get magnitude
Phase = angle(X);          % Get phase in radians not unwrapped
Phase1 = 4*Phase;          % Modify phase spectrum
%
am = Mag.*cos(Phase1);     % Calculate new real coefficients
bm = Mag.*sin(Phase1);     % Calculate new imaginary coefficients
Y = am + j*bm;             % New complex spectrum
y = ifft(Y);               % Take modified inverse Fourier transform.
%
subplot(1,2,2); hold on;   % Plot modified and original signal
plot(t,x1);                % Signal from unmodified spectrum
plot(t,y,':');             % Signal from modified spectrum
   .......labels and axis.......
```

Results: The original signal, Figure 3.27A, is the same as that used in Example 3.11 except that the Morlet wavelet occurs at the beginning of the signal. The signal reconstructed from the unmodified complex signal is identical to the original, Figure 3.27B (solid line). The signal constructed from the modified phase spectrum has the same shape as the original, Figure 3.27B (dotted line), but has been shifted in time approximately 5 s. This is expected because increasing the slope of the phase trajectory is equivalent to a time shift.

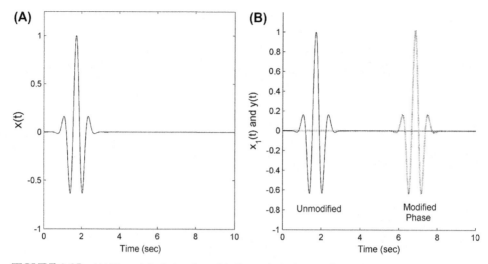

FIGURE 3.27   (A) The original signal used in Example 3.13 is similar to that used in Example 3.11 except for the position of the wavelet waveform. (B) The signal reconstructed from the unmodified Fourier transform of the signal in A is the same. The signal reconstructed from the Fourier transform after the phase shift has been modified by increasing its slope is the same but shifted in time.

There are many signal processing algorithms that follow the strategy used in this example; they move signals to the frequency domain, process the resulting spectral curves, and then move the signals back into the time domain. Although this may seem convoluted, the speed of the FFT and inverse Fourier transform makes this approach attractive and often faster than operations in the time domain.

3.5 SUMMARY

The sinusoid (i.e., $A \cos (\omega t + \theta)$) is a unique signal with a number of special properties. A sinusoid can be completely defined by three values: its amplitude A, its phase θ, and its frequency ω (or $2\pi f$). Any periodic signal can be broken down into a series of harmonically related sinusoids, although that series might have to be infinite in length. Reversing that logic, a periodic signal can be reconstructed from a series of sinusoids so a periodic signal can be equivalently represented by a sinusoidal series. A sinusoid is also a pure signal in that it has energy at only one frequency, the only waveform to have this property. Since they have such a simple frequency domain representation, sinusoids are useful intermediaries between the time and frequency domain representation of signals.

The technique for determining the sinusoidal series representation of a periodic signal is known as Fourier series analysis. To determine the Fourier series, the signal is represented by an equivalent series of sinusoids that have frequencies harmonically related to the signal. The amplitude and phase of this series is found by correlating the signal with each sinusoid in the series. The resulting amplitude and phase of these sinusoids can be plotted against the associated frequency to construct the frequency domain representation of the signal. As harmonically related sinusoids are orthogonal, the components of the Fourier series have no influence on one another. Fourier series analysis is often described and implemented using the complex representation of a sinusoid.

If the signal is not periodic, but exists for a finite time period, Fourier decomposition is still possible by assuming that this aperiodic signal is in fact periodic, but that the period is infinite. This approach leads to the true Fourier transform where the correlation is now between the signal and infinite number of sinusoids having continuously varying frequencies. The frequency plots then become continuous curves. The inverse Fourier transform also applies to continuous frequency.

Fourier decomposition applied to digitized data is known as the discrete Fourier series or the DFT. In the real world (as opposed to the textbook world), Fourier analysis is always done on a computer using a high-speed algorithm known as the fast Fourier transform or FFT. The discrete equations follow the same pattern as those developed for the continuous time signal except integration becomes summation and both the sinusoidal and signal variables are discrete. The digitized signal is assumed to be periodic even if it is not. Cases where this assumption causes problems, and possible solutions, are described in the next chapter. There is a discrete version of the continuous Fourier transform known as the discrete time Fourier transform, but this equation and its inverse are of theoretical interest only.

The FFT works best on signals that have a length that is a power of 2. They produce a complex spectrum that is the same length as the signal, N, and ranges in frequencies between 0 and f_s Hz. But the second half of the spectrum is redundant: it is the mirror image of the first half so only the components between and 0 and $f_s/2$ Hz are of interest. Taking the absolute value and angle of the complex spectrum produces the magnitude and phase spectra (Equations 3.23 and 3.24). Each of these spectra is N points long, but only $N/2$ points are unique, so the total number of unique points in the two spectra is equal to the number of points in the signal.

The FFT implements the Fourier analysis equations while the IFFT implements the Fourier synthesis equations. A number of signal processing operations involve converting a signal to the frequency domain using the FFT, operating on the signal while in the frequency domain, then converting back to a time domain signal using the IFFT. The great speed of the FFT and IFFT not only makes such involved operations practical, but also greatly enhances the value and practicality of time–frequency conversion.

PROBLEMS

1. Find the Fourier series of the square wave below using analytical methods (i.e., Equations 3.8 and 3.9). (Hint: Take advantage of symmetries to simplify the calculation.)

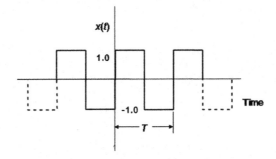

2. Find the Fourier series of the waveform below using analytical methods. The period, T, is 1 s.

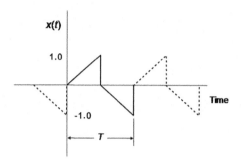

3. Find the Fourier series of the half-wave rectified sinusoidal waveform below using analytical methods. Take advantage of symmetry properties.

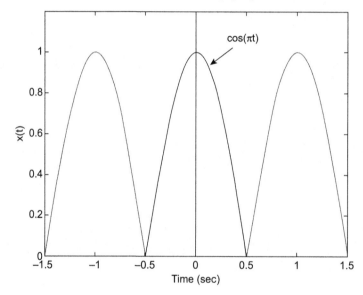

4. Find the Fourier series of the "sawtooth waveform" below using analytical methods.

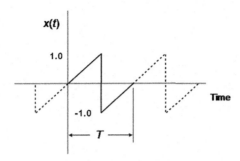

5. Find the Fourier series of the periodic exponential waveform shown below where $x(t) = e^{-2t}$ for $0 < t \leq 2$ (i.e., $T = 2$ s). Use the complex form of the Fourier series in Equation 3.20.

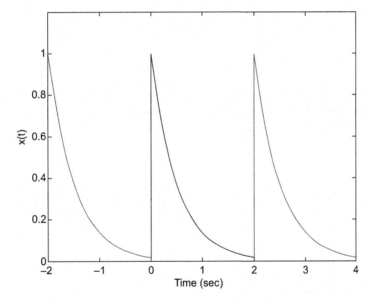

6. Find the continuous Fourier transform of an aperiodic pulse signal given in Example 3.5 using the complex equation, Equation 3.25 (or Equation 3.35).

7. Find the continuous Fourier transform of the aperiodic signal shown below using the complex equation, Equation 3.25.

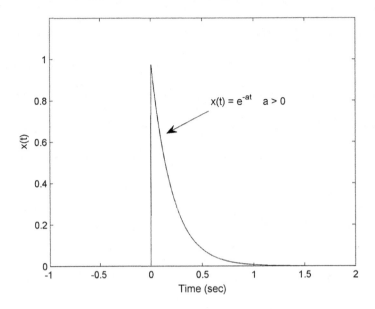

8. Find the continuous Fourier transform of the aperiodic signal shown below. This is easier to solve using the noncomplex form, Equation 3.26, in conjunction with symmetry considerations.

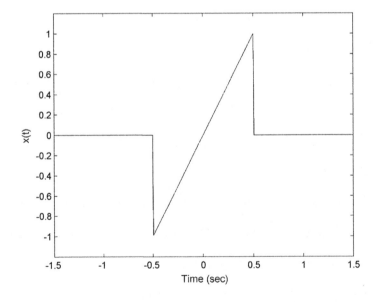

9. Use the MATLAB Fourier transform routine to find the spectrum of a waveform consisting of two sinusoids with frequencies of 200 and 400 Hz. Make $N = 512$ and $f_s = 1$ kHz. Take the Fourier transform of the waveform and plot the magnitude of the full spectrum (i.e., 512 points) as in Example 3.10. Generate a frequency vector so the spectrum plot has a properly scaled horizontal axis. (Hint: To generate the waveform, first construct a time vector `t`, then generate the signal using the code: `x = sin(2*pi*f1*t) + sin (2*pi*f2*t)`where `f1` $= 200$ and `f2` $= 400$.)

10. Use the routine `sig_noise` to generate a waveform containing 200 and 400 Hz sine waves as in Problem 9, but add noise so that the signal to noise ratio (SNR) is -8 dB, i.e., `x = sig_noise([200 400], -8,N)` where `N` $= 512$. Recall `sig_noise` assumes $f_s = 1$ kHz. Plot the magnitude spectrum, but only plot the nonredundant points (2 to $N/2$) and do not plot the DC term, which again is zero. Repeat for an SNR of -16 dB. Use `subplot` to plot the two spectra side by side for easy comparison and scale the frequency axis correctly. Note that the two sinusoids are hard to distinguish at the higher (-16 dB) noise level.

11. Use the routine `sig_noise` to generate a waveform containing 200 and 400 Hz sine waves as in Problems 9 and 10 with an SNR of -12 dB with 1000 points. Plot only the nonredundant points (no DC term) in the magnitude spectrum. Repeat for the same SNR, but for a signal with only 200 points. Use subplot to plot the two spectra side by side for easy comparison and scale the frequency axis correctly.

    Note that the two sinusoids are hard to distinguish with the smaller data sample. Taken together, Problems 10 and 11 indicate that both data length and noise level are important when detecting sinusoids (also known as "narrowband signals") in noise.

12. Generate the signal shown in Problem 2 and use MATLAB to find both the magnitude and phase spectrum of this signal. Assume that the period, T, is 1 sec and $f_s = 500$ Hz (hence, $N = 500$). Plot the time domain signal, $x(t)$, then calculate and plot the spectrum. The spectrum of this curve falls off rapidly, so plot only the first 20 points plus the DC term and plot them as discrete points, not as a line plot. As always scale the frequency axis correctly. Note that every other frequency point is zero as predicted by the symmetry of this signal.

13. Generate the signal used in Problem 4 with two different time shifts. One version should be as shown and the other shifted by $T/2$. For both signals, take the Fourier transform using `fft` and plot only the first 20 values (plus the DC term) as discrete points and use `subplot` to place the magnitude and phase plots side by side (you could put all four plots together). Unwrap the phase plot and scale by $360/(2\pi)$ so it is in degrees. Note the similarities and differences between the two frequency domain representations. (Hint: The hardest part of this problem is constructing the signals, particularly the first signal. To make the first signal, the one shown in Problem 4, merge two linear segments using MATLAB brackets. Care must be taken

when defining the intercepts of the second segment. The second signal is easier, as it is just a straight line over the entire period with a slope of 2 and an intercept of −1.)

14. Generate two versions of the signal used in Problems 4 and 13. One version should be as shown and the other its inverse. For both signals, take the Fourier transform using `fft` and plot only the first 20 values (plus the DC term) as discrete points. Use `subplot` to place the magnitude and phase plots side by side (you could put all four plots together). Scale the phase plot by $360/(2\pi)$ so it is in degrees. The two signals have very similar frequency domain characteristics, but there is a difference. Can you find the difference?

15. Use the `fft` routine to find the spectrum of a waveform consisting of two sinusoids at 100 and 200 Hz similar to Problem 9. Again generate the signal with $N = 512$ and $f_s = 1$ kHz. Plot the full magnitude spectrum (all N points). Repeat using sinusoidal frequencies of 200 and 900 Hz and plot. Use `subplot` to plot the two spectra side by side for easy comparison. Can you explain the results? If not, no problem, as we discuss this phenomenon in the next chapter.

16. Find the DFT for the signal shown below using the manual approach of Example 3.7. The dashed lines indicate one period.

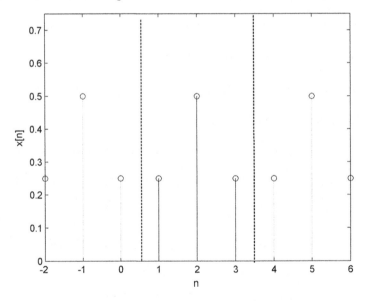

17. Find the DFT for the signal shown below using the manual approach of Example 3.7. The dashed lines indicate one period. (Hint: You can just ignore the zeros on either side except for the calculation of N.)

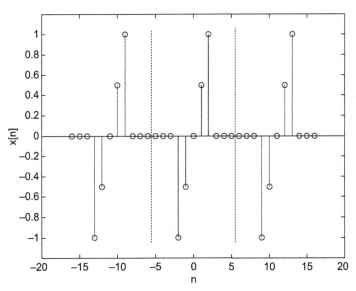

18. Plot the magnitude and phase components of the ECG signal found as variable ecg in file ECG.mat. Plot only the nonredundant points and do not plot the DC component (i.e., the first point in the Fourier series.) Also plot the time function and correctly label and scale the time and frequency axes. The sampling frequency was 125 Hz. Based on the magnitude plot, what is the bandwidth of the ECG signal, i.e., the range of frequencies that have some energy?

19. The data file pulses.mat contains three signals: x1, x2, and x3. These signals are all 1 s in length and were sampled at 500 Hz. Plot the three signals and show that each contains a single 40-ms pulse, but at three different delays: 0, 100, and 200 ms. Calculate and plot the spectra for the three signals superimposed on a single magnitude and single phase plot. Plot only the first 20 points as discrete points plus the DC term using a different color for each signal's spectra. Apply the unwrap routine to the phase data and plot in degrees. Note that the three magnitude plots are identical, whereas the phase plots are all straight lines but with radically different slopes.

20. Load the file chirp.mat, which contains a sinusoidal signal, x, that increases its frequency linearly over time. The sampling frequency of this signal is 5000 Hz. This type of signal is called a "chirp signal" because of the sound it makes when played through an audio system. If you have an audio system, you can listen to this signal after loading the file using the MATLAB command: sound(x,5000);. Take the Fourier transform of this signal and plot magnitude and phase (no DC term). Note that the magnitude spectrum shows the range of frequencies that are present but there is no information on the timing of those frequencies. Actually, information on signal timing

is contained in the phase plot but, as you can see, this plot is not easy to interpret. Advanced signal processing methods known as "time–frequency methods" are necessary to recover the timing information. These methods are briefly covered in the next chapter.

21. Load the file ECG_1min.mat that contains 1 min of ECG data in variable ecg. Take the Fourier transform. Plot both the magnitude and phase (unwrapped) spectrum up to 20 Hz and do not include the DC term. Find the average heart rate using the strategy found in Example 3.12. The sample frequency is 250 Hz. [Note: The spectrum will have a number of peaks; the largest low-frequency peak (which is also the largest overall) corresponds to the cardiac cycle.]

22. Construct two waveforms: the waveform in Problem 7 (e^{-at}) and a short pulse. For the first waveform make $a = 2$ and the period equal 2 s. Make the pulse as short as possible, i.e., one point. Make both waveforms 1024 points long and assume $f_s = 400$ Hz. Take the Fourier transform of both waveforms. Use ifft to reconstruct the Problem 7 waveform. Then multiply the two complex spectra together (point by point) and use ifft to get the time domain response. The two waveforms should look the same. Can you explain why the multiplication of the two complex spectra had no effect? If not, no worries, as we revisit the issue in Chapter 5.

23. Construct two waveforms. For the first, use sig_noise to produce a 2 Hz sinusoid, a signal to noise ratio of −3 dB, and $N = 2048$ (sig_noise(2,-3,N);). Recall sig_noise assumes an f_s of 1 kHz. The other signal should be a pulse having the same length as the noisy signal and a width of 10 samples. Take the Fourier transform of both waveforms. As in Problem 22, multiply the two complex spectra together (point by point) and use ifft to get the time domain response of the result. Plot the time domain of the original noisy signal and the signal produced by the frequency domain multiplication. Note that the latter has less noise. Again we revisit the concepts behind the results in Chapter 5. In the meantime you know one way of reducing noise in a signal.

Signal Analysis in the Frequency Domain—Implications and Applications

4.1 GOALS OF THIS CHAPTER

We are now quite skilled in transforming data into, and out of the frequency domain. We understand how this transformation works by correlating our signal with a series of harmonically related sinusoids. Once we have an equivalent sinusoidal series, it is easy to map the sinusoidal correlation coefficients to magnitude and phase in the frequency domain. In this chapter, we use our expertise in frequency domain signals to explore some important features of signals, particularly the consequences of signal digitization.

To convert a real-world continuous analog signal to discrete data, we know that two major steps are involved: slicing the signal in amplitude and sampling the signal in time. But we do not know how the resulting sliced and diced digitized signal corresponds to the original. Since all of our signal processing tools are applied to discrete signals, it is very important to determine if these signals closely reflect the original analog signals. In addition, it is often necessary to analyze only a portion of the original signal due to computer memory limitations, and we would like to understand how this signal truncation affects our analysis. Frequency domain representation helps us to understand both these limitations.

In signal analysis, usually only the magnitude spectrum of a signal is of interest. This is because the phase spectrum, while necessary to reconstruct the original signal, is difficult to interpret. In such cases, we often use the "power spectrum" which shows the spectral distribution of signal energy. The power spectrum is directly determined from the magnitude spectrum. Although the original signal cannot be reconstructed from the power spectrum, it is very easy to interpret.

To summarize, in this chapter we will:

- Use frequency domain methods to analyze the influence of time sampling on the digitized waveform.
- Define the power spectrum.

- Describe the effect of data truncation and introduce the concept of nonrectangular windowing of a signal.
- Define the bandwidth of a signal.
- Show how spectral averaging can be used to emphasize the broadband characteristics of a signal.

4.2 DATA ACQUISITION AND STORAGE

As mentioned, converting a continuous signal to discrete format involves amplitude slicing (quantization) and time sampling. In addition, it may be necessary to convert only a portion of the signal because of memory constraints. We look at each of these operations in turn and determine their influence on the discrete signal stored in memory.

4.2.1 Data Sampling: The Sampling Theorem

Slicing the signal into discrete time intervals (usually evenly spaced) is a process known as sampling and is described in Chapter 1 (see Figures 1.5 and 1.6). Sampling is a nonlinear process and has some peculiar effects on the signal's spectrum. Figure 4.1A shows an example of a magnitude frequency spectrum of a hypothetical 1.0-second periodic, continuous signal, as determined using continuous Fourier series analysis. The fundamental frequency is $1/T = 1\,Hz$ and the first 10 harmonics are plotted out to 10 Hz. For this particular signal, there is little energy above 7.0 Hz.

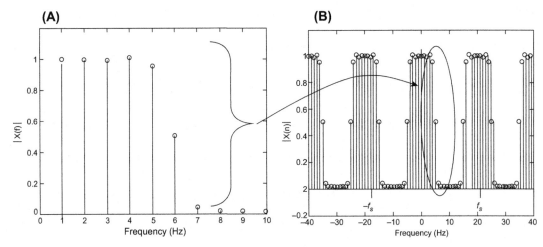

FIGURE 4.1   (A) The spectrum of a continuous signal. (B) The spectrum of this signal after being sampled at $f_s = 20\,Hz$. Sampling produces a larger number of frequency components not in the original spectrum, even components having negative frequency. The sampled signal has a spectrum that is periodic at the sampling frequency (20 Hz) and has an even symmetry about 0.0 Hz, as well as symmetry about the sampling frequency, f_s. Since the sampled spectrum is periodic, it goes on forever and only a portion of it can be shown. The spectrum is just a series of points; the vertical lines are drawn to improve visualization.

If we apply the mathematics describing the sampling to the Fourier transform, we find that the sampling process produces many additional frequencies that were not in the original signal. After sampling at 20 Hz, the sampled magnitude spectrum, now plotted over a larger frequency range, is shown in Figure 4.1B. Only a portion can be shown because the sampled spectrum is, theoretically, infinite. The new spectrum is itself periodic, with a period equal to the sample frequency, f_s (in this case 20 Hz). The spectrum also contains negative frequencies (it is, after all, a theoretical spectrum). It has even symmetry about $f = 0.0$ Hz as well as about both positive and negative multiples of f_s. Finally, the portion between 0 and 20 Hz also has even symmetry about the center frequency, $f_s/2$ (in this case 10 Hz).

The spectrum of the sampled signal is certainly bizarre, but comes directly out of the mathematics of sampling as described in Chapter 5 (Section 5.7.1). When we sample a continuous signal, we effectively multiply the original signal by a periodic impulse function (with period f_s) and that multiplication process produces all those additional frequencies. Even though the negative frequencies are mathematical constructs, their effects are felt because they are responsible for the symmetrical frequencies above $f_s/2$ as clearly noted in Figure 3.21. If the sampled signal's spectrum is different from the original signal's spectrum, it stands to reason that the sampled signal is different from the original. If the sampled signal is not the same as the original and we cannot somehow link the two, then digital signal processing is a lost cause. We would be processing something unrelated to the original signal. The critical question is: given that the sampled signal is different from the original, can we find some way to reconstruct the original signal from the sampled signal? The frequency domain version of that question is: can we reconstruct the unsampled spectrum from the sampled spectrum? The definitive answer is: maybe, but it depends on things we can understand and measure.

Figure 4.2 shows just one period of the spectrum shown in Figure 4.1B, the period between 0 and f_s Hz. In fact, this is the only portion of the spectrum that can be calculated by the discrete Fourier transform (DFT); all the other frequencies shown in Figure 4.1B are theoretical (but not inconsequential). Comparing this spectrum to the spectrum of the original signal, Figure 4.1A, we see that the two are the same for the first half of the spectrum, that is, up to $f_s/2$. The second half is just the mirror image of the first half. These mirror image frequencies are just the negative frequencies reflected back from f_s.

The mirror image frequencies, those above $f_s/2$, were not part of the original signal. But the lower frequencies, those below $f_s/2$, are in the original spectrum. So if we somehow got rid of all frequencies above $f_s/2$, we would have our original spectrum. We can get rid of the frequencies above $f_s/2$ by filtering them out. Just knowing that it is possible to get back to the original spectrum is sufficient to justify our sampled computer data; we just ignore the frequencies above $f_s/2$. The frequency $f_s/2$ is so important it has its own name: the "Nyquist[1] frequency."

This strategy of just ignoring all frequencies above the Nyquist frequency ($f_s/2$) works well and is the approach that is commonly adopted. But it can be used only if the original signal does not have spectral components at or above $f_s/2$. Consider a situation in which four sinusoids with respective frequencies of 100, 200, 300, and 400 Hz are sampled at a frequency of

---

[1]Nyquist was one of many prominent engineers to hone his skills at the former Bell Laboratories during the first half of the 20th century. He was born in Sweden, but received his education in the United States.

FIGURE 4.2 A portion of the spectrum of the sampled signal whose unsampled spectrum is shown in Figure 4.1A. There are more frequencies in this sampled spectrum than in the original, but they are distinct from and do not overlap the original frequencies. Separating out these unwanted spectral components through some sort of filtering should be possible.

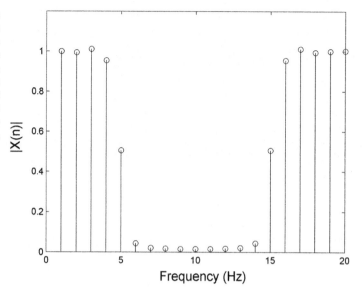

1000 Hz. The spectrum produced after sampling actually contains eight frequencies, Figure 4.3A: the four original frequencies plus the four mirror image frequencies reflected about $f_s/2 = 500$ Hz. As long as we know, in advance, that the sampled signal does not contain any frequencies above the Nyquist frequency (500 Hz), we do not have a problem: we know that the first four frequencies are those of the signal and the second four, above the Nyquist frequency, are the reflections, which can be ignored. However, a problem occurs if the signal contains frequencies higher than the Nyquist frequency. The reflections of these high-frequency components will be reflected back into the lower half of the spectrum. This is shown in Figure 4.3B where the signal now contains two additional frequencies at 650 and 850 Hz. These frequency components have their reflections in the lower half of the spectrum: at 350 and 150 Hz, respectively. It is now no longer possible to determine if the 350 and 150 Hz signals are part of the true spectrum of the signal (i.e., the spectrum of the signal before it was sampled) or whether these are reflections of signals with frequency components greater than $f_s/2$ (which in fact they are). Both halves of the spectrum now contain mixtures of frequencies above and below the Nyquist frequency, and it is impossible to know where they really belong. This confusing condition is known as "aliasing." The only way to resolve this ambiguity is to ensure that all frequencies in the original signal are less than the Nyquist frequency.

If the original signal contains frequencies above the Nyquist frequency, then you cannot determine the original spectrum from what you have in the computer and you cannot reconstruct the original analog signal from the one in the computer. The frequencies above the Nyquist frequency have hopelessly corrupted the signal stored in the computer. Fortunately, the converse is also true. If there are no corrupting frequency components in the original signal (i.e., the signal contains no frequencies above half the sampling frequency), the spectrum in the computer can be adjusted to match the original signal's spectrum if we eliminate

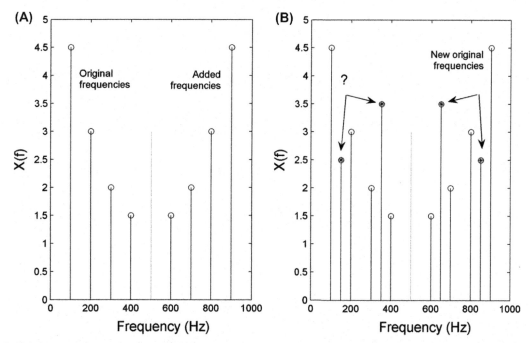

FIGURE 4.3 (A) Four sine waves between 100 and 400 Hz are sampled at 1 kHz. Sampling essentially produces new frequencies not in the original signal. The additional frequencies are a mirror image reflection around $f_s/2$, the Nyquist frequency. As long as the frequency components of the sampled signal are all below the Nyquist frequency as shown here, the upper frequencies do not interfere with the lower spectrum and can simply be ignored. (B) If the sampled signal contains frequencies above the Nyquist frequency, they are reflected into the lower half of the spectrum (filled circles). It is no longer possible to determine which frequencies belong where, an example of aliasing.

or disregard the frequencies above the Nyquist frequency. (Elimination of frequencies above the Nyquist frequency can be achieved by low-pass filtering, and the original signal can be reconstructed.) This leads to the famous "Sampling Theorem" of Shannon: the original signal can be recovered from a sampled signal provided the sampling frequency is more than twice the maximum frequency[2] contained in the original:

$$f_s > 2f_{max} \qquad (4.1)$$

Usually the sampling frequency is under software control, and it is up to the biomedical engineer doing the sampling to ensure that f_s is high enough. To make elimination of the unwanted higher frequencies easier, it is common to sample at three to five times f_{max}. This increases the spacing between the frequencies in the original signal and those generated by the sampling process, Figure 4.4. The temptation to set f_s higher than is really necessary is strong, and it is a strategy often pursued. However, excessive sampling frequencies lead to large data storage and processing requirements that needlessly overtax the computer system.

---

[2]The signal must not contain any frequencies above the Nyquist frequency. In other words, any signal frequencies above the sampling frequency, including those from noise, must have negligible energy.

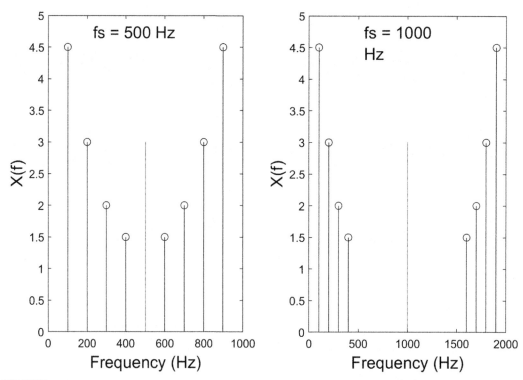

FIGURE 4.4 The same signal is sampled at two different sampling frequencies. The higher sampling frequency provides much greater separation between the original spectral components and those produced by the sampling process.

The concepts behind sampling and the sampling theorem can also be described in the time domain. Consider a single sinusoid. (Since all periodic waveforms can be broken into sinusoids, the influence of sampling on a single sinusoid can be extended to cover any general waveform.) In the time domain, Shannon's sampling theorem states that a sinusoid can be accurately reconstructed as long as two or more (evenly spaced) samples are taken over its period. This is equivalent to saying that f_s must greater than $2f_{sinusoid}$. Figure 4.5 shows a main sine wave (solid line) defined by two samples per cycle (black circles). The Shannon sampling theorem states that no other sinusoids of a lower frequency can pass through both these points, so these two samples uniquely define this sine wave. This spacing would also uniquely define any sine wave that was of a lower frequency. However, there are many higher frequency sine waves that can pass cleanly through these two points, two of which are shown in Figure 4.5 as dashed and dotted lines. The two higher frequency sine waves shown are second and third harmonics of the main sine wave. In fact, all the higher harmonics of the main sine wave would pass though these two points, so there are an infinite number of higher frequencies defined by the 2 samples. These higher frequency sine waves give rise to the added points in the sample spectrum as shown in Figure 4.1; they are the source of the additional frequencies.[3]

[3]The fact that higher frequency sinusoids are all harmonics explains why the added frequencies are themselves periodic.

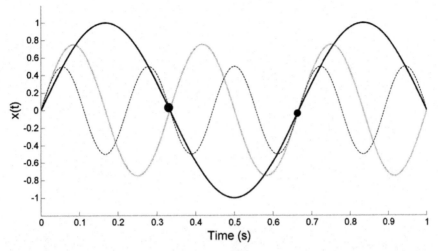

FIGURE 4.5   A sine wave (*solid line*) is sampled at two locations (*black circles*) within one period. The time domain interpretation of Shannon's sampling theorem states that no other sine wave of a lower frequency can pass though these two points. This sine wave is uniquely defined. However, an infinite number of higher frequency sine waves can pass though those two points (all the harmonics), two of which are shown. These higher frequency sine waves contribute the additional points shown on the spectrum of Figure 4.1.

Figure 4.6 illustrates aliasing using an undersampled sine wave. A 5 Hz sine wave is sampled at 7 samples/s, so $f_s/2 = 3.5$ Hz. A 2 Hz sine wave (*dotted line*) also passes through the seven points. This is predicted by aliasing: the 5 Hz sine reflected about f_s would be $7 - 5 = 2$ Hz.

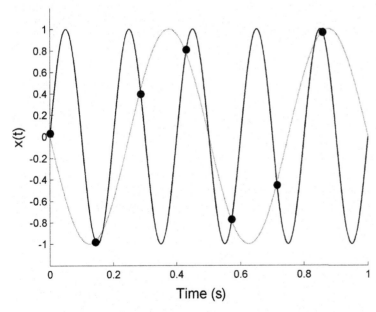

FIGURE 4.6   A 5 Hz sine wave (*solid line*) is sampled at 7 Hz (seven samples over a 1-s period). The seven samples also fit a 2 Hz sine wave, as is predicted by aliasing.

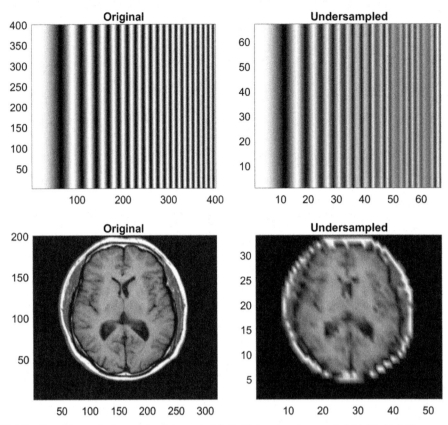

FIGURE 4.7 Two images that are correctly sampled (left side) and undersampled (right side). The upper pattern is a sine wave that increases in spatial frequencies from left to right. The undersampled image shows additional sinusoidal frequencies folded into the original pattern because of aliasing. The lower image is an MR image of the brain and the undersampled image has jagged diagonals and a moiré pattern, both characteristics of undersampled images.

Aliasing can also be observed in images. Figure 4.7 shows two image pairs: the left images are correctly sampled and the right images are undersampled. The upper images are of a sine wave that increases in spatial frequency going from left to right. This is the image version of a "chirp" signal that increases in frequency over time. The left image shows the expected smooth progression of sinusoidal bars. The right image is fine for the lower spatial frequencies, but as spatial frequency increases, additional frequencies are created because of aliasing, disrupting the progression. The lower images are MR images of the brain and the undersampled image shows jagged diagonals and a type of moiré pattern that is characteristic of undersampled images.

EXAMPLE 4.1

Construct a signal consisting of three sine waves at 100, 200, and 350 Hz. Find and plot the magnitude spectrum of this signal assuming two different sampling frequencies: $f_s = 800$ and 500 Hz. Use an N of 512 points and label and scale the frequency axes. Plot only valid (i.e., $N/2$) points.

Solution: Use a loop to operate on the two sampling frequencies. Construct a time vector and a frequency vector with the appropriate sampling frequency. Use the time vector to construct the signal as the sum of three sine waves and the frequency vector for plotting. Find the magnitude spectrum and plot only $N/2$ points.

```
% Example 4.1 Example of aliasing.
%
N = 512;                          % Number of points, N
N2 = N/2;                         % Half N
fs = [800 500];                   % Sample frequencies
for k = 1:2
  t = (0:N-1)/fs(k);              % Time vector
  f = (1:N)*fs(k)/N;              % Frequency vector
  x = sin(2*pi*100*t) + sin(2*pi*200*t) + sin(2*pi*350*t);    % Signal
  Xmag = abs(fft(x));             % Magnitude spectrum
  subplot(1,2,k);
  plot(f(1:N2),Xmag(1:N2),'k'); % Plot magnitude spectrum
  .......labels and title........
end
```

Results: The two spectra are shown in Figure 4.8. In both cases, f_{max} is 350 Hz. When $f_s > 2 f_{max}$ (i.e., 700 Hz), the three peaks are found at the correct frequencies: 100, 200, and 350 Hz, Figure 4.8A. When the sampling frequency is reduced to 500 Hz, f_s is now $< 2 f_{max}$. Consequently, the peak at 350 Hz manifests as a false peak at 150 Hz, Figure 4.8B. (Note: $500 - 350 = 150$ Hz). This is a classic example of aliasing.

4.2.2 Amplitude Slicing—Quantization

By selecting an appropriate sampling frequency, it is possible to circumvent problems associated with time slicing, but what about amplitude slicing, i.e., quantization? In Chapter 1, we note that amplitude resolution is given in terms of the number of bits in the binary output with the assumption that the least significant bit in the output is accurate (which is not always true). Typical analog-to-digital converters (ADCs) feature 8-, 12-, and 16-bit outputs, with 12 bits presenting a good compromise between conversion resolution and cost. In fact, most biological signals do not have sufficient signal-to-noise ratio (SNR) to justify a higher resolution; you are simply obtaining a more accurate conversion of the noise.

The number of bits used for conversion sets an upper limit on the resolution, and determines the quantization error. The more bits used to represent the signal, the finer the resolution of the digitized signal and the smaller the quantization error. The quantization error is

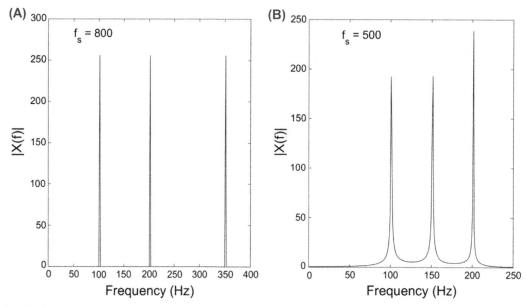

FIGURE 4.8 Magnitude spectra of a signal consisting of three sine waves at 100, 200, and 350 Hz sampled at two different frequencies: 800 and 500 Hz. (A) When $f_s = >2f_{max}$, three peaks are found at the correct frequencies. (B) When $f_s = <2f_{max}$, the peak that was greater than $f_s/2$ (350 Hz) appears folded back as a false peak at 150 Hz (Note: $f_s/2 - f_{max} = 500 - 350 = 150$ Hz.)

the difference between the original continuous signal value and the digital representation of that level after sampling, Figure 4.9. This error can be thought of as some sort of noise superimposed on the original signal. If a sufficient number of quantization levels exist (say $N > 64$, equivalent to seven bits), the distortion produced by quantization error may be modeled as additive, independent white noise with zero mean and a variance determined by the quantization step size, q. As described in Example 1.1, the quantization step size, q, is the maximum voltage the ADC can convert, divided by the number of quantization levels, which is $2^N - 1$; hence, $q = V_{MAX}/2^N - 1$. The variance or mean square error can be determined using the expectation function from basic statistics:

$$s^2 \equiv \sigma^2 = \overline{\sigma^2} = \int_{-\infty}^{\infty} e^2 PDF(e)de \tag{4.2}$$

where $PDF(e)$ is the uniform probability density function and e is the error voltage (the bottom trace of Figure 4.8). The $PDF(e)$ for a uniform probability distribution that ranges between $-q/2$ to $+q/2$ is simply $1/q$. Substituting $1/q$ for $PDF(e)$ in Equation 4.2:

$$\sigma^2 = \int_{-q/2}^{q^6/2} e^2 \left(\frac{1}{q}\right) de = \left. \frac{\left(\frac{1}{q}\right)e^3}{3} \right|_{-q/2}^{q/2} = \frac{q^2}{12} = \frac{V_{MAX}^2}{12(2^N - 1)^2} \tag{4.3}$$

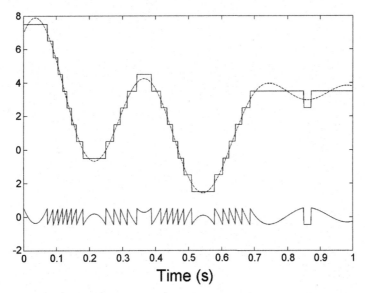

FIGURE 4.9 Quantization (amplitude slicing) of a continuous waveform. The lower trace shows the error between the quantized signal and the input.

where V_{MAX} is the maximum voltage the ADC can convert and N is the number of bits out of the ADC. Assuming a uniform distribution with zero mean for the quantization noise, the root mean square (RMS) value of the noise (Equation 2.3) would be approximately equal to the standard deviation, σ (Equation 2.12).

EXAMPLE 4.2

What is the equivalent quantization noise (RMS) of a 12-bit ADC that has an input voltage range of ± 5.0 volts? Assume the converter is perfectly accurate.

Solution: Apply Equation 4.3. Note that V_{max} would be 10 volts since the range of the ADC is from -5 to $+5$ volts.

$$\sigma^2 = \frac{V_{MAX}^2}{12(2^N - 1)^2} = \frac{10^2}{12(2^{12} - 1)^2} = \frac{100}{12(4095)^2} = 4.97 \times 10^{-7}$$

$$V_{RMS} \cong \sigma = \sqrt{4.97 \times 10^{-7}} = 0.705 \text{ mV}$$

4.2.3 Data Truncation

Given the finite memory capabilities of computers, it is usually necessary to digitize a shortened version of the signal. For example, only a small segment of a subject's ongoing electroencephalography (EEG) signal can be captured for analysis on the computer. The length of this shortened data segment is determined during the data acquisition process and is usually a compromise between the desire to acquire a large sample of the signal and the need to limit computer memory usage or data acquisition time. Taking just a segment of a much longer signal involves truncation of the signal, and this has an influence on the spectral resolution of the stored signal. As described next, both the number of data points and the manner in which the signal is truncated will alter the spectral resolution.

4.2.3.1 Data Length and Spectral Resolution

As discussed in Chapter 3, all digitized signals are assumed to be periodic. The period, T, is the time length of the data segment. Recall that the Fourier series frequency components depend on this period, specifically:

$$f_m = \frac{m}{T} = mf_1 \tag{4.4}$$

where m is the harmonic number, T is the period, and f_1 equals $1/T$.

For a discrete signal of N samples, the equivalent time duration, T, is N times the sample interval, T_s or N divided by the sample frequency, f_s:

$$T = \frac{N}{f_s} \tag{4.5}$$

and the equivalent frequency of a given harmonic number, m, is:

$$f_m = \frac{m}{T} = \frac{m}{\dfrac{N}{f_s}} = \frac{mf_s}{N} \tag{4.6}$$

The last data point from the DFT has a frequency value of f_s since:

$$f_{\text{Last}} = f|_{m=N} = \frac{Nf_s}{N} = f_s \tag{4.7}$$

The frequency resolution of a spectral plot is the difference in frequency between harmonics, which according to Equation 4.6[4] is f_s/N:

$$f_{\text{Resolution}} = \frac{f_s}{N} \tag{4.8}$$

Just as with the continuous Fourier transform, frequency resolution of the DFT depends on the period (i.e., time length) of the data. For the DFT, the resolution is equal to f_s/N, from Equation 4.8. So, for a given sampling frequency, the more samples (N) in the signal, the smaller the frequency increment between successive DFT data points. The more points sampled, the higher the spectral resolution. Of course you could also reduce f_s, but this strategy is limited by the sampling theorem, Equation 4.1.

---

[4]This is essentially the same equation we use in MATLAB programs to generate the horizontal axis used in frequency plots (i.e., in MATLAB, f = (0:N−1)*fs/N).

Once the data have been acquired, it would seem that the number of points representing the data, N, is fixed, but there is a trick that can be used to increase the data length post hoc. We can increase N simply by tacking on constant values, usually zeros. We used this approach, zero padding, to extend signals when computing cross-correlation in Chapter 2. Using it here to extend the length of a signal may sound like cheating, but we could argue that we cannot really know what the signal was doing outside of the data segment we have, and maybe it really was zero. Other, more complicated padding techniques can be used, such as extending the last values of the signal on either end, but zero padding is by far the most common strategy for extending data.

Zero padding gives the appearance of a spectrum with higher resolution. Example 4.3 shows that zero padding certainly appears to improve the resolution of the resulting spectrum, producing a spectrum with more points that better illustrate the spectrum's details. Of course, artificially extending the period with zeros does not increase the information in the spectrum and the spectral resolution is really not any better. But zero padding provides an interpolation of the points that were in the unpadded signal: it fills in the gaps of the original spectrum using an estimation process. Overstating the value of zero padding is a common mistake of practicing engineers. Nonetheless, the interpolated spectrum certainly looks better when plotted.

EXAMPLE 4.3

Generate a waveform with a period $T = 1.0$ s that is a triangle wave for the first 0.5 s and zero for the rest of the period. Assume $f_s = 100$ Hz so $N = 100$ points. Calculate and plot the magnitude spectrum. Zero pad the signal to extend the period to 2 and 6 s and recalculate and plot the power spectrum. Limit the spectral plot to a range of 0–20 Hz.

Solution: Generate the 1.0 triangle waveform. Since $f_s = 100$ Hz, the original signal should be padded by 100 and 500 additional points to extend the period to 2 and 6 s. Calculate the spectrum using fft. Use a loop to calculate and plot the magnitude spectra of the three signals.

```
% Example 4.3  Example of zero padding on apparent spectral resolution
%
fs = 100;                        % Sampling frequencies
N1 = [0 100  500];               % Zero padding numbers
x =[ (0:25) (24:-1:0),zeros(1,49) ];   % Generate triangle signal, 100 pts
for k = 1:3
  y = [x zeros(1,N1(k))];        % Zero pad signal with 0, 100, and 500
  N = length(y);                 % Get data length
  t = (0:N-1)/fs;                 % Construct time vector for plotting
  f = (0:N-1)*fs/N;              % Construct frequency vector for plotting
  subplot(3,2,k*2-1);
  plot(t,y,'k');                  % Plot the signal
  ....... Labels and titles.........
  subplot(3,2,k*2);
  Y = abs(fft(y));                % Calculate the magnitude spectrum
  plot(f, Y,'.k');                % Plot spectrum using individual points
  .......Labels and titles.......
end
```

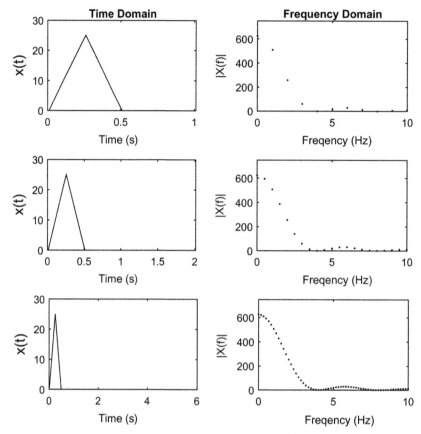

FIGURE 4.10 A waveform having an actual period of 1.0 s (upper left) and its associated frequency spectrum (upper right). Extending the period to 2 and 6 s by adding zeros decreases the spacing between the frequency points, producing smoother looking frequency curves (middle and lower plots).

Results: The time and magnitude spectrum plots are shown in Figure 4.10. The spectral plots all have the same shape, but the points are more closely spaced with the zero-padded data. As shown in Figure 4.10, simply adding zeros to the original signal produces a better looking curve, which explains the popularity of zero padding (even if no additional information is produced). MATLAB makes zero padding during spectral analysis easy, as the second argument of the fft routine (i.e., fft(x,N)) adds zeros to the original signal if the original signal length is less than N. (If N is less than the data length, the data are truncated to that length.)

Taking the DFT implicitly assumes that the signal is periodic with a period of $T = N/f_s$ (Equation 4.5). The frequency spectrum produced is a set of numbers spaced f_s/N apart (Equation 4.8). However, if we make the assumption that our signal is actually aperiodic (i.e., that $T \to \infty$), then the spacing between points goes to zero and the points become a continuous line. This is the assumption that is commonly made when we plot the spectrum

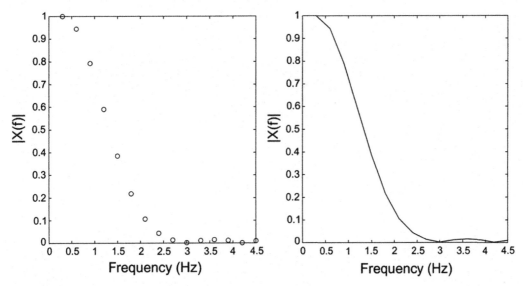

FIGURE 4.11   (A) The discrete Fourier transform assumes the signal is periodic and computes a spectrum consisting of discrete points. (B) Making the assumption that the signal really is aperiodic with an infinite period justifies connecting the discrete points forming a smooth curve. This assumption is commonly made when plotting frequency spectra.

not as a series of points, but as a smooth curve, Figure 4.11 (right side). Although this is commonly done when plotting spectra, you should be aware of the underlying truth: an implicit assumption is being made that the signal is aperiodic and, although the calculated spectrum is really a series of discrete points, they are joined together because of this assumption.

4.2.4 Data Truncation—Window Functions

The third limitation of signal digitization is the usual need to truncate the original real-world signal because of memory constrains. Truncation can be thought of as a "windowing" process whereby the original data are multiplied by a finite-length function that limits the data length. For example, often the original signal is simply cut at two time points to extract the digitized signal. We could think of this extraction as multiplying the original signal by a rectangular waveform that is 1.0 over the length of the digitized signal and 0.0 everywhere else, Figure 4.12. Note that this approach usually produces abrupt changes or discontinuities at the endpoints.

The discontinuities at the endpoints can produce artifacts in the spectrum, particularly if the data length is small. One way to reduce these discontinuities would be to impose a window function of our own invention that tapers the data at the ends to zero. No need to invert such a window as many such tapering window shapes have been developed. An example of one such tapering window is the "Hamming window" shown in Figure 4.13. This window is often used in MATLAB routines that are likely to involve short data sets. The Hamming

FIGURE 4.12 Data truncation or shortening can be thought of mathematically as multiplying the original data, $x(t)$ (*upper curve*), by a rectangular window function, $w(t)$ (*middle curve*), to produce the truncated data that are in the computer, $x'(t)$ (*lower curve*). The window function has a value of 1.0 over the length of the truncated data and 0.0 everywhere else. Note that this window function generally results in abrupt changes at the endpoints.

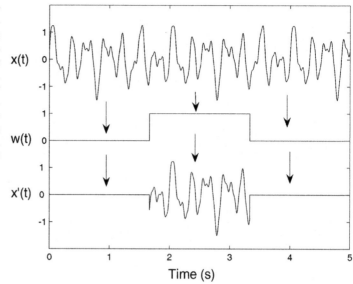

FIGURE 4.13 The application of a Hamming window to shape the data truncated by a rectangular window in Figure 4.12. This window is similar to a half sine function and, when multiplied with the data set, tapers the two ends of the signal toward zero. The application of such a tapering window decreases the resultant spectral resolution, but also reduces the influence of noise on the spectrum.

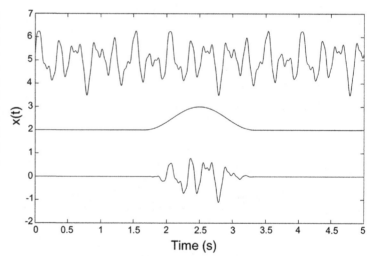

window has the shape of a raised half sine wave, which when applied to the truncated signal reduces endpoint discontinuities:

$$w[n] = 0.5 - 0.46 \cos\left(\frac{2\pi n}{N}\right) \tag{4.9}$$

where $w[n]$ is the Hamming window function and N is the window (and data) length. The variable n ranges between 1 and $N + 1$.

Multiplying your signal by a function like that shown in Figure 4.13 may seem a bit extreme and it is controversial in the signal processing community. Moreover, the benefit of windows such as the Hamming window is generally quite subtle. Except for very short data segments, a rectangular window (i.e., simple truncation and no additional windowing) is usually the best option. Applying the Hamming window to the data will result in a spectrum with slightly better noise immunity, but overall frequency resolution is reduced. Figure 4.14 shows the magnitude spectra of two closely spaced sinusoids with noise that were obtained using the FFT after application of two different windows: a rectangular window (i.e., simple truncation) and a Hamming window. The spectra are really quite similar, although the spectrum produced after applying a Hamming window to the data shows a slight loss in resolution as the two frequency peaks are not as sharp. The peaks due to background noise are also somewhat reduced by the Hamming window.

If the data set is fairly long (perhaps 256 points or more), the benefits of a nonrectangular window are slight. Figure 4.15 shows the spectra obtained from a signal containing two closely spaced sinusoids (100 and 120 Hz) in noise (SNR = −12 dB) with and without the Hamming window. The resulting spectra are seen to be nearly the same except for a scale difference produced by the Hamming window. MATLAB's Signal Processing Toolbox contains routines to generate 17 different window functions of varying complexity. Semmlow

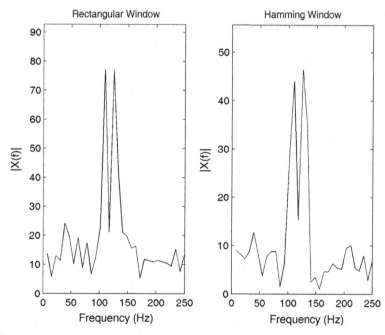

FIGURE 4.14 The spectrum from a short signal ($N = 128$) containing two closely spaced sinusoids (100 and 120 Hz) with added noise. A rectangular and a Hamming window have been applied to the signal to produce the two spectra. The spectrum produced after application of the Hamming window shows a slight loss of spectral resolution, as the two peaks are not as sharp; however, background spectral peaks due to noise have been somewhat reduced.

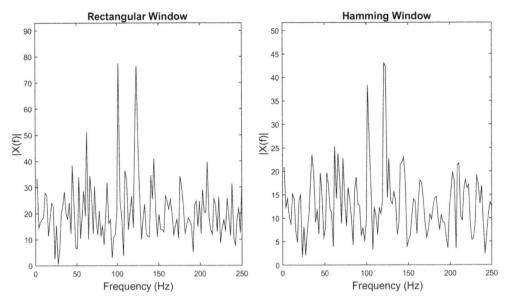

FIGURE 4.15   A spectrum from a relatively long signal ($N = 512$) containing two closely spaced sinusoids (100 and 120 Hz) in noise (SNR $= -12$ dB). Rectangular and Hamming windows have been applied to the signal. When the data set is long, the window function has little effect on the resultant spectrum and you might as well use a rectangular window, i.e., simple truncation.

(2014) provides a thorough discussion of the influence of various nonrectangular window functions, but the bottom line is that they typically have little effect.

4.3 POWER SPECTRUM

The power spectrum is commonly defined as the Fourier transform of the autocorrelation function. In continuous and discrete notations the power spectrum equation becomes:

$$PS(f) = \frac{1}{T} \int_0^T r_{xx}(t)e^{-j2\pi mf_1 t}dt \qquad m = 0, 1, 2, 3\ldots \tag{4.10}$$

$$PS[m] = \sum_{n=1}^{N} r_{xx}[n]e^{-\frac{j2\pi mn}{N}} \quad m = 0, 1, 2, 3\ldots N \tag{4.11}$$

where $r_{xx}(t)$ and $r_{xx}[n]$ are autocorrelation functions as described in Chapter 2. Since the autocorrelation function has even symmetry, the sine terms of the Fourier series are all zero (see Table 3.1), and two equations can be simplified to include only real cosine terms:

$$PS[m] = \sum_{n=0}^{N-1} r_{xx}[n]\cos\left(\frac{2\pi nm}{N}\right) \quad m = 0, 1, 2, 3\ldots N \tag{4.12}$$

$$PS(f) = \frac{1}{T} \int_0^T r_{xx}(t)\cos(2\pi mft)dt \quad m = 0, 1, 2, 3, \ldots \tag{4.13}$$

Equations 4.12 and 4.13 are sometimes referred to as "cosine transforms."

Nowadays, these equations are principally of theoretical, or perhaps historical, value. We now use the direct approach to calculate the power spectrum. This approach is motivated by the fact that the energy contained in an analog signal, $x(t)$, is related to the magnitude of the signal squared integrated over time:

$$E = \int_{-\infty}^{\infty} |x(t)|^2 dt \tag{4.14}$$

By an extension of a theorem attributed to Parseval, it can be shown that:

$$\int_{-\infty}^{\infty} |x(t)|^2 dt = \int_{-\infty}^{\infty} |X(f)|^2 df \tag{4.15}$$

Equations 4.14 and 4.15 show that the magnitude spectrum squared, $|X(f)|^2$, has the same energy as the time signal. Accordingly the magnitude spectrum squared is referred to as the "energy spectral density," or more commonly, the "power spectral density" or simply the "power spectrum" (PS). So in the direct approach, the power spectrum is just the magnitude squared of the Fourier transform (or Fourier series):

$$PS(f) = |X(f)|^2 \tag{4.16}$$

This direct approach of Equation 4.16 has displaced the cosine transform for determining the power spectrum because of the efficiency of the fast Fourier transform.[5] One of the problems at the end of this chapter compares the power spectrum obtained using the direct approach of Equation 4.16 with the traditional cosine transform method represented by Equation 4.11 and, if done correctly, shows them to be identical.

Unlike the Fourier transform, the power spectrum does not contain phase information, so the power spectrum is not an invertible transformation: it is not possible to reconstruct the signal from the power spectrum. However, the power spectrum has a wider range of applicability and can be defined for some signals that do not have a meaningful Fourier transform such as signals resulting from random processes. Since the power spectrum does not contain phase information, it is applied in situations in which the phase is not considered useful or to data that contain a lot of noise, since phase information is easily corrupted by noise.

An example of the descriptive properties of the power spectrum is given using the heart rate data shown in Example 2.19. The heart rates showed major differences in the mean and standard deviation between meditative and normal states. Applying the autocovariance to the meditative heart rate data suggested a possible repetitive structure for the variation in

---

[5]A variation of the cosine transform approach is still used in some advanced signal processing techniques involving simultaneous time and frequency transformation.

heart rate (Problem 29 in Chapter 2). The next example uses the power spectrum to search for structure in the frequency characteristics of both normal and meditative heart rate data.

EXAMPLE 4.4

Determine and plot the power spectra of heart rate variability (HRV) during both normal and meditative states.

Solution: The power spectrum can be determined using the direct method given in Equation 4.16. However, the heart rate data should first be converted to evenly sampled time data, and this is a bit tricky. Nonetheless, it is important to learn how to deal with such data, as unevenly sampled data are often found in studies of cardiac and neural function.

The data set obtained by a download from the PhysioNet data base provides the heart rate in beats per minute at unevenly spaced times, whenever a heartbeat occurred. The (uneven) sample times are provided as a second vector. These rate data need to be rearranged into evenly spaced time samples. This process, known as "resampling," constructs a vector of the heart rate at some selected sample interval. This evenly timed sampled vector is constructed through interpolation using MATLAB's interp1 routine. This routine takes in the unevenly spaced x-y pairs as two vectors along with a vector containing the desired evenly spaced x values. The routine then uses linear interpolation (other options are possible) to approximate the y values that match the evenly spaced x values. Details can be found in the MATLAB help file for interp1.

In the following program, the uneven x-y pairs for normal conditions are in time vector t_pre and heart rate vector hr_pre, both found in the MATLAB file Hr_pre. For meditative conditions, the vectors are t_med and hr_med in file Hr_med. To convert the heart rate data to a sequence of evenly spaced points in time, a time vector, xi, is first created. This vector increases in increments of 0.01 s $(t_s = 1/f_s = 1/100)$ between the lowest and highest values of time (rounded appropriately) in the original data. A 100 Hz resampling frequency was chosen because this is common in HRV studies that use certain nonlinear methods, but in this example, a wide range of resampling frequencies give the same result. Evenly spaced data are produced in vector yi using the MATLAB interpolation routine interp1. Since we want the power spectrum of HRV, not heart rate per se, we subtract out the average heart rate before evaluating the power spectrum. (In Example 2.19, we used autocovariance, which automatically subtracts the mean.)

After interpolation and removal of the mean heart rate, the power spectrum is determined using fft and taking the square of the magnitude component. The frequency plots are limited to a range between 0.0 and 0.15 Hz since this is where most of the spectral energy is to be found.

```
% Example 4.4
% Frequency analysis of heart rate data in the normal and meditative state
%
fs = 100;              % Sample frequency (100 Hz)
Ts = 1/fs;             % Sample interval
load Hr_pre;           % Load normal and meditative data
%
% Convert to evenly-spaced time data using interpolation; i.e., resampling
%  First generate evenly space time vectors having one second
%  intervals and extending over the time range of the data
```

```
%
Tmin = ceil(t_pre(1));                    % Initial time (rounded upward)
Tmax = floor(t_pre(end));                 % Final time (rounded downward);
t = (Tmax:Ts:Tmin);                       % Evenly-spaced time vector
yi = interp1(t_pre,hr_pre,xi');           % Interpolate
yi = yi - mean(yi);                       % Remove average
N2 = round(length(yi)/2);
f = (1:N2)*fs/N2;                         % Vector for plotting
%
% Now determine the Power Spectrum
YI = abs((fft(yi)).^2);                   % Direct approach (Eq. 4.16)
subplot(1,2,1);
plot(f,YI(2:N2+1,'k');                    % Plot spectrum, but not DC value
axis([0 .15 0 max(YI)*1.25]);             % Limit frequency axis to 0.15 Hz
   .......label and axis.......
%
% Repeat for meditative data
```

Results: The power spectrum of normal HRV is low and decreases with frequency, showing little energy above 0.1 Hz, Figure 4.16A. The meditative state, Figure 4.16B, shows large peaks at around 0.1−0.12 Hz, indicating that some resonant process is active at these frequencies, frequencies corresponding to a time frame of around 10 s. Feel free to speculate on the cause of this change in the heart rate rhythm.

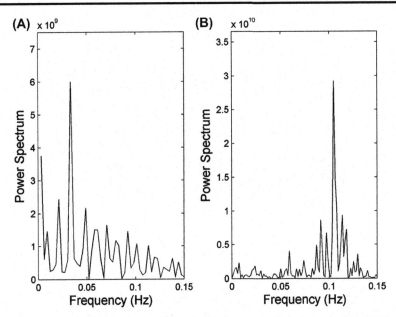

FIGURE 4.16   Power spectra of heart rate variability in the frequency range between 0 to 0.15 Hz where most of the energy is located. (A) Power spectrum of heart rate variability under normal conditions. The power decreases with frequency. (B) Power spectrum of heart rate variability during meditation. Strong peaks in power around 0.12 Hz are seen, indicating that much of the variation in heart rate is organized around these frequencies. Note the larger scale of the meditative power spectrum.

Analysis of HRV is an area of considerable current research. Spectral analysis is a commonly used tool in HRV studies. Many physiological processes influence the heart rate and the feedback nature of these processes can produce unique oscillations that can be seen in the HRV spectrum. HRV has been shown to be of diagnostic value in myocardial infarction and other heat diseases, and also in diabetes, neural disorders, sepsis, sudden infant death syndrome, depression, and other psychological disorders.

4.4 SPECTRAL AVERAGING

Although the power spectrum is usually calculated using the entire waveform, it can also be applied to isolated segments of the data. The power spectra determined from each of these segments can then be averaged to produce a spectrum that better represents the broadband, or global features of the spectrum. This approach is popular when the available waveform is only a sample of a longer signal. In such situations, spectral analysis is necessarily an estimation process, and averaging improves the statistical properties of the result. When the power spectrum is based on a direct application of the Fourier transform followed by averaging, it is referred to as an average "periodogram."

Averaging is usually achieved by dividing the waveform into a number of segments, possibly overlapping, and evaluating the power spectrum of each of these segments, Figure 4.17. The final spectrum is constructed from the ensemble average of the power

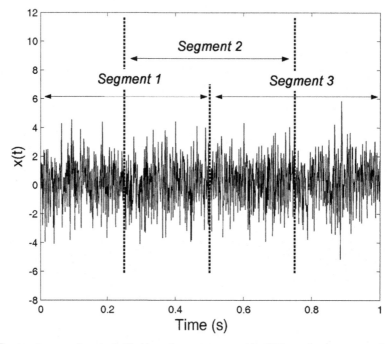

FIGURE 4.17 A noisy waveform is divided into three segments with a 50% overlap between each segment. In the Welch method of spectral analysis, the power spectrum of each segment is computed separately and an average of the three transforms gives the final spectrum.

spectra obtained from each segment.[6] Note that this averaging approach can only be applied to the power spectrum or magnitude spectrum because these spectra are not sensitive to time translation as shown in Example 3.11. Applying this averaging technique to the standard Fourier transform would not make sense because the phase spectrum would be sensitive to segment position. Averaging phases obtained for different time positions would be meaningless.

One of the most popular procedures to evaluate the average periodogram is attributed to Welch and is a modification of the segmentation scheme originally developed by Bartlett. In this approach, overlapping segments are used and a shaping window (i.e., a nonrectangular window) is sometimes applied to each segment. Periodograms obtained from noisy data traditionally average spectra from half-overlapping segments: segments that overlap by 50%. Higher amounts of overlap have been recommended in applications when computing time is not a factor. Maximum overlap occurs when the segment is shifted by only one sample.

If we segment a signal, the length of each segment is obviously less than that of the original unsegmented signal. Each segment will have fewer samples, N. Equation 4.8 states that frequency resolution is proportional to f_s/N where N is now the number of samples in a segment. So, averaging produces a trade-off between spectral resolution, which is reduced by averaging, and increased statistical reliability. Choosing a short segment length (a small N) will provide more segments for averaging and improve the reliability of the spectral estimate, but will also decrease frequency resolution. This trade-off is explored in the next example.

EXAMPLE 4.5

Evaluate the influence of averaging on power spectrum estimation as applied to a signal consisting of broadband and narrowband signals with added noise. The broadband signal was constructed by taking white noise and filtering it to remove frequencies above 300 Hz; then two closely spaced sinusoids were added at 390 and 410 Hz. Determine both the averaged and unaveraged spectra. The signal is in vector x in file broadband1.mat and $f_s = 1 \text{ kHz}$.

Solution: Load the file broadband1 containing the signal with narrowband and broadband components. First, calculate and display the unaveraged power spectrum. Then apply power spectrum averaging using an averaging routine. Use a segment length of 128 points and the maximum overlap of 127 points. To implement averaging, write a routine called welch that takes in the data, segment size, and the number of overlapping points and produces the averaged power spectrum. The main routine is:

```
% Example 4.5 Investigation of the use of averaging to improve
%   broadband spectral characteristics in the power spectrum.
%
load broadband1;              % Load data (variable x)
fs = 1000;                    % Sampling frequency
N = length(x);                % Find signal length
```

[6]Section 2.2.4 describes ensemble averaging to reduce noise in the time domain.

```
N2 = round(N/2);              % Half data length
nfft = 128;                   % Segment size for averaging
f = (1:N)*fs/N;               % Frequency vector for plotting
PS = abs((fft(x)).^2)/N;      % Calculate un-averaged PS
subplot(1,2,1)
plot(f(1:N2),PS(1:N2));       % Plot un-averaged Power Spectrum
  ........labels and title........
%
[PS_avg,f_avg]=welch(x,nfft, nfft-1,fs);       % Calculate periodogram, max. overlap
%
subplot(1,2,2)
  plot(f_avg,PS_avg);         % Plot periodogram
  .......labels and title.......
```

The welch routine takes in the sampling frequency, an optional parameter, which is used to generate a frequency vector useful for plotting. MATLAB power spectrum routines generally include this feature and we need to make our programs as good as MATLAB's. This program outputs only the nonredundant points of the power spectrum (i.e., up to $f_s/2$) along with the frequency vector. Finally, this routine checks the number of arguments and uses defaults if necessary, another common MATLAB feature. The program begins with a list and description of input and output parameters, then sets up defaults as needed. This is the text that is printed when someone types in "help welch."

```
function [PS,f] = welch(x,nfft,noverlap,fs);
  ......descriptive comments...........
%
N = length(x);          % Get data length
nfft = round(nfft);     % Make sure nfft is an interger
nfft2 = round(nfft/2);  % Half segment length
if nargin < 4           % Check arguments
  fs = 2*pi;            % Default fs
end
if nargin < 3 || isempty(noverlap) == 1
  noverlap = nfft2;     % Set default overlap at 50%
end
noverlap = round(noverlap);        % Make sure noverlap is an interger
%
%  Defaults complete. The routine now calculates the appropriate number of points
%  to shift the window and the number of averages that can be done given
%  the data length (N), window size (nff) and overlap (noverlap).
%
f = (1:nfft2)* fs/(nfft);   % Calculate frequency vector
increment = nfft - noverlap;     % Calculate window shift
nu_avgs = round(N/increment);    % Determine the number of segments
%
% Now shift the segment window and calculate the PS using Eq. 4.17
for k = 1:nu_avgs                    % Calculate spectra for each data point
```

```
first_point = 1 + (k - 1) * increment;
if (first_point + nfft -1) > N          % Check for possible overflow
  first_point = N - nfft + 1;           % Shift last segment to prevent overflow
end
data = x(first_point:first_point+nfft-1); % Get data segment
  % MATLAB routines would add a non-rectangular window here, the default being
  % a Hamming window. This is left as an exercise in one of the problems
if k == 1
  PS = abs((fft(data)).^2);             % Calculate PS, first segment
else
  PS = PS + abs((fft(data)).^2);        % Calculate PS, add to average
end
end
% Scale average and remove redundant points. Also do not include DC term
PS = PS(2:half_segment+1)/(nu_avgs*nfft2);
```

Analysis: This routine first checks if the sampling frequency and desired overlap are specified. If not, the routine sets f_s to 2π and the overlap to a default value of 50% (i.e., half the segment length: nfft2). These are the values that MATLAB uses in a similar routine called pwelch found in the Signal Processing Toolbox. Argument names nfft and noverlap are also lifted from the pwelch routine. Next, a frequency vector, f, is generated from 1 to fs/2 (or π). The number of segments to be averaged is determined based on the segment size (nfft) and the overlap (noverlap). A loop is used to calculate the power spectrum using the direct method of Equation 4.16, and individual spectra are summed. Finally, the power spectrum is shortened to eliminate redundant points and the DC term, and then normalized by both the segment length and the number of spectra to generate an average spectrum.

This example determines both the unaveraged and averaged power spectra. For the averaged spectrum or periodogram, a segment length of 128 (a power of two) is specified as is the maximal overlap (an overlap equal to the segment length $N - 1$). In practice, the selection of segment length and averaging strategy is usually based on experimentation with the data.

Results: In the unaveraged power spectrum Figure 4.18A, the two sinusoids at 390 and 410 Hz are clearly seen; however, the broadband signal is noisy and poorly defined. The periodogram in Figure 4.18B is much smoother, better reflecting the constant energy in white noise, but the loss in frequency resolution is apparent as the two sinusoids are hardly visible. This demonstrates one of those all-so-common engineering compromises. Spectral techniques that produce a good representation of global characteristics such as broadband features are not good at resolving narrowband or local features such as sinusoids and vice versa.

EXAMPLE 4.6

Determine and plot the frequency characteristics of HRV during both normal and meditative states using spectral averaging. Divide the time data into eight segments and use a 50% overlap (the default). Plot using both a linear vertical axis and an axis scaled in decibel (dB).

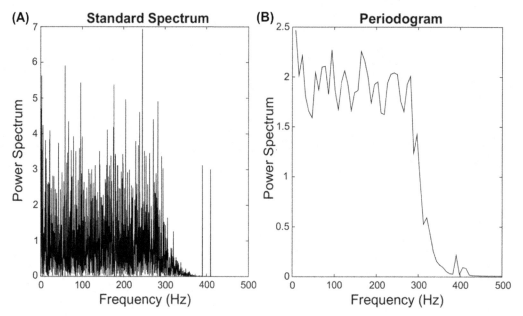

FIGURE 4.18   Power spectra obtained from a waveform consisting of a broadband signal having constant energy between 0 and 300 Hz and two sine waves at 390 and 410 Hz. (A) The unaveraged spectrum clearly shows the 390 and 410 Hz components, but the features of the broadband signal are unclear. (B) The averaged spectrum barely shows the two sinusoidal components, but produces a smoother estimate of the broadband spectrum, which is expected to be flat up to around 300 Hz. In the next example, averaging is used to estimate the power spectrum of the normal and meditative heart rate variability data.

Solution: Construct the time data by resampling at 100 Hz as was done in Example 4.4. Find the segment length by dividing the total length by eight and round this length downward. Use the welch routine with the default 50% overlap to calculate the average power spectrum and plot.

```
% Example 4.6 Influence of spectral averaging on heart rate data.
...... Data loading and reorganization as in Example 4.4 .......
%
nfft = floor(length(yi)/8);              % Calculate segment length
[PS_avg,f] = welch(yi,nfft,[ ],fs);      % Periodogram, 50% overlap
PS_avg_dB = 20*log10(PS_avg);            % Put in dB
   .......plot, label and axis........
   ....... Repeat for meditative data ........
```

Results: The welch routine developed in the last example is used to calculate the average power spectrum. The sample segment length with a 50% overlap is chosen empirically, as it produces smooth spectra without losing the peak characteristic of the meditative state.

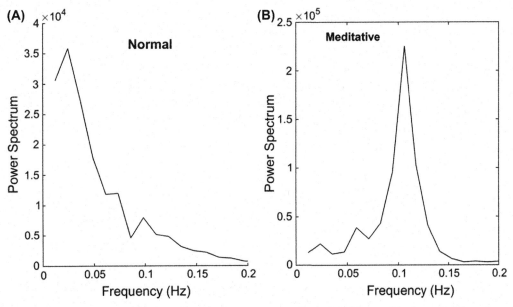

FIGURE 4.19 Power spectra taken from the same heart rate variability data used to determine the spectra in Figure 4.16, but constructed using an averaging process. The spectra produced by averaging are considerably smoother. (A) Normal conditions; (B) meditative conditions.

The results in Figure 4.19 show much smoother spectra than those of Figure 4.16, but they also lose some of the detail. The power spectrum of HRV taken under normal conditions now appears to decrease smoothly with frequency. It is common to plot power spectra in dB. Since we are dealing with power, we apply a variation of Equation 2.5: $PS_{dB} = 10 \log (PS)$. (We use 10 log instead of 20 log because the square of the magnitude spectrum was taken in the calculation of the power spectrum; see Equation 4.16.) Replotting in dB shows that the normal spectrum actually decreases linearly with frequency (Figure 4.20A). The principle feature of the dB plot of meditative data is the large peak at 0.12 Hz (Figure 4.20B). Note that this peak is also present, although much reduced, in the normal data (Figure 4.20A). This suggests that meditation enhances a process that is present in the normal state as well, some sort of feedback process with an 8-s delay (1/0.12). Note that the linear decrease in power with frequency and the small peak at 0.12 Hz are not apparent in the previous spectral plots of these EEG data. This demonstrates the importance of presenting your data appropriately.

4.5 SIGNAL BANDWIDTH

The concept of representing a signal in the frequency domain brings with it additional concepts related to a signal's spectral characteristics. One of the most important of these new concepts is signal bandwidth. In Chapter 6, we extend the concept of bandwidth to systems as well as signals and find that the concepts are the same. The modern everyday use of

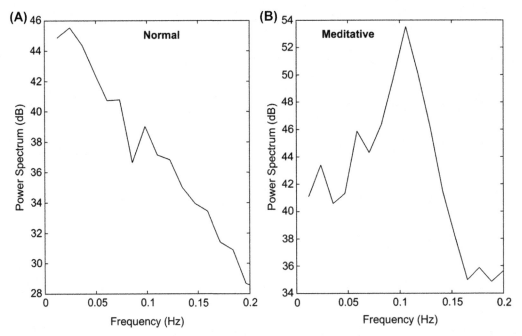

FIGURE 4.20 Normal and meditative spectra of heart rate variability data replotted in dB by taking 10 log of the power spectral curves in Figure 4.19. (A) The spectrum obtained under normal conditions is seen to decrease linearly with frequency. (B) The spectrum obtained under meditative conditions shows a large peak at 0.12 Hz. A reduced version of this peak is also seen in the normal data.

the word "bandwidth" relates to a signal's general ability to carry information. More specifically, we engineers define the term with regard to the range of frequencies found in a signal.

Figure 4.21A shows the spectrum of a hypothetical signal that contains equal energy at all frequencies up to 200 Hz, the frequency labeled f_c. Above this frequency, the signal contains no energy. We would say that this signal has a "flat" spectrum up to frequency f_c, and no energy above that frequency. That critical frequency, f_c, is called the "cutoff frequency." Since bandwidth is defined in terms of a signal's energy range and this signal contains energy only between 0 and f_c Hz, the bandwidth would be defined as from 0 to 200 Hz, or simply 200 Hz. The frequency range below 200 Hz is called the "passband," whereas the frequency range above 200 Hz is call the "stopband," as labeled in Figure 4.21A.

Although some real signals can be quite flat over selected frequency ranges, they are unlikely to show such an abrupt cessation of energy above a given frequency as seen in Figure 4.21A. Figure 4.21B shows the spectrum of a more realistic signal, where the energy decreases gradually. When the decrease in signal energy takes place gradually, as in Figure 4.21B, defining the signal bandwidth is problematic. If we want to attribute a single bandwidth value for this signal, we need to define a "cutoff frequency": a frequency defining a boundary between the region of substantial energy and the region of minimal energy. Such a boundary frequency has been arbitrarily defined as the frequency when the signal's value has declined by 3 dB with respect to its average unattenuated RMS value. (In Figure 4.21, the

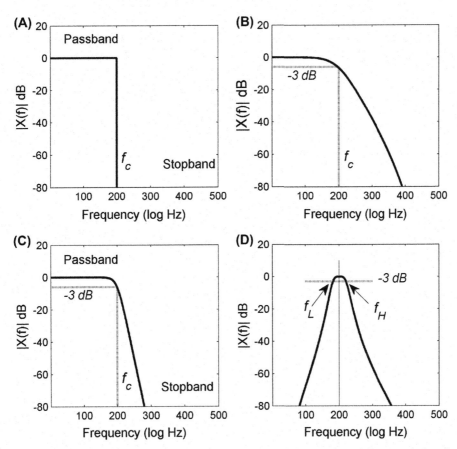

FIGURE 4.21 Frequency characteristics of ideal and realistic signals. The magnitude frequency plots shown here are plotted in dB. (A) A signal with an idealized, well-defined frequency range: 0–200 Hz. (B) A more realistic signal where signal energy decreases gradually at higher frequencies. (C) A realistic signal where signal energy decreases more sharply at higher frequencies. (D) A realistic signal where the signal energy decreases both above and below a "center frequency" of 200 Hz.

nominal unattenuated value for all signals is normalized 0 dB). The negative 3 dB boundary is not entirely arbitrary. If we convert −3 dB to linear units using Equation 2.9 we get: $10^{(-3\text{dB}/20)} = 0.707$. Thus the amplitude of a signal reduced by 3 dB is 0.707 of its unattenuated value. A signal's power is proportional to the amplitude squared (Equation 2.7), so at this attenuation, the signal's power is $0.707^2 = 0.5$. When the amplitude of the signal is attenuated by 3 dB from its nominal value, its amplitude is reduced by 0.707 and it power is halved. (Since the dB scale is logarithmic, −3 dB means a reduction of 3 dB; see Table 2.3.)

When the signal magnitude spectrum (the magnitude spectrum, not the power spectrum) is reduced by 3 dB, the power in the signal is half the nominal value, so this boundary frequency is also known as the "half-power point." The terms "−3 dB point," "3 dB down point," and "half-power point" are all used synonymously to define the boundary frequency,

f_c. In Figure 4.21B, the signal again has a bandwidth of 0.0–200 Hz, or just 200 Hz. The signal in Figure 4.21C has a sharper decline in energy, referred to as the "rolloff," but it still has a bandwidth of 200 Hz based on the −3 dB point.

It is possible that a signal "rolls off" or "attenuates" at both the low-frequency and high-frequency ends as shown in Figure 4.21D. In this case the signal has two cutoff frequencies, one labeled f_L and the other f_H. For such signals, the bandwidth is defined as the range between the two cutoff frequencies (or −3 dB points), that is, $BW = f_H - f_L$ Hz.

EXAMPLE 4.7

Find the effective bandwidth of the noisy signal, x, in file Ex4_7_data.mat. $f_s = 500$ Hz.

Solution: To find the bandwidth we first need the spectrum. Since the signal is noisy, we use spectral averaging to produce a cleaner spectrum. As explained in the results, test runs indicated that the Welch method using a segment length of 256 samples with maximal overlap works well. We then use MATLAB's find routine to search the magnitude spectrum for the first and last points that are greater than 0.5 (since we are using the power spectrum).

```
% Example 4.7 Find the effective bandwidth of the signal x in file Ex4_7_data.mat
%
load Ex4_7_data.mat;              % Load the data file. Data in x
fs = 500;                        % Sampling frequency (given)
[PS1,f1]  =  welch(x,length(x),0,fs); % Determine  and  plot  the  un-averaged
spectrum
    ......plot, axes labels and new figure.......
%
nfft = 256;                      % Power spectrum window size
[PS,f] = welch(x,nfft,nfft-1,fs);  % Compute avg. spectrum, maximum overlap
PS = PS/max(PS);                 % Normalize peak spectrum to 1.0
plot(f,PS); hold on;             % Plot the normalized, average spectrum
.......axes labels.......
%
i_fl = find(PS > .5, 1, 'first'); % Find index of low freq. cutoff
i_fh = find(PS > .5, 1, 'last');  % Find index of high freq. cutoff
f_low = f(in_fl);                % Convert low index to cutoff freq.
f_high = f(in_fh);               % Convert high index to cutoff freq.
    .......plot markers.......
BW = f_high - f_low;             % Calculate bandwidth
title(['Bandwidth: ',num2str(BW,3),' Hz']); % Display bandwidth in title
```

Results: The unaveraged power spectrum shown in Figure 4.22 was determined using the welch routine, but with the segment size equal to the data length. Yes, we could have taken the square of the magnitude Fourier transform, but the welch routine provides the frequency vector, making life easier. The unaveraged power spectrum is quite noisy, as seen in Figure 4.22, so we do want to use spectral averaging. Using the welch routine with a window size of 256 samples produces a spectrum that is fairly smooth, but still has a reasonable spectral resolution, Figure 4.23. (The effect of changing window size is explored in one of the problems.)

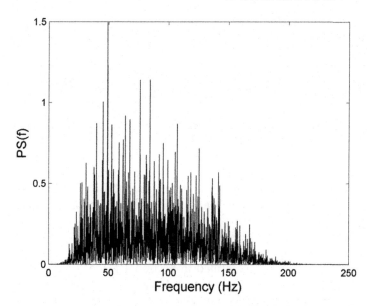

FIGURE 4.22 The power spectrum of the noisy signal used in Example 4.7. The difficulty of finding the general shape of this curve suggests that spectral averaging is in order. The averaged power spectrum is shown in Figure 4.23.

The `find` routine with the proper options gives us the indices of the first and last points above 0.5, and these are plotted superimposed on the spectral plot (large dots in Figure 4.23). These high and low sample points can be converted to frequencies using the frequency vector. The difference between the two cutoff frequencies is the bandwidth and is 111 Hz as displayed in the title of the

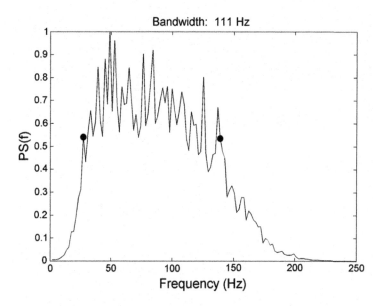

FIGURE 4.23 The averaged power spectrum of the signal used in Example 4.7. The window size is selected empirically to be 256 samples with maximal overlap. This size appears to give a smooth spectrum while maintaining good resolution. The cutoff frequency is found by searching for the half-power points. MATLAB's `find` routine is used to locate these points, which are identified as large dots on the spectrum. The frequency vector is used to convert the index of these points to equivalent frequencies and the calculated bandwidth is shown in the figure title.

spectral plot, in Figure 4.23. The estimation of the high and low cutoff frequencies could have been improved by interpolating between the spectral frequencies on either side of the 0.5 values. This is done in one of the problems.

4.6 TIME–FREQUENCY ANALYSIS

All Fourier analyses are done on a block of data, so the result represents a signal's spectrum over a finite period of time. An underlying and essential assumption is that the data have consistent statistical properties over the length of the data set being analyzed; that is, the signal is stationary over the analysis period (see Section 1.4.2.1). There is a formal mathematical definition of stationary based on consistency of the data's probability distribution function, but for our purposes, a stationary signal is one that does not change its mean, standard deviation, or autocorrelation function over time.

Most biological signals are not stationary, so to use Fourier analyses and many other signal processing techniques, some additional data transformation may be needed as mentioned in Chapter 1. Taking a page from spectral averaging described in Section 4.4, we could try to break up a signal into stationary segments and then perform a Fourier transform on each segment. This approach is known as "time–frequency" analysis; each segment produces a spectrum that covers a specific time period of the original signal. For example, if a 1-h data set was divided into sixty 1-min segments and a Fourier analysis performed on each segment, the resulting data would be 60 spectra, each describing a different time period.

There are many different approaches to performing time–frequency analyses, but the one based on the Fourier series is the best understood and the most popular. Since it is based on dividing a data set into shorter segments, sometimes quite short, it is called the "short-term Fourier transform" or "STFT." Since the data segments are usually short, a window function such as the Hamming window is usually applied to each segment before the Fourier transform is taken.

The STFT is very easy to implement in MATLAB: it is just like spectral averaging except, rather than averaged, each spectrum is plotted separately. Usually the resulting spectra are displayed on a three-dimensional plot with the spectra's magnitude on the vertical axis, and time and frequency on the other axes. The next example illustrates how easy it is to evaluate the STFT in MATLAB.

EXAMPLE 4.8

Load the respiratory signal found as variable `resp` in file `Resp_long.mat`. This signal is similar to the one used in Example 3.12 except that it is 1 h long. The sampling frequency is 125 Hz. Find the power spectra for sixty 1-min segments that overlap by 30 s (i.e., 50%). Plot these spectra in three dimensions using MATLAB's `mesh` routine. Limit the frequency range to 2 HZ and do not plot the DC components.

Solution. Simply modify the `welch` routine so that it saves the individual spectra in a matrix rather than computing an average spectrum. Also add code to compute the relative time corresponding to each segment. The main program is given below followed by the modified `welch` routine now called `stft` for the STFT.

```
% Example 4.8 Example of the STFT time-frequency analysis.
%
load Resp;                          % Get respiratory data
N = length(resp);                   % Determine data length
fs = 125;                           % Sampling frequency
nfft = fs*60;                       % 1 min of samples
noverlap = round(nfft/2);           % Use 50% overlap
m_plot = round(2/(fs/nfft));        % Find m for 2 Hz
[PS,f,t] = stft(resp',nfft,noverlap,fs);     % Calculate the STFT
PS1 = PS(:,2:m_plot);               % Resize power spectra to range to 2 Hz
f1 = f(1:m_plot-1);                 % Resize frequency vector for plotting
mesh(f1,t,PS1);                     % Plot in 3D
view([17 30]);                      % Adjust view for best perspective
  ......labels.......
```

The modified welch routine is:

```
function [PS,f,t] = stft(x,nfft,noverlap,fs);
% Function to calculate short term Fourier Transform
%
  .....same as welch.m up to the calculation of the power spectra.......
% Data have been windowed
  PS_temp = abs((fft(data)).^2)/N1;     % Calculate PS (normalized)
  PS(k,:) = PS_temp(2:half_segment+1)/(nfft/2);   % Remove redundant points
  t(k) = (k - 1)*nfft/fs;               % Construct time vector
end                                     % Program ends here
```

Results: The stft routine generates a matrix, PS, that contains the spectra in rows. The frequency vector, f, is constructed as in welch and not shown here. There is also a time vector, t, that shows the time when each segment starts.

The results from the STFT are shown in Figure 4.24. The view was initially adjusted interactively to find a perspective that showed the progression of power spectra. In Figure 4.24, each time frame shows a spectrum that is similar to the spectrum in Figure 3.24, with a large peak around the respiratory rate.

4.7 SUMMARY

Converting a real-world continuous signal to a digital signal requires three operations: time sampling, amplitude slicing, and, usually, truncation. The DFT can be used to understand the relationship between a continuous time signal and the sampled version of that signal. This frequency-based analysis shows that the original, unsampled signal can be recovered if the sampling frequency is more than twice the highest frequency component in the unsampled signal. Quantization is viewed as noise added to the signal and can be quantitatively determined from the amplitude and quantization level (i.e., number of bits) of the

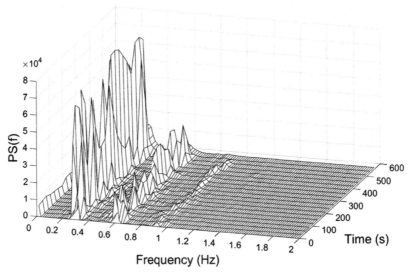

FIGURE 4.24   Time–frequency plot of a 1-h respiratory signal. This plot was constructed by segmenting the 1-h signal into 60-sec increments and taking the power spectrum of each increment separately rather than averaging as in the Welch method. Segments overlap by 50%.

ADC. Data truncation can produce discontinuities at the endpoints that may result in artifacts in the DFT, particularly for short data segments. Tapering window functions can be imposed on the digitized data to reduce these artifacts, but for reasonable signal lengths (>256 samples), they are not needed.

The Fourier transform can be used to construct the power spectrum of a signal by taking the square of the magnitude spectrum. The power spectral curve shows signal power as a function of frequency. The power spectrum is particularly useful for noisy or random data where phase characteristics have little meaning. By dividing the signal into a number of possibly overlapping segments and averaging the power spectrum obtained from each segment, a smoothed power spectrum can be obtained.[7] The resulting averaged power spectrum curve will emphasize the broadband or general characteristics of a signal's spectrum, but at the cost of fine detail.

Bandwidth is a term used commonly to describe the capability of a communication channel to carry information. Bandwidth is specifically defined as the range of frequencies included in a signal. To be included, the energy associated with a given frequency must be greater than half that of the signal's nominal values. So for frequencies to be included, the signal amplitude must be greater that 0.707 of the nominal value or, equivalently, attenuated no more than −3 dB from these values.

Nonstationarity is a common problem in biological signals. For spectral analysis, the STFT is a common approach to dealing with such signals. The idea is to divide the signal into short

---

[7]Although averaging is generally applied to the power spectrum, the magnitude spectrum can also be used. Averaging cannot be applied to the phase spectrum because the phase is altered by the time shifting inherent in averaging.

segments that are, we hope, stationary, and then apply the Fourier transform to each separate segment. This results in a number of spectra that represent a specific time frame of the overall signal, hence the term "time—frequency analysis." These spectra are usually displayed on three-dimensional plot of time, frequency, and spectral magnitude (or power spectrum). Alternatively, time—frequency information could be represented in a two-dimensional plot of time and frequency where color or shading maps the magnitude value. In principal, phase information could also be displayed this way, but the sensitivity of phase to time shifts would make such displays difficult to interpret.

PROBLEMS

1. This problem demonstrates aliasing similar to Example 4.1. Generate a 512-point waveform consisting of two sinusoids at 200 and 400 Hz. Assume a sampling frequency of 1 kHz. Generate another waveform containing frequencies at 200 and 900 Hz. Take the Fourier transform of both waveforms and plot the magnitude of the spectrum up to $f_s/2$. Plot the two spectra superimposed, but in different colors to highlight the additional peak due to aliasing at 100 Hz. [Hint: Follow the approach in Example 4.1 to generate the sine waves.]

2. Load the chirp signal, x, in file `chirp.mat` and plot the magnitude spectrum. The sample frequency is 5 kHz. Now decrease the sampling frequency by a factor of two by removing every other point, and recalculate and replot the magnitude spectrum. Note the distortion of the spectrum produced by aliasing. Do not forget to recalculate the new frequency vector based on the new data length and sampling frequency. (The new sampling frequency is half that of the original signal.) Decreasing the sampling frequency of an existing signal is termed "downsampling" and can be done easily in MATLAB. To downsample by a factor of two use: `x1 = x(1:2:end)`.

3. Load the file `sample_rate.mat`, which contains signals x and y sampled at 1 kHz. Is either of these signals likely to be undersampled (i.e., $f_s/2 \leq f_{max}$)? Alternatively, could the sampling rate of either signal be safely reduced? Justify your answer.

4. Repeat Problem 3 for the data in `sample_rate1.mat`, which contains signals x and y, also sampled at 1 kHz. Is either of these signals undersampled or could the sampling rate of either signal be safely reduced? Again, justify your answer.

5. Generate a 1024-point waveform consisting of four sinusoids at 100, 200, 300, and 400 Hz. Make $f_s = 1$ kHz. Generate another waveform containing frequencies at 600, 700, 800, and 900 Hz. Take the Fourier transform of both waveforms and plot the magnitude of the full magnitude spectrum (i.e., 1024 points) on separate plots. (Plotting one above the other using `subplot` makes a nice comparison.)

6. The file labeled `quantization.mat` contains three vector variables: x, y, and z, all sampled at 1 kHz. This file also contains a time vector, t, useful for plotting. Variable x represents an original signal. Variable y is the same signal after it has been sampled by a 6-bit ADC, and variable z is the original signal after sampling by a 4-bit ADC. Both converted signals have been scaled to have the same range as x, 0—5 volts. On one plot, show the three variables superimposed and on another plot, show the error

signals $x - y$ and $x - z$. Then calculate the RMS error between x and y and x and z. Also calculate the theoretical error for the two ADCs based on Equation 4.3.

7. The file short.mat contains a very short signal of 32 samples, $f_s = 40$ Hz. Plot the nonredundant magnitude spectrum as discrete points obtained with and without zero padding. Zero pad out to a total of 256 points.

8. Generate a short signal consisting of a short impulse function: x = [1 0 0 0 0 0];, ($f_s = 50$ Hz). As in Problem 7, find the magnitude transform with no padding and zero padding to 256 points. Plot the nonredundant spectrum.

   Repeat for a double impulse function of the same length: x = [1 0 0 1 0 0];. Note how information about the impulse function is not really improved by padding, but really makes a difference for the double impulse.

9. The variable x in file prob4_9_data contains 200 and 300 Hz sine waves with SNRs of −6 dB. The sampling frequency is 1000 Hz and the data segment is fairly short; 64 samples. Plot the magnitude spectrum obtained with and without a Hamming window. Use Equation 4.9 to generate a window function 64 points long and multiply point by point with the signal variable x. Note that the difference is slight, but could be significant in certain situations.

10. Use sig_noise to generate a 256-point waveform consisting of a 300 Hz sine wave with an SNR of −12 dB (x = sig_noise(300,−12,256);). (Recall that sig_noise assumes $f_s = 1$ kHz.) Calculate and plot the power spectrum using two different approaches. In the first approach, use the direct approach: take the Fourier transform and square the magnitude function. In the second approach, use the traditional method defined by Equation 4.11: take the Fourier transform of the autocorrelation function. Calculate the autocorrelation function using crosscorr(x,x), then take the absolute value of the fft of the autocorrelation function. You should only use the second half of the autocorrelation function (those values corresponding to positive lags). Plot the power spectrum derived from both techniques. The scales will be different because the MATLAB fft routine does not normalize the output. You may find a very slight difference between results because of the difference in computer round-off errors.

11. Construct two arrays of white noise using randn: one 128 points in length and the other 1024 points in length. ($f_s = 1$ kHz.) Take the power spectrum of both. Eliminate the first point—the average or DC term—when you plot the spectra and plot only nonredundant points. Does increasing the length improve the spectral estimate of white noise, which should be flat? This demonstrates the benefit of spectral averaging when examining broadband features.

12. Use MATLAB routine sig_noise to generate two arrays, one 128 points long and the other 512 points long. Include two closely spaced sinusoids having frequencies of 320 and 340 Hz with an SNR of −12 dB. The MATLAB call should be:

   x = sig_noise([320 340],-12,N); where N = either 128 or 512.

   Calculate and plot the (unaveraged) power spectrum. Repeat the execution of the program several times and note the variability of the results indicating that noise is noisy. (Remember sig_noise assumes $f_s = 1$ kHz.)

13. Use sig_noise to generate a 128-point array containing 320 and 340 Hz sinusoids as in Problem 12. Calculate and plot the unaveraged power spectrum of this signal for an

SNR of -10, -14, and -16 dB. Plot all three on the same graph using subplot and execute the program several times and observe that variability. How does the presence of noise affect the ability to detect and distinguish between the two sinusoids? (Remember $f_s = 1$ kHz.)

14. Load the file `broadband2`, which contains variable x, a broadband signal with added noise. ($f_s = 1$ kHz.) Calculate the averaged power spectrum using the `welch` routine. Evaluate the influence of segment length using segment lengths of N/4 and N/16, where N is the length of the date of variable, x. Use the default overlap.

15. Load the file `eeg_data.mat` that contains EEG data. ($f_s = 50$ Hz.) Analyze these data using the unaveraged power spectral technique and an averaging technique using the `welch` routine. Find a segment length that smooths the background spectrum, but still retains any important spectral peaks. Use a 99% overlap.

16. Modify the `welch` routine to create a routine `welch_win` that applies a "Blackman window" to the data before taking the Fourier transform. The Blackman window is another of many window functions. It adds another cosine term to the equation of the Hamming window giving:

$$w[n] = 0.41 + 0.5 \cos\left(\frac{2\pi n}{N}\right) + 0.08 \cos\left(\frac{4\pi n}{N}\right)$$

Load the file `broadband3.mat`, which contains a broadband signal and a sinusoid at 400 Hz in variable x ($f_s = 1$ kHz). Analyze the broadband/narrowband signal using both `welch` and `welch_win` with segment lengths of $N/4$ and $N/16$, where N is the length of the data of variable, x. Use the default overlap. Note that for the shorter segment ($N/16$), the Blackman window provides slightly more smoothing that the rectangular window used by `welch`.

17. Load the file broadband3.mat, which contains a broadband signal and a sinusoid at 400 Hz in variable x ($f_s = 1$ kHz). Analyze the broadband/narrowband signal using `welch` with two different overlaps: 50% (the default) and maximum. As in Problem 16, use segment lengths of $N/4$ and $N/16$, where N is the length of the data of variable, x.

18. Find the effective bandwidth of the signal x in file `broadband2.mat` ($f_s = 1$ kHz). Load the file `broadband2.mat` and find the bandwidth using the methods of Example 4.7. Find the power spectrum using `welch` and a window size ranging between 50 to 200 points and the maximum overlap. Find the window size that appears to give the best estimate of bandwidth. (Note: window sizes that produce the best looking power spectrum may not result in the most accurate estimate of bandwidth. Can you explain?)

19. Load the file `broadband2.mat` and find the bandwidth using the methods of Example 4.7, but this time using interpolation ($f_s = 1$ kHz). Determine the power spectrum using `welch` with a window size of 100 and the maximum overlap. This will give a highly smoothed estimate of the power spectrum but a poor estimate of bandwidth because of the small number of points. Interpolation can be used to improve the bandwidth estimate from the smoothed power spectrum. The easiest way to implement interpolation in this problem is to increase the number of points in the power spectrum using MATLAB's `interp` routine. (For example, PS1 = interp(PS,5) increases the number of points in PS by a factor of 5.) Estimate the bandwidth using

the approach of Example 4.7 before and after expanding the power spectrum by a factor of five. Be sure to expand the frequency vector (f in Example 4.7) by the same amount.

20. Load the file ECG_1hr, which contains 1 h of electrocardiogram signal in variable ecg.mat. Plot the time—frequency spectra of this signal using the STFT approach shown in Example 4.8 (stft.m can be found on the associated files). Limit the frequency range to 3 Hz and do not include the DC term. The sample frequency is 250 Hz. [Hint: You can improve the appearance of the higher frequency components by limiting the z axis using ylim and the color axis using caxis.]

21. Repeat Problem 20, but change the parameters to emphasize a different feature of this signal: the influence of respiration and other factors on heat rate. Make the segment length 30 s and the overlap 20 s (2/3). Also change the upper limit on the frequency range to 8 Hz, but again do not include the DC term ($f_s = 250$ Hz).

    Plot the resulting spectra as a three-dimensional "heat map"-type display. Use pcolor with shading set to interp to plot the spectra and correctly label and scale the two axes. Limit the color range to 200. You will see a light-colored wavy line around the heart rate frequency (approximately 1 Hz) that reflects the variation in heart rate due to respiration. Now lower the upper limit of the color range to around 20 and note the chaotic nature of the higher frequency components. [Hint: If you make the lower limit of the color range slightly negative, it lightens the background and improves the image.]

PART II

SYSTEMS

Linear Systems Analysis in the Time Domain—Convolution

5.1 GOALS OF THIS CHAPTER

You now have expertise in the fundamentals of signal analysis, including basic and advanced time-domain measurements (mean, root mean square, standard deviation, variance, and correlations, including auto- and cross-correlation) and frequency-domain analysis involving time—frequency transformations with emphasis on signal spectrum. Now it is time to move on to systems: processes that moderate, influence, and/or produce signals.

Ultimately, we want to know how a given physiological system does what it does, but we start with a more modest goal: to have a definitive understanding of what the system does. So the primary objective of "linear system analysis" is to determine how a system responds to any input, no matter how complicated the system or the input. This input/output determination can be done in either the time domain or frequency domain. In both approaches, we start with a definition of the system itself. These representations are quite different in the two domains, but they achieve the same end: a mathematical function that allows us to predict the output of the system to any input. In this chapter, we look at the time domain representation of a system, whereas Chapter 6 takes the frequency-domain approach.

Specific topics include:

- Basic goals of linear signal analysis, system constraints, superposition
- Defining a system in the time domain: the impulse function (i.e., signal) and impulse response
- Finding a system's output to any specific input signal in the time domain: convolution

5.2 LINEAR SYSTEMS ANALYSIS—AN OVERVIEW

Systems act on signals. The objective of systems analysis is to describe how a system modifies a signal: any (linear) system and any signal(s). To be able to predict the behavior of complex processes in response to complex stimuli, we usually impose rather severe simplifications and/or assumptions. The most common assumptions are: (1) that the process responds linearly to all inputs, and (2) that the basic characteristics do not change over time. In Section 1.4 we note that these two assumptions define an LTI (linear, time invariant) system. These assumptions allow us to apply a powerful array of mathematical tools known as linear systems analysis.

Of course, living systems change over time, they are adaptive and often nonlinear. But the power of linear systems analysis is sufficiently seductive that assumptions or approximations are made so that these tools can be used. Linearity can be approximated by using small-signal conditions since, when a system's range of action is restricted, it often behaves more or less linearly. Alternatively, piecewise linear approaches can be used where the analysis is confined to operating ranges over which the system behaves linearly. In Chapter 9, we learn computer simulation techniques that allow us to analyze systems with certain types of nonlinearity. To deal with processes that change over time, we can use the same approach underlying the short-term Fourier transform: restrict the analysis time frame to a period when the system can be considered time invariant.

5.2.1 Superposition and Linearity

Finding the output of any system to any input is a tall order and the only way we can achieve it is to break up the signal. For the time-domain analysis described here, we break up the signal into little slices of time. In frequency-domain analyses as described in the next chapter, we break up the signal into frequency components (i.e., sinusoids). Either way, we determine how the system responds to each of these components as if it was acting alone. To get the output signal, we put those individual output components back together. For this divide-and-conquer approach, we rely on an important concept known as super-position; for superposition we need linearity.

Superposition states that when multiple influences act on some process, the resultant behavior is the sum of the process's response to each influence acting alone. This means that when a system receives two or more signals, a valid solution can be obtained by solving for the response to each signal in isolation, and then algebraically summing these partial solutions. The sources could be anywhere in the system or even from different locations. This is exactly what we need for our decomposition technique to work.

The superposition principle is a consequence of linearity and applies only to LTI systems. It makes all signal processing tools more powerful since it enables decomposition strategies, either in the frequency or the time domain. Examples using the principle of superposition are found throughout this book.

Two time-domain approaches can be used to find the output of an LTI system to almost any input. Both decompose the signal into slices of time. Slicing the input signal into small, or even infinitesimal, segments uses the approach of integral calculus where a function is divided up into infinitely small segments, the solution to each segment is determined, and these solutions are summed. This approach is termed "convolution," which applies this time-slice trick to the entire system in one operation. For continuous signals, the signal segments can be infinitesimal and integration used to sum up the contributions of each segment. If digital signals are involved, then the minimum segment size is one sample (i.e., one data point) so the size of the time slice depends on the sample interval.

The second time-domain approach is a very powerful computer-based method called "simulation" and is described in Chapter 9. In this inherently digital approach, the response of each element to the first time slice is found element by element. This process is repeated for each subsequent time slice and including the last time slice in the input signal. It is as computationally intensive as it sounds, but it gives access to internal signals as well as the output signal and can be used on systems that contain nonlinearities.

5.3 A SLICE IN TIME: THE IMPULSE SIGNAL

The convolution approach takes advantage of the fact that any linear system responds in a characteristic way to sufficiently short time[1] slices, or pulse signals, irrespective of pulse amplitude. Pulse amplitude simply scales this characteristic response up or down, Figure 5.1. A pulse signal having the smallest possible time slice is called the "impulse function," or "impulse signal," and the system's response to this signal is called the "impulse response."

Since the shape of the impulse response is unique to a given system, it can be used as the time-domain representation of the system. For a unique representation, we need to agree on standard impulse amplitude. As we show, it is better to define a standard impulse function area as having a given value rather than a given amplitude. By convention, we set the area of a standard impulse to 1.0. For a digital impulse signal this is no problem: the area is the amplitude multiplied by one sample interval, T_s, so the amplitude should be $1/T_s$ to produce an area of 1.0. For a continuous signal, the width of an impulse function, Δt, becomes

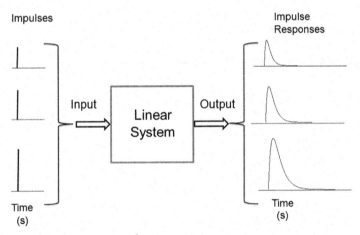

FIGURE 5.1   A sufficiently short input signal[1] is called the impulse function. For all linear systems, the response to an impulse has a characteristic shape determined by the system. This shape is the same regardless of the amplitude of the impulse. This characteristic response is called the impulse response and essentially defines the system in the time domain.

[1]By sufficiently short, we mean infinitely short for continuous systems, and one sample interval (i.e., T_s) for discrete systems. For real-world systems, we can determine the maximum slice size empirically as shown in Example 5.1.

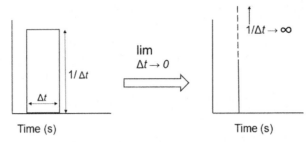

FIGURE 5.2 A continuous impulse function is a theoretical construct that has a width that goes to zero, an amplitude that goes to infinity, but an area of 1.0. This function is called a delta(δ)-function.

infinitely small, $\Delta t \to 0$, so the amplitude, $1/\Delta t$ must become infinite as $\Delta t \to 0$, at least in theory, to maintain an area of 1.0, Figure 5.2. (The need for infinite amplitude to keep the area of an infinitely short pulse equal to 1.0 is the reason we define area rather than amplitude.) This theoretical impulse input is called the "delta function" and is notated $\delta(t)$.

5.3.1 Real-World Impulse Signals

A signal that is infinitely short with infinite amplitude is not compatible with real-world situations. If we want to generate a real impulse function, we use a pulse that is sufficiently short and has enough amplitude to produce a reasonable response signal. (In real-world situations there is always noise present and we would like the impulse response to be much larger than the noise.) How short is sufficiently short? This depends on the dynamics of the system: it should be much shorter than the fastest system response. This may still sound vague, but there is a simple way to set the maximum limit of a practical impulse input empirically as shown in our first example.

EXAMPLE 5.1

An unknown system is represented by the MATLAB routine:

`y = unknown_sys5_1(x).m where x` is the input to the system and `y` is its output. Although we do not know any details of the system, we suspect that it responds to stimulus changes in around a tenth of a second (0.1 s). Find the maximum pulse width that can be considered sufficiently short to be taken as an impulse input to this system.

Solution: An impulse should produce a characteristic response from any given system. If we make it shorter the response amplitude will decrease, but the shape will stay the same. So if we compare responses to two pulse signals, one shorter than the other, and they both give the same-shaped responses, then both can be considered short enough to be impulse functions. Since we know the system in question responds on the order of 0.1 s, we compare the system's response to pulses having widths ranging from 100 ms down to 2.0 ms. Specifically, we compare the following pairs of pulse widths: 50 and 100 ms; 10 and 25 ms; 5.0 and 10 ms; and 2.0 and 5.0 ms. These are just guesses, so we write the MATLAB code so that pulse width can be easily changed. To make the comparison easier, we normalize the output responses to a maximum value of 1.0.

We need to set a sample interval and data length. To get a 2.0-ms pulse, we need to have a sample frequency of at least 500 Hz, so we select f_s to be 1000 to be conservative. As to data length, given responses take around 0.1 s we start with a data length of 0.2 s. Again we can always make the time frame shorter or longer if need be.

```
% Example 5.1  Example to evaluate the responses of an unknown
% system to pulses of various widths.
%
PW = [50 10 5 5; 100 25 10 2];          % Pulse widths in msec
fs = 1000;                              % Sample frequency
N = round(.2 *fs);                      % Data length for 0.2 sec
t = (0:N-1)/fs;                         % Time vector for plotting
%
for k = 1:4                             % Do four comparison plots
  subplot(2,2,k); hold on;              % Plot two by two
  x = [ones(1,PW(1,k)) zeros(1,N-PW(1,k))];  % Generate pulse signal
  y = unknown_sys5_1(x);                % Stimulate system, get response
  y = y/max(y);                         % Normalize peak to 1.0
  plot(t,y,'k'); hold on;               % Plot pulse response
  x = [ones(1,PW(2,k)) zeros(1,N-PW(2,k))];  % Generate pulse signal
  y = unknown_sys5_1(x);                % Stimulate system, get response
  y = y/max(y);                         % Normalize peak to 1.0
  plot(t,y,'k');                        % Plot pulse response
        .......labels and text.......
end
```

Results: The four comparison pulse responses are shown in Figure 5.3. All responses are scaled to a maximum amplitude of 1.0. (Since the different pulse widths have different energies, they produce

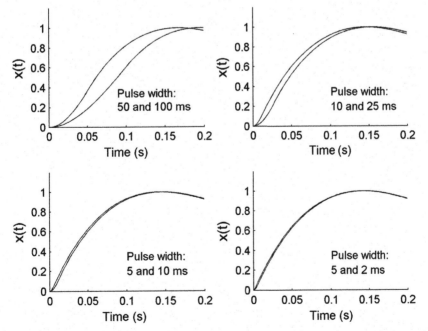

FIGURE 5.3 Pulse responses of an unknown system to different combinations of pulse width. These responses have been normalized to a maximizing value of 1.0. When the inputs have widths of either 2.0 or 5.0 ms, the normalized responses are nearly identical, indicating that both can be considered impulse signals with respect to this system.

different response amplitudes, irrespective of shape.) Figure 5.3 shows that as the pulses reduce in width, the responses become more similar despite differences in the pulse width of the input signals. There is very little difference between the responses generated by the 5- and 2.5-ms pulses, so that a 5-ms pulse (or less) is close enough to a true impulse for this system. This example is a realistic simulation of the kind of experiment one might do to determine the impulse response empirically, provided the physical system and some type of pulse stimulator are available. This also assumes the system can be stimulated with pulse waveforms and the response monitored, not always the case with biological systems.

5.3.2 The Impulse Signal in the Frequency Domain

The impulse signal has a very special frequency-domain representation. Again, we show this by example. In the next example, we find the magnitude spectra for two of the pulse signals used in Example 5.1 and the magnitude spectrum of a true discrete impulse signal: a signal that has a value of 1.0 for one the first sample and zero everywhere else.

EXAMPLE 5.2

Find the magnitude spectra of two pulse signals having widths of 5 and 2 ms and a true discrete impulse signal. The true impulse signal should have a value of 1.0 for the first sample and zeros everywhere else. Assume a sample rate of 1 kHz as in the last example, but make the signal length 1000 samples. Plot the spectra superimposed, but scale the maximum value of each magnitude spectrum to 1.0 for easy comparison. Label the three spectra. As always, plot only the valid spectral points and label the plots.

Solution: We can borrow the same code used in Example 5.1. Since $f_s = 1$ kHz, the pulse width vector has only three entries of 5, 2, and 1 representing the three pulse widths in milliseconds. Instead of using the pulse signals as inputs to an unknown system, we find their magnitude spectra using the fft routine. We normalize the magnitude spectra to have a maximum value of 1.0 then plot the three spectra superimposed.

```
% Example 5.2 Find the magnitude spectra of two pulses having a width of
%   5 and 2 msec and a true discrete impulse signal.
%
PW = [5, 2, 1];          % Pulse widths in msec
fs = 1000;               % Sample frequency (given)
N = 1000;                % Data length (given)
N2 = 500;                % Valid spectral points
f = (1:N)*fs/N;          % Frequency vector for plotting
%
for k = 1:3              % Do 3 spectral plots
   x = [ones(1,PW(k)) zeros(1,N-PW(k))];    % Generate pulse signal
   X = abs(fft(x));      % Find magnitude spectrum
   X = X/max(X);         % Normalize the spectra
   plot(f(1:N2),X(1:N2),'k'); hold on;      % Plot pulse spectra
   ......label spectral curves.......
end
```

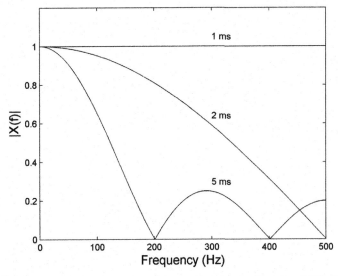

FIGURE 5.4 The magnitude spectra of three pulse signals. Since $f_s = 1$ kHz, the 1.0-ms pulse width signal is a true discrete impulse signal. Although the longer signals decrease in frequency (as $|\sin(x)/x|$, see Example 3.3), the impulse signal's spectrum is a constant over all valid frequencies, 0 to $f_s/2$.

Results: The three spectral curves are shown in Figure 5.4. As shown in Example 3.3, the magnitude spectrum of a pulse signal has a shape given by $|\sin(x)/x|$ (see Figure 3.12). This is seen for the 2- and 5-ms pulses, where the shorter pulse produces a spectrum that falls off less rapidly with increasing frequency, but still goes to zero at the Nyquist frequency, $f_s/2$. The true impulse has a much different magnitude spectrum. It is a constant value across all frequencies between 0 and $f_s/2$ Hz. Its phase spectrum is also a constant. As shown in one of the problems, the phase angle is 0.0 degree over the frequency range of 0 to $f_s/2$ Hz.

The spectrum of the true impulse is quite different and rather remarkable. It contains an equal amount of energy for all the valid frequencies in the signal, a property that can be very useful in exploring the frequency characteristics of a system. Just as a signal can have a spectrum, so can a system.[2] A system's spectrum shows how that system attenuates or enhances an input signal as a function of frequency. The impulse response can be used to find a system's spectrum. Here is the rationale: if the input signal in the frequency domain is a constant across all frequencies, the output frequencies show how the system modifies signals as a function of frequency. In other words, if the impulse has a constant spectrum, the spectrum of the impulse response must be identical to the spectrum of the system. So to find the spectrum of a system, we only need to convert the impulse response to the frequency domain using the Fourier transform.

---

[2]The spectrum of a system is mathematically embodied in the "transfer function" as described extensively in the next chapter.

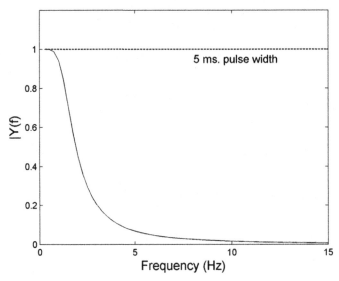

FIGURE 5.5   The magnitude spectrum of the unknown system from Example 5.1. The response of this system decreases rapidly for frequencies above 2.0 Hz. In the frequency domain it looks like a low-pass filter, although its actual function is unknown. The dashed line shows the magnitude spectrum of a 5.0-ms pulse signal, which appears to be constant over this limited frequency range. This explains the finding in Example 5.1 that such a pulse can serve as an impulse signal.

We can use the impulse signal to find the frequency characteristics of the unknown system used in Example 5.1. When the input is effectively an impulse, the spectrum of the output is shown in Figure 5.5 (solid line). (Note the expanded frequency scale ranges between 0 and 15 Hz.) The system is seen to decrease for frequencies that are above 2.0 Hz. This system, whatever its real purpose, acts like a low-pass filter. It can respond to signals having low-frequency energy, but for input signals much above 5 Hz there is little response. The dashed line in Figure 5.5 shows the magnitude spectrum of a 5.0-ms pulse to be almost constant over this limited frequency range. That is why it acts like an impulse in Example 5.1. For the limited range of frequencies to which the unknown system is capable of responding, the 5.0-ms pulse signal looks like an impulse.

This illustrates another way to determine whether a short pulse can be considered an impulse if you know the frequency characteristics of the system. Just compare the system's spectrum with that of the short pulse. If the pulse spectrum is more or less flat over the range of the system's nonzero spectrum, it can be considered an impulse. This idea is presented in one of the problems. In the next two chapters, we become deeply involved in the frequency characteristics of the system and we will return to the impulse input with its unique spectral properties.

5.4 USING THE IMPULSE RESPONSE TO FIND A SYSTEMS OUTPUT TO ANY INPUT—CONVOLUTION

Any signal can be chopped up into time slices; any signal can be represented as a sequence of impulse signals. Each pulse will produce its own impulse response. All these individual

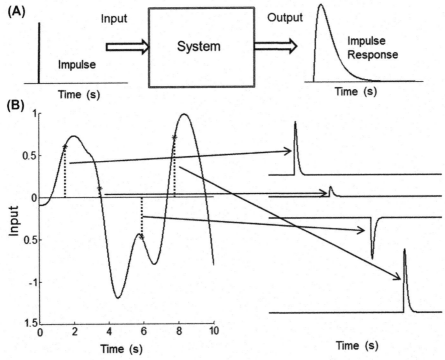

FIGURE 5.6 (A) An impulse input produces a characteristic response known as the impulse response. (B) A general signal can be represented as a number (possibly infinite) of impulses in sequence. Each impulse produces its own response and all responses have the same shape, a shape characteristic of the system. The only difference is the amplitude of the impulse response, which depends on the amplitude of the signal.

impulse responses have the same shape; the only difference is amplitude, which is scaled by the input signal's amplitude during a given time slice, Figure 5.6. The scaled impulse response is also shifted to correspond to its time slot in the input signal, Figure 5.6B. Since LTI systems are, by definition, time invariant, time shifting the signal does not alter the impulse response. The output signal is the sum of all those scaled and shifted impulse responses. This approach to finding the output of a system given the input is called "convolution."

When implementing convolution, we reverse the input signal, because it is the input signal's smallest time value that produces the initial output. From a graphical perspective, the left-hand side of a time plot is actually the low-time side and it is this segment that enters the system first. So, from the point of view of the system, the left side of the input signal is encountered first and the leftmost segment produces the first impulse response, Figure 5.7. The input signal then proceeds through the system backward from the way it is normally plotted.

Since the segments are infinitesimal, there are an infinite number of impulse responses so the summation becomes integration. Standard calculus is well equipped to describe this operation, although implementation is another matter. Scaling and summing the time-shifted impulse responses generated by a reversed input signal leads to the "convolution integral" equation.

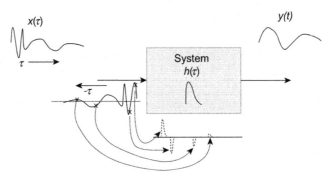

FIGURE 5.7 An input signal first enters its target system at its smallest (i.e., most negative) time value. From a graphical perspective, the first impulse response generated by the input signal comes from the left side of the signal as it is normally plotted. As it proceeds in time through the system backward, it generates a series of impulse responses that are scaled by the value of the input signal at any given time.

$$y(t) = \int_{-\infty}^{\infty} h(\tau)x(t - \tau)d\tau \tag{5.1}$$

where x is the input signal, $h(\tau)$ is the system's impulse response, and y is the output. Although the limits of integrations are $\pm\infty$, the actual limits are set by the length of the signals.

Convolution is the same as the cross-correlation equation (Equation 2.39) except for two alterations. First, the role of t and τ are switched so that t is now the time-shift variable and the second time variable, τ, is used for the input and impulse response functions. This modification is a matter of convention and is trivial. Second, there is a change in the sign between the time variables of x: $x(t + \tau)$ changes to $x(t - \tau)$. This is because the input signal, $x(\tau)$, must be reversed as described earlier, and the -τ accounts for this reversal. A third difference not reflected in the equations themselves is the way the two equations are used: cross-correlation is used to compare two signals, whereas convolution is used to find the output of an LTI system to any input.

As with cross-correlation, it is just as valid to shift the impulse response instead of the input signal. This leads to an equivalent representation of the convolution integral equation:

$$y(t) = \int_{-\infty}^{\infty} x(\tau)h(t - \tau)d\tau \tag{5.2}$$

The solution of Equation 5.1 or 5.2 requires multiple integrations, one integration for every value of t. Since t is continuous, that works out to be an infinite number of integrations. Fortunately, in real-world situations we apply convolution only in the discrete domain. In the discrete domain, t takes on discrete values at sample intervals T_s and integration becomes summation, so solving the convolution equation becomes a matter of performing a large number of summations. This would still be pretty tedious except that, as usual, MATLAB does all the work.

An intuitive feel for convolution is provided in Figures 5.8—5.10 as more and more impulse responses are added to the convolution process. Figure 5.8 shows the impulse response of a system in the left plot and the input to that system in the right plot. Note the different time scales: the impulse response is much shorter than the input signal, as is generally the case.

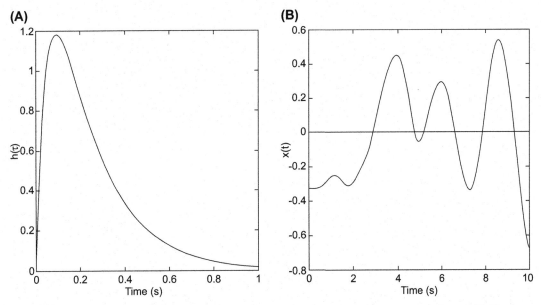

FIGURE 5.8 (A) The impulse response of a hypothetical system used to illustrate convolution. (B) The input signal to a system having the impulse response shown in (A). The impulse response has a much shorter duration than the input signal: 1 s versus 10 s. The impulse response is usually much shorter than the input signal.

In Figure 5.9A, the impulse responses to four signal segments at 2, 4, 6, and 8 s are shown. Each segment produces an impulse response that is shifted to the same time slot as the impulse, and the impulse response is scaled by the amplitude of the input signal at that time. Note that the input signal has been reversed as explained earlier. Some responses are larger,

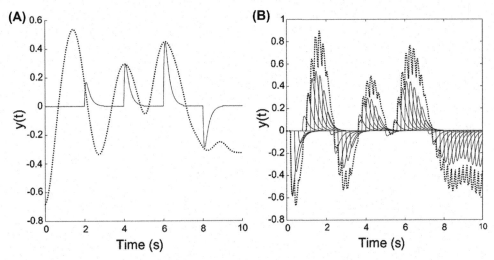

FIGURE 5.9 (A) The output response produced by four time slices of the input signal shown in Figure 5.6B. Impulse responses from input segments at 2, 4, 6, and 8 s are shown (*solid curve*) along with the reversed input signal, $x(-\tau)$ in Equation 5.2 (*dotted curve*). (B) Fifty impulse responses (*solid curves*) for the same reversed signal shown in (A). The sum of these signals is also shown (*dotted curve*).

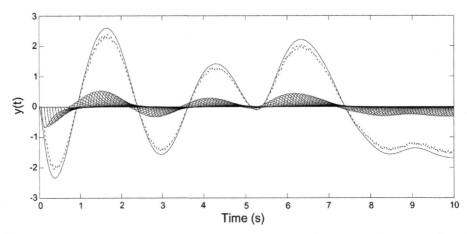

FIGURE 5.10 The summation of 150 impulse responses (*dashed curve*) from input segments spaced evenly over the input signal now closely resembles the actual output curve (*solid line*). The actual output is obtained from implementation of the digital convolution equation (Equation 5.3).

and one is scaled negatively, but they all have the basic shape as the impulse response shown in Figure 5.8A. A larger number (50) of these scaled impulse responses is shown in Figure 5.9B.

The 50 impulse responses in Figure 5.9 begin to suggest the shape of the output signal. A better picture is given in Figure 5.10, which shows 150 impulse responses, the summation of those responses (dashed line), and for comparison the actual output response (solid line) obtained by convolution. The summation of 150 impulse responses looks very similar to the actual output. (The actual output signal is scaled down to aid comparison.) As mentioned earlier, convolution of a continuous signal requires an infinite number of segments, but the digital version is limited to one for each data sample as discussed later. (The input signal used in the figures has 1000 samples.)

If the impulse response extends over a long period of time at high values, then a large contribution comes from past segments of the input signal. Such a system is said to have more memory, as the output signal is based on more of the input signal's past. If the impulse response is short, then very little comes from the past and more of the output is shaped by the instantaneous input at a given time. In the extreme case, if the impulse response is itself an impulse, then nothing comes from past inputs—all of the output comes from the current instantaneous input, and the output looks just like the input. The only difference would be the scale of the output, which would alter the height of the output impulse. Such a system is memoryless and would be called a "gain element."

For discrete signals, the integration becomes a summation and the discrete version of the convolution integral becomes the "convolution sum":

$$y[n] = \sum_{k=-\infty}^{\infty} x[k]h[n-k] = \sum_{k=-\infty}^{\infty} h[k]x[n-k] \qquad (5.3)$$

where n represents the time-shift variable and may extend over only the shorter of the two functions or, with zero padding, over all possible signal combinations. Again, it does not matter whether $h[k]$ or $x[k]$ is shifted; the net effect is the same. Since both $h[k]$ and $x[k]$ must be finite (they are stored in finite memory), the summation is also finite. The continuous and discrete convolution operations may also be abbreviated by using an "*" between the two signals:

$$y(t) = \int_{-\infty}^{\infty} h(\tau)x(t - \tau)d\tau \equiv h(t) * x(t) \tag{5.4}$$

or:

$$y[n] = \sum_{n=0}^{N-1} x[k]h[n - k] \equiv x[n] * h[n] \tag{5.5}$$

Unfortunately the * symbol is broadly used to represent multiplication in many computer languages, including MATLAB, so its use to represent the convolution operation can be a source of confusion and we avoid it here.

Superposition is a fundamental assumption of convolution. The impulse response describes how a system responds to a small (or infinitely small) signal segment. Each of these small (or infinitesimal) segments of an input signal generates its own little impulse response scaled by the amplitude of the segment and shifted in time to correspond with the segment's time slot. The output is obtained by summing (or integrating) the multitude of impulse responses, but this is valid only if superposition holds. Since convolution invokes superposition, it can be applied only to LTI systems where superposition is valid.

In the real world, convolution is done only on a computer, but in the textbook world, manual convolution can be found, in both the continuous and digital domains, as demonstrations. Sometimes these exercises provide insight and, in that hope, two examples of manual convolution are presented here. There are also a few in the problems.

EXAMPLE 5.3

Find the result of convolving the two continuous signals shown in Figure 5.11A and B. Use direct application of Equation 5.2. In this and the next example, one of the signals is assumed to be an impulse response $h(\tau)$ as is often the case, but of course convolution can be applied to any two signals.

The equations for the two signals can be determined from Figure 5.11:

$$x(\tau) = \begin{cases} 1 & 0 < \tau < 1.5 \\ 0 & \text{otherwise} \end{cases} ; \quad h(\tau) = e^{-t} \quad 0 \le t < 1.5$$

Solution: First, one of the signals has to be flipped, then a sliding integration can be performed. In this example, we flip the impulse response as shown in Figure 7.6 C. So the basic signals in Equation 5.2 are the equation for x above and:

$$h(t - \tau) = \begin{cases} e^{-(t-\tau)} & 0 < \tau < 1.5 \\ 0 & \text{otherwise} \end{cases}$$

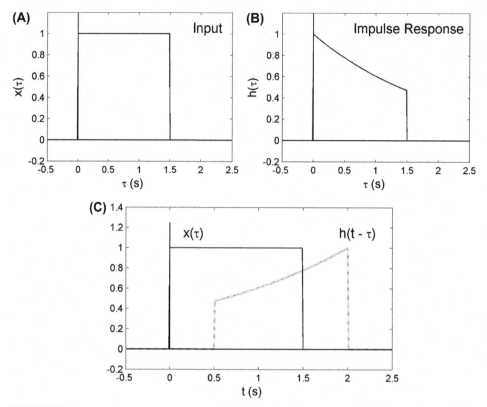

FIGURE 5.11   (A) and (B) Two signals to be convolved manually in Example 5.2. (C) To convolve the two signals, one signal, $h(\tau)$ is reversed. This reverse signal then slides across the other signal and the area of overlap is determined.

To solve the integral equation we use the classic calculus trick of dividing the problem into segments where the relationship between the two signals is consistent. For these two signals, there are three time periods during which the mathematical statement for $x(t)\,h(t - \tau)$ remains consistent, Figure 5.12.

In the first region, for shifts where $t < 0$, there is no overlap between the two signals, Figure 5.12A. So their product is zero and the integral is also 0:

$$y(t) = 0 \quad \text{for} \quad t < 0$$

When the time shift $t > 0\,\text{s}$, the functions overlap to an extent determined by the impulse response, $h(t - \tau)$, as shown in Figure 5.12B.

$$y(t) = \int_0^\infty x(\tau)h(t - \tau)d\tau = \int_0^t 1e^{-(t-\tau)}d\tau = e^{-t}\left(e^\tau\big|_0^t\right) = e^{-t}e^t - e^{-t}e^0 = 1 - e^{-t} \quad 0 \le t < 1.5$$

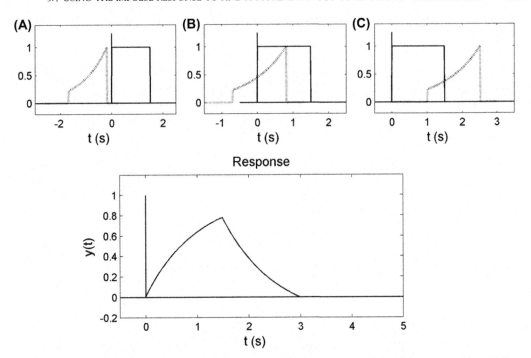

FIGURE 5.12   (A) For this relative position, where the shift time is $t < 0$, two signals have no overlap. (B) For the time shift where $0 \leq t < 1.5$ s, the overlap depends on $h(t - \tau)$. (C) When $t > 1.5$ s, the overlap depends on the pulse function. (D) The result of the manual convolution of the signals shown in Figure 5.11.

This relationship continues until $t > 1.5$ s when the overlap between the two signals is determined by the pulse. For this region the integral is the same, except that the upper limit of integration is 1.5 s:

$$y(t) = \int_0^{1.5} 1e^{-(t-\tau)}d\tau = e^{-t}\left(e^{\tau}|_0^{1.5}\right) = e^{-t}e^{1.5} - e^{-t} = \left(e^{1.5} - 1\right)e - t \quad t \geq 1.5$$

Combining these three solutions gives:

$$y(t) = \begin{cases} 0 & t < 0 \\ 1 - e^{-t} & 0 \leq t < 1.5 \\ \left(e^{1.5} - 1\right)e^{-t} & t \geq 1.5 \end{cases}$$

A time plot of the signal resulting from this convolution is shown in Figure 5.12D. An example of manual convolution with a step function that is 0 for $t < 0$ and a constant for all $t > 0$ is given in the problems.

Manual convolution of digital signals is comparatively easy; it is only a matter of keeping track of the shifts. The next example applies convolution to a digital signal.

EXAMPLE 5.4

Find the result of convolving the two discrete signals shown in Figure 5.13A and B. Use the direct application of the convolution sum in Equation 5.3. It is worth paying close attention to this example: if you understand what is going on in this example, you understand the basics of discrete convolution.

Solution: This example is similar to the digital cross-correlation shown in Example 2.15 except that one of the signals is flipped. The signal values for $x[k]$ and $h[k]$ are given in Figure 5.13. After reversing $h[k]$, we slide it across $x[k]$ one point at a time. At each position, we take the product of all the overlapping points and add them up. To handle the end points we pad $x[n]$ with two zeros on each end.

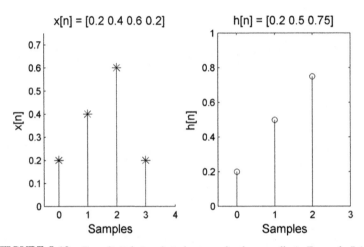

FIGURE 5.13   Two digital signals to be convolved manually in Example 5.4.

The operation is best illustrated graphically, Figure 5.14.

We begin with the flipped impulse response, $(h[-k]$, the "o" points) on the left side of the signal $(x[k]$, the "*" points): The first two "o" points overlap padded zeros; only the third point contributes to the convolution sum. So the first output point is

$x[k]$	0.0	0.0	0.2	0.4	0.6	0.2	0.0	0.0
$h[-k]$	0.75	.05	0.2					

$$y[1] = 0.2(0.2) = 0.04$$

After sliding the "o" points one position to the right, we now have two points overlapping nonzero values of the "*" points, giving the convolutions sum:

$x[k]$	0.0	0.0	0.2	0.4	0.6	0.2	0.0	0.0
$h[-k]$		.75	0.5	0.2				

$$y[2] = 0.5(0.2) + 0.2(0.4) = 0.18$$

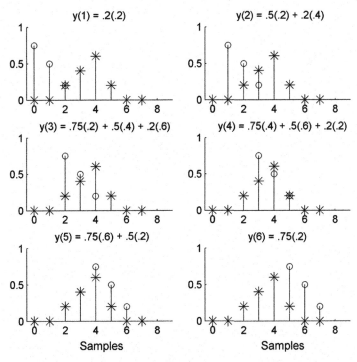

FIGURE 5.14 A step-by-step example of the manual convolution of the two digital signals shown in Figure 5.13. Note how $h[-k]$, the "o" points, slides across $x[k]$, the "*" points, producing one output point, $y[n]$, at each position.

For the next two positions, all three of the "o" points overlap nonzero "*" points, giving sums of:

$x[k]$	0.0	0.0	0.2	0.4	0.6	0.2	0.0	0.0
$h[-k]$			0.75	0.5	0.2			

$y[3] = 0.75(0.2) + 0.5(0.4) + 0.2(0.6) = 0.47$

$x[k]$	0.0	0.0	0.2	0.4	0.6	0.2	0.0	0.0
$h[-k]$				0.75	0.5	0.2		

$y[4] = 0.75(0.4) + 0.5(0.6) + 0.2(0.2) = 0.64$

For the last two positions of "o" points, first two points, then only one point, overlaps nonzero "*" points:

$x[k]$	0.0	0.0	0.2	0.4	0.6	0.2	0.0	0.0
$h[-k]$					0.75	0.5	0.2	

$y[5] = 0.75(0.6) + 0.5(0.2) = 0.55$

$x[k]$	0.0	0.0	0.2	0.4	0.6	0.2	0.0	0.0
$h[-k]$						0.75	0.5	0.2

$y[5] = 0.75(0.6) + 0.5(0.2) = 0.55$

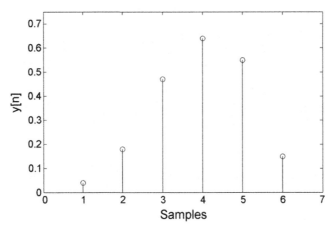

FIGURE 5.15   The result of the manual convolution of the two signals shown in Figure 5.11.

Results: The output signal $y[n]$ is shown in Figure 5.15. Although this may seem like a simple example since the number of points in both signals is small, it fully reflects the operations performed in all discrete convolutions. Being able to visualize this sliding summation (as $h[-k]$ slides to the right) is also helpful in understanding the operation of digital filters described in Chapter 8.

5.4.1 Finding the Impulse Response

Although convolution can be applied to any two signals, it is most commonly used to evaluate the output of a system to any input given the system's impulse response. Convolution is often used in signal processing to modify a signal in some controlled way. A common example is signal filtering where a system, the filter, is applied to a signal to remove unwanted frequencies. In such cases, the impulse response can be calculated based on the filter's desired frequency characteristics. We learn how to construct such impulse responses in Chapter 8.

Alternatively, if the system is physically available, as is often the case in the real world, the impulse response can be determined empirically by monitoring the response of the system to an impulse. Of course, a mathematically correct impulse function that is infinitely short but still has an area of 1.0 is impossible to produce in the real world, but an appropriately short pulse will serve just as well. Just what constitutes an appropriately short pulse depends on the dynamics of the system, and an example of how to determine if a pulse is a suitable surrogate for an impulse function was given in Example 5.1.

5.4.2 MATLAB Implementation

The convolution integral can, in principle, be evaluated analytically using standard calculus operations, but it quickly becomes tedious or impossible for complicated inputs. The discrete form of the convolution integral (Equation 5.3) is easy to implement on a computer. In fact, the major application of convolution is in digital signal processing, where it is frequently used to apply filtering to signals.

It is not difficult to write a program to implement Equation 5.3, but, of course, MATLAB has a routine to compute convolution, the `conv` routine:

```
y = conv(x,h,'option') ;        % The convolution sum (Equation 5.2)
```

where x and h are vectors containing the waveforms to be convolved, and y is the output signal. The options are given in Table 5.1.

If `'option'` is not given (or given as `'full'`), the length of the output signal is longer than the input signal (as in Example 5.4). When having a longer output signal is a problem, the `'same'` option computes all possible combinations, but outputs only the central portion of the result the same size as signal x. The option `'valid'` limits the summations to time shifts where x and h completely overlap returning a signal that is shorter than x.

TABLE 5.1 Options for the MATLAB `conv` routine

Option	Operation	Output Signal Length
None or `'full'`	Uses all possible relative positions Zero pads as necessary	`length(x) + length(h) -1`
`'same'`	Uses all possible relative positions with zero padding, but returns only the central sums	`length(x)`
`'valid'`	Sums over only relative positions possible without zero padding	`length(x) - length(h) -1`

An alternative routine, `filter`, can also be used to implement convolution. This routine always produces the same number of output samples as in x. When used for convolution, the calling structure is:

```
y = filter(h,1,x);      % Convolution sum (output same length as x)
```

where the variables are the same as described earlier and the second variable is always 1 as used here (more advanced applications of this routine are found in Chapter 8). In the following examples, the `conv` routine is used with the `same` option. You can try experimenting with using `filter` where appropriate in the problems. To get the proper output signal amplitude, both `conv` and `filter` require that the output signal be divided by the sampling frequency, f_s.

EXAMPLE 5.5

Find and plot the convolution sum for discrete versions of the two continuous signals used in Example 5.3.

Solution: Compared with the analytical solution in Example 5.3, this example is frightfully easy. We simply define the two signals in the digital domain and then apply the conv routine. With this routine, we need to normalize the output by f_s to get the proper scale. Since we use discrete versions of the continuous signals in Example 5.3, we do need to select a sample frequency. Figure 5.12D

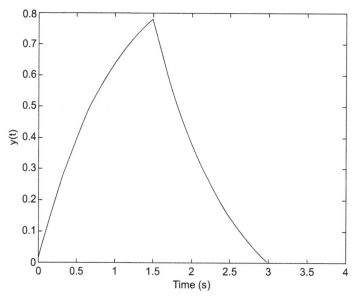

FIGURE 5.16   The convolution of the two signals in Figure 5.11 using MATLAB's `conv` routine. The resulting signal is the same as that found by manual calculation and plotted in Figure 5.12D.

shows that the output signal was 3 s long so if we choose an $f_s = 100$ Hz, our output signal will contain 300 samples, adequate for a decent plot. If we define h and x to be 1.5 s and use `conv` with the default option, the output will be approximately 3 s.

```
% Ex 5.5 Find the convolution of the two signals used in Example 5.3
%
fs = 100;                    % Sampling frequency
N = 150;                     % Number of samples in 1.5 sec
x = [ones(1,N)];             % Define pulse signal
t = (0:N-1)/fs;             % Time vector for the exponential signal
h = exp(-t);                % Define exponential signal
y = conv(x,h)/fs;           % Convolve and normalize
t = (0:length(y)-1)/fs;     % Time vector for plotting
      ........plot, label, and extend to 4 sec.......
```

Results: The result is shown in Figure 5.16 plotted out to 4 s. Using the default `conv` option outputs all possible points to give us the full response. The result is identical to the analytical solution found in Example 5.3 and shown in Figure 5.12D.

5.5 APPLIED CONVOLUTION—BASIC FILTERS

As mentioned, one of the most common uses of convolution is in signal processing to filter signals. The popularity of using convolution to implement signal filtering explains why MATLAB calls one of its convolution routines filter. The basic idea is to construct impulse responses that when applied to a signal remove or reduce undesired frequencies. Again, we discover how to construct such impulse responses given the desired frequency characteristic in Chapter 9. The next example gives a taste of filtering using convolution.

EXAMPLE 5.6

Plot an electrocardiogram (ECG) signal before and after application of a filter (or system) consisting of a three-point running average. The ECG signal can be found as variable `ecg` in file `ecg_noise.mat`. $f_s = 125$ Hz. Also determine and plot the magnitude spectrum of the three-point running average system in decibels.

Solution: This example covers a lot of territory as it introduces the basic concepts behind a large class of digital filters: averaging across a series of sequential signal samples to construct the filtered signal. In this example of a three-point running average, we average three consecutive points, but in other filters we often apply a different weighting to each of the sequential points. We also get the spectral characteristics of this filter with the Fourier transform (and a lot of zero padding) as in Example 5.2.

A three-point running average (also known as a three-point moving average) constructs a filtered signal by taking three consecutive points from the unfiltered signal, averaging them to produce a new point in the filtered signal. This averaging process repeats after shifting one sample over the unfiltered data, Figure 5.17. The result is a smoother signal due to averaging.

We could easily write a program to perform a three-point running average. But looking back at the operations in Figure 5.14, a three-point running average is very similar to the sliding operation in discrete convolution. Convolution of the signal with an impulse response that consisted of three equal values of 1/3 each would effectively implement a three-point running average. In other words, a three-point moving average is equivalent to convolving a signal with an impulse response of $h = [1/3 \ 1/3 \ 1/3]$. When used in a filter, the individual samples of the impulse response are called the filter "weights," or filter "coefficients." Filters usually have short impulse responses.

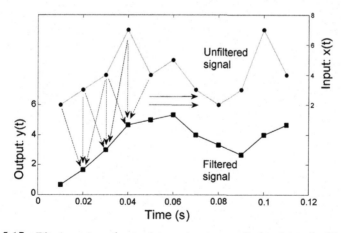

FIGURE 5.17 Filtering using a three-point running average. Each point on the filtered (lower) signal is constructed from the average of three consecutive points from the unfiltered signal. Endpoints are treated using zero padding. The three-point running average is identical to convolving the unfiltered signal with an impulse response having three equal values; i.e., $h = [1/3 \ 1/3 \ 1/3]$.

So after loading the data, we define the impulse response, carry out the convolution, and plot the results.

To find the equivalent spectrum of the three-point running average, note that in Section 5.3.2 we argued that a system's spectrum is the Fourier transform of the impulse response. Our three-point moving average is actually a system and its spectral characteristics are given by the Fourier transform of its impulse response, [1/3, 1/3, 1/3]. With only three points, the Fourier transform would be meaningless, so it is zero padding to the rescue. We will zero pad our short impulse response out to a large number of points to produce a smooth spectrum.

```
%  Example 5.6 Find the output of the ECG signal after filtering by a
%  filter that preforms a 3-point running average.
%
load ecg_noise;              % ECG signal in variable eeg
N = length(ecg);             % Number of data samples
fs = 125;                    % Sample frequency 125 Hz (given)
t = (1:N)/fs;                % Construct time vector for plotting
h = [1 1 1]/3;               % Define h(t) (3-point running avg)
out = conv(ecg,h,'same');    % Perform convolution
   ........ plot the filtered and unfiltered signal, new figure........
%
H = 20*log10(abs(fft(h,N)));  % Fourier transform of filter (heavily zero-padded)
N2 = round(N/2);             % Plot only non-redundant points
f = (1:N)*fs/N;              % Construct frequency vector for plotting
     ......plot the filter's spectrum........
```

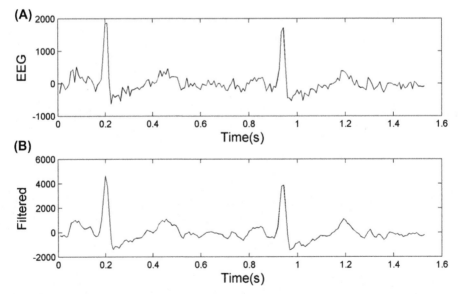

FIGURE 5.18 (A) A somewhat noisy unfiltered electrocardiogram (ECG) signal. (B) The ECG signal after filtering with the three-point running average filter.

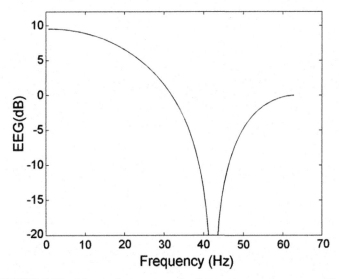

FIGURE 5.19 Magnitude spectrum of a three-point moving average filter.

Results: The ECG signal is shown before and after filtering in Figure 5.18. Note that the three-point running average smooths the signal slightly.

The filter's spectrum as obtained from the Fourier transform of the impulse response is shown in Figure 5.19. It has the form of a low-pass filter with a cutoff frequency of around 20 Hz. Reducing the higher frequencies reduces the spikey noise seen in the unfiltered ECG signal, Figure 5.18A.

You might wonder what would happen if we increased the number of sequential points that are averaged. What about a 10-point running average, or even a 20-point moving average? What would happen if not all the points in the impulse response had the same value? The not very informative answer is that then you would have a different filter. These questions are explored a bit in the problems and extensively in Chapter 8.

Just as signals and systems can be represented in either the time or frequency domain, the time-domain operation of convolution has a frequency-domain equivalent. To find out what the convolution operation would look like in the frequency domain, we use the same approach that we use to convert signals to the frequency domain: we apply the Fourier transform. We use the continuous representation of both convolution and the Fourier transform and will substitute ω for $2\pi f$ to make the equations cleaner. Starting with the convolution operation:

$$\int_{-\infty}^{\infty} h(\tau)x(t-\tau)d\tau = h(t) * x(t)$$

and taking the Fourier transform:

$$FT[h(t) * x(t)] = FT\left[\int_{-\infty}^{\infty} h(\tau)x(t-\tau)d\tau\right] = \int_{-\infty}^{\infty}\left[\int_{-\infty}^{\infty} h(\tau)x(t-\tau)d\tau\right] e^{-j\omega t}dt$$

Rearranging the order of integration:

$$= \int_{-\infty}^{\infty} h(\tau) \left[\int_{-\infty}^{\infty} x(t - \tau) e^{-j\omega t} dt \right] d\tau$$

The integral within the brackets is the Fourier transform of $x(t - \tau)$ a time-shifted version of $x(t)$. From the time-shift equation in Chapter 3 (Equation 3.41), that integral in brackets equals $X(\omega) e^{-j\omega \tau}$ (where $X(\omega)$ is the Fourier transform of $x(t)$). So the equation for convolution in the frequency domain becomes:

$$FT[h(t) * x(t)] = \int_{-\infty}^{\infty} h(\tau) X(\omega) e^{-j\omega \tau} d\tau = X(\omega) \int_{-\infty}^{\infty} h(\tau) e^{-j\omega \tau} d\tau = X(\omega) H(\omega) \qquad (5.6)$$

Equation 5.6 shows that convolution of two signals in the time domain is the same as multiplying their frequency-domain representations. For example, if we had an impulse response of a system and wanted to find the output to a given signal, but for some reason we did not want to use convolution, we could multiply their frequency-domain representations instead:

1. Covert the impulse response and signal to the frequency domain by taking their Fourier transform: $H(\omega) = FT[h(t)]$ and $X(\omega) = FT[x(t)]$.
2. Multiply both: $Y(\omega) = H(\omega) X(\omega)$.
3. Take the inverse Fourier transform to get the time-domain output: $y(t) = FT^{-1}[Y(\omega)]$.

In the next example, we try out this strategy and compare it with straightforward convolution using MATLAB.[3]

---

[3]Insider tip: The roundabout way of performing convolution may seem needlessly complicated (one might even say convoluted), but it is the approach MATLAB uses in the conv routine! MATLAB uses this approach because the FFT algorithm and its inverse are so fast that it is actually faster to do convolution in the frequency domain going back and forth between the two domains than to execute all those time domain multiplications and additions called for by Equation 5.3.

EXAMPLE 5.7

Given the system impulse response in the equation below and shown in Figure 5.20B, find the output of the system to a periodic sawtooth wave having a frequency of 20 rad/s = 3.2 Hz, Figure 5.20A. Use both convolution and frequency-domain approaches. Since the sawtooth has a frequency of 16 Hz ($100/2\pi$), we use a sampling frequency that is 10 times greater: $f_s = 160$ Hz.

$$h(t) = 5.0 \, e^{-5t} \sin(100t)$$

Solution: First construct the impulse response function given the equation above using a time vector of 1.0 s, Figure 5.20B. Finding the response using convolution requires only the application of MATLAB's conv. To find the output using the frequency-domain approach follow these steps: convert the impulse response and input signal to the frequency domain using the Fourier transform, multiply the two frequency-domain functions, and take the inverse Fourier transform.

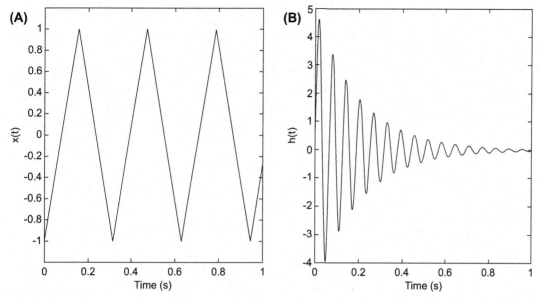

FIGURE 5.20 (A) Input signal to the system used in Example 5.7. (B) The impulse response of that system.

```
% Example 5.7  System output found two different ways.
%
fs = 160;                    % Sample frequency
T = 1;                       % Time in sec
t = (0:T)/fs;                % Time vector
N = length(t);               % Determine length of time vector
x = sawtooth(20*t,.5);       % Construct input signal
h = 5* exp(-5*t).*sin(100*t);  % Construct the impulse response
%
y = conv(x,h);               % Calculate output using convolution
subplot(2,1,1)
plot(t,y(1:N));              % Plot output from convolution
   .......label and title.......
% Now calculate in the frequency domain.
TF = fft(h);                 % Get TF from impulse response (Step 1)
X = fft(x);                  % Take Fourier transform of input (Step 1)
Y = X.*TF;                   % Take product in the frequency domain (Step 2)
y1 = ifft(Y);                % Take the inverse FT to get y(t) (Step 3)
subplot(2,1,2)
plot(t,y1);                  % Plot output from frequency domain
    .......labels and title.......
```

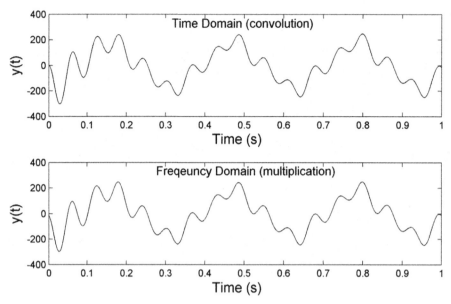

FIGURE 5.21   Output signals from a system defined by the impulse response in Figure 5.20B. These signals were computed using standard convolution (top) and multiplication of their frequency-domain representations (bottom).

Results: As shown in Figure 5.21, the output waveforms look like sinusoids riding on the input signal, and are the same using either time- or frequency-domain analysis.

5.6 CONVOLUTION IN THE FREQUENCY DOMAIN

Equation 5.6 states that convolution in the time domain is the same as multiplication in the frequency domain. The equations leading to Equation 5.6 can be rearranged to show that the reverse is also true: convolution in the frequency domain is the same as multiplication in the time domain.

$$X_1(\omega) * X_2(\omega) = x_1(t) \, x_2(t) \tag{5.7}$$

There are several important operations that involve time domain multiplication. In data sampling, a continuous signal is multiplied by a series of pulses spaced T_s apart (see Sections 4.2.1 and 1.2.3.2). Signal truncation is equivalent to multiplying the signal by a window function (see Section 4.2.3). Time domain multiplication is also found in amplitude modulation, where a signal is multiplied by a sinusoid or square wave. In the next section, we reexamine the sampling process specifically with respect to its influence on the signal's spectrum.

5.6.1 Data Sampling Revisited

In Chapter 4, we found that data sampling produces additional frequencies, in fact an infinite number of additional frequencies. The sampling process is like multiplying a continuous time function by a series of impulse functions. Figure 5.22A shows a continuous sinusoidal signal that is multiplied by a series of impulse functions spaced 50 ms apart. The result shown in Figure 5.22B is the same as a signal that is sampled at a frequency of $1/T_s = 1/0.05$ or $f_s = 20$ Hz. (A repeating series of pulses is referred to as a "pulse train.") So the sampling process can be viewed as a time domain multiplication of a continuous signal by a pulse train of impulses having a period equal to the sample interval, T_s.

Since multiplication in the time domain is equivalent to convolution in the frequency domain Equation 5.7), we should be able to determine the effect of sampling in a signal's spectrum using convolution. We just convolve the sampled signal's original spectrum with the spectrum of the pulse train. We get the former by applying the Fourier transform to the original signal. For the latter, the spectrum of a periodic series of impulse functions,

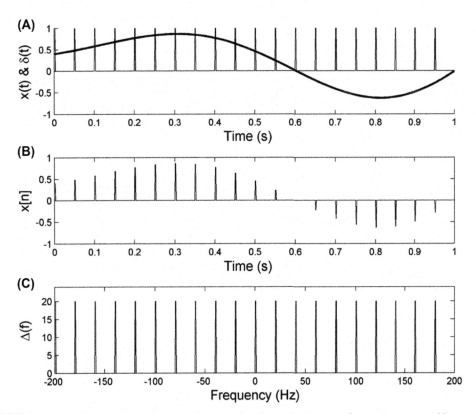

FIGURE 5.22 (A) The sampling process can be viewed as the multiplication of a continuous signal by a periodic series of impulse functions. In this figure the pulse train's period is 0.05 s, equivalent to the sample interval, T_s, or $1/f_s$. (B) The sampled signal, $x[n]$, where $f_s = 1/0.05 = 20$ Hz. (C) The frequency spectrum of a periodic impulse function is itself a series of impulse functions (Equation 5.10).

we use the Fourier series analysis, since the functions are periodic. In Example 3.3, the complex form of the Fourier series equation is used to find the Fourier coefficients of a periodic series of ordinary pulses having a pulse width W:

$$X_m = \left(\frac{V_p W}{T}\right) \frac{\sin(\pi m f_1 W)}{\pi m f_1 W} \qquad m = 0, \pm 1, \pm 2, \pm 3, \ldots \tag{5.8}$$

For an impulse function $W \to 0$, but the area, $V_p W$, remains equal to 1.0. Also for small angles, $\sin(x) \approx x$, so as $W \to 0$ and $V_p W \to 1$, Equation 5.8 becomes:

$$X_m = \lim_{\substack{WV_p \to 1 \\ W \to 0}} \left| \left(\frac{V_p W}{T}\right) \frac{\sin(\pi m f_1 W)}{\pi m f_1 W} \right| = \lim_{\substack{WV_p \to 1 \\ W \to 0}} \left| \left(\frac{V_p W}{T}\right) \frac{\pi m f_1 W}{\pi m f_1 W} \right|$$

$$= \lim_{WV_p \to 1} \left| \frac{V_p W}{T} \right| = \frac{1}{T} \tag{5.9}$$

Since the period of the impulse train is $T = T_s$, the spectrum of this impulse train interval is:

$$C_m = 1/T = 1/T_s = f_s \tag{5.10}$$

So the spectrum of the impulse function is itself a pulse train at frequencies of mf_1, where $f_1 = 1/T = f_s$ as shown in Figure 5.22C. Since the complex version of the Fourier series equation was used, the resultant spectrum has both positive and negative frequency values. Recall the impulse function is denoted by $\delta(t)$ so its spectrum is $\Delta(f)$ and is given as:

$$\Delta(f) = \frac{1}{T} \sum_{k=-\infty}^{\infty} \delta(f_s - k f_s) = f_s \sum_{k=-\infty}^{\infty} \delta(f_s - k f_s) \tag{5.11}$$

Now back to frequency-domain convolution. If the spectrum of the signal being sampled is $X(f)$ and the frequency spectrum of the impulse train is $\Delta(f)$, given by Equation 5.11, then the spectrum of the sampled signal is the convolution of the two. Applying the discrete convolution sum, Equation 5.3, and replacing time with frequency terms:

$$X_{samp}(f) = \Delta(f) * X(f) = \sum_{n=-\infty}^{\infty} \Delta(f) X(f - nf)$$

Substituting in Equation 5.11 for $\Delta(f)$, the spectrum of the sampled signal is:

$$X_{samp}(f) = \sum_{n=-\infty}^{\infty} \left(\frac{1}{T} \sum_{k=-\infty}^{\infty} \delta(f_s - k f_s) \right) X(f - nf) = \sum_{n=-\infty}^{\infty} \left(f_s \sum_{k=-\infty}^{\infty} \delta(f_s - k f_s) \right) X(f - nf)$$

$$\tag{5.12}$$

After we rearrange the sums and simplify by combining similar frequencies, Equation 5.12 becomes:

$$XSamp = fs \sum_{k=-\infty}^{\infty} X(f - kfs) \qquad (5.13)$$

This application of convolution in the frequency domain shows that the spectrum of a sampled signal is an infinite sum of shifted versions of the original spectrum. This is stated, but unproved, in Chapter 4.

5.7 SUMMARY

The impulse function, $\delta(t)$, is a very short pulse; in theory it is infinitely short, but also infinitely tall, so its area remains equal to 1.0. In the real world, whether a pulse can be considered an impulse depends on the response characteristics of the system it is used to evaluate. A pulse input is effectively an impulse if it produces a response that does not change in shape for incremental changes in pulse width. The response of such a short pulse is called the impulse response and it has a unique shape that does not depend on the amplitude of the pulse: it simply scales up or down proportionally with pulse amplitude. Since the shape of the impulse response is always the same for a particular system, it can be used as a descriptor of that system. Impulse responses can be determined empirically if the system is available by monitoring the system's response to a sufficiently short pulse. The most useful feature of the impulse and impulse response is that it can be used in a time-slicing approach to determine the output of a system to any input using convolution.

Convolution employs time slicing to determine the response of a system to any input signal. All you need is the system's impulse response, which can be considered the system's response to an infinitesimally short segment of the input. If the system is linear (LTI) and superposition holds, the impulse response from each input segment can be summed to produce the system's output. The convolution integral and convolution sum (Equations. 5.1, 5.2, and 5.3) are basically running correlations between the input signal and the impulse response. This integration can be cumbersome to compute analytically, but is easy to program on a computer. MATLAB provides two routines, `conv` and `filter`, to perform convolution.[4]

Convolution is commonly used in signal processing to implement digital filtering. In these applications, the impulse response of the desired filter is first determined analytically using techniques described in Chapter 8. The filtered signal is obtained by applying the filter's impulse response to the signal using convolution.

Convolution performs the same function in the time domain as multiplication in the frequency domain. The reverse is also true: convolution in the frequency domain is like

---

[4]The `conv` routine actually operates in the frequency domain: it converts both the impulse response and input signal to the frequency domain, multiplies the two, then converts back to the time domain. It is actually faster than the `filter` routine.

multiplication in the time domain. Frequency-domain convolution is useful in explaining time-domain operations such as sampling. Specifically, frequency-domain convolution explains the additional frequencies found in the spectra of time sampled signals. Recall that if the frequencies of a digitized signal exceed the Nyquist frequency, $f_s/2$, these added frequencies produce aliasing hopelessly corrupting the signal's spectrum.

PROBLEMS

1. Modify the code in Example 5.1 to find the maximum length of a pulse that will serve as an impulse to the unknown system, `unknown_sys5_2`. Note this system has quite different response characteristics and the maximum pulse width will also be quite different. (Hint: You need to look at complete impulse response to determine the maximum pulse width. This system is much slower than that in Example 5.1, so you need to substantially increase the time length of the plots to capture the full impulse response.)

2. Using the pulse width obtained for an unknown_sys5_2 in Problem 1, show that the impulse responses to a wide range of pulse amplitudes have exactly the same shape. Use three different pulse inputs ranging in amplitude by a factor of 100. (Hint: Scale the responses to be the same maximum amplitude and show that they are identical.)

3. A pulse having a width of 40 ms is known to be sufficiently short to be an impulse input for the system `unknown_sys5_3`. Plot the impulse response in the time domain and be sure to use a long enough time period to accurately capture the complete impulse response.

   Convert the system's impulse response and the 40-ms pulse to the frequency domain. Plot the magnitude spectra of the two superimposed to show that the 40-ms pulse has an nearly flat magnitude spectrum over the frequencies to which the system is capable of responding, i.e., the frequencies in the impulse response. Limit the frequency axis to a range that clearly shows the impulse response frequencies. If you plot the two spectra on the same graph you need to scale one or both to have similar ranges.

4. The `file unknown_impulse_response.mat` contains the impulse response of system in variable `i_r`. The impulse response was sampled at $T_s = 0.01$ s ($f_s = 100$ Hz.) Use this impulse response to find the magnitude spectrum of the unknown system. Then find the longest pulse that will still serve as an impulse to that system based on the pulse spectra. When generating the pulse, use the same data length as the impulse response. (Hint: The magnitude spectrum of the pulse should be relatively flat in the nonzero region of the system. You can plot the spectra of several pulses superimposed to speed up the search for maximum pulse width. When plotting the spectra, restrict the frequency axis to the region where the impulse response spectrum is nonzero. Finally, in reporting the maximum pulse width, note the sample interval and report the maximum pulse width in s.)

5. Use the basic convolution equation (Equation 5.1) to find the output of a system having an impulse response of $h(t) = 2\,e^{-10t}$ to an input that is a unit step function; i.e., $x(t) = 1$. $t > 0$. (Hint: If you reverse the step input, then for $t < 0$ there is no overlap

and the output is 0. For $t > 0$, the integration limit is determined only by t, the leading edge of the reversed step function, so only a single integration, with limits from 0 to t, is required. See the following figure.)

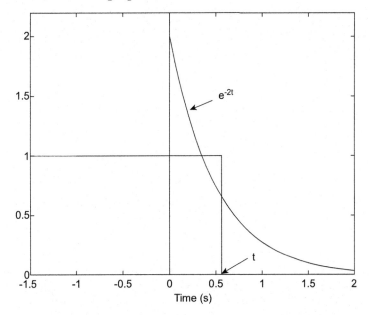

6. Use the basic convolution equation (Equation 5.1) to find the output of a system with an impulse response, $h(t) = 2(1 - t)$, to a 1-sec pulse having an amplitude of 1.

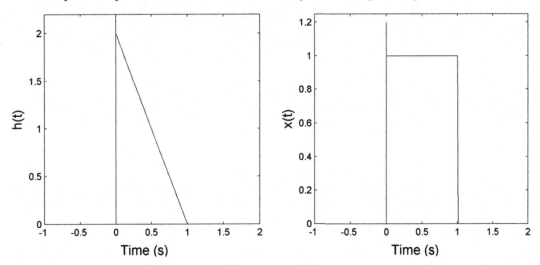

II. SYSTEMS

7. Use the basic convolution equation (Equation 5.1) to find the output of a system with an impulse response shown below to a step input with an amplitude of 5; i.e., $x(t) = 5$ for $t > 0$.

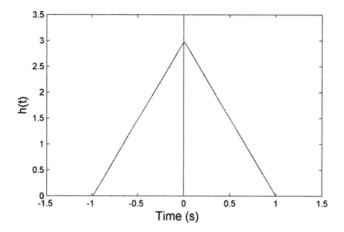

8. Find the convolution sum (Equation 5.3) for the discrete impulse response and discrete input signal shown in the following figure.

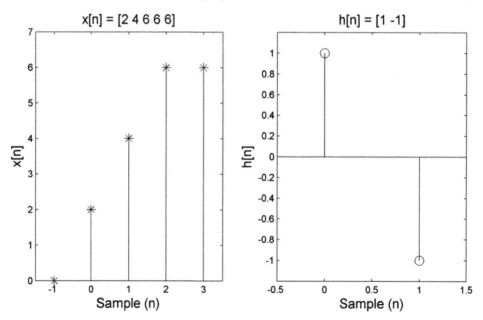

9. Find the convolution sum (Equation 5.3) for the discrete impulse response and discrete input signal shown in the following figure.

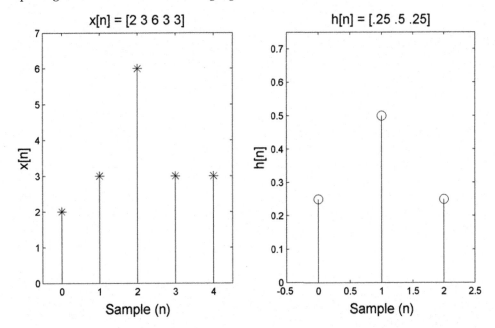

10. Systems known as filters are designed to alter the spectral characteristics of signals. In Example 5.6, we saw how a filter represented by an impulse response of three equal weights smoothed a noisy signal. What if we used more weights? What if we used weights that were not all the same? These questions are addressed (at least in part) by this and the next four problems.

Compare the frequency characteristic of three impulse responses that could be used as filters. The first impulse response consists of three equal weights used in Example 5.6. The second consists of five equal weights of 1/5. (Thus it computes a five-point running average.) The third filter consists of five different weights: h3 = [0.0264 0.1405 0.3331 0.3331 0.1405 0.0264]. To find the filter's spectrum, take the Fourier transform of these impulse responses. Be sure to sufficiently zero pad (and make N a power of 2 for a faster FFT while you are at it). As always, plot only the valid points and be sure the frequency axis is correctly scaled.

You will see that they are all low-pass filters and that, although the two equal-weight filters cut off more sharply, the unequal impulse response filter gives a much smoother frequency characteristic. We will learn how to design these filter impulse responses in Chapter 8.

11. There is more than one way to identify the magnitude spectrum of a system. The approach described in Section 5.3.2 and used in Problem 10 is to convert the impulse response to the frequency domain. Another way is to use sinusoidal inputs to the

system: multiple sinusoids that cover the frequency range of this system. Use convolution to find the system's output to these sinusoids, then plot the output amplitude divided by the input amplitude as a function of frequency. In the next three problems we use both approaches.

In this problem, we examine the four-coefficient "Daubechies filter":

$$h[n] = [0.683 \quad 1.183 \quad 0.3169 \quad -.0183]$$

A. Determine and plot magnitude frequency characteristics of this filter by taking the Fourier transform of the impulse response. Use an $f_s = 100$ Hz.[5] Be sure to pad sufficiently and plot frequencies between 0 and $f_s/2$.

B. Set up a loop to evaluate the output of this filter to 50 sinusoids varying in frequency from 1 to 50 Hz in 1.0-Hz intervals; i.e., from 1 to $f_s/2$ Hz. At each of the 50 frequencies, generate a sine wave (suggested N of 256), convolve it with $h[n]$ above, and take the root mean square (rms) value of the output. Plot that value at the frequency of the sine wave. (If you make the peak-to-peak amplitude of the sine wave input 1.414, it will have an rms value of 1.0 and you will not need to divide the output by the input.)

Use subplot to plot the two magnitude spectra. If done correctly, they should be nearly identical. Note that this is a low-pass filter with very gradual, smooth attenuation. This filter has some very special properties and is one member of a family of filters used in Wavelet analysis.

12. Repeat Problem 11 using one of the simplest of all filters, the "Haar" filter. This filter has an impulse response consisting of only two coefficients each with a value of 1/2:

$$h[n] = \begin{bmatrix} \dfrac{1}{2} & \dfrac{1}{2} \end{bmatrix}$$

Although the Haar filter has a gradual attenuation and is not a very strong filter, it is used to demonstrate some basic principles of Wavelet analysis.

13. Load the respiratory signal resp in file resp_noise.mat ($f_s = 125$ Hz). Apply a 3-point moving average filter and a 12-point moving average filter. For variety, use the MATLAB filter routine:

$(y = $ filter(h,1,x);), which produces a filtered signal that is the same length as the input signal. Display the results from the two filters in two plots using subplot. Each plot should compare the unfiltered signal with one of the filtered signals. Title the plots and offset the filtered signal (by 1.0) for better viewing. Note the improved noise reduction with the 12-point moving average filter.

14. So far we have examined only the magnitude spectrum of our filters. Sometimes it is important to know the phase characteristic of the filter. The phase spectrum of any system can be obtained directly from the Fourier transform of the impulse response

---

[5]It does not matter what f_s you use, you get the exact same spectral shape if you plot the entire range of valid frequencies, 0 to $f_s/2$. This is because f_s cancels out in the DFT equation. See Equation 3.30.

using the MATLAB `angle` routine. Usually you also have to apply MATLAB's `unwrap` routine to get the correct phase spectrum.

Show both the magnitude and phase spectra of two filters, one a four-point moving average and the other the four-coefficient "Daubechies" filter used in Problem 11. Use $f_s = 200$ Hz and adequate padding. Plot phase in degrees and plot only valid points. Use subplot to bring the four plots together.

Note that, although the four-point moving average has a sharper cutoff, the Daubechies filter has a smoother phase characteristic.

15. Load the MATLAB file `filter1.mat`, which contains the impulse response, `h`, of a mystery filter. Also load the signal in `cardiac_press.mat`, which holds one cycle of a cardiac pressure wave in variable `c_press` $f_s = 200$ Hz. Apply the impulse response, `h`, to the pressure signal using convolution. Plot the cardiac pressure signal before and after filtering on separate plots using subplot. What does this mystery filter do?

16. Repeat Problem 11 using the impulse response, `h`, in file `bp_filter.mat`. This is the impulse response of a band-pass filter. This is a narrowband filter, so to get an accurate spectrum using sinusoids you should make the frequency increments 0.5 Hz instead of 1.0 Hz. Then you need to increase the number of sine waves to 100 to get the full valid spectrum.

17. Repeat Problem 11 using the impulse response, `h`, in file `x_impulse.mat`. This is a resonant system with a sharp peak at low frequencies. To get an accurate spectrum using sine waves, decrease the frequency increment to 0.1 Hz, but use only 50 sine waves covering the range of 0–5 Hz. When taking the Fourier transform, you may have to increase the padding to get the correct magnitude spectrum.

18. This problem demonstrates that the spectrum of the sampling process, essentially a pulse train, is itself a pulse train. To simulate the sampling process, construct a 16384-point signal (2^{14}) of zeros then add a pulse of 1.0 at every 16th position (i.e., `pt(1:16:end) = 1`). Take the Fourier transform and plot. If you do not scale the horizontal axis, you should see a peak at every 1024 points. Assuming the horizontal axis is in Hz, this corresponds to an $f_s = 1024$. Now simulate halving the sampling frequency by setting every 32nd point of the sampling signal to 1.0. You will now see peaks every 512 Hz reflecting the lower sampling frequency. Although you do not need to scale the axes, they should still be labeled. Combine the plots using `subplot` and title them for clarity.

19. This problem simulates the effect of sampling frequency on a spectrum. The spectrum can be found in file `unknown_spectrum.mat` as variable `Spec`. This spectrum represents the true spectrum of a signal before sampling. Plot this spectrum without scaling the frequency axis. Simulate the Fourier transform of a sampling process where $f_s = 2000$ Hz. Construct a 10,000-point array and make every 2000th point 1.0 (i.e., `pt(1:2000:end) = 1;`). Convolve this array with the spectrum `Spec` and plot the results. This is the spectrum you would get when sampling the signal at 2000 Hz. Now lower the sampling frequency to 1000 Hz by making every 1000th point of the sampling spectrum 1.0. Convolve and plot the results. Note the confused spectra that result from this lower sampling frequency, a condition described in Chapter 4 known as aliasing. Although you do not need to scale the axes, they should still be labeled. Combine the plots using subplot and title them for clarity.

6

Linear Systems in the Frequency Domain: The Transfer Function

6.1 GOALS OF THIS CHAPTER

Just as signals have spectral properties, systems have spectral properties. The difference is that a signal's spectrum describes how energy is distributed over the frequency range of the signal; a system's spectrum (also given as magnitude and phase) shows how a signal's energy distribution is modified as it passes through the system. In other words, the system's spectrum describes the relationship of the output signal's spectrum to the input signal's spectrum. The difference between the output and input signal spectrum is due to the system and is defined by the system's spectrum. Given the input signal's spectrum and the system's spectrum, we can determine the output signal's spectrum. We know there is a direct bilateral relationship between a signal's spectrum and its time-domain representation through the Fourier transform and its inverse, so the output time-domain signal can be constructed by taking the inverse Fourier transform of the output signal's spectrum. So if we know a system's spectrum, we can achieve in the frequency domain the same result that convolution achieves in the time domain: the output of the system to any input.

Why do we need another approach when we already have something that works, and works well? Convolution is a powerful technique for determining the behavior of any system that can be described by an impulse response, but it does not give much insight into the inner workings of these systems. Most systems can be viewed as collections of fundamental elements. In fact, all linear systems can be represented (i.e., modeled) using just a few basic element types. A system's spectrum provides information on the type of elements that make up the system.

We begin this chapter by describing the spectra associated with the different element types. We then show how, when such elements are combined into a system, the system's spectrum can be determined from the spectra of the individual elements. Once we have the system's spectra, we can determine its output to any input. We can also work the other way around: we can estimate the system's spectrum if we have both the input and output signals spectra. Discovering a system's spectra using the input and output signals is known as "system identification."

Topics covered in this chapter include:

- The basic concept behind systems models and the various types of model elements.
- The response of system elements to sinusoidal stimuli determined using complex sinusoids, an approach called "phasor analysis."
- How and why to construct the system's transfer function (introduced in Section 1.4.5.2).
- The spectral characteristics of basic system elements.
- How to derive the system's frequency response (i.e., spectrum) from the transfer function using an approach termed "Bode plots."
- Linking the transfer function and Fourier transform to find the output to any input signal.
- How to determine the system spectrum given the input and output signals.
- How to estimate the transfer function from the system's spectrum.

6.2 SYSTEMS ANALYSIS MODELS

A systems model is a process-oriented representation that emphasizes the influences, or flow, of information between model elements. A systems model describes how processes interact and what operations these processes perform, but it does not go into details as to how these processes are implemented. The basic element of the system model is a block enclosing a transfer function. Some very general systems-type models are based solely on descriptive (i.e., nonquantitative) elements. The purpose of such models is to show the relationships, or flow of influence, between processes, but they are not amenable to mathematical analysis.

Systems model elements are sometimes referred to as "black boxes" because they are defined only by the transfer function, which is not concerned with, and does not reveal, the inner working of the box. Both the greatest strength and the greatest weakness of system models are found in their capacity to ignore the details of mechanism and emphasize the interactions between the elements.

A mathematical expression defines the input—output relationship of a system element. Since a systems element is just a mathematical relationship, it is represented graphically by a geometrical shape, usually a rectangle, or sometimes a circle when an arithmetic process is involved. Two typical system elements are shown in Figure 6.1. The circle is an arithmetic element that subtracts signal x_2 from signal x_1 as indicated by the plus and minus signs next to the element (If both signs were positive the two signals would be summed.). The rectangular element could represent any general mathematical process. The inputs and outputs of

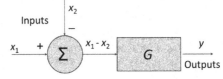

FIGURE 6.1   Typical graphical representation of systems elements. Input and output signals are shown as *arrows* and possible associated variable names.

all elements are signals with a well-defined direction, or flow, or influence. These signals and their direction of influence are shown by lines and arrows connecting the system elements.

The letter G in the right-hand element of Figure 6.1 is a stand-in for the mathematical description of the element: the mathematical operation that converts the input signal into an output signal. Stated mathematically:

$$Output = G(Input) \tag{6.1}$$

Equation 6.1 represents a very general concept: G could be any linear relationship and the terms *Input* and *Output* could be any signal of any complexity. It could represent a single element or an entire system composed of many elements. If it is a black box, how can we know? All we really know is the mathematical relationship. Capital letters are commonly used to represent the mathematical relationship, whereas lower case letters represent the signal.

Rearranging this basic equation, G can be expressed as the ratio of output to input:

$$G = \frac{Output}{Input} \tag{6.2}$$

Equation 6.2 emphasizes that G relates the output of an element to its input; it can be thought of as transferring information from the input to the output giving rise to the term "transfer function." Again, the input–output relationship is the only concern of a system element so the transfer function is a complete description of the element (or an entire system for that matter). Although the transfer function concept is sometimes used very generally, in linear systems analysis it is only an algebraic term (although possibly/probably having complex notation). This implies that linear system elements can be represented as algebraic functions and indeed we show how systems can be represented algebraically even if they contain calculus operators.

Making the transfer function an algebraic function greatly simplifies the analysis of even very complicated systems. For example, if two elements (or systems) are connected together as in Figure 6.2, the transfer function of the paired systems is just the product of the two individual transfer functions:

$$Out_2 = G_2 \times In_2 \text{ but } Out_1 \equiv In_2 = G_1 \times In_1,$$

then combining:

$$Out_2 = G_2 \times (G_1 \times In_1) = G_2 G_1 In_1. \tag{6.3}$$

and the overall transfer function for the two series elements is:

$$\frac{Out_2}{In_1} = G_1 G_2 \tag{6.4}$$

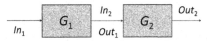

FIGURE 6.2 Two systems or elements connected together with transfer functions G_1 and G_2, respectively. Simple algebra shows that the transfer function of the combined systems is $Out_2/In_1 = G_1 G_2$.

Note that the overall transfer function of the combined systems in Figure 6.2 would be the same even if the order of the two systems was reversed. This is a property termed "associativity." We can extend this concept to any number of system elements in series:

$$Output = Input \prod_i G_i \tag{6.5}$$

And the overall transfer function is just the product of the individual series element transfer functions:

$$\frac{Output}{Input} = \prod_i G_i \tag{6.6}$$

The transfer function concept is very powerful and makes determining the input—output characteristics of even the most complex systems easy. In practice, the trick is determining the transfer function of the process of interest. Finding the transfer function of most biological systems is challenging and usually involves extensive empirical observation of the input/output relationship. However, once the relationships for the individual elements have been determined, finding the input/output relationship of the overall system is straightforward.

EXAMPLE 6.1

Find the transfer function for the systems model in Figure 6.3. The mathematical description of each element is either an algebraic term or simply an arithmetic operation. In this system, the element with the Σ identifier does subtraction since one of the input signals has a negative sign. G and H are assumed to be linear algebraic functions so they produce an output that is the product of the function times the input (i.e., Equation 6.2). The system shown is a classic feedback system because the output is coupled back to the input via the lower pathway. In this system, the upper pathway is called the "feedforward pathway" because it moves the signal toward the output. The lower pathway is called the "feedback pathway"; it takes a signal from further along the forward pathway and feeds it back as an input to an element at, or nearer, the system input. Note that the signal in this feedback pathway becomes mixed with the real input signal through the summation operator (actually subtraction, but we still generally call the is summation element). To finish off the terminology lecture, G in the configuration is called the "feedforward gain" and H is called the "feedback gain."

Solution: Generate an algebraic equation based on the configuration of the system and the fact that the output of each process is the input multiplied by the associated gain term (Equation 6.2 $Output = G \times Input$).

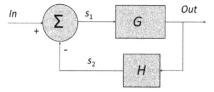

FIGURE 6.3   Systems model used in Example 6.1. This is an example of a classic feedback control system.

For the upper box: $G = \frac{Out}{s_1}$

and for the lower box: $H = \frac{s_2}{Out}$

Rearranging the two equations: $s_1 = \frac{Out}{G}$ and $s_2 = Out\ (H)$

Since: $s_1 = In - s_2$; Substituting: $\frac{Out}{G} = In - Out(H)$

Rearranging: $Out = In(G) - Out(GH)$; $Out(1 + GH) = In(G)$

$$TF = \frac{Out}{In} = \frac{G}{1 + GH} \tag{6.7}$$

Analysis: The solution given in Equation 6.7 is known as the "feedback equation" and will be used in subsequent analyses of more complex systems. In this example, the two elements, G and H, could be anything as long as they can be treated algebraically and we show later how all elements, even calculus operations, can be transformed into algebraic functions.

Figures 6.1 and 6.2 show another important property of systems models. The influence of one process on another is explicitly stated and indicated by the line connecting two processes. This line has a direction usually indicated by an arrow, which implies that the influence or information flow travels only in that direction. If there is also a reverse flow of information, such as in feedback systems, this must be explicitly stated in the form of an additional connecting line showing information flow in the reverse direction.

As described in Chapter 1, analog models represent physiological processes using elements that are, to some degree, analogous to those in the actual processes. Good analog models can represent systems at a lower level, and in greater detail, than system models. Analog models provide better representation of secondary features such as energy use, which is usually similar between analog elements and the components they represent. However, in an analog model, the interaction between components may not be obvious from inspection of the model.

System models emphasize component interaction, particularly information flow. They also provide a clear illustration of a system's overall organization. This can be of great benefit in clarifying the control structure of a complex system. These models explain what an element does with its input or stimulus, but they give no clue as to how it does it. Perhaps the most significant advantage of the systems approach is what it does not represent: it allows the behaviors of biological processors to be quantitatively described without requiring the modeler to know the details of the underlying physiological mechanism. Given our lack of understanding of the details of some biological processes, this can be a considerable blessing.

6.3 THE RESPONSE OF SYSTEM ELEMENTS TO SINUSOIDAL INPUTS: PHASOR ANALYSIS

If the signals or variables in a system are sinusoidal or are converted to sinusoids using the Fourier transform, then a technique known as phasor analysis can be used to convert calculus operations into algebraic operations. As used here, the term "phasor analysis" is considerably more mundane than the name implies: the analysis of phasors such as those used on Star Trek is, unfortunately, beyond the scope of this text. The phasor analysis we will learn about here combines complex representation of sinusoids with the fact that calculus operations (integrations and differentiation) change only the magnitude and phase of a sinusoid. This analysis assumes the signals are in sinusoidal steady state; that is, they always have been and always will be sinusoidal.

There are four basic systems elements: arithmetic operators, scale operation, differentiators, and integrators. Since sinusoidal frequency is not altered by these operations, if the input(s) to a system are sinusoids, then all signals in the system will be sinusoidal at the same frequency (sinusoidal Property 6 in Chapter 3). Then all signals in a linear, time invariant (LTI) system can be described by the same general equation:

$$x(t) = A\cos(\omega t + \theta) = A\cos(2\pi f t + \theta) \tag{6.8}$$

where the values of A and θ can be modified by the system elements, but the value of ω (or f) will be the same throughout the system.

Sinusoids require just three variables for complete description, amplitude, phase, and frequency (sinusoidal Property 2 in Chapter 3). But if the frequency is always the same, then we really only need to keep track of two variables: amplitude[1] and phase. Here is where complex notation sounds like it could be useful since a single complex variable is actually two variables rolled into one (i.e., $a + jb$). A single complex number or variable is all that is needed to describe the amplitude and phase of a sinusoid.

To find how to represent a sinusoid by a single complex variable, we return to the complex representation of sinusoids given by Euler's equation:

$$e^{jx} = \cos x + j\sin x \tag{6.9}$$

or for sinusoidal signals such as Equation 6.8:

$$Ae^{j(\omega t + \theta)} = A\cos(\omega t + \theta) + jA\sin(\omega t + \theta) \tag{6.10}$$

Comparing the basic sinusoid equation, Equation 6.8, with Equation 6.10 indicates that only the real part of e^{jx} is needed to represent a sinusoid:

$$A\cos(\omega t + \theta) = \mathrm{Re}Ae^{j(\omega t + \theta)} = \mathrm{Re}Ae^{j\theta}e^{j\omega t} \tag{6.11}$$

If all variables in an equation contain the real part of the complex sinusoid, the real terms can be dropped. Consider,

If, $\mathrm{Re}Ae^{j\theta_1}e^{j\omega t} = \mathrm{Re}Be^{j\theta_2}e^{j\omega t}$ for all t,

Then, $A = B$ and[2]

$$Ae^{j\theta_1}e^{j\omega t} = Be^{j\theta_2}e^{j\omega t} \tag{6.12}$$

---

[1]Some engineers use the word "amplitude" when referring to the peak value of the sinusoid and the word "magnitude" when referring to the rms value. I use these words interchangeably.

[2]In general, A does not necessarily equal B when Re A = Re B, but the only way that Re $Ae^{j\omega t}$ can equal the Re $Be^{j\omega t}$ at all values of t is when $A = B$. Appendix E presents a review of complex arithmetic.

Since all variables in a sinusoidally driven LTI system are the same except for amplitude and phase, they will all contain the "Re" operator, and these terms can be removed from the equations as was done in Equation 6.12. They do not actually cancel; they are just unnecessary since the equality stands just as well without them. Similarly, since all variables will be at the same frequency, identical $e^{j\omega t}$ terms will appear in each variable and will cancel once the Re's are dropped. Therefore, the two defining sinusoidal variables in Equation 6.8, A and θ, can be represented by a single complex variable:

$$A \cos(\omega t + \theta) \Leftrightarrow A e^{j\theta} \tag{6.13}$$

where $A e^{j\theta}$ is the "phasor" representation of a sinusoid.

Equation 6.13 does not indicate a mathematical equivalence, but a transformation from the standard sinusoidal representation to a complex exponential representation without loss of information. In the phasor representation, the frequency, ω, is not explicitly stated, but is understood to be associated with every variable in the system (sort of a virtual variable). Note that the phasor variable, $A e^{j\theta}$, is in polar form as opposed to the rectangular form $(a + jb)$ and you may need to convert it to the rectangular representation in some calculations. Since the phasor, $A e^{j\theta}$, is defined in terms of the cosine (Equation 6.11), sinusoids defined in terms of sine waves must be converted to cosine waves when using this analysis.

If phasors offered only a more succinct representation of a sinusoid, their usefulness would be limited. It is their calculus-friendly behavior that endears them to engineers. To determine the derivative of the phasor representation of a sinusoid, we return to the original complex definition of a sinusoid (i.e., Re $A e^{j\theta} e^{j\omega t}$):

$$\frac{d(\mathrm{Re} A e^{j\theta} e^{j\omega t})}{dt} = \left(\mathrm{Re} j\omega A e^{j\theta} e^{j\omega t}\right) \tag{6.14}$$

The derivative of a sinusoid in complex notation is the original complex variable, but multiplied by $j\omega$. After the Re operators are dropped, the $e^{j\omega t}$'s cancel, and the derivative of a phasor becomes multiplication by $j\omega$. For phasors, the derivative operation reduces to a simple arithmetic operation:

$$\frac{d}{dt} \Leftrightarrow j\omega \tag{6.15}$$

Similarly, integration using the complex definition of a sinusoid:

$$\int \mathrm{Re} A e^{j\theta} e^{j\omega t} dt = \mathrm{Re} \frac{A e^{j\theta} e^{j\omega t}}{j\omega} \tag{6.16}$$

Again, integration of a sinusoid in complex notation gives rise to the same complex variable except divided by $j\omega$. Integration applied to a phasor is accomplished by dividing by $j\omega$, again an arithmetic operation:

$$\int dt \Leftrightarrow \frac{1}{j\omega} \tag{6.17}$$

The basic rules of complex arithmetic are covered in Appendix E; however, a few properties of the complex operator j are important enough to be repeated here. Note that $1/j = -j$, since:

$$\frac{1}{j} = \frac{1}{\sqrt{-1}} = \frac{-\sqrt{-1}}{(-\sqrt{-1})(-\sqrt{-1})} = \frac{-\sqrt{-1}}{-(-1)} = -\sqrt{-1} = -j \qquad (6.18)$$

So Equation 6.17 could also be written as:

$$\int dt \Leftrightarrow -j/\omega \qquad (6.19)$$

Multiplying by j in complex arithmetic is the same as shifting the phase by 90 degrees, which follows directly from Euler's equation:

$$je^{jx} = j(\cos x + j \sin x) = j \cos x - \sin x = -\sin x + j \cos x$$

Substituting in $\cos(x + 90)$ for $-\sin x$, and $\sin(x + 90)$ for $\cos x$, je^{jx} becomes[3]:

$$je^{jx} = \cos(x + 90) + j \sin(x + 90)$$

This is the same as e^{jx+90}, which can also be written as:

$$je^{jx} = e^{jx}e^{90} \qquad (6.20)$$

Similarly, dividing by j is the equivalent of shifting the phase by -90 degrees:

$$\frac{e^{jx}}{j} = \frac{\cos x + j \sin x}{j} = \frac{\cos x}{j} + \sin x = -j \cos x + \sin x$$

Substituting in $\cos(x - 90)$ for $\sin x$, and $\sin(x - 90)$ for $-\cos x$:

$$\frac{e^{jx}}{j} = \cos(x - 90) + j \sin(x - 90) = e^{jx}e^{-90} \qquad (6.21)$$

Equations 6.15, 6.17, and 6.19 demonstrate the benefit of representing sinusoids by phasors: the calculus operations of differentiation and integration become the algebraic operations of multiplication and division. Moreover, the transformation used to convert a time-domain sinusoid to a phasor (Equation 6.13), or vice versa, is super easy.

---

[3]Note that we are using degrees in these equations, although the mathematics would normally be carried out in radians. I feel it is easier to visualize the term "90 degrees" than "$\pi/2$." In any case, the conclusions reached by these equations are the same.

EXAMPLE 6.2

Find the derivative of $x(t) = 10 \cos(2t + 20)$ using phasor analysis.

Solution: Convert $x(t)$ to a phasor (represented as $x(j\omega)$). To take the derivative, multiply by $j\omega$, then take convert the resulting phasor back to a sinusoid:

$$10 \cos(2t + 20) \Leftrightarrow 10e^{j20}$$

$$\frac{dx(j\omega)}{dt} \Leftrightarrow j\omega(10e^{j20}) = j2(10e^{j20}) = j20e^{j20}$$

$$j20e^{j20} = 20e^{j20}e^{j90} \Leftrightarrow 20 \cos(2t + 20 + 90) = 20 \cos(2t + 110)$$

Since $\cos(x) = \sin(x + 90) = -\sin(x - 90)$, this can also be written as $-20 \sin(2t + 20)$, which is what you would get from standard differentiation. Note that the frequency ($\omega = 2$) is not explicitly stated in the phasor solution (i.e., $20e^{j20}e^{j90}$), but is reinserted when converting back to sinusoidal representation. Again, an explicit representation of frequency is unnecessary since frequency is the same for all elements in the system.

A shorthand notation is common for the phasor description of a sinusoid. Rather than write $Ve^{j\theta}$, we simply write $V \angle \theta$ (stated as "V at an angle of θ"). When a time variable such as $v(t)$ is converted to a phasor variable, it is common to write it as a function of ω using capital letters for the amplitude, i.e., $V(\omega)$. This acknowledges the fact that phasors represent sinusoids at specific frequencies, even though the sinusoidal term, $e^{j\omega t}$, is not explicit in the phasor itself. Also, in phasor analysis, it is common to represent frequency as rad/s (i.e., ω) rather than as Hz, even though Hz is generally used in engineering settings. Putting these conventions together, the time-to-phasor transformation for variable $v(t)$ can be stated as:

$$v(t) \Leftrightarrow V(\omega) = V \angle \theta \tag{6.22}$$

In this notation, the phasor representation of $20 \cos(2t + 110)$ would be written as $20 \angle 110$ rather than $20 \, e^{j110}$. Sometimes, the phasor representation of a sinusoid expresses the amplitude of the sinusoid in root mean square (rms) values rather than peak-to-peak values, in which case the phasor representation of $20 \cos(2t + 110)$ would be written as $(0.707) \, 20 \angle 110 = 14.14 \angle 110$. Here we use peak-to-peak values, but it really does not matter as long as we are consistent.

The phasor approach is an excellent method for simplifying the mathematics of LTI systems. It can be applied to all systems that are driven by sinusoids and, in conjunction with the Fourier transform, systems driven by periodic, or aperiodic, signals.

6.4 THE TRANSFER FUNCTION

The transfer function mathematically transfers the input to an output. The transfer function concept is so compelling that it has been generalized to include many different types of processes or systems with different types of inputs and outputs. In the phasor domain, the transfer function is a minor modification of Equation 6.2:

$$\text{Transfer Function}(\omega) = \frac{Output(\omega)}{Input(\omega)} \tag{6.23}$$

Frequently, both *Input(ω)* and *Output(ω)* are signals measured in volts and represented by symbols such as $v_{in}(\omega)$ and $v_{out}(\omega)$, but for now we continue to use general terms. By strict definition, the transfer function should always be a function of ω, f, or, as shown in the next chapter, the Laplace variable, s, but the idea of expressing the behavior of a process by its transfer function is so powerful that it is sometimes used in a nonmathematical, conceptual sense.

EXAMPLE 6.3

The system shown in Figure 6.8 is a simplified, linearized model of the Guyton–Coleman body fluid balance system presented by Ridout (1991). This model describes how the arterial blood pressure P_A in mmHg responds to a small change in fluid intake, F_{IN}, in mL/min. Find the transfer function for this system, $P_A(\omega)/F_{IN}(\omega)$ using phasor analysis.

Solution: The two elements in the upper path can be combined into a single element using Equation 6.4, which states that the transfer function of the combined element is the product of the two individual transfer functions:

$$G(\omega) = G_1(\omega)G_2(\omega) = \frac{0.06}{j\omega}(16.67) = \frac{1}{j\omega}$$

The resulting system has the same configuration as the feedback system in Figure 6.3, which has already been solved in Example 6.1 (Equation 6.7). We could go through the same algebraic process, but it is easier to use the fact that the system in Figure 6.4 has the same configuration as the feedback system in Figure 6.3, where the transfer function of G is $1/j\omega$ and the feedback gain, H, is 0.05.

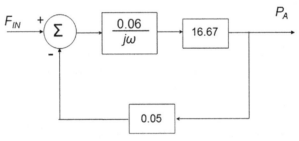

FIGURE 6.4 A simplified linear version of the Guyton–Coleman fluid balance system that relates arterial blood pressure, P_A, to small changes in fluid intake, F_{IN}. The transfer function of this model is developed in Example 6.3.

Substituting the expressions for G and H into the feedback equation, Equation 6.7, gives the transfer function of this system:

$$\frac{Out(\omega)}{In(\omega)} = \frac{G}{1+GH} = \frac{\dfrac{1}{j\omega}}{1+0.05\left(\dfrac{1}{j\omega}\right)} = \frac{1}{0.05+j\omega} = \frac{20}{1+j20\omega}\frac{\text{mmHg}}{\text{mL/min}}$$

Analysis: This transfer function applies to any input signal at any frequency, ω, as long as it is sinusoidal steady state. Note that the denominator consists of a real and an imaginary part, and that the constant term is normalized to 1.0. This is the common format for transfer function equations in the frequency domain: the lowest power of ω, usually a constant term, is normalized to 1.0.

Systems having transfer functions with a $1 + jk\omega$ term in the denominator, where k is a constant, are called "first-order" systems and are discussed, along with other common systems, later in this chapter. The output of this system to a specific sinusoidal input is determined in the next example.

Many of the transfer functions encountered in this text have the same units for the numerator and denominator terms (e.g., volts/volts), so these transfer functions are dimensionless. However, when this approach is used to represent physiological systems, the numerator and denominator often have different units as is the case here (i.e., the transfer function has the dimensions of mmHg/mm/min).

EXAMPLE 6.4

Find the output of the system in Figure 6.4 if the input signal is $F_{IN}(t) = 0.6 \sin(0.3t + 20)$ mL/min.

Solution: Since phasors are based on the cosine, we first need to convert $F_{IN}(t)$ to a cosine wave. From Appendix C, Equation 6.4: $\sin(\omega t) = \cos(\omega t - 90)$, so $F_{IN}(t) = 0.6 \cos(0.3t - 70)$ mL/min.

In phasor notation the input signal is:

$$F_{IN}(t) = 0.6 \cos(0.3t - 70) \Leftrightarrow F_{IN}(\omega) = 0.5 \angle -70 \text{ mL/min}$$

In Example 6.3, we found the transfer function as:

$$\frac{Out(\omega)}{In(\omega)} = \frac{20}{1 + j20\omega}$$

Solving for $Out(\omega)$ and then substituting in $0.5 \angle -70$ for $In(\omega)$ and letting $\omega = 0.3$

$$Out(\omega) = \frac{20}{1 + j20\omega} In(\omega) = \frac{20}{1 + j20(0.3)\omega}(0.5 \angle -70) = \frac{10 \angle -70}{1 + j6\omega}$$

The rest of the problem is just working out the complex arithmetic. To perform division (or multiplication), it is easiest to have a complex number in polar form. The number representing the input, which is also the numerator, is already in polar form ($6 \angle -70$), so the denominator needs to be converted to polar form:

$$1 + j6 = \sqrt{1^2 + 6^2} \angle \tan^{-1}\left(\frac{6.1}{1}\right)\left(\frac{360}{2\pi}\right) = 6.08 \angle 80$$

Result: Note that $\tan^{-1}(6.1)$ has been converted to degrees by multiplying by $360/2\pi$. Substituting and solving:

$$P_A(\omega) = \frac{10 \angle -70}{6.08 \angle 80} = 1.64 \angle -150 \text{ mmHg}$$

Converting to the time domain:

$$P_A(t) = 1.64 \cos(0.3t - 150) \text{ mmHg}$$

So the arterial pressure response to this input is a sinusoidal variation in blood pressure of 1.64 mmHg. That variation has a phase of -150 degrees, which is shifted 80 degrees from the input phase of −70 degrees (using the cosine form of the input signal). The corresponding time delay between the stimulus and response sinusoid can be found from Equation 2.21:

$$t_d = \frac{\theta}{360f} = \frac{80}{360\left(\frac{0.3}{2\pi}\right)} = 4.65 \text{ min}$$

So the response is delayed 4.65 min from the stimulus because of the long delays in transferring fluid into the blood. As with many physiological systems, generating a sinusoidal stimulus, in this case a sinusoidal variation in fluid intake, is difficult. The next example shows the application of phase analysis to a more complicated system.

EXAMPLE 6.5

Find the transfer function of the system shown in Figure 6.5. Find the time-domain output if the input is $10 \cos (20t + 30)$.

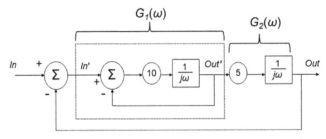

FIGURE 6.5   The system used in Example 6.5.

Solution: Since we do not already have a transfer function for this configuration of elements, we need to find the transfer function from the element equations and then use this transfer function to solve for the output in terms of the input. In all of these problems, the best strategy is to find the equivalent feedforward gain, G, and then apply the feedback equation to find the transfer function. To facilitate this strategy, some internal signals are labeled for reference. We observe that the elements inside the dashed rectangle constitute an internal feedback system with input In' and output Out'. So we can use the basic feedback equation, Equation 6.7, to get the transfer function of this internal subsystem. For this subsystem, the feedforward gain is a combination of the scalar and integrator:

$G' = 10\left(\frac{1}{j\omega}\right)$ and $H = 1$. So the transfer function for the subsystem becomes:

$$G_1 = \frac{Out'}{In'} = \frac{\frac{10}{j\omega}}{1 + \frac{10}{j\omega}} = \frac{10}{j\omega + 10} = \frac{1}{1 + j0.1\omega}$$

Following convention, the denominator is normalized so the constant term (or lowest power of ω) is 1. The rest of the feedforward path consists of a scalar and integrator, which can be represented by transfer function G_2.

$$G_2 = 5\left(\frac{1}{j\omega}\right) = \frac{5}{j\omega}$$

The transfer function of the feedforward gain is the product of G_1 and G_2:

$$G = G_1 G_2 = \left(\frac{5}{j\omega}\right)\left(\frac{1}{1 + j0.1\omega}\right) = \frac{5}{-0.1\omega^2 + j\omega}$$

We apply the feedback equation yet again, this time to find the transfer function of the overall system. Substituting G above into the feedback equation and letting $H = 1$:

$$\frac{Out}{In} = \frac{G}{1 + GH} = \frac{\dfrac{5}{-0.1\omega^2 + j\omega}}{1 + \dfrac{5}{-0.1\omega^2 + j\omega}} = \frac{5}{-0.1\omega^2 + j\omega + 5} = \frac{1}{1 - 0.1\omega^2/5 + j\omega/5}$$

Rearranging the denominator to collect the real and imaginary part and normalizing the constant term to 1:

$$\frac{Out}{In} = = \frac{1}{1 - 0.02\omega^2 + j0.2\omega}$$

This type of transfer function is known as a "second-order function," as the polynomial in the denominator includes a second-order frequency term, ω^2.

Finding the output given the input is a matter of working through some complex algebra. In phasor notation the input signal is $10\angle 30$ with $\omega = 20$. The output in phasor notation becomes:

$$Out = TF\ In = \frac{10\angle 30}{1 - 8 + j4} = \frac{10\angle 30}{-7 + j4} = \frac{10\angle 30}{8\angle 150} = 1.24\angle - 120$$

Converting to the time domain: $out(t) = 1.24\cos(20t - 120)$.

6.4.1 The Spectrum of a Transfer Function

One of the best ways to examine transfer functions, and the behavior of the systems they represent, follows the approach used for signals: find the spectral characteristics. The transfer function specifies how a system transforms the input signals into output signals as a function of frequency. Specifically, the transfer function shows how the amplitudes and phases of sinusoids at different frequencies are modified as they pass through the system. So plotting the magnitude and phase of the complex transfer function as a function of frequency provides a complete description of how a system modifies the signal passing through it.

The signal spectrum describes a signal's magnitude and phase characteristics as a function of frequency. The system spectrum describes how the system changes signal magnitude and phase as a function of frequency. For example, Figure 6.6 shows the magnitude and phase spectra of some hypothetical system. At the lower frequencies, below around 80 Hz, the magnitude spectrum is 1.0. This means that sinusoids from 0.0 to around 80 Hz pass through the system without a change in amplitude: they emerge from the system at the same amplitude as when they went in. This does not mean that the output will contain energy at those frequencies if there is no energy in the input signal, but if frequencies in that range exist in the input they will be unchanged in the output. At the higher frequencies the system magnitude spectrum diminishes, so any input signal frequencies in that range will be reduced at the output. For example, at around 100 Hz the transfer function has a magnitude value of around

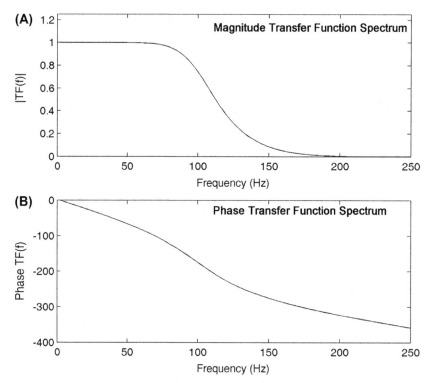

FIGURE 6.6 The magnitude (A) and phase (B) spectra of a hypothetical system. These spectra show the changes over frequency in the magnitude and phase of sinusoids as they pass (are transformed) from input to output.

0.707.[4] That means that the amplitude of a 100 Hz sinusoid will be reduced by a factor of 0.707 as it goes through the system. Moreover, the energy in the output signal will be reduced by half. (Remember energy is proportional to magnitude squared so when the magnitude is reduced by 0.707, the energy is reduced by 0.707^2 or 0.5.)

Following the same logic, unlike the phase spectrum of a signal, the phase plot does not show the phase angle of the system's output; rather it shows the change in the phase angle induced by the system. For this system, sinusoids would emerge from the system with a more negative phase angle, that angle becoming increasingly more negative at higher frequencies.

If the entire spectral range of a system is explored, then the transfer function spectrum provides a complete description of the way in which the system alters all possible input frequencies. Transfer function properties of systems are usually represented as frequency-domain plots because it is easier to conceptualize the operation of a system from spectral plots as in Figure 6.6 than as a complex mathematical function. It is easy to generate the frequency characteristics of any transfer function using MATLAB, as shown in the next two examples.

---

[4]It may be hard to tell from the figure, but the system really has a cutoff frequency of 100 Hz. The definitions for bandwidth in Section 4.6 apply directly to systems.

EXAMPLE 6.6

Use MATLAB to find and plot the magnitude and phase spectra of the transfer function below (also a second-order transfer function similar to that in Example 6.5):

$$\frac{Out}{In} = = \frac{1}{1 - 0.03\omega^2 + j0.01\omega}$$

Solution: After defining a frequency vector, w, the transfer function can be coded directly in MATLAB. The magnitude and phase are determined by applying the abs and angle routines. We have to guess the frequency range, and increments for w. We want a frequency range that will show the most interesting portion of the spectrum. A range of 0.1−1000 rad/s in increments of 0.1 rad/s worked well.

```
% Example 6.6 Use MATLAB to plot a second-order transfer function
%
w = .1:.1:1000;                      % Select frequency range an increment
TF = 1./(1 - .03*w.^2 + .01*j*w);    % Define transfer function
Mag = 20*log10(abs(TF));             % Magnitude spectrum in dB
Phase = unwrap(angle(TF))*360/(2*pi); % Phase spectrum in deg
%
semilogx(w,Mag);
..............label and grid, repeat for phase spectrum......
```

Results: Note how easy it is to define the transfer function in MATLAB. It is common to plot the system magnitude spectrum on a log-log scale, that is, in dB (a log scale) versus log frequency (using semilogx). The phase spectrum is plotted in degrees versus log frequency. Figure 6.7 shows

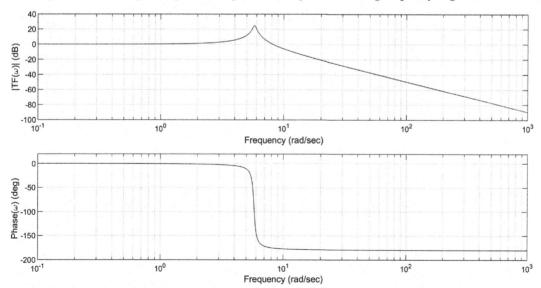

FIGURE 6.7 The magnitude and phase spectra of a second-order system. The magnitude is in dB, the phase in deg, and the frequency in log radians. Note that inputs with frequency components around 6 rad/s are enhanced in the output by 20 dB (i.e., a factor of 10).

the resulting magnitude and phase spectra. Here the plot is in radians, but it would have been easy to convert to frequency in Hz. The magnitude spectrum shows that the amplitude of input sinusoids having frequencies around 6 rad/s will be enhanced by 20 dB (a factor of 10 on a linear scale) in the output. The phase of output sinusoids will be roughly the same as that of input sinusoids up to that frequency, but as input sinusoidal frequencies increase, the output shows a −180-degree phase shift.

EXAMPLE 6.7

The transfer function in this example represents the relationship between the applied external pressure and airway flow for a person on a respirator in an intensive care unit. This transfer function was derived by Koo (2000) for typical lung parameters, but could be modified for specific patients. The original model is in the Laplace domain (described in the next chapter), but has been modified to the frequency domain for this example. We want to find the magnitude and phase spectrum of this airway system. We use MATLAB to plot the magnitude in dB and phase spectrum in deg, both against log frequency, this time in Hz.

$$\frac{Q(\omega)}{P(\omega)} = = \frac{9.52j\omega(1 + j0.0024\omega)}{1 - 0.00025\omega^2 + j0.155\omega} \frac{\text{L/min}}{\text{mmHg}}$$

where $Q(\omega)$ is airway flow in L/min and $P(\omega)$ is the pressure applied by the respirator in mmHg. Traditional engineering systems usually have the same units for the input(s) and output(s), for example, and the electronic system would have the input and the output in volts. These input and output units cancel, so the transfer function is dimensionless. However, transfer functions for biological systems often have inputs and outputs in different units, as is the case here, where the input is a pressure in mmHg and the output is a flow in L/min. In this case the transfer function has a dimension: L/(min mmHg).

Solution: To plot the spectra we follow the same procedure used in the last example. For the frequency vector, we originally tried a range of 0.1 to 1000 rad/s in steps of 0.1 rad, but then extended it to range between 0.001 and 10,000 in steps of 0.001 to better show the lower frequencies. Note that in this model, time is in minutes not seconds, so the plotting frequency needs to be modified accordingly. Since there are 60 s in a minute, the plotting frequency should be divided by 60 (as well as by 2π to convert from rad to Hz).

```
% Example 6.7 Use MATLAB to plot the given transfer function
%
w = .001:0.001:1000;                    % Construct frequency vector
TF = (9.62*j*w.*(1+j*w/420))./(1 - (w.^2)/4000 + .166*j*w); % Define TF
Mag = 20*log10(abs(TF));                % Magnitude of TF in dB
Phase = unwrap(angle(TF))*360/(2*pi);   % Phase of TF in deg
semilogx(w/(60*2*pi),Mag);              % Plot |TF| in log freq in Hz.
..... labels, grid, and repeat for phase spectrum
```

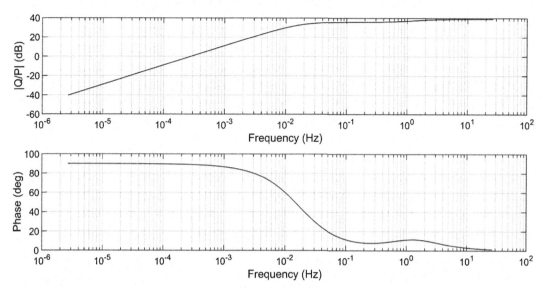

FIGURE 6.8 Magnitude and phase plot of a model that links airway flow to the air pressure in a respirator. The relationship between airway flow and respirator pressure is fairly constant over the range of normal respiratory rates (around 0.18 Hz).

Results: The spectral plots for this respiratory-airway system are shown in Figure 6.8. The flow is constant for a given external pressure for frequencies above 1 Hz (corresponding to 60 breaths/min), but decreases for slower breathing. However, in the normal respiratory range (11 breaths/min = 0.18 Hz), the magnitude spectrum is also constant and nearly the same. This shows that the relationship between airway flow and respiratory flow remains about the same for physiological respiratory rates.

Relegating the problem to the computer is easy but more insight is found by working it out manually. To really understand the frequency characteristics of transfer functions, it is necessary to examine the typical components of a general transfer function. In so doing, we learn how to plot transfer functions without the aid of a computer. More importantly, by examining the component structure of a typical transfer function, we also gain insight into what the transfer function actually represents. This knowledge is often sufficient to allow us to examine a process strictly within the frequency domain and learn enough about the system that we need not bother with time-domain responses.

6.5 THE SPECTRUM OF SYSTEM ELEMENTS: THE BODE PLOT

To dig into the meaning of transfer functions and be able to determine their spectral plots directly (i.e., without using MATLAB), we start by identifying the spectral characteristics of the four basic elements. These spectral characteristics are termed "Bode plot primitives." We then look at a couple of popular configurations of these elements. The arithmetic element (summation and subtraction) does not really have a spectrum, as it performs its operations the same way at all frequencies. Learning the spectra of the other elements is not difficult, but combining the various spectral elements into a spectrum that represents the overall system can be a bit tedious. For engineers, this depth is necessary as it results in increased understanding of systems and their design.

6.5.1 Constant Gain Element

The constant gain element shown in Figure 6.9 has already been introduced in previous examples. Since this element simply scales the input to produce the output, the transfer function for this element is a constant called the "gain." When the gain is greater than 1, the output is larger than the input and vice versa. In electronic systems, a gain element is called an "amplifier."

The transfer function for this element is:

$$TF(\omega) = G \tag{6.24}$$

The output of a gain element is $y(t) = G\,x(t)$ and hence only depends on the instantaneous and current value of t. Since the output does not depend on past values of time, it is called a "memoryless" element.

In the spectrum plot, the transfer function magnitude is usually plotted in dB so the transfer function magnitude equation is:

$$|TF(\omega)|_{dB} = |20 \log TF(\omega)| = 20 \log G \tag{6.25}$$

The magnitude spectrum of this element is a horizontal line at $20 \log G$. If there are other elements contributing to the spectral plot, it is easiest simply to rescale the vertical axis so that the former zero line equals $20 \log G$. This rescaling will be shown in the later examples.

The phase of this element is zero since the phase angle of a real constant is zero. The phase spectrum of this element plots as a horizontal line at a level of zero and makes no contribution to the phase plot.

$$\angle TF(\omega) = \angle G = 0 \tag{6.26}$$

6.5.2 Derivative Element

The derivative element, Figure 6.10A, has somewhat more interesting frequency characteristics than the constant gain element. The output of a derivative element depends on both the

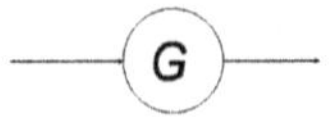

FIGURE 6.9   A constant gain element.

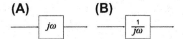

FIGURE 6.10   (A) A derivative element. (B) An integrator element.

current and past input values, so it is an example of an element with memory. This basic element is sometimes referred to as an "isolated zero" for reasons that become apparent later. The transfer function of this element is:

$$TF(\omega) = j\omega \qquad (6.27)$$

The magnitude of this transfer function in dB is $20 \log(|j\omega|) = 20 \log(\omega)$, which is a logarithmic function of ω that plots as a curve on a linear frequency axis, but as a straight line against log frequency. This is another reason to use dB against log frequency in spectral plots. To find the intercept, note that when $\omega = 1$, $20 \log(\omega = 1)$ equals 0. So the magnitude plot of this transfer function is a straight line that intersects the 0 dB line at $\omega = 1$ rad/s.

To find the slope of this line, note that when $\omega = 10$ the transfer function in dB equals 20 log (10) = 20 dB, and when $\omega = 100$ it equals 20 log (100) = 40 dB. So for every order of magnitude increase in frequency, there is a 20-dB increase in the value of Equation 6.27. This leads to unusual dimensions for the slope; specifically, 20 dB/decade. These units are due to the logarithmic scaling of the horizontal and vertical axes. The magnitude spectrum plot is shown in Figure 6.11 along with that of the integrator element described next.

The angle of $j\omega$ is +90 degrees irrespective of the value of ω:

$$\angle j\omega = 90 \text{ degrees.} \qquad (6.28)$$

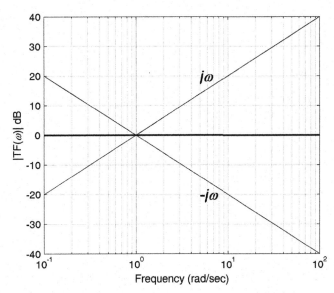

FIGURE 6.11   The upward sloping line is the magnitude spectrum of a derivative element, whereas the downward sloping line is the magnitude spectrum of an integrator element.

So the phase spectrum of this transfer function is just a straight line at $+90$ degrees. In constructing spectral plots manually, this term is usually not plotted, but is just used to rescale the phase plot after the phase characteristics of the other elements are plotted. (The same approach as used to deal with a gain element in the magnitude plot.)

6.5.3 Integrator Element

The last of the basic elements is the integrator element, Figure 6.10B, with a transfer function that is the inverse of the derivative element:

$$TF(\omega) = \frac{1}{j\omega} \tag{6.29}$$

This element also depends on the current and past values of the input, so is an element with memory. The integrator element is sometimes referred to as an "isolated pole," explained later. The magnitude spectrum in dB is just $20 \log \left| \frac{1}{j\omega} \right| = -20 \log |j\omega| = -20 \log$ (ω), which plots as a straight line when the frequency axis is in log ω. The line intercepts 0 dB at $\omega = 1$ since $-20 \log(1) = 0$ dB. This is similar to the derivative element described earlier, but with the opposite slope: -20 dB/decade. The magnitude plot of this transfer function is shown in Figure 6.11.

The phase spectrum of this transfer function is:

$$\angle \frac{1}{j\omega} = -90 \text{ degrees.} \tag{6.30}$$

Again this is usually not plotted since it is a straight line; rather the phase axis is rescaled after the plot is complete.

6.5.4 First-Order Element

We can now plot both the magnitude and phase spectra of four basic elements, and none of them is particularly exciting. However, if we start putting some of these elements together, we can construct more interesting, and useful, spectra. The "first-order element" can be constructed by placing a negative feedback path around a gain term and an integrator, Figure 6.12A. The gain term, which is just a constant, is given the symbol ω_1 for reasons that will be apparent.

The transfer function for a first-order element can be found easily from the feedback equation, Equation 6.7.

$$TF(\omega) = \frac{G}{1 + GH} = \frac{\omega_1 \left(\dfrac{1}{j\omega} \right)}{1 + \omega_1 \left(\dfrac{1}{j\omega} \right)} = \frac{\omega_1}{\omega_1 + j\omega} = \frac{1}{1 + j\dfrac{\omega}{\omega_1}} \tag{6.31}$$

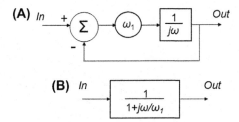

FIGURE 6.12 Constructing a first-order element for a gain term and integrator element in a feedback path. (A) A first-order term as a collection of elements. (B) The transfer function symbol for a first-order element.

This combination is usually presented as a single element as shown in Figure 6.12B. It is called a first-order element because the denominator of its transfer function is a first-order polynomial of the frequency variable, ω. This is also true of the integrator element (Equation 6.32), but it already has a name. Since the first-order element contains an integrator and an integrator element has memory, the first-order element also has memory; its output depends on current and past values of the input.

Finding the magnitude spectrum of the first- and higher-order elements uses graphical techniques based on transfer function asymptotes, an approach originally developed by Hendrik Bode in the 1930s, so the resulting spectral plots are called "Bode plots." First, the high-frequency asymptote and low-frequency asymptotes are plotted. Here high frequency means frequencies where $\omega \gg \omega_1$ and low frequency means frequencies where $\omega \ll \omega_1$. To find the low-frequency asymptote in dB, take the limit as $\omega \ll \omega_1$ of the absolute value of 20 log of Equation 6.31:

$$|TF(\omega_{low})| = \lim_{\omega \ll \omega_1} 20 \log \left| \frac{1}{1 + j\dfrac{\omega}{\omega_1}} \right| = 20 \log \left(\frac{1}{1 + j0} \right) = 0 \text{ dB} \tag{6.32}$$

The low-frequency asymptote given by Equation 6.32 plots as a horizontal line at 0 dB. The high-frequency asymptote is obtained when $\omega \gg \omega_1$:

$$|TF(\omega_{high})| = \lim_{\omega \gg \omega_1} \left[20 \log \left| \frac{1}{1 + j\dfrac{\omega}{\omega_1}} \right| \right] = 20 \log \left| \frac{1}{j\dfrac{\omega}{\omega_1}} \right| = -20 \log \left(\frac{\omega_1}{\omega} \right) \text{ dB} \tag{6.33}$$

The high-frequency asymptote, $-20 \log(\omega_1/\omega)$, is a logarithmic function of ω and plots as a straight line when frequency is plotted against log frequency (again, it is $\log(\omega_1/\omega)$ against $\log(\omega)$). It has the same slope as the integrator element: -20 dB/decade. This line intersects the 0 dB line at $\omega = \omega_1$ since $-20 \log(\omega_1/\omega_1) = -20 \log(1) = 0$ dB.

Often just plotting the asymptotes is enough to give us a general picture of an element's spectrum. Errors between the actual curve and the asymptotes occur when the asymptotic assumptions are no longer true; that is, when the frequency, ω, is neither much greater than, nor much less than, ω_1. The biggest error occurs when ω exactly equals ω_1. At that frequency the magnitude value is:

$$|TF(\omega = \omega_1)| = 20 \log \left| \frac{1}{1 + \dfrac{j\omega_1}{\omega_1}} \right| = 20 \log \left| \frac{1}{1 + j} \right| = -20 \log \left(\sqrt{2} \right) = -3 \text{ dB} \qquad (6.34)$$

From our definition of bandwidth in Chapter 4, recall that the "cutoff frequency" or "break frequency" occurs when the nominal spectral value is reduced by 3 dB. To approximate the spectrum of the first-order element, we first plot the high- and low-frequency asymptotes as shown in Figure 6.13 (dashed lines). Then we identify the cutoff frequency or −3 dB point, which occurs at $\omega = \omega_1$. Finally, we draw a freehand curve from asymptote to asymptote through the −3 dB, Figure 6.13, solid line. In Figure 6.13, this curve was actually plotted using MATLAB, but you could do almost as well freehand. The deviation of this curve from the two asymptotes is quite small.

The phase plot of the first-order element can be estimated using the same approach: taking the asymptotes and the worst case point, $\omega = \omega_1$. The low-frequency asymptote is:

$$\angle TF(\omega_{low}) = \lim_{\omega \ll \omega_1} \left[\angle \left(\frac{1}{1 + j \dfrac{\omega}{\omega_1}} \right) \right] = \angle -1 = 0 \text{ degrees.} \qquad (6.35)$$

This is a straight line at 0 degrees. In this case, the assumption "much much less than" is taken to mean one order of magnitude less, so that the low-frequency asymptote is assumed

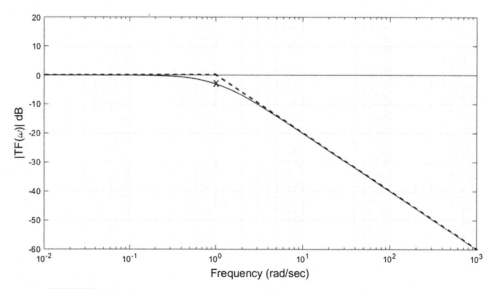

FIGURE 6.13   The magnitude spectrum of a first-order element where $\omega_1 = 1$ rad/s.

to be valid from $0.1\omega_1$ down to zero frequency. The high-frequency asymptote is determined to be -90 degrees in Equation 6.36, and by the same reasoning is assumed to be valid from $10\omega_1$ to all higher frequencies.

$$\angle TF(\omega_{high}) = \lim_{\omega \gg \omega_1} \left[\angle \left(\frac{1}{1 + j\frac{\omega}{\omega_1}} \right) \right] = \angle \left(\frac{1}{j\frac{\omega}{\omega_1}} \right) = -90 \text{ degrees.} \quad (6.36)$$

Again the greatest difference between the asymptotes and the actual curve is when ω equals ω_1:

$$\angle (TF(\omega = \omega_1)) = \angle \left(\frac{1}{1 + j\frac{\omega_1}{\omega_1}} \right) = \angle \left(\frac{1}{1 + j} \right) = -45 \text{ degrees.} \quad (6.37)$$

This value, -45 degrees, is exactly halfway between the high- and low-frequency asymptotes. Usually, a straight line is drawn between the high end of the low-frequency asymptote at $0.1\omega_1$ and the low end of the high-frequency asymptote at $10\omega_1$ passing through 45 degrees. Although the phase curve is nonlinear in this range, the error induced by a straight line approximation is small as shown in Figure 6.14. To help you to plot these spectral curves by hand, the associated files contain semilog.pdf, which can be printed out to produce semilog graph paper.

There is a variation of the transfer function described in Equation 6.31 that, although not as common, is sometimes found as part of the transfer function of other elements. One such variation has the first-order polynomial in the numerator but nothing in the denominator:

$$TF(\omega) = 1 + j\frac{\omega}{\omega_1} \quad (6.38)$$

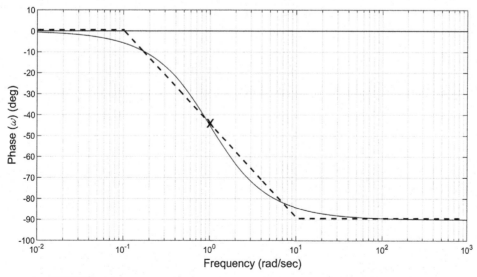

FIGURE 6.14 The phase spectrum of a first-order element where $\omega_1 = 1$ rad/s.

The asymptotes for this transfer function are very similar to that of Equation 6.31; in fact, the low-frequency asymptote is the same.

$$|TF(\omega_{low})| = \lim_{\omega \ll \omega_1} 20 \log\left|1 + j\frac{\omega}{\omega_1}\right| = 20 \log(1) = 0 \text{ dB} \tag{6.39}$$

The high-frequency asymptote has the same intercept, but the slope is positive, not negative:

$$|TF(\omega_{high})| = \lim_{\omega \gg \omega_1}\left[20 \log\left|1 + j\frac{\omega}{\omega_1}\right|\right] = 20 \log\left|j\frac{\omega}{\omega_1}\right| = 20 \log\left(\frac{\omega}{\omega_1}\right) \text{ dB} \tag{6.40}$$

Similarly the value of the transfer function when $\omega = \omega_1$ is $+3$ dB instead of -3 dB:

$$|TF(\omega = \omega_1)| = 20 \log\left|1 + \frac{j\omega_1}{\omega_1}\right| = 20 \log|1 + j| = 20 \log\left(\sqrt{2}\right) = 3 \text{ dB} \tag{6.41}$$

The phase plots are also the same except that the change in angle with frequency is positive rather than negative. Demonstration of this is left as an exercise in the problems at the end of this chapter. The magnitude and phase spectra of the transfer function given in Equation 6.38 is shown in Figure 6.15.

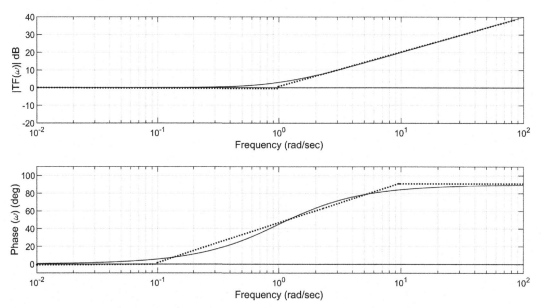

FIGURE 6.15   The magnitude and phase spectra of the transfer function given in Equation 6.38. These spectra are similar to those of the standard first-order process, Figures 6.13 and 6.14, except both curves go up instead of down at the higher frequencies. $\omega_1 = 1$ rad/s.

EXAMPLE 6.8

Plot the magnitude and phase spectrum of the following transfer function:

$$TF(\omega) = \frac{100}{(1 + j0.1\omega)}$$

Solution: This transfer function is actually a combination of two elements: a gain term of 100 and a first-order term where ω_1 is equal to $1/0.1 = 10$. The easiest way to plot the spectrum of this transfer function is to plot the first-order term first, then rescale the vertical axis of the magnitude spectrum so that 0 dB is equal to $20 \log(100) = 40$ dB. The rescaling accounts for the gain term. The gain term has no influence on the phase plot.

Results: Figure 6.16 shows the resulting magnitude and phase spectral plots. For the magnitude plot the spectrum is drawn freehand through the -3 dB point. For both plots, the asymptotes alone provide a sufficiently accurate picture of the system's spectra.

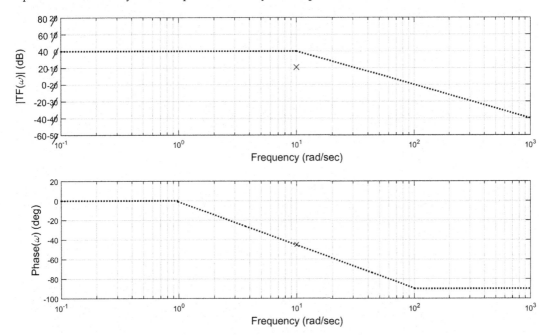

FIGURE 6.16 The magnitude and phase spectra of the system described in Example 6.8. Only the asymptotes are plotted and they provide an adequate picture of the system's spectra.

6.5.5 Second-Order Element

A second-order element has, as you might expect, a transfer function that includes a second-order polynomial of ω. A second-order element contains two integrators so it also has memory. Again that means that its output at any given time depends on both current and past values of the input. One way to construct a second-order system is to put two first-order systems in series, but other configurations also produce a second-order equation as long as they contain two integrators. For example, the system shown in Figure 6.6, and the transfer function plotted in Example 6.6, have a second-order ω terms in the denominator. The equation for a general second-order system has the format:

$$TF(\omega) = \cfrac{1}{1 - \left(\cfrac{\omega}{\omega_n}\right)^2 + j\cfrac{2\delta\omega}{\omega_n}} \tag{6.42}$$

where ω_n and δ are constants associated with the second-order polynomial. The format of this second-order equation may seem strange, but the two constants, ω_n and δ, have direct relationships to both the time behavior and the spectrum. The parameter δ is called the "damping factor" and ω_n the "undamped natural frequency." The rational for these strange names comes from the time-domain behavior and is described in Chapter 7. But ω_n and δ also relate directly to a system's spectra; in fact, sometimes they are all we need for an overview of the system spectral characteristics. Determining these constants from a typical second-order equation is the motivation of the next example.

EXAMPLE 6.9

Find the constants ω_n and δ in the following second-order transfer function:

$$TF(\omega) = \frac{1}{1 - .03\omega^2 + j.01\omega} \tag{6.43}$$

Solution/Results: We obtain the values of constants ω_n and δ by equating coefficients between the denominator terms of Equations 6.42 and 6.43:

$$\left(\frac{1}{\omega_n}\right)^2 = 0.03; \quad \omega_n = \frac{1}{\sqrt{0.03}} = 5.77 \text{ rad/s}$$

$$\frac{2\delta}{\omega_n} = 0.1; \quad \delta = \frac{0.01\omega_n}{2} = \frac{(5.77)0.01}{2} = 0.029$$

If the roots of the denominator polynomial are real, it can be factored into two first-order terms, then separated into two first-order transfer functions using partial fraction expansion. This is done in Chapter 7, but for spectral plotting the second-order term will be dealt with as is.

Not surprisingly, the second-order element is more challenging to plot than the previous transfer functions. Basically the same strategy is used as in the first-order function except special care must be taken with the worst case error at frequency $\omega = \omega_n$. We begin by finding the high- and low-frequency asymptotes. These now occur when ω is either much greater or much less than ω_n. When $\omega << \omega_n$, both the ω and the ω^2 terms go to 0 and the denominator goes to 1.0:

$$|TF(\omega_{low})| = \lim_{\omega \ll \omega_n} \left[20 \log \left| \cfrac{1}{1 - \left(\cfrac{\omega}{\omega_n}\right)^2 + j\cfrac{2\delta\omega}{\omega_n}} \right| \right] = 20 \log(1) = 0 \text{ dB} \tag{6.44}$$

The low-frequency asymptote is the same as for the first-order element, namely, a horizontal line at 0 dB.

The high-frequency asymptote occurs when $\omega >> \omega_n$ where the ω^2 term in the denominator dominates:

$$|TF(\omega_{high})| = \lim_{\omega \gg \omega_n} \left[20 \log \left| \frac{1}{1 - \left(\frac{\omega}{\omega_n}\right)^2 + j\frac{2\delta\omega}{\omega_n}} \right| \right] = 20 \log \left(\frac{1}{\frac{\omega}{\omega_n}}\right)^2 = -40 \log \left(\frac{\omega_n}{\omega}\right) \text{ dB}$$

(6.45)

The high-frequency asymptote is also similar to that of the first-order element, but with double the downward slope: -40 dB/decade instead of -20 dB/decade.

A major difference between the first- and second-order terms occurs when $\omega = \omega_n$:

$$|TF(\omega = \omega_n)| = 20 \log \left| \frac{1}{1 - \left(\frac{\omega}{\omega_n}\right)^2 + j\frac{2\delta\omega}{\omega_n}} \right| = 20 \log \left| \frac{1}{j2\delta} \right| = -20 \log(2\delta) \text{ dB} \qquad (6.46)$$

The magnitude spectrum at $\omega = \omega_n$ is not a constant, but depends on δ: specifically, $-20 \log(2\delta)$. In fact, the value of δ can radically alter the shape of the magnitude curve and must be taken into account when plotting. If δ is less than 0.5, then $\log(2\delta)$ will be negative and the transfer function at ω_n will be positive. Usually the magnitude plot is determined by first plotting the asymptotes, then calculating and plotting the TF's value at $\omega = \omega_n$ using Equation 6.46, and finally drawing a curve freehand through the ω_n point, converging smoothly with the high- and low-frequency asymptotes above and below ω_n. The freehand curve can sometimes take some artistic skill. Nonetheless, the spectrum is wholly determined by the two constants, ω_n and δ. The second-order magnitude plot is shown in Figure 6.17 for several values of δ.

The phase plot of a second-order system is also approached using the asymptote method. For phase angle the high- and low-frequency asymptotes are given as:

$$\angle TF(\omega_{low}) = \lim_{\omega \ll \omega_n} \angle \left[\frac{1}{1 - \left(\frac{\omega}{\omega_n}\right)^2 + j\frac{2\delta\omega}{\omega_n}} \right] = \angle 1 = 0 \text{ degrees.} \qquad (6.47)$$

$$\angle TF(\omega_{high}) = \lim_{\omega \gg \omega_n} \angle \left[\frac{1}{1 - \left(\frac{\omega}{\omega_n}\right)^2 + j\frac{2\delta\omega}{\omega_n}} \right] = \angle \left(\frac{1}{-\omega^2}\right) = -180 \text{ degrees.} \qquad (6.48)$$

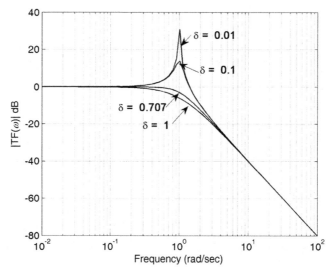

FIGURE 6.17   Magnitude spectra of second-order systems with different values of δ. For all spectra $\omega_n = 1.0$ rad/s.

This is similar to the asymptotes of the first-order process except the high-frequency asymptote is at -180 degrees instead of -90 degrees. The phase angle when $\omega = \omega_n$ can easily be determined:

$$\angle TF(\omega_n) = \angle \left(\frac{1}{1 - (\omega_n/\omega_n)^2 + j\dfrac{2\delta\omega_n}{\omega_n}} \right) = \angle \left(\frac{1}{j2\delta} \right) = -90 \text{ degrees.} \qquad (6.49)$$

So the phase at $\omega = \omega_n$ is -90 degrees, halfway between the two asymptotes. Unfortunately, the shape of the phase curve between $0.1\omega_n$ and $10\omega_n$ is a function of δ and can no longer be approximated as a straight line except at larger values of δ. Phase curves are shown in Figure 6.18 for the same range of values of δ used in Figure 6.17. The curves for low values of δ have steep transitions between 0 and -180 degrees, whereas the curves for high values of δ have gradual slopes approximating the phase characteristics of a first-order element (except for the 180-degree phase change). Hence if δ is 2.0 or more, a straight line between the low-frequency asymptote at $0.1\omega_n$ and the high-frequency asymptote at $10\omega_n$ works well. If δ is much less than 2.0, the best that can be done using this manual method is to approximate, freehand, the appropriate curve (or approximately appropriate curve) in Figure 6.18.

EXAMPLE 6.10

Plot the magnitude and phase spectra of the second-order transfer function used in Example 6.9 (i.e., Equation 6.43). Confirm the results obtained manually with those obtained from MATLAB.[5]

Solution: The equivalent values of ω_n and δ are found in Example 6.9 to be 5.77 rad/s and 0.029, respectively. The magnitude and phase spectra can be plotted directly using the asymptotes, noting that the magnitude spectrum equals $-20 \log(2\delta) = +24.7$ dB when $\omega = \omega_n$.

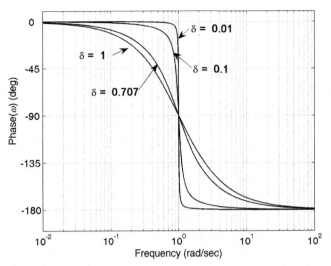

FIGURE 6.18   Phase spectra of second-order systems with different values of δ. The values of δ are the same as used in Figure 6.17 and $\omega_n = 1.0$ rad/s.

To plot the transfer function using MATLAB, first construct the frequency vector. Given that $\omega_n = 6.77$ rad/s, use the frequency vector range between $0.01\omega_n = 0.0677$ and $100\omega_n = 677$ rad/s in increments of, say, 0.1 rad/s. Define the function in MATLAB and for the magnitude spectrum, take the 20 times log magnitude and for the phase spectrum take the angle in degrees. Plot these functions against frequency in radians using a semilog plot.

```
% Example 6.10  Use MATLAB to plot the transfer function given in Equation 6.42
%
w = .067:.1:667;                          % Define frequency vector
wn = 5.77;                                % Define wn
delta = 0.029;                            % Define delta
TF = 1./(1 - (w/wn).^2 + j*2*delta*w/wn); % Transfer function
Mag = 20*log10(abs(TF));                  % Magnitude in dB
Phase = angle(TF)*360/(2*pi);             % Phase in deg
subplot(2,1,1);
  semilogx(w,Mag,'k');                    % Plot as log frequency
  .......labels.......
subplot(2,1,2);
  semilogx(w,Phase,'k');
  .......labels........
```

Results: The asymptotes, the point where $\omega = \omega_n$, and freehand spectra are shown in Figure 6.19, and the results of the MATLAB code are shown in Figure 6.20. The ease of plotting the transfer function using MATLAB is seductive, but the manual method provides more insight into the system. More importantly, it also allows us to go in the reverse direction: from a spectrum to an equivalent transfer function as shown later. This also aids in the design of systems having particular spectral characteristics.

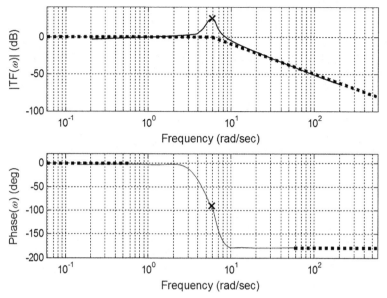

FIGURE 6.19 Magnitude and phase spectra of the system defined by the transfer function equation, Equation 6.42. The *solid line* was drawn freehand using the asymptotes (*dotted lines*) and the points where $\omega = \omega_n$.

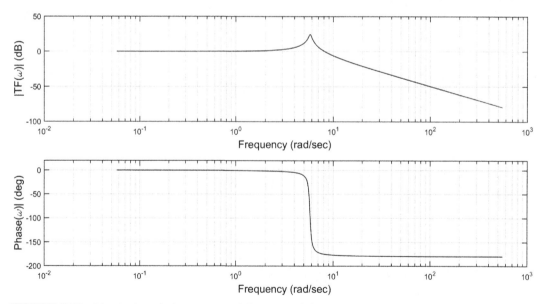

FIGURE 6.20 Magnitude and phase spectra of the system defined by the transfer function equation, Equation 6.43, as determined by MATLAB. These spectra approximately match those constructed manually shown in Figure 6.19.

[5]MATLAB has a function, `bode.m`, that plots the spectral characteristics of any linear system. The system is defined by its transfer using another MATLAB routine, `tf.m`. Both these routines are in the Control System Toolbox. However, as shown here, it is easy to generate Bode plots using standard MATLAB.

Occasionally the second-order term is found in the numerator, usually as part of a more complex transfer function. In this case, both the magnitude and phase spectra are inverted so the phase spectrum goes from 0 to +180 degrees passing through +90 degrees when $\omega = \omega_n$. Modifying the equations to demonstrate this is an exercise in the problems.

Table 6.1 summarizes the various elements, which along with their magnitude and phase characteristics, are known as "Bode plot primitives." This table also lists alternate names for these elements where applicable. Table 6.2 brings together the magnitude and phase characteristics of these Bode plot primitives.

6.6 BODE PLOTS COMBINING MULTIPLE ELEMENTS

We now know how to find and plot the spectra of some basic, and not so basic, systems even without a computer. But what of more complicated systems? We can deal with these using an extension of what we already learned: first use our Bode plot techniques to plot the spectra of individual elements (or systems), then combine those spectra graphically.

As always, we start with the transfer function, then assume it can be factored into combinations describing the elements we have already covered (i.e., the Bode plot primitives summarized in Table 6.2). Under this assumption, the transfer function of any system, no matter how complicated, can be written as:

$$TF(\omega) = \frac{Gj\omega\left(1+j\dfrac{\omega}{\omega_1}\right)\left(1-\left(\dfrac{\omega}{\omega_{n1}}\right)^2+j\dfrac{2\delta_1\omega}{\omega_{n1}}\right)\cdots}{j\omega\left(1+j\dfrac{\omega}{\omega_2}\right)\left(1-\left(\dfrac{\omega}{\omega_{n2}}\right)^2+j\dfrac{2\delta_2\omega}{\omega_{n2}}\right)\cdots} \tag{6.50}$$

TABLE 6.1 Transfer Function Elements

Basic Equation	Name(s)	
	Numerator	Denominator
G	Constant or gain	—
$j\omega$	Differentiator[a] Isolated zero	Integrator[a] Isolated pole
$1 + j\omega/\omega_1$	Real zero or just zero Lead element	First-order element[a] Real pole or just pole Lag element
$1 - (\omega/\omega_n)^2 + j2\delta\omega/\omega_n$	Complex zeros[b]	Complex poles[b] Second-order element[a]

[a]*Name most commonly used in this text.*
[b]*Depends on the values of* δ.

TABLE 6.2   Bode Plot Primitives

Denominator Term	Magnitude Plot	Phase Plot
Constant Gain	20 log G	—
$j\omega$		-90 degrees

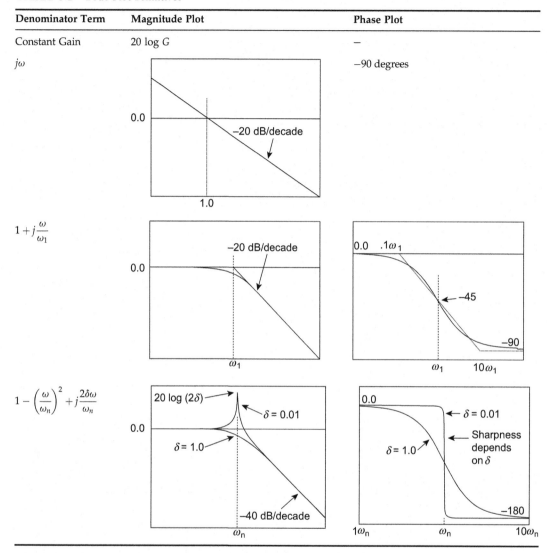

Admittedly, polynomial factoring can be tedious, but it is easily accomplished using MATLAB's `roots` routine.[6] An example involving a fourth-order denominator polynomial is found at the end of the next chapter. The constants for first- and second-order elements

[6]Why would we first use MATLAB to factor a higher-order polynomial, then proceed manually with our graphical approach? Why not just plot the system's spectrum straight away in MATLAB? Plotting the transfer function would be much easier, but designing a system to duplicate that transfer function would be much more difficult without knowing the individual elements. You would be unaware of these more basic elements embedded in that higher-order transfer function.

were described earlier. Note that this is a general equation illustrating the various possible element combinations; if the same elements actually appear in both the numerator and denominator (e.g., the $j\omega$ term), of course they would cancel.

To convert Equation 6.50 to magnitude in dB, we take 20 log of the absolute value:

$$TF(\omega) = 20\log\left|\frac{Gj\omega\left(1+j\dfrac{\omega}{\omega_1}\right)\left(1-\left(\dfrac{\omega}{\omega_{n1}}\right)^2+j\dfrac{2\delta_1\omega}{\omega_{n1}}\right)\cdots}{j\omega\left(1+j\dfrac{\omega}{\omega_2}\right)\left(1-\left(\dfrac{\omega}{\omega_{n2}}\right)^2+j\dfrac{2\delta_2\omega}{\omega_{n2}}\right)\cdots}\right| \tag{6.51}$$

Note that multiplication and division in Equation 6.51 become addition and subtraction after taking the log, so in dB Equation 6.51 can be expanded to:

$$|TF(\omega)|_{dB} = \overbrace{20\log(G)} + \overbrace{20\log|j\omega|} + \overbrace{20\log|1+j\omega/\omega_1|} + \overbrace{20\log\left|1-\left(\omega/\omega_{n1}\right)^2+j2\delta_1\omega/\omega_{n1}\right|} + \ldots$$
$$\underbrace{-\,20\log|j\omega|} \underbrace{-\,20\log|1+j\omega/\omega_2|} \underbrace{-\,20\log\left|1-\left(\omega/\omega_{n2}\right)^2+j2\delta_2\omega/\omega_{n2}\right|} + \ldots. \tag{6.52}$$

where the first line in Equation 6.52 is the expanded version of the numerator and the second line is the expanded version of the denominator. Aside from the constant term, the first and second lines in Equation 6.52 have the same form except for the sign: numerator terms are positive and denominator terms are negative. Each term in the summation is one of the element types described in the last section as emphasized by the horizontal brackets. This shows that the magnitude spectrum of any transfer function can be plotted by plotting each individual element and then adding the spectral curves graphically.

Adding the individual spectra is not as difficult as it first appears. Usually only the asymptotes and a few other important points (such as the value of a second-order term at $\omega = \omega_n$) are plotted, then the overall curve is completed freehand by connecting the asymptotes and critical points. Aiding this procedure is the fact that most real transfer functions do not contain a large number of elements. Although the resulting Bode plot is only approximate and often somewhat crude, it is usually sufficient to represent the general spectral characteristics of the transfer function.

The phase portion of the transfer function can also be dissected into individual components:

$$\angle TF(\omega) = \frac{\angle\left[Gj\omega\left(1+j\dfrac{\omega}{\omega_1}\right)\left(1+\left(\dfrac{\omega}{\omega_{n1}}\right)^2+j\dfrac{2\delta_1\omega}{\omega_{n1}}\right)\right]}{\angle\left[j\omega\left(1+j\dfrac{\omega}{\omega_2}\right)\left(1+\left(\dfrac{\omega}{\omega_{n2}}\right)^2+j\dfrac{2\delta_2\omega}{\omega_{n2}}\right)\right]} \tag{6.53}$$

By the rules of complex arithmetic, the angles of the individual elements simply add if they are in the numerator or subtract if they are in the denominator:

$$\angle TF(\omega) = \angle G + \angle j\omega + \angle\left(1 + \frac{j\omega}{\omega_1}\right) + \angle\left(1 + \left(\frac{\omega}{\omega_{n1}}\right)^2 + j2\delta_1\frac{\omega}{\omega_{n1}}\right)$$
$$- \angle j\omega - \angle\left(1 + j\frac{\omega}{\omega_2}\right) - \angle\left(1 + \left(\frac{\omega}{\omega_{n2}}\right)^2 + j2\delta_2\frac{\omega}{\omega_{n2}}\right) \tag{6.54}$$

So the phase transfer function also consists of individual components (emphasized by the brackets) that correspond to the Bode plot primitives shown in Table 6.2. Again these components add or subtract depending on whether they are in the numerator or denominator of the transfer function. The phase spectrum of any general transfer function can be constructed manually by plotting the spectrum of each component and adding the curves graphically. Again, we usually start with the asymptotes and fill in freehand; like the magnitude spectrum the result is a sometimes crude, but usually adequate, approximation of the phase spectrum. Often only the magnitude plot is of interest and it is not necessary to construct the phase plot.

The next two examples demonstrate the Bode plot approach for transfer functions of increasing complexity.

EXAMPLE 6.11

Plot the magnitude and phase spectra using Bode plot methods for the transfer function:

$$TF(\omega) = \frac{100j\omega}{(1 + j1\omega)(1 + j.1\omega)} \tag{6.55}$$

Solution: The transfer function contains four elements: a constant, an isolated zero (i.e., $j\omega$ in the numerator), and two first-order terms in the denominator. For the magnitude curve, plot the asymptotes for all but the constant term, then add these asymptotes together graphically to get an overall asymptote. Finally, use the constant term to scale the value of the vertical axis. For the phase plot, construct the asymptotes for the two first-order denominator elements, then rescale the axis by $+90$ degrees to account for the $j\omega$ in the numerator. Recall that the constant term does not contribute to the phase plot.

The general form for the two first-order elements is $\frac{1}{1 + j\frac{\omega}{\omega_1}}$ where ω_1 is the point where the high-

and low-frequency asymptotes intersect. In this transfer function, ω_1 and ω_2 are $1/1 = 1.0$ and $1/0.1 = 10$ rad/s. Figure 6.21 shows the asymptotes obtained for the two first-order primitives plus the $j\omega$ primitive. Note that the asymptotes can be determined directly from the transfer function; it is not necessary to rewrite the equation into the format of Equation 6.54.

Results: Graphically adding the three asymptotes shown in Figure 6.21 gives the curve consisting of three straight lines shown in Figure 6.22. Note that the numerator and denominator asymptotes cancel out at $\omega = 1.0$ rad/s, so the overall asymptote is flat until the additional downward asymptote comes in at $\omega = 10$ rad/s. The actual magnitude transfer function is also shown in Figure 6.23 and closely follows the overall asymptote. A small error (3 dB) is seen at the two breakpoints: $\omega = 1.0$ and $\omega = 10$. A final step in constructing the magnitude curve is to rescale the

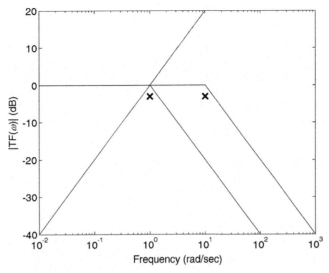

FIGURE 6.21 The magnitude spectrum asymptotes for three of the elements in the transfer function equation, Equation 6.55. Also shown are the two breakpoints where $\omega = \omega_1$ and $\omega = \omega_2$. The next step is to add these three curves graphically resulting in the overall spectrum shown in Figure 6.22.

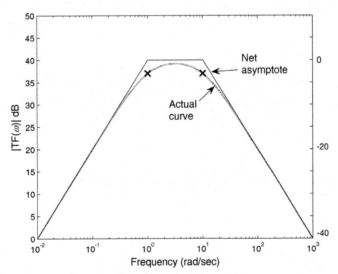

FIGURE 6.22 The *solid line* is the algebraic summation of asymptotes shown in Figure 6.23. The *dotted line* is the actual magnitude spectrum of the transfer function given by Equation 6.5. The right-hand axis shows the original scaling and the left-hand axis shows the rescaling due to the gain term. The actual curve was drawn by MATLAB, but does not differ significantly from the asymptotes. A hand-drawn curve would likely have been just as good.

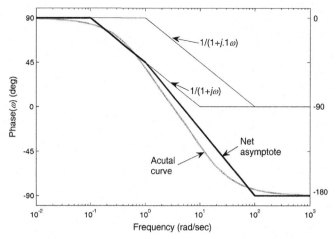

FIGURE 6.23 The *light lines* show the phase spectrum asymptotes for two elements found in the transfer function Equation 6.52. The *heavy line* is the summation of these two asymptotes and the *light curve* is the actual phase spectrum.

vertical axis so that 0 dB corresponds to 20 log (100) = 40 dB. (The original axis values are shown on the right vertical axis, whereas the rescaled values are shown on the left.)

The asymptotes of the phase curve are shown in Figure 6.23 along with the overall asymptote that is obtained by graphical addition. Also shown is the actual phase curve, which, as with the magnitude curve, closely follows the overall asymptote. As a final step the vertical axis of this plot has been rescaled by +90 degrees on the right side to account for the $j\omega$ term in the numerator.

In both the magnitude and phase plots, the actual curves follow the overall asymptote fairly closely for transfer functions that have terms no higher than first order. Tracing freehand through the −3 dB points further improves the match, but often the asymptotes are sufficient. As we see in the next example, this is not true for transfer functions that contain second-order terms, at least when the damping factor, δ, is small.

EXAMPLE 6.12

Find the magnitude and phase spectra using both Bode plot primitives and MATLAB code for the transfer function.

$$TF(\omega) = \frac{10(1 + j2\omega)}{j\omega(1 - 0.04\omega^2 + j0.04\omega)} \qquad (6.56)$$

Solution: This transfer function contains four elements: a gain constant, a numerator first-order term, an isolated pole (i.e. $j\omega$) in the denominator, and a second-order term in the denominator. For the magnitude curve, plot the asymptotes for all primitives except the constant, then add these up graphically. Lastly, use the constant term to scale the value of the vertical axis. For the phase curve, plot the asymptotes of the first-order and second-order terms, then rescale the vertical axis by

90 degrees to account for the $j\omega$ term in the denominator. To plot the magnitude asymptotes, it is first necessary to determine ω_1, ω_n, and δ from the associated coefficients:

$$\omega_1 : \frac{1}{\omega_1} = 2; \quad \omega_1 = 0.5 \text{ rad/s}$$

$$\omega_n : \frac{1}{\omega_n^2} = 0.04; \quad \omega_n = \frac{1}{\sqrt{0.04}} = 5 \text{ rad/s}$$

$$\delta : \frac{2\delta}{\omega_n} = 0.04; \quad \delta = \frac{0.04\omega_n}{2} = \frac{0.04(5)}{2} = 0.1$$

$$-20 \log(2\delta) = 14 \text{ dB}$$

Results: Note that the second-order term is positive 14 dB when $\omega = \omega_n$. This is because 2δ is less than 1 and so the log is negative and the two negatives make a positive. Using these values and including the asymptotes leads to the magnitude plot shown in Figure 6.24.

The log frequency scale can be a little confusing (particularly without grid lines, which are left off to improve visibility), so the positions of 0.5 and 5.0 rad/s are indicated by arrows in Figure 6.24. The overall asymptote and the actual curve for the magnitude transfer function are shown in Figure 6.25. The vertical axis has been rescaled by 20 dB to account for the constant term in the numerator. Note that the actual magnitude spectrum (light dashed line) goes higher than +14 dB because of the contribution of the first-order numerator term. The net asymptote (heavy dashed line) at $\omega = \omega_n$ is approximately 5 dB, so the actual curve, not including the rescaling, peaks around $14 + 5 = 19$ dB. Adding in the constant 20 dB brings the peak up to 39 dB.

The individual phase asymptotes are shown in Figure 6.26, whereas the overall asymptote and the actual phase curve are shown in Figure 6.27. Note that the actual phase curve is quite different from the overall asymptote, again because the second-order term has a small value for δ. This low

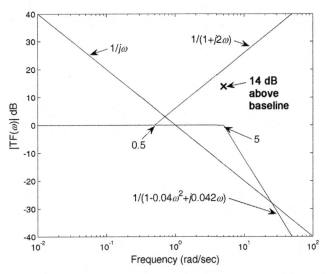

FIGURE 6.24 The magnitude spectrum asymptotes for three of the elements in the transfer function Equation 6.56. Also shown is the point where $\omega = \omega_n$, the undamped natural frequency of the second-order element. The next step is to add these three curves graphically resulting in the overall spectrum shown in Figure 6.25.

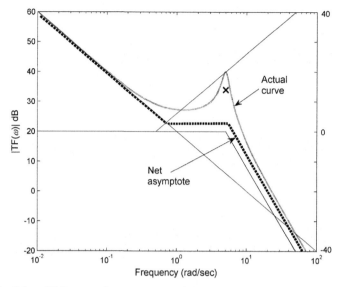

FIGURE 6.25   The *light solid lines* are the asymptotes of the magnitude spectra of the individual elements of Equation 6.55. The *heavy dashed line* is the algebraic sum of the individual asymptotes and the *light dashed line* is the actual magnitude spectrum. Because of the second-order term, the magnitude spectrum deviates significantly from the net asymptote in the region of $\omega = \omega_n$. The left vertical axis has been rescaled by 20 dB to account for the constant gain term.

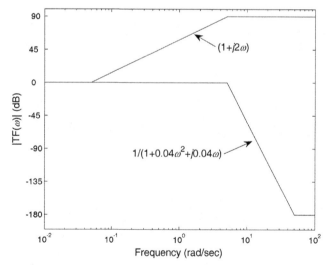

FIGURE 6.26   The asymptotes of the phase spectra of two of the individual elements of Equation 6.56. The algebraic summation of these asymptotes is shown in Figure 6.27.

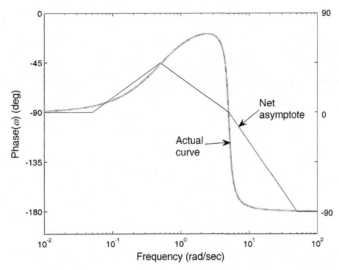

FIGURE 6.27 The *solid line* is the algebraic sum of the individual asymptotes shown in Figure 6.26 and the *light dashed line* is the actual magnitude spectrum. Because of the second-order term, the magnitude spectrum deviates significantly from the net asymptote in the region of $\omega = \omega_n$.

value of δ means the phase curve has a very sharp transition between 0 and -180 degrees and will deviate substantially from the asymptote. The vertical axis is rescaled by -90 degrees to account for the $j\omega$ term in the denominator. The original axis is shown on the right side.

Results, MATLAB: The MATLAB code required for this example is a minor variation of that required in previous examples. Only the range of the frequency vector and the MATLAB statement defining the transfer function have been changed. Note that the point-by-point operators, .* and ./, are required whenever vector terms are multiplied. The plots produced by this code are shown as light dashed lines in Figures 6.26 and 6.27.

```
% Example 6.12  Use MATLAB to plot the transfer function given in this example
%
w = .005:.1:500;
w1 = 0.5;                % First-order cutoff freq (rad)
wn = 5;                  % Second-order undamped resonant freq. (rad)
delta = 0.1;             % Damping factor
TF  =  10*(1+j*w/w1)./(j*w.*(1  -  (w/wn).2  +  j*2*delta*w/wn));   % Transfer
function

......the rest of the code is the same as Example 6.7.......
```

The code in this example could easily be modified to plot a transfer function of any complexity as some of the problems demonstrate. The Bode plot approach may seem like a lot of effort to achieve a graph that could be done better with a couple of lines of MATLAB code. However, these techniques based on Bode plot primitives provide us with a mental map that links the transfer function to the system's spectrum. These primitives will guide us when we go the other way, from spectrum (perhaps derived experimentally) to transfer function.

6.6.1 Constructing the Transfer Function From the System Spectrum

Bode plot primitives can also be used to derive the transfer function given a desired frequency response as illustrated in the next example.

EXAMPLE 6.13

Find the transfer function of a system that has the magnitude spectrum shown in Figure 6.28.

Solution: We start by identifying some basic features of the spectrum. From our knowledge of Bode plot primitives we note that the high-frequency portion of the curve looks like a second-order element with an undamped natural frequency (i.e., ω_n) of 100 rad/s. The baseline is 20 dB so $G = 10$ and the peak rises 20 dB above the baseline (from 20 to 40 dB), so the damping factor is:

$$-20 \log(2\delta) = 20 \text{ dB}; \quad -\log(2\delta) = 1.0; \quad 1/(2\delta) = 10^1; \quad \delta = 1/20 = 0.05$$

So a partial transfer function that accounts for the high-frequency spectrum is:

$$TF(\omega) = \frac{10}{1 - \dfrac{\omega^2}{10^4} + j\dfrac{0.1\omega}{100}}$$

The low-frequency spectrum looks like the lower frequencies of the spectrum in Example 6.11. In that example, the low-frequency curve was constructed by a combination of a $j\omega$ in the numerator and a $(1 + j\omega/\omega_1)$ in the denominator where ω_1 is the cutoff or break frequency. In Figure 6.28, the break frequency appears to be around 0.1 rad/s. This gives the partial transfer function:

$$TF(\omega) = \frac{\dfrac{j\omega}{0.1}}{1 + j\dfrac{\omega}{0.1}}$$

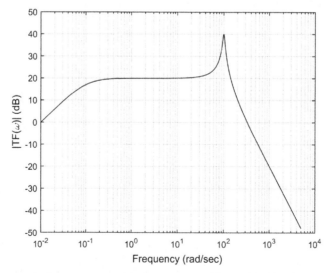

FIGURE 6.28   The magnitude spectrum used in Example 6.13. The goal in this example is to derive a transfer function that will give us this magnitude spectrum.

Result: Combining the two partial transfer functions gives:

$$TF(\omega) = \frac{100j\omega}{(1 + j10\omega)(1 - 10^{-4}\omega^2 + j10^{-3}\omega)} \tag{6.57}$$

Plotting this function will give the spectrum shown in Figure 6.28 as confirmed in one of the problems. Once we derive the transfer function, it is only one more step to design a system that has this transfer function, a step that is touched on later in this book. More relevant to us as bioengineers is that, if we can determine the frequency characteristics of a physiological process experimentally, we can use Bode plot primitives to determine its transfer function. From this transfer function, we can develop a quantitative model for this system. Moreover, we can predict the response of this biosystem to a wide range of input stimuli by combining the transfer function with Fourier decompositions. Examples of this approach are given in the next section.

6.7 THE TRANSFER FUNCTION AND THE FOURIER TRANSFORM

Combining the transfer function with the Fourier transform is a no-brainer: it allows you to predict the output to any signal that can be decomposed into a sinusoidal series. The approach is straightforward: (1) decompose the input signal into a Fourier series using the complex representation; (2) at each frequency, find the output signal by multiplying the input sinusoid by the transfer function; and (3) construct the output waveform from the output sinusoidal series using the inverse Fourier transform. Although this seems like a roundabout approach, it is really easy to do in MATLAB. In Step 1, we keep the complex sinusoidal representation as real and imaginary components rather than converting to magnitude and phase. Then in Step 2, we simply multiply the complex spectrum by the complex transfer function to get a representation of the output sinusoidal series, also complex. For Step 3, we apply the inverse Fourier transform directly to the output of Step 2 to get our output signal. This approach is illustrated in Figure 6.29 and implemented in the next example. More examples are explored in later chapters. Note that this decomposition approach invokes the principle of superposition since it assumes the system responds the same way to each sinusoidal component whether presented in isolation or as a combined signal.

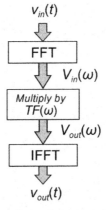

FIGURE 6.29   The three steps involved in predicting the output signal, $v_{out}(t)$, of any system, to any input signal, $v_{in}(t)$, given the transfer function of the system, $TF(\omega)$.

EXAMPLE 6.14

Find the output of the system having the following transfer function when the input is the electroencephalography (EEG) signal shown Figure 1.7 and found as variable eeg in file eeg data.mat. Plot both input and output signals in both the time and the frequency domains.

$$TF(\omega) = \frac{V_{out}(\omega)}{V_{in}(\omega)} = \frac{1}{1 - .05\omega^2 + j.1\omega} \tag{6.58}$$

Solution: In MATLAB, Step 1 applies the fft command directly to the EEG signal, essentially giving us $V_{in}(\omega)$ as real and imaginary components. Step 2 finds the output from these sinusoids by multiplying their complex representation with the transfer function: $V_{out}(\omega) = TF(\omega) \, V_{in}(\omega)$. In Step 3, this frequency-domain output (still represented as real and imaginary components) is converted back into the time domain using the inverse Fourier transform command, ifft. The program plots the original time-domain signal and the calculated system output in both time and frequency domains. It also plots the magnitude spectrum of transfer function, Equation 6.58.

```
% Example 6.14 Apply a specific transfer function to EEG data
%
load eeg_data;                   % Load EEG data (in variable eeg)
N = length(eeg);                 % Get data length
fs = 50;                         % Sample frequency is 60 Hz
t = (1:N)/fs;                    % Construct time vector
f = (0:N-1)*fs/N;                %  and frequency vector for plotting
Vin = (fft(eeg));                % Step 1: decompose data
TF = 1./(1 - .002*(2*pi*f).^2 + j*.003*2*pi*f);    % Define TF
Vout = Vin .* TF                 % Step 2, get output
vout = ifft(Vout);               % Step 3 Convert output to time domain
    ........plot and label system input and output in the time and frequency
domains.......
    .......plot magnitude spectrum of transfer function.........
```

The program loads the data, constructs time and frequency vectors based on the sampling frequency, and plots the time data. Next, the fft routine converts the EEG time data to the frequency domain, the transfer function is defined, then multiplied with input signal frequency components. The output of the product of transfer function and the signal (in the frequency domain) is converted back into the time domain using the inverse Fourier transform. Note that each of the steps in Figure 6.29 requires only one line of MATLAB code. The last section plots the magnitude spectra and time data of the input and output signals and the magnitude of the transfer function.

Results: The plots generated by this program are shown in Figures 6.30–6.32. The transfer function indicates that the system is of second order, but the magnitude spectrum, Figure 6.32, shows that it is acting as a narrowband band-pass filter with a peak frequency between 3 and 4 Hz. This operation is evident in the frequency-domain plots of the input and output signals, Figure 6.31, which shows the system reduces both high- and low-frequency components. This emphasizes EEG activity around 3 to 4 Hz and, as shown in the time-domain plots, Figure 6.30, appreciably alters the appearance of the EEG signal.

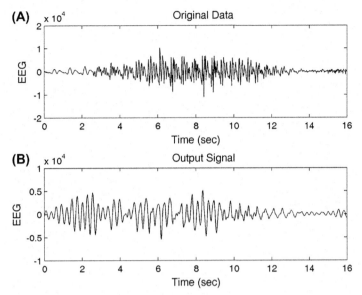

FIGURE 6.30 (A) The electroencephalography signal that is used as input to the system defined by the transfer function in Equation 6.58. (B) The output signal after passing through the system. This signal clearly contains a smaller range of frequencies.

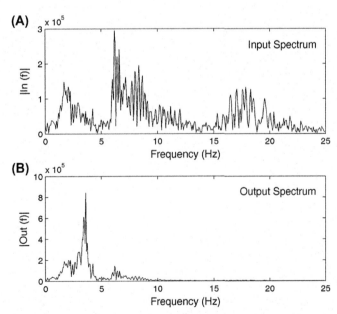

FIGURE 6.31 The magnitude spectra of the input (A) and output signals (B). The action of a filter that reduces the high- and low-frequency components is evident.

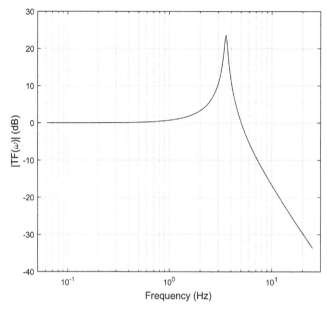

FIGURE 6.32   The magnitude spectrum of the system used in Example 6.14 and defined by the transfer function in Equation 6.58. The system acts as a narrowband filter with a peak frequency between 3 and 4 Hz. The bandwidth of this filter is approximately 0.4 Hz.

6.8 SUMMARY

Linear systems can be represented by differential equations and contain combinations of only four unique elements. If the input signals can be restricted to steady-state sinusoids, then phasor techniques can represent these elements using algebraic equations that are functions only of frequency. Phasor techniques represent steady-state sinusoids as a single complex number. With this representation, the calculus operation of differentiation can be implemented in algebra simply as multiplication by $j\omega$, where $j = \sqrt{-1}$ and ω is frequency in radians. Similarly, integration becomes division by $j\omega$.

Since system elements can be represented by algebraic operations, they can be combined into a single equation termed the transfer function. The transfer function relates the output of a system to the input. As presented here, the transfer function can only deal with signals that are sinusoidal or can be decomposed into sinusoids, but the transfer function concept is extended to a wider class of signals in the next chapter.

The transfer function not only offers a succinct representation of the system, it also gives a direct link to the frequency spectrum of the system. Frequency plots can be constructed directly from the transfer function using MATLAB, or they can be drawn by hand using a graphical technique termed Bode plot methods. With the aid of Bode plot methods, we can also go the other way, from a frequency plot to the corresponding transfer function.

Using the Bode plot approach, we are able to estimate the transfer function of a system given its magnitude spectrum. The transfer function can then be used to predict the system's behavior to other input signals, including signals that might be difficult to generate experimentally. These signals must be suitable for Fourier decomposition, but that applies to all digital signals stored in finite memory. The transfer function concept, coupled with Bode primitives, is a powerful tool for determining the transfer function representation of a wide range of physiological systems and for predicting their behavior to an equally broad range of stimuli.

If a real system can be stimulated and the resultant response measured, the system's spectral response can be used to estimate its transfer function. Alternatively, we can use the system's own natural stimulus as long as we can measure that stimulus and it contains energy over the spectral frequencies of interest. Determining a system's transfer function–based input/output data is an area of ongoing development known as "systems identification," and is briefly covered in the next chapter.

PROBLEMS

1. Assume that the feedback control system presented in Example 6.1 is in a steady-state or static condition. If $G = 100$ and $H = 1$ (i.e., a unity gain feedback control system), find the output if the input equals 1. Find the output if the input is increased to 10. Find the output if the input is 10 and G is increased to 1000. Note how the output is proportional to the input, which accounts for why this system (having the configuration shown) is sometimes termed a "proportional control system."

2. In the system given in Problem 1 with $G = 100$, the input is changed to a signal that smoothly goes from 0.0 to 5.0 in 10 s (i.e., $In(t) = 0.5\ t$ s). What does the output look like? Give the equation for the output. (Note G and H are simple constants so Equation 6.7 still holds.)

3. Convert the following to phasor representation:

   A. $10 \cos (10t)$
   B. $5 \sin (5t)$
   C. $6 \sin (2t + 60)$
   D. $2 \cos (5t) + 4 \sin (5t)$
   E. $\int 5 \cos(20t)dt$
   F. $\dfrac{d(2 \cos(20t + 30))}{dt}$

4. Add the following real, imaginary, and complex numbers to form a single complex number:

$$6, j10,\ 5 + j12,\ 8 + j3,\ 10\angle 0,\ 5\angle -60,\ 1/(j0.1)$$

5. Evaluate the following expressions:

   A. $(10{+}j6) + 10\angle{-30} - 10\angle 30$
   B. $6\angle -120 + \dfrac{5 - j10}{j4}$
   C. $\dfrac{10 + j5}{16 - j6} - \dfrac{8 - j8}{12 + j4}$
   D. $\dfrac{j(6 + j5)(3 - j4)}{(8 + j3)10\angle 260}$

6. Find the transfer function, $Out(\omega)/In(\omega)$, using phasor notation for the following system.

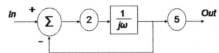

7. Find the transfer function, $Out(\omega)/In(\omega)$, for the following system. Find the time function, $out(t)$, if $in(t) = 5\cos(5t - 30)$. (Hint: You could use the feedback equation to find this transfer function, but the feedforward term contains an arithmetic operator; specifically, a summation operation.)

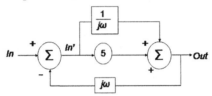

8. Using the transfer function of the respirator-airway pathway given in Example 6.7, find the airway pressure, $Q(\omega)$, if the respirator produces an external sinusoidal pressure of 2.0 mmHg at 11 breaths per minute. (Hint: Generate the input function $P(\omega)$, then solve using phasor analysis. Recall the transfer function is in minutes, which needs to be taken into account for the calculation of $P(\omega)$.)

9. Find the transfer function of the modified Guyton–Coleman fluid balance system shown below. Find the time function $P_A(t)$ if $F_{IN} = 5\sin(5t-30)$.

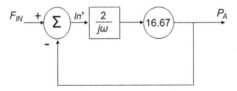

10. Find the transfer function of the following circuit. Find the time function, $out(t)$, if $in(t) = 5\sin(5t - 30)$. (Hint: This system contains a feedback system in the feedforward path and the feedback equation must be applied twice.)

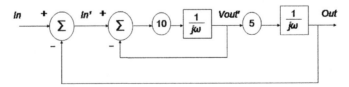

11. Plot the magnitude spectra of the transfer functions in Problems 6 and 7. What is the difference between the two?

12. Modify Equations 6.36 and 6.37 to find the equations for the asymptotes and worst case angle for the phase spectrum of the transfer function: $TF(\omega) = 1 + j\frac{\omega}{\omega_1}$.

13. Modify Equations 6.44−6.49 to show that when the second-order term appears in the numerator, the magnitude and phase spectra are inverted. Use MATLAB to plot the magnitude and phase spectra of the second-order transfer function used in Example 6.9 but with the second-order term in the numerator (i.e., $TF(\omega) = 1 - 0.03\omega^2 + j0.1\omega$).

14. Plot the Bode plot (magnitude and phase) for the following transfer function using graphical techniques.

$$TF(\omega) = \frac{100(1 + j.05\omega)}{(1 + j.01\omega)}$$

15. Plot the Bode Plot (magnitude and phase) for the following transfer function using graphical techniques:

$$TF(\omega) = \frac{100(1 + j\omega)}{(1 + j0.005\omega)(1 + j0.0002\omega)}$$

16. Plot the Bode plot (magnitude and phase) of the following transfer function using graphical techniques.

$$TF(\omega) = \frac{10j\omega}{(1 - 0.0001\omega^2 + j0.002\omega)}$$

17. Plot the Bode plot for the transfer function of the respirator-airway system given in Example 6.7 using graphical methods. Compare with the spectrum generated by MATLAB given in Figure 6.8.

18. Use MATLAB to plot the transfer functions (magnitude and phase) given in Problems 14 and 16.

19. Plot the Bode plot (magnitude and phase) of the following transfer function using graphical techniques.

$$TF(\omega) = \frac{100(1 + 0.1j\omega)}{(1 - 0.0004\omega^2 + j0.028\omega)}$$

20. Use MATLAB to plot the transfer functions (magnitude and phase) given in Problems 16 and 19.

21. In the system used in Problem 10, find the values of the ω_n and δ.

22. For the respirator-airway system of Example 6.7, find the values of ω_n and δ. Remember that time for this model is in minutes (not seconds) so this will have an effect on the value of ω_n. Convert ω_n to Hz and determine the breaths per minute that correspond to that frequency.

23. In the following system, find the value of k that makes $\omega_n = 10$ rad/s.

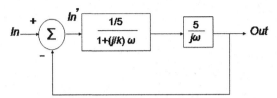

24. In the following system, find the maximum value of k so that the roots of the transfer function are real. As we show in the next chapter, the TF has real roots when the value of k results in $\delta \geq 1$.

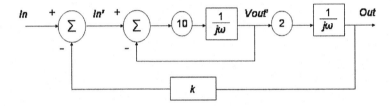

25. Estimate the transfer function that produces the following magnitude spectrum curve. Use MATLAB to plot the phase curve graphically of your estimated transfer function.

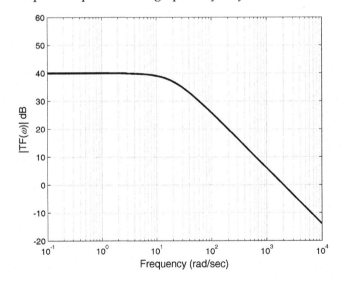

26. Find the transfer function that produces the following magnitude frequency curve. Plot the magnitude spectrum of your estimated transfer function using MATLAB and compare. Also plot the phase spectrum using MATLAB.

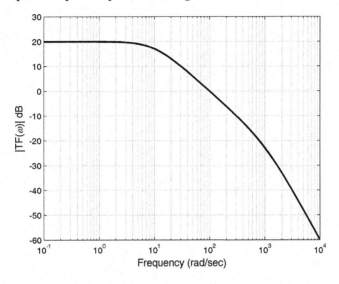

27. Find the transfer function that produces the following magnitude frequency curve. Plot the magnitude spectrum of your estimated transfer function using MATLAB and compare. Also use MATLAB to plot the phase spectrum.

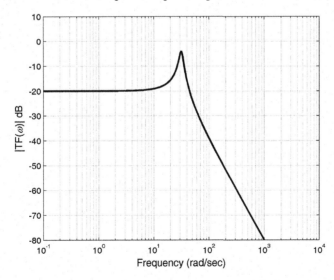

28. Find the transfer function that produces the following magnitude frequency curve. Plot the magnitude spectrum of your estimated transfer function using MATLAB and compare. Also use MATLAB to plot the phase spectrum.

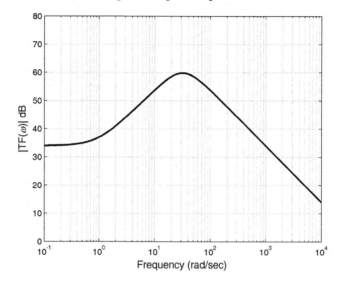

Linear Systems in the Complex Frequency Domain: The Laplace Transform

7.1 GOALS OF THIS CHAPTER

We now have two techniques to find the response of a system to any input… well, almost any input: it has to be either periodic or aperiodic. In Chapter 5, we divided the input into short time segments and used convolution to sum up the resultant impulse responses. In Chapter 6, we used the Fourier transform to divide the input into sinusoids, found the output to each sinusoid by multiplying it with the transfer function, and then summed the output sinusoids using the inverse Fourier transform. We should always keep in mind that both of these methods involve the principle of superposition and time invariance so they apply only to linear, time invariant (LTI) systems. Even so, neither of these approaches can handle a third class of signals: waveforms that suddenly change and never return to a baseline level (recall Section 1.4.2). A classic example is the "step function," which changes from one value to another at one instant of time (often taken as $t = 0$) and remains at the new value for all eternity, Figure 7.1.

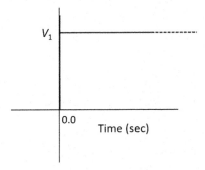

FIGURE 7.1 Time plot of a step function that changes from 0 to value V_1 at $t = 0.0$ s. It remains at this value for all time.

Circuits, Signals, and Systems for Bioengineers
https://doi.org/10.1016/B978-0-12-809395-5.00007-2

One-time changes, or changes that do not return to some baseline level, are common in nature and even in everyday life. For example, this text should leave the reader with a lasting change in his or her knowledge of biosystems and biosignals, one that does not return to the baseline, precourse level. However, unlike a step function, this change is not expected to occur instantaneously. An input signal could change a number of times, either upward or downward, and might even show a partial return to its original starting point. But if the signal never returns to some original level, it cannot be either periodic or aperiodic and we do not have the tools to analyze how a system responds to these signals. As noted in Chapter 1, signals that change and never return to baseline are sometimes referred to as "transient" signals. Of course all signals vary in time and therefore could technically be called transient signals, but this term is often reserved for signals that have one-time or step-like changes. This is another linguistic issue in engineering where context can vary the meaning.

The goal of this chapter is to master the Laplace transform, a technique that allows us to analyze the response of LTI systems to transient or step-like inputs. It also enables us to analyze systems that have initial conditions. The downside is that this is a purely analytical technique and cannot be applied to discrete signals. So this rules out computer applications. Moreover, think back on the difficulty of working out the convolution integral analytically as in Example 5.3. Maybe we should avoid the Laplace transform and try working around the signal limitations of our previous methods. Could we not approximate a step function as a pulse having a very long pulse width? You could, and that is very likely what you would do in many situations. But we still cannot avoid the Laplace transform; it is just too important a concept in systems analysis.

The importance of the Laplace transform lies in its theoretical implications, although it does have some practical applications. Like the Fourier transform, it converts signals or functions from the time domain to something we call the Laplace domain. The Laplace domain is actually more general than the Fourier transform and, since it is easy to go from the Laplace domain to the frequency domain (but not vice versa), it is the preferred domain for transfer function equations. So most systems are described in the Laplace domain even if they do not use Laplace transforms for evaluation. MATLAB's powerful simulation language Simulink® described in Chapter 9 uses the Laplace notation to define system elements. If the input signals are periodic or aperiodic, we can easy slip back into the frequency domain and apply all the computer-friendly methods described in Chapters 5 and 6.

Topics in this chapter reflect many of those in Chapter 6 and include:

- The development of the Laplace transform as an extension of the Fourier transform and the introduction of the concept of complex frequency;
- How to determine the Laplace transform of a signal;
- The transfer function in the Laplace domain;
- How to find the output to any input in the Laplace domain (the output signal is found in the same way as in Chapter 6, by multiplying the signal by the transfer function, but this time in the Laplace domain);
- How to take inverse Laplace transform to get the time domain signal (this may involve additional algebraic tools such as partial fraction expansion);
- The relationships between the various methods for defining systems and describing their behavior.

7.2 THE LAPLACE TRANSFORM

The Laplace transform is used to define systems using transfer functions similar to the frequency domain transfer functions used in the last chapter.[1] But the Laplace domain can also represent signals that do not return to a baseline level. With this tool, we can extend all of the techniques developed in the last chapter to systems exposed to this wider class of signals. The Laplace domain can also be used to analyze systems that begin with nonzero initial conditions, a situation we have managed to avoid thus far, but one that does occur in the real world.

7.2.1 Definition of the Laplace Transform

Transfer functions written in Laplace notation differ only slightly from the frequency domain transfer function used in the last chapter. The reason that frequency domain transfer functions cannot be used when signals change in a step-like manner is simply that these functions cannot be decomposed into sinusoidal components using either Fourier series analysis or the Fourier transform. Consider a function similar to that shown in Figure 7.1, a function that is 0.0 for a $t \leq 0$ and 1.0 for all $t > 0$:

$$x(t) = \begin{cases} 0 & t \leq 0 \\ 1 & t > 0 \end{cases} \tag{7.1}$$

The function defined in Equation 7.1 is known as the "unit step function" since it begins at zero and jumps to 1.0 at $t = 0$, but a generic step function could begin at any level and jump to any other level. Try to find the Fourier transform of this function and you get:

$$FT(\omega) = \int_0^\infty x(t)e^{-j\omega t}dt = \int_0^\infty 1e^{-j\omega t}dt \Rightarrow \infty \tag{7.2}$$

The problem is that because $x(t)$ does not return to its baseline level (zero in this case), the limits of the integration must be infinite. Since the sinusoidal function $e^{-j\omega t}$ has nonzero values out to infinity, the integral becomes infinite. In the past, our input signals had a finite life span and the Fourier transform integral need only be taken over that finite time period. But for transient signals this integral cannot be computed.

The trick used to solve this infinite integral problem is to modify the exponential function so that it converges to zero at large values of t even if the signal, $x(t)$, does not. This can be accomplished by multiplying the sinusoidal term, $e^{-j\omega t}$, by a decaying exponential such as $e^{-\sigma t}$ where σ is some positive real variable. In this case the sinusoid term in the Fourier transform becomes a complex sinusoid, or rather a sinusoid with a complex frequency:

$$e^{-j\omega t}e^{-\sigma t} = e^{-(\sigma + j\omega)t} = e^{-st} \tag{7.3}$$

---

[1] In fact, some purists believe that only transfer functions written in Laplace notation are worthy of the term "transfer function."

where $s = \sigma + j\omega$ and is termed the "complex frequency" because it is a complex variable, but has the same role as frequency, ω, in the Fourier transform exponential. The complex variable, s, is also known as the "Laplace variable" since it plays a critical role in the Laplace transform (another example of multiple names for the same thing). A modified version of the Fourier transform can now be constructed using complex frequency in place of regular frequency; that is, s (which is $\sigma + j\omega$) instead of just $j\omega$. This modified transform is termed the Laplace transform:

$$X(s) = \mathcal{L}x(t) = \int_0^\infty x(t)e^{-st}dt \tag{7.4}$$

where the script $\mathcal{L}$ indicates the Laplace transformation.

So the trick is to use complex frequency, with its decaying exponential component, to cause convergence for functions that would not otherwise converge. For any general signal, $x(t)$, the product of $x(t)e^{-st} = x(t)e^{-(\sigma+j\omega)t}$ may not necessarily converge to zero as $t \to \infty$, in which case the Laplace transform does not exist.[2] Some advanced signal processing gurus have spent a lot of time worrying about such functions: which functions converge, their ranges of convergence, or how to get them to converge. Fortunately, such matters need not concern us since most common real-world signals, including the step function, do converge for some values of σ and so have a Laplace transform. The range of σ's over which a given product of $x(t)$ and $e^{-(\sigma+j\omega)t}$ converges is another occupation of signal processing theoreticians, but again is not something bioengineering signal processors need worry about. If a signal has a Laplace transform (and all the signals we use do), then the product $x(t)e^{-st}$ converges as $t \to \infty$.

The Laplace transform is a purely analytical technique so it cannot be solved on a computer and also has two other issues: it cannot be applied to functions for negative values of t, and it is difficult to evaluate using Equation 7.4 for any but the simplest of functions. The restriction against representing functions for negative time values comes from the fact that e^{-st} becomes a positive exponential and will go to infinity as t goes to large negative values. (For negative t, the real part of the exponential becomes $e^{+\sigma t}$ and does exactly the opposite of what we want it to do: it forces divergence rather than convergence.) The only way around this is to limit our analyses to $t > 0$, but this is usually not a serious restriction. The other problem, the difficulty in evaluating Equation 7.4, stems from the fact that s is complex, so although the integral in Equation 7.4 does not look so complicated, the complex integration becomes very involved for all but a few simple functions of $x(t)$. To get around this problem, we use tables that give us the Laplace transforms of frequently used functions. Such a table is given in Appendix B, and a more extensive list can easily be found on the internet. A Laplace transform table is used both to determine the Laplace transform of a signal and, using it in reverse, the inverse Laplace transform. The only difficulty in finding the inverse Laplace transform is rearranging the output Laplace function into one of the formats found in the table.

---

[2]For example, the function $x(t) = e^{t^2}$ will not converge as $t \to \infty$ even when multiplied by e^{-st}, so it does not have a Laplace transform.

EXAMPLE 7.1

Find the Laplace transform of the step function in Equation 7.1.

Solution: The step function is one of the few functions that can be evaluated easily using the basic defining equation of the Laplace transform, Equation 7.4.

$$X(s) = \int_0^\infty x(t)e^{-st}dt = \int_0^\infty 1e^{-st}dt = -\frac{e^{-st}}{s}\Big|_0^\infty = 0 - \left(-\frac{1}{s}\right)$$

$$X(s) = \frac{1}{s}$$

7.2.2 Calculus Operations in the Laplace Domain

Just as in the frequency domain, the calculus operations of differentiation and integration can be reduced to algebraic operations in the Laplace domain. The Laplace transform of the derivative operation can be determined from the defining equation, Equation 7.4.

$$\mathcal{L}\frac{dx(t)}{dt} = \int_0^\infty \frac{dx(t)}{dt}e^{-st}dt$$

Integrating by parts:

$$\mathcal{L}\frac{dx(t)}{dt} = x(t)e^{-st}\Big|_0^\infty + s\int_0^\infty x(t)e^{-st}dt$$

From the definition of the Laplace transform, $\left(\int_0^\infty x(t)e^{-st}dt = X(s)\right)$, the right-most term in the summation is $sX(s)$, and the equation becomes:

$$\mathcal{L}\frac{dx(t)}{dt} = x(\infty)e^{-\infty} - x(0)e^{-0} + sX(s)$$

$$\mathcal{L}\frac{dx(t)}{dt} = sX(s) - x(0-) \tag{7.5}$$

Equation 7.5 shows that in the Laplace domain, differentiation becomes multiplication by the Laplace variable s with the additional subtraction of the value of the function at $t = 0$. The value of the function at $t = 0$ is known as the "initial condition." This value can be used, in effect, to account for all negative time history of $x(t)$. In other words, all of the behavior of $x(t)$ when t was negative can be lumped together as a single initial value at $t = 0$. This trick allows us to include some aspects of the system's behavior over negative values of t even if the Laplace transform does not itself apply to negative time. If the initial condition is zero, as is frequently the case, then differentiation in the Laplace domain is simply multiplication by s.

Multiple derivatives can also be taken in the Laplace domain, although this is not such a common operation. Multiple derivatives involve multiplication by s n-times, where n is the number of derivative operations, and taking the derivatives of the initial conditions:

$$\mathcal{L}\frac{d^n x(t)}{dt^n} = s^n X(s) - s^{n-1}x(0-) - s^{n-2}\frac{dx(0-)}{dt}\cdots\frac{d^{n-1}x(0-)}{dt} \tag{7.6}$$

Again, if there are no initial conditions, taking n derivatives becomes just a matter of multiplying $x(t)$ by s^n.

If differentiation is multiplication by s in the Laplace domain, then integration is simply a matter of dividing by s. Again, a second term account for the initial conditions:

$$\mathcal{L}\left[\int_0^T x(t)dt\right] = \frac{1}{s}X(s) + \frac{1}{s}\int_{-\infty}^0 x(t)dt \tag{7.7}$$

If there are no initial conditions, then integration is accomplished by simply dividing by s. The second term of Equation 7.7 is a direct integral that accounts for the initial conditions and is again a way of accounting for the negative time history of the system.

7.2.3 Sources—Common Signals in the Laplace Domain

In the Laplace domain, both signals and systems are represented by functions of s. As mentioned earlier, the Laplace domain representation of signals is determined from a table. Although it might seem that there could be a large variety of such signals, which would require a very large table, in practice only a few signal types commonly occur in systems analysis. The signal most frequently encountered in Laplace analysis is the step function shown in Figure 7.1, or its more constrained version, the unit step function given in Equation 7.1 and repeated here:

$$x(t) = u(t) = \begin{cases} 0 & t \le 0 \\ 1 & t > 0 \end{cases} \tag{7.8}$$

The symbol u is used frequently to represent the unit step function. The Laplace transform of the step function was found in Example 7.1 and repeated here:

$$X(x) = U(s) = \mathcal{L}u(t) = \frac{1}{s} \tag{7.9}$$

As with the Fourier transform, it is common to use capital letters to represent the Laplace transform of a time function. Two functions closely related to the unit step function are the ramp and impulse functions, Figure 7.2.

These functions are related to the step function by differentiation and integration. The unit ramp function is a straight line with slope of 1.0.

$$r(t) = \begin{cases} t & t > 0 \\ 0 & t \le 0 \end{cases} \tag{7.10}$$

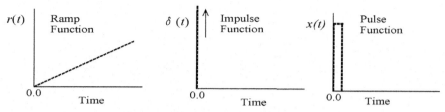

FIGURE 7.2 The ramp and impulse are two signals related to the step function and are commonly encountered in Laplace analysis. As described in Chapter 5, an ideal impulse occurs at $t = 0$ and is infinitely narrow and infinitely tall. Real-world impulse signals are approximated by short pulses (see Example 5.1).

Since the unit ramp function is the integral of the unit step function, its Laplace transform will be that of the step function divided by s:

$$R(s) = \mathcal{L}r(t) = \frac{1}{s}\left(\frac{1}{s}\right) = \frac{1}{s^2} \tag{7.11}$$

The impulse function is the derivative of a unit step, which leads to one of those mathematical fantasies: a function that becomes infinitely short, but as it does its amplitude becomes infinite so the area under the function remains 1.0 as shown in Equation 7.12.

$$x(t) = \delta(t) = \lim_{a \to 0} \frac{1}{a} \quad \frac{-a}{2} \le t \le \frac{a}{2} \tag{7.12}$$

In practice, a short pulse is used as an impulse function and an approach to finding the appropriate pulse width for any given system is described in Example 5.1.

Since the impulse response is the derivative of the unit step function, its Laplace transfer function is that of a unit step multiplied by s:

$$\Delta(s) = \mathcal{L}\delta(t) = s\left(\frac{1}{s}\right) = 1 \tag{7.13}$$

Hence the Laplace transform of an impulse function is a constant, and if it is a unit impulse (the derivative of a unit step) then that constant is 1. As you might guess, this fact will be especially useful in the analysis of Laplace transfer functions. The Laplace transforms of other common signal functions are given in a table in Appendix B.

7.2.4 Converting the Laplace Transform to the Frequency Domain

Remember that s is a complex frequency and equals a real term plus the standard imaginary frequency term: $s = \alpha + j\omega$. To convert a Laplace function to the frequency domain, we simply substitute $j\omega$ for s. When we do this, we are agreeing to restrict ourselves to sinusoidal steady-state signals, or those that can be decomposed into such signals, so the real component of s with its convergence properties is no longer needed. Converting the Laplace transform function to the frequency domain allows us to determine the frequency characteristics of that function using Bode plot techniques or computer-based methods. This approach is used in the next section to evaluate the frequency characteristics of a time delay process.

TABLE 7.1   Transformations Between the Time, Frequency, and Laplace Domains.

From	To	Transformation
Time	Frequency	Fourier transform
Frequency	Time	Inverse Fourier transform
Time	Laplace	Laplace transform
Laplace	Time	Inverse Laplace transform
Laplace	Frequency	$s \rightarrow j\omega$
		Assumes sinusoidal steady state
Frequency	Laplace	$j\omega \rightarrow s$
		Assumes no initial conditions

7.2.5 The Inverse Laplace Transform

Working in the Laplace domain is pretty much the same as working in the frequency domain: the math is still algebra. The transfer function of system elements will be in Laplace notation as described later and summarized in Table 7.1. First, the input signal is converted to the Laplace domain using the table in Appendix B. Then multiplying the input signal with the transfer function provides the output response, but again as a function of s, not $j\omega$. Sometimes the Laplace representation of the solution or even just the Laplace transfer function is sufficient, but if a time domain solution is desired, then the inverse Laplace transform must be determined.

The equation for the inverse Laplace transform is given as:

$$x(t) = \mathcal{L}^{-1}X(x) = \frac{1}{2\pi} \int_{\sigma-j\infty}^{\sigma+j\infty} X(s)e^{st}ds \tag{7.14}$$

Unlike the inverse Fourier transform, this equation is quite difficult to solve even for simple functions. So to evaluate the inverse Laplace transform, we use the Laplace transform table in Appendix B in the reverse direction: find a function (on the right side of the table) that matches your Laplace output and convert it to the equivalent time domain signal. The difficulty is usually in rearranging the Laplace output function to conform to one of the formats given in the table. Methods for doing this are described next and specific examples given.

7.3 THE LAPLACE DOMAIN TRANSFER FUNCTION

The analysis of systems using the Laplace transform is no more difficult than in the frequency domain, except that there may be the added task of accounting for initial conditions. In addition, to go from the Laplace domain back to the time domain may require some algebraic manipulation of the Laplace output function.

The transfer function introduced in Chapter 6 is ideally suited to Laplace domain analysis, particularly when there are no initial conditions. In the frequency domain, the transfer function is used primarily to determine the spectrum of a system or system element, but it can also be used to determine the system's response to any input provided that input can be expressed as a sinusoidal series or is aperiodic. In the Laplace domain, the transfer function can be used to determine a system's output to a broader class of input signals. The Laplace domain transfer function is similar to its cousin in the frequency domain, except the frequency variable, ω, is replaced by the complex frequency variable, s:

$$TF(s) = \frac{Output(s)}{Input(s)} \tag{7.15}$$

Like its frequency domain cousin, the general Laplace domain transfer function consists of a series of polynomials. The general transfer function in the frequency domain was given in Equation 6.50 and repeated here:

$$TF(\omega) = \frac{Gj\omega\left(1 + j\frac{\omega}{\omega_1}\right)\left(1 - \left(\frac{\omega}{\omega_{n1}}\right)^2 + j\frac{2\delta_1\omega}{\omega_{n1}}\right)\cdots}{j\omega\left(1 + j\frac{\omega}{\omega_2}\right)\left(1 - \left(\frac{\omega}{\omega_{n2}}\right)^2 + j\frac{2\delta_2\omega}{\omega_{n2}}\right)\cdots} \tag{7.16}$$

In the Laplace domain, the general form of the transfer function differs in three ways: (1) the frequency variable $j\omega$ is replaced by the complex frequency variable s; (2) the polynomials are normalized so that the highest order of the frequency variable is normalized to 1.0 (as opposed to the constant term as in Equation 7.16); and (3) the order of the frequency terms is reversed with the highest-order term (now normalized to 1) coming first:

$$TF(s) = \frac{Gs(s + \omega_1)(s^2 + 2\delta_1\omega_{n1}s + \omega_{n1}^2)\cdots}{s(s + \omega_2)(s^2 + 2\delta_2\omega_{n2}s + \omega_{n2}^2)\cdots} \tag{7.17}$$

The constant terms, δ, ω, ω_n, etc. have the same meaning as in Equation 6.50. Even the terminology used to describe the elements is the same. For example, $\frac{1}{s+\omega_1}$, which is related to $\frac{1}{1+\frac{j\omega}{\omega_1}}$, is called a first-order element, and $\frac{1}{s^2+2\delta\omega_n s+\omega_n^2}$ is a second-order element. Note that in Equation 7.17, the constants δ and ω_n have a more orderly arrangement in the transfer function equation.

As described in Section 7.1.4, it is possible to go from the Laplace domain transfer function to the frequency representation simply by substituting $j\omega$ for s, but you should also rearrange the coefficients and frequency variable to fit the frequency domain format. As in the frequency domain, the assumption is that any higher-order polynomial (third-order or above) can be factored into the first- and second-order terms.

As with the frequency domain transfer function, we cover each of the element types separately: gain and first-order elements followed by the intriguing behavior of the second-order element. But first we introduce a new element, the "time delay" element, commonly found in physiological systems.

7.3.1 Time Delay Element: The Time Delay Theorem

Many physiological processes have a delay before they begin to respond to a stimulus. Such physiological delays are termed "reaction time," or "response latency," or simply "response delay." In these processes, there is a period of time between the onset of the stimulus and the beginning of the response. In systems involving neurological control, this delay is due to processing delays in the brain. The Laplace domain has an element to represent such delays.

The Time Delay theorem can be derived from the defining equation of the Laplace transform, Equation 7.4. Assume a signal $x(t)$ that is zero for negative time, but normally changes at $t = 0$ s. If this signal is delayed by T seconds, then the delayed function would be $x(t - T)$. From the defining equation (Equation 7.4), the Laplace transform of such a delayed function would be:

$$\mathcal{L}[x(t - T)] = \int_0^\infty x(t - T)e^{-st}dt$$

Defining a new variable: $\gamma = t - T$

$$\mathcal{L}[x(t - T)] = \int_0^\infty x(\gamma)e^{-s(T+\gamma)}d\gamma = \int_0^\infty x(\gamma)e^{-sT}e^{-s\gamma}d\gamma = e^{-sT}\int_0^\infty x(\gamma)e^{-s\gamma}d\gamma$$

But the integral in the right-hand term is the same as Laplace transform of the function, $x(t)$, it just has a different variable for time. So the right-hand integral is the Laplace transform of an unshifted $x(t)$; i.e., it is $\mathcal{L}[x(t)]$. Hence the Laplace transform of the shifted function becomes:

$$\mathcal{L}[x(t - T)] = e^{-sT}\mathcal{L}[x(t)] \tag{7.18}$$

Equation 7.18 is the Time Delay theorem, which can also be used to construct an element that represents a pure time delay, an element with a transfer function:

$$TF(s) = e^{-sT} \tag{7.19}$$

where T equals the delay, usually in seconds. So an element with a transfer function $TF(s) = e^{-sT}$ is a time delay of T seconds. This element is often found in models of neurological control where it represents neural processing delays.

EXAMPLE 7.2

Find the spectrum of a system time delay element for two values of delay: 0.5 and 2.0 s. Use MATLAB to plot the spectral characteristics of this element.

Solution: The Laplace transfer function of a pure time delay of 2.0 s would be:

$$TF(s) = e^{-2s}$$

To determine the spectrum of this element, we need to convert the Laplace transfer function to the frequency domain. This is done simply by substituting $j\omega$ for s in the transfer function equation:

$$TF(\omega) = e^{-j2\omega}$$

The magnitude of this frequency domain transfer function is:

$$|TF(\omega)| = \left|e^{-j2\omega}\right| = 1 \tag{7.20}$$

for all values of ω since changing ω (or the constant 2 for that matter) only changes the angle of the imaginary portion of the exponential, not its magnitude. The phase of this transfer function is:

$$\angle TF(\omega) = \angle e^{-j2\omega} = -2\omega \text{ rad.} \tag{7.21}$$

Since the magnitude of the transfer function is 1.0, the input and output signals have the same magnitude at all frequencies. The time delay process only changes the phase of the output signal by making it more negative with increasing frequency (i.e., adding -2ω to the phase of the input signal).

For a 0.5-s delay, $\angle TF(\omega) = \angle e^{-0.5j\omega}$, so the phase curve would be -0.5ω. A pure time delay decreases the phase component of a signal's spectrum in a linear manner, as shown in Figure 7.3. A time delay process is also explored in one of the problems.

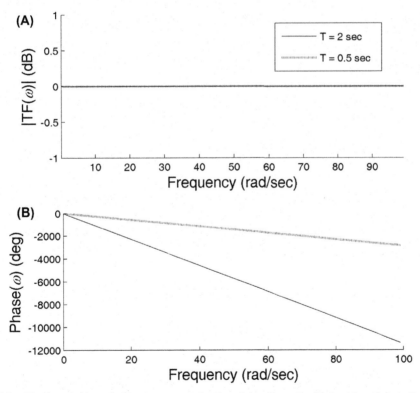

FIGURE 7.3 The magnitude and phase spectrum of a time delay element with two time delays: 2 and 0.5 s. In both cases the magnitude spectrum is a constant 0 dB (output equal input) as expected and the phase spectrum decreases linearly with frequency. Note that when $\omega = 100$ rad/s the phase angle is $-12,000$ degrees for the 2-s time delay as predicted by Equation 7.21 ($-2*100*360/2\pi = -12,000$ degrees).

The MATLAB code used to generate the magnitude and phase spectrum is a minor variation of Example 6.7. The plotting is done using linear frequency rather than log, to emphasize that the phase spectrum is a linear function of frequency.

```
% Example 7.2 Use MATLAB to plot the transfer function of a time delay
%
T = 2;                                    % Time delay in sec.
w = .1:1:100;                             % Frequency vector
TF = exp(-j*T*w);                        % Transfer function
Mag = 20*log10(abs(TF));                 % Calculate magnitude spectrum
Phase = unwrap(angle(TF))*360/(2*pi);            % Calculate phase spectrum
  ....... Repeat for T = 0.5 and plot and label......
```

Results: The time delay element increases the phase linearly with frequency as shown in Figure 7.3.

7.2.2 Constant Gain Element

The gain element is not a function of frequency, so its transfer function is the same irrespective of whether standard or complex frequency is involved:

$$TF(s) = G \tag{7.22}$$

The system representation for a gain element is the same as in the frequency domain representation, Figure 6.9.

7.3.3 Derivative Element

Equation 7.5 shows that the derivative operation in the Laplace domain is implemented simply by multiplication with the Laplace variable s. This is analogous to this operation in the frequency domain, where differentiation is accomplished by multiplying by $j\omega$. So in the absence of initial conditions, the Laplace transfer function for a derivative element is:

$$TF(s) = s \tag{7.23}$$

The Laplace system representation of this element is shown in Figure 7.4A

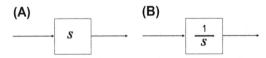

FIGURE 7.4   (A) Laplace system representation of a derivative element. (B) Laplace system representation of an integrator element.

7.3.4 Integrator Element

As shown in Equation 7.7, integration in the Laplace domain is accomplished by dividing by s. So in the absence of initial conditions, the system representation of integration in the Laplace domain has a transfer function of

$$TF(s) = 1/s \tag{7.24}$$

This parallels the representation in the frequency domain where $1/s$ becomes $1/j\omega$. The representation of this element is shown in Figure 7.4B.

EXAMPLE 7.3

Find the Laplace and frequency domain transfer function of the system in Figure 7.5. The gain term k is a constant.

Solution: The approach to finding the transfer function of a system in the Laplace domain is exactly the same as in the frequency domain used in Chapter 6. Here we are asked to determine both the Laplace and frequency domain transfer function. First we find the Laplace transfer function, then substitute $j\omega$ for s, and rearrange the format for the frequency domain transfer function.

We could solve this several different ways, but this is clearly a feedback system so the easiest solution is to use the feedback equation, Equation 6.7. All we need to do is find the equivalent feedforward gain function, $G(s)$, and the equivalent feedback function, $H(s)$. The feedback function is $H(s) = 1$ since all of the output feeds back to the input in this system.[3] The feedforward gain is the product of the two elements:

$$G(s) = k\frac{1}{s}$$

Substituting $G(s)$ and $H(s)$ into the feedback equation:

$$TF(s) = \frac{G(s)}{1 + G(s)H(s)} = \frac{\dfrac{k}{s}}{1 + \dfrac{k}{s}} = \frac{k}{s + k}$$

Result: Again, the Laplace domain transfer function has the highest coefficient of complex frequency, in this case s, normalized to 1. Substituting $s = j\omega$ and rearranging the normalization:

$$TF(\omega) = \frac{k}{j\omega + k} = \frac{1}{1 + j\dfrac{\omega}{k}}$$

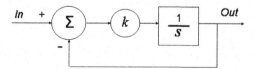

FIGURE 7.5   Laplace representation of a system used in Example 7.3.

This has the same form as a first-order transfer function given in Equation 6.31 and shown in Figure 6.12B, where the constant term ω_1 is k.

[3]Such systems are known as "unity gain feedback systems" since the feedback gain, $H(s)$, is 1.0 (unity).

7.3.5 First-Order Element

First-order processes contain an $s + \omega_1$ term in the denominator (or possibly only an s). Later we find that these systems contain a single energy storage device. The Laplace transfer function for a first-order process is:

$$TF(s) = \frac{1}{s + \omega_1} = \frac{\frac{1}{\tau}}{s + \frac{1}{\tau}} \tag{7.25}$$

In this transfer function equation, a new variable is introduced, τ, which is termed the "time constant" for reasons shown in Example 7.4. As seen in Equation 7.25, this time constant variable, τ, is simply the inverse of the frequency constant, ω_1:

$$\tau = \frac{1}{\omega_1} \tag{7.26}$$

In Chapter 6, we found that the frequency constant ω_1 is where the magnitude spectrum is -3 dB (Equation 6.34) and the phase spectrum is -45 degrees (Equation 6.37). In the next example, we show that the time constant τ has a direct relationship to the time behavior of a first-order element.

Now that the transfer function is in the Laplace domain, we can explore the response of these two similarly behaving systems using input signals other than sinusoids. Two popular signals that are used are the step function, shown in Figure 7.1, and the impulse function, shown in Figure 7.2. Typical impulse and step responses of a first-order system are found in the next example.

EXAMPLE 7.4

Find the arterial pressure response of the linearized model of the Guyton–Coleman body fluid balance system (Ridout, 1991) to a step increase of fluid intake of 0.5 mL/min. The frequency domain version of this model is shown in Figure 6.4. Also find the pressure response to fluid intake of 250 mL administered as an impulse.[4] This is equivalent to drinking approximately 1 cup of fluid quickly. After finding the analytical solution, use MATLAB to plot the outputs in the time domain.

Solution, step response: The first step is to convert to Laplace notation from the frequency notation used in Figure 6.4. The feedforward gain becomes $G(s) = 16.67\left(\frac{0.06}{s}\right) = \frac{1}{s}$ and the feedback gain becomes $H(s) = 0.05$. Again applying the feedback equation (Equation 6.7), the transfer function is:

$$TF(s) = \frac{P_A(x)}{F_{IN}(s)} = \frac{G(s)}{1 + G(s)H(s)} = \frac{\frac{1}{s}}{1 + 0.05\frac{1}{s}} = \frac{1}{s + 0.05}$$

where P_A is arterial blood pressure in mmHg and F_{IN} is a small change in fluid intake in mL/min.

To find the step response, multiply the Laplace input signal by the transfer function:

$$P_A(s) = F_{IN}(s)TF(s) = \frac{1}{s}TF(s) \tag{7.27}$$

Substituting in for $TF(s)$, Equation 7.27 becomes:

$$P_A(s) = \frac{1}{(s+0.05)}\left(\frac{1}{s}\right) = \frac{1}{s(s+0.05)} \tag{7.28}$$

To take the inverse Laplace transform and get the time domain output, we need to rearrange the resulting Laplace output function so that it has the same form as one of the functions in the Laplace Transform Table of Appendix B. Often this can be the most difficult part of the problem. Fortunately, in this problem, we see that the right-hand term of Equation 7.28 matches the Laplace function in entry number 4 of the table.

$$\frac{1}{s(s+0.05)} \text{ has the form}: \quad \frac{\alpha}{s(s+\alpha)} \Leftrightarrow (1-e^{-\alpha t})/\alpha$$

where $\alpha = 0.05$. Hence the step response in the time domain for this system is an exponential:

$$p_A(t) = \frac{1}{0.05}\left(1 - e^{-0.05}\right) = 20\left(1 - e^{-0.05}\right) \text{ mmHg}$$

Solution, impulse response: Solving for the impulse response is even easier since for the impulse function, $F_{IN}(s) = 1$. So the impulse response of a system in the Laplace domain is the transfer function itself. Therefore, the impulse response in the time domain is the inverse Laplace transform of the transfer function:

In this example the impulse input was 250 mL, so $F_{IN}(s) = 250$, and $P_A(s) = 250\ TF(s)$:

$$P_A(s) = \frac{250}{s + 0.05} \text{ mmHg}$$

The Laplace output function is matched by entry 3 in the Laplace Transform Table except for the constant term. From the definition of the Laplace transform in Equation 7.4, any constant term can be removed from the integral, so the Laplace transform of a constant times a function is the constant times the Laplace transform of the function. Similarly, the inverse Laplace transform of a constant times a Laplace function is the constant times the inverse Laplace transform. Stating these two characteristics formally:

$$\mathcal{L}[kx(t)] = k\mathcal{L}x(t) \tag{7.29}$$

$$\mathcal{L}^{-1}[kX(s)] = k\mathcal{L}^{-1}[X(s)] \tag{7.30}$$

So the time domain solution for $P_A(s)$ is obtained:

$$V_{out}(s) = 250\left(\frac{1}{s+0.05}\right) \text{ which has the same form as}: \quad k\frac{1}{s+\alpha} \Leftrightarrow ke^{-\alpha t}$$

$$v_{out}(t) = 250e^{-0.05} \text{ mmHg}$$

Plotting this result with the help of MATLAB is easy as shown.

```
% Example 7.4 First-order system impulse and step response for two time constants
%
t = 0:.1:100;                    % Define time vector: 0 to 100 min
x = 250*exp(-0.05);              % Impulse response
plot(t,x,'k');                   % Plot impulse response
   ........labels, title, and other text......
x = 20*(1 - exp(-0.05));         % Step response.
plot(t,x,'k');                   % Plot step response
   ........plot, labels, title, and other text......
```

Results: The system responses are shown in Figure 7.6. Both the impulse and step responses are exponential with time constants of $1/0.05 = 20$ min. When time equals the time constant (i.e., $t = \tau$), the value of the exponential becomes: $e^{-t/\tau} = e^{-1} = 0.37$. Hence at time $t = \tau$, the exponential is within 37% of its final value. In other words, the exponential has attained 73% of its final value after one time constant.

The time constant makes a good measure of the relative speed of an exponential response. As a rule of thumb, an exponential is considered to have reached its final value when $t > 5\tau$, although in theory the final value of an exponential is reached only when $t = \infty$.

[4]Logically, the response to a step change in input is called the "step response," just as the response due to an impulse is termed the "impulse response."

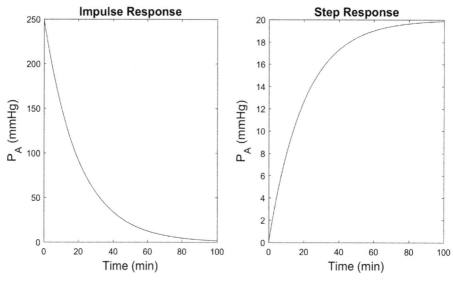

FIGURE 7.6   Response of a linearized model of the Guyton–Coleman body fluid balance system. The graph shows the change in arterial pressure to an increase of fluid intake, either taken in a step-like manner (right) or as an impulse (left).

There are many other possible input waveforms in addition to the step and impulse function. As long as the waveform has a Laplace representation, the response to any signal can be found from the transfer function.[5] However, the step and/or impulse responses usually provide the most insight into the general behavior of the system.

7.3.5.1 The Characteristic Equation

First-order transfer functions can be slightly different than that of Equation 7.25. There can be a single s or even an $(s + 1/\tau)$ in the numerator. What bonds all first-order systems is the denominator term, $(s + 1/\tau)$, or equivalently $(s + \omega_1)$. All first-order systems will contain a s in the denominator irrespective of what is in the numerator. In the next section, we find that all second-order systems contain an s^2 in the denominator of the transfer function, usually as a quadratic polynomial of s. Again the numerator can be a variety of polynomials of s.

The characteristics of the transfer function are primarily determined by the denominator, which explains why the denominator is called the "characteristic equation." The characteristic equation describes the general behavior of the system. Often, only the characteristic equation is required to determine the salient features of a response: a time domain solution is not needed. For example, the characteristic equation $s + 3$ tells us that the system's response will include an exponential having a time constant of $1/3$ or 0.33 s. Second-order characteristic equations are even more informative, as the next section shows.

EXAMPLE 7.5

Find the transfer function of the system shown in Figure 7.7.

Solution: As in Example 7.3 and previous examples, we use the feedback equation. We just need to find $G(s)$ and $H(s)$. Again $H(s) = 1$: another unity gain feedback system. The feedforward gain function is just the product of the two feedforward elements:

$$G(s) = \frac{1}{s + 7}\left(\frac{5}{s}\right) = \frac{5}{s^2 + 7s}$$

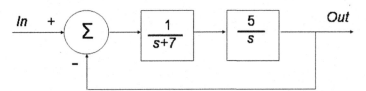

FIGURE 7.7 System used in Example 7.5. The transfer function of this system is to be determined.

---

[5]Even a sine wave has a Laplace transform, but it is complicated. It is much easier to use the frequency domain techniques developed in Chapter 6 for sine waves as long as they are in steady state. If the sine wave is not in steady state, but starts at some particular time, say $t = 0$, then Laplace techniques and the Laplace transform of a sine wave must be used.

Result: Substituting into the feedback equation gives the transfer function:

$$TF(s) = \frac{G(s)}{1 + G(s)H(s)} = \frac{\dfrac{5}{s^2 + 7s}}{1 + \dfrac{5}{s^2 + 7s}} = \frac{5}{s^2 + 7s + 5}$$

The s^2 in the characteristic equation indicates that this is a second-order system. These elements are covered in the next section.

7.3.6 Second-Order Element

Second-order processes have an s^2 in the denominator of the transfer function, usually as a quadratic polynomial:

$$TF(s) = \frac{1}{s^2 + 2\delta\omega_n s + \omega_n^2} = \frac{1}{s^2 + bs + c} \tag{7.31}$$

The denominator of the right-hand term is the familiar notation for a standard quadratic equation and can be factored using Equation 7.32. As we find later, second-order systems must contain two energy storage devices.

One method for dealing with second-order terms would be to factor them into first-order terms. We could then use partial fraction expansion to expand the two factors into two terms of the form $s + \alpha$. This approach is perfectly satisfactory if the factors, the roots of the quadratic equation, are real. Examination of the classic quadratic equation demonstrates when this approach will work. Since the coefficient of the s^2 term is always normalized to 1.0, the a coefficient in the quadratic is always 1.0 and the roots of the quadratic equation become:

$$r_1, r_2 = \frac{-b}{2} \pm \frac{1}{2}\sqrt{b^2 - 4c} \tag{7.32}$$

If $b^2 \geq 4c$ then the roots will be real and the quadratic can be factored into two first-order terms: $s - r_1$ and $s - r_2$. However, if $b^2 < 4c$, both roots will be complex and have real and imaginary parts:

$$r_1 = \frac{-b}{2} + j\frac{1}{2}\sqrt{4c - b^2} \text{ and } r_2 = \frac{-b}{2} - j\frac{1}{2}\sqrt{4c - b^2} \tag{7.33}$$

If the roots are complex, they both have the same real part $(-b/2)$, whereas the imaginary parts also have the same values but with opposite signs. Complex number pairs that feature this relationship, the same real part but oppositely signed imaginary parts, are called "complex conjugates."

Whether or not the roots of a second-order characteristic equation are real or complex has important consequences in the behavior of the system. Sometimes all we need to know about a second-order system is whether the roots in the characteristic equation are real or imaginary. This saves the effort of finding the inverse Laplace transform.

The second-order transfer function variables δ and ω_n were introduced in Chapter 6. Recall that the parameter δ is called the "damping factor," whereas ω_n is called the "undamped natural frequency." As with the term "time constant," applied to first-order systems, these names relate to the step and impulse response behavior of the second-order systems. As is shown later, second-order systems with low damping factors will respond with an exponentially decaying oscillation, and the smaller the damping factor, the slower the decay. The rate of oscillation is related to ω_n. Specifically, the rate of oscillation, ω_d, of these underdamped systems is:

$$\omega_d = \omega_n \sqrt{1 - \delta^2} \tag{7.34}$$

So as δ becomes smaller and smaller, the oscillation frequency ω_d approaches ω_n. When δ equals 0.0, the system is "undamped" and the oscillation continues forever at frequency ω_n. Hence the term "undamped natural frequency" for ω_n: it is the frequency at which the system oscillates if it has no damping; i.e., $\delta \to 0$.

Here we equate the variables ω_n and δ to coefficients a and b in Equation 7.31 and insert them into the solution to the quadratic equation:

$$r_1, r_2 = \frac{-2\delta\omega_n}{2} \pm \frac{\sqrt{4\delta^2\omega_n^2 - 4\omega_n^2}}{2} = -\delta\omega_n \pm \omega_n \sqrt{\delta^2 - 1} = -\delta\omega_n \pm \omega_d \tag{7.35}$$

From this equation, we see that the damping factor δ alone determines if the roots will be real or complex. Specifically, if $\delta > 1$ then the constant under the square root is positive and the roots will be real. Conversely, if $\delta < 1$ the square root will be a negative number and the roots will be complex. If $\delta = 1$, the two roots are also real, but are the same: both roots equal $-\omega_n$.

Again, the behavior of the system is quite different if the roots are real or imaginary, and the form of the inverse Laplace transform is also different. Accordingly, it is best to examine the behavior of a second-order system with real roots and complex roots separately: they act as two different animals.

7.3.6.1 Second-Order Elements With Real Roots

If the roots are real (i.e., $\delta > 1$), the system is said to be "overdamped" because it does not exhibit oscillatory behavior in response to a step or impulse input. Such systems have responses consisting of double exponentials.

The best way to analyze overdamped second-order systems is to factor the quadratic equation into two first-order terms each with its own time constant, τ_1 and τ_2:

$$TF(s) = \frac{1}{s^2 + 2\delta\omega_n s + \omega_n^2} = \frac{1}{\left(s + \dfrac{1}{\tau_1}\right)\left(s + \dfrac{1}{\tau_2}\right)} \tag{7.36}$$

These two time constants are the negative of the inverted roots of the quadratic equation as given by Equation 7.32. In other words, $\tau_1 = -1/r_1$ and $\tau_2 = -1/r_2$. The second-order transfer function can also have numerator terms other than 1, but the essential behavior does not change and the analysis strategy does not change. Typical overdamped impulse and step responses will be shown in the following example.

After the quadratic equation is factored, the next step is either to separate this function into two individual first-order terms $\frac{k_1}{s+1/\tau_1} + \frac{k_2}{s+1/\tau_2}$ using partial fraction expansion (see later discussion), or find a Laplace transform that generally matches the unfactored equation in Equation 7.36. If the numerator is a constant, then entry 9 in the Laplace Transform Table (Appendix B) matches the unfactored form. If the numerator is more complicated, a match may not be found and partial fraction expansion will be necessary.

Figure 7.8 shows the response of a second-order system to a unit step input. The step responses shown are for an element with four different combinations of the two time constants: τ_1, $\tau_2 = 1$, 0.2; 1, 2; 4, 0.2; and 4, 2 s. The time it takes for the output of this element to reach its final value depends on both time constants, but is primarily a factor of the slowest time constant. The next example illustrates the use of Laplace transform analysis to determine the impulse and step responses of a typical second-order overdamped system.

7.3.6.2 Partial Fraction Expansion—Manual Methods

Partial fraction expansion is the opposite of finding the common denominator: instead of trying to combine fractions into a single fraction, we are trying to break them apart. We want to determine what series of simple fractions will add up to the fraction we are trying to decompose. The technique described here is a simplified version of partial fraction expansion that deals with distinct linear factors, that is, denominator components of the form $(s - p)$. Moreover, this analysis will be concerned only with single components, not multiple components such as $(s - p)^2$.

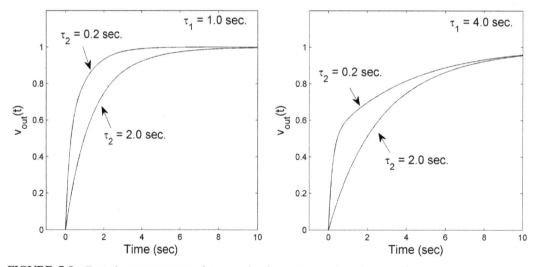

FIGURE 7.8　Typical step responses of a second-order system with real roots. These responses are termed overdamped relating to the exponential-like behavior of the response. Four different combinations of τ_1 and τ_2 are shown. The speed of the response depends on both time constants, but is dominated by the slower of the two.

Under these restrictions: the partial fraction expansion can be defined as:

$$TF(s) = \frac{N(s)}{(s - p_1)(s - p_2)(s - p_3)\ldots} = \frac{k_1}{s - p_1} + \frac{k_2}{s - p_2} + \frac{k_3}{s - p_{31}} + \ldots \tag{7.40}$$

$$k_n = (s - p_n)TF(s)|_{s=p_n} \tag{7.41}$$

Since the constants in the Laplace equation denominator will always be positive, the values of p will always be negative. The next example uses partial fraction expansion to find the step response of a second-order system.

EXAMPLE 7.6

Find the impulse and step responses of the following second-order transfer function. Use Laplace analysis to find the time functions and MATLAB to plot the two time responses.

$$TF(s) = \frac{25}{s^2 + 12.5s + 25} \tag{7.37}$$

Solution, General: First find the values of δ and ω_n by equating coefficients with the basic equation shown in Equation 7.31:

$$\omega_n^2 = 25; \quad \omega_n = 5$$

$$2\delta\omega_n = 12.5; \quad \delta = \frac{12.5}{2\omega_n} = \frac{12.5}{10} = 1.25$$

Since $\delta > 1$, the roots are real. In this case, the next step is to factor the denominator using the quadratic equation, Equation 7.32[6]:

$$r_1, r_2 = \frac{-b}{2} \pm \frac{1}{2}\sqrt{b^2 - 4c} = \frac{-12.5}{2} \pm \frac{1}{2}\sqrt{12.5^2 - 4(25)} = -6.25 \pm 3.75$$

$$r_1, r_2 = -10.0 \text{ and } -2.5$$

Solution, Impulse Response: For an impulse function input, the output $V_{out}(s)$ is the same as the transfer function since the Laplace transform of the input is $V_{in}(s) = 1$:

$$V_{out}(s) = \frac{25}{(s + 10)(s + 2.5)} = \left(\frac{25}{7.5}\right)\frac{7.5}{(s + 10)(s + 2.5)}$$

The right hand term of $V_{out}(s)$ has been rearranged to match entry #9 in the Laplace Transform Table:

$$\frac{\gamma - \alpha}{(s + \alpha)(s + \gamma)} \Leftrightarrow e^{-\alpha t} - e^{-\gamma t} \quad \text{where } \alpha = 2.5 \text{ and } \gamma = 10$$

So $v_{out}(t)$ becomes:

$$v_{out}(t) = \left(\frac{25}{7.5}\right)\left(e^{-2.5t} - e^{-10t}\right) = 3.33\left(e^{-2.5t} - e^{-10t}\right) \tag{7.38}$$

Solution, Step response: To find $V_{out}(s)$, multiply the transfer function by the Laplace transform of the step function, $1/s$:

$$V_{out}(s) = \left(\frac{1}{s}\right)\frac{25}{(s+2.5)(s+10)} = \frac{25}{s(s+2.5)(s+10)} \tag{7.39}$$

With the extra s added to the denominator, this function no longer matches any in the Laplace Transform Table. However, we can expand this function using partial fraction expansion into the form:

$$V_{out}(s) = \frac{k_1}{s} + \frac{k_2}{s+2.5} + \frac{k_3}{s+10}$$

Applying partial fraction expansion to the Laplace function of Equation 7.39, the values for p_1, p_2, and p_3 are, respectively: -0, -2.5, and -10, which produces the numerator terms k_1, k_2, and k_3:

$$k_1 = (s+0)\frac{25}{s(s+2.5)(s+10)}\Bigg|_{S=-0} = \frac{25}{2.5(10)} = 1.0$$

$$k_2 = (s+2.5)\frac{25}{s(s+2.5)(s+10)}\Bigg|_{S=-2.5} = \frac{25}{-2.5(-2.5+10)} = -1.33$$

$$k_3 = (s+10)\frac{25}{s(s+2.5)(s+10)}\Bigg|_{S=-10} = \frac{25}{-10(2.5-10)} = 0.33$$

This gives rise to the expanded version of Equation 7.39:

$$V_{out}(s) = \frac{25}{s(s+2.5)(s+10)} = \frac{1}{s} - \frac{1.33}{s+2.5} + \frac{0.33}{s+10} \tag{7.42}$$

Each of the terms in Equation 7.42 has an entry in the Laplace Table. Taking the inverse Laplace transform of each of these terms separately gives:

$$v_{out}(t) = 1.0 - 1.33e^{-2.5t} + 0.33e^{-10t} \tag{7.43}$$

MATLAB is used as in the Example 7.4 to plot the two resulting time domain equations, Equations 7.38 and 7.43 in Figure 7.9.

The next example reiterates the role of partial fraction expansion in the solution of the inverse Laplace transform.

---

[6]If you are challenged by basic arithmetic you can easily factor any polynomial using the MATLAB `roots` routine. An example of using this routine to factor a fourth-order polynomial is given in Example 7.9.

7.3.6.3 Partial Fraction Expansion—MATLAB

Not surprisingly, MATLAB has a routine to perform partial fraction expansion. The routine also finds the roots of the denominator. (A MATLAB routine that only calculates the roots of a polynomial is described in the next section.) To expand a ratio of polynomials, you first define the numerator and denominator polynomials using two vectors: one that specifies the numerator coefficients, the other the denominator coefficients. For example, for the transfer function:

$$TF(s) = \frac{s^2 + 5s + 15}{s(s^2 + 12.5s + 25)} = \frac{s^2 + 5s + 15}{s^3 + 12.5s^2 + 25s}$$

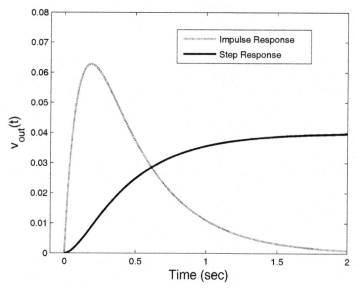

FIGURE 7.9 The impulse (*dashed curve*) and step (*solid curve*) responses of the transfer function used in Example 7.7.

The vector defining the numerator would be b = [1, 5, 15];, whereas the denominator would be defined as a = [1, 12.5, 25 0];. Note that since the constant term is missing in the denominator, a zero must be added to the vector definition. Once these vectors are defined, the partial fraction expansion terms are found using:

```
[numerators, roots, const] = residue(b,a)  % Partial fraction expansion
```

This routine gives the outputs:

```
numerators =
  0.8667
 -0.4667
  0.6000

roots =
 -10.0000
 -2.5000
  0

const =
  [ ]
```

and leads to the expanded transfer function.

$$TF(s) = \frac{0.6}{s} - \frac{0.47}{s + 2.5} + \frac{0.87}{s + 10}$$

The roots are the same as in Example 7.9 because the denominator equation is the same. The basic components of the time domain solution found from the inverse Laplace transform will also be the same. The numerator influences only the amplitude of these components. This again shows why the denominator polynomial is called the "characteristic equation" (Section 7.2.5.1).

EXAMPLE 7.7

Find the step response of the system having the following transfer function. Determine the solution analytically and also use MATLAB's residue routine.

$$TF(s) = \frac{\frac{s}{4}}{s^2 + 15s + 50}$$

Solution, Analytical: Since $In(s) = 1/s$, the output in Laplace notation becomes:

$$Out(s) = \left(\frac{1}{s}\right)\frac{\frac{s}{4}}{s^2 + 15s + 50} = \frac{.25}{s^2 + 15s + 50} \tag{7.44}$$

Next, we evaluate the value of δ by equating coefficients:

$$2\delta\omega_n = 15; \quad \delta = \frac{15}{2\omega_n} = \frac{15}{2\sqrt{50}} = 1.06$$

Since $\delta = 1.07$, the roots will be real and the system will be overdamped. The next step is to factor the roots using the quadratic equation, Equation 7.30:

$$r_1, r_2 = \frac{-15}{2} \pm \frac{1}{2}\sqrt{15^2 - 4(50)} = -7.5 \pm 2.5 = -10.0, -5.0$$

and the output becomes: $Out(s) = \frac{.25}{(s + 10)(s + 5)}$

The inverse Laplace transform for this equation can be found in Appendix B (#9) where $\gamma = 10$ and $\alpha = 5$. To get the numerator constants to match, multiply top and bottom by $10 - 5 = 5$:

$$Out(s) = \left(\frac{0.25}{5}\right)\frac{5}{(s + 10)(s + 5)} = (0.05)\frac{5}{(s + 10)(s + 5)}$$

From the Laplace Transform Table we get the time function:

$$out(t) = 0.05(e^{-5t} - e^{-10t}) \tag{7.45}$$

Solution, MATLAB: For those of us who are arithmetically challenged, this problem could be solved using the MATLAB residue routine described in Section 7.2.6.3. First define the numerator and denominator vectors: b = 0.25; and a = [1, 15, 50];. The MATLAB command:

```
[numerators, roots, const] = residue(b,a)    % Partial fraction expansion

Gives:

Numerators =
  -0.0500
   0.0500

Roots =
  -10
  -5

Const =
  [ ]
```

This gives the expanded equation:

$$Out(s) = \frac{0.05}{s+5} - \frac{0.05}{s+10}$$

Result: Taking the inverse Laplace transform of the two first-order terms leads to the same time function found using the unfactored transfer function. Figure 7.10 shows *out(t)* plotted using MATLAB code.

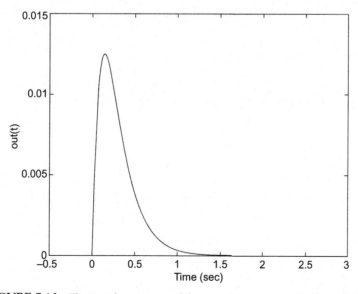

FIGURE 7.10 The impulse response of the second-order system in Example 7.7.

7.3.6.4 Second-Order Processes With Complex Roots

If the damping factor, δ, of a second-order transfer function is <1, then the roots of the characteristic (i.e., denominator) equation are complex and the step and pulse responses have the behavior of a damped sinusoid: a sinusoid that decreases in amplitude exponentially with time (i.e., of the general form $e^{-\alpha t}\sin\omega_n t$). Second-order systems that have complex roots are said to be "underdamped," a term relating to the oscillatory behavior of such systems. This terminology can be a bit confusing. Underdamped systems respond with damped sinusoidal behavior, whereas overdamped responses have double exponential responses and no oscillatory behavior.

If the roots are complex, the quadratic equation is not factored as the inverse Laplace transform can usually be determined directly from the Table. Inverse Laplace transforms for second-order underdamped responses are provided in the Table in terms of ω_n and δ and in terms of general coefficients (Transforms #13—17). Usually, the only difficulty in finding the inverse Laplace transform to these systems is in matching coefficients and scaling the transfer function to match the constants in the Table. The next example demonstrates the solution of a second-order underdamped system.

EXAMPLE 7.8

In a simple biomechanics experiment, a subject grips a rotating handle that is connected through an opaque screen to a mechanical system that can generate an impulse of torque. To create the torque impulse, a pendulum behind the screen strikes a lever arm attached to the handle. The subject is unaware of when the strike and the torque impulse it generates will occur. The resulting rotation of the wrist in degrees is measured under relaxed conditions. The rotation can be represented by the following second-order equation (Equation 7.46). As shown in Chapter 14, such an equation is typical of mechanical systems that contains a mass, elasticity, and friction or viscosity. In this example, we solve for the angular rotation, $\theta(t)$.

$$TF(s) = \frac{\theta(s)}{T(s)} = \frac{150}{s^2 + 6s + 310} \tag{7.46}$$

where $T(s)$ is the torque input and $\theta(s)$ is the rotation of the wrist in radians. The impulse function has a value of s 2×10^{-2} newton-meters.

Solutions: The values for δ and ω_n are found from the quadratic equation's coefficients.

$$\omega_n = \sqrt{310} = 17.6 \quad \text{and } \delta \text{ is}: \ 2\delta\omega_n = 6; \ \delta = \frac{6}{2\omega_n} = \frac{6}{2(17.6)} = 0.17$$

Since $\delta < 1$, the roots are complex.

The input $T(s)$ was an impulse of torque with a value of 2×10^{-2} newton-meters, so in the Laplace domain: $T(s) = 2 \times 10^{-2}$. $\theta(s)$ becomes:

$$\theta(s) = \left(2 \times 10^{-2}\right)\frac{150}{s^2 + 6s + 310} = \frac{3}{s^2 + 6s + 310} \tag{7.47}$$

Two of the transforms given in the Laplace tables will work (#13 and #15): entry #13 is used here with $b = 0$ (although #15 is more direct):

$$e^{-\alpha t}\left(\frac{c - b\alpha}{\beta}\sin(\beta t) + b\cos(\beta t)\right) \Leftrightarrow \frac{bs + c}{s^2 + 2\alpha s + \alpha^2 + \beta^2}$$

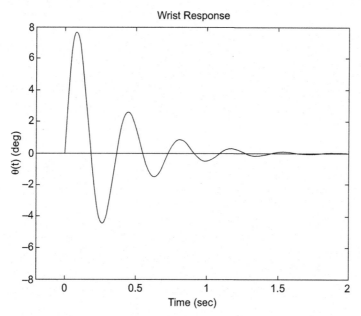

FIGURE 7.11 The response of the relaxed human wrist to an impulse of torque. This system is of second order, underdamped so the impulse response is a damped sinusoid.

Equating coefficients with Equation 7.47 we get:

$$c = 3; \quad b = 0; \quad \alpha = \frac{6}{2} = 3; \quad \alpha^2 + \beta^2 = 310; \quad \beta = \sqrt{310 - \alpha^2} = \sqrt{310 - 9} = 17.3$$

Substituting these values into the time domain equivalent on the left side of the Transform Table gives $\theta(t)$:

$$\theta(t) = e^{-3t}(0.173 \sin(17.3t)) = 10e^{-3t}(\sin(17.3t)) \; rad$$

Result: This response, scaled to degrees, is plotted in Figure 7.11 using MATLAB code.

7.3.7 Higher-Order Transfer Functions

Transfer functions that are higher than second-order can usually be factored into multiple second-order terms, or a combination of second- and first-order terms. The math would be tedious, so we turn to MATLAB for assistance. The MATLAB routine `roots` computes the roots of a polynomial. The polynomial equation is defined by a vector, c, that contains the polynomial's coefficients beginning with the highest power in the polynomial and continuing down to the constant. For example, the roots of the polynomial:

$$s^4 + 14s^3 + 25s^2 + 10s + 35$$

are found using: `roots([1, 14, 25, 10, 35]);`.

If a root is real, it is represented by a first-order term. Complex roots are given as complex conjugates and should be combined into second-order terms. This is achieved by multiplying the roots together. For example, the polynomial above has four roots: two are real and two are complex conjugates, $0.1431 \pm j1.1122$. We could combine these by multiplying $(s + 0.1431 + j1.1122)\,(s + 0.1431 - j1.1122)$. If we want to make our life easier and avoid doing the complex arithmetic, we could use the MATLAB routine poly, which performs the inverse of roots. So the command:

```
poly([0.1431 + j*1.1122, 0.1431 - j*1.1122]);
```

gives the polynomial coefficients: 1.0, −0.2862, 1.2575 corresponding to the second-order polynomial $s^2 - 0.2862s + 1.2575$.

The use of roots and poly to factor a fourth-order transfer function is illustrated in the next example. After factoring, we find the magnitude and phase spectrum. This example uses MATLAB to generate the plot, but the plotting could also have been done using the graphical Bode plot methods developed in Chapter 6.

EXAMPLE 7.9

Factor the transfer function shown below into first- and second-order terms. Write out the factored transfer function and plot the Bode plot (magnitude and phase). Use roots to factor the numerator and denominator and use poly to rearrange complex congregate roots into second-order terms. Generate the Bode plot from the factored transfer function.

$$TF(s) = \frac{s^3 + 5s^2 + 3s + 10}{s^4 + 12s^3 + 20s^2 + 14s + 10}$$

Solution: The numerator contains a third-order term and the denominator contains a fourth-order term. We use the MATLAB roots routine to factor these two polynomials and rearrange them into first- and second-order terms, then substitute $s = j\omega$ and plot. Of course, we could plot the transfer function directly without factoring, but the factors provide some insight into the system. The first part of the code factors the numerator and denominator.

```
% Example 7.9 Find the Bode plot of a higher order transfer function.
%
num = [1 5 3 10];          % Define numerator polynomial
den = [1 12 20 14 10];     % Define denominator polynomial
n_root = roots(num)        % Factor numerator and display
d_root = roots(den)        % Factor denominator and display
```

The results of this program are:

```
n_root = -4.8087
         -0.0957 + 1.4389i
         -0.0957 - 1.4389i
```

```
d_root = -10.1571
          -1.4283
          -0.2073 + 0.8039i
          -0.2073 - 0.8039i
```

The numerator factors into a single root and a complex pair indicating a second-order term. The denominator factors into two roots and one complex pair. To combine two complex roots into a single second-order term we use `poly`:

```
nun_sec_ord = poly([-0.0957+j*1.4389,-0.0957-j*1.4389])     % 2nd-order numerator
den_sec_order = poly([-0.2073+j*0.8039,-0.2073-j*0.8039])   % 2nd-order denominator
```

This gives the output:

```
nun_sec_ord = 1.0000  0.1914  2.0797
den_sec_order = 1.0000  0.4147  0.7893
```

Results: Combining these first- and second-order roots, the factored transfer function is written as:

$$TF(s) = \frac{(s + 4.81)(s^2 + 0.19s + 2.08)}{(s + 10.2)(s + 1.43)(s^2 + 0.41s + 0.69)}$$

The Bode plot can be obtained by substituting $j\omega$ for s in this equation and evaluating over ω where now each of the terms is identifiable as one of the Bode plot elements presented in Chapter 6. As mentioned, although we could use MATLAB to plot the unfactored transfer function, factoring shows us the Bode plot primitives (see Table 6.2), providing insight into what the elements make up the system. In this example, we use MATLAB to plot this function so there is no need to rearrange this equation into the standard Bode plot format (i.e., normalized so that the constant terms are equal to 1).

```
% Plot the Bode Plot
w = 0.1:.05:100;                         % Define the frequency vector
TF = (j*w + 4.81).*(-w.^2+j*0.19*w + 2.08)...
   ./((j*w +10.2).*(j*w+1.43).*(-w.^2+j*0.41*w+0.79));     % Define TF
subplot(2,1,1);
semilogx(w,20*log10(abs(TF)));           % Plot magnitude spectrum in dB
  .......labels.......
subplot(2,1,2);
  semilogx(w,angle(TF)*360/(2*pi));      % Plot the phase spectrum
  .......labels........
```

The resulting spectrum plot is shown in Figure 7.12.

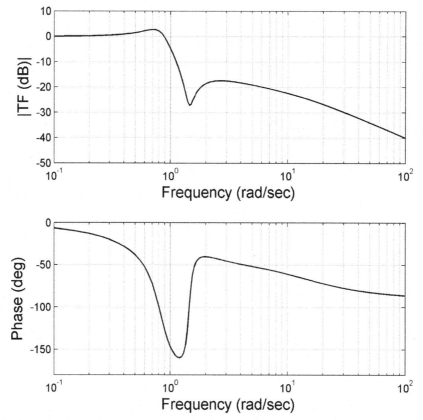

FIGURE 7.12   The magnitude and phase frequency characteristic (i.e., Bode plot) of the higher-order transfer function used in Example 7.9.

7.4 NONZERO INITIAL CONDITIONS—INITIAL AND FINAL VALUE THEOREMS

7.4.1 Nonzero Initial Conditions

Although the Laplace analysis method cannot deal with negative values of time, it can handle systems that have nonzero conditions at $t = 0$. So one way of dealing with systems that have a history for $t < 0$ is to summarize that history as an initial condition at $t = 0$. To evaluate systems with initial conditions, the full Laplace domain equations for differentiation and integration must be used. These equations, Equations 7.5 and 7.7 are repeated here.

$$\mathcal{L}\frac{dx(t)}{dt} = sX(s) - x(0-) \tag{7.48}$$

$$\mathcal{L}\left[\int_0^T x(t)dt\right] = \frac{1}{s}X(s) + \frac{1}{s}\int_{-\infty}^0 x(t)dt \tag{7.49}$$

In both cases, we add a term to the standard Laplace operator. In taking the derivative, the term is a constant: the negative of the initial value of the variable; that is, $-x(0-)$. With integration, the term is a constant divided by s where the constant is the integral of the past history of the variable from minus infinity to zero. This is because the current state of an integrative process is the result of integration over all past values.

So for systems with nonzero initial conditions, it is only necessary to add the appropriate terms in Equations 7.48 and 7.49. An example is given in the solution of a one-compartment diffusion model. This model is a simplification of diffusion in biological compartments such as the cardiovascular system. The one-compartment model would apply to large molecules in the blood that cannot diffuse into tissue and are only slowly eliminated. The compartment is assumed to provide perfect mixing; that is, the mixing of blood and substance is instantaneous and complete.

EXAMPLE 7.10

Find the concentration, $c(t)$, of a large molecule solute delivered as a step input to the blood compartment. Assume an initial concentration of $c(0)$ and a diffusion coefficient, K.

Solution: The kinetics of a one-compartment system with no outflow is given by the differential equation:

$$V\frac{dc(t)}{dt} = F_{in} - Kc(t)$$

where V is the volume of the compartment, K is a diffusion coefficient, and F_{in} is the input, which is assumed to be a step change of a given amount A: $F_{in} = A\mu(t)$. Converting this to the Laplace domain, $F_{in}(s) = A/s$. Next we assume that the initial concentration of solute is $c(0)$. Substituting A/s for $F_{in}(s)$, converting to Laplace notation, and applying the derivative operation with the initial condition (Equation 7.48):

$$V(sC(s) - c(0)) = \frac{A}{s} - KC(s)$$

Solving for the concentration, $C(s)$:

$$C(s)(Vs + K) = Vc(0) + \frac{A}{s}$$

$$C(s) = \frac{A}{s(Vs + K)} + \frac{Vc(0)}{Vs + K} = \frac{\frac{A}{V}}{s\left(s + \frac{K}{V}\right)} + \frac{c(0)}{s + \frac{K}{V}}$$

Taking the inverse Laplace transform:

$$c(t) = \frac{\frac{A}{V}}{\frac{K}{V}}\left(1 - e^{-tK/V}\right) + c(0)e^{-tK/V} = \left(\frac{A}{K}\right)\left(1 - e^{-tK/V}\right) + c(0)e^{-tK/V}$$

Results: We can interpret the solution as consisting of two components: the exponentially increasing concentration from an initial value of $c(0)$ to A/K resulting from the step input and an

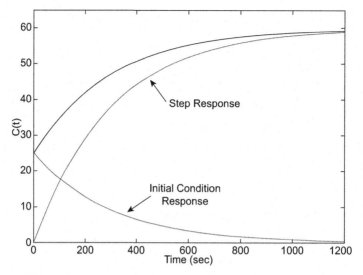

FIGURE 7.13   The diffusion of a large molecule solute delivered as a step input into the blood compartment. The molecule had an initial concentration of 25 g/mL. The solution (*dark curve*) can be viewed as having two components (*light curves*): the exponentially increasing concentration due to the step input and the exponential decay of the initial concentration. Perfect mixing in the blood compartment is assumed.

exponential decay of the initial value to zero. This second component is what we would see if the step input is not present. Using typical values from Rideout (1991) of $A = 0.7$ g, $V = 3.0$ mL, $K = 0.01$, $c(0) = 25.0$ g/mL, leads to the results for both components and their sum as plotted in Figure 7.13.

Nonzero initial conditions also occur in circuits where preexisting voltages exist or in mechanical systems that have nonzero initial velocities, so you can look forward to more examples of systems with nonzero initial conditions in Chapter 13.

7.4.2 Initial and Final Value Theorems

The time representation of a Laplace function is obtained by taking the inverse Laplace transform using tables such as that found in Appendix B. But sometimes we are only looking for the value of the function at the very beginning of the stimulus, $t = 0$, or at its very end, $t \to \infty$. Two useful theorems can supply us with this information without the need to take the inverse Laplace transform: the "Initial and Final Value Theorems." These theorems give us the initial and the final output values directly from the Laplace transform equation.

Recall that frequency and time are inversely related: $f = 1/T$. For example, as the period of a sine wave increases, its frequency decreases and vice versa. The Laplace variable, s, is complex frequency so we might expect to find the value of a Laplace function at $t = 0$ by letting $s \to \infty$. The Initial Value Theorem supports this idea, but we must first multiply $X(s)$ by s, then let $s \to \infty$:

$$x(0+) = \lim_{t \to 0} x(t) = \lim_{s \to \infty} sX(s) \tag{7.50}$$

The Final Value Theorem follows the same logic, but now since $t \to \infty$, it is s that goes to 0. The Final Value Theorem states:

$$x(\infty) = \lim_{t \to \infty} x(t) = \lim_{s \to 0} sX(s) \tag{7.51}$$

The application of either theorem is straightforward as is shown in the following example.

EXAMPLE 7.11

Use the Final Value Theorem to find the final value of $x(t)$ to a step input for the system whose transfer function is given below. Also find the final value the hard way: by determining $x(t)$ from the inverse Laplace transform, then letting $t \to \infty$.

$$TF(s) = \frac{.28s + .23}{s^2 + 0.3s + 2} \tag{7.52}$$

Solution, Inverse Laplace Transform: First find the output Laplace function $X(s)$ by multiplying $TF(s)$ by the step input function in the Laplace domain:

$$X(s) = \frac{1}{s}TF(s) = \frac{.28s + 0.92}{s(s^2 + 0.03s + 2)} \tag{7.53}$$

Next we find the full expression for $x(t)$ from the inverse Laplace transform. We need to examine δ to determine if the roots are real or complex:

$$2\delta\omega_n = 0.3; \quad \delta = \frac{0.3}{2\omega n} = \frac{0.3}{2\sqrt{2}} = 0.106 < 1.$$

So the system is underdamped and one of the transfer functions #13–17 in the Laplace Transform Table should be used to find the inverse. The function $X(s)$ matches entry #14 in the Laplace Transform Table, but requires some rescaling to match the numerator.

$$1 - e^{-\alpha t}\left(\frac{\alpha - b}{\beta}\right)\sin \beta t + b \cos \beta t \Leftrightarrow \frac{bs + \alpha^2 + \beta^2}{s(s^2 + 2\alpha s + \alpha^2 + \beta^2)}$$

Considering only the denominator, the sum $\alpha^2 + \beta^2$ should equal 2, but then the numerator needs to be rescaled so the numerator constant is also 2 (i.e., $\alpha^2 + \beta^2 = 2$). To make the numerator constants match we need to multiply the numerator in Equation 7.53 by: $2/0.92 = 2.17$. Multiplying top and bottom by 2.17, the rescaled Laplace function becomes:

$$X(s) = \left(\frac{1}{2.17}\right)\frac{0.61s + 2}{s(s^2 + 0.3s + 2)} = \frac{0.46(0.61s + 2)}{s(s^2 + 0.3s + 2)}$$

Now equating coefficients with entry #14:

$$b = .61; \quad \alpha = \frac{0.3}{2} = 0.15; \quad \alpha^2 + \beta^2 = 2; \quad \beta^2 = 2 - 0.15^2; \quad \beta = \sqrt{1.98} = 1.41$$

The inverse Laplace Transform becomes:

$$x(t) = 0.46\left[1 - e^{-.15t}\left(\frac{.15 - .61}{1.41}\sin(1.41t) + \cos(1.41t)\right)\right]$$

$$x(t) = 0.46\left(1 - e^{-.15t}(-0.33\sin(1.41t) + \cos(1.41t))\right)$$

Now letting $t \to \infty$, the exponential term goes to zero and the final value becomes:

$$x(t) = 0.46$$

Solution, Final Value Theorem: This approach is much easier. Substitute the output Laplace function given in Equation 7.52 into Equation 7.51:

$$\lim_{s \to 0} sX(s) = \lim_{s \to 0} s \left[\frac{0.281s + 0.92}{s(s^2 + 0.3s + 2)} \right] = \lim_{s \to 0} \left[\frac{0.28s + 0.92}{s^2 + 0.3s + 2} \right] = \frac{0.921}{2} = 0.46$$

This is the same value that is obtained when letting $t \to \infty$ in the time solution $x(t)$.

An example of the application of the Initial Value Theorem is found in the problems.

7.5 THE LAPLACE DOMAIN, THE FREQUENCY DOMAIN, AND THE TIME DOMAIN

We know how to move between the Laplace domain and the time domain: take the Laplace transform or its inverse. To move between the frequency domain and the time domain: take the Fourier transform or its inverse. To move from the Laplace domain to the frequency domain, we note that s is complex frequency, $\sigma + j\omega$, so we just do away with the real part, σ, and substitute $j\omega$ for s. (We should also renormalize the coefficients so the constant term equals 1, particularly if we are planning to use Bode plot techniques.) To make this transformation, we must assume that we are dealing with periodic steady-state signals only. Since we usually work with systems in the Laplace domain, converting the other way, from the frequency domain to the Laplace domain ($j\omega \to s$), is not common and we would also need to assume zero initial conditions. These transformations are summarized in Table 7.1.

Some of the relationships between the Laplace transfer function and the frequency domain characteristics have already been mentioned, and these depend largely on the characteristic equation. A first-order characteristic equation gives rise to first-order frequency characteristics such as those shown in Figures 6.13 and 6.14.

Second-order frequency characteristics, like second-order time responses, are highly dependent on the value of the damping coefficient, δ. As shown in Figure 6.17, the frequency curve shows a peak for values of $\delta < 1$, and the height of that peak increases as δ decreases. This peak occurs at the undamped natural frequency, ω_n.

The peaks in the frequency domain have dramatic correlates in the time domain. As illustrated in the next example, when $\delta < 1$ the response overshoots the final value, oscillating around this value at frequency ω_d. This oscillation frequency is not quite the same as the undamped frequency, ω_n, although for small values of δ the oscillation frequency approaches ω_n (Equation 7.34). The undamped natural frequency, ω_n, is the frequency that the system would like to oscillate at if there was no damping, that is, if it could oscillate unimpeded. But the decay in the oscillation caused by the damping lowers the actual oscillation frequency to ω_d. As the damping factor, δ, decreases, its influence on oscillation frequency, ω_d, is reduced, so it approaches the undamped natural frequency, ω_n, as described in Equation 7.34.

The next example uses MATLAB to compare the frequency and time characteristics of a second-order system for various values of damping.

EXAMPLE 7.12

A system with the following transfer function has an undamped natural frequency of 10,000 rad/s and can have three different values of the damping factor: 0.05, 0.1, and 0.5.

$$TF(s) = \frac{1}{s^2 + 200\delta s + 10^4}$$

Plot the frequency spectrum (i.e., Bode plot) of the time domain transfer function and the step response of the system for the three damping factors. Use the Laplace transform to solve for the time response and MATLAB for calculation and plotting.

Solution, Time Domain: The step response can be obtained by multiplying the transfer function by $1/s$, then determining the inverse Laplace transform leaving δ as a variable:

$$V_{out}(s) = \frac{1}{s(s^2 + 2 * 10^2 \delta s + 10^4)}$$

This matches entry #17 for $\delta < 1$, although a minor rescaling is required:

$$V_{out}(s) = \left(\frac{1}{10^4}\right) \frac{10^4}{s(s^2 + 200\delta s + 10^4)}$$

The time domain solution is:

$$v_{out}(t) = \left(\frac{1}{10^4}\right)\left(1 - \frac{e^{-\delta\omega_n t}}{\sqrt{1-\delta^2}}\sin\left(\omega_n\sqrt{1-\delta^2}\,t + \theta\right)\right); \quad \theta = \tan^{-1}\left(\frac{\sqrt{1-\delta^2}}{\delta}\right)$$

where $\omega_n^2 = 10^4$, $\omega_n = 100$.

The equivalent time response will be programmed directly into the MATLAB code.

Solution: Frequency Domain: To find the frequency response, convert the Laplace transfer function to a frequency domain transfer function by substituting $j\omega$ for s and rearranging into frequency domain format where the constant is normalized to 1.0:

$$TF(\omega) = \frac{1}{(j\omega)^2 + 200\delta j\omega + 10^4} = \frac{1}{1 - 10^{-4}\omega^2 + 0.02\delta j\omega}$$

This equation can also be programmed directly into MATLAB. The following is the resulting program.

```
% Example 7.12 Comparison of time and frequency characteristics of a second-order
system
%  with different damping factors.
%
w = 1:1:10000;              % Frequency vector: 1 to 10,000 rad/sec
t = 0:.0001:.4;            % Time vector: 0 to 0.4 sec.
delta = [0.5 0.1 0.05];    % Damping factors
% Calculate spectra
hold on;                   % Plot superimposed
for k = 1:length(delta)    % Repeat for each damping factor
```

```
TF = 1./(1 - 10.^-4*w.^2 + j*delta(k)*.02*w);  % Transfer function
Mag = 20*log10(abs(TF));                        % Take magnitude in dB
semilogx(w,Mag);                                % Plot semilog (log frequency)
end
........labels and title........
% Calculate and plot time characteristics
figure ; hold on;                               % Plot superimposed
wn = 100;                                        % Undamped natural frequency
for k = 1:length(delta)
  c = sqrt(1-delta(k)^2);                        % Useful constant
  theta = atan(c/delta(k));                      % Calculate theta
  xt = (1 - (1/c)*(exp(-delta(k)*wn*t)).*(sin(wn*c*t + theta)))/1000;  % Time
function
  plot(t,xt);
end
.......label, title, and text.......
```

Results: The results are shown in Figures 7.14 and 7.15. The correspondence between the frequency characteristics and the time responses is evident. When a system's spectrum has a peak, the system's time response has an overshoot. Note that the frequency peak associated with a δ of 0.5 is modest, but the time domain response still has some overshoot. As shown here and in one of the problems, the larger the spectral peak, the greater the overshoot. The minimum δ for no overshoot in the response is left as an exercise in one of the problems.

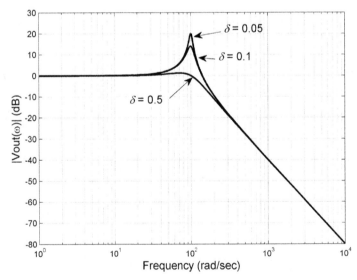

FIGURE 7.14 Comparison of magnitude spectra of the second-order system used in Example 7.12 having three different damping coefficients: $\delta = 0.05, 0.1$, and 0.05.

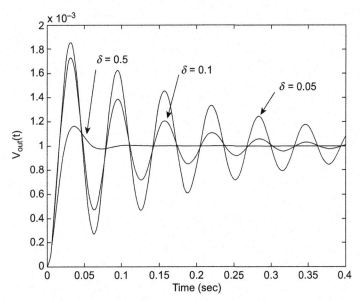

FIGURE 7.15 Comparison of time domain step responses for a second-order system with three different damping factors. The magnitude spectrum for this system is shown in Figure 7.14.

7.6 SYSTEM IDENTIFICATION

Bioengineers often face complex systems with unknown internal components. In such cases, it is usually impossible to develop equations for system behavior to construct the frequency characteristics. However, if you can control the stimulus to the system, and measure its response, you should be able to determine the system's spectrum experimentally, provided that the system is linear or can be taken as linear. Once you determine the system's spectrum, its transfer function can be estimated using Bode plot methods from Chapter 6. Then the system's response to any input can easily be computed. Finding a system's transfer function from external behavior is called "system identification."

The are several approaches to identifying a system if we can control, or at least have access to, its input. If we can generate an impulse input, we can determine its spectrum from the impulse response by taking the Fourier transform of the impulse response (see Section 5.3.2). If the inputs are sinusoidal, or are decomposed into sinusoids, then we can estimate the frequency characteristics by taking the ratio of output amplitude to input amplitude at each frequency. That is, we transform the input (if needed) and output signals to the frequency domain and divide $Output(\omega)$ by $Input(\omega)$ to get $TF(\omega)$, as shown in Equation 6.23. This approach works if we can measure the input signal as long as that signal contains energy over the frequency range of the system.

Another powerful method that works as long as we can measure both input and output signals is model-based simulation. In this approach, we construct a model we believe represents the system, then adjust the model parameters (or elements) until the inputs and outputs match our data. The result is not just a transfer function that matches our system, but a representation of possible internal biological elements. Simulation approaches are explored in Chapter 9.

Here we use frequency-based methods to find the spectrum of several unknown systems, then apply Bode plot primitives to estimate the transfer function. In the following three examples of system identification, the first uses Fourier decomposition, the second uses sinusoidal and impulse inputs, and the last example uses measurements of the input and output signals.

EXAMPLE 7.13

Use white noise to estimate the magnitude spectrum of a system represented by MATLAB routine $y = \text{unknown\_sys7\_1(x).m}$. The input argument, x, is taken as the input signal and the output argument, y, is the output signal. Plot the magnitude spectrum, then apply Bode plot primitives to estimate the transfer function.

Solution: Since we have complete control of the input signal (not often the case in dealing with biological systems), we use an input signal that contains energy over a broad range of frequencies. As discussed in Chapter 1, a random white noise signal has equal energy over all frequencies or, for digital signals, equal energy up to $f_s/2$ (see Figure 1.13). With a random signal as our input, we take the Fourier transform of both this input signal and the system's output. We divide the two frequency domain signals and plot the magnitude (in dB) as the spectrum of our transfer function. As always, we only plot the meaningful spectral points below $f_s/2$.

We initially select a sampling frequency of 1000 Hz. This gives us a spectrum that ranges between 0 and 500 Hz. As we have no knowledge of the system's transfer function, we do not know if this range is sufficient to cover the frequencies of interest. We should be prepared to adjust f_s to cover other frequencies if they are needed to define the transfer function. We use a signal with a large number of points, $N = 10{,}000$, to improve the resolution.

```
% Example 7.13 Use MATLAB to find the transfer function of an unknown
% system from its input and output signals.
%
fs = 1000;                          % Assumed sample rate
N = 10000;                          % Number of points
nf = round(N/2);                    % Number of valid spectral points
f = (0:N-1)*fs/N;                   % Frequency vector for plotting
t = (1:N)/fs;                       % Time vector for plotting
x = rand(1,N);                      % Input signal
y = unknown_sys7_1(x);              % Unknown system
X = fft(x);                         % Convert input to freq. domain
Y = fft(y);                         % Convert output to freq. domain
TF = Y./X;                          % Find TF
Mag = 20*log10(abs(TF));            % TF in dB
semilogx(f(1:nf),Mag(1:nf),'k');    % Plot magnitude, valid points only
```

Results: The spectrum produced by this code is shown in Figure 7.16. To convert these frequency characteristics into transfer functions, we need to rely on the skills developed in the last chapter. The spectrum looks like a combination of three Bode plot primitives: a second-order underdamped

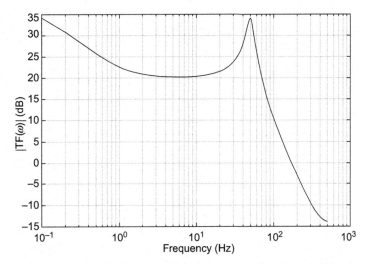

FIGURE 7.16 The spectrum of a signal from the unknown system used in Example 7.13 when the input is white noise. Since white noise has energy at all frequencies, this spectrum is the same as the system's spectrum and can be used to estimate the transfer function.

element, an integrator (i.e., $1/j\omega$), and an inverted low-pass filter (a numerator $1 + j\omega/\omega_1$). In fact, it looks a lot like the spectrum in Figure 6.25, which was derived from a transfer function having these three elements. Thus the transfer function of this system has the general form:

$$TF(\omega) = K \frac{1 + j\dfrac{\omega}{\omega_1}}{j\omega\left(1 + \left(\dfrac{\omega}{\omega_n}\right)^2 + j2\delta\dfrac{\omega}{\omega_n}\right)}$$

We can get the parameters for ω_1, ω_n, and δ from the spectral plot. The peak occurs at 50 Hz, so $\omega_n = 2\pi f = 2\pi(50) = 314$ rad sec. and the descending low frequency slope is within 3 dB of leveling off at about 2 Hz so $\omega_1 = 2\pi f = 2\pi(2) = 12.6$ rad/sec. To find the baseline gain, K, note that at $f = 0$ Hz, the integrator element has a gain of 0.0 (Figure 6.11) as do the other elements, but the spectrum has a gain of 20 dB. So K must be 10 (i.e., 20 dB). To find δ, note that the peak is about 14 dB above the baseline, so:

$$-20\log(2\delta) = 14dB; \quad -\log(2\delta) = 0.7; \quad 1/(2\delta) = 10^{-0.7}; \quad \delta = .1999/2 = 0.1$$

This gives the complete transfer function as:

$$TF(\omega) = 10 \frac{1 + j.08\omega}{j\omega\left(1 - (0.0032\omega)^2 + j0.00064\omega\right)}$$

If we ignore initial conditions: we can rewrite this transfer function in Laplace notation:

$$TF(s) = \frac{0.8s\left(s + \dfrac{1}{0.08}\right)}{0.0032^2 s\left(s^2 + \dfrac{0.00064s}{0.0032^2} + \dfrac{1}{0.0032^2}\right)} = \frac{78125(s + 12.6)}{s(s^2 + 62.5s + 97656)}$$

White noise is not an easy stimulus to induce in most biological systems. Another way to determine frequency response experimentally is to take advantage of the fact that a sinusoidal stimulus into a linear system will produce a sinusoidal response at the same frequency. By stimulating the biological system with sinusoids over a range of frequencies and measuring the change in amplitude and phase of the response, we can construct a plot of the frequency characteristics by simply combining all the individual measurements. This approach is illustrated in the next example.

EXAMPLE 7.14

Find the magnitude of spectral characteristics of the process represented by `unknown_sys7_2(x).m`. Use sinusoids to identify the spectrum and Bode plot primitives to estimate the transfer function. Also determine the system spectrum from the impulse response and compare. The actual magnitude spectrum in dB can be found as the second output argument of `unknown_sys7_2(x).m`.

Solution: Generate a sinusoid with an RMS value of 1.0. This requires the amplitude to be 1.414. Input this sinusoid to the unknown process, and measure the RMS value of the output. The RMS value is usually a more accurate measurement of a signal value than the peak-to-peak amplitude as it is less susceptible to noise-induced error. Repeat this protocol for increasing frequencies until the output falls to very low levels (in this case up to 400 Hz). Plot the results in dB against log frequency. Also construct an impulse signal and input it to the system. Take the Fourier transform of the impulse response and plot in dB against log frequency. Finally, plot the actual system spectrum and compare it with the two experimentally obtained spectra.

To find the transfer function from the spectrum, use the sinusoidal responses as they are likely to be more accurate. Estimate asymptotes if need be and apply Bode techniques to determine the transfer function.

```
% Example 7.14 Identify an unknown system unknown_sys7_2.
%
N = 1000;                        % Input signal length
t = (0:N-1)/N;                   % Time vector ( 1 sec)
for k = 1:400                    % Try frequencies up to 400 Hz
  f(k) = k;                      % Freq = k = 1-400Hz
  x = 1.414*cos(2*pi*f(k)*t);    % Sinusoid with RMS = 1.0
  y = unknown_sys7_2(x);         % Input stimulus to process
  Y(k) = sqrt(mean(y.^2));       % Calculate RMS value as magnitude
end
Y = 20*log10(Y);                 % Convert to dB
% Now use the impulse response for system identification
x1 = [1.0, zeros(1,N-1)];        % Generate impulse response
[y1, spec] = unknown_sys7_2(x1);       % Get impulse response and true spectrum
Mag = abs(fft(y1));              % Get magnitude spectrum
Y1 = 20*log10(Mag(1:400));       % Convert to dB
semilogx(f,Y,'k','LineWidth',1); hold on;    % Plot magnitude as dB/log f
semilogx(f,Y1,':k','LineWidth',2);           % Plot as dB/log f
semilogx(f,spec(1:400),'--k','LineWidth',2); % Plot actual spectrum
   .......labels.......
```

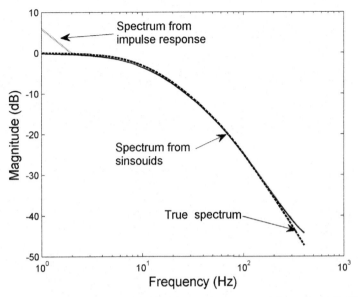

FIGURE 7.17 Three spectra generated in Example 7.14: one from sinusoidal simulation (*solid line*), another from the impulse response (*dashed line*), and a third showing the true spectrum (*dotted line*). Except at the lowest frequencies, the three spectra overlap and are hard to distinguish.

Results: The three spectra are plotted superimposed in Figure 7.17 with labels. Note that the spectrum generated from sinusoids (solid line) closely matches the true spectrum (dotted line). The spectrum determined from the impulse response (dashed line) deviates slightly at the higher frequencies. This is likely due to computational error at the low output amplitudes.

Based on the spectrum from sinusoidal stimulation, the process appears overdamped, as there are no spectral peaks (Figure 7.18). At the higher frequencies, the slope is 40 dB/decade, indicating a second-order system. The lower frequencies show a slope of 20 dB/decade. Applying Bode plot techniques, we fit the curve with lines of 20 and 40 dB/decade (Figure 7.18 dashed lines).

From the spectrum of Figure 7.18, it looks like this system consists of two first-order elements with cutoff frequencies somewhere around 10 and 80 Hz (or 63 and 503 rad/sec). The gain is around 0 dB or 1.0. So an estimate of the frequency domain transfer function of this system would be:

$$TF(\omega) = \frac{1.0}{\left(1 + \dfrac{j\omega}{63}\right)\left(1 + \dfrac{j\omega}{503}\right)} = \frac{1.0}{(1 + j0.0159\omega)(1 + j0.002j\omega)}$$

Assuming no initial conditions, the transfer function in Laplace notation becomes:

$$TF(s) = \frac{1.0}{(0.0159)(0.002)\left(s + \dfrac{1}{0.0159}\right)\left(s + \dfrac{1}{0.002}\right)} = \frac{3.14 \times 10^4}{(s + 63)(s + 503)}$$

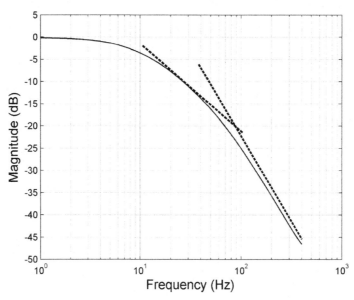

FIGURE 7.18 The magnitude spectrum of an unknown system that is represented in the routine `unknown_sys7_2.m`. The spectrum was determined by stimulating the system with sine waves ranging in frequency from 1 to 400 Hz. The stimulus sine waves all had root mean square (RMS) values of 1.0, so the RMS values of the output indicate the magnitude spectrum of the unknown system.

When it is not practical to simulate biological systems, the system's natural input can be used as long as it contains energy covering the range of the system's spectrum. To estimate the Fourier transform, you take the Fourier transform of the output signal and divide it by the Fourier transform of the input signal. The next example embodies this approach to identify a biological system.

EXAMPLE 7.15

The data file `bio_sys.mat` contains the input and output signals of a biological system in variables x and y, respectively. $f_s = 150$ Hz. This file also contains the true spectrum of the system in variable `spec`. Determine if the input signal can be used to accurately determine the system's spectrum. If so, estimate that spectrum and then use Bode plot primitives to find the related transfer function.

Solution: Use the Fourier transform to calculate both the input and output magnitude spectra, then plot. Check to see if the input spectrum has energies out to a frequency range where the output spectrum is considerably attenuated. In other words, ensure that a decrease in the output spectrum is due to the system and not insufficient energy in the input spectrum. Then divide the magnitude spectrum of the output by the input to get an estimate of the system's spectrum. Use Bode plot methods to estimate the transfer function.

```
% Example 7.15 Identify a biological system from input/output data.
%
load bio_sys;
fs = 150;                    % Sample frequency
N = length(x);               % Signal length
N_2 = round(N/2);            % Half signal length for fft
f = (1:N)*fs/N;              % Frequency vector for plotting
X = abs(fft(x));             % Fourier transform of the input signal
Y = abs(fft(y));             % Fourier transform of the output signal
.......linear plot of x and y, label, new figure........
TF = Y./X;                   % Calculate magnitude transfer function
TF_dB = 20*log10(TF);        % in dB
.......semilog plot, label........
```

Results: Figure 7.19 shows the input and output spectra plotted as linear functions. Although the energy in the input spectrum falls off at the higher frequencies, it does appear to have energy over the range of system output frequencies except possibly at 30 and 60 Hz.

Since the input spectrum appears to contain sufficient energy over the frequency range of interest, the ratio of output spectrum to input spectrum should give a reasonable estimate of the system's spectrum. The result of dividing the output spectrum by the input spectrum produces the system's spectrum estimate shown in Figure 7.20 (solid line). This estimated spectrum closely follows the actual spectrum (dashed line) except at the two frequency extremes. Despite the deviations at the high and low frequencies, the estimated spectrum is sufficient to determine the transfer function using Bode plot methods.

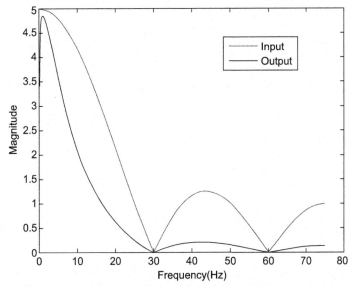

FIGURE 7.19 The magnitude spectra of input and output signals associated with an unknown biological system in Example 7.15. The input spectrum is seen to decrease with increasing frequency, but still exceeds that of the output spectrum except possibly at 30 and 60 Hz. This indicates that the attenuation seen in the output spectrum is not a result of insufficient energy in the input spectrum.

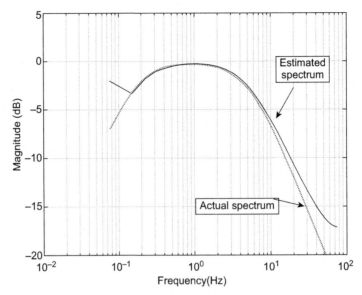

FIGURE 7.20   An estimate of system spectrum obtained by dividing the output spectrum by the input spectrum (*solid line*) and the system's actual spectrum (*dotted line*).

The system spectrum in Figure 7.20 has the shape of a band-pass filter with cutoff frequencies around 0.2 and 5 Hz corresponding to 1.26 and 31.4 rad/sec. The slope on either side appears to be 20 dB/decade and the midrange gain is near 0 dB or 1.0. Applying Bode plot techniques with this type of curve (see Example 6.11 and Equation 6.56) with these parameters leads to an estimated transfer function.

$$TF(\omega) = \frac{j\omega}{\left(1 + \frac{j\omega}{\omega_1}\right)\left(1 + \frac{j\omega}{\omega_2}\right)} = \frac{j\omega}{\left(1 + \frac{j\omega}{1.26}\right)\left(1 + \frac{j\omega}{31.4}\right)} = \frac{j\omega}{(1 + j0.794\omega)(1 + j0.032\omega)}$$

Again assuming no initial conditions, the Laplace transfer function is:

$$TF(s) = \frac{s}{(0.794)(0.032)(s + 1.26)(s + 31.4)} = \frac{39.4\,s}{(s + 1.26)(s + 31.4)}$$

If you can control, or at least measure, the stimulus and response of a system, the approaches used here can be very useful. Variations of both impulse and frequency response methods have been used to estimate the transfer function of the extraocular muscles, the iris, and lens muscles in the eye, and the response of chemoreceptors in the respiratory system and numerous other biosystems.

7.7 SUMMARY

With the Laplace transform, all of the analysis tools developed in Chapter 6 can be applied to systems exposed to a broader class of signals. Transfer functions written in terms of the

Laplace variable s (i.e., complex frequency) serve the same function as frequency domain transfer functions, but now include transient signals such as the step function. Here only the response to the step and impulse signals is used in examples because these are the two stimuli that are most commonly used in practice. Their popularity stems from the fact that they provide a great deal of insight into system behavior, and they are usually easy to generate in practical situations. However, responses to other signals such as ramps or exponentials, or any signal that has a Laplace transform, can be analyzed using these techniques. Laplace transform methods can also be extended to systems with nonzero initial conditions, a useful feature explored later.

The Laplace transform can be viewed as an extension of the Fourier transform where complex frequency, s, is used instead of imaginary frequency, $j\omega$. With this in mind, it is easy to convert from the Laplace domain to the frequency domain by substituting $j\omega$ for s in the Laplace transfer functions. Bode plot techniques can be applied to these converted transforms to construct the magnitude and phase spectra. Thus the Laplace transform serves as a gateway into both the frequency domain and the time domain through the inverse Laplace transform.

Determining a system's transfer function from external behavior is a broad and expanding area of signal processing known as systems identification. If it is practical to generate a sinusoidal or impulse input to a system and measure the response, we should be able to determine the spectral characteristics of that system. Alternatively, we can use the system's own natural stimulus as long as we can measure that stimulus and it contains energy of the spectral frequencies of interest. From the system's spectrum, we can use Bode plot methods developed in Chapter 6 to reconstruct the system's transfer function.

PROBLEMS

1. Find the Laplace transform of the following time functions:
   a.

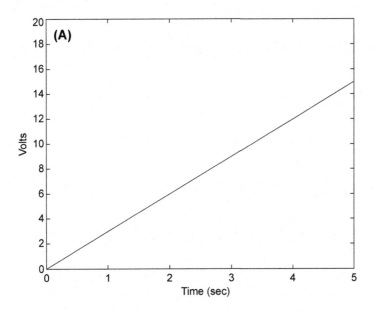

b.

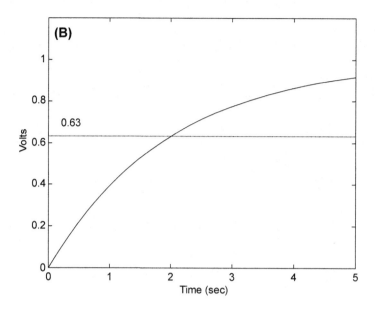

c. $(e^{-2t} - e^{-5t})$

d. $2e^{-3t} - 4e^{-6t}$

e. $5 + 3e^{-10t}$

2. Find the inverse Laplace transform of the following Laplace functions:

a. $\dfrac{10}{s+5}$

b. $\dfrac{10}{s(s+5)}$

c. $\dfrac{5s+4}{s^2+5s+20}$ (Hint: Check roots.)

d. $\dfrac{5s+4}{s(s^2+5s+20)}$

3. Use partial fraction expansion to find the inverse Laplace transform of these functions:

a. $\dfrac{s+4}{s^2+10s+10}$

b. $\dfrac{10}{s(s^2+4s+3)}$

4. Find the time response of the following system if the input is a step from 0 to 5:

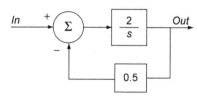

5. Find the time response of the following system to a unit step function. Use Laplace methods to solve for the time response as a function of k. Then use MATLAB to plot the time function for $k = 0.1$, 1, and 10.

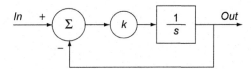

6. Find the time response of the following two systems if the input is a step from 0 to 8. Use MATLAB to plot the time responses. Plot superimposed.

(A)   **(B)**

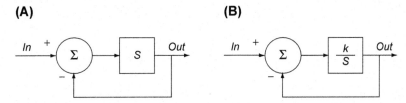

7. Solve for the Laplace transfer function of the following system where $k = 1$. Find the time response to a step from 0 to 4 and an impulse having a value of 4. (Hint: You can apply the feedback equation twice.)

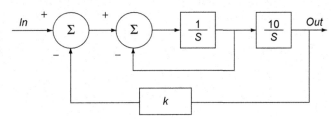

8. Find the time response of the system with the following transfer function if the input is a unit step. Also find the response to a unit impulse function. Assume $k = 5$.

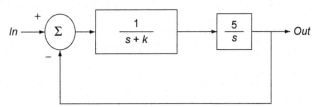

9. Repeat Problem 8 assuming $k = 0.2$.
10. Find the unit impulse response to the system in Problem 7 for $k = 1$ as on Problem 7 and $k = 0.1$. Use MATLAB to plot the two impulse responses. How does decreasing k change the output behavior?
11. Find the impulse response to the following system for $k = 1$ and $k = 0.1$. Use MATLAB to plot the two impulse responses. How does decreasing k change the output behavior? Compare with the results in Problem 10 to the same values of k.

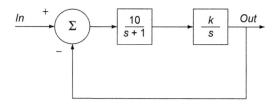

12. Use MATLAB to plot the magnitude spectra to the two systems shown in Problem 6. Plot superimposed.
13. Use MATLAB to plot the magnitude spectra of the system in Problem 10 to the two values of k. Repeat for the system in Problem 11.
14. Use Laplace analysis to find the transfer function of the following system that contains a time delay of 0.01 sec $(e^{-0.01s})$. Find the Laplace domain response of a step from 0 to 10. Then convert to the frequency domain and use MATLAB to find the magnitude and phase spectrum of the response. Note that such a time delay is typical in biological systems. Plot the system spectrum over a frequency range of 1 to 200 rad/sec. Owing to the delay, the phase curve will exceed -180 deg and will wraparound, so use the MATLAB unwrap routine.

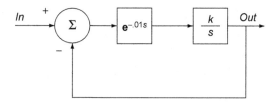

15. Demonstrate the effect of a 0.2-sec delay on the frequency characteristics of a second-order system. The system should have an ω_n of 10 rad/sec and a δ of 0.7. Plot the magnitude and phase with and without the delay. (Hint: Use MATLAB to plot the spectrum of the second-order system by substituting $j\omega$ for s. Then replot adding an $e^{-0.2s}$ $(=e^{-j0.2\omega})$ to the transfer function.) Plot for a frequency range of 1 to 100 rad/sec. Again the unwrap routine should be used since the phase plot will exceed -180 deg.

16. Find the time function of the following higher-order Laplace function. Use `roots` to factor the denominator (and `poly` if needed). Then apply partial fraction expansion to separate out the denominator terms and find the inverse Laplace transform.

$$TF(s) = \frac{s^3 + 17s^2 + 80s + 100}{s^4 + 10s^3 + 45s^2 + 110s + 104}$$

17. Find the time function of the following higher-order Laplace function. Use `roots` to factor the denominator (and `poly` if needed). Then apply partial fraction expansion to separate out the denominator terms and find the inverse Laplace transform. Alternatively, use MATLAB's residue to find the partial fractions directly,

$$TF(s) = \frac{3(s + 5)}{s^3 + 6s^2 + 11s + 6}$$

18. The impulse response of a first-order system is:

$$V_{out}(s) = \left(\frac{1}{\tau}\right)\left(\frac{1}{s + \frac{1}{\tau}}\right)$$

Use the Initial Value Theorem to find the filter output's value at $t = 0$ (i.e. $v_{out}(0)$) for the filter.

19. The transfer function of an electronic system has been determined as:

$$TF(s) = \frac{5s + 4}{s^2 + 5s + 20}$$

Use the Final Value Theorem to find the value of this system's output for $t \rightarrow \infty$ if the input is a step function that jumps from 0 to 5 at $t = 0$.

20. The MATLAB file `unknown_sys7_4(x).m` found on the associated files represents a linear system as in Examples 7.13 and 7.14. The input is x and the output is the output argument, i.e., `y = unknown_sys7_4(x);`. Assume a sampling frequency of 1.0 kHz and use 40,000 points to get good spectral resolution. Determine the magnitude spectrum for this unknown process using a random input as in Example 7.13. Estimate the transfer function of this system based on Bode plot primitives.

21. Find the magnitude spectrum of the unknown system as represented by `unknown_sys7_5.m` using sinusoids as in Example 7.14. Vary the range of frequencies of the sine wave between 0 and 400 Hz in 1.0 Hz intervals. Make $N = 1000$ and assume $f_s = 1$ kHz. Estimate the transfer function for this system based on Bode plot primitives.

22. Use the impulse response to find the magnitude spectrum of the system represented by `unknown_sys7_6(x).m`. Estimate the transfer function for this based on Bode plot primitives.

Use an impulse input of 1000 points and assume they are spaced 1.0 msec apart. As always, be sure to plot only valid spectral points.

23. The file `bio_sys1.mat` contains the input and output signals of a biological system sampled at 1000 Hz. Follow the approach used in Example 7.15 to find the magnitude transfer function, then use Bode plot primitives to estimate the transfer function. Ignore obvious artifacts.

Analysis of Discrete Linear Systems—The z-Transform and Applications to Filters

8.1 GOALS OF THIS CHAPTER

The transfer function, in either the frequency or complex frequency (Laplace) domain, provides a complete description of the behavior of any continuous linear, time-invariant (LTI) system. Given the transfer function, or equivalently the impulse response, we can evaluate the response of the system to any stimulus. All physiological systems are continuous, and to the extent that they can be considered linear and time invariant, the techniques of Chapters 5–7 can be used to analyze these systems.

We have the tools to describe all continuous linear systems, but not discrete systems. The only discrete systems you are likely to encounter live in computers,[1] and of these, the only systems likely to be important to you are "digital filters." But they are important. Every signal you measure contains some noise and most would benefit from filtering. So you need to understand the basics of discrete systems analysis to intelligently use, and possibly design digital filters.

The digital filters described here are all linear, time-invariant, discrete-time systems, abbreviated "LTID." LTID systems can be analyzed using the discrete Fourier transform, but restrictions apply. Recall that to apply the Fourier transform methods to (continuous) LTI systems, we need to assume steady-state signals and that the system has no initial conditions. If these conditions are not met, the more general Laplace transform must be used. Additional requirements apply when using the discrete Fourier transform to analyze a discrete system: all signals must be summable (that is, they cannot be growing with time), and the system must meet certain stability requirements. Biomedical signals and digital filters are likely to meet these criteria, so the good old Fourier transform, specifically the fast Fourier transform (FFT), will be our primary tool for analyzing digital filters. Nonetheless, there is a more general approach to discrete system analysis, known as the "z-transform," and you need at least

---

[1] Or in computer-like devices such as field programmable gate arrays known as FPGAs.

a passing facility with this approach as z-transform notation is used for describing all discrete systems, including digital filters.

The z-transform is nothing more than a discrete version of the Laplace transform. It can be developed as an extension of the discrete Fourier transform in a manner that parallels the development of the Laplace transform from the continuous Fourier transform (Equations 7.3 and 7.4). With the z-transform, we have a formidable array of system and signal processing techniques. Table 8.1 summarizes these analytical tools and shows how the z-transform fits in with these other techniques.

This chapter covers the development of the z-transform and its application to digital filters, the only discrete systems you are likely to encounter. Specific topics include:

- Development of the z-transform equation, which closely parallels the development of the Laplace transform.
- Definition of the digital (i.e., discrete) transfer function. This z-transform plays the same role as the Laplace transfer function for discrete systems. It is a function of the complex discrete frequency variable, z.
- Difference equations: discrete-time representation of discrete systems.
- Spectrum of a digital transfer function. How to use the FFT to plot the spectrum given a digital transfer function.
- General properties associated with all filters: bandwidth and roll-off or slope characteristics.

TABLE 8.1   Summary of System and Signal Analysis Techniques

Analysis	Continuous	Discrete	Constraints
Signals Time domain	Autocorrelation Cross-correlation	Autocorrelation Cross-correlation	Finite signals
Signals Frequency domain	Fourier series analysis Fourier transform	Discrete Fourier series Discrete Fourier transform	Periodic or aperiodic Dirichlet conditions
Systems Time domain	Impulse response, $h(t)$ Convolution	Impulse response, $h[n]$ Convolution	Linear, time invariant (LTI) ...or for discrete systems Linear, time invariant discrete-time (LTID)
Systems, restricted Frequency domain	Fourier transform Transfer function, $TF(\omega)$	Discrete Fourier transform Transform function, TF $[\omega]$	LTI, steady state, no transient signals ...or for discrete systems LTID, stable systems, summable signals (not growing with time)
Systems, general Complex frequency domain	Laplace transform Laplace transfer function, $TF(s)$	z-transform z-transfer function, TF $[z]$	LTI ...or for discrete systems LTID

- Definition and properties of digital filters, including finite impulse response (FIR) filters and infinite impulse response (IIR) filters.
- Designing FIR filters with and without MATLAB. Finding the filter coefficients (or filter weights) from the desired frequency characteristics for FIR filters.
- Designing IIR filter with MATLAB. Finding the filter coefficients (or filter weights) from the desired frequency characteristics for IIR filters using a MATLAB Toolbox.

8.2 THE Z-TRANSFORM

The z-transform is a discrete domain analogy to the Laplace transform. In fact, it can be viewed as just a different version of the Laplace transform. The transform provides us with a discrete transfer function, which is, again, analogous to the Laplace transfer function. The discrete transfer function gives us the same analytical power for discrete systems as the Laplace transform provides for continuous systems. Specifically, the transfer function is a complete description of the behavior (the input/output characteristics) of the system.

To develop a digital version of the Laplace transform, we recap the approach that was used to develop the Laplace transform. Recall the Laplace transform is just an extension of the Fourier transform that uses complex frequency s (where $s = \sigma + j\omega$) instead of ordinary frequency $j\omega$. Replacing the regular frequency term (i.e., $j\omega$) in the Fourier transform with complex frequency (i.e., s) leads to the Laplace transform, repeated here:

$$X(\sigma, \omega) = \int_0^\infty x(t)e^{-\sigma t}e^{-j\omega t}dt = \int_0^\infty x(t)e^{-st}dt \qquad (8.1)$$

The addition of the $e^{-\sigma t}$ term in the Laplace transform ensures convergence for a wider range of signals that includes step-like signals. These signals do not converge in the Fourier transform equation, but the additional $e^{-\sigma}$ term added to the Laplace equation provides convergence provided σ takes an appropriate value.

To develop a discrete version of the Laplace transform, we start by putting the Laplace equation into discrete notation. This conversion is achieved by replacing the integral with a summation and substituting a sample number, n, for the time variable t. (Actually $t = nT_s$, but we assume that T_s is normalized to 1.0 for this development.) These modifications give:

$$X(\sigma, \omega) = \sum_{n=-\infty}^{\infty} x[n]e^{-\sigma n}e^{-j\omega n} \qquad (8.2)$$

Next, define a new variable:

$$r = e^{-\sigma} \qquad (8.3)$$

Substituting r for $e^{-\sigma}$ into Equation 8.2 gives:

$$X(r, \omega) = \sum_{n=-\infty}^{\infty} x[n]r^n e^{-j\omega n} \qquad (8.4)$$

Equation 8.4 is a valid form for the z-transform, but commonly, the equation is simplified by defining another new complex variable, z. This is essentially the idea used in the Laplace transform where the complex variable s is introduced to represent $\sigma + j\omega$. The new variable, z, is defined as:

$$z = r\,e^{j\omega} = |z|e^{j\omega} \tag{8.5}$$

where r is now simply the magnitude of the new complex variable z. Substituting z into Equation 8.4 gives:

$$X(z) = Z[x[n]] = \sum_{n=-\infty}^{\infty} x[n]z^{-n} \tag{8.6}$$

This is the defining equation for the z-transform of a digital signal $x[n]$ and is notated as $Z[x[n]]$. This is called the "bilateral z-transform" as the summation limits extend to $\pm\infty$. As defined in Section 1.4.3, causal signals are those that exist only for $t \geq 0$. For discrete causal signals, this translates to $n \geq 0$. Thus for discrete causal signals, the summation in Equation 8.6 only needs to be from $n = 0$ to ∞. This gives rise to the "unilateral z-transform" that is commonly used in digital systems analysis:

$$X(z) = Z[x[n]] = \sum_{n=0}^{\infty} x[n]z^{-n} \tag{8.7}$$

In any real application, the limit of the summation is finite, usually the length of the signal $x[n]$.

As with the Laplace transform, the z-transform is based on a complex variable: in this case the arbitrary complex variable z, which equals $|z|e^{j\omega}$. Analogous to Laplace variable s, the complex frequency, z, is the "discrete complex frequency." As with the Laplace variable s, it is possible to substitute $e^{j\omega}$ for z to perform a strictly sinusoidal analysis.[2] For us, this is a very useful property, as it allows us to easily determine the frequency characteristic of a z-transform transfer function using only the Fourier transform.

8.2.1 The Unit Delay

Multiplying a data sequence by z^{-n} in the z-domain has the useful effect of shifting the sequence by n data samples in the time domain. For example, assume you have a data sequence $x[n]$ that has a z-transform $X[z]$. Define another data sequence, $y[n]$, that is the same sequence but shifted over one position: $y[n] = x[n-1]$. The z-transform of this shifted sequence is by reference to Equation 8.7 is:

$$Y(z) = \sum_{n=0}^{\infty} y[n]z^{-n} = \sum_{n=0}^{\infty} x[n-1]z^{-n}$$

---

[2]If $|z|$ is set to 1, then from Equation 8.5, $z = e^{j\omega}$. This is called evaluating z on the "unit circle" because $|z| = 1$ describes a circle of radius one when z is plotted in polar coordinates. Substituting $e^{j\omega}$ for z is a useful operation because in digital filter analysis, our primary interest is the filter's spectrum.

If $k = n - 1$, then $x[n - 1] = x[k]$ and $Y(z)$ becomes:

$$Y(z) = \sum_{k=0}^{\infty} x[k]z^{-(k+1)} = \sum_{k=0}^{\infty} x[k]z^{-k}z^{-1} = z^{-1}\sum_{k=0}^{\infty} x[k]z^{-k}$$

But the summation:

$$\sum_{k=0}^{\infty} x[k]z^{-k} = X(z)$$

So $Y(z)$ due to a shift in x of one sample becomes:

$$Y(z) = z^{-1}X(z) \tag{8.8}$$

This time-shifting property generalizes to higher powers of z, so multiplication by z^k in the z-domain is equivalent to a shift of k time samples:

$$Z[x[n - k]] = z^{-k}Z[x[n]] \tag{8.9}$$

For a discrete variable, $x[n] = [x_1, x_2, x_3,\dots x_N]$, the power of z in the z-domain can be used to define a sample's position in the sequence. This is useful in understanding the z-domain transfer function.

8.2.2 The Digital Transfer Function

As in Laplace transform analysis, the z-transform allows us to define a digital transfer function. By analogy to previous transfer functions, the digital transfer function is defined as:

$$H(z) = \frac{Y(z)}{X(z)} = \frac{Z[y[n]]}{Z[x[n]]} \tag{8.10}$$

where $X(z)$ is the z-transform of the input signal, $x[n]$, and $Y(z)$ is the z-transform of the system's output, $y[n]$. As an example, a simple linear element known as a "unit delay" is shown in Figure 8.1. In this system, the time shifting characteristic of z^{-n} (Equations 8.8 and 8.9) is used to define processes where the output is the same as the input but shifted (or delayed) by one data sample. The z-transfer function for this process in Figure 8.1 is $H(z) = z^{-1}$.

The unit delay also generalizes to a higher power of n so if $H(z) = z^{-3}$ then $y[n] = x[n - 3]$. The z-transform of the delayed version of $x[n]$ is z raised to the power of the delay and vice versa:

$$x[n - k] \Leftrightarrow z^{-k} \tag{8.11}$$

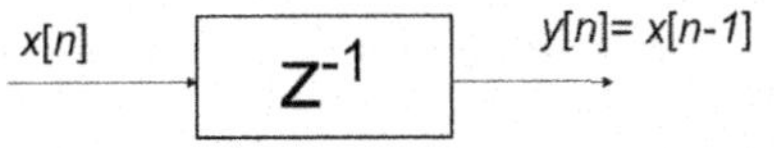

FIGURE 8.1   A unit delay is a linear system that shifts the input by one data sample. Other powers of z can be used to provide larger shifts. The transfer function for this system is $Y(z) = z^{-1}$.

Most transfer functions are more complicated than that of Figure 8.1 and can include polynomials of z in both the numerator and denominator, just as analog transfer functions contain polynomials of s in the numerator and denominator:

$$H(z) = \frac{Y(z)}{X(z)} = \frac{b[0] + b[1]z^{-1} + b[2]z^{-2} + \ldots + b[K]z^{-K}}{1 + a[1]z^{-1} + a[2]z^{-2} + \ldots + a[L]z^{-L}} \qquad (8.12)$$

where the $b[k]$'s are constant coefficients of the numerator and the $a[l]$'s are coefficients of the denominator.[3] Although $H(z)$ has a structure similar to the Laplace domain transfer function $H(s)$, there is no simple relationship between them.[4] For example, unlike analog systems, the order of the numerator, K, need not be less than, or equal to, the order of the denominator, L, for stability. In fact, systems that have a denominator order of 1 (i.e., such as FIR filters) are more stable than those that have higher-order denominators. Remember that in the time domain, the powers of z indicate the position of the sample in a data sequence.

But just as in the Laplace transform, an LTID system is completely defined by the denominator, a, and numerator, b, coefficients. Equation 8.12 can be more succinctly written as:

$$H(z) = \frac{\sum_{k=0}^{K} b[k]z^{-k}}{\sum_{l=0}^{L} a[l]z^{-l}} \qquad (8.13)$$

where $a[0] = 1.0$. From the digital transfer function, $H(z)$, it is possible to determine the output given any input:

$$Y(z) = X(z)H(z), \quad y(n) = Y^{-1}(z) \qquad (8.14)$$

where $Y^{-1}(z)$ is the inverse z-transform. Just as with the Laplace transform, z-transform tables can be used to convert from the (discrete) time domain to the z-domain. Extensive z-transform tables can easily be found on the web.

The z-transform transfer function (Equations 8.13) can be used in the same manner as the Laplace transfer function to find the output of an LTID system to a given input. Multiply the transfer function by the z-transform of the input and rearrange the resulting equation to match an entry in the z-transform table. You may have to go through a few of the same algebraic manipulations such as partial fraction expansion, but the basic idea is the same.

---

[3]This equation can be found in a number of different formats. A few authors reverse the role of the a and b coefficients with a's being the numerator coefficients and b's being the denominator coefficients. Another variation is to reverse the coefficient order, with $b[0]$ representing the highest power of z. Sometimes, the denominator coefficients are written with negative signs changing the nominal sign of the a coefficients. We will follow the format used by MATLAB with the exception that coefficient series begins with $b[0]$ and $a[1]$ so the coefficient index is the same as the power of z. Since MATLAB does not allow an index of 0, this equation is formatted with the series index beginning with $b[1]$ and $a[2]$ (so $a[1]$ is always 1.0 in MATLAB).

[4]Nonetheless, the z-transfer function borrows from the Laplace terminology, so the term "pole" is sometimes used for denominator coefficients and the term "zeros" for numerator coefficients.

The inverse z-transform method is tedious, not amenable to computer analysis (except possibly for finding the roots of a polynomial), and, fortunately, not necessary for filter analysis. With digital filters, we are interested only in the filter's spectrum. When we determine the spectrum of a system, any system, if we assume steady-state conditions. This means we can avoid solving for the inverse z-transform. We simply substitute $e^{j\omega}$ for z in the transfer function and evaluate the result using the Fourier transform as shown in the next section.[5]

8.2.3 Transformation From the z-Domain to the Frequency Domain

To transform a z-transform to the (discrete) frequency domain, we substitute $e^{j\omega}$ for z. This is analogous to substituting $e^{j\omega}$ for s to convert a Laplace transfer function to the frequency domain. With this substitution, Equation 8.13 becomes:

$$H(m) = \frac{\sum_{k=0}^{K-1} b[k]e^{-j\omega k}}{\sum_{l=0}^{L-1} a[l]e^{-j\omega l}} = \frac{\sum_{k=0}^{K-1} b[k]e^{-\frac{j2\pi mk}{K}}}{\sum_{l=0}^{L-1} a[l]e^{-\frac{j2\pi ml}{L}}} = \frac{FT\{b[k]\}}{FT\{a[l]\}} \tag{8.15}$$

where K is the number of b (numerator) coefficients and L is the number of a (denominator) coefficients. As in all Fourier transforms, the actual frequency can be obtained from the variable m by multiplying m by f_s/N or, equivalently, $1/(NT_s)$. In practice, both the numerator and denominator coefficients are zero padded to be the same length, a length large enough to generate a smooth spectrum. Equation 8.15 is easily implemented in MATLAB as shown in Example 8.1.

EXAMPLE 8.1

Find and plot the frequency spectrum (magnitude and phase) of a system having the following digital transfer function. Assume this system is used with signals that are sampled at 1000 Hz.

$$H(z) = \frac{0.2 + 0.5z^{-1}}{1 - 0.2z^{-1} + 0.8z^{-2}} \tag{8.16}$$

Solution: First identify the a and b coefficients from the digital transfer function. From Equation 8.16, the numerator coefficients are $b = [0.2, 0.5]$ and the denominator coefficients are $a = [1.0, -0.2, 0.8]$. Then solve Equation 8.15 using these coefficients. Zero pad both coefficients to the same large number of samples to get a smooth spectrum. (Here we use $N = 512$, which is more than sufficient.) When plotting, use a frequency vector: $f = mf_s/N$.

```
%  Example 8.1  Plot the Frequency characteristics of an LTID system defined
%  by the z-transfer function, Equation 8.16.
%
%
fs = 1000;              % Assumed sampling frequency
```

---

[5]For an extensive discussion of the inverse z-transform, see, for example, Lathi, 2005.

```
N = 512;                              % Number of points (arbitrary)
%
% Define a and b coefficients based digital transfer function
a = [1 -.2 .8];                       % Denominator coefficients
b = [.2 .5];                          % Numerator coefficients
%
H = fft(b,N)./fft(a,N);               % Equation 8.15 with zero padding

Hm = 20*log10(abs(H));                % Get magnitude in dB
Theta = (angle(H))*360/(2*pi);        % and phase in deg.
f = (1:N/2) *fs/N;                    % Frequency vector for plotting
     .......plot magnitude and phase transfer function with grid and labels.......
```

Result: The spectrum of the system defined in Equation 8.16 is presented in Figure 8.2. The spectrum has a hump at around 240 Hz, indicating that the system is underdamped with an undamped resonant frequency (ω_n) of approximately 240 Hz.

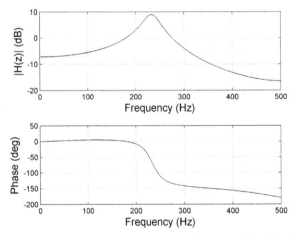

FIGURE 8.2   Plot of the frequency characteristic (magnitude and phase) of the digital transfer function given in Equation 8.16. The spectrum describes an underdamped system with a resonant frequency of approximately 240 Hz.

This example demonstrates how easy it is to determine the time and frequency behavior of a system defined by a z-transform transfer function. The approach used here can readily be extended to any transfer function of any complexity. Other examples of the use of the z-transform are given in the problems.

The next question you might ask is, how do we implement a digital filter, or any digital system, given its z-transfer function? That is, how do we apply the digital filter to a digital signal? Again we would like to avoid having to take the inverse z-transform and implement the system on a computer. For this, we turn to the time domain representation of digital systems, a representation that uses "difference equations."

8.3 DIFFERENCE EQUATIONS

Difference equations are the discrete-time equivalent of continuous differential equations. A simple first-order differential equation and its discrete-time counterpart illustrate this relationship. Given a typical first-order differential equation:

$$\frac{dy}{dt} + ky(t) = x(t) \tag{8.17}$$

If $x(t)$ is converted to a discrete signal by sampling at T_s seconds, then Equation 8.17 becomes:

$$\lim_{T_s \to 0} \frac{y[n] - y[n-1]}{T_s} + ky[n] = x[n]$$

Assuming T_s is small, but not 0, multiplying through by T_s and combining $y[n]$'s yields:

$$(1 + kT_s)y[n] - y[n-1] = x[n]T_s \tag{8.18}$$

Normally the coefficient of $y[n]$ is normalized to 1:

$$y[n] - \left(\frac{1}{1+kT_s}\right)y[n-1] = \left(\frac{T_s}{1+kT_s}\right)x[n] \tag{8.19}$$

Equation 8.19 is a difference equation, so-called because it is based on the difference between samples (i.e., $y[n]$ and $y[n-1]$). Equation 8.19 is the discrete-time equivalent of the continuous differential equation given in Equation 8.17. The highest order difference in either x or y indicates the order of the difference equation and corresponds to the order of the equivalent differential equation. In Equation 8.18, the highest difference is 1.0 (between the n and $n-1$ terms), so this is a first-order difference equation. This is expected since it is the digital version of a first-order differential equation, Equation 8.17. A second-order difference equation would have an $n-2$ term (in either x or y).

Although the input–output relationship can be determined from equations in the form of Equation 8.18 or 8.19, it is more common to use difference equations that are arranged to be discrete versions of convolution. To convert from the z-domain transfer function of Equation 8.13, we begin with the expanded version of Equation 8.13 given in Equation 8.12 and repeated here:

$$H(z) = \frac{Y(z)}{X(z)} = \frac{Z[y[n]]}{Z[x[n]]} = \frac{b[0] + b[1]z^{-1} + b[2]z^{-2} + \dots + b[K]z^{-K}}{1 + a[1]z^{-1} + a[2]z^{-2} + \dots + a[L]z^{-L}}$$

Clearing the fraction in Equation 8.12 by multiplying both sides by both denominators gives:

$$
\begin{aligned}
&Z[y[n]]1 + Z[y[n]]a[1]z^{-1} + Z[y[n]]a[2]z^{-2} + \dots + Z[y[n]]a[L]z^{-L} = \\
&Z[x[n]]b[0] + Z[x[n]]b[1]z^{-1} + Z[x[n]]b[2]z^{-2} + \dots + Z[x[n]]b[K]z^{-K}
\end{aligned}
\tag{8.20}
$$

Now we apply the time shift interpretation of z^{-k} given in Equation 8.11 and repeated here:

$$z^{-k} \Leftrightarrow x[n-k] \quad (\text{or } z^{-k} \Leftrightarrow y[n-k])$$

Taking the inverse z-transform of each term in Equation 8.20 using this time shift interpretation:

$$
\begin{aligned}
y[n] + y[n-1]b[1] + y[n-2]b[2] + \ldots + y[n-L]b[L] = \\
x[n]b[0] + x[n-1]b[1] + x[n-2]b[2] + \ldots + x[n-K]b[K]
\end{aligned}
$$

Note that $a[0]$ is assumed to be 1.0 (Equation 8.13) so $y[n]a[0] = y[n]$. Solving for $y[n]$ in terms of $x[n]$ and past values of $y[n]$:

$$
\begin{aligned}
y[n] = x[n]b[0] + x[n-1]b[1] + x[n-2]b[2] + \ldots + x[n-K]b[K] \\
-y[n-1]b[1] - y[n-2]b[2] - \ldots - y[n-L]b[L]
\end{aligned}
$$

Combining the sums:

$$y[n] = \sum_{k=0}^{K-1} b[k]x[n-k] - \sum_{l=1}^{L-1} a[l]y[n-l] \tag{8.21}$$

This difference equation defines any general LTID system in the discrete-time domain. It contains two summations: a summation involving the b coefficients applied to the input signal and a summation of the a coefficients applied to delayed samples of the output signal. Note that the summation of a coefficients begins at $l = 1$ (since $a[0] = 1$), so only past values of the output signal, y, contribute to the current output signal. Nonetheless, the a coefficients >1 are former outputs and they contribute, in part, to the current output ($y[n]$), so they are called "recursive coefficients."

The difference equation (Equation 8.21) also provides us with a method to find the output, $y[n]$, to any input, $x[n]$, without having to deal with the inverse z-transform. Compare the two terms in Equation 8.21 with the convolution equation introduced in Chapter 5 (Equation 5.3) and repeated here:

$$y[n] = \sum_{k=-\infty}^{\infty} h[k]x[n-k] \tag{8.22}$$

The two summations in Equation 8.21 have the same form as the convolutions equation. In fact, if $h[k] = b[k]$ the first summation of Equation 8.21 is identical to the convolution equation, Equation 8.22. So the output to any discrete system can be obtained by a double convolution, convolving the input with the b coefficients and past values of the output with the a coefficients. Implementation of Equation 8.21 in MATLAB is straightforward as shown in the next example.

EXAMPLE 8.2

Find the output of the system defined by Equation 8.16 in Example 8.1 to a unit step input. Use the same values for f_s and make the signal length 256 samples.

Solution: Define the a and b coefficients as vectors then generate the unit step. The first summation in Equation 8.21 can be found by direct convolution of b with x using MATLAB's conv routine. Use the 'same' option with x as the first parameter so the output signal is the same length as the input signal x.

Evaluating the second term is a bit more complicated, as it is recursive: the output depends on past values of itself. We must write our own code for this convolution. Since there are three a coefficients, we need to pad the output with two zeros and the first output term. Since $a[0] = 1$, the first output term is just the first term of the convolution of x and b. We then use a loop to carry out the recursive convolution of the second summation. After the convolution is completed, we left shift the output one position to account for the leading zero padding. The result is plotted against a time vector based on f_s. Only the first 60 points are plotted to emphasize the transient response.

```
% Example 8.2 Compute the step response of the system defined by Equation 8.21
fs = 1000;                       % Sampling frequency (from Example 8.1)
N = 256;                         % Number of points (arbitrary)
% Define a and b coefficients (from Example 8.1)
a = [1 -.2 0.8];
b = [.2 .5];
% Compute the Step Response
x = [0, ones(1,N-1)];            % Generate a step function
b_conv = conv(x,b,'same');       % Generate first convolution
y = [0 0 b_conv(1)];             % Initialize y with padding and first term
for k = 1:N
  y(k+2) = b_conv(k) - (y(k+1)*a(2) + y(k)*a(3));  % Equation 8.21, second conv
end
y = y(2:N+1);                    % Shift y to account for padding
t = (1:N)/fs;                    % Time vector for plotting
  ..... plot and label.......
```

Results: The time response generated by this program is shown in Figure 8.3. The step response is clearly that of an underdamped system showing a decaying oscillation. We can estimate the time difference between oscillatory peaks as roughly 0.004 s, which is equivalent to 250 Hz. This is close to the resonant peak shown in Figure 8.2 estimated to be approximately 240 Hz.[6] From the time plot we could estimate the damping factor of this system, but that is easier to do using the spectrum in Figure 8.2.

Equation 8.21 is very useful for implement digital filters. Once we know the appropriate filter coefficients, any signal could be filtered using the code in Example 8.2. However, this is unnecessary because Equation 8.21 is so useful that MATLAB has a routine that implements this equation with one statement:

```
y = filter(b,a,x);              % Find the output of a discrete system
```

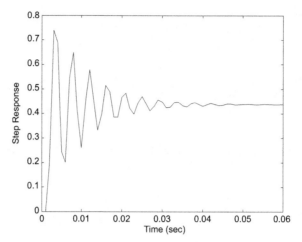

FIGURE 8.3   The step response of the discrete system defined by the transfer function in Equation 8.16. The spectrum of this system was determined in Example 8.1 and shown in Figure 8.2. The step response is clearly that of an underdamped system as is expected from its spectrum. The oscillatory peaks are spaced about 0.004 s apart, suggesting a damped resonance around 240 Hz in rough agreement with the spectrum in Figure 8.2.

where b is a vector of b coefficients, a is a vector of a coefficients, and x is the input signal. The filter routine is much faster than the code in Example 8.2. In a test run with more samples, the code in Example 8.2 required 6.316 ms, whereas filter required 0.309 ms, more than 20 times faster. It is hard to outperform MATLAB.

[6]The resonant oscillations in Figure 8.3 reflect the damped resonant frequency ω_d, whereas the spectral peak is at the undamped natural frequency, ω_n. As given by Equation 7.34, the damped frequency is slightly less than the undamped natural frequency: $\omega_d A \omega_n \sqrt{1 A \delta^2}$. But in this system δ is quite small: approximately 0.07 based on the height of the resonance peak in Figure 8.2. So ω_d should be very close to ω_n. More than likely, the difference (222 vs. 240 Hz) is due to the difficulty accurately reading the time and spectral plots.

We can now find the output of any LTID system given its transfer function, or only the coefficients of that transfer function. To apply a digital filter to any waveform, we use Equation 8.21 embodied in MATLAB as filter (b,a,x). But we still need to find the a and b coefficients that give the filtering we desire. In other words, we need to find coefficients that define a system having a spectrum that attenuates the noise in our signal, but not the frequencies of interest. Finding these coefficients is called "filter design" or "filter synthesis", and is the subject of the remainder of this chapter.

8.4 LINEAR FILTERS—INTRODUCTION

Filtering is a process that alters a signal in some desired manner and is usually described in terms of its spectrum. Filters exist in both the analog or digital domain. The former consists of analog electronics such as those described in Chapter 15. Because of the problem of aliasing due to sampling (see Section 4.1.1), most biomeasurement systems have an analog low-pass filter, an "anti-aliasing filter," that is applied to the signal before it is digitized. Once the signal is digitized, additional filtering can be done using digital filters.

Digital filters can be adaptive, modifying their coefficients to respond to certain features of the signal, or they can be fixed with constant filter coefficients. Here we concentrate on fixed filters implemented by discrete linear (LTID) systems.

While we implement filters in the time domain using Equation 8.21, we think about filters in the frequency domain; specifically, how a filter reshapes the signal's spectrum. A frequent goal of filtering is to reduce noise. Most noise occurs over a wide range of frequencies (noise is broadband), whereas most signals have a limited frequency range (signals are narrowband or bandlimited). Reducing the energy of frequencies outside the signal range reduces noise, but not signal; the signal to noise ratio (SNR) is improved. Such filtering is usually done using a fixed digital filter: a specially designed LTID system that reshapes a signal's spectrum in some well-defined, and we hope, beneficial manner.

Chapter 5 touched on the use of simple filters. Example 5.6 used a three-point running average, implemented through convolution, to reduce noise in an electrocardiogram (ECG) signal (Figure 5.18). The spectrum of this system was found using the Fourier transform to be that of a low-pass filter (Figure 5.19). Putting the three-point average in context of what we have learned thus far, it was a digital filter with three b coefficients of 0.33 each and had no a coefficients except $a[0] = 1$. The z-transform transfer function of this filter would be:

$$H(z) = \frac{0.33 + 0.33z^{-1} + 0.33z^{-2}}{1} \tag{8.23}$$

Discrete systems for filtering can be classified into two groups depending on their impulse response characteristics: those that have short well-defined impulse responses and those that produce impulse responses that theoretically go on forever. The former are termed finite impulse response ("FIR") filters and the latter infinite impulse response ("IIR") filters.

FIR filters have only one a coefficient, $a[0] = 1$, but any number of b coefficients. Since they do not have additional a coefficients, the second, recursive summation in Equation 8.21 does not occur, so FIR filters are nonrecursive. As shown in Chapter 5, their spectrum can be found by taking the Fourier transform of the b coefficients. This is in agreement with Equation 8.15 because the Fourier transform of a single a coefficient with a value of 1.0, would be simply 1.0, making the denominator 1.0 and the numerator FT{$b[k]$}. The three-point moving average defined by the transfer function in Equation 8.23 is an example of an FIR filter. FIR filters can be implemented using convolution or the `filter` routine where a = 1.

IIR filters have both a and b coefficients. They include the recursive operation defined by the second term in Equation 8.21. They are implemented by the `filter` routine, which, again, embodies Equation 8.21. Real-world IIR filters do not really have infinite impulse responses, but they do have longer impulse responses than FIR filters having the same number of total coefficients. Figure 8.4 shows the impulse response of an IIR filter with 5 a coefficients and 5 b coefficients and an FIR filter with 10 b coefficients. As shown in Figure 8.4, the IIR filter has an impulse response that is longer than the FIR filter with the same number of total coefficients. Both filter types are easy to implement in basic MATLAB; however, the design of IIR filters is best done using the MATLAB Signal Processing Toolbox. The advantages and disadvantages of each of the filter types are summarized in the next section.

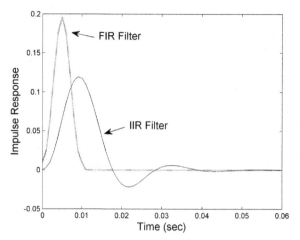

FIGURE 8.4   The impulse response of infinite impulse response (IIR) and finite impulse response (FIR) filters having the same number of total coefficients. The IIR filter has 5 a coefficient and 5 b coefficients, whereas the FIR filter has 10 b coefficients. The impulse response of the IIR filter is longer.

8.4.1 Filter Properties

Although you might think that a filter could have any imaginable magnitude and phase curve, in practice the variety of spectral characteristics of most filters is limited. Filter spectra can be defined by three basic properties: basic filter type, bandwidth, and attenuation slope.

8.4.1.1 Filter Bandwidth

Bandwidth was defined in Section 4.5 in terms of signals, but the same definitions apply to systems. The magnitude spectra of an ideal and a real filter are shown in Figure 4.21 (Chapter 4). The cutoff frequency of a filter is taken where the spectrum has decreased by 3 dB from its unattenuated level, usually the level in the passband (see Figure 4.21B–D).

For a filter, the nominal gain region, the frequency region where the signal is not attenuated more than 3 dB, is termed the "passband," and the gain of the filter in this region is called the "passband gain." The passband gain of all filters in Figure 4.21 is 1.0. The frequency region where the signal is attenuated more than 3 dB is termed the "stopband."

8.4.1.2 Filter Type

Filters are usually classed based on the range of frequencies they do not suppress, that is, the frequencies they pass. A "low-pass" filter allows low frequencies to pass with minimum attenuation, whereas higher frequencies are attenuated. Conversely, a "high-pass" filter permits high frequencies to pass, but attenuates low frequencies. "Band-pass" filters allow a range of frequencies through, whereas frequencies above and below this range are attenuated. An exception to this terminology is the "band-stop" filter, which passes frequencies on either side of a range of attenuated frequencies: it is the inverse of the band-pass filter. The frequency characteristics of the four filter types are shown in Figure 8.5.

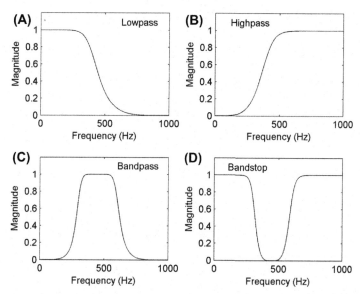

FIGURE 8.5 Magnitude spectral curves of the four basic filter types. (A) Low pass $f_c = 400$ Hz; (B) high pass $f_c = 400$ Hz; (C) band pass $f_{c1} = 300$ Hz, $f_{c2} = 600$ Hz; and (D) band-stop $f_{c1} = 300$ Hz, $f_{c2} = 600$ Hz.

8.4.1.3 Filter Attenuation Slope—Filter Order

Filters are also defined by the sharpness with which they increase or decrease attenuation as frequency varies. The attenuation slope is sometimes referred to as the filter's "rolloff." Spectral sharpness is specified in two ways: as an initial sharpness in the region where attenuation first begins and as a slope further along the attenuation curve.

For FIR filters, the attenuation slope is directly related to the length of the impulse response, i.e., the number of b coefficients. In FIR filters, the filter order is defined as one less than the number of b coefficients. Figure 8.6 shows the spectral characteristics of four FIR filters that have the same cutoff frequency but increasing filter orders, and the commensurate increase in attenuation slope is apparent.

The attenuation slope of IIR filters follows the pattern found for transfer functions analyzed in Chapter 6. The magnitude of the slope is determined by the order of the denominator polynomial: a first-order system has a slope of 20 dB/decade, whereas a second-order system has a slope of 40 dB/decade. This generalizes to higher-order IIR filters so that an nth-order IIR filter has an attenuation slope of $20n$ dB/decade. In Chapter 15, we examine analog circuits and find that the order of a circuit's denominator polynomial is related to the complexity of the electrical circuit, specifically the number of energy storage elements in the circuit.

8.4.1.4 Filter Initial Sharpness

For both FIR and IIR filters, the attenuation slope increases with the filter order, although for FIR filters, it does not follow the neat 20 dB/decade per filter order slope of IIR filters. For IIR filters, it is possible to increase the initial sharpness of the filter's attenuation characteristics without increasing the order of the filter, if you are willing to accept some unevenness, or

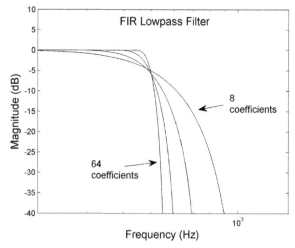

FIGURE 8.6   The magnitude spectral characteristics of four finite impulse response filters that have the same cutoff frequency, $f_c = 500$ Hz, but increasing filter orders of 8, 16, 32, and 64. The higher the filter order, the steeper the attenuation slope.

"ripple," in the passband. Figure 8.7 shows two low-pass, fourth-order IIR filters, differing in the initial sharpness of the attenuation. The one marked Butterworth has a smooth passband, but the initial attenuation is not as sharp as the one marked Chebyshev, which has a pass-band that contains ripples. This figure also shows why the Butterworth filter is popular in biomedical engineering applications: it has the greatest initial sharpness while maintaining a smooth passband. For most biomedical engineering applications, a smooth passband is needed.

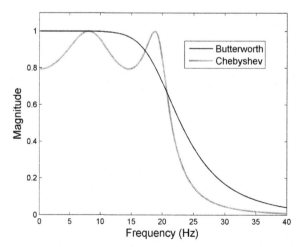

FIGURE 8.7   Two fourth-order infinite impulse response filters with the same cutoff frequency of 20 Hz, but differing in the sharpness of the initial slope. The filter marked "Chebyshev" has a much steeper initial slope, but contains ripple in the passband. The final slope also looks steeper, but would not be if it were plotted on a log scale.

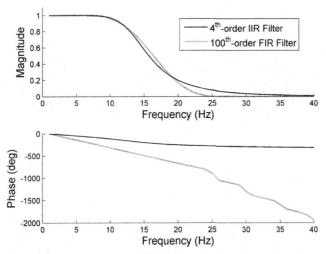

FIGURE 8.8 Comparison of the magnitude and phase spectra of a 100th-order finite impulse response (FIR) filter and a 4th-order infinite impulse response (IIR) filter, both having a cutoff frequency, f_c, of 14 Hz. (Upper graph) The magnitude spectra show approximately the same attenuation slope, but the FIR filter requires 10 times the number of coefficients (a fourth-order IIR filter has 10 coefficients, including $a[0] = 1.0$). (Lower graph) The phase spectra show that the FIR filter has a greater phase change, but that change is linear within the passband. This turns out to be an important feature of FIR filters.

8.4.2 Finite Impulse Response Versus Infinite Impulse Response Filter Characteristics

There are several important behavioral differences between FIR and IIR filters aside from the characteristics of their impulse responses. FIR filters require more coefficients to achieve the same filter slope. Figure 8.8A compares the magnitude spectra of a 4th-order IIR filter with a 100th-order FIR filter having the same cutoff frequency of 14 Hz. They both have approximately the same slope, but the FIR filter requires more than 10 times the number of coefficients. (A 4th-order IIR filter has 10 coefficients and a 100th-order IIR filter has 125 coefficients). In general, FIR filters do not perform as well for low cutoff frequencies as is the case here, and the discrepancy in filter coefficients is less if the cutoff frequency is increased. This behavior is explored in the problem set. Another difference is the phase characteristics as shown for the two filters in Figure 8.8B. The FIR has a larger change in phase, but the phase change is linear within the passband region of 0—14 Hz.

Figure 8.9 shows the application of the two filters to a fetal ECG signal that contains noise. Both low-pass filters do a good job of reducing the high-frequency noise, but, again, the FIR filter has 10 times the number of coefficients, and the computing time required to apply this filter is correspondingly longer. (For a 10,000 sample signal, the IIR filter required 0.56 ms, whereas the FIR filter required 7.04 ms on a typical personal computer.) The filtered data are also shifted to the right in comparison with the original signal, more so for the FIR filtered data. This is because both filters are causal (see Section 1.4.3) and the output of these filters depends on past values of the input. Causal and noncausal filtering are discussed in the next section (Section 8.4.3).

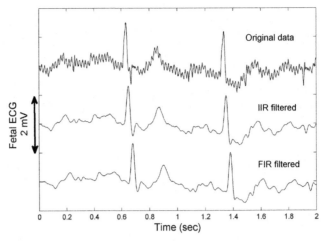

FIGURE 8.9   A fetal electrocardiography signal recorded from the abdomen of the mother. The unfiltered (upper) signal is filtered by the infinite impulse response (middle) and finite impulse response (FIR) (lower) filters whose spectra are shown in Figure 8.8. The 60-Hz noise in the original is considerably reduced in both filtered signals, but, again, the FIR filter requires 10 times the data samples and a proportionally longer computation time. Both filters induce a time shift because they are causal filters as described later.

The linear phase property of FIR filters shown in Figure 8.8 makes FIR filters much easier to design since we need only be concerned with the magnitude spectrum of the desired filter. FIR filters are also more stable than IIR filters. Since IIR filters are recursive, they can produce unstable oscillatory outputs for certain combinations of filter coefficients. This is not a problem if the coefficients are fixed as in standard filters, but in adaptive filters, where filter coefficients are modified on the fly, severe problems can occur. For this reason, adaptive filters employ FIR filters. IIR filters are also unsuitable for filtering two-dimensional data, so FIR filters are used exclusively for filtering images. A summary of the benefits and appropriate applications of the two filter types is given in Table 8.2.

TABLE 8.2   Finite Impulse Response (FIR) Versus Infinite Impulse Response (IIR) Filters: Features and Applications

Filter Type	Features	Applications
FIR	Easy to design Stable Applicable to two-dimensional data (i.e., images)	Fixed, one-dimensional filters Adaptive filters Image filtering
IIR	Require fewer coefficients for the same attenuation slope, but can become unstable Particularly good for low cutoff frequencies. Mimic analog (circuit) filters	Fixed, one-dimensional filters, particularly at low cutoff frequencies Real-time applications where speed is important

8.4.3 Causal and Noncausal Filters

If a filter uses only the current and past data samples, it is a "causal" filter. As noted previously, all real-world systems must be causal since they do not have access to the future; they have no choice but to operate on current and past values of a signal. However, if the data are already stored in a computer, it is possible to use future signal values along with current and past values to compute an output signal, that is, future data with respect to any given data sample. Filters (or systems) that use future values of a signal in their computation are noncausal.

The motivation for using future values in filter calculations is provided in Figure 8.10. The upper curve in Figure 8.10A is the response of the eyes to a target that jumps inward in a step-like manner. (The curve is actually the difference in the angle of the two eyes with respect to the straight-ahead position.) These eye movement responses are corrupted by 60-Hz noise riding on top of the signal, a ubiquitous problem in the acquisition of biological signals. A simple FIR filter consisting of 10 equal coefficients of 0.1 (i.e., a 10-point moving average) was applied to the noisy data and this filter does a good job of removing the noise as shown in the lower two curves. The filter was applied in two ways: as a causal filter using only current and past values of the eye movement data, (output shown in lower curve of Figure 8.7A), and as a noncausal filter using an equal number of past and future values, (output shown in

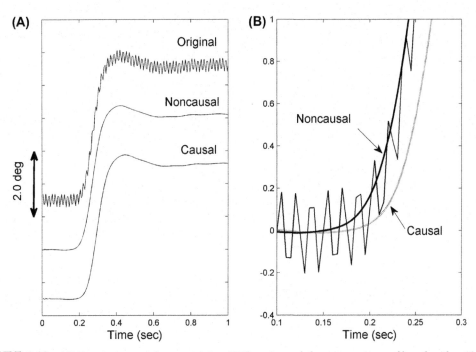

FIGURE 8.10 (A) Eye movement data containing 60-Hz noise and the same response filtered with a 10-point moving average filter applied in a causal and noncausal mode. (B) Detail of the initial response of the eye movement showing the causal and noncausal filtered data superimposed. The noncausal filter overlays the original data, whereas the causal filter produces a time shift in the filtered response.

middle curve of Figure 8.7A). The causal filter was implemented using MATLAB's `conv` routine, and so was the noncausal filter, but with the option `'same'` i.e.:

```
y = conv(x,b,'same');     % Noncausal filtering
```

where x is the signal and b is the filter coefficients. When the `'same'` option is invoked, the convolution algorithm returns only the center section of output, effectively shifting future values into the present. In Chapter 5, Figure 5.17 shows the noncausal implementation of a three-point moving average, as one of the values in the average is taken from a sample ahead (i.e., in the future) of the output sample.

Both filters do a good job of reducing the 60-cycle noise, but the causal filter has a slight delay. In Figure 8.7B, the initial responses of the two filters are plotted superimposed over the original data and the delay in the causal filter is apparent. Eliminating the delay or time shift inherent in causal filters is the primary motivation for using noncausal filters. In Figure 8.10, the time shift in the output is small, but can be much larger if filters with longer impulse responses are used or if the data are passed through multiple filters. However, in many applications a time shift does not matter, so causal filter implementation is adequate.

8.5 DESIGN OF FINITE IMPULSE RESPONSE FILTERS

For FIR filters, the impulse response is the filter coefficients, $b[k]$. Since there are no a coefficients except $a[0]$, FIR filters are implemented using convolution and in MATLAB, either the `conv` or `filter` routine can be used. Referring to Equation 8.15, since the Fourier transform of the a coefficient is 1.0, the FIR filter spectrum is just the Fourier transform of the b coefficients.

To design an FIR filter, we start with the desired spectrum. Since the desired spectrum is the Fourier transform of the b coefficients, to find these coefficients, all we need do is take the inverse Fourier transform of that spectrum. We need only the magnitude spectrum because FIR filters have a linear phase, so the phase is uniquely determined by the magnitude spectrum. Occasionally we might want some specialized frequency characteristic, but usually we just want to separate out a selected frequency range from everything else, for example, the signal frequency range from everything else. Also, we usually want that separation to be as sharp as possible. In other words, we usually prefer an ideal filter such as that shown in Figure 4.21A with a particular cutoff frequency. The spectrum of an ideal low-pass filter has a shape like a rectangular window, so sometimes the filters are called "rectangular window" filters.[7]

This spectrum of an ideal filter is a fairly simple function, so we ought to be able to find the inverse Fourier transform analytically from the defining equation given in Chapter 3 (a rare situation in which we are not totally dependent on MATLAB). When using the complex form, we must include the negative frequencies, so the desired filter's frequency characteristic is as

---

[7]This filter is sometimes just called a "window filter," but the term "rectangular window filter" is used here so as not to confuse such a filter with a "window function" as described in Chapter 4. This can be particularly confusing since, as shown below, rectangular window filters use window functions, but the two words (window filter and window function) mean two very different things!

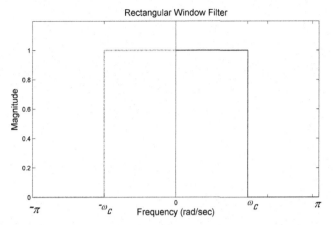

FIGURE 8.11 The magnitude spectrum of a rectangular window filter: an ideal filter. The negative frequency portion of this filter is also shown because it is needed in the computation of the complex inverse Fourier transform. The frequency axis is in radians per second and the cutoff frequency is ω_c.

shown in Figure 8.11. Frequency is shown in radians per seconds to simplify the derivation of the rectangular window impulse response and the cutoff frequency is ω_c.

Since we are trying to solve for the inverse Fourier transform analytically, we use the continuous form given in Equation 3.27, to solve for a continuous series of coefficients $b(t)$, and then convert the result to discrete numbers, $b[k]$.

$$b(t) = \frac{1}{2\pi} \int_{-\infty}^{\infty} B(\omega)e^{j\omega t}d\omega = \frac{1}{2\pi} \int_{-\omega_c}^{\omega_c} 1e^{j\omega t}d\omega \qquad (8.24)$$

since the window function is 1.0 between $\pm\omega_c$ and zero elsewhere. Integrating and putting in the limits:

$$b[t] = \frac{1}{2\pi} \frac{e^{j\omega t}}{jt}\Big|_{-\omega_c}^{\omega_c} = \frac{1}{\pi t} \frac{e^{j\omega_c t} - e^{-j\omega_c t}}{2j} \qquad (8.25)$$

The term $\frac{e^{j\omega_c t} - e^{-j\omega_c t}}{2j}$ is the exponential definition of the sine function and equals $\sin(\omega_c t)$. So the impulse response of a rectangular window is:

$$b(t) = \frac{\sin(\omega_c t)}{\pi t} = \frac{\sin(2\pi f_c t)}{\pi t} \qquad (8.26)$$

The impulse response of a rectangular window filter has the general form of a "sinc" function: $\sin(x)/x$. The filter coefficients, $b[k]$, can be obtained from Equation 8.26 by taking discrete values, k, for the continuous time variable t. The resulting coefficients are shown for 64 values of k and two values of f_c (or $\omega_c/2\pi$) in Figure 8.12. These cutoff frequencies are relative to the sampling frequency as explained in the following discussion.

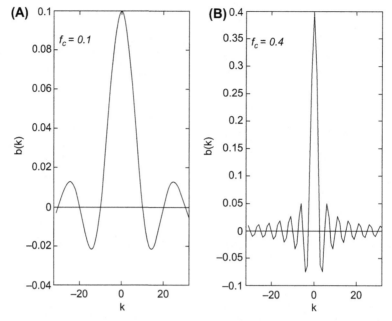

FIGURE 8.12   The impulse response of a rectangular window filter for 64 coefficients determined from Equation 8.26. The cutoff frequencies are given relative to the sampling frequency, f_s. (A) Low-pass filter with a relative cutoff frequency of 0.1 Hz. (B) Low-pass filter with a higher relative cutoff frequency of 0.4 Hz.

From Figure 8.11, the cutoff frequency, ω_c, is relative to a maximum frequency of π. In the spectrum of a discrete signal, the maximum frequency in Hz is the sampling frequency, f_s. So to convert the relative frequencies of Equation 8.26 to the actual frequency, we need merely to multiply f_c by f_s.

$$f_{actual} = f_c f_s \tag{8.27}$$

(Note that MATLAB filter design routines included in the Signal Processing Toolbox use a cutoff frequency relative to $f_s/2$ so $f_{actual} = f_c\,(f_s/2)$.

The symmetrical impulse responses shown in Figure 8.12 have both positive and negative values of k. Since MATLAB requires indexes to be positive, we need to shift the index, k, on the right side of Equation 8.26 to half the total length of the filter. To be compatible with MATLAB, the discrete version of Equation 8.26 becomes:

$$b[k] = \frac{\sin\left(\omega_c\left(k - \dfrac{L}{2}\right)\right)}{\pi\left(k - \dfrac{L}{2}\right)} = \frac{\sin\left(2\pi f_c\left(k - \dfrac{L}{2}\right)\right)}{\pi\left(k - \dfrac{L}{2}\right)} \tag{8.28}$$

where f_c is the cutoff frequency relative to f_s (Equation 8.27) and L is the length of the filter. If L is odd, then an adjustment is made so that $L/2$ is an integer as shown in the next example.

When Equation 8.28 is implemented, a problem occurs when the denominator goes to zero at $k = L/2$. The actual value of the function for $k = L/2$ can be obtained by applying the limits and noting that $\sin(x) \to x$ as x becomes small:

$$b[L/2] = \lim_{(k-L/2)} \left| \frac{\sin[2\pi f_c(k - L/2)]}{\pi(k - L/2)} \right| = \lim_{(k-L/2)} \left| \frac{2\pi f_c(k - L/2)}{\pi(k - L/2)} \right| = 2f_c \qquad (8.29)$$

If frequency is in radians per seconds, the value of $b[L/2]$ is:

$$b[L/2] = \lim_{(k-L/2)} \left| \frac{\sin[\omega_c(k - L/2)]}{\pi(k - L/2)} \right| = \lim_{(k-L/2)} \left| \frac{\omega_c(k - L/2)}{\pi(k - L/2)} \right| = \frac{\omega_c}{2} \qquad (8.30)$$

There is one serious problem with the logic used thus far: the FIR coefficient equation in Equation 8.26 is infinite. That is, Equation 8.26 (or Equation 8.28) produces nonzero values for all finite values of t or k. Since we are designing filters with a finite number of coefficients, we need to truncate the functions produced by these equations. You might suspect that truncating the filter coefficient limits the filter's performance: we might not get the rectangular cutoff we desire. In fact, truncation has two adverse effects: the filter no longer has an infinitely sharp cutoff and oscillations are produced in the filter's spectrum. These adverse effects are demonstrated in the next example, where we show the spectrum of a rectangular window filter truncated to two different lengths.

EXAMPLE 8.3

Find the magnitude spectrum of the rectangular window filter given by Equation 8.28 for two different coefficient lengths: $L = 17$ and $L = 65$. Use a cutoff frequency of 300 Hz assuming a sampling frequency of 1 kHz (i.e., a relative cutoff frequency of $f_c = 0.3$).

Solution: First generate the filter's impulse response, $b[k]$, by direct implementation of Equation 8.28. Note that the two filter lengths are both odd, so to make the shift a whole number, we reduce L by 1 and then shift by $k - (L-1)/2$. The coefficient $b[L/2]$ should be set to $2f_c$ as given by Equation 8.29. After calculating the coefficients, find the spectrum by taking the Fourier transform of the response. Plot only the magnitude spectrum.

```
%Example 8.3  Generate the coefficients of two rectangular window filters and
%  find and plot their magnitude spectra.
%
%
N = 256;                % Padding for Fourier transform (arbitrary)
fs = 1000;              % Sampling frequency (assumed)
f = (1:N)*fs/N;         % Frequency vector for plotting
fc = 300/fs;            % Cutoff frequency (normalized to fs)
L = [17 65];            % Filter lengths (filter order + 1)
for m = 1:2             % Loop for the two filter lengths

  for k = 1:L(m)
    n = k-(L-1)/2 ;     % Whole number shift
    if n == 0           % Generate sin(n)/n function.  Use Equation 8.28.
```

```
    b(k) = 2*fc;                    % Case where denominator is zero.
  else
    b(k) = sin(2*pi*fc*n)/(pi*n);   % Filter impulse response
  end
 end
% Now plot the filter's spectrum
 H = fft(b,N);                      % Calculate spectrum
 subplot(1,2,m);                    % Plot magnitude spectrum
 plot(f(1:N/2),abs(H(1:N/2)),'k');
 .......labels and title.......
end
```

Results: The spectrum of this filter is shown in Figure 8.13 to have two artifacts associated with finite length. The oscillation in the magnitude spectrum is another example of Gibbs artifact first encountered in Chapter 3 and is due to the truncation of the filter coefficients given by Equation 8.28. In addition, the slope is less steep when the filter's impulse response is shorter.

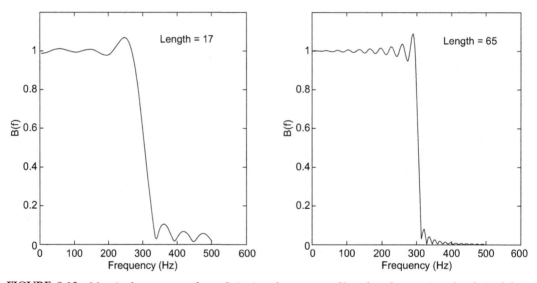

FIGURE 8.13   Magnitude spectrum of two finite impulse response filters based on an impulse derived from Equation 8.28. The impulse responses are abruptly truncated at 17 and 65 coefficients. The low-pass cutoff frequency is 300 Hz for both filters with an assumed sample frequency of 1 kHz. The oscillations seen are Gibbs artifacts and are due to the abrupt truncation of what should be an infinite series. Like the Gibbs artifacts seen in Chapter 3, they do not diminish with increasing filter coefficient length, but do increase in frequency.

We can probably live with the less-than-ideal slope (we should never expect to get an ideal anything in the real world), but the oscillations in the spectrum are serious problems. Since Gibbs artifacts are due to truncation of an infinite function, we might reduce them if the function were tapered toward zero rather than abruptly truncated. In Chapter 4, Section 4.2.4, window functions, such as the Hamming window, are used to improve the spectra obtained from short data sets. These tapering window functions could be used to reduce the abrupt truncation. There are many different window functions, but the two most popular and most useful for FIR impulse responses are the Hamming and Blackman windows. The Hamming window equation is given in Equation 4.9 and repeated here:

$$w[n] = 0.5 - 0.46 \cos\left(\frac{2\pi n}{N}\right) \tag{8.31}$$

where N is the length of the window, which should be the same length as the data. The Blackman window is more complicated, but like the Hamming window, is still easy to program in MATLAB.

$$w[n] = 0.42 - 0.5 \cos\left(\frac{2\pi n}{N}\right) + 0.16 \cos\left(\frac{4\pi n}{N}\right) \tag{8.32}$$

The following example presents the MATLAB code for the Blackman window[8] and applies it to the filters of Example 8.3.

EXAMPLE 8.4

Apply a Blackman window to the rectangular window filters used in Example 8.3. (Note that the word "window" is used in two completely different contexts in the last sentence as you were warned about in footnote 7.) Calculate and display the magnitude spectrum of the impulse functions after they have been windowed.

Solution: Write a MATLAB function, blackman(N), to generate a Blackman window of length N. Apply it to the filter impulse responses using point-by-point multiplication (.* operator). Modify the last example by applying the window to the filter's impulse response before taking the Fourier transform.

```
% Example 8.4 Apply the Blackman window to the rectangular window impulse
responses
% developed in Example 8.3.
%
% Generate the filter coefficients
..... same code as in Example 8.3.......
    if n == 0              % Generate sin(n)/n function.  Use Equation 8.28.
        b(k) = 2*fc;    % Case where denominator is zero.
    else
        b(k) = sin(2*pi*fc*n)/(pi*n);    % Filter impulse response
    end
    w = blackman(L);    % Get Blackman window of length L
    b = b .* w ;        % Apply window to impulse response
    H = fft(b,N);       % Calculate spectrum
......same code as in Example 8.3, plot and label.......
```

[8]This equation is written assuming the popular value for constant α of 0.16.

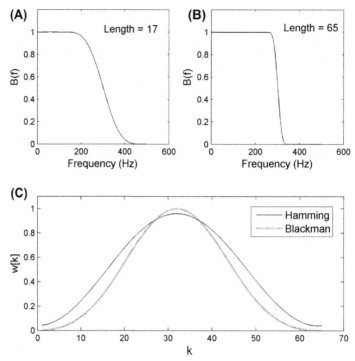

FIGURE 8.14 (A) Magnitude spectrum produced by the 17-coefficient finite impulse response (FIR) filter in Figure 8.13 except a Blackman window was applied to the filter coefficients. The Gibbs oscillations seen in Figure 8.13 are no longer visible. (B) Magnitude spectrum produced by the 65-coefficient FIR filter in Figure 8.13 also after application of the Blackman. (C) Plot of the Blackman and Hamming window functions. Both have cosine-like appearances.

```
    end

    function w = blackman(L)
    % Function to calculate a Blackman window L samples long
    %
    n = (1:L);          % Generate vector for window function
    w = 0.42 - 0.5*cos(2*pi*n/(L-1)) + 0.08*cos(4*pi*n/(L-1));
```

Results: The Blackman window is easy to generate in MATLAB, and when applied to the impulse responses of Example 8.3, substantially reduces the oscillations as shown in Figure 8.14. The filter rolloff is still not that of an ideal filter, but becomes steeper for a longer filter length. Of course, increasing the length of the filter increases the computation time required to apply the filter to a data set, a typical engineering compromise. Figure 8.14C shows a plot of the Blackman and Hamming windows. Both of these popular windows are quite similar since they use raised cosines to taper the filter coefficients.

The next example applies a rectangular window low-pass filter to a signal of human respiration that was obtained from a respiratory monitor.

EXAMPLE 8.5

Apply a low-pass rectangular window filter to the 10-min respiration signal shown in the top trace of Figure 8.15. This signal is found as variable resp in file Resp.mat. $f_s = 12.5$ Hz. The sampling frequency is low because the respiratory signal has a very low bandwidth. Use a cutoff frequency of 1.0 Hz and a filter length of 65. Use the Blackman window to truncate the filter's impulse response. Plot the original and filtered signal.

Solution: Reuse the code in Examples 8.3 and 8.4 and apply the filter to the respiratory signal using convolution. MATLAB's conv routine is invoked with noncausal filtering by using the option 'same' to avoid a time shift in the output.

```
% Example 8.5  Application of a rectangular window low-pass filter to a respiratory
signal.
%
load Resp;                    % Get data
fs = 12.5;                    % Sampling frequency
N = length(resp);            % Get data length
t = (1:N)/fs;                % Time vector for plotting
L = 65;                      % Filter lengths
fc = 1/fs;                   % Cutoff frequency: 1.0 Hz
plot(t,resp+1,'k'); hold on;       % Plot original data (offset for clarity)
for k = 1:L                        % Generate sin(n)/n function symmetrically
  n = k-(L-1)/2 ;                  % Same code as in previous examples
  if n == 0
    b(k) = 2*fc;                   % Case where denominator is zero.
  else
    b(k) = (sin(2*pi*fc*n))/(pi*n); % Filter impulse response
  end
end
b = b.*blackman(L);                % Apply Blackman window
%
y = conv(resp,b,'same');           % Apply filter
plot(t,y,'k');
```

Results: The filtered respiratory signal is shown in the lower trace of Figure 8.15. The filtered signal is smoother than the original signal as the noise riding on the original signal has been eliminated.

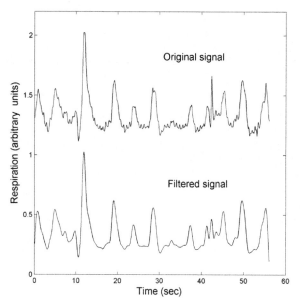

FIGURE 8.15   The output of a respiratory monitor is shown in the upper trace to have some higher frequency noise. In the lower trace, the signal has been filtered with a finite impulse response rectangular low-pass filter. The filter has a cutoff frequency of 1.0 Hz and a filter length of 65. *Original data from PhysioNet, Goldberger, A.L., Amaral, L.A.N., Glass, L., Hausdorff, J.M., Ivanov, P.C., Mark, R.G., Mietus, J.E., Moody, G.B., Peng, C.-K., Stanley, H.E., 2000. PhysioBank, PhysioToolkit, and PhysioNet: components of a new research resource for complex physiologic signals. Circulation 101 (23), e215–e220. Circulation Electronic Pages. http://circ.ahajournals.org/cgi/content/full/101/23/e215.*

The FIR filter coefficients for high-pass, band-pass, and band-stop filters can also be derived by applying an inverse Fourier transform to rectangular spectra having the appropriate associated shape. These equations have the same general form as Equation 8.28 except they may include additional terms:

$$b[k] = \begin{cases} -\dfrac{\sin[2\pi f_c(k - L/2)]}{\pi(k - L/2)} & k \neq \dfrac{L}{2} \\[2ex] 1 - 2f_c & k = \dfrac{L}{2} \end{cases} \quad \text{Highpass} \qquad (8.33)$$

$$b[k] = \begin{cases} \dfrac{\sin[2\pi f_h(k - L/2)]}{\pi(k - L/2)} - \dfrac{\sin[2\pi f_l(k - L/2)]}{\pi(k - L/2)} & k \neq \dfrac{L}{2} \\[2ex] 2(f_h - f_l) & k = \dfrac{L}{2} \end{cases} \quad \text{Bandpass} \qquad (8.34)$$

$$b[k] = \begin{cases} \dfrac{\sin[2\pi f_l(k - L/2)]}{\pi(k - L/2)} - \dfrac{\sin[2\pi f_h(k - L/2)]}{\pi(k - L/2)} & k \neq \dfrac{L}{2} \\[3ex] 1 - 2(f_h - f_l) & k = \dfrac{L}{2} \end{cases} \quad \text{Bandstop} \quad (8.35)$$

The order of high-pass and band-stop filters should always be even, so the number of co-efficients in these filters should be odd. The next example applies a band-pass filter to the electroencephalogram (EEG) signal introduced in Chapter 1.

EXAMPLE 8.6

Apply a band-pass filter to the EEG data in file ECG.mat. Use a lower cutoff frequency of 6 Hz and an upper cutoff frequency of 12 Hz. Use a Blackman window to truncate the filter's impulse to 129 coefficients. Plot the data before and after band-pass filtering. Also plot the spectrum of the original signal and superimpose the spectrum of the band-pass filter.

Solution: Construct the filter's impulse response using the band-pass equation shown in Equation 8.34. Note the special case where the denominator in the equation goes to zero (i.e., when $k = L/2$). The application of limits and the small angle approximation ($\sin(x) \to x$) gives a coefficient value of $b[L/2] = 2f_h - 2f_l$, as shown in Equation 8.33. Recall that the sampling frequency of the EEG signal is 100 Hz.

After applying the filter and plotting the resulting signal, compute the signal spectrum using the Fourier transform and plot. The filter's spectrum is found by taking the Fourier transform of the impulse response ($b[k]$) and is plotted superimposed on the signal spectrum.

```
% Example 8.6 Apply a band-pass filter to the EEG data in file ECG.mat.
%
load EEG;                  % Get data
N = length(eeg);           % Get data length
fs = 100;                  % Sample frequency
fh = 12/fs;                % Set highpass and
fl = 6/fs;                 % lowpass cutoff frequencies
L = 129;                   % Set number of weights
%
for k = 1:L                % Generate bandpass filter coefficients

   n = k - (L-1)/2 ;       % Make symmetrical
   if n == 0
     b(k) = 2*fh - 2*fl;   % Case where denominator is zero
   else
     b(k) = sin(2*pi*fh*n)/(pi*n) - sin(2*pi*fl*n)/(pi*n) ; % Filter coefficients
   end
end
b = b .* blackman(L);      % Apply Blackman window
y = conv(eeg,b,'same');    % Filter the data using convolution
.......plot eeg before and after filtering; plot eeg and filter spectrum......
```

Results: Band-pass filtering the EEG signal between 6 and 12 Hz reveals a strong higher-frequency oscillatory signal that is washed out by lower frequency components in the original

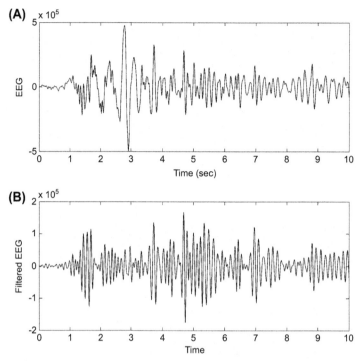

FIGURE 8.16   Electroencephalogram signal before (A) and after (B) band-pass filtering between 6 and 12 Hz. A fairly regular oscillation is seen in the filtered signal.

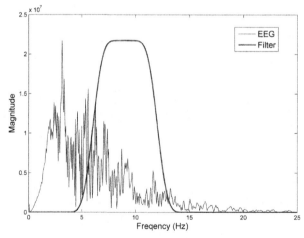

FIGURE 8.17   The magnitude spectrum of the electroencephalography signal used in Example 8.5 along with the spectrum of a band-pass filter based on Equation 8.12. The band-pass filter range is designed to be between 6 and 12 Hz.

signal as shown in Figure 8.16B. This figure shows that filtering can significantly alter the appearance and interpretation of biomedical data. The band-pass spectrum shown in Figure 8.17 has the desired cutoff frequencies and, when compared with the EEG spectrum, is shown to reduce the high- and low-frequency components of the EEG signal.

Implementation of other FIR filter types is found in the problem set. A variety of FIR filters exist that use strategies other than the rectangular window to construct the filter coefficients, and some of these are explored in the section on MATLAB implementation. One FIR filter of particular interest is used to construct the derivative of a waveform, since the derivative is often of interest in the analysis of biosignals. The next section explores a popular filter for this operation.

8.5.1 Derivative Filters—The Two-Point Central Difference Algorithm

The derivative is a common operation in signal processing, and is particularly useful in analyzing certain physiological signals. Digital differentiation is defined as $dx[n]/dn$ and can be calculated directly from the slope of $x[n]$ by taking differences:

$$\frac{dx[n]}{dn} = \frac{\Delta x[n]}{T_s} = \frac{x[n+1] - x[n]}{T_s} \tag{8.36}$$

This equation can be implemented by MATLAB's `diff` routine. This routine uses no padding, so the output is one sample shorter than the input. As shown in Chapter 6 (Section 6.5.2), the frequency characteristic of the derivative operation increases linearly with frequency, so differentiation enhances higher frequency signal components (see Figure 6.11). Since the higher frequencies frequently contain a greater percentage of noise, this operation tends to produce a noisy derivative curve. The upper curve of Figure 8.18A is a fairly clean physiological motor response, an eye movement response similar to that used in Figure 8.10. The lower curve of Figure 8.18A is the velocity of the movement obtained by calculating the derivative using MATLAB's `diff` routine which implements Equation 8.36. Considering the relative smoothness of the original signal, the velocity curve obtained using Equation 8.36 is quite noisy.

In the context of FIR filters, Equation 8.36 is equivalent to a two-coefficient filter: $b[k] = [+1/T_s, -1/T_s]$. (Note that the positive and negative coefficients are reversed by convolution, so they are sequenced in reverse order in the impulse response.) A better approach to differentiation is to construct a filter that approximates the derivative at lower frequencies but attenuates higher frequencies that are likely to be only noise. The "two-point central difference algorithm" achieves such an effect, acting as a differentiator at lower frequencies and a low-pass filter at higher frequencies. Figure 8.18B shows the same responses when this algorithm is used to estimate the derivative. The result is a much cleaner velocity signal that still captures the peak velocity of the response.

The two-point central difference algorithm still subtracts two points to get a slope but the two points are no longer adjacent, rather, they may be spaced some distance apart. Putting this in FIR filter terms, the algorithm is based on an impulse function containing two coefficients of equal but opposite sign, spaced L points apart. The equation defining the filter coefficients of this differentiator is:

$$\frac{dx[n]}{dn} = \frac{x(n+L) - x(n-L)}{2LT_s} \tag{8.37}$$

where L is now called the "skip factor" that defines the distance between the points used to calculate the slope, and T_s is the sample interval. The skip factor, L, influences the effective

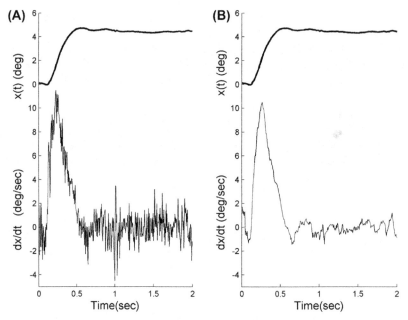

FIGURE 8.18   An eye movement response to a step change in target depth is shown in the upper trace, and its velocity (i.e., derivative) is shown in the lower trace. (A) The derivative was calculated by taking the difference in adjacent points and scaling by the sample interval, following Equation 8.36. The velocity signal is noisy, even though the original signal is fairly smooth. (B) The derivative was computed using the two-point central difference algorithm with a skip factor of four. This filter results in a much cleaner derivative signal.

bandwidth of the filter as shown below. Implemented as an FIR filter, Equation 8.37 leads to coefficients:

$$b[k] = \begin{cases} 1/2LT_s & k = -L \\ -1/2LT_s & k = +L \\ 0 & k \neq \pm L \end{cases} \tag{8.38}$$

Again, note that the $+L$ coefficient is negative and the $-L$ coefficient is positive since the convolution operation reverses the order of $b[k]$. As with all FIR filters, the frequency response of this filter algorithm can be determined by taking the Fourier transform of $b[k]$. Since this function is fairly simple, it is not difficult to take the Fourier transform analytically (trying to break from MATLAB wherever possible) as well as in the usual manner using MATLAB. Both methods are presented in the following example.

EXAMPLE 8.7

(A) Determine the magnitude spectrum of the two-point central difference algorithm analytically, then (B) use MATLAB to determine the spectrum and apply it to the EEG signal.

Analytical Solution: Starting with the equation for the discrete Fourier transform (Equation 3.34) substituting k for n:

$$X[m] = \sum_{k=0}^{N-1} b[k]\, e^{-j2\pi mk/N}$$

Since $b[k]$ is nonzero only for $k = \pm L$, the Fourier transform, after the summation limits are adjusted for a symmetrical coefficient function with positive and negative n, becomes:

$$X(m) = \sum_{k=-L}^{L} b[k]e^{-j2\pi mk/N} = \frac{1}{2LT_s}e^{-j2\pi m(-L)/N} - \frac{1}{2LT_s}e^{-j2\pi mL/N}$$

$$X(m) = \frac{e^{-j2\pi m(-L)/N} - e^{-j2\pi mL/N}}{2LT_s} = \frac{-j\,\sin(2\pi mL/N)}{LT_s}$$

where L is the skip factor and N is the number of samples in the waveform. To put this equation in terms of frequency, note that $f = m/(N\,T_s)$, hence $m = f\,N\,T_s$.

To find $|X(f)|$, substitute $f\,N\,T_s$ for m and take the magnitude.

$$|X(f)| = \left| -j\frac{\sin(2\pi f LT_s)}{L\,T_s} \right| = \frac{|\sin(2\pi f LT_s)|}{L\,T_s}$$

This equation shows that the magnitude spectrum, $|X(f)|$, is a sine function that goes to zero at $f = 1/(LT_s) = f_s/L$. Figure 8.19 shows the frequency characteristics of the two-point central difference algorithm for two different skip factors: $L = 2$ and $L = 6$.

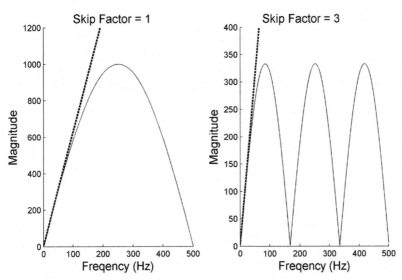

FIGURE 8.19 The frequency response of the two-point central difference algorithm using two different skip factors: (A) $L = 2$; (B) $L = 6$. The *dashed line* shows the frequency characteristic of a simple differencing operation. The sample frequency is 1.0 kHz.

MATLAB Solution: Finding the spectrum using MATLAB is straightforward once the coefficient vector is constructed. An easy way to construct the impulse response is to use brackets and concatenate zeros between the two end values, essentially following Equation 8.38 directly. Again, the initial filter coefficient is positive and the final coefficient negative to account for the reversal produced by convolution.

```
% Example 8.7 Determine the frequency response of
%  the two point central difference algorithm used for differentiation.
%
Ts = .001;                    % Assume a Ts of 1 msec. (i.e., fs = 1 kHz)
N = 1000;                     % Number of data points (time = I sec)
Ln = [1 3];                   % Define two different skip factors
for skip = 1:2                % Repeat for each skip factor
  L = Ln(skip);               % Set skip factor
  b = [1/(2*L*Ts) zeros(1,2*L-1) -1/(2*L*Ts)];   % Filter impulse response
  H = abs(fft(b,N));          % Calculate magnitude spectrum
  ..... plot and label spectrum; plot straight line for comparison....
end
```

Result: The result of both this program and the analysis are shown in Figure 8.19. A true derivative has a linear change with frequency: a line with a slope proportional to f as shown by the dashed lines in Figure 8.19. The two-point central difference spectrum approximates a true derivative over the lower frequencies, but has the characteristic of a low-pass filter for higher frequencies. Increasing the skip factor, L, has the effect of lowering the frequency range over which the filter acts like a derivative operator as well as lowering the low-pass filter range. Note that for skip factors >2, the response curve repeats above $f = 1/(LT_s)$. Usually the assumption is made that the signal does not contain frequencies in this range. If this is not true, then these higher frequencies can be removed by an additional low-pass filter as shown in one of the problems. It is also possible to combine the difference equation (Equation 8.15) with a low-pass filter and this is also explored in one of the problems.

8.5.2 Determining Cutoff Frequency and Skip Factor

Determining the appropriate cutoff frequency of a filter or the skip factor for the two-point central difference algorithm can be somewhat of an art (meaning there is no definitive approach to a solution). If the frequency ranges of the signal and noise are known, setting cutoff frequencies is straightforward. But this knowledge is not usually available in biomedical engineering applications. In most cases, filter parameters such as filter order and cutoff frequencies are set empirically based on the data. In one trial-and-error scenario, the signal bandwidth is progressively reduced until some desirable feature of the signal is lost or compromised. Although it is not possible to establish definitive rules because of the task-dependent nature of filtering, the next example gives an idea about how these decisions are approached.

When taking derivatives, we often strive to remove the most noise while still preserving the derivative peaks. For example, if the signal represents a movement, then its derivative is velocity, and we frequently want a good measure of peak velocity. In the next example, we return to the eye movement signal shown in Figure 8.18 to evaluate several different skip factors to find the one that gives the best reduction of noise without reducing the accuracy of the velocity trace. We use a strictly empirical approach: we increase the skip factor until the measurement of peak velocity is reduced.

EXAMPLE 8.8

Use the two-point central difference algorithm to compute velocity traces of the eye movement step response in file `eye.mat`. Use four different skip factors (1, 2, 5, and 10) to find the skip factor that best reduces noise without substantially reducing the peak velocity of the movement.

Solution: Load the file and use a loop to calculate and plot the velocity determined with the two-point central difference algorithm using the four different skip factors. Find the maximum value of the velocity trace for each derivative evaluation and display on the associated plot. The original eye movement shown in Figure 8.18 (upper traces), is in degrees so the velocity trace is in degrees per second.

```
% Example 8.8  To evaluate the two-point central difference algorithm using
%   different derivative skip factors
%
load eye;                          % Response in vector eye_move
fs = 200;                          % Sampling frequency
Ts = 1/fs;                         % Calculate Ts
t = (1:length(eye_move))/fs;       % Time vector for plotting
L = [1 2 5 10];                    % Filter skip factors
for skip = 1:4                     % Loop for different skip factors
   b = [1/(2*L(skip)*Ts) zeros(1,2*L(skip)-1) -1/(2*L(skip)*Ts)]; % Construct
filter
   der = conv(eye_move,b,'same');  % Apply filter
   subplot(2,2,skip);
   plot(t,der,'k');                % Plot velocity curve
.....text, labels, and axis.......
end
```

Results: The results of this program are shown in Figure 8.20. As the skip factor increases, the noise decreases, and the velocity trace becomes quite smooth at a skip factor of 10. But our measurement of peak velocity also decreases at the higher skip factors. Deciding on the best skip factor requires that we exercise some judgment. Examining the curve with the lowest skip factor ($L = 1$) suggests that the peak velocity shown, 23 degrees/s, may be augmented a bit by noise. A peak velocity of 21 degrees/s is found for both skip factors of 2 and 5. This peak velocity appears to be reflective of the true peak velocity. The peak found using a skip factor of 10 is clearly reduced. A skip factor of around 5 seems appropriate, but this is a judgment call. In fact, judgment and intuition are all too frequently involved in signal processing and signal analysis tasks, a

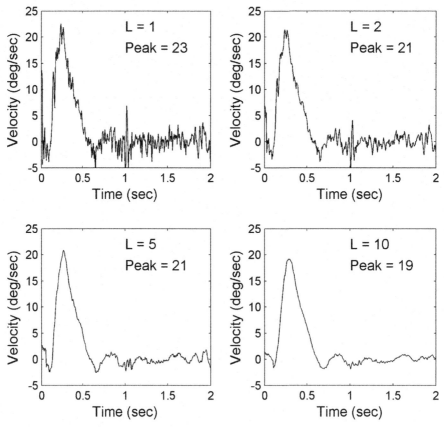

FIGURE 8.20   Velocity traces for the eye movement shown in Figure 8.18, calculated by the two-point central difference algorithm for different values of skip factor as shown. The peak velocities are shown in degrees per second.

reflection that signal processing is still an art. Other examples that require empirical evaluation are given in the problem set.

8.6 FINITE IMPULSE RESPONSE AND INFINITE IMPULSE RESPONSE FILTER DESIGN USING THE SIGNAL PROCESSING TOOLBOX

Unlike FIR filters, IIR filters are quite difficult to design from first principles. MATLAB again comes to our rescue with the Signal Processing Toolbox. This toolbox offers considerable support for the design and evaluation of both FIR and IIR filters. The remainder of this chapter is devoted to using this toolbox to design both simple and more complicated filters.

Within the MATLAB environment, filter design and application occur in either one or two stages, each stage executed by separate but related routines. In the two-stage protocol, the user supplies information regarding the filter type and desired attenuation characteristics, but not the filter order. The first-stage routines determine the appropriate order as well as other parameters required by the second-stage routines. The second-stage routines then generate the filter coefficients, $b[k]$, based on the arguments produced by the first-stage routines, including the filter order.

It is possible to bypass the first stage routines if you already know, or can guess, the filter order. Only the second-stage routines are needed. In this situation the user supplies the filter order along with other filter specifications. In practical situations, trial and error is often required to determine the filter order and first stage routines are not helpful. For this reason, we ignore first-stage routines, and discuss only the second-stage filter design routines.

Other tools that we will not describe include an interactive filter design package called FDATool (for Filter Design and Analysis Tool) that uses a graphical user interface to design filters with highly specific or demanding spectral characteristics. Yet another Signal Processing Toolbox package, the SPTool (Signal Processing Tool), is useful for analyzing filters and generating spectra of both signals and filters. The following basic routines should cover all your filtering needs, but you can always go to MATLAB help for detailed information on these and other packages.

Irrespective of how we get there, the net result is a set of b and a coefficients that uniquely define the filter's spectrum. From there, we already know how to apply the filter to a signal using either the conv or filter routine. Alternatively the Signal Processing Toolbox contains a third routine, filtfilt that, like conv with the 'same' option, employs noncausal methods to implement filters that have no time delay. However, filtfilt can also be used to implement IIR filters. The calling structure is exactly the same as filter:

```
y = filtfilt(b,a,x);     % Noncausal IIR filter applied to signal x
```

One of the problems dramatically illustrates the difference between the use of filter and filtfilt.

As a brief aside, there is a useful Signal Processing Toolbox routine that determines the frequency response of a filter given the coefficients. We already know how to do this, we did it in Example 8.1 using the Fourier transform (i.e., Equation 8.15). But the MATLAB routine freqz also includes frequency scaling and plotting, making it quite convenient.

```
[H,f] = freqz (b,a,n,fs);
```

where b and a are the filter coefficients and n is optional and specifies the number of points in the desired frequency spectra. For FIR filters, the value of a is set to 1.0 as in the filter routine. The input argument, fs, is also optional and specifies the sampling frequency. Both output arguments are also optional and are usually not given. If freqz is called without the output arguments, the magnitude and phase plots are produced automatically. If the output arguments are specified, the output vector H is the complex frequency response of the filter (the same variable produced by fft). The second optional output f is a frequency vector useful in plotting. If fs is given, f is in hertz and ranges between 0 and $f_s/2$; otherwise a less useful f is provided in radians per sample and ranges between 0 and π.

8.6.1 Finite Impulse Response Filter design

Although we already know how to design the basic rectangular window FIR filters, the Signal Processing Toolbox makes it so easy that it is hard to resist. The toolbox supports all rectangular window FIR filters: low pass, high pass, band pass, and band stop (Equations 8.28, 8.33−8.35). The calling structure is:

```
b = fir1(N,Wn,'type');    % Design a rectangular window filter
```

where b is a vector containing the coefficients, N is the filter order, and Wn[9] is the cutoff frequency (with respect to $f_s/2$, not f_s, so when wn = 1, $f_c = f_s/2$). For a low-pass filter, the optional variable 'type' is absent. If you want a band-pass filter, simply make Wn a two-element vector that contains low and high cutoff frequencies, again relative to $f_s/2$. If you want a high-pass filter, make the third variable 'high' and use with a single Wn to specify the cutoff frequency. Finally for a band-stop filter, again make Wn a two-element variable specifying the lower and upper cutoff frequencies and make the third variable 'stop.' An optional fourth variable can be used to specify a tapering window (which must be N + 1 points long), but if it is omitted, the filter coefficients are tapered by the ever popular Hamming window. Use help fir1 to find other features you are unlikely to need.

In some rare cases, there may be a need for a filter with a more exotic spectrum for which we turn to fir2. This routine produces an FIR filter with a spectrum of almost any shape. The command structure for fir2 is:

```
b = fir2(N,F,A)
```

where N is the filter order, F is a vector of normalized frequencies in ascending order, and A is the desired gain of the filter at the corresponding frequency in vector F. In other words, plot(F,A) would show the desired magnitude frequency curve. Clearly Fand A must be the same length, but duplicate frequency points are allowed corresponding to step changes in the frequency response. In addition, the first value of f must be 0.0 and the last value 1.0 (equal to $f_s/2$). Some additional optional input arguments are mentioned in the MATLAB help file. The next example shows the use of fir1 and the flexibility of fir2.

EXAMPLE 8.9

Use fir1 to design a band-pass filter with a passband between 50 and 100 Hz. Use fir2 to design a second filter, a double band-pass filter with one passband also between 50 and 100 Hz and a second passband between 150 and 200 Hz. Make the order 65 for both filters. Apply this filter to a signal containing sinusoids at 75, and 175 Hz buried in 20 dB of noise (SNR = −20 dB). Use routine sig_noise (Section 3.4.3) to generate a 2000-sample signal. (Recall sig_noise assumes $f_s = 1$ kHz.)

[9]Normally, we use f for frequency in hertz and ω for frequency in radians. But here Wn is frequency in hertz, although it is a relative frequency (relative to $f_s/2$). It is confusing to use Wn for frequency in hertz, but MATLAB uses it in most of their filter routines, so we stick with it to be consistent with MATLAB. Yet in routine fir2, described next, MATLAB uses F for relative frequency in hertz. So much for consistency.

Plot the magnitude spectrum of both filters and the magnitude spectrum of the signal before and after filtering with both filters.

Solution: Constructing a single band-pass filter using `fir1` is straightforward. We could construct a double band-pass filter by putting two such band-pass filters in series, but instead we can use `fir2` to construct a single FIR filter having the desired filter characteristics.

For the double band-pass filter, we first specify the desired frequency characteristics in terms of a frequency and a gain vector. The frequency vector must begin with 0.0 and end with 1.0, but can have duplicate frequency entries to allow for step changes in gain. We can apply these filters in a causal manner using either `filter` or `conv`. In this example, we use `filter` for a little variety. Then, the magnitude spectra of the filtered and unfiltered waveforms are determined using the Fourier transform.

```
% Example 8.9 Design a double bandpass filter.
fs = 1000;               % Sample frequency
N = 2000;                % Number of points
Nf = 65;                 % Filter order
% Cutoff frequencies of single bandpass filter
fl1 = 50/(fs/2);         % First peak low cutoff freq.
fh1 = 100/(fs/2);        % First peak high cutoff freq.
fl2 = 150/(fs/2);        % Second peak low freq. cutoff
fh2 = 200/(fs/2);        % Second peak high freq. cutoff
%
x = sig_noise([75 175],-20,N);         % Generate noise waveform
%
b1 = fir1(Nf,[fl1,fh1]);               % Design single bandpass filter.
% Design double bandpass filter
F = [0 fl1 fl1 fh1 fh1 fl2 fl2 fh2 fh2 1]; % Construct desired
A = [0 0 1 1 0 0 1 1 0 0];             %  frequency/gain characteristic
b2 = fir2(Nf,F,A);
[H1,f1] = freqz(b1,1,512,fs);          % Calculate filter1 frequency response
H2 = freqz(b2,1,512,fs);               % Calculate filter2 frequency response
%
y1 = filter(b1,1,x);                   % Apply single bandpass filter
y2 = filter(b2,1,x);                   % Apply double bandpass filter
Xf = abs(fft(x));                      % Get signal spectrum
Yf1 = abs(fft(y1));                    % Get spectrum of filtered signals
Yf2 = abs(fft(y2));
.......plot and label magnitude spectra.......
```

Results: Figure 8.21 is a plot of the magnitude spectrum of the noisy signal. It looks noisy with no hint of the buried sinusoids. Figure 8.22 shows the magnitude spectra (upper curves) of the single and double band-pass filters. The lower curves of Figure 8.22 show the spectra of the noisy signal after filtering by the two filters, and the signal peaks are clearly visible. These peaks are not any larger than they were in Figure 8.21, but are much more evident because the surrounding energy, energy outside the various passbands, has been attenuated.

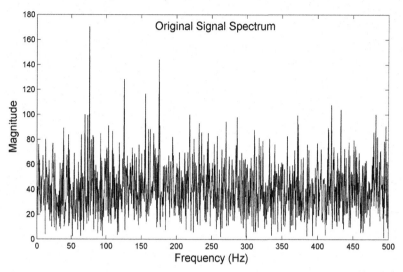

FIGURE 8.21 The magnitude spectra of the noisy signal used in Example 8.9. There are two sinusoids in this signal buried in 20 dB of noise; that is, the signal to noise ratio is −20 dB.

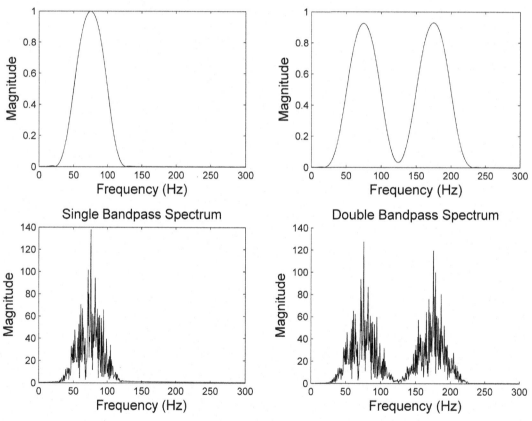

FIGURE 8.22 The spectra (upper curves) of the single and double band-pass filters used in Example 8.9. The lower curves show the spectra of the noisy signal after filtering. Although the signal peaks are not really any larger than they are in Figure 8.21, they are much more apparent because of the attenuation of the surrounding noise.

Although the effect of filtering in this example is quite dramatic, we set the filter's frequency ranges to center on the signal frequencies. This can be done only if we know the frequencies we are trying to isolate ahead of time.

8.6.3 Designing Infinite Impulse Response Filters

IIR filters can achieve sharp rolloffs with far fewer filter coefficients than FIR filters (e.g., see Figure 8.8). For standard noise filtering of one-dimensional signals, the IIR filter is usually your go-to filter. The design of IIR filters is not as straightforward as that for FIR filters and whole books have been written on this topic. But with the MATLAB Signal Processing Toolbox, anyone can do it. Although you need the Signal Processing Toolbox to design IIR filters, you do not need it to apply these filters. IIR filters are totally specified by their a and b coefficients, so once you have these coefficients, you can apply the filters using the standard `filter` routine, or with any other implementation of Equation 8.21. You could use another programming language to implement these filters; you do not need MATLAB.

IIR filters are analogous to some popular analog filters and originally were designed using tools developed for analog filters. In an analog filter, there is a direct relationship between the number of independent energy storage elements in the system and the filter's rolloff slope: each energy storage element increases the downward (or upward) slope by 20 dB/decade. In IIR filters, the first a coefficient, $a[0]$, always equals 1.0, but each additional a coefficient adds 20 dB/decade to the attenuation slope. So, an eighth-order IIR filter with nine a coefficients has the same attenuation characteristics as an eighth-order analog filter.[10] Since the slope of an IIR filter increases by 20 dB/decade for each filter order, determining the filter order needed for a given desired attenuation is straightforward.

IIR filter design under MATLAB follows the same procedures as FIR filter design, only the names of the routines are different. In the MATLAB Signal Processing Toolbox, the two-stage design process is supported for most of the IIR filter types. But as with FIR design, a single-stage design process can be used provided you specify the filter order. Again, we stick to single-stage design as is commonly used practice.

The Yule-Walker recursive filter is the IIR equivalent of the `fir2` FIR filter routine in that it allows for the specification of a flexible frequency characteristic. The calling structure is also very similar to that of `fir2`.

```
[b,a] = yulewalk(N,F,A);
```

where N is the filter order, and F and A specify the desired frequency characteristic in the same manner as `fir2`: A is a vector of the desired filter gains at the frequencies specified in F. The F and A vectors follow the same rules as in `fir2`: frequencies are relative to $f_s/2$, the

---

[10]In the analog world, one rarely finds a filter higher than eighth-order. Even using integrated circuits, the circuitry becomes overly complex. Moreover, an eighth-order filter is usually sufficient, even for demanding filtering tasks, such as antialiasing. In fact, lower order filters such as fourth-order analog filters are often used in many analog filtering applications.

first point in F must be 0, the last point 1.0, and duplicate frequency points are allowed. Duplicate frequency points enable step changes in the frequency response. This routine is used in the next example.

EXAMPLE 8.10

Design the double band-pass filter that is used in Example 8.9 with one passband, also between 50 and 100 Hz and a second passband between 150 and 200 Hz. Use a 12th-order IIR filter and compare the results with the 65th-order FIR filter used in Example 8.9. Plot the frequency spectra of both filters superimposed for easy comparison.

Solution: Modify Example 8.9 to add the Yule-Walker filter, determine its spectrum using freqz, and plot superimposed on the spectrum of the double band-pass FIR filter. Remove the code relating to the single band-pass filter and the signal filtering as it is not needed in this example.

```
% Example 8.10 Design a double bandpass IIR filter and compare with a similar
FIR filter
%
    ..... same initial code as in Example 8.9......
N1 = 65;                % FIR Filter order
N2 = 12;                % IIR Filter order
fl1 = 50/(fs/2);        % First peak low cutoff freq.
fh1 = 100/(fs/2);       % First peak high cutoff freq.
fl2 = 150/(fs/2);       % Second peak low freq. cutoff
fh2 = 200/(fs/2);       % Second peak high freq. cutoff
%
% Design filter
F = [0 fl1 fl1 fh1 fh1 fl2
fl2 fh2 fh2 1];                       % Construct desired
A = [0 0 1 1 0 0 1 1 0 0];           % frequency characteristic
b1 = fir2(N1,F,A);                    % Construct FIR filter
[b2 a2] = yulewalk(N2,F,A);          % Construct IIR filter
[H1,f1] = freqz(b1,1,512,fs);        % Calculate FIR frequency response
[H2 f2] = freqz(b2,a2,512,fs);       % Calculate IIR frequency response
            .......plots and labels........ .
```

Results: The spectral results from two filters are shown in Figure 8.23. The magnitude spectra of both filters look quite similar despite the fact that the IIR filter has far fewer coefficients. (The FIR filter has 66 b coefficients, whereas the IIR filter has 13 b coefficients and 13 a coefficients.)

In an extension of Example 8.10, the FIR and IIR filters are applied to a random signal of 10,000 data points, and MATLAB's tic and toc are used to evaluate the time required for each filter operation.

```
% Example 8.10 (continued) Time required to apply the two filters in Example 8.10

%  to a random array of 10,000 samples.
```

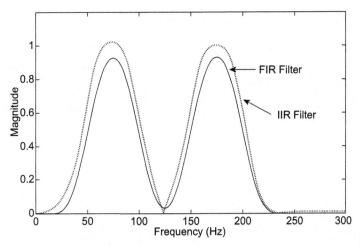

FIGURE 8.23 The filter spectra of two double band-pass filters, a 65th-order finite impulse response (FIR) used in Example 8.9, and a 12th-order infinite impulse response (IIR) filter. The two spectra are quite similar despite the large difference in the number of filter coefficients. (The FIR filter has 66 coefficients, whereas the IIR filter has 13 coefficients.)

```
%
x = rand(1,10000);        % Generate random data
tic                       % Start clock
y = filter(b2,a2,x);      % Filter data using the IIR filter
toc                       % Get IIR filter operation time
clear y;
tic                       % Restart clock
y = filter(b1,1,x);       % Filter using the FIR filter
toc                       % Get FIR filter operation time
```

Surprisingly, despite the differences in the number of coefficients, the two filters require around 1.2 msec for implementation. However, if conv is used to implement the FIR filter, then it takes three times longer.

As mentioned earlier, some well-known analog filter types can be duplicated as IIR filters. Specifically, analog filters known as Butterworth, Chebyshev types I and II, and Elliptic (or Cauer) designs can be implemented as IIR digital filters and are supported by the MATLAB Signal Processing Toolbox. Butterworth filters provide a frequency response that is maximally flat in the passband and monotonic overall. To achieve this characteristic, Butterworth filters sacrifice rolloff steepness; hence, the Butterworth filter has a less sharp initial attenuation characteristic than other filters. The other filters achieve a faster rolloff than Butterworth filters, but have ripple in the passband. The Chebyshev type II filter has ripple only in the stopband, and its passband is monotonic, but it does not rolloff as sharply as type I which has ripple in the passband. The ripple produced by Chebyshev filters is termed equiripple since it is of constant amplitude across all frequencies. Finally, Elliptic filters have steeper rolloff than any of the above-mentioned filters, but have equiripple in both the passband and stopband. Although the sharper initial rolloff is a desirable feature as it provides a more definitive boundary between passband and stopband, most biomedical engineering applications require a smooth passband. This makes the Butterworth the filter of choice and the most commonly used in practice.

The filter coefficients for a Butterworth IIR filter can be determined using the MATLAB routine:

```
[b,a] = butter(order,wn,'ftype') ;   % Design Butterworth filter
```

where `order` and `wn` are the order and cutoff frequencies, respectively. (Of course `wn` is relative to $f_s/2$.) The other arguments are similar to those in the FIR filter `fir1`. For a low-pass filter, `wn` is scalar and the `'ftype'` argument is missing. For a band-pass filter, `wn` is a two-element vector, `[w1 w2]`, where `w1` is the low cutoff frequency and `w2` is the high cutoff frequency and again there is no `'ftype'`. If a high-pass filter is desired, then `wn` should be scalar and `'ftype'` should be `'high.'` For a stopband filter, `wn` is a two-element vector indicating the frequency ranges of the stop band and `'ftype'` should be `'stop'`. The outputs of `butter` are the `b` and `a` coefficients.

Although the Butterworth filter is the only IIR filter you are likely to use, the other filters are easily designed using the associated MATLAB routine. The Chebyshev type I and II filters are designed with similar routines except that an additional parameter is needed to specify the allowable ripple:

```
[b,a] = cheby1(order,rp,wn,'ftype');   % Design Chebyshev Type I
```

where the arguments are the same as in `butter`, except for the additional argument, `rp`, which specifies the maximum desired passband ripple in dB. The type II Chebyshev filter is designed using:

```
[b,a] = cheby2(n,rs, wn,'ftype');   % Design Chebyshev Type II
```

where again the arguments are the same, except `rs` specifies the stopband ripple, again in dB, but with respect to the passband gain. In other words, a value of 40 dB means that the ripple will not exceed -40 dB where the passband gain is 0.0 dB. In effect, this value specifies the minimum attenuation in the stopband.

The Elliptic filter includes both stopband and passband ripple values:

```
[b,a] = ellip(n,rp,rs,wn,'ftype');   % Design Elliptic filter
```

where the arguments presented are in the same manner as described earlier, with rp specifying the passband gain in dB and rs specifying the stopband ripple relative to the passband gain.

The following example uses these routines to compare the frequency response of the four IIR filters discussed earlier.

EXAMPLE 8.11

Plot the frequency response curves (in dB) obtained from an eighth-order low-pass filter using the Butterworth, Chebyshev types I and II, and Elliptic filters. Use a cutoff frequency of 200 Hz and assume a sampling frequency of 2 kHz. For all filters, the ripple or maximum attenuation should be less than 3 dB in the passband, and the stopband attenuation should be at least 60 dB.

Solution: Use the MATLAB IIR design routines to determine the a and b coefficients, use freqz to calculate the complex frequency spectrum, take the absolute value of the spectra and convert to dB ($20*\log(abs(H))$), then plot using semilogx. Repeat this procedure for the four filters.

```
% Example 8.11 Frequency response of four IIR 8th-order lowpass filters
%
fs = 2000;                      % Sampling filter
n = 8;                          % Filter order
wn = 200/1000;                  % Filter cutoff frequency
rp = 3;                         % Maximum passband ripple
rs = 60;                        % Maximum stopband ripple
% Determine filter coefficients
[b,a] = butter(n,wn);           % Butterworth filter coefficients
[H,f] = freqz(b,a,256,fs);      % Calculate complex spectrum
H = 20*log10(abs(H));           % Convert to magnitude in dB
subplot(2,2,1);
semilogx(f,H,'k');              % Plot spectrum in dB vs log freq.
   .......labels and title.......
   ...... repeat for the other 3 IIR filters.......
[b,a] = cheby1(n,rp,wn);        % Chebyshev Type I filter coefficients
[b,a] = cheby2(n,rs,wn);        % Chebyshev Type II filter coefficients
[b,a] = ellip(n,rp,rs,wn);      % Elliptic filter coefficients
   .......use freqz, plot, labels, and titles.......
```

The spectra of the four filters are shown in Figure 8.24. As described earlier, the Butterworth is the only filter that has smooth frequency characteristics in both the passband and stopband; it is this feature that makes it popular in biomedical signal processing both in its analog and digital incarnations. The Chebyshev type II filter also has a smooth passband and a slightly steeper initial slope than the Butterworth, but it does have ripple in the stopband, which can be problematic in some situations. The Chebyshev type I has an even sharper initial slope, but also has ripple in the passband, which limits its usefulness. The sharpest initial slope is provided by the Elliptic filter, but

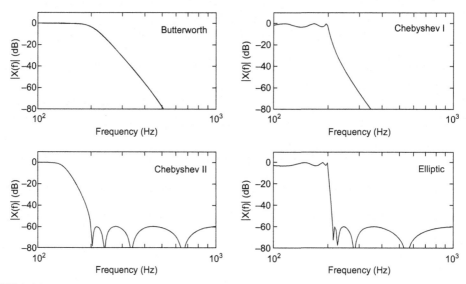

FIGURE 8.24   The spectral characteristics of four different eighth-order infinite impulse response (IIR) low-pass filters with a cutoff frequency of 200 Hz. The assumed sampling frequency is 2 kHz. It is possible to increase the initial sharpness of an IIR filter if various types of ripple in the spectral curve can be tolerated, not usually the case in bioengineering applications.

ripple is found in both the passband and stopband. Reducing the ripple of these last three filters is possible through the design routine, but this also reduces the filter's initial sharpness, yet another engineering compromise.

Other types of filters exist in both FIR and IIR forms; however, the filters described here are the most common and are the most useful in bioengineering. Given the seeming superiority of IIR filters, you might wonder why we even need FIR filters. There are two important applications where IIR do not work: image filtering and adaptive filtering. The former are described in Chapter 11, whereas the latter are left for more advanced signal processing textbooks (see Semmlow and Griffel, 2014).

8.7 SUMMARY

A digital version of the Laplace transform, called the z-transform, can be used to analyze discrete-time systems. These systems are sometimes abbreviated as LTID, which stands for linear time-invariant discrete systems. The z-transform, so-called because it uses the complex variable z in place of the Laplace variable, s, can be used to construct discrete transfer functions, $H(z)$. These transfer functions consist of numerator and denominator polynomials, just like Laplace transfer functions, except that these polynomials feature powers of z rather than powers of s (Equation 8.13). The z transfer function can be used to find the output to any input through a process similar to that used with the Laplace transfer function: take the

z-transform of the input signal, multiply it by $H(z)$, then take the inverse z-transform of the result. The z-transform and its inverse are found using tables as with the Laplace transform. This approach can be tedious and does not lend itself to computer analysis.

If the signals are assumed to be in steady state and the system has zero initial conditions, the sinusoidal variable, $e^{-j\omega}$, can be substituted for z. With this substitution, the spectrum of the system can be obtained from $H(z)$ simply by taking the Fourier transform of the numerator divided by the Fourier transform of the denominator (Equation 8.15). This operation can be done on a computer. The only discrete systems we biomedical engineers are likely to encounter are digital filters where we can assume a steady state with zero initial conditions. Hence, our interest in the z transfer function is limited to its ability to provide the system's spectrum through the Fourier transform.

Discrete systems, specifically digital filters, can be implemented using difference equations. Difference equations are discrete versions of differential equations and use the time-delay feature for the z variable. These equations consist of two terms: convolution between the signal and filter coefficients and convolution between the delayed output and additional filter coefficients (Equation 8.21). Difference equations are easy to solve using a computer and several MATLAB routines implement digital filters by solving these equations.

Filters are used to alter the spectrum of a signal, usually to eliminate frequency ranges that include noise or to enhance frequencies of interest. Filters vary in frequency range (i.e., bandwidth), basic type (low pass, high pass, band pass, and band stop), and characteristics of attenuations (rolloff and initial sharpness). Filters that have very sharp cutoffs can have ripple in the band-pass region, making them unsuitable for most biomedical engineering applications.

Designing a digital filter is a matter of determining the filter coefficients that give you the desired spectrum. Digital filters come in two basic versions: FIR and IIR. FIR filters are essentially moving average operations with the averages weighted by filter coefficients. The filter coefficients are identical to the system's impulse response. FIR filters are implemented using standard convolution. They have linear phase characteristics, but their spectra do not cut off as sharply as IIR filters of the same complexity. They can be applied to images where filter coefficients consist of a two-dimensional matrix (see Chapter 11). Their inherent stability makes them popular in adaptive filtering where the filter coefficients are continuously modified based on the signal. FIR filters can be designed by taking the inverse Fourier transform of an ideal filter, one with a rectangular frequency window. The inverse Fourier transform gives the desired filter's impulse response, which is identical to the filter coefficients. Unfortunately, in real applications, these filter coefficients must be truncated and this creates artifacts, including oscillations, in the frequency response. Appling a tapering window, such as the Hamming window, to the truncated coefficients reduces these artifacts. Although the filter coefficients of FIR rectangular window filters can be determined from basic concepts, MATLAB has routines that simplify their design.

IIR filters consist of two sets of filter coefficients, one applied to the input signal through standard convolution and the other to a delayed version of the output, and also using convolution. They can produce greater attenuation slopes and some types can produce very sharp initial cutoffs at the expense of some ripple in the band-pass region. An IIR filter known as the Butterworth filter produces the sharpest initial cutoff without band-pass ripple and is the most commonly used in biomedical applications. The design of IIR filters is more complicated than that of FIR filters, and is best achieved using MATLAB routines.

PROBLEMS

1. Find the magnitude and phase characteristic of a unit delay. (Hint: As shown in Figure 8.1, the transfer function of a unit delay is $H(z) = z^{-1}$. Refer to Equation 8.12 to help determine the a and b coefficients.) Compare the resultant magnitude and phase spectra with that of the time delay element analyzed in Example 7.2.

2. Find the spectrum (magnitude and phase) of the system represented by the z-transform:

$$H(z) = \frac{0.06 - 0.24z^{-1} + 0.37z^{-2} - 0.24z^{-3} + 0.06z^{-4}}{1 - 1.18z^{-1} + 1.61z^{-2} - 0.93z^{-3} + 0.78z^{-4}}$$

   For plotting purposes, assume $f_s = 500$ Hz. Be sure to pad the Fourier transforms sufficiently and, as always, plot only the nonredundant points.

3. Use the difference equation in Equation 8.21 to find the step response of the LTID system described by the z transfer function given in Problem 2. Use the same f_s as in Problem 2 and plot 0.25 sec of the response. (Hint: It is easiest to implement Equation 8.21 using the MATLAB `filter` routine.)

4. Use `sig_noise` to generate a 20-Hz sine wave in 5 dB of noise (i.e., SNR = −5 dB) and apply two moving average filters using the MATLAB `filter` routine: a 3-point moving average and a 10-point moving average. Plot the time characteristics of the two outputs. (Use `subplot` to combine the two plots). Use a data length (N) of 200 and remember that `sig_noise` assumes a sample frequency of 1 kHz.

5. Find the magnitude spectrum of an FIR filter with a coefficients of $b = [1\ 1\ 1\ 1\ 1\ ]/5$ in two ways: (a) apply the Fourier transform with padding to the filter coefficients and plot the magnitude spectrum; (b) pass white noise through the filter using `conv` and plot the magnitude spectra of the output. Since white noise has, theoretically, a flat spectrum, the spectrum of the filter's output to white noise should be the spectrum of the filter. In the second method, use a 20,000-point noise array, i.e., y = conv (b, randn(20000,1)). Use the Welch averaging method described in Section 4.4 to smooth the spectrum. For the Welch method, use a suggested segment size of 128 points and a 50% segment overlap. Since the `welch` routine (Example 4.5) produces the power spectrum, you need to take the square root to get the magnitude spectrum for comparison with the Fourier transform method. The two methods use different scaling, so the vertical axes are slightly different. (Use `subplot` to combine the two spectral plots). Assume a sampling frequency of 200 Hz for plotting the spectra.

6. Use sig_noise to construct a 512-point array consisting of two closely spaced sinusoids of 200 and 230 Hz with an SNR of −14 dB. Plot the magnitude spectrum using the Fourier transform. Generate a 24th-order (i.e., 25 coefficients) rectangular window band-pass filter using Equation 8.34 to modify the approach in Example 8.4. Set the low cutoff frequency to 180 Hz and the high cutoff frequency to 250 Hz. Apply a Blackman window (Equation 8.32, and Example 8.5) to the filter coefficients, then filter the data using MATLAB's `filter` routine. Plot the magnitude spectra before and after filtering. (Use `subplot` to combine the two spectral plots). (Hint: You can modify a section of the code in Example 8.4 to generate the band-pass filter.)

7. Write a program using Equation 8.28 to construct the coefficients of a 15th-order low-pass rectangular window filter (i.e., 16 coefficients). Assume $f_s = 1000$ and make the cutoff frequency 200 Hz. Apply the appropriate Hamming (Equation 8.31) and Blackman (Equation 8.32) windows to the filter coefficient. Find and plot the spectra of the filter without a window and with the two windows. Plot the spectra superimposed with different line types to aid comparison and pad the coefficients to $N = 256$ when determining the spectra. As always, do not plot redundant points. You can use the blackman routine for the Blackman window, but you need to write your own code for the Hamming window. (Suggestion: Simply modify the Blackman routine appropriately.)

8. Comparison of Blackman and Hamming windows. Generate the filter coefficients of a 127th-order low-pass rectangular window filter. Apply the Blackman and Hamming windows to the coefficients. Apply these filters to an impulse function using the MATLAB filter routine. The impulse input should consist of a 1 followed by 255 zeros. The impulse responses will look nearly identical, so take the Fourier transform of each the two responses and plot the magnitude and phase. You need to use MATLAB's unwrap routine on the phase data before plotting. (Use subplot to combine the magnitude and phase spectral plots). Is there any difference in the spectra of the impulses produced by the two filters? Look carefully.

9. Load file ECG_9.mat, which contains 9 sec of ECG data in variable x. These ECG data have been sampled at 250 Hz. The data have a low frequency signal superimposed over the ECG signal, possibly because of respiration artifact. Filter the data with a high-pass filter constructed using Equation 8.33. Use 65 coefficients (63rd-order) and a cutoff frequency of 8 Hz. Apply the Blackman window to the coefficients. Plot the spectrum of the filter to confirm the correct type and cutoff frequency. Also plot the filtered and unfiltered ECG data using subplot. (Hint: You can modify a section of the code in Example 8.4 to generate the high-pass filter.)

10. ECG data are often used to determine the heart rate by measuring the time interval between the peaks that occur in each cycle known at the "R wave." This is termed the R-R interval. To determine the position of this peak accurately, ECG data are first prefiltered with a band-pass filter that enhances the QRS complex, the spike-like section of the ECG. Load file ECG_noise.mat, which contains 10 s of noisy ECG data in variable ecg. Filter the data with a 64th-order FIR band-pass filter based on Equation 8.34 to best enhance the R-wave peaks. Determine the low and high cutoff frequencies empirically. (They will both be somewhere between 2 and 30 Hz.) The sampling frequency is 250 Hz. Plot the unfiltered and filtered signals on the same plot.

11. Load the variable x found in impulse_resp1.mat. Take the derivative of these data using the two-point central difference algorithm with a skip factor of 6, implemented using the filter routine. Now add an additional low-pass filter using a rectangular window filter with 65 coefficients (64th-order) and a cutoff frequency 25 Hz (Equation 8.28). Apply the filter using the conv routine with option 'same' to eliminate the added delay that would be induced by the filter. Plot the original time data, the result of the two-point central difference algorithm, and the low-pass filtered derivative data. (Use subplot to combine the three plots). Also plot the spectrum of the two-point central difference algorithm, the low-pass filter, and the combined spectrum again combining

with `subplot`. The combined spectrum can be obtained simply by multiplying the two-point central difference spectrum point by point with the low-pass filter spectrum.

12. Load the variable x found in `impulse_resp1.mat`. These data were sampled at 250 Hz. Use these data to compare the two-point central difference algorithm with a differencer (Equation 8.36) combined with a low-pass filter. Use a skip factor of 8 for the two-point central difference algorithm and a 54th-order rectangular window low-pass filter (Equation 8.28) with a cutoff frequency of 20 Hz. Use MATLAB's `diff.m` to produce the difference output; then, after you divide by T_s, filter this output with a low-pass filter. Note that the derivative taken this way is smooth, but has low-frequency noise.

13. Use `sig_noise` to construct a 512-point array consisting of two widely separated sinusoids: 150 and 350 Hz, both with SNR of −14 dB ($f_s = 1$ kHz). Use MATLAB's `fir2` to design a 65th-order FIR filter having a spectrum with a double band pass. The bandwidth of the two band-pass regions should be ±10 Hz centered about the two peaks. Plot the filter's spectrum superimposed on the desired spectrum. (Hint: To plot the desired spectrum, note that based on the calling structure of `fir2`, you can plot m versus f, but you need to rescale the frequency vector, f, since it is in relative frequency (relative to $f_s/2$.) Also plot the signal's spectrum before and after filtering. (Use `subplot` to combine the two signal spectra plots.)

14. The file `ECG60.mat` contains an ECG signal in variable x that was sampled at 250 Hz and has been corrupted by 60-Hz noise. The 60-Hz noise is at a high frequency compared with the ECG signal, so it may appear as a thick line superimposed on the signal. Construct a 126th-order FIR rectangular band-stop filter with a center frequency of 60 Hz and a bandwidth of 10 Hz and apply it to the noisy signal. Implement the filter using either `filter` or `conv` with the `'same'` option, but note the time shift if you use the former. Plot the signal before and after filtering and plot the filter spectrum to ensure that the filter is correct. You should combine all three plots using `subplot`. (Hint: You can modify a portion of the code in Example 8.5 to implement the band-stop filter.)

15. We have a unique opportunity to check a MATLAB routine. Construct the filter used in Problem 10: a 64th-order FIR band-pass filter based on Equation 8.34. Make the low and high cutoff frequencies 0.1 f_s and 0.5 f_s. Apply the Blackman window to this filter. Also construct a similar filter using `fir1`. (Remember that with MATLAB filters the cutoff frequencies are based on $f_s/2$.) Take the Fourier transform of both filter coefficients and plot the magnitude spectra superimposed. Note the slight differences perhaps because `fir1` uses a Hamming window. (Suggestion: Make $N = 256$ and arbitrarily select an f_s.)

16. Comparison of causal and noncausal FIR filter implementation. Generate a 64th-order low-pass rectangular window filter using either Equation 8.28 (Blackman window) or `fir1`. Make the cutoff frequency 200 Hz. Then apply the filter to the sawtooth wave, x, in file `sawth.mat`. This waveform was sampled at $f_s = 1000$ Hz. Implement the filter in two ways. Use the causal routine `filter` and the noncausal routine `conv` with option `'same'`. Plot the two waveforms along with the original superimposed for comparison. Note the differences.

17. Given the advantage of a noncausal filter with regard to the time shift shown in Problem 16, why not use a noncausal filter routinely? This problem shows the downsides of noncausal FIR filtering. Generate a 33rd-order low-pass rectangular window filter

using either Equation 8.28 (Blackman window) or `fir1`. Make the cutoff frequency 100 Hz and assume $f_s = 1$ kHz. Generate an impulse function consisting of a 1 followed by 255 zeros. Now apply the filter to the impulse function in two ways: using the MATLAB `filter` routine (causal), and the `conv` routine with the `'same'` option (noncausal). The latter generates a noncausal filter since it performs symmetrical convolution. Plot the two time responses separately limiting the x axis to 0–0.05 s to better visualize the responses. Then take the Fourier transform of each output and plot the magnitude and phase in degrees. Use the MATLAB unwrap routine on the phase data before plotting.

Note the strange spectrum produced by the noncausal filter (i.e., `conv` with the `'same'` option). This is because the implementation of the noncausal filter truncates the initial portion of the impulse response.

To confirm this, rerun the program using an impulse that is delayed by 10 sample intervals (i.e., `impulse = [zeros(1,10) 1 zeros(1,245)];`). Note that the magnitude spectra of the two filters are now the same. The phase spectrum of the noncausal filter shows less maximum phase shift with frequency as would be expected. This problem demonstrates that noncausal filters can create an artifact with the initial portion of an input signal because of the way they compensate for the time shift.

18. Differentiate the variable `x` in the file `impulse_resp1.mat` using the two-point central difference operator with a skip factor of 10. Construct another differentiator using a 16th-order least square IIR filter implemented in MATLAB's `yulewalk` routine. The filter should perform a modified differentiator operation by having a spectrum that has a constant upward slope until some frequency f_c, and then a rapid attenuation to zero. Adjust f_c to minimize noise and still maintain derivative peaks. (A relative frequency around 0.1 is a good place to start.) To maintain the proper slope, the desired gain at the 0.0 Hz should be 0.0 and the gain at $f = f_c$ should be $\frac{f_c f_s \pi}{2}$. Plot the original data and derivative for each method side by side. The derivative should be scaled for reasonable viewing. Also plot the new filter's magnitude spectrum for the value of f_c you selected. Note the cleaner response given by this new, but somewhat more complicated derivative method because it now contains a low-pass component.

19. Compare the step response of an eighth-order Butterworth filter and a 64th-order rectangular window filter, both having a cutoff frequency of $0.2\ f_s$. Assume a sampling frequency of 2 kHz for plotting. Use `fir1` to generate the FIR filter coefficients and implement both filters using MATLAB's `filter` routine. Use a step input of 256 samples but delay the step by 20 samples for better visualization (i.e., the step change should occur at the 20th sample). Plot the time responses of both filters. Also plot the spectra of both filters. You can combine all the plots using `subplot`. Note the oscillations induced by filtering and also note how these oscillations differ between the two filters.

20. Repeat Problem 14, but use an eighth-order Butterworth band-stop filter to remove the 60-Hz noise. Implement the filter using either `filter` or `filtfilt`, but note the time shift if you use the former. Plot the signal before and after filtering; also plot the filter spectrum to ensure the filter is correct. You can combine all three plots using `subplot`.

21. This problem demonstrates a comparison of a causal and a noncausal IIR filter implementation. Load file `Resp_noise1.mat` containing a noisy respiration signal in variable

`resp_noise1`. Assume a sample frequency of 125 Hz. Construct a 14th-order Butterworth filter with a cutoff frequency of 0.15 $f_s/2$. Filter the respiratory signal using both `filter` and `filtfilt` and plot the original and both filtered signals. Plot the signals offset on the same graph to allow for easy comparison. Finally, plot the noise-free signal found as variable `resp` in file `Resp_noise1.mat` below the other signals. Note how the original signal compares with the two filtered signals in terms of the restoration of features in the original signal and the time shift.

22. The downsides of noncausal filtering revisited. This problem is similar to Problem 16 except that it involves an IIR filter. Generate the filter coefficients of an eighth-order Butterworth filter with a cutoff frequency of 100 Hz assuming $f_s = 1$ kHz. Generate an impulse function consisting of a 1 followed by 255 zeros. Now apply the filter to the impulse function using both the MATLAB `filter` routine and the `filtfilt` routine. The latter generates a noncausal filter. Plot the two time responses separately, limiting the x axis to 0—0.05 s to better visualize the responses.

    Then take the Fourier transform of each output and plot the magnitude and phase. Use the MATLAB `unwrap` routine on the phase data before plotting. Note the differences in the magnitude spectra. The noncausal filter (i.e., `filtfilt`) has ripple in the passband. Again, this is because the noncausal filter has truncated the initial portion of the impulse response.

    To confirm that the artifact is due to the initial period, rerun the program using an impulse that is delayed by 20 sample intervals (i.e., `impulse = [zeros(1,20) 1 zeros(1,235)];`). Note that the magnitude spectra of the two filters are now the same. The phase spectrum of the noncausal filter shows reduced phase shift with frequency as would be expected.

23. Load the data file ensemble_data.mat. Filter the average with a 12th-order Butterworth filter. Select a cutoff frequency that removes most of the noise, but does not unduly distort the response dynamics. Implement the Butterworth filter using `filter` and plot the data before and after filtering. Implement the same filter using `filtfilt` and plot the resultant filter data. Compare the two implementations of the Butterworth filter. For this signal, the noncausal filter works well because the interesting part is not near the edges. Use MATLAB's `text` command to display the cutoff frequency on the plot containing the filtered data.

24. FIR—IIR filter comparison. Construct a 12th-order Butterworth high-pass filter with a cutoff frequency of 80 Hz assuming $f_s = 300$ Hz. Use `fir1` to construct a FIR high-pass filter having the same cutoff frequency. Plot the spectra of both filters and adjust the order of the FIR filter to approximately match the slope of the IIR filter. Compare the number of a and b coefficients in the IIR filter with the number of coefficients in the FIR filter.

25. Find the power spectrum of an LTID system four ways: (a) use white noise as the input and take the Fourier transform of the output; (b) use white noise as an input and take the Fourier transform of the autocorrelation function of the output; (c) use white noise as an input and take the Fourier transform of the cross-correlation of the output with the input, and (d) apply Equation 8.15 to the a and b coefficients. The third approach works even if the input is not white noise.

    As a sample LTID system, use a fourth-order Butterworth band-pass filter with cutoff frequencies of 150 and 300 Hz. For the first three methods, use a random input signal with 20,000 samples. Use `crosscorr` from Chapter 2 to calculate the auto- and cross-correlation and `welch` to calculate the power spectrum. For the `welch` routine, use a window of 128 points and a 50% overlap. (Owing to the large number of samples, this program may take 60 s or more to run so you may want to debug it with fewer samples initially.)

26. Write the z-transform equation for a fourth-order Butterworth high-pass filter with a relative cutoff frequency of $0.3 f_s/2$. (Hint: Get the coefficients from MATLAB's `butter` routine.)

System Simulation and Simulink

9.1 GOALS OF THIS CHAPTER

We create models for a variety of reasons and with a range of complexity, from a general qualitative model on which to hang our thoughts to a detailed quantitative model that reflects our deepest understanding of a system. In this chapter, we are interested in quantitative systems models with elements defined by Laplace transfer functions. One of the great advantages of quantitative models is that they can be used to predict a system's response to any input. In Chapter 6, we developed frequency domain methods that could be implemented on a computer, but with the limitations that signals are periodic in steady state (or aperiodic) and system elements have no initial conditions. Considering that many real-world signals are transient (i.e., steplike) and zero initial conditions are rare, these are serious limitations. In Chapter 7, we introduced Laplace techniques that can deal with both these limitations, but the system responses were solved analytically and were not open to computer solution. Here we introduce a method that lets us have it all: a computer-based analysis of systems with nonzero initial conditions, which applies to any class of signals. This approach even lets us relax the linear and time invariant requirement. Selected nonlinear elements and elements that change their properties over time can be included in the system model. In this most useful approach, a continuous system operating in the continuous domain is simulated on a digital computer.

Digital simulation is an outgrowth of an early computation device known as an "electrical analog computer." Such computers used continuous variations in voltage to represent signal levels in a system. Electrical components were used to represent system processes and were capable of performing linear operations such as summation, scaling, and integration.[1] As they operated in the continuous domain, they did not suffer from quantization error, but did suffer from electronic noise (see Section 1.3.2.1). Programming these computers required manually connecting the electronic components together with patch cables (flexible wires with plugs at both ends), a time-consuming but rewarding effort.

Digital simulation is sort of a mix between the discrete and continuous domains. It is implemented on a computer, so the fundamental calculations are digital, but it mimics a

---

[1]A circuit that does summations of signals is described in Chapter 15 (Section 15.8). Integrator circuits use capacitors that produce voltages that are time integrals of their current (Chapter 12, Section 12.4.1.2).

continuous system. Like the Laplace transfer function, it is used to describe the behavior of a continuous system, but with much less effort. Simulation programs such as MATLAB's Simulink are basically digital computer representations of the now-extinct analog computer.

In this chapter, we explore the capabilities of MATLAB's simulation program, specifically we perform the following:

- Explain the principles of simulating continuous systems on a digital computer.
- Learn the basics of Simulink including some of its more useful options.
- Explore some of Simulink's extensive features that are particularly relevant to biomedical engineers.
- Apply Simulink to several biological and nonlinear systems.

9.2 DIGITAL SIMULATION OF CONTINUOUS SYSTEMS

Conceptually, digital simulation of a continuous system is straightforward. Information flows through the simulated system one instant (i.e., time step) at a time. Both time and amplitude are still quantized, but the time and amplitudes steps are made small enough to appear continuous (for all practical purposes). The basic idea is similar to that used in convolution: calculate responses to a number of small time slices of the input signal. The main difference is that the response of each element in the system is determined individually. Given the element's input(s) for a specific time slice and the operation performed by the element, its output is calculated. Then, for that same time slice, the element's response becomes the input to any element(s) connected to it. This proceeds until the outputs of all the elements, and the system itself, have been determined, and then begins again for the next time slice.

Figure 9.1 illustrates the digital simulation process for two elements assumed to be a subsection of a larger model. Five time slices are shown. The process begins at the first time slice

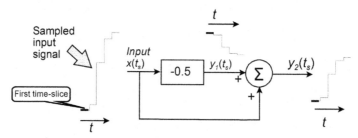

FIGURE 9.1 Two elements that might be part of a larger system showing hypothetical inputs and resultant outputs. The first time slice is shown as darker than the subsequent four time slices. Simulation software takes the value of the input to this subsystem, $x(t_s)$ during the first time slice and applies the specified scaling operation of the first element, in this case multiplying by −0.5. This produces an output $y_1(t_s)$, which feeds the summation element. The summation element adds $y_1(t_s)$ to $x(t_s)$ to produce the output $y_2(t_s)$, again, all during the first time slice. These steps are then repeated for all the elements of the system. For the next time slice, a new input value, $x(t_s)$, arrives at the input and the calculations proceed through the system. This process repeats until the last input time slice at the end of the input signal. If this were part of a larger system, $y_2(t_s)$ would feed additional elements.

(Figure 9.1, shown in bold) when a signal value is taken from the subsystem input and becomes the input to the first element. It is also the input to the summation element. This input comes from the output of another subsystem or a simulated signal source. The signal value at the first time slice is $x(t_s)$, at $t_s = 0$. One of the elements receiving this signal is a scaling element that multiples the input by -0.5 and produces $y_1(t_s=0)$, also shown in bold. The signal $x(t_s=0)$ is also sent to a summation element, which adds it to the output of the scalor. This summation element produces an algebraic sum of the two values $(x(t_s=0) + y_1(t_s=0))$ as its output, $y_2(t_s= 0)$.

The progression of the input to output of the signal continues through other system elements (not shown) until it reaches the last element in the system. The response of this last element is the output of the system and is usually displayed or recorded. This whole process is then repeated for the second time slice, the one immediately following the bold slice. After the second time slice propagates through all the elements, the cycle is repeated for successive time slices until the input signal ends or some other stopping criterion is met. During simulation, the output signal and the response of each element evolve over time just as in a real continuous system (admittedly in discrete steps, but these should be small enough that the operation appears continuous). Generally, it is possible to view the response of any element in the system, as this evolution takes place.

At the heart of continuous simulation, analog or digital, is the integrator. In the Laplace transfer function, there is integration behind every occurrence of $1/s$ (or $1/j\omega$ for a frequency domain transfer function). For example, a $1/j\omega$ by itself is just an integrator (Equation 6.29) and a $1/(1+j\omega)$ is an integrator in a feedback loop (e.g., Example 6.3). A second-order transfer function represents a system with two integrators (e.g., Example 6.5). Thus to simulate the behavior of a system, you need to perform integration, and on a computer, you need to perform integration digitally.

Previously, when we transformed a continuous equation to a discrete version, if it contained an integral we replaced it with summation. Summation is the digital equivalent of integration. But simulation programs want to mimic continuous systems, so they use sophisticated algorithms that behave more like true integrators. Simulink has a number of different integrator algorithms. These algorithms trade off speed, stability, and accuracy, but, as is often the case, the default integrator works well for most problems. Nonetheless, we will write a simple simulation routine using summation as an integrator to simulate the response of a first-order system to a step function.

EXAMPLE 9.1

Write a program to simulate the response of the continuous first-order system shown in Figure 9.2 to a step input.

In Chapter 7 we showed that this system has the transfer function:

$$TF(s) = \frac{1}{s + \omega_1} = \frac{1/\tau}{s + 1/\tau} \tag{9.1}$$

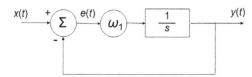

FIGURE 9.2 Simple first-order system whose output to a step function is found in Example 9.1 and later in Example 9.2.

where ω_1 is the cutoff frequency and equals $1/\tau$ (Equation 7.26). Make the cutoff frequency 5 rad/ s ($\tau = 0.2$ s) and show the first 1.0 s of the response. Assume that the initial condition of the integrator is 0.0.

Solution

We use summation for the integration process. We divide the simulation time of 1.0 s into 100 time slices, so each time slice represents 10 ms. For the first step, the output of the integrator is the initial condition (which in this case is 0.0, but it need not be). After that, it produces the output of the next step, which is the summation of its last input with its last output.

```
% Example 9.1 Crude digital simulation of a first-order system

%
Ts = 0.010;                        % Step size 10 ms
N = 100;                           % Number of steps (Tt = 1 s)
w1 = 5*Ts;                         % Cutoff frequency of first-order system
out(1) = 0;                        % Initial condition for integrator

x = [0, ones(1,N-1)];             % Input (step function)

y = zeros(N); % Output array

for k = 1:N-1

   e(k) = x(k) - out(k);          % Error signal
   out(k+1) = w1*e(k) + out(k);   % Integrator summation
end
t = (1:N)*Ts;                      % Time vector for plotting
      ........plotting and labels........
```

Results

Both the system output, $y(t)$, and the integrator input, $e(t)$, are shown in Figure 9.3. As expected from our analytical solutions (see Example 7.3), the output is a smooth exponential with a time constant of 0.2 s (i.e., the response is 0.63 of its final value at 0.2 s). Note that it is easy to add in a nonzero initial condition for the integrator: just make y(1) equal to the desired initial condition. Similarly, you can think of ways to embed nonlinear elements into this program. An example is found in the problems.

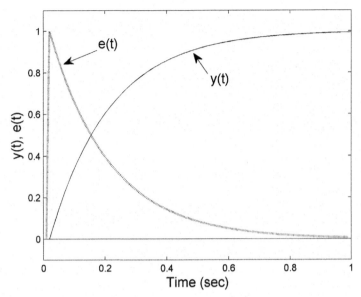

FIGURE 9.3 Output from the simulation of the first-order system. Both system output, $y(t)$, and the difference signal, $e(t)$ (for "error" signal), are shown.

MATLAB's simulation program, Simulink, is bit more sophisticated than the approach taken in Example 9.1. For starters, the system is defined with the aid of a Graphical User Interface (GUI) that not only simplifies model specification but also provides a nice diagram of the systems (suitable for publication). Unless instructed otherwise, Simulink uses time slices that vary in width depending on the dynamics of the system. Unlike signal sampling, there is no reason to keep step size constant, and Simulink varies the size of the time slice to match the dynamic properties of system elements. When the signals are changing quickly, the step size decreases to improve accuracy, but when signals change slowly the step size increases. Not only does this decrease computation time but by reducing the number of steps required during the simulation also reduces error propagation.[2] Finally, as mentioned above, Simulink uses sophisticated integration algorithms that behave more like real-world continuous integrators.

9.3 INTRODUCTION TO SIMULINK

All linear systems can be decomposed (at least in theory) into elements that perform addition or subtraction, scaling (i.e. gain elements), differentiation, and integration. In fact, it is

---

[2]Roundoff errors, although small for any given calculation, build up over the many steps involved in a simulation that requires short time slices over an extended time. It was this problem of error propagation that occurred in so-called "wide bandwidth" simulations (i.e., fast dynamics with long simulation times) that kept analog computers alive long after digital simulation became popular.

possible to rearrange systems to eliminate derivative elements, but this is not necessary when you use Simulink. Any simulation program must perform a number of tasks: (1) it must allow you to define the system you want to simulate including initial conditions, (2) generate the input(s) to the system and set up signals for display, and (3) calculate and track the responses of the individual elements as the signals flow through the system and display and/or record the system's output. These steps are detailed below:

Step (1) Setting up the model and initial conditions. A systems model is a collection of interconnected elements, so to set up a model you need to specify the elements and their interconnections. In systems models, interactions are unidirectional and explicitly indicated, usually by an arrow. In Simulink, model setup is done graphically: elements are selected from a number of libraries and the connections between them specified by dragging lines between the various elements. Those elements that have initial conditions can be adjusted after they have been specified. A step-by-step procedure is given in the next example.

Step (2) Generating input waveforms and setting up the output display. In Simulink, the input signal usually comes from a waveform-generating element, which is chosen from the library just like all the other elements. Simulink has a wide range of waveform generators stored in a library called Sources. It is even possible to generate a waveform in a standard MATLAB program and pass it to your model as an input signal. Outputs are also just elements selected from the library called Sinks. [3] Output elements can be attached to any element in the system and include time displays, outputs to a MATLAB routine, or outputs to a file.

Step (3) Tracking the responses of the individual elements and displaying the output. This is handled by Simulink so that once the simulation is initiated, it runs automatically for the requested duration. Simulink goes to great lengths to optimize step size by using special integration algorithms. In most simulations, including all of those covered here, we need not be concerned with integration techniques and step size manipulation. If needed, we can alter the simulation parameters, including the algorithm used for integration, and sometimes these alterations are useful to create smoother response curves. In addition we may want to request evenly spaced step sizes if we want to compare simulation results directly with time sampled data.

9.3.1 Model Specification and Simulation

The best way to learn Simulink is by doing, or through example. While the examples below provide a rather basic introduction to Simulink's extensive simulation capabilities, they cover the range of applications that most bioengineers need. As always, detailed instructions can be found in MATLAB's Help. [4] A simulation starts by entering simulink in MATLAB's main window. This opens a window that shows the libraries of possible elements (Figure 9.4). A list of the most important element libraries (for us) and the types of elements they include are given in Table 9.1. The opening window File tab allows saved

---

[3] Output and display devices are called "sinks" because the output signal flows into, and ends at, these elements just as water flows into a sink and disappears.

[4] Click on "Help" in the main MATLAB window, then select "Product Help" in the drop-down menu, and then scroll down the list on the left side and select "Simulink." The main Simulink help page will open and a large variety of helpful topics will be available.

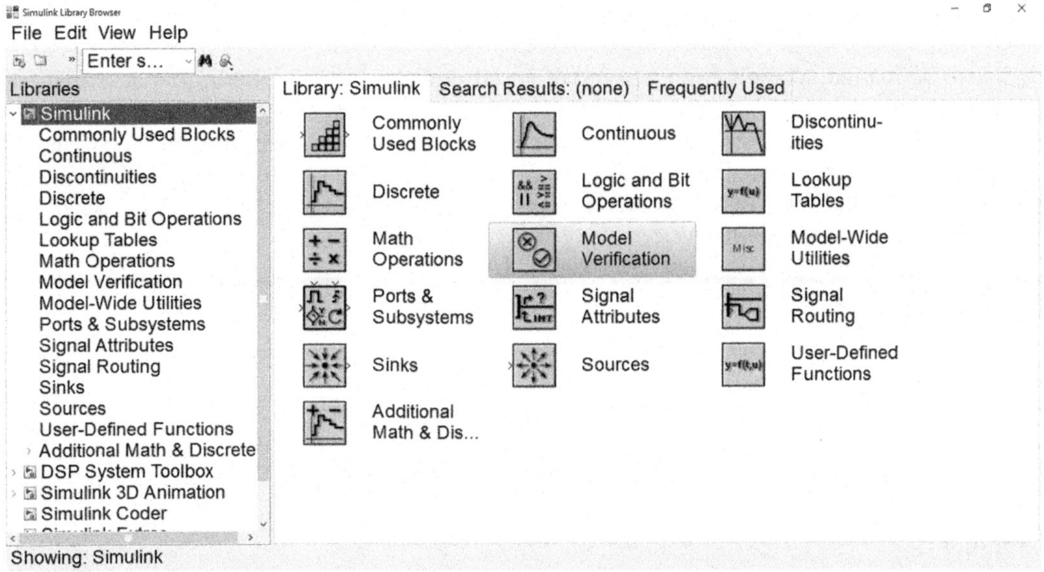

FIGURE 9.4 The initial Simulink window showing a list and icons for the various element libraries. The libraries most important to bioengineers are presented in Table 9.1 along with their most useful elements.

TABLE 9.1 Useful Simulink Libraries

Library	Element Name	Function
Sources (23 elements)	Constant	Constant output with adjustable value.
	Step	Step output with adjustable onset time, initial and final values.
	Ramp	Ramp output with adjustable onset time, slope, and initial value.
	Sine	Sine wave with adjustable frequency, phase, and amplitude.
	Pulse Generator	Pulse generator with adjustable pulse rate, width, and amplitude.
	Chirp	Chirp with adjustable start and stop frequencies and duration.
	Random	Gaussian random number with adjustable variance and sample time.
	Repeating Sequence	Outputs an adjustable repeating sequence of levels set by table.
Sinks (9 elements)	Scope[a]	Display signal with axes set interactively.
	To File	Sends signal to a specified file as a time series
	Out1[a]	or other.
	To Workspace	Sends signal as to a MATLAB program.
		Sends signal as time signal or other to workspace as specified variable.
Continuous (13 elements)	Integrator[a]	Output is integral of input.
	Transfer Fcn	General transfer function with any number of adjustable numerator
	Delay	and denominator coefficients.
		Adjustable time delay.
Math operations (37 elements)	Add	Adds or subtracts any number of signals.
	Sum[a]	Same as add, but with circular symbol
	Gain[a]	Multiplies signal by adjustable constant.
	Product[a]	Takes the produce of two signals.
	Math Function	Performs various math operations, including log, power, and exponential.
Discontinuous (12 elements)	Saturation[a]	Limits signal amplitude to adjustable minimum and maximum.
	Rate Limiter	Limits signal rate of change to adjustable minimum and maximum.
	Dead Zone	Outputs zero for signals within adjustable dead zone.

[a]Also found in the "Commonly Used Blocks" library

models to be loaded. Models can be saved at any time from the model window as explained below.

Example 9.2 provides a step-by-step demonstration of its use to determine the response of a second-order system to a step input.[5] Similar instructions can be found in MATLAB's help documentation following the links *Simulink*, *Getting Started*, *Creating a Simulink Model*, and ending up at *Creating a Simple Model*.

EXAMPLE 9.2

Use Simulink find the step response of the simple first-order system used in Example 7.1.

$$TF(s) = \frac{1}{s + \omega_1} = \frac{1/\tau}{s + 1/\tau}$$

where $\omega_1 = 5$ rad/s.

Solution

Although Simulink provides a way to represent this first-order system as a single element, here we use an integrator and gain element in a feedback loop and shown in Figure 9.2. The Simulink graphical interface is used to set up the system model using mouse manipulations similar to those of other graphic programs such as PowerPoint. A model window is opened first, and then elements are installed in the window and moved by dragging. They can also be duplicated using Copy and Paste.

Step (1) Set up the model window. To create a system model, open Simulink by typing simulink in the MATLAB command window. The window that is opened by this command is shown in Figure 9.4, which also contains the Simulink Library Browser window. Use the File drop-down menu and select New and then Model. This opens a blank window called "untitled1." If you save this window as a model file (name.slx)[6], the name you choose appears in the title. The model window can be saved, reopened, and resaved just as with any other MATLAB program file.

Step (2) The next step is to populate the model window with the desired elements. We have a diagram of the system in Figure 9.2, which shows it contains three elements: a summation, a gain term, and an integrator. We also need the step input and some sort of output. Normally, we would add a Scope element for displaying the output. (For publication purposes, we actually output the data to MATLAB workspace then plot, but the Scope element is easier to use and gives us the signal immediately.) Table 9.1 indicates that all of these can be found in the Commonly Used Blocks library except for the step input, which is in the Sources library. To open a library, double-click on the library icon. Double-clicking on the Commonly Used Blocks library opens the window shown in Figure 9.5.

Next we drag the desired element(s) into the model window creating a duplicate of that element in the model window. Dragging the Sum, Gain, Integrator, and Scope elements onto our model window gives Figure 9.6. The elements could be placed anywhere, but it is reasonable to put them in approximately the same general position as in Figure 9.2.

For the step input signal, we turn to the Sources library, Figure 9.7, and drag out the Step icon found in the lower left corner.

---

[5]The Simulink examples shown here use MATLAB release R2013a, version 8.1.

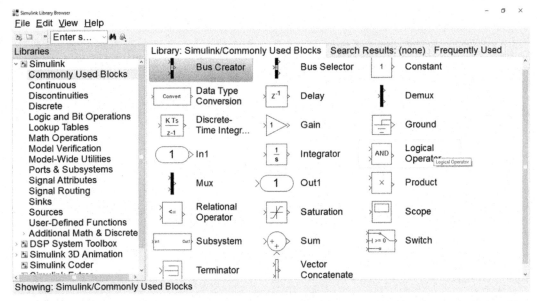

FIGURE 9.5 The Commonly Used Blocks library. All of the elements required in Example 9.2 can be found here except the step input.

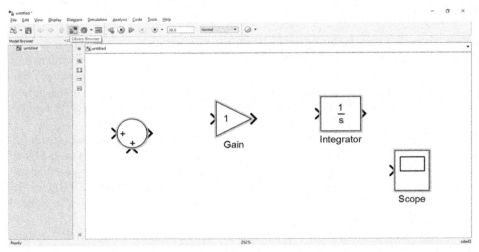

FIGURE 9.6 Three of the elements needed for the model used in Example 9.2 placed in approximately the same position as in the system diagram of Figure 9.2.

Step (3) Adjust element parameters. Most elements have parameters. In some case the default values are fine, but often they need to be modified. In our model, the Integrator element has an initial value parameter, the Gain element has a gain value, the Step element has initial and final values, and the Sum element parameters specify the signs and number of inputs. For the Integrator element, the default initial condition value is zero, which is appropriate for our system. As seen in

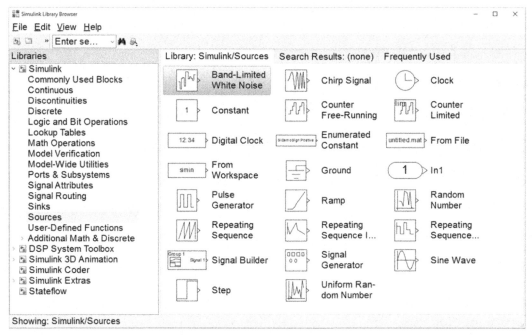

FIGURE 9.7   The Sources library window. The Step element is in the lower left corner.

Figure 9.6, the default signs for the Sum element are two positive inputs as indicated by the two "+" signs on the Sum icon. We need an element that takes one positive and one negative input, Figure 9.2. To change the "+" to "−" we double-click on the Sum icon, which opens the element's parameter window as shown in Figure 9.8A. The List of signs: line shows two "+" symbols. We simply backspace over the second "+" symbol and add a "−" symbol (Figure 9.8B). This makes the second (i.e., lower) input negative as shown in the final model in Figure 9.10.

The Step element parameter window is shown in Figure 9.9A. The default initial value is 0.0 and the final value is 1.0, which is fine for our simulation. However, the default Step time is 1.0 and we change it to 0.01 to match the small onset delay of our homegrown simulation program used in Example 9.1 The Gain element also needs adjustment as the desired gain value is 5.0 and the default value is 1.0 as seen in Figure 9.6. The Gain element parameter window is shown in Figure 9.9B where the Gain value has already been changed to 5.0.

Step (4) Connect the elements. This is done graphically by clicking on an element output and dragging the line to the desired element input, or vice versa. The complete, connected system is now in the model window as shown in Figure 9.10. A second Scope has been added to display the error signal. (Elements can be duplicated using copy-paste as in any graphics program.) The model can be saved any time using Save As... command under the File pull-down menu. (See Footnote 6 regarding computability issues for saved models.)

Step (5) Run the simulation. You can start the simulation simply by clicking the run icon found on the second bar of the model window shown in Figure 9.11. You can also use the Simulation pull-down menu found on the top bar of the model window and which displays a similar run icon.

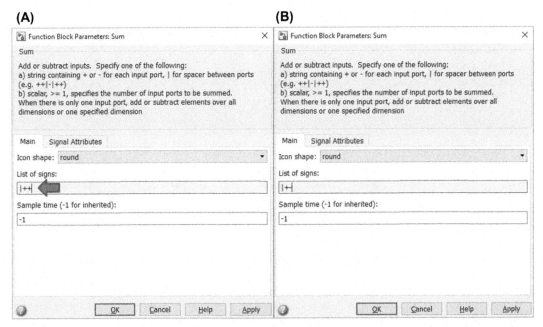

FIGURE 9.8 The parameter window of the Sum element. A) The List of signs: line (arrow) shows two '+' symbols, the default for this element. B) The second sign has been changed to a '−' sign which will make the second input a negative.

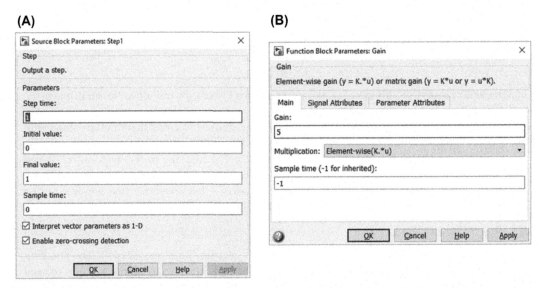

FIGURE 9.9 A) The Step element parameter window shows the default values for initial and final step valued to be 0.0 and 1.0, respectively. These are okay for our simulation; however, the Step time needs to be changed to 0.1. B) The Gain element parameter window where the Gain value has already been changed to the desired value of 5.0.

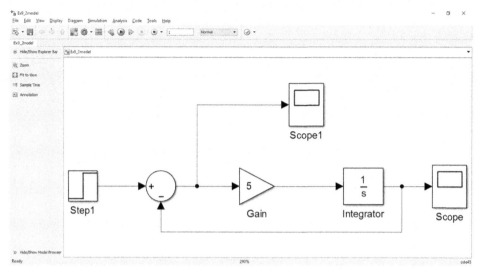

FIGURE 9.10 The completed model used in Example 9.2. A second Scope has been added to display the error signal. Note that the model layout is similar to that of the original system presented in Figure 9.2.

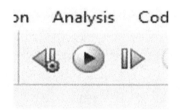

FIGURE 9.11 The Run icon found on the second bar of the model window shown in Figure 9.10. Clicking this button initiates the simulation.

In older versions of MATLAB there is no run icon, but the Simulation pull-down menu displays the word "Run."

The default time of the simulation is 10 (seconds).[7] To change the simulation run time along with many other simulation options, you can open the Model Configuration Parameters window also found on the Simulation pull-down menu (Figure 9.12). The first line of this window is labeled Simulation Time and has entries for start and stop time. In Figure 9.13, the stop time has been changed to 1.0 s. Alternatively, you can change the number that appears in the window on the second line of the simulation window (barely visible in Figure 9.10).

Results

The output of the model can be viewed by double-clicking the Scope element. The resulting window is a plot of the signal to which the scope is attached. The display features a light colored line against a dark background. As this does not reproduce well, the graph shown in Figure 9.13 is constructed from variables placed in MATLAB workspace using the simout element. This approach produces standard MATLAB plots and is fully described in the next example.

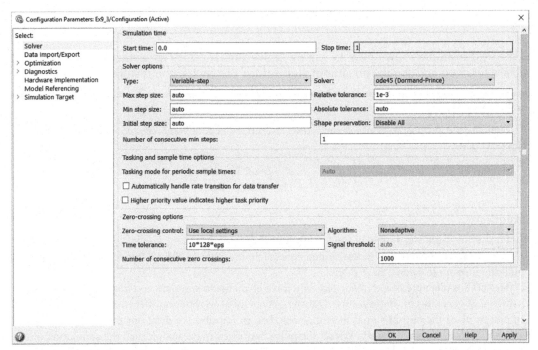

FIGURE 9.12 The Model Configuration Parameters window found on the Simulation pull-down menu can be used to set the start and stop time of the simulation along with other simulation parameters.

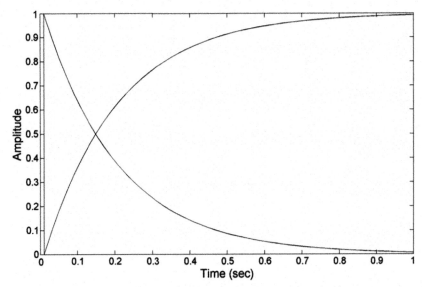

FIGURE 9.13 The output and error signal produced by simulation in Example 9.2. The two signals are very similar to those of Figure 9.3 that were produced by our homebrewed simulation program in Example 9.1.

[6]For MATLAB versions 2012b and earlier, models were stored differently under the .mdl extension. Newer versions of Simulink can load .mdl files and save them in this older format using Export Model in the File pull-down menu. However, earlier versions of Simulink cannot read the newer .slx files.

[7]Simulink time units are generic: they reflect the units used in the model parameters. We generally work in seconds, so Simulink time units can usually be taken as seconds. Some very slow biological processes such as the glucose response simulated in Example 7.5 use hours as the basic time unit.

The next example shows how to use the powerful Transfer Fcn element to simulate any transfer function and the useful simout element for displaying the results.

EXAMPLE 9.3

Simulate the impulse response of the fourth-order transfer function analyzed in Example 7.9 and repeated here:

$$TF(s) = \frac{s^3 + 5s^2 + 3s + 10}{s^4 + 12s^3 + 20s^2 + 14s + 10} \tag{9.2}$$

Solution

Follow the same steps used in the last example.

Step (1) Set up the model window. Enter simulink in the main MATLAB window and use the File pull-down menu to open a new model.

Step (2) Populate the model. We need only three elements to simulate and display this system: a Source that generates an impulse, the Transfer Fcn element to represent the transfer function, and simout to place the output signal in workspace. For an impulse we drag out the Pulse Generator element from the Sources library. We set the pulse frequency to be greater than the simulation time so we get only one pulse. We set the pulse width to be very short, as we want to mimic an impulse. The Transfer Fcn element is taken from the Continuous library,[8] and the simout element from the Sinks library. The resultant model is shown with connections in Figure 9.14

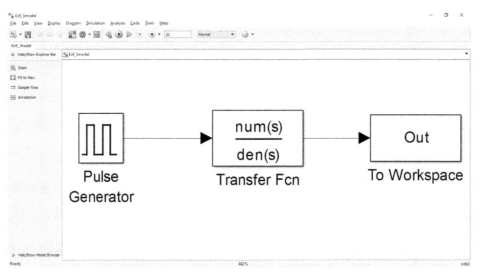

FIGURE 9.14   The model used in Example 9.3 to simulate the impulse response of a fourth-order transfer function.

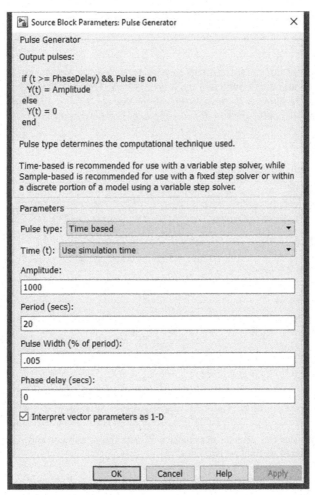

FIGURE 9.15 The Pulse Generator parameter window showing the new values; Pulse Width (% of period) is .005%, Amplitude is 1000, and Period (seconds) is 20.

Step (3) Adjust element parameters. The Pulse Generator window is shown in Figure 9.15. Initially, the default simulation time of 10 s was used, but a longer time period was needed to show the complete response. The simulation time was changed to 20 s and the Period (seconds) is set to 20, so only one pulse would be produced. Using the default Phase delay (seconds) of 0 ensures that the single pulse occurs at $t = 0$. The Pulse Width (% of period) is set to 0.005% to produce a short pulse. As the period is 20 s, this produces a 1.0 ms pulse. To determine if the pulse input realistically represents an impulse, we used the trick described in Example 5.1. We shortened the impulse and noted minimal change in the shape of the response, indicating that our 1.0 ms pulse is acting like an impulse to this system. For a unit impulse the area under the pulse should be 1.0 so we set the amplitude to 1000.

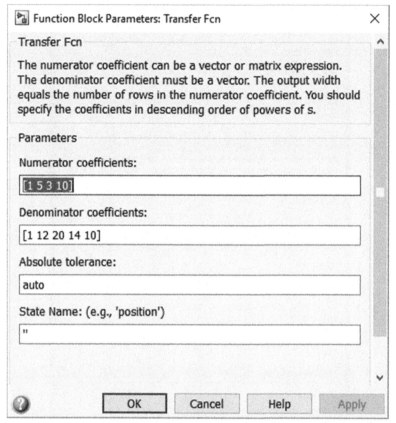

FIGURE 9.16 The Transfer Fcn element parameter window. The numerator and denominator coefficients of Equation 9.2 have been entered on the first two lines.

The Transfer Fcn element parameter window must be set to represent the transfer function in Equation 9.2. This is accomplished by entering the numerator and denominator coefficients as vectors in the parameter window. The numerator coefficients array is [1 5 3 10] and the denominator array is [1 12 20 14 10]. These coefficients have been entered in the Transfer Fcn parameter window shown in Figure 9.16.

The To Workspace element parameter window must also be modified to place the data in the workspace in the appropriate format. Specifically, we need to modify the Save format: pull-down menu near the bottom of the parameter window (Figure 9.17). This menu is used to change the format from the default Timeseries to Array. This element then produces two standard MATLAB arrays: tout, which is the time array, and Out, which is the response array. To plot the response, we simply enter plot(tout,Out); in the main MATLAB window.

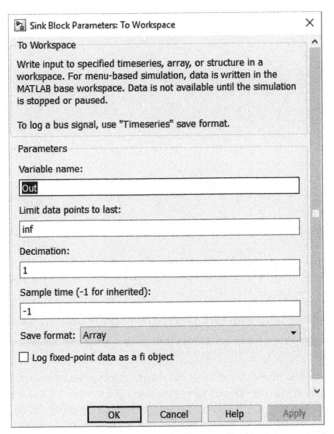

FIGURE 9.17 The To Workspace parameter window. The Save format has been changed from the default Timeseries to the Array format. The name given to the output variable in the workspace is "Out" as indicated by the Variable name: parameter. (Often we just use the default name of "simout.")

Step (4) Connect the elements. Because the model consists of only three elements, this is performed in Step 2 when the model window is populated.

Step (5) Run the simulation. After setting the simulation time to 20 s, we use the run icon to start the simulation.

Results

After the simulation is run, the output can be found as vector Out in the workspace along with the time vector tout. As mentioned above, the response can now be plotted using the command plot(tout,Out);. Axis labels were added resulting in Figure 9.18.

[8]As linear systems deal with signals that are theoretically continuous in time, simulation of LTI systems is sometimes referred to as "continuous simulation" although the simulation approach is still based on discrete time slices. This continuous simulation terminology is the basis of the name MATLAB used for this library.

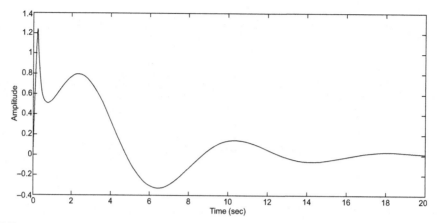

FIGURE 9.18   The impulse response of the system defined by Equation 9.2 obtained through simulation.

As we have the impulse response in the workspace, we can find the spectrum of this system by taking Fourier transform of this response. This is done in the next example.

EXAMPLE 9.4

Find the magnitude and phase spectrum of the system represented by the transfer function given in Equation 9.2. To conform to the spectral plots generated in Example 7.9 and shown in Figure 7.12, plot the magnitude spectrum in dB versus log and the phase spectrum in deg, both versus log frequency. Also, limit the frequency axis to between 0.1 and 100 Hz.

Solution

We have a simulation of the impulse response, so we should, in principle, be able to determine the spectrum from the Fourier transform of this response. However, in the default mode, Simulink does not use equal time spacing: the time spacing is optimized based on the model dynamics. To set Simulink to produce signals with a fixed time interval, we use the Model Configuration Parameters window found in the Simulation pull-down menu on the model window. This window is shown in its default configuration in Figure 9.12 (except for Stop Time, which was set to 1). The first entry below Type shows Variable-step, the default step mode. Use this pull-down menu to select Fixed-step (the only other option) as shown in Figure 9.19. This opens a line titled Fixed-step size (fundamental sample time): which we set to 0.001 s making $f_s = 1$ kHz.

Because we have changed to a fixed-step time, we also need to change the Pulse Width (% of period) value. We want to ensure that we have a true digital impulse; that the Pulse Generator has a nonzero output for only the first step. The duration of one sample is 0.001 s, so the pulse width should be 0.00005 (20) = 0.001. Because the pulse width variable is in percent, we should make Pulse Width (% of period) entry 100 (0.00005) = 0.0050. (Note: to check this, use simout to put the output of the Pulse Generator in the workspace and make sure only the first term is nonzero. Also

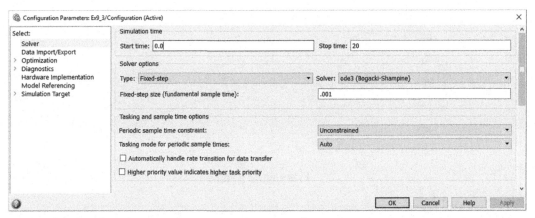

FIGURE 9.19 The Model Configuration Parameters window after modification to produce a simulation with a fixed-step size of 0.001 s ($f_s = 1$ kHz).

when using the fixed-step mode, Simulink requires that the pulse width should be an integer value of the step size: in this case the integer is 1.0.)

After these modifications, we run the model using the Run button (Figure 9.11), as before. This produces an impulse response as variable Out in the workspace. We then write a short MATLAB program to take the Fourier transform of this variable and calculate the magnitude and phase spectra, and plot.

```
% Example 9.3 Find the magnitude and phase spectrum of the transfer function
% given in Equation 9.2. Assumes data in vector Out
%
fs = 1000;                     % Sample frequency
N = length(Out);               % Get data length
N_2 = round(N/2);              % N/2 (valid spectral points)
f = (1:N)*fs/N;                % Frequency vector for plotting
X = fft(Out);
Mag = 20*log10(abs(X));        % Magnitude spectrum in dB
Phase = angle(X)*320/(2*pi);   % Phase spectrum
subplot(2,1,1);
semilogx(f(1:N_2),Mag(1:N_2),'k','LineWidth',1); % Plot magnitude spectrum
.......label and repeat for phase spectrum.......
```

Result

This program produces the plots shown in Figure 9.20. Comparing these plots to those of Figure 7.12, we see the same general shape with dips in both magnitude and phase curves around 1.5 rad/s. The curves found by simulation are not as smooth as those found in Example 7.9. This is to be expected because the former are developed indirectly from a simulation of the time-domain response, whereas the latter were found directly from the frequency domain transfer function. However, as shown later, when a system contains nonlinear elements, simulation can often be used to determine both the time and frequency domain behavior where all our linear methods fail.

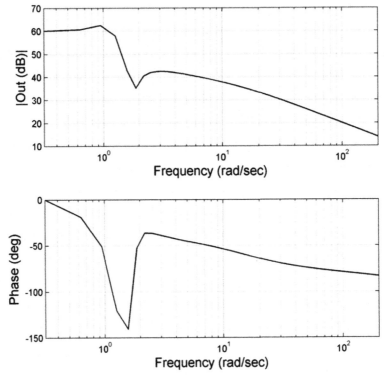

FIGURE 9.20 Spectra of the system defined in Equation 9.2 found by applying the Fourier transform to an impulse response generated by simulation.

9.3.2 Complex System Simulations

Even very complex systems are easy to analyze using Simulink. The next example applies the same methodology described in Example 9.2 and 9.3 to a somewhat more complicated system.

EXAMPLE 9.5

Simulate the step response of the system shown in Figure 9.2. Also simulate the response to a 200-ms pulse. Then determine the pulse width that represents an impulse input. Display not only the output signal but also the contributions of the upper and lower pathways.

Solution

Construct the model shown in Figure 9.21, adding a step input and displays. After simulating the step response, replace the step function element with a pulse generator element. Adjust the pulse width so that it appears as an impulse to the system using the strategy used in Example 9.3.

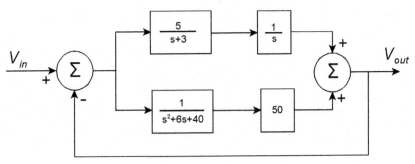

FIGURE 9.21   System simulated in Example 9.5

Step (1,2) From the Continuous library, we drag out three elements: two Transfer Fcn elements and an Integrator. From the Commonly Used Blocks library, we drag out two Sum elements (one will be changed to a subtraction element), a Gain element, and three display elements from the Sinks library. Again, we use simout to place the output signals in the workspace. Then we go to the Sources library to pick up the Step element. The resulting model window appears as shown in Figure 9.22, where the elements have been dragged into the window more or less randomly. The element parameters have already been modified in this figure.

Step (3) Next we modify the element parameters. As in Example 9.3, the transfer function elements must be set with the coefficients of the numerator and denominator equations. The element in the upper path in Figure 9.21 has a transfer function of $5/(s + 3)$, giving rise to a numerator term of 5 (this is a scalar, so no brackets are needed) and a denominator term of $[1, 3]$. The transfer function in the lower branch, $1/(s^2 + 6s + 40)$, has a numerator term of 1 and a denominator term of $[1, 6, 40]$. The gain element parameter, Gain:, should be set to 50. The Sum elements have a textbox labeled List of signs: in one of the two elements the '+' symbols should be changed to a '−' sign, so the

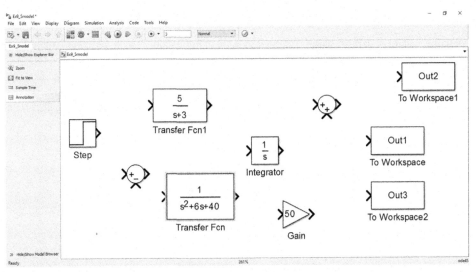

FIGURE 9.22   Elements used to simulate the model in Figure 9.21. The element parameters have already been modified as reflected by the text in each element.

element performs subtraction. The element's icon shows which input is negative. The Integrator block does not need to be changed. The Step element was modified, so the step begins 90 ms after the start of the simulation. This provides us with a better view of the initial response.

Step (4) The elements are then arranged in a configuration with the signals going from left to right (except for the feedback signal) and connected. There are shortcuts to connecting these elements, but the technique of going for input back to output is effective. The second connection does not have to be an element output terminal; it can be just another line which allows for 'T' connections. Right angles are made by momentarily lifting up on the mouse button as with other graphics programs. The three simout elements are connected to the output and upper and lower pathways. The resulting model window is shown in Figure 9.23.

Step (5) We run the simulation setting the Stop time to 5 s. With Simulink, we can monitor any element in the system; in this simulation we use three simout elements to monitor the upper, lower, and output signal paths. The three signals are plotted from the workspace and are shown in Figure 9.24. The upper pathway contributes a smooth exponential-like signal (actually it is a double exponential that continues to increase). The lower pathway contributes a transient component that not only serves to bring the response more quickly toward its final value but also produces the inflection in the combined response.

To determine the pulse response of the system, replace the Step element with the Pulse Generator element from the Sources library. Setting the Period textbox of the Pulse Generator to 10 s will ensure that the pulse occurs only once during the 5-s simulation period. Then setting the Pulse Width (% of Period) to 2% produces a 0.2-s pulse as shown in Figure 9.25 along with other pulse responses.

To estimate the pulse width that approximates an impulse, simulate the system using a range of pulse widths. Shorten the pulse to 0.1 s (i.e., 1%), then to 0.05, 0.25, and finally 0.01 s to produce the responses shown in Figure 9.25. These responses are normalized to the same amplitude to aid comparison. The responses produced by either the 10-ms or 25-ms pulse widths are very similar, so

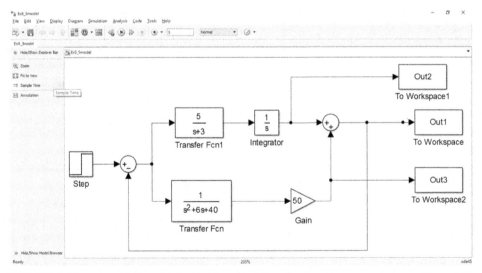

FIGURE 9.23 Assembled model used in Example 9.5 to simulate the system shown in Figure 9.21.

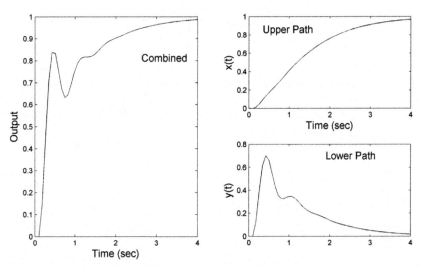

FIGURE 9.24   Three signals from the system shown in Figure 9.23.

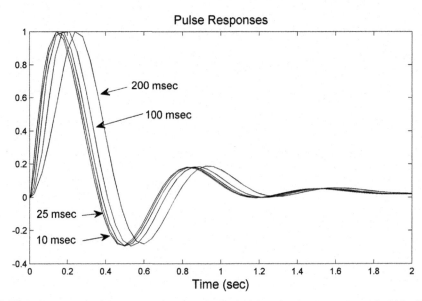

FIGURE 9.25   The response of the system shown in Figure 9.21 to pulses having a range of widths. Responses to pulses having widths of 10 and 25 ms produce nearly the same response indicating that pulses less than 25 ms can be considered impulse inputs with respect to the dynamics of this system.

a pulse having a width of 25 ms or less can be taken as an impulse input with respect to the dynamics of this system. A unit impulse function should have an area of 1.0, so the amplitude of the 0.025-s pulse should be set to $1/0.025 = 40$, whereas the amplitude of the 0.01-s pulse should be 100 to produce the correct response amplitude.

9.4 IMPROVING CONTROL SYSTEM PERFORMANCE: THE PID CONTROLLER

Simulink can be used to show how the performance of real system can be improved. The classic control system consists of a controller subsystem, which directs the actions of an effector subsystem (Figure 1.26). As stated in Chapter 1 (Section 1.4.5.2), the effector subsystem is sometimes referred to as the "plant." Often the controller uses feedback from the plant to modify its control signals to the plant. For example, in the human body, various components of the central nervous system serve as controllers for muscle subsystems. Vision and proprioception provide feedback to the neural controllers. In this section, we use Simulink to study a classic feedback control system and show how to improve its performance using different control strategies. Again we introduce these control strategies through an example.

Our system consists of controller and plant as shown in Figure 1.26. As a model plant, we use a second-order overdamped system, but the controller strategies we investigate are general and work to improve the performance of a wide range of plants. The transfer function of our model plant is

$$TF(s) = \frac{50}{s^2 + 20s + 50} \tag{9.3}$$

In this subsystem, $\omega_n^2 = 50$, so $\omega_n = 7.07$, and $2\delta\omega_n = 20$, so $\delta = 1.4$. The plant is an overdamped second-order system. The roots of the denominator are -17.07 and 2.9, so we could factor Equation 9.3 into two first-order elements, but as Simulink can represent elements of any order, why bother.

Our first controller will be just a gain term that operates on the difference between the input signal and the feedback system from the plant (Figure 9.26). We will investigate the response of the controller–plant system to a step input with the goal of improving the speed of response and the response accuracy.

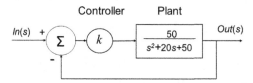

FIGURE 9.26 Classic feedback control system with a second-order, overdamped plant and a constant gain controller. This system is simulated in Example 9.6.

EXAMPLE 9.6

Find the step response of the system in Figure 9.26 to various values of controller gain k. Find the gain that results in the fastest step response without causing overshoot.

Solution

Simulate the step response of this system to a unit step input. Plot the response to several values of gain and find the largest gain that does not produce overshoot. Because we have no idea what the controller gain, k, should be, we start with a value of 1.0 and work our way up (or down if that produces overshoot). The model is laid out with a Step element, Sum element, Gain element, and Transfer Fcn element. The Step element parameter, Step time, is set to 0.1 (the default is 1.0) to clearly show the onset of the movement. In addition, one of the signs on the sum element is made negative, and the Transfer Fcn numerator and denominator coefficients are set to represent the transfer function of Equation 9.3. The output is sent to a To Workspace element, so the results can be plotted. This element's Save format: parameter was set to Array so the output would be a standard MATLAB array. (The default format is a structure.) This results in the model shown in Figure 9.27. The simulation time is reduced to 2.0 s, and the elements are renamed to reflect their role in the model.

Results

Initial simulations used gains of 0.1, 1.0, and 10.0. With a gain of 10, a noticeable overshoot was observed in the response. The gain was reduced to 2.0 and a slight overshoot was still observed. The system's step responses to gains of 0.8, 1.2, and 2.0 are shown in Figure 9.28. A gain of 1 produces a response with very little overshoot, but a gain of 2.0 produces noticeable overshoot. Increasing the controller gain, k, provides two benefits: the step response becomes faster and the final value of the

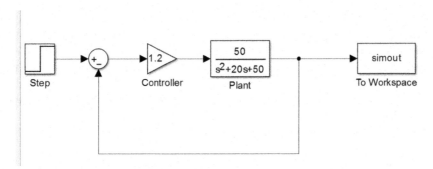

FIGURE 9.27   The Simulink representation of the system shown in Figure 9.26. The elements have been modified to reflect the model parameters. In this model version, the gain, k, was set to 1.2.

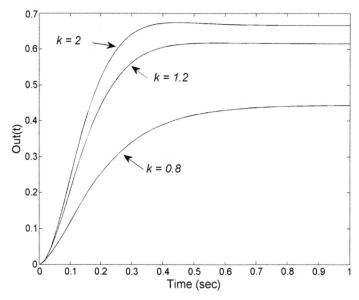

FIGURE 9.28   Simulation of the system in Figure 9.26 to three different values of controller gain, k. As k increases, the response becomes faster and the final value becomes closer to the input final value of 1.0; however, higher values of k also lead to overshoot. A steady-state error between the input and output is inherent in this type of controller as discussed below. More advanced controllers can eliminate this error as shown in the next example.

response comes closer to the step input amplitude of 1.0. However, even at a gain of 2.0 there is still a large error between the output final value of approximately 0.66 and the input final value which is 1.0.

The controller used in the feedback control system shown in Figure 9.26 is known as a "proportional controller" because its output is proportional to the error signal where the constant of proportionality is k. As the plant is driven by error, the error of such a system can never be zero. In fact, we do not need Simulink to tell us there will be an error, or what that error will be. Using the tools of Chapters 6 and 7, we can calculate this final error as a function of k analytically. Applying the feedback equation from Chapter 6 to the system in Figure 9.26 gives the overall transfer function:

$$TF(s) = \frac{Out(s)}{In(s)} = \frac{G(s)}{1 + G(s)H(s)}$$

where $G(s) = (k)\dfrac{50}{s^2 + 20s + 50}$ and $H(s) = 1$. So the overall transfer function is

$$TF(s) = \frac{\dfrac{50k}{s^2 + 20s + 50}}{1 + \dfrac{50k}{s^2 + 20s + 50}} = \frac{50k}{s^2 + 20s + 50 + 50k} \tag{9.4}$$

TABLE 9.2 Output Error as a Function of Controller Gain

Controller Gain, k	Final Output Value	Percent Error
1.2	0.546	45%
1.6	0.615	39%
2.0	0.667	33%

The Laplace domain output to a unit step, $1/s$, is

$$Out(s) = TF(s)\left(\frac{1}{s}\right) = \frac{50k}{s^2 + 20s + 50 + 50k}\left(\frac{1}{s}\right)$$

To find the final value, we turn to the Final Value Theorem (Equation 7.51, Section 7.4.2), which states that $x(t \to \infty) = \lim_{s \to 0} sX(s)$. Thus the final output value as a function of time is

$$Out(t \to \infty) = \lim_{s \to 0}\left[\frac{50ks}{s^2 + 20s + 50 + 50k}\left(\frac{1}{s}\right)\right] = \frac{50k}{50 + 50k} = \frac{k}{1 + k} \quad (9.5)$$

As the final value of the input is 1.0 (the input is a unit step), the final value given by Equation 9.5 shows us the error. This is summarized for three values of k in Table 9.2. The final values appear to be the same as those found in the simulation results shown in Figure 9.28. Our analytical approach (*sans* MATLAB) gives us more information regarding final values than simulation because it shows exactly how error changes with controller gain.

We could also use Equation 9.4 to determine the maximum value of k before response oscillation occurs by finding the value of k that makes the damping factor, δ, equal to 1.0. Recall δ can be determined from the characteristic equation (the denominator equation) of Equation 9.4. As Equation 9.5 shows, as the controller gain, k, increases the error is reduced, but the error can never be zero unless the gain becomes infinite.

The problem with this controller is that its drive is produced solely by the error; without some error there can be no signal to drive to the plant. Perhaps we could reduce this error with a more intelligent controller. For example, if the controller included an integrator, some of the controller's drive signal would come from integration of the error signal. This signal component should continue to increase until the error is finally zero. Such a combined proportional and integral controller is investigated in the next example.

EXAMPLE 9.7

Add an integrator into the feedback controller of the system shown in Figure 9.26. Adjust the gain of the integrator pathway to decrease the final error between input and output values, but still produce a response with no overshoot.

Solution

Modify the model shown in Figures 9.26 and 9.27 to include a second controller pathway containing an integrator and gain term. Such a modification is shown in Figure 9.29. We leave the

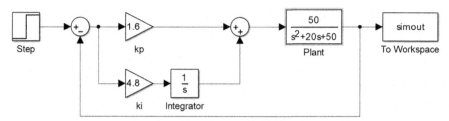

FIGURE 9.29 The system shown in Figure 9.26 with an improved controller that includes an additional pathway containing an integrator. One component of the controller signal should continue to increase (or decrease) as long as the error is nonzero. This system is simulated in Example 9.7.

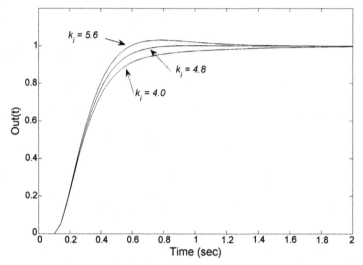

FIGURE 9.30 The response of a system containing both integral and proportional control for three different values of integrator gain, k_i. An integrator gain of 4.8 produces the fastest response without an overshoot. The proportional gain was set at 1.6, which was found to be the largest possible without overshoot in the last example. The timescale of this figure is twice as long as that of Figure 9.28, so these responses are actually a little slower, but they go to the correct final value.

proportional gain set at 1.6, the largest value possible before an overshoot occurs as found in the previous example. We then increase the gain of the integrator pathway until we get response oscillation as in the previous example.

Results

The simulated output of the system with this modified controller is shown in Figure 9.30. Our modified controller was quite successful. The response now goes to 1.0 indicating zero error between the final values of input and output. An integrator gain, k_i, of 4.8 produces a rapid response without overshoot. Note that the time frame of this plot is 2.0 s, twice as long as that of Figure 9.28, so these responses are actually a little slower, but they do go to the correct final value.

Modifying the controller to include an integrator was very successful in improving performance. But why stop there? We have eliminated the final error in the response, but maybe we could make it still faster without inducing overshoot. To boost the response speed, we might try adding a derivative component to the controller signal. This would produce an initial signal in response to the step input, but this signal would fade away as the response progressed and would be zero in the late steady-state section of the response. We will try this added component in the next example.

EXAMPLE 9.8

Modify the controller of the system simulated in the last example to include a derivative component. Increase the gain of this signal component to produce the fastest signal possible without inducing response overshoot.

Solution

Add a third pathway to the controller that includes a derivative operator and a gain term. Although Simulink has a derivative element, it does not work well in simulations with step changes, as the element produces an extremely large, spikelike output. It is better to use a realistic derivative operator consisting of a derivative and a first-order element that acts as a low-pass filter.

$$TF(s) = \frac{\omega_1 s}{s + \omega_1} \tag{9.6}$$

The low-pass element functions to limit the derivative output to more reasonable levels. Equation 9.6 is also the transfer function of a realistic derivative operator, one you would get if you used analog hardware to build a derivative operator. It is only necessary that the low-pass cutoff frequency, ω_1, be much greater than that of the plant, so the derivative controller is still faster than the plant. From the plant's transfer function, Equation 9.3, the undamped natural frequency, $\omega_n = \sqrt{50} = 7.07$ rad/s, so we should make the derivative controller much faster, say, with a cutoff frequency, ω_1, of 50 rad/s. This results in the model shown in Figure 9.31.

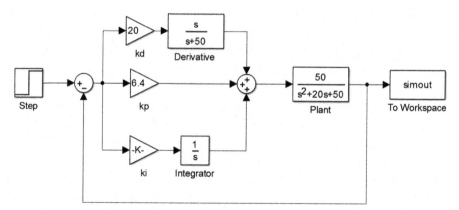

FIGURE 9.31 The system of Figure 9.26 with a more sophisticated controller containing proportional, integral, and derivative components. This system is simulated in Example 9.8.

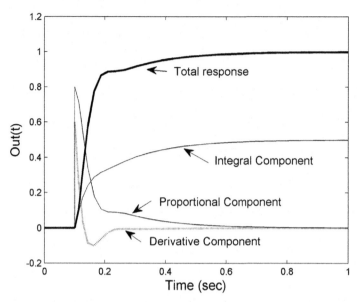

FIGURE 9.32   The response of a feedback control system having a plant defined by Equation 9.3 and a three-component controller. The three component signals are shown along with the total response. The component gains that produced the fastest response with no overshoot and no steady-state error are $k_p = 6.4$; $k_i = 14.4$; and $k_d = 20$.

Adjusting the gains for an optimal response is much more difficult when all three control components are present. Various strategies exist, including commercial software packages that guide parameter adjustment. The approach here is to increase the gains of both proportional and integral elements by a factor of 2 or 3, and then increase the derivative gain until the overshoot is eliminated. This requires multiple simulations and the simultaneous adjustment of all three parameters. A similar empirical approach is often used when actual hardware is involved.

Results

The approximately optimal step response of this system with a three-component controller is shown in Figure 9.32. The proportional, integral, and derivative component gains that produce this response are $k_p = 6.4$; $k_i = 14.4$; and $k_d = 20$. The three signal components that combine to drive the plant are also shown scaled to have about the same amplitude.

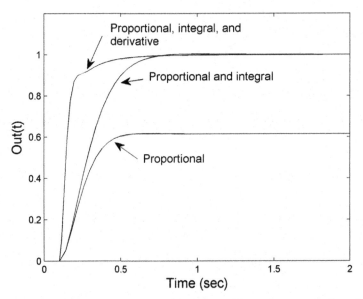

FIGURE 9.33 A comparison of the response of the same plant to three different controllers. The improvement in performance in terms of speed and accuracy of controllers that include proportional, integral, and derivative signals is obvious. Such controllers are called PID controllers, and a range of these controllers is offered commercially

Figure 9.33 compares the responses of the same plant (Equation 9.3) when driven by increasingly sophisticated controllers. The improvement in output error achieved by adding an integral component is striking as is the improvement in speed due to the derivative component. In many practical situations the dynamics of the plant are difficult to modify, but the controller (which is usually a computer) is easy to extend and to change.[9] Because of the obvious performance enhancement, three-component controllers are very common in feedback control systems. Such controllers are called "PID" controllers. They are found in many systems where the system output must track a desired level or simply maintain a desired level. A variety of commercial hardware PID controllers are available that will drive plants ranging from small motors to very large machines.

9.5 BIOLOGICAL EXAMPLES

Many biomedical models use Simulink, and some are quite detailed involving more than 50 elements. For a comprehensive Simulink model of the cardiovascular system, check out the model by Zhe Hu, downloadable from MATLAB Central at http://www.mathworks.com/matlabcentral/fileexchange/818-hsp. This model, with over 90 elements, is far too large to be used here.

---

[9]Think of the cruise control in an automobile. The plant dynamics (the way the car responds to increased throttle) depend on the weight of the car, the size of the engine, and unknown variables such as road conditions. In such cases, a control system that can respond quickly and accurately to keep speed at the requested level is essential.

Example 9.9 presents a realistic physiological model on a much smaller scale: a model of glucose and insulin balance in extracellular tissue. This model demonstrates the ability of Simulink to handle nonlinear elements.

9.5.1 Stolwijk–Hardy Model of Glucose–Insulin Concentrations

EXAMPLE 9.9

Simulate a simplified version of the Stolwijk–Hardy model for glucose–insulin concentrations in extracellular space.

Solution

The Stolwijk–Hardy model consists of two integrators involved in a complex feedback system. The concentration of glucose or insulin in the extracellular fluid is determined by integration of the net flow into each fluid compartment divided by the effective volumes of the fluid compartment. The Simulink version of the Stolwijk–Hardy model is shown in Figure 9.34. Concentrations of glucose and insulin are represented by the outputs of integrators labeled "Glucose" and "Insulin."

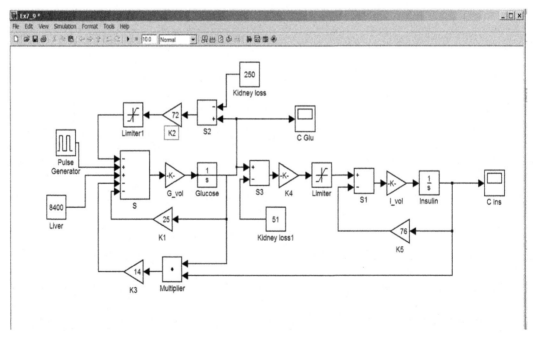

FIGURE 9.34 Simulink representation of the Stolwijk–Hardy model, which represents the concentration of glucose and insulin in the extracellular space.

The insulin integrator receives a negative feedback signal from its output modified by gain K5 and also receives a contribution from the glucose concentration signal. The glucose signal is modified by gain term, K4, and loss through the kidney is represented by a constant negative signal from Kidney loss1. This glucose signal also has an upper limit or saturation imposed by element Limiter.

The glucose integrator receives two feedback signals from its output: one direct through gain K1 and the other modified by Kidney loss and gain K2. This kidney feedback signal saturates: it has a maximum output, imposed by Limiter1. The glucose integrator also receives feedback from the insulin signal modified by its own output signal through the multiplier and gain K4. The glucose integrator receives two input signals: a constant input from the liver and one representing glucose intake. The latter is implemented by a pulse generator set to produce a single 0.5-h pulse after a 1-h delay. (Owing to the slow response of this system, the assumed time unit is in hours for this simulation.) Both integrators are preceded by gain terms that reduce the signal in proportion to the effective volumes of the extracellular fluid for glucose and insulin. This converts the glucose and insulin signals to concentrations. Both integrators also have initial values incorporated into the integrator. Alternatively, a longer delay between the start of the simulation and the onset of the pulse can be used to allow the integrators to obtain appropriate initial values. A problem at the end of this chapter demonstrates this approach.

The limiters and multiplier elements make the equations for this system nonlinear so that they cannot be solved using the linear techniques presented in Chapters 6 and 7. However, Simulink has no difficulty incorporating many different nonlinear elements in a model.

The various model parameters, from a description of the Stolwijk—Hardy model in Rideout (1991), are shown in Table 9.3.

TABLE 9.3 Parameter Values for the Stolwijk—Hardy Model of Glucose and Insulin Concentrations

Parameter	Value	Description	Parameter	Value	Description
C_{G_init}	81 (%mg/mg)	Glucose integrator initial value	G_vol	1/150 (% mL)	Inverse effective volume of glucose extracellular
C_{I_init}	5.65 (%mg/mg)	Insulin isntegrator initial value	I_vol	1/150 (% mL)	Inverse effective volume of insulin extracellular
Liver	8400 (mg/h)	Glucose inflow from liver	Limiter	10000	Glucose saturation driving insulin
K1	25	Gain of glucose feedback	Limiter1	10000	Glucose saturation in feedback signal
K2	72	Gain of glucose feedback from kidney loss	Kidney loss	200 (mg/h)	Renal spill in feedback signal
K3	14	Gain of insulin feedback to glucose	Kidney loss1	51 (mg/h)	Kidney loss in glucose feedback
K4	14	Gain of glucose drive to insulin	K5	76	Gain of insulin feedback

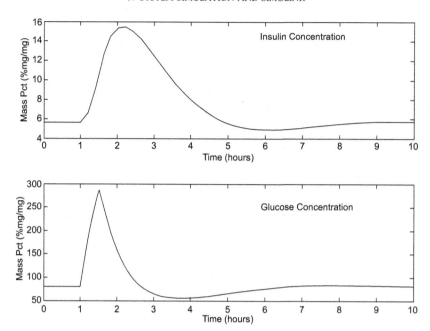

FIGURE 9.35 Simulated responses of glucose and insulin concentrations in the extracellular fluid after infusion of glucose. The 10-h simulation shows that, following a 1/2-h pulse infusion, glucose rises sharply, then falls off undershooting the baseline level. Insulin also increases, but more slowly, and undershoots only slightly before returning to baseline levels.

Results

The results of a 10-h simulation of this model are shown in Figure 9.35. The influx of glucose, modeled as a half-hour pulse of glucose, results in a sharp rise in glucose concentration and a slightly slower rise of insulin. When the glucose input stops, glucose concentration immediately begins to fall and undershoots before returning to the baseline level several hours later. Insulin concentration falls much more slowly and shows only a slight undershoot but does not return to baseline levels until about 7 h after the stimulus.

9.5.2 Model of the Neuromuscular Motor Reflex

Example 9.10 describes a model of the neuromuscular motor reflex. In this model, introduced by Khoo (2000), the mechanical properties of the limb and its muscles are modeled as a second-order system. The model addresses an experiment in which a sudden additional weight is used to exert torque on the forearm joint. The muscle spindles detect the stretching of arm muscles and generate a counterforce though the spinal reflex arc. The degree to which this counterforce compensates for the added load on the muscle can be determined by measuring the change in the angle of the elbow joint. The model consists of a representation of the forearm muscle mechanism in conjunction with the spindle feedback system. The input to the model is a mechanical torque applied by a steplike increase in the load on the arm; the output is forearm position measured at the elbow in degrees.

EXAMPLE 9.10

Model the neuromuscular reflex of the forearm to a suddenly applied load.

Solution

To generate a model of the forearm reflex, we need a representation of the mechanics of the forearm at the elbow joint and a representation of the stretch reflex. The transfer function of mechanical models is covered in Chapter 13, where the function for the forearm joint is derived from basic mechanical elements. For now, we accept that the transfer function of the forearm muscle mechanism is given by a second-order equation. Using typical limb parameters, the Laplace domain transfer function is

$$TF(s) = \frac{V(s)}{M(s)} = \frac{250}{s2 + 25s + 500} \tag{9.7}$$

where $V(s)$ is the velocity of the change in angle of the elbow joint and $M(s)$ is the applied torque. As position is the integral of velocity, Equation 9.7 should be integrated to find position. In the Laplace domain, integration is just divided by s:

$$TF(s) = \frac{\theta(s)}{M(s)} = \frac{1}{s}\frac{V(s)}{M(s)} = \frac{250}{s(s2 + 25s + 500)} \tag{9.8}$$

The transfer function of Equation 9.8 represents the effector element in this system, i.e., the plant. The muscle spindles make up the controller of this reflex but are located in the feedback pathway. Khoo (2000) shows that the muscle spindle system can be approximated by the transfer function:

$$TF(s) = \frac{M(s)}{\theta(s)} = \frac{s + 60}{s + 300} \tag{9.9}$$

where $M(s)$ is the effective counter torque produced by the spindle reflex and $\theta(s)$ is the angle of the elbow joint. In addition, there is a Gain term in the spindle reflex nominally set to 50 and a time delay (i.e., e^{-sT}) of 20 ms. The spindles provide feedback, so these elements are placed in the feedback pathway. The overall model is shown in Figure 9.36.

Results

Simulation of this model in response to added load of 5 kg applied as step input is shown in Figure 9.37. Simulations are performed for two values of feedback gain, 50 and 100. At the higher feedback gain, the reflex response shows some oscillation, but the net displacement produced by the added weight is reduced. If the goal of the stretch reflex is to compensate for the change in load on a limb, the increase of feedback gain improves the performance of the reflex, as it reduces the change

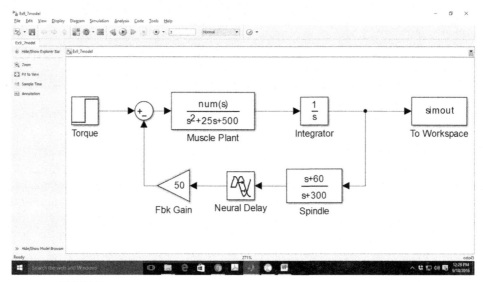

FIGURE 9.36　Model of the neuromuscular reflex developed by Khoo (2000).

produced by the added load. However, feedback systems with high loop gains tend to oscillate, a behavior that is exacerbated when there is a time delay in the loop. The effects of gain and delay on this oscillatory behavior are examined in a couple of problems based on this model.

Example 9.11, an extension of Example 9.10, finds the magnitude and phase spectra of the neuromuscular model used in that example. The approach is the same as in Example 9.4.

EXAMPLE 9.11

Find the magnitude and phase spectrum of the neuromuscular reflex model used in Example 9.10. Use a feedback gain of 90.

Solution

As in Example 9.4, we simulate the impulse response, then take the Fourier transform, and plot the magnitude and phase spectra. Again, to generate the impulse response, we shorten the width of a pulse until the responses generated by a pair of short pulses are the same. To find this pulse width, we replace the step generator in Figure 9.36 with a pulse generator. Simulations to pulses having widths less than 2 ms produce the same response dynamics, indicating that a 2-ms pulse width can be taken as an impulse. The response to a 2-ms pulse with an amplitude of 5 kg is shown in Figure 9.38. For all practical purposes, this is the impulse response.

The Model Configuration Parameters window was modified, so the simulator used a fixed step of 1.0 ms as in Example 9.4 ($f_s = 1$kHz). Because a 2-ms pulse acts as an impulse to this system, any sample interval ≤ 2 ms would work. After running the simulation, the same code developed in Example 9.4 was used to calculate and plot the magnitude and phase spectra

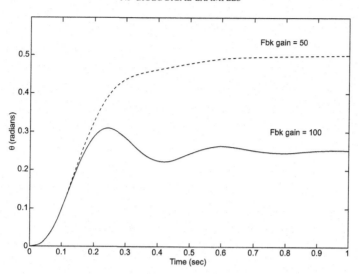

FIGURE 9.37 Model responses of the neuromuscular reflex to a steplike increase on load. Simulated responses are shown for two values of feedback gain.

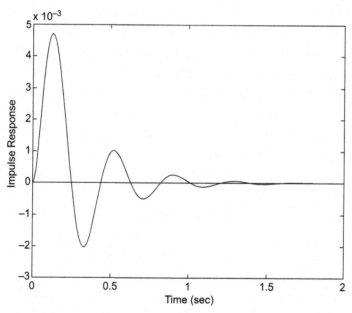

FIGURE 9.38 The response of the neuromuscular system shown in Figure 9.28 to a 2-ms pulse. Reducing pulse width does not change the shape of the response indicating that the 2-ms pulse is acting as an impulse for this system.

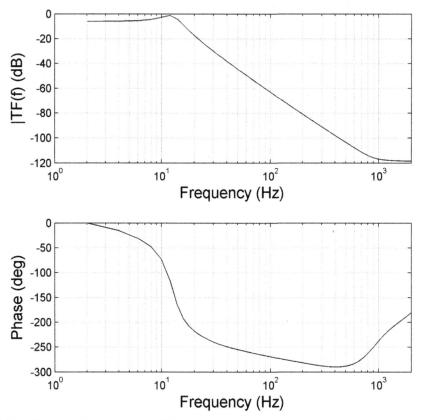

FIGURE 9.39   The spectral characteristics of the neuromuscular reflex shown in Figure 9.38. These spectra were determined by taking the Fourier transform of the impulse response.

Results

These two frequency components are plotted against log frequency in Hz in Figure 9.39. The magnitude curve shows a low pass characteristic with a cutoff frequency of approximately 15 Hz and an attenuation of 60 dB/decade as expected from a third-order system. Determining the spectral characteristics of a system using simulation is explored in a couple of the problems.

9.5.3 The Makay and Glass Model of Neutrophil Density

The last example features a highly nonlinear model and shows how to construct a model, given a differential equation for the system. This model represents white blood cell counts, or neutrophils, in patients with chronic myeloid leukemia. It was originally developed as a differential equation by Mackay and Glass in 1977 and is described in Khoo (2000). The equation defining neutrophil density, x, is given as

$$\frac{dx}{dt} = (\beta\theta^n)\left(\frac{x(t-T_D)}{\theta^n + x(t-T_D)^n}\right) - \gamma x(t) \tag{9.10}$$

where t is time in days (the time frame of this system is quite long), γ is the neutrophil extinction rate, θ and n describe the relationship between the neutrophil production rate and the past neutrophil density, T_D is the delay in neutrophil production, and β is a scale factor. Values of γ, θ, and n suggested by Khoo are 0.1, 1.0, and 10.0, and the value for β is around 0.3. The delay T_D varies between 2 and 20 days. In addition, we need an initial value for neutrophil density $x(0)$ and use 0.1. Equation 9.10 is highly nonlinear and would be difficult to solve analytically, but simulating the solution in MATLAB is straightforward.

EXAMPLE 9.12

Generate a Simulink model based on Equation 9.10 and simulate the response for neutrophil maturation delays of 2, 8, and 20 days.

Solution

The major hurdle in this example is generating the model from the equation. In general, it is easiest to start with the derivative dx/dt (or the highest derivative) and use an integrator to generate x. Then we work from x to produce the right side of the equation.

The $-\gamma x$ term is generated by applying a gain term gamma to x, then feeding this back negatively to the derivative input (Figure 9.40). Sending signal x through a transport delay produces the

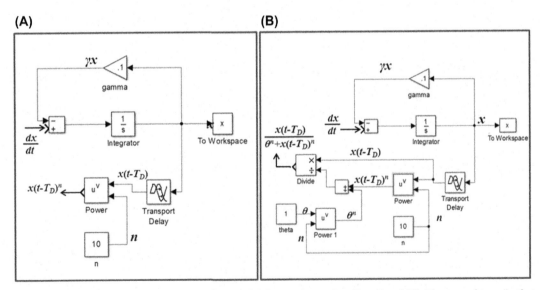

FIGURE 9.40   A) Partial model representation of differential equation (Equation 9.10). The second term in that equation, γx, is produced by applying a gain term to signal x. This is fedback negatively to be one component of the derivative signal. Two component signals of the first term, $x(t - T_D)$ and $x(t - T_D)^n$, are also generated. B) The constant θ^n is generated and with other signals is used to construct a signal representing the first term of Equation 9.10, missing only the two multiplier terms, θ^n and β.

signal $x(t - T_D)$ (Figure 9.40A). To raise this signal to power n, use the Math Function block found in the Math Operations library. This block supports a wide variety of mathematical operation, including exponentiation, natural and common logs, squaring, and raising to a power. We use the Function Block Parameter window to set Function to pow. This raises the signal in the upper input to the power of the signal (or constant, in this case) in the lower input. The output of this block, labeled Power in Figure 9.40A, is $x(t - T_D)^n$.

The Math Function block set to pow is also used to generate the θ^n term as seen in Figure 9.40B. Adding this constant term to signal $x(t - T_D)^n$ gives $\theta^n + x(t - T_D)^n$, the denominator of the fraction in Equation 9.10. As we already have signal $x(t - T_D)$, we use a divider block, also from the Math Operations library to produce a signal equivalent to $\left(\frac{x(t-T_D)}{\theta^n+x(t-T_D)^n} \right)$ as shown in Figure 9.40B. To finish the first component, all we need to do is to multiply this signal by θ^n, then scale by β. The resultant signal is added to the $-\gamma x$ term to complete the derivative, $\frac{dx}{dt}$. The final model is shown in Figure 9.41.

In essence, the system shown in Figure 9.41 can be thought of as an integrator that received two signals from two feedback pathways. One is a negative feedback pathway with linear gain of γ. The other is a positive feedback pathway that is highly nonlinear. The default Simulink solver showed slight instabilities, so an alternative solver was selected from the Model Configuration Parameters window. Under the Solver: tab, ode15 was selected on a trial and error basis simply because it produced stable simulations with short run times.

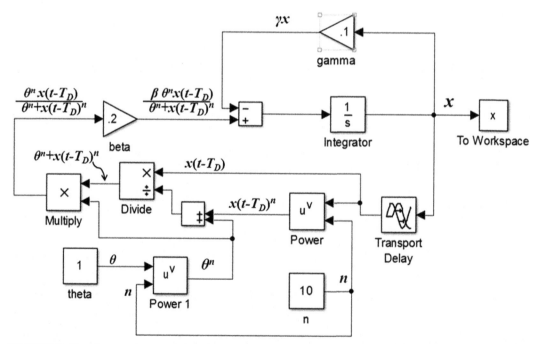

FIGURE 9.41 The completed model used in simulate Equation 9.10. This model is highly nonlinear. Adapted from Khoo (2000).

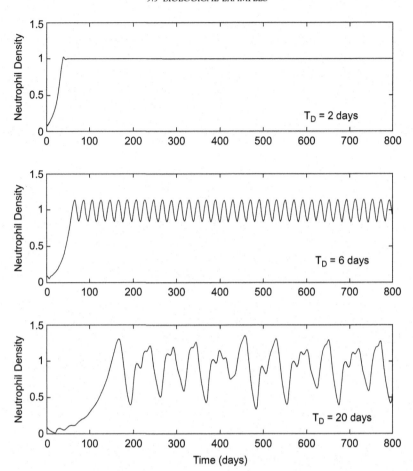

FIGURE 9.42 The modification in neutrophil density found for three maturation times: 2, 6, and 20 days. The 6-day delay produces a sustained oscillation of approximately 20 days, whereas the 20-day delay produces a response that oscillates randomly. The 20-day delay signal shows chaotic behavior: an apparent randomness in what is a completely deterministic system.

Results

Simulations were performed using the parameter specified above for three values of TD: 2, 8, and 20 days, and the results are shown in Figure 4.42.

Recall that the transport delay T_D represents the maturation time, the time required to produce mature neutrophils in the bone marrow. The two day period is normal and the simulation shows an initial rise to a constant neutrophil density that is sustained. The 6-day delay leads to an oscillatory response and is known as "cyclical neutropenia." When maturation is extended to 20 days, cyclical behavior now has a random component, a pattern sometimes found in "chronic myeloid leukemia." This random component is surprising, as the system is completely deterministic; there is no added noise in the system. This is an example of a chaotic system, a deterministic system governed by fixed, well-defined equations, which shows apparently random behavior. It occurs in some nonlinear systems with positive feedback. (Linear systems with positive feedback may show sustained oscillations, but not chaotic behavior.)

One distinguishing feature of a chaotic system is high sensitivity to changes in initial conditions. Thus we might expect that if the initial condition, which is 0.1 for all the responses shown in Figure 9.42, were changed slightly, the response might be quite different. This assessment is found in one of the problems.

9.6 SUMMARY

Simulation is a powerful time-domain tool for exploring and designing complex systems. Simulation operates by automatically time-slicing the input signal and determining the response of each element in turn until the final output element is reached. This analysis is repeated time slice by time slice until a set time limit is reached. As the response of each element is analyzed for each time slice, it is possible to examine the behavior of each element, as it processes its local input as well as overall system behavior. Simulation also allows for incorporation of nonlinear elements such as the divider and exponential elements found in the model used in Example 9.12. MATLAB's Simulink is a powerful simulation tool that provides a wide range of elements and is easy to use.

In classic feedback control models, the overall system is divided into two subsystems: an effector subsystem, often called the plant, which implements the system's output, and a controller that combines information from the input and feedback from the output to produce signals that control the plant. For instance, in a home heating system the plant contains the furnaces, radiators, and home thermal properties, whereas the controller is the thermostat. Often the dynamic characteristics of the plant are difficult to modify, or may change over time, or even be unknown, whereas those of the controller, which is usually computer based, can be easily changed. Improvements in the computer algorithm can often lead to significant improvements in performance of the overall system. One such control subsystem is the PID controller, which uses the error between the actual and desired response, along with the derivative and integral of that error, to drive the plant. Adjusting the three signal gains can be tricky but can significantly enhance system performance. Some advanced controllers are adaptive and can change their parameters, for example, the gains of PID control, in response to changes in the plant.

Simulation techniques are well-suited to analyzing biological systems, as they can represent the nonlinear properties found in these systems. Examples of glucose–insulin balance and the stretch reflex are given, but many more can be found on the Web.

PROBLEMS

1. Write your own program following the approach used in Example 9.1 to simulate the system shown below to a unit step input. Note the first-order element can be simulated using an integrator in a feedback loop. (Using the feedback equation, Equation 7.6,

putting $\frac{0.5}{s}$ in a feedback loop gives $TF(s) = \frac{G}{1+GH} = \frac{0.5/s}{1 + 0.5/s} = \frac{0.5}{s+0.5}$. Thus you need to add another, external, feedback loop to the one in Example 9.1.)

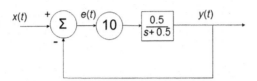

2. It is easy to add nonlinearity to a simulation, even when coding it yourself as in Problem 1. Write a program to simulate the response of the system below to a unit step. The system is the same as the one in Problem 1 except for saturation nonlinearity in the controller. The saturation nonlinearity restricts the output signal to fall within some maximum and minimum limits, in this case ± 0.3. (Hint. If you use MATLAB's min and max routines you only need one additional line of code.)

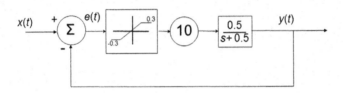

3. Simulate the system in Problem 2 using Simulink. Simulate the response to a unit step and to a 10 rad/s sine wave with an amplitude of 1.0. For the latter, increase the simulation time from 2.0 to 5.0 s. (To submit the MATLAB plot, you can use the To Workspace block in the Sinks library. Remember to set the Save format option to array and select a variable name or just use the default "simout".)
4. Use Simulink to find the step response of the system shown in Problem 7 of Chapter 7. Use values of 0.5 and 1 for k. Simulate a 10-s response. Use To Workspace to get the data in the workspace and plot the two responses on a single graph.
5. Use the approach shown in Example 9.3 to find the maximum pulse width that can still be considered an impulse. Use Simulink to generate the pulse response of the system used in Problem 4 with the value of $k = 0.5$. To determine when responses are essentially the same dynamically, use the scope to measure carefully some dynamic characteristic of the response such as the time of the first peak. To get an accurate estimate of the time when the first peak occurs, use the magnifying feature on the scope. When two different pulse widths produce responses with the same peak time, the input pulse can be taken effectively as an impulse. Use To Workspace to plot this impulse response.

6. Simulate 10-s of the response of the system below to a 0.1 s (i.e., 1%) pulse. Make $k = 1$ and 10 and plot the responses on the same graph. Can you explain the response?

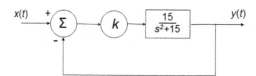

7. Find the magnitude spectrum of a system using white noise. Use Simulink to simulate the fourth-order transfer function of Equation 9.2 used in Example 9.4. Replace the Pulse Generator with a Band-Limited White Nosie generator also found in the Sources library. Theoretically, white noise contains energy at all frequencies, so the Fourier transform of the white noise response of a system can be used to find the magnitude spectrum (but not the phase spectrum because the phase of white noise is undefined). As in Example 9.4, open the Model Configuration Parameters window (under the pull-down Simulation tab) to set the Solver options/Type to Fixed-step and set the Fixed-step size to 0.001 s (1.0 ms). As we are dealing with noise input, it is better to have more data, so increase the simulation time to 200 s. Plot the magnitude in dB against log frequency. The resulting magnitude spectrum will be noisy, particularly at high frequencies, but should have the same shape as the magnitude curve in Figure 9.20. A filter can be applied to the spectrum to smooth the curve.

8. Develop a PID control to improve the step response of a feedback control system having a plant defined by the transfer function. Use a 2.0-s simulation time.

$$TF(s) = \frac{4}{s^2 + 8s + 8}$$

First set the proportional gain so as to get no overshoot in the response. Plot this response. As in Example 9.6, use the feedback equation in conjunction with the final value of the plant to calculate the steady-state error. Compare this error with the final value from the simulation. Next add integral control to eliminate the steady-state error. Plot this response. Finally, add derivative control and try to get the fastest response. You will find you can set the derivative gain very high and increase the other gains by four or five times to get a very rapid response. Plot your best response and include the PID parameter values in the title.

9. Use the optimized model from Problem 8 and replace the step input with a Ramp element. Set the Slope parameter of this input element to 2.0 (units/s). Although there was no steady-state error in this system's step response, there will be a steady-state error to the ramp input signal. This error can best be observed by plotting the error signal that feeds the three PID controller elements as shown in the figure.

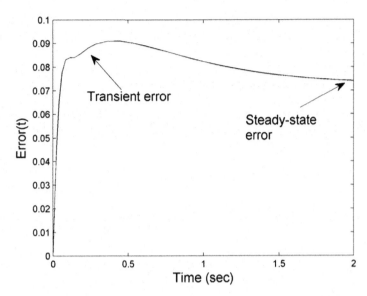

Although all three of the PID controller signals affect the transient error, only one has an effect on the steady-state error (see figure). Which PID signal influences steady-state error? Scale up this parameter until the transient error just begins to show undershoot (in addition to the initial overshoot.) Record the steady-state error at this parameter value and at the values used in Problem 8. Plot the final error signal as in the figure and include the original and final steady-state errors in the title.

10. Can you do better than the human central nervous system? The variable verg_movement in file vergence_resp.mat is the response of the eyes to an inward moving target. The response, called a vergence eye movement, shows the angle between the two lines-of-sight in deg. The response covers the first 2 s after a step change in target distance and is sampled at 200 Hz. Your task, given the transfer function of the eye movement plant below, is to develop a PID controller that does as well. Use a fixed-step simulation with a Fixed-step size... of 0.005 s, so you can plot the actual and simulated response superimposed.

Several highly accurate models of the neuroocular plant have been developed based on the mechanics of the eye and its muscles. These sophisticated models feature fourth-order dynamics and nonlinear elements, but a reasonable second-order approximation is given by the transfer function:

$$TF(s) = \left(\frac{62}{s+62}\right)\left(\frac{4.2}{s+4.2}\right)$$

You should be able to generate a simulated response that comes quite close to the actual response. Plot the simulation to a 4.0 (deg) step input and actual response superimposed with the correct timescale. Note the vertical axis will be in deg. Provide the PID parameter values in the title of the plot. (You may want to store the simulation output for use in the next problem.)

11. You can actually make the response much faster without oscillation, provided you are willing to give up feedback. This means that there is some error in the final position; nonetheless the nervous system actually adapts this "open-loop" strategy to produce an ultrafast movement called "saccadic eye movements." (Adaptive processes are used by the nervous system to reduce systematic movement errors over the long term.) To generate this fast movement (the fastest in the human body), the nervous system generates a pulse that moves the eyes quickly to the final position. A step signal is also generated to sustain the eyes at the desired final position. There is neurophysiological evidence that the step is generated by neural integration of the pulse. This leads to the control model shown below.

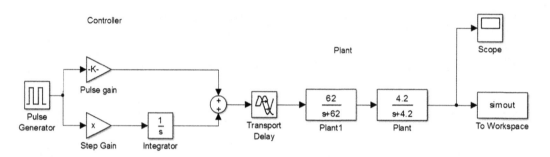

Adjust the step gain to produce a 4-deg response and then adjust the pulse gain to give the fastest response without overshooting. The result should be a response that is many times faster than that achievable with the feedback control system used in Problem 10. Plot the results of your saccadic simulation superimposed with the results of the vergence simulation found in Problem 10. Note the pulse and step gains in the title of the graph.

12. Use Simulink and the neuromuscular reflex model developed by Khoo (2000) and presented in Example 9.10 to investigate the effect of changing feedback gain on final value. Set the feedback gain to values: 9, 20, 40, 60, 90, 130, and 160. (Recall that the final value is a measure of the position error induced by the added force on the arm.) Measure the final error induced by the steplike addition of the 5 kg force: the same force used in Example 9.10. Lengthen the simulation time to 3 s and use the scope output to find the value of the angle at the end of the simulation period. Plot the final angle (i.e., force-induced error) versus the feedback gain using standard MATLAB. Label the axes correctly.

13. Use Simulink and the neuromuscular reflex model developed by Khoo (2000) and presented in Example 9.10 to investigate the effect of changing feedback gain on response overshoot. Set the feedback gain to values: 40, 60, 90, 130, and 160. Use the scope to measure the response overshoot in percent as the maximum overshoot divided by the final angle for three values of delay: 0, 20, and 30 ms. If there is no overshoot, record it as 0%. Plot percent overshoot versus feedback gain for the various delays using standard MATLAB. Label the axes correctly.

14. The figure below shows a physiologically realistic model of the vergence eye movement control system developed by Krishnan and Start in the 1960s. (Vergence eye movements are used and defined in Problem 10.) The controller of this model contains two elements of a PID controller: the upper path is the derivative control and the lower path the proportional control. (Note that the plant is a first-order approximation of that used in Problem 10.) Transcribe the model into Simulink and use simulation to find the first 2 s of the response to a 4.0-deg step input. (Use a fixed step of 0.005 s so you can compare the simulated response with actual data found as verg_movement in file vergence_resp.mat.) Increase the value of k until the response just begins to overshoot and note the improved speed of response. Plot the responses to your optimal value of derivative gain, k, and without derivative control; i.e., $k = 0$. On the same graph, plot the actual data. You will note the simulated data falls short of the actual response in two ways: there is a fairly large steady-state error, and the dynamics are slightly slower. An added integral controller could solve both these problems as shown in Problem 10.

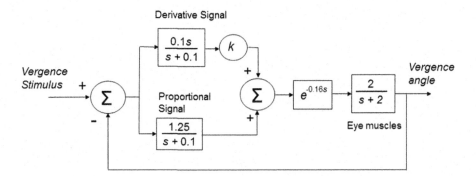

15. The model shown below is a two-compartment circulatory pharmacokinetic model developed by Richard Upton of the University of Adelaide in Australia. This model has interacting compartments and can become unstable if the rate constants become imbalanced. Simulate the model using Simulink for the following parameter values: $k21 = k21a = 0.2$; $k12a = 0.5$; Gain = 0.5; Gain1 = 0.1. Then reduce the parameter $k21a$ to 0.15 and note that the model becomes unstable with an exponential increase in compartment concentrations. Find the lowest value of $k21a$ for which the model is stable. (Note: The vertical black bar combines the two input signal into an array so that the scope displays both signals. If you use a To Workspace output element with this signal, plot(tout,simout) plots both signals superimposed.)

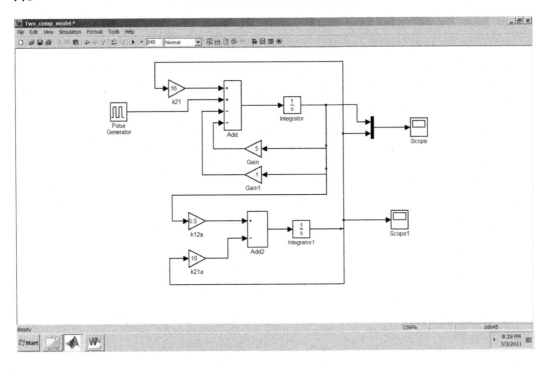

16. Evaluate the Mackey and Glass model of neutrophils shown in Figure 9.41 for sensitivity to initial conditions. Use the parameters specified in Table 9.4 with a maturation delay, T_D, of 20 days. Plot the response when $IC = 0.10$ and superimpose the response when the $IC = 0.095$. Note the differences in the later response caused by this minor change in initial condition.

17. Construct a model based on the nonlinear van der Pol oscillator equation:

$$\frac{d^2x}{dt^2} = c\left(1 - x^2\right)\frac{dx}{dt} - x$$

TABLE 9.4   Parameters used in the Mackey and Glass model of Neutrophil Density

Parameter	Value	Description
γ	0.1	Neutrophil extinction
θ	1.0	Neutrophil production as function of past density
n	10	
β	0.2	Scale factor
T_D	20	Maturation time
IC	0.1	Initial condition

Plot the response of the system with an initial condition for x of 0.1 and values of c of 2.0 and 5.0. Make the simulation time 100 s. Note the sustained oscillations and the change in frequency with c. (Hint. Start with d^2x/dt^2 and integrate to get dx/dt and integrate again to get signal $x(t)$. Put the initial condition into the second integrator. Use x^2, gain c, and a constant 1.0 to construct $c(1 - x^2)$. You can use the Math Function with the option "square" to generate x^2. Then multiply this signal by dx/dt. Sum the resultant signal with $-x$ to get d^2x/dt^2, the input to the first integrator.) Be sure to save this model for possible use in Problem 9 of Chapter 10.

10

Stochastic, Nonstationary, and Nonlinear Systems and Signals

10.1 GOALS OF THIS CHAPTER

In this chapter, we treat some of the thornier challenges of real biological systems: nonstationarity and nonlinearity. Except for a few simulations done in Chapter 9, all of the systems we have studied thus far have been linear and time invariant (i.e., stationary) (LTI) systems. Linear systems produce linear signals, and we now have a powerful array of tools to analyze both linear signals and systems. Many linear models have been quite successful in representing biological systems even though these systems contain some nonlinearity. We also know how to use a simulation to analyze systems containing certain well-defined nonlinear processes. For example, the Stolwijk–Hardy model for extracellular glucose–insulin concentrations (Example 9.9) contains two rate-limiting nonlinearities. However, biological systems sometimes contain major nonlinearities and often change over time; i.e., they are nonstationary. The latter may be due to well-defined adaptive processes that can be added to a model, or they may be due to less well-defined degenerative processes that are hard to represent. Finally all nonlinear signals contain noise, either due to the measurement instrumentation or inherent in the signals themselves, and this makes the identification of nonlinearities particularly difficult.

Nonlinear systems produce nonlinear signals, and these nonlinear properties can provide important information about the system. Despite considerable work to date, none of the signal analysis techniques can reliably tell if a signal contains nonlinear features. Gross nonlinearities such as signal saturation can usually be detected, but identifying subtle signal nonlinearities is an ongoing challenge.

Chapters 1 and 2 introduced the basic properties of signals, including linearity versus nonlinearity, stochastic versus deterministic, and stationary versus nonstationary (see Table 2.1). Chapter 2 also introduced some of the basic measurements that can be applied to all signals in the time domain, including mean, variance, and correlations. In this chapter we expand on some of these concepts; specifically, we perform the following:

- Delve deeper into the properties of random data, including the concepts of stationarity and ergodicity.

Circuits, Signals, and Systems for Bioengineers
https://doi.org/10.1016/B978-0-12-809395-5.00010-2

- Explore methods to analyze nonstationary signals.
- Describe ongoing work to identify and quantify subtle nonlinear signal features.
- Show how nonlinear feature detection can be applied to biological signals, particularly heart rate variability.

10.2 STOCHASTIC PROCESSES: STATIONARITY AND ERGODICITY

When dealing with signals that display stochastic (i.e., random) behavior, one of the first things we would like to know is if the signal is stationary. As briefly mentioned in Chapter 1 (Section 1.4.2), a stationary signal does not change its statistical properties over time. In other words, if we measure a statistical property, such as the mean, the value found for one segment of the data will be the same as for any other segment. If we measure the mean of a large number of segments (or an infinite number), the variance of those mean values will converge to 0. If this is the case, the signal is said to be "ergodic in the mean." For such a signal, the measurement of the mean over only one segment would be sufficient to estimate the signal's true mean (also known as the "expectation" of the mean).

The concept of ergodicity extends to other common statistical measurements including higher orders such as variance. For a system to be considered completely ergodic (without qualifiers such as "in the mean" or "in variance"), both the mean and the autocorrelation must be the same for all signal segments, or, alternatively, the first four moments of the signal. The first four moments are the mean, variance, "skewness," and "kurtosis." The mean and variance are defined in Chapter 2, and skewness and kurtosis are defined in Equations 10.1 and 10.2. If a signal satisfies these conditions, it greatly simplifies the analysis of its statistical properties because statistical measurements need to be made on only one segment of the waveform. Ergodic signals are necessarily stationary. If a signal is ergodic in the mean, it is sometimes called "weakly stationary," and in many engineering cases these signals are also ergodic. Unfortunately, bioengineering signals such as the EEG can be ergodic in the mean but still nonstationary.

To determine if a signal is ergodic, multiple measurements must be made over different time periods, and this usually requires a large amount of data. Example 10.1 compares multiple measurements of mean and variance on a band-limited Gaussian noise signal and an EEG signal. Gaussian noise signals are known to be ergodic, so we should get similar values for our measurements.

EXAMPLE 10.1

Evaluate the band-limited Gaussian signal found in the file `Bandlimit_gauss.mat` and the EEG signal in the file `EEG10.mat` to determine if they are ergodic. Both signals are stored as MATLAB variable x and have been normalized to have an overall RMS of 1.0. The effective time length of both signals is 10 min, where $f_s = 256\,\text{Hz}$.

Solution: First we divide the data into a number of segments so that we can check the consistency of measurement between segments. We could either check the consistency of the mean and autocorrelation functions, or the first four moments. We have MATLAB routines for the first two moments (mean and variance) and can easily write a routine that calculates the third and fourth moments.

The equation for the third moment, skewness, is:

$$Skew = \frac{1}{\sigma^3 N} \sum_{n=0}^{N-1} (x_n - \bar{x})^3 \tag{10.1}$$

The fourth moment is called kurtosis and is determined as

$$Kurtosis = \frac{1}{\sigma^4 N} \sum_{n=0}^{N-1} (x_n - \bar{x})^4 \tag{10.2}$$

In both these equations, the standard deviation, σ, is normalized by $1/N$ instead of $1/(N-1)$. This is achieved in MATLAB by making the second argument of standard deviation as 1 (i.e., std(x,1);). These equations are implemented in function skew_kurt(x) as follows:

```
function [skew, kurt] = skew_kurt(x)
% Function to calculate skew and kurtosis of data x
%
N = length(x);          % Data length
x = x - mean(x);        % Remove mean
sd = std(x,1);          % Standard deviation (normalized by 1/N)
skew = (1/(N*sd^3))*mean(x.^3);   % Calculate skew
kurt = (1/(N*sd^4))*mean(x.^4);   % Calculate kurtosis
```

We then apply these statistical measures along with mean and variance to the two data sets. From the 10 min of data, we extract five segments, each being 30 s long. Segment length and the number of segments are somewhat arbitrary, but if the data are ergodic, we should get the same statistical values for all segments. If we are uncertain of the results, we can alter these parameters. This strategy leads to the following code:

```
% Example 10.1 Program to check for ergodicity of two signals.

%
load Bandlimit_gauss;    % Bandlimited Gaussian noise
% load EEG10;            % Alternate data set: 10 min EEG
fs = 254;                % Sample frequency
N = 30 * fs;             % Segment size: 30 sec intervals
K = 5;                   % Divide signal into K, 30 sec intervals
for k = 1:K              % Isolate and plot 5 segments of EEG
  ix = N*(k-1) + 1;
  y(k,:) = x(ix:ix+N-1);
  avg(k) = mean(y(k,:));    % Calculate mean
  va(k) = var(y(k,:));      % Second moment
  [skew(k), kurt(k)] = skew_kurt(y(k,:));  % Other two moments.
end
output = [avg' va' skew' kurt']    % Display individual segment results
variance = var(output)             % Display variance of individual results
```

TABLE 10.1   Value of First Four Moments of Five Band-Limited Gaussian
             Signals

Segment	Mean	Variance	Skewness	Kurtosis
1	0.0650	0.9853	0.0000	0.0004
2	0.0235	1.0109	−0.0000	0.0004
3	0.0182	0.9632	0.0000	−0.0004
4	0.0166	1.0042	−0.0000	0.0004
5	0.0223	1.0302	−0.0000	0.0004
Variance	0.0004	0.0007	0.0000	0.0000

Results: The values of the four moments and the associated variances are shown for the band-limited Gaussian signals in Table 10.1. The variance of all four moments is very close to 0 indicating that this signal is ergodic. Note that the third- and fourth-order moments are, themselves, also near 0 for the individual segments. This indicates that the probability distribution function for each segment is symmetrical as expected from Gaussian data.

Rerunning the program and loading the file EEG10.mat gives the results shown in Table 10.2. For this signal, the means of the segments are the same (and near 0) but not the variances. Therefore the EEG signal might be called ergodic in the mean, but it is not ergodic in variance. Again, the third- and fourth-order moments are near 0, showing that although the probability distribution functions are different for each segment, they are all symmetrical.

Example 10.1 shows how to test for ergodicity, but it requires multiple segments. If your signal is long enough, you can always slice it into multiple segments. Although the formality of testing for ergodicity is all very nice, often all you really need to do is to look at the signal. Figure 10.1A shows an EEG signal that is clearly nonstationary: the mean is changing and the

TABLE 10.2   Value of First Four Moments of Five EEG Signals

Segment	Mean	Variance	Skewness	Kurtosis
1	−0.0037	1.6631	0.0001	0.0014
2	0.0055	0.7442	0.0004	0.0041
3	0.0008	1.2456	0.0003	0.0040
4	−0.0026	0.4931	0.0003	0.0024
5	−0.0032	0.3694	0.0002	0.0020
Variance	0.0004	0.2933	0.0002	0.0020

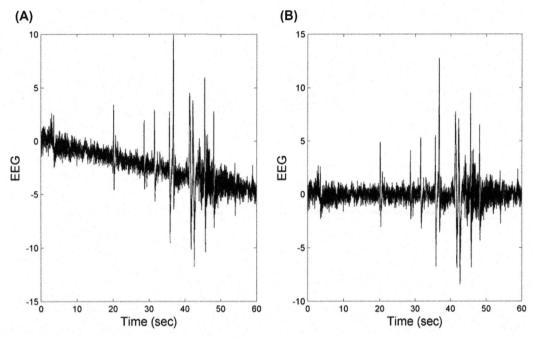

FIGURE 10.1 (A) A nonstationary EEG signal where the mean steadily decreases with time and where the variance is greater in the period between 30 and 50 s. (B) The same EEG signal where the variation in mean has been corrected using detrending, but the variation in variance remains.

variance is larger toward the end of the movement. Baseline variation can be sometimes corrected by taking the derivative, high-pass filtering, or "detrending" the data. In the later approach, baseline wander is identified and then subtracted out. The MATLAB routine detrend(x) subtracts out a linear, or piecewise linear, trend from the data. Detrending is described in Section 10.3.3 and the MATLAB routine is used in Example 10.9. The effect of this routine is seen in Figure 10.1B where the changing baseline has been removed. The signal is still nonstationary because the variance changes over the course of the response. Other methods for compensating for a wandering baseline are explored in the problems.

If you are unable to remove the nonstationarity, then you need to work with it. This means you have to apply analyses to short segments of the signal, segments that can be assumed to be stationary. Sometimes the nonstationarities, themselves, are of special interest. Because each signal has different features of interest, the particular analysis of a nonstationary signal is unique to each signal, but a common thread is signal segmentation. The segments may be chosen arbitrarily, but sometimes the underlying physiology suggests how the signal should be segmented. For example, in analyzing ECG signals, it is usually assumed that each beat is a stationary process but that beat-to-beat variations may be nonstationary. Thus segmenting the signal on a beat-to-beat basis makes sense. With the EEG signal, the segmentation strategy is less obvious. Nonstationarities may be due to a variety of factors including changes in mental state, but the timing of these changes may not be known in advance. In such cases, segments may be selected arbitrarily, but the appropriateness of these segments can be confirmed during the analysis. The next two examples illustrate both these strategies of signal segmentation.

The next example of signal segmentation again involves the EEG signal, a signal usually interpreted in terms of rhythmic waves in the frequency domain. The Fourier transform is used to convert the time domain EEG signal to the frequency domain. Because the EEG is nonstationary, a time–frequency analysis such as the short-term Fourier transform (STFT) is ideal (see Section 4.6 and Example 4.9). In this example, rather than apply the STFT directly, we use a band-pass filter to track the activity of a specific frequency range over time. The best known EEG rhythm is the alpha rhythm that ranges between 8 and 13 Hz. Currently these waves are thought to come from areas of the cortex that are not in use, or they may play a role in neural coordination. This example determines the alpha wave activity as it varies over a 1-h period.

EXAMPLE 10.2

Determine the time-dependent activity of a 1-h EEG signal within a frequency band of 8–13 Hz. The signal is found in the file EEG_FP1_FP7.mat. Segment the signal into 10-s intervals using a 50% overlap (i.e., 5 s). Then filter these isolated segments with a band-pass filter having a passband between 8 and 13 Hz. Use a fourth-order Butterworth band-pass filter. Next, take the RMS value of the filtered signal. To account for possible changes in signal gain, normalize this RMS value by the RMS value of the unfiltered EEG signal. Plot this normalized RMS value as a function of the window's time position. The one-hour EEG signal is a variable, val, in the file EEG and $f_s = 256$ Hz. (These data are from the PhysioBank data collecTion of physionet.org, Golberger, 2000. Record number chb03, electrode position FP1-F7.)

Solution: The analysis is straightforward. Load the data and define the parameters (f_s, filter frequencies, window size, overlap, and number of total iterations). In a loop, isolate the data using the band-pass filter, and then calculate the RMS of the filtered and unfiltered data. Store the ratio of filtered to unfiltered data and the window time. Plot the results.

```
% Example 10.2 Example to quantify the alpha activity in a 1 hr EEG signal
%
load EEG_FP1_FP7;              % Get data
fs = 256;                     % Sample frequency (given)
N = length(val);              % EEG length
wl = 8 *2/fs;                 % Define bandpass filter cutoff freq.
wh = 13*2/fs;
[b,a] = butter(4,[wl,wh])     % Define 4th-order bandpass filter
window_size = 10 * fs;        % 10 sec window size
overlap = 5 * fs;             % and 5 sec overlap
incr = window_size - overlap; % Window increment
K = round(N/incr) - 2;        % Number of windows to analyze
for k = 1:K
  i_st = incr*k;                            % Define window indices
  i_end = i_st + window_size;               % beginning and end
  rms_wind = sqrt(mean(val(i_st:i_end).‘2)); % Overall rms
```

```
alpha = filter(b,a,val(i_st:i_end));       % Filter segment
rms(k) = sqrt(mean(alpha.^2))/rms_wind;     % Compute normalized RMS
t(k) = mean([i_st,i_end])/fs;               % Save window center time
end
plot(t/60,rms,'k');                         % Plot results
  ......... labels..........
```

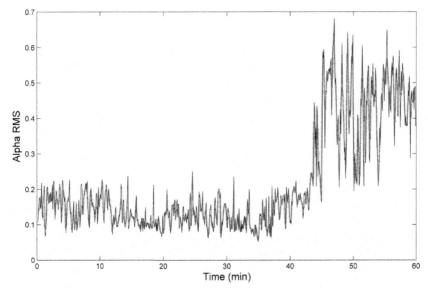

FIGURE 10.2 The alpha wave activity in a 1-h recording of EEG activity. The signal was windowed into 10-s segments with a 50% overlap. The energy of each segment between 8 and 13 Hz was determined by first isolating this frequency range with a band-pass filter and then taking the RMS signal of the result. This RMS value was normalized by the RMS of the unfiltered signal. Results obtained using a 20-s window (also 50% overlap) are superimposed over the original record as a dotted line but are difficult to see as they are quite similar indicating that exact window size is not critical.

Results: Figure 10.2 shows the alpha wave activity over the 1-h period. This activity is seen to be low initially but to increase markedly after around 45 min. We might wonder if the interval size was chosen appropriately. Perhaps a larger or smaller window would give different results. One way to check this empirically is to modify the window size and see if it changes the results. The window size here was doubled to 20 s with a 10-s overlap, and the result was plotted as a dashed line superimposed over the original results. This second line is difficult to see in Figure 10.2 because it closely follows the original results. One of the problems explores the effect of larger window sizes.

It might be interesting to see the activity of other EEG rhythms and of other electrode positions, and these are examined in the problems.

When analyzing an ECG signal, a common first step is to isolate the signal into segments that represent single cardiac cycles. Fortunately the electrical activity of the heart produces a sharp-peaked waveform known as the QRS complex. Such peaks are seen in a 4-s ECG

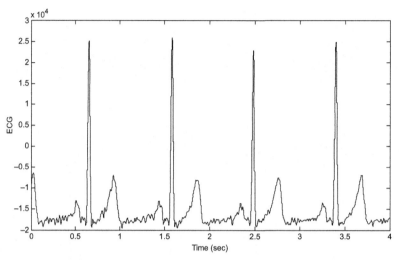

FIGURE 10.3   Portion of the 10 min ECG signal used in Example 10.3. This signal was obtained from ECG lead I.

segment in Figure 10.3. There are a number of hardware and software devices dedicated to identifying the QRS complex including the advanced Pan-Tompkins QRS detector available in MATLAB code at https://www.mathworks.com/matlabcentral/fileexchange/45840-complete-pan-tompkins-implementation-ecg-qrs-detector. However, in the next example, a simple home-brew QRS detector will be used to illustrate the basic principles.

After identifying and isolating the electrical activity associated with each beat, a typical ECG analysis might involve measuring specific features to find electrical patterns that are indicative of a disease. This step is called "feature detection" where the features are used to "classify" the beats in normal and various abnormal[1] classes. (Using features to classify data is part of a broader analysis strategy known as "pattern recognition.") Isolated abnormal electrical patterns would be further analyzed to determine their diagnostic implications. The next example gives the flavor of feature extraction and classification of beat-to-beat ECG activity.

EXAMPLE 10.3

Load the 10 min ECG signal found as variable `ECG` in the file `ECG_10 min.mat` ($f_s = 100$ Hz). Detect the "R-wave," the peak of the QRS complex. Then measure two features: the mean and RMS values, of the waveform on either side of the R-wave peak. Use signal segments that begin 0.25 s before the R-wave and end 0.5 s after. Plot each feature against the other as individual points, a plot known as a "scatter plot" or "scattergram." See if you can find clusters of feature points that might indicate different types of patterns. If more than one type is found, plot examples of each waveform.

---

[1]Abnormal electrical patterns are associated with so-called "ectopic beats." Common examples are "premature ventricular contractions" and "premature atrial contractions". In most cases ectopic beats are benign, but some ectopic beats may indicate a greater risk of developing serious arrhythmias.

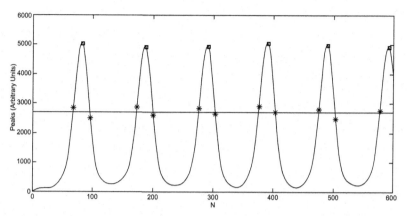

FIGURE 10.4 The curve and markers produced by the QRS detection routine `qrs_detect.m`. The smooth curve is the result of applying three filters: a band-pass filter, a derivative filter applied to the absolute value, and a moving average smoothing filter. The horizontal line is a threshold, halfway between the minimum and maximum of the smooth curve. The * points indicate where the curve crosses the threshold in the upward and downward direction and the *square points* indicate the peaks. The location of these peaks is the same as the location of the R-wave peak except for a small constant shift due to the phase characteristic of the filters.

Solution, QRS detector: Probably the most challenging component of this example is identifying the R-wave peaks. The routine `qrs_detect.m` begins with a filter strategy developed by Pan-Tompkins: a triple filter consisting of an initial band-pass filter, then a derivative filter applied to the absolute value of the signal, and finishing up with a moving average low-pass filter. (Because units are arbitrary, we do not bother to scale the derivative coefficients by $1/T_s$.) The result when applied to a typical section of ECG signal is the smooth curve shown in Figure 10.4. The rest of the routine identifies the location of the R-wave peaks. (There is a small time shift due to the phase properties of the various filters.) Finding these peaks could be done in a number of ways. Here routine `find_peaks.m` (not shown) detects when the filtered curve crosses a threshold chosen midway between the maximum and minimum values, the * points, Figure 10.4. The routine then finds the location of the maximum value between these points, shown as square points in Figure 10.4. The location of these peaks after correcting for the shift due to filtering is the output of the routine. This fairly simple approach works well on the relatively noise-free data used in this example.

```
function qrs_peaks = qrs_detect(ECG_data)
% Function to locate R-wave peaks in the ECG data
%
fs = 100;
% Initialize filter coefficients
wn = [8 16]/(fs/2);              % QRS bandpass filter freq
order = 4;                        % QRS bandpass filter order
[bn an] = butter(order,wn);      % Bandpass QRS filter coefficients
ma = ones(20,1)/20;              % Moving average filter coefficients
b = [1 0 0 0 0 0 -1];           % Derivative filter coefficients.  Skip = 3
% Triple filter ECG data
```

```
temp = filtfilt(bn,an,ECG_data);      % Bandpass filter
temp = abs(filter(b,1,temp));         % Absolute value of derivative
peaks = filtfilt(ma,1,temp);          % Moving average over 200 msec
thresh = (max(peaks) + min(peaks))/2;    % Threshold to find peaks
qrs_peaks = find_peaks(peaks,thresh);    % Find peaks using threshold crossings
qrs_peaks = qrs_peaks - 5;            % Apply correction factor.
```

The main program isolates the activity of each beat using the R-wave peak and the time parameters specified: 0.25 s before and 0.5 s after the R-wave peak. Then the two features, RMS and mean, are easily determined for each isolated segment and stored. After all the beats are analyzed, the two features are displayed, one against the other. This plot is examined for clusters of feature points indicating qualitatively different behaviors.

```
% Example to demonstrate use and analysis of ECG signals
%
load ECG_10min;                % Load data
fs = 100;                      % Sampling frequency
N = length(ECG);               % Data length
qrs = qrs_detect(ECG);         % Detect QRS peaks
N_qrs = length(qrs);           % Number of QRS peaks in data
%
% Isolate single beats and measure two features
P_wave = round(0.25*fs);       % P wave sample range: 0.25 sec
qrst_wave= round(0.5*fs);      % QRS-T sample range: 0.5 sec
for k = 1:N_qrs
  ix = qrs(k);                 % Get peak location
  i1 = max([1,ix_wave]);       % Limit indices to be
  i2 = min([N,ix+qrst_wave]);  % between 1 - N
  y(k,:) = ECG(i1:i2);         % Isolate a beat
  f1(k) = sqrt(mean(y(k,:).^2)); % Feature 1; RMS
  f2(k) = mean(y(k,:));        % Feature 2; mean
end
plot(f1,f2,'k*');              % Plot features
    ........label axes........
```

Results: The scatter plot of the two features (one feature of a beat plotted against the other) is shown in Figure 10.5. It appears that there are two primary classes of beats: those with relatively small RMS values and those with a wide range of RMS values and generally higher mean values. We will call the former "cluster 1" and the latter "cluster 2." It is likely that cluster 1, which includes the majority of beats, represents normal beats, whereas cluster 2 represents some types of unusual or ectopic beats.

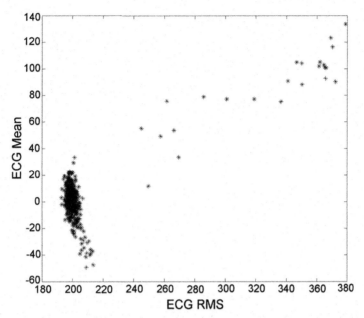

FIGURE 10.5   The scatter plot of two features measured on each beat of a 10 min ECG signal. The majority of data points cluster in the lower left corner (cluster 1), but a few are distributed across a range of higher mean and RMS values (cluster 2). Examples of beats from these two ranges are shown in Figure 10.6.

Because we saved the isolated segments, it is easy to plot some examples from the two groups. We can separate the two groups using only the RMS feature, for example, above and below an RMS of 230. The following code plots eight examples from each group superimposed using different line styles.

```
figure; hold on;              % New figure
n1 = 1;                       % Counter of group 1 plots
n2 = 1;                       % Counter of group 2 plots
t = (1:length(y(1,:)))/fs;    % Time vector for plotting
%
for k = 100:N_qrs             % Go through all records
  if f1(k) < 230 && n1 < 8    % plot only first 8 records
    plot(t,y(k,:),'k');       % Plot group 1
    n1 = n1 + 1;              % Increment group number counter
  end
  if f1(k) > 230 && n2 < 8
    plot(t,y(k,:),':k','LineWidth',2);    % Plot group 2
    n2 = n2 + 1;             % Increment group number counter
  end
end
      ........label axes........
```

The examples are shown in Figure 10.6. The examples from Cluster 1 (closely grouped lines) are very similar and follow the typical ECG pattern. Beats classified as Cluster 2 (widely varying lines) show a very different electrical pattern with a smaller QRS followed by a large, somewhat slower

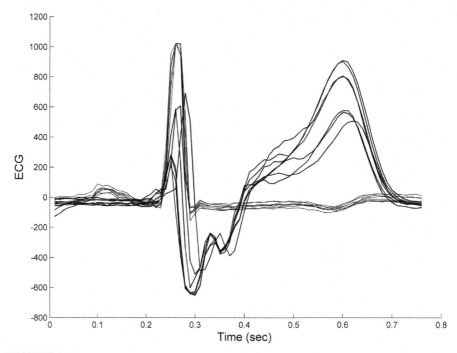

FIGURE 10.6 Examples from the two clusters found in the scatter plot of Figure 10.5. Cluster 1 samples (*closely spaced lines*) are very similar and follow a typical ECG pattern. Cluster 2 samples (*varying lines*) are more variable and show a large wave following the QRS complex.

wave. Again these unusual patterns are termed ectopic beats and are often the most diagnostically useful.

Example 10.3 touched on several important topics including QRS detection, feature extraction (a branch of pattern recognition), and classification based on feature clusters (known as "cluster analysis"). Each of these topics has been extensively developed and has made, and will continue to make, valuable contributions to biomedical engineering. Although more sophisticated approaches exist for each of these topics, Example 10.3 demonstrates the basic concepts in a real-world application. Other applications and expansion of these three topics are explored in the problems.

10.3 SIGNAL NONLINEARITY

Because nonlinear systems generate nonlinear signals and all biological systems contain some nonlinearity, you might expect all biological signals to contain some nonlinearity.

Often the nonlinearity is small and so unimportant that it can be ignored, but sometimes the most important diagnostic information is in the nonlinear behavior. Tests for nonlinearities search for two general features related to signal complexity: fractal behavior and long-term interactions. The diagnostic utility of these features drives development of new methods to identify these behaviors quickly and reliably, as they are often clouded by physiological and measurement noise.

Fractal behavior or fractal scaling methods identify similarities within a signal when viewed at different timescales or resolutions. Fractal behavior reflects the activity of multiple biological processes influencing a signal. Each process is likely to have a different timescale that leads to fractal-like behavior. For example, human heart rate is influenced by several different biological feedback loops, all having a different delay. Biological processes with long feedback loops lead to long-term signal interactions and those with short delays lead to short-term signal modifications. The health of these various physiological processes can be assessed by analyzing their impact on the signal. Fractal properties and long-term interactions have been observed in ECG and EEG signals as well as recordings of human movement such as walking gait, running gait, standing posture, and eye movements.

There are a number of popular approaches to identifying fractal behavior and long-term interaction. "Correlation dimension" and the "Lyapunov exponent" analyze a signal construct known as the "phase trajectory" in "state space" or "phase space". These approaches attempt to determine if the state space trajectory has a fractional dimension. Entropy-based methods that measure a signal's information content are used to identify both long- and short-term interactions. An approach termed "multiscale entropy" (MSE) estimates signal entropy over different timescales. These entropy estimates can be used to identify fractal scaling but can also reflect long- and short-term correlations. Another popular nonlinear method is "detrended fluctuation analysis" (DFA) that removes trends over varying timescales to differentiate long- and short-term interactions. There are an ever-increasing number of methods that search for signal nonlinearities and correlations. The three methods covered here, correlation dimension, MSE, and DFA, give a good introduction to the challenges and techniques of measuring signal complexity and nonlinearity.

10.3.1 Fractal Dimension

Fractal data show similar behavior over a wide range of scales. The classic example is a coastline that presents more or less the same random pattern over a wide range of scales, i.e., similar patterns are found when it is viewed from a high altitude, a low altitude, sea level, or even microscopically. Fractal dimension methods search for fractal behavior using a seemingly bizarre concept that a trajectory can have a fractional dimension. Although a plane is two dimensional and a sphere is three dimensional, there are trajectories that can

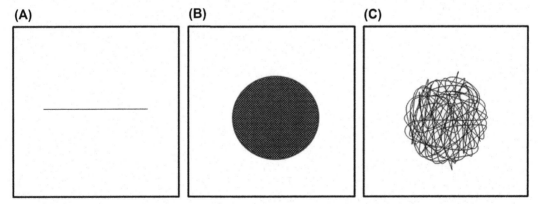

FIGURE 10.7   Three objects used to illustrate the concept of fractional dimension. (A) A one-dimensional line. (B) A two-dimensional disk. (C) A one-dimensional line but one that is so complicated that it fills almost as much space as a two-dimensional disk. To quantify this space-filling characteristic, we could extend the concept of dimension and say this object has a fractional dimension, somewhere between 1 and 2.

be, say, 2.5 dimensional. For example, the line in Figure 10.7A is a one-dimensional object (at least, in theory), whereas the disk in Figure 10.7B is clearly a two-dimensional object. But the object in Figure 10.7C is also just a line, so it is technically one dimensional, yet it has almost the appearance of a two-dimensional object (you have to use a bit of imagination here). The concept of fractional dimension is an extension of the traditional definition of dimension to include the idea of "space filling." Clearly the line in Figure 10.7C fills more space than the line in Figure 10.7A and to quantify this space-filling characteristic, we might assign the object in Figure 10.7C a fractional dimension, say, somewhere between 1 and 2.

Given this concept of fractional or fractal dimension, the question becomes what sort of objects would benefit from quantifying their space-filling characteristics. This brings us to the next new concepts of state variables and state space.

10.3.1.1 *State Variables and State Space*

In Chapter 5 we learned how to predict the behavior of any LTI system using convolution and the impulse response. "State variables" provide an alternative time-domain description of system behavior. State variables are internal to the system and can be used to describe internal as well as external behavior. They are more general in that they can be used with nonlinear systems and systems that are not time invariant. State variables are not unique; they are just some minimum number of variables that are required to describe the system completely. "State space," also called "phase space," is the space that defines the state variables as they move through time.

If you have a model of the system, the state variables can be derived from the model description, either based on internal model transfer functions or electrical or mechanical components as described in Chapter 12. Of course, if you have such a description of the system, you could use simulation (Chapter 9) to find the internal signals and would not need state variables. Simulation also works for systems that are neither linear nor time invariant. So although we will not actually solve for system behaviors using the state space approach, the concept is important in nonlinear signal analysis because it is the state space curves that are analyzed for their space-filling characteristics.

Although state space variables are based on a system's internal variables, they can also be found from external signals. Because the state variables are not unique, all we need is a set of variables that completely describe the system. In general, when we use external signals, the state space variables we come up with are not the same as those we would find internally. The important point is only that they relate to, in some way, the internal signals. This is the key to nonlinear signal analysis using state space concepts: if there is some internal nonlinearity whose presence is buried in the output signal, we might be able to find it by searching for special characteristics in the state space variables. This applies even when the state space trajectory is reconstructed solely from the output signal. The special characteristics we look for are fractal dimensions in the system's state space.

If the system is second order (or less), the state space variables can be obtained directly from the output signal: they are the output signal itself and its first derivative. In this case, the state space, or phase space, is two dimensional and is called the "phase plane." It is easy to plot the phase plane for any second-order system as shown in the next example.

EXAMPLE 10.4

Plot the time course and phase plane of a second-order system having the transfer function $\frac{100}{s^2+6s+100}$ in response to a unit step response. The step response of this transfer function is

$$X(s) = \frac{1}{s}TF(s) = \frac{100}{s(s^2 + 6s + 100)} \tag{10.4}$$

Solution: Applying the methods developed in Chapter 7, we note that from the transfer function, $\omega_n{}^2 = 100$, so $\omega_n = 10$, and $2\delta\omega_n = 6$, so $\delta = 6/20 = 0.3$. Since $\delta < 1$, the system is underdamped. Referring to the table of Laplace transforms, the solution to the transfer function equation for complex roots is as follows:

$$x(t) = 1 - \left[\frac{e^{-\delta\omega_n t}}{\sqrt{1 - \delta^2}} \sin\left(\omega_n\sqrt{1 - \delta^2}\,t + \theta\right)\right] \quad \theta = \tan^{-1}\left(\frac{\sqrt{1 - \delta^2}}{\delta}\right)$$

$$x(t) = 1 - \left[\frac{e^{-3t}}{0.95} \sin(9.5\,t + \theta)\right] \quad \theta = \tan^{-1}\left(\frac{0.95}{0.3}\right)$$

We solve this equation using MATLAB and plot dx/dt versus x. Because the signal is computer generated and is noise free, we can use MATLAB's `diff` routine to take the derivative. To get the proper derivative scale, we divide by T_s (or multiply the result by f_s). (Because the output of `diff` is one sample shorter than the input, we zero-pad one sample.) For the solution, assume $f_s = 1$ kHz and solve for t from 0 to 10 s. This leads to the following code:

```
% Example 10.4 Plot the time course and phase plane of a second-order system
%
fs = 1000;               % Sampling frequency
N = fs * 10;             % Solve for 10 sec
t = (1:N)/fs;            % Time vector
theta = atan2(.95,.3);   % Solve for theta
x = 1 - (exp(-3*t).*sin(9.5*t+theta))/0.95;    % Solve equation
dx_dt = [diff(x) 0];     % Take derivative
subplot(1,2,1);
  plot(t,x,);
...... labels and axis limits......
subplot(1,2,2);
  plot(x,dx_dt);
...... labels, zero line, and limits.......
```

Results: The time plot generated by this code is shown in Figure 10.8A and the phase plane in Figure 10.8B. The inward spiraling curve in Figure 10.8B is typical of second-order overdamped systems. Note that this curve, like all state space curves, shows how the variables evolve over time but, unlike Figure 10.8A, is not itself an explicit function of time. The spiral winds down to a point at values $dx/dt = 0.0$ and $x = 1.0$. It is as if the phase trajectory is attracted to this point. For this reason, the point (which is the final position in this case) is called an "attractor." The phase trajectory of this system will be attracted to this point regardless of the initial condition. This is a "fixed point" attractor, but attractors can also be lines, surfaces, or even regions of phase space. For some

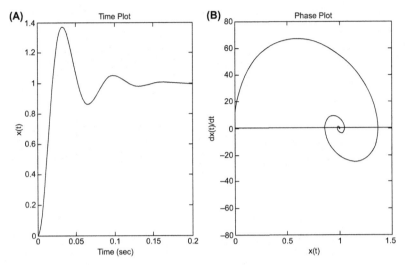

FIGURE 10.8   (A) The time plot generated by an underdamped, second-order system. (B) The phase plane produced by the time response in A. The inward spiral is typical of all such underdamped systems. The final point at $(1, 0)$ is called the system's attractor. All trajectories for this system end up at this point, irrespective of their starting point.

systems, the phase trajectory never settles but continues to orbit; a so-called "limit cycle" attractor. An example of this is the van der Pol oscillator that was simulated in the last chapter. Its phase trajectory is explored in one of the problems. An attractor can even have a fractal structure, and it is called a "strange attractor." Strange attractors are often associated with chaotic systems. The phase trajectories developed by these attractors circle around the attractor endlessly but never repeat the same path. For an awesome demonstration of a strange attractor's phase trajectory, type `lorenz` in MATLAB.[2]

[2]Entering lorenz in MATLAB calls up an animated demo. Once the start button is pushed, the demo traces out the Lorenz phase space trajectory in real time. It has no inputs or outputs.

We now know how to find and plot the state variables of a second-order system, but what about higher-order systems? Finding the state variables of a higher-order system from only the output response uses a trick called "delay embedding" or just "embedding" and is the topic of the next section.

10.3.1.2 Delay Embedding

If all the internal state variables are coupled, they will all have an influence on the output. Delay embedding is a technique for constructing a set of likely state variables from just the output. In delay embedding, delayed versions of the measured signal are used as substitutes for the hidden internal signals. The phase space can be constructed by plotting these delayed signals against the original signal. From a single signal, a collection of signals is created, each being a delayed version of the original.

$$y(t, \tau) = [x(t); x(t - \tau); x(t - 2\tau); \cdots x(t - m\tau)] \tag{10.5}$$

where τ is the constant that sets the basic delay and m is the number of additional signals created to represent the state variables. The constant m is known as the embedding dimension because the phase trajectory, which is constructed by plotting these signals against one another, now requires m dimensions.

A classic theorem by Takens states that delay embedding accurately generates the state variables, provided τ and m are appropriately chosen. Although there is some guidance in choosing these two parameters, it usually comes down to trial and error. For us, delay embedding is just a means to an end. We use it to construct a state space trajectory that is then used in conjunction with correlation dimension analysis to determine if the trajectory has a fractal dimension. A fractal dimension is a characteristic of a strange attractor and of a fractal signal. The two embedding parameters may have to be adjusted to get good results in the subsequent correlation dimension analysis.

When it comes to setting the basic delay, τ, Takens originally stated that any value would do, but with real data, there is a preferred range. If τ is too small, the constructed signals are too much alike, and if it is too large, they have no relationship to one another. Biological signals, in particular, usually have a timescale over which the samples are relevant. For example, if we assume that each beat in the ECG is independent, then a delay corresponding to several cardiac cycles would be appropriate. One method to estimate a value for τ is through the autocorrelation (Section 2.4.5), specifically, selecting the time lag where the autocorrelation function has its first minimum or reaches zero. Another approach uses mutual information, a similar function, in the same manner. Here autocorrelation is used, keeping in mind that it offers only a best guess.

Takens has come up with a rule for setting the embedding dimension, m, specifically

$$m \leq 2D + 1 \tag{10.6}$$

where m is the embedding dimension and D is the dimension of the system, in our case the system order. The problem is that we rarely know the system order or dimension. Of course we could figure that out from a mathematical description or model of the system, but if we had that we could just look at the system to see if it had any nonlinear components. Some tests have been developed such as the false nearest-neighbor analysis or single value decomposition, but these usually fail when applied to real signals. The maximum embedding dimension, m, is often limited by the length of the data, so a reasonable range of values for m can be evaluated by trial and error. The next example reconstructs the phase space of the electrical activity of the heart using the ECG signal.

EXAMPLE 10.5

Construct the phase space trajectory for the heart rate data obtained from a 30 min ECG signal. To get this signal, the peak of the R-wave was found using routine `qrs_detect` as in Example 10.3. The difference between R-wave peak indices was determined using MATLAB's `diff` routine, then divided by f_s to convert these sample differences to time intervals in seconds. The interpolation approach presented in Example 4.4 was then applied to the RR-interval data to transform it into evenly spaced time samples at 10 Hz. The heart rate signal is found as variable `hr` in the file `hr_data`. We would like to set the delay, τ, such that the delayed signals are more or less independent; a delay equivalent to three to five cardiac cycles should be sufficient. Because $f_s = 10$ Hz, we choose a delay of 40 samples (a typical cardiac cycle is around 1 s), but, in fact, a wide range of delays would work. Pragmatically, we make $m = 3$, so we can plot the trajectory on a 3D plot. In fact, previous work in heart rate analysis has suggested 3 is an appropriate value for m.

Solution: We first write a routine to perform embedding, then plot the resulting signals. The routine delay_em.m takes in the original signal along with constants τ and m. It outputs an array containing the delayed signals. We make the length of the output signals as long as possible, which is the input signal length minus the maximum shift, $\tau(m - 1)$.

```
function y = delay_emb(x,m,tau)
% Function to perform delay embedding on input signal x.
%
L = length(x)-(m-1)*tau;    % Length of output signals
y = zeros(m,L);             % Initialize output array for speed
for k = 1:m
    a = (1+tau*(k-1));      % Calculate first
    b = (a+L-1);           % and last indices
    y(k,1:L) = x(a:b);     % Build array in m dimensions
end
```

The main program loads the data, embeds it in a 3D space, and then plots the trajectory in 3D using MATLAB's plot3.

```
% Example 10.5 Program to calculate and display the phase trajectory
% of an ECG signal
%
load hr_data;              % Get data
hr = hr-mean(hr);          % Remove mean
m = 3;                     % Define parameters
tau = 90;
y = delay_emb(x, m, tau);            % Embed signal
plot3(y(1,:), y(2,:), y(3,:));       % Plot trajectory
```

Results: The phase trajectory based on state variables reconstructed from the heart rate signal is shown in Figure 10.9. It almost fills the three dimensions showing that the order or dimension of the heart rate phase space trajectory is close to three. It is difficult to tell from this trajectory if nonlinearity is involved, but then this reconstruction is just one step in the evaluation of nonlinearity. Further analysis of the trajectory is required to determine if the system contains nonlinearities. Two popular methods that analyze the phase trajectory for nonlinear behavior are the Lyapunov exponent and correlation dimension analyses. The latter is covered in the next section.

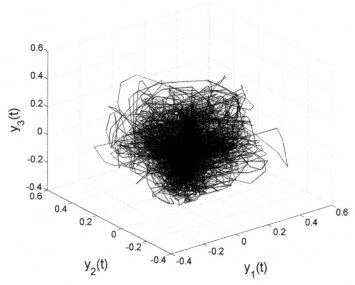

FIGURE 10.9 The phase trajectory obtained by embedding a 30 min segment of heart rate data into three dimensions. The trajectory densely fills the three dimensions.

10.3.1.3 *Correlation Dimension*

The correlation dimension describes the space-filling characteristics of the phase trajectory. Nonlinear chaotic systems usually have phase trajectories that fill phase space, (i.e., state space) such that the correlation dimension is fractional. A fractional correlation dimension suggests that the trajectory's attractor is fractal (i.e., a strange attractor) and that the system is chaotic.

The most popular method for estimating the correlation dimension is the correlation sum (or correlation integral if the data are continuous) that is based on the algorithm developed by Grassberger and Procaccia. Basically the Grassberger–Procaccia algorithm measures the relative relationship between the points of the phase space trajectory and describes the probability that any two points are close together. The proximity of phase space points is clearly related to how extensively the trajectory fills the space. To accomplish this measurement, the Grassberger–Procaccia algorithm measures how the number of points within a certain distance, r, varies as r increases. Specifically, it finds the distance between all two-point pairs on the trajectory and counts the number of pairs that are closer than r. The distance between any two points in a multidimensional phase space is as follows:

$$d = \sqrt{\left(\overline{x_i^2} + \overline{x_j^2} \right)} = \|x_i + x_j\| \tag{10.7}$$

where $\overline{x_i}$ and $\overline{x_j}$ are two vectors in a multidimensional space that point to the location of the two trajectory points, x_i and x_j. Measuring the distance between two points in a multidimensional space can be easily done using the matrix "norm" operator that is indicated by the double vertical bars, $\| \ \|$. (Of course, MATLAB has a routine that performs the norm operation.) The distance, d, in Equation 10.7 is known as the "Euclidian distance."

To count up how many pairs of points fall within a given distance, r, we subtract the distance between two points, d, from r and count the number of subtractions that are positive using the Heaviside operator, $\Theta(x)$. The Heaviside operator is 1 for $x > 0$ and 0 otherwise. If we then take the sum over all point pairs, we get a count of the number of point pairs that were closer than r.

$$\text{Number closer than } r = \sum_{i=1}^{N} \sum_{j=1, j \neq i}^{N} \Theta(r - d) = \sum_{i=1}^{N} \sum_{j=1, j \neq i}^{N} \Theta(r - \|x[i] - x[j]\|)$$

For the Grassberger–Procaccia algorithm, we actually want the ratio of point pairs that are closer than r with respect to the total number of point pairs, so we need to normalize the aforementioned equation by $2/N(N-1)$. This gives the correlation sum equation:

$$C(r) = \frac{2}{N(N-1)} \sum_{i=1}^{N} \sum_{j=1, j \neq i}^{N} \Theta(r - \|x[i] - x[j]\|) \tag{10.8}$$

To estimate the correlation dimension from the correlation sum (Equation 10.8), we should examine how $C(r)$ changes as r gets very small. If the decrease follows a power law, i.e., $C(r) \backsim r^D$, then D is the correlation dimension. Stated mathematically, the correlation dimension, D, is defined as follows:

$$D = \lim_{r \to 0} \frac{\ln C(r)}{\ln r} \tag{10.9}$$

The problem is that as r gets very small the number of point pairs decreases and the estimate is not very accurate. It would be better if more of the $C(r)$ function was involved in estimating the correlation dimension. So what is commonly done is to plot $C(r)$ against r on a log–log plot and to fit a straight line to the segment of the curve where r is small. D is then the slope of this line. Practically, Equation 10.9 is applied without the limit operator. A plot of this equation is searched for a linear region known as the "scaling region" because it determines the dimension. The next example takes a single signal, embeds it in multiple dimensions, finds the correlation sum, and then plots $\ln C(r)$ against $\ln r$ (Equation 10.9) and searches for a scaling region to estimate the correlation dimension.

EXAMPLE 10.6

Estimate the correlation dimension of the signal x in the file `ex10_6_data.mat`. This signal is obtained from one axis of the Lorenz phase space trajectory (see footnote 2). Because the correlation dimension analysis does not involve time, we do not need to know the sampling frequency. We can assume that the dimension of the system that produced this signal was ≤ 3.[3]

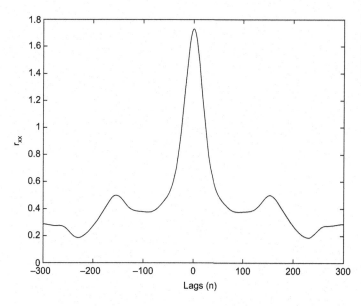

FIGURE 10.10 Autocorrelation function of the signal x before embedding. The first minimum occurs at a lags of ±100. We use that value for the delay in the delay embedding operation.

Solution: After we load the signal, the first step is to embed the signal in three dimensions using the routine `delay_emb`. To select the value for τ, we take the autocorrelation function of our signal (Figure 10.10).

The first minimum occurs at lags of ±100, so we use that value for the delay, τ, in the delay embedding routine. The original signal is embedded in three dimensions, as we are using a priori knowledge that the system is less than third order. (Otherwise we would have to try different embedding dimensions and search for values that had led to a linear scaling range in the correlation sum.)

We then plot out the phase space trajectory, as this helps us determine a range of radius values, r. The correlation sum is computationally intensive, so evaluating the sum over a large number of radius values can be quite time consuming. The phase space trajectory is shown in Figure 10.11 and

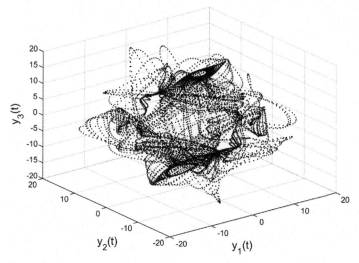

FIGURE 10.11 The phase trajectory of the signal under evaluation after being embedded in three dimensions. The trajectory ranges between ±20 in all three dimensions.

ranges between ± 20 in all three dimensions. To estimate the system dimension, we need to probe values of r that are small (recall, ideally $r \to 0$). Given the scale of the phase space, we should evaluate radii much less than 1.0. Accordingly, we select a range of radii between 0.05 and 0.5.[4] If we make the increment 0.01, this leads to 46 different values for r and 46 correlation sums that require 10–15 min on a typical personal computer. The code up to this point is as follows:

```
% Example 10.6.  Program to calculate the correlation dimension of a
%  system from a single signal.
%
load ex10_6_data.mat            % Load data (Lorenz attractor)
[rxx,lags] = crosscorr(x,x);    % Calculate autocorrelation function
plot(lags,rxx);                 % and plot
xlim([-300 300]);               % Limit x-axis of plot
tau = input('Input value for tau: ');      % tau = 100, from autocorrelation
m = 3;                          % Embedding dimension (given)
y = delay_emb(x,m,tau);         % Construct the phase space trajectory
figure;
plot3(y(:,1),y(:,2),y(:,3));    % Plot trajectory
    ........labels and grid.......
r = (0.05:0.01:.5);             % Define range of r
```

The next step is to compute the correlation sum. This is done in the following routine corr_dim.

```
function Cr = corr_dim(y,r)
%
N = length(y(:,1));        % Number points per dimension
for k1=1:length(r)         % For each r count number of close neighbors
  for k=1:(N-1);           % For number of points count close neighbors
    repl = repmat (y(k,:),N,1);      % Copy y(k,:) into N rows
    dist = sqrt(sum((y-repl).^2,2)); % Distance between y(k,:) and all other y
    dist(1:k) = [];                  % Eliminate self-match
    nu_inside(k)= sum(dist < r(k1)); % Sum number of inside points
  end
  Cr(k1)=sum(nu_inside);             % Total inside points for this r
end
Cr=2*CR/(N*(N-1));                   % Normalize count
```

The correlation sum is calculated using two loops. The outer loop iterates the test radius r, whereas the inner loop totals the number of point pairs closer than r. For any given point, the distance to all other points is calculated using a MATLAB trick to improve speed. The three variables that define the test point in 3D space are duplicated into N rows using the routine repmat. Now the number of test points is the same as the total number of points on the trajectory, and the distance between all the point pairs can be determined in a single vectorized command. Distances less than r are counted and the inner loop repeats this calculation for all trajectory points. This strategy results in an additional self-match, which is eliminated. The counts for each point are then totaled and normalized as in Equation 10.8.

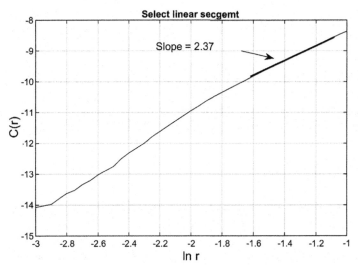

FIGURE 10.12 A plot of ln(C), the correlation sum versus ln(r), the test radius. The segment of this curve identified as most linear is shown in bold. The slope of this linear segment is an estimate of the correlation dimension: 2.37. This fractional dimension indicates that the attractor in this system is fractal and that the system is chaotic.

Returning to the main program, the ratio of ln(C) to ln(r) is plotted and examined for a scaling region, a section of the plot that is linear. A linear segment is identified interactively as a thin line superimposed on the plot. The slope of the linear region is displayed as the estimate of correlation dimension. The remaining code of Example 10.6 is as follows:

```
r = (0.05:0.01:.5);              % Define range of r
Cr = corr_dim(y,r);              % Calculate correlation sum
plot(log(r),log(Cr),'k'); hold on;  % Plot correlation sum
 ........labels, grid, and title.......
[x,y] = ginput(2);               % Get linear segment interactively
plot(x,y,'k','LineWidth',2);     % Plot linear segment as bold line
slope = (y(2) - y(1))/(x(2) - x(1))  % Output correlation dimension
```

Results: The plot ln(C) to ln(r) is shown in Figure 10.12. The entire curve is fairly linear, but the segment identified as most linear is shown in bold. In some of the problems, we take the derivative of ln(C) to help identify the flattest region. The slope of this segment is 2.37, a fractional dimension. This suggests the system has a strange attractor that is fractal and that the system behavior is chaotic. This is in the same range as that found in previous studies of this system known as the Lorenz attractor (see footnotes 2 and 3).

[3]This signal is known to be nonlinear because it is a component of the Lorenz phase space trajectory. The Lorenz phase space features a strange attractor. The Lorenz attractor is a well-studied system that is defined by three coupled first-order differential equations. The Lorenz attractor is known to have a fractional dimension between two and three as we find in this example.
[4]Because the natural log of the correlation sum is plotted against the natural log of r (Equation 10.9), it is common to specify r in terms of an exponential. For example, exp(−3:−0.5:−1) would give approximately the same range of r as in the current program. The approach used in this example, defining r as a standard sequence of numbers, is easier to appreciate.

The correlation dimension approach to identifying nonlinear behavior, specifically chaotic behavior, has a number of problems. The embedding dimension often must be chosen without much guidance. Identifying the scaling region, the linear segment of the correlation sum versus radius curve is somewhat subjective; however, this could be made more rigorous by quantifying the fit to a straight line. However, the biggest concern is that it requires relatively noise-free data, as it can easily be misled by noise. The influence of noise is illustrated in one of the problems. In the next section, we turn our attention to approaches based on information theory that attempt to estimate the entropy of a signal and how this entropy changes with scale.

10.3.2 Information-Based Methods

In this section, we examine approaches to detect nonlinearities and correlations in a signal based on the concept of entropy. In physical systems, entropy is a measure of disorder. In 1948, Claude Shannon applied this concept to signals, translating system disorder in physics into measures of signal unpredictability in signal analysis. Unpredictability is related to the information in a signal: a signal that is very regular, hence predictable, has less information than one that is unpredictable. For example, a sine wave of constant amplitude and frequency is completely predictable and carries no information.[5] Unfortunately, this relationship between entropy and information content breaks down for noise: noise is completely unpredictable, so it should have high entropy, yet like a perfectly predictable signal it carries no information.

Signal entropy has a formal definition based on probabilities, but basically entropy is related to the number of states a signal can have: signals that are very regular have a limited number of states and therefore have low entropy. When applied to signal analysis, efforts to quantify entropy search for, or describe, patterns in the signal. Here we discuss two such methods: "sample entropy" and "multiscale entropy".

10.3.2.1 Sample Entropy

Sample entropy attempts to estimate the amount of entropy, or uncertainty, of information in a signal by searching for, and measuring, repeating patterns in the signal. Sample entropy is one of the more popular methods to measure the uncertainty; it provides a single measurement based on the number of repetitions of a pattern. The basic idea is that signals that have a lot of repeated patterns are more predictable than signals that have very few repeated patterns. Sample entropy uses counts of patterns that more or less match to quantify uncertainty.

To count the number of matching patterns, we first identify a sequence of data samples to use as a "template." These template sequences are generally short, usually two or three samples. We then compare this template with all other data sequences of the same length. If the template and a sequence match, they are added to the count of matches. To count as a match, the difference between the template samples and the sequence samples must be less than some maximum error. In Figure 10.13 a two-sample template is shown that matches two other two-sample sequences in the data set.

---

[5]You could turn the sine wave off and on to transmit information, but then it would not have constant amplitude or frequency.

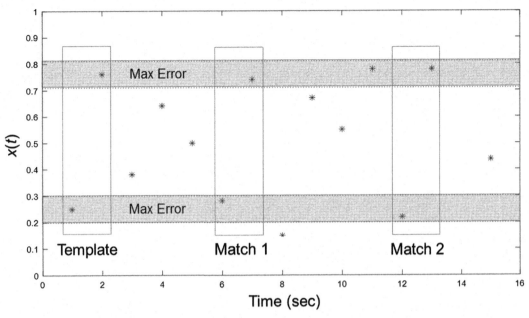

FIGURE 10.13 A two-point template on the left matches two other two-point sequences in this data set as indicated by *rectangles*. The maximum error for the data points to be considered a match in this example is ±0.05.

In the determination of sample entropy, two match counts are determined: one for a template m samples long and the other for a template $m + 1$ samples long. The sample entropy is then defined as the negative log of the ratio of $m + 1$ point matches to m point matches. It can be stated formally as follows:

$$SampEn = -\log \frac{A}{B} \tag{10.10}$$

where A is the number of $(m + 1)$-point matches and B is the number of m-point matches. Usually $m = 2$ so that A is the number of three-point matches and B is the number of two-point matches. The calculation of sample entropy is shown in the next example.

EXAMPLE 10.7

Find the sample entropy of a data set consisting of 20,000 Gaussian points

Solution: We first develop a routine to count the number of two- and three-point matches to a template. For a template, we use every two- and three-point sequence in the data set. The code for this routine is straightforward. We use an outer loop to establish an index of the template and an inner loop to index the test sequences. For each combination, we find the absolute difference between the first and second sets of samples. Because the signal is one dimensional, we can evaluate the points by subtraction and do not have to use the norm operator (i.e., $\| \ \|$). When counting three-sample matches, we also test the third set of corresponding points. If all two, or three, comparisons produce an absolute difference less than the error, we increment the match counter by 1.0. A special routine is used to count 2 or 3 point matches.

```
function matches = num_matches(x,m,max_err)
% Function to determine number of matches in data set of length m
%
N = length(x);
matches = 0;          % Initialize match counter
for k = 1:N-m         % Template loop
   for j = 1:N-m      % Test point(s) loop
      if k ~= j       % Avoid self matches
        if abs(x(k) - x(j)) < max_err                  % Test first points
          if abs(x(k+1) - x(j+1)) < max_err      % Test second points
            if m == 2 || abs(x(k+2) - x(j+2)) < max_err      % Test third points if
m = 3
              matches = matches + 1;               % Increment match counter
            end
          end
        end
      end
   end
end
```

The routine num_matches is then used by the main program to calculate the sample entropy. After defining the random number data set, the main program calls this routine twice: once to determine two-point matches and again to determine three-point matches. The data are normalized to have a standard deviation of 1.0 and the maximum error is set to 0.15, a commonly used value.

```
% Example 10.7 Calculation of sample entropy on random data
%
N = 20000;            % Number of data points
x = randn(1,N);       % Generate data
max_err = 0.15;       % Maximum error between points
x = (x-mean(x))/std(x);    % Subtract mean, normalize so std = 1
%
B = num_matches(x,2,max_err);      % Determine two-component matches
A = num_matches(x,3,max_err);      % Determine three-component matches
samp_entropy = -log(A/B);          % Calculate sample entropy, Equation 10.10
disp(samp_entropy)                 % and display
```

Result: The output of this program is a single number. For these data, the sample entropy is 2.4727.

Sample entropy gives us a single number that can be used to compare the information content of different signals. Changes in entropy can be of diagnostic value in certain diseases, but often of greater importance are changes in long-term processes that incorporate physiological feedback loops. To identify long- and short-term interactions, we can evaluate sample entropy over different timescales. High entropy levels suggest the influence of a process, so the variation in entropy over different timescales highlights processes that act over different time periods. Measuring entropy at different scales can also show fractal or "self-similar" behavior. Signal scaling is done essentially by "downsampling" the signal in an approach also known as "coarse graining."

10.3.2.2 *Multiscale Entropy*

MSE was developed to analyze the signal properties at different scales (Costa et al, 2005). In MSE, the sample entropy is determined over a range of scales, that is, at different signal resolutions. The sample entropy is then plotted as a function of scale. To view a signal or data set at different scales, MSE uses a technique called "coarse graining" to change the resolution of the signal.

10.3.2.2.1 DATA SCALING THROUGH COARSE GRAINING

Coarse graining uses downsampling, usually in conjunction with an antialiasing low-pass filter, to reduce the resolution of the signal. Reducing signal resolution illuminates processes that act over longer timescales. The finest scale is set by the resolution of the sampled signal; once it is sampled, we cannot examine the signal at a finer scale. However, we can examine the data on coarser, lower resolution scales simply by downsampling the data. Figure 10.14 shows a signal at its finest scale and at lower resolutions achieved by downsampling the original signal. Decreasing the scale enhances the slow (or low-frequency) behavior of the signal, but unlike low-pass filtering, we are not simply attenuating high frequencies; we are actually eliminating high frequency information.

Data scaling or coarse graining uses downsampling but with the addition of a low-pass filter to prevent aliasing. Recall that with discrete data, frequencies greater than $f_s/2$ are folded back and confounded (i.e., they are added in with) the lower frequencies (see Section 4.1.1), a condition known as aliasing. If the data are properly sampled, they will not contain frequencies greater than $f_s/2$. However, once we start downsampling, for example, by removing every other data point, we effectively reduce the sample frequency, so that aliasing can occur. The solution is to low-pass filter the data before downsampling.

Adding a filter before downsampling is not difficult, but the trick is to properly define the filter's cutoff frequency so as to remove potential aliasing frequencies. The cutoff frequency of the filter depends on the amount of downsampling. If we retain every other data sample, we are downsampling by a factor of two and halving the effective f_s. Setting the cutoff frequency to $f_s/4$ attenuates, at least partially, frequencies above the new $f_s/2$. If we keep every third point, we reduce f_s by 2/3, so we now want the cutoff frequency to be $f_s/6$. In general, we want the low-pass filter cutoff frequency to be $f_s/2n$, where n is the downsampling factor. Fortunately, MATLAB filter routines specify the cutoff frequency relative to $f_s/2$ (see Chapter 8 and Section 8.6), so to get the proper cutoff frequency we make $f_c = 1/n$. In the routine to perform coarse graining, we use a sixth-order Butterworth filter and, after filtering, downsample by n.

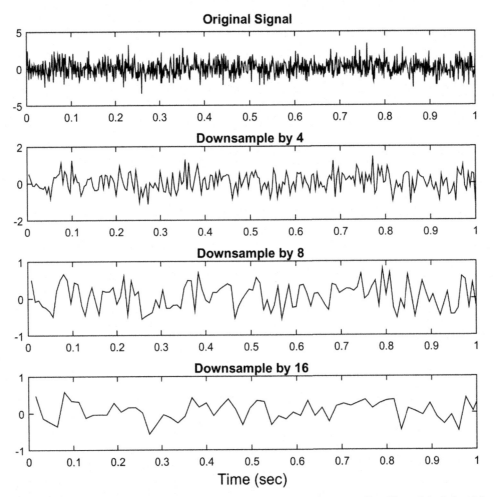

FIGURE 10.14 A signal shown at four different scales achieved by downsampling. The original signal (upper plot) has the finest resolution, whereas the downsampled signals show behavior over longer time frames. In the downsampled signals, the higher frequency information is not just attenuated; it is removed from the signal.

```
function y = coarse_graining(x,n)

% Function performs coarse graining
%
[B,A] = butter(6,1/n);        % Remove freq > fs/2n
y = filtfilt(B,A,x);          % Filter the data
y = y(1:n:end);               % resample by n
```

In the next example, we combine measurements of sample entropy with coarse graining to find the MSE of a random signal.

EXAMPLE 10.8

Calculate the MSE of Gaussian random noise used in Example 10.7. Use the same parameters as in Example 10.7 and calculate the sample entropy for 20 different scales. Plot sample entropy as a function of scale.

Solution: Modify the program in Example 10.7 so that the sample entropy calculation is placed in a loop where each iteration uses data at a different scale. The error tolerance, r, is set at 0.15 times the standard deviation of the original data after the mean is removed.[6] The first iteration should act on the unmodified data. Subsequent iterations should use coarse graining to alter the scale.

```
% Example 10.8 Evaluate MSE on Gaussian random data
%
N = 10000;                % Data length
msf = 20;                 % Maximum scale factor
%
x = randn(1,N);           % Construct random data set
r = 0.15 * std(x);        % Determine error tolerance
for k = 1:msf
  if k == 1
    x_scaled = x;         % Original (unscaled) data
  else
    x_scaled = coarse_graining(x,k);   % Scale data
  end
  B = num_matches(x_scaled,2,r);       % Determine two-component matches
  A = num_matches(x_scaled,3,r);       % Determine three-component matches
  samp_entropy(k) = -log(A/B);         % Calculate sample entropy
  disp(['Scale factor: ',num2str(k)])  % Indicates progress
  end
    plot(samp_entropy);                % Plot result
      .......labels.........
```

Result: Sample entropy as a function of scale is shown in Figure 10.15. An exponential decrease in entropy is seen, indicating that the data become less uncertain as the resolution is decreased. This is understandable because the original Gaussian random data are highly unpredictable and therefore have high entropy. Increasing the timescale reduces the noise and hence the uncertainty.

---

[6]Some researchers believe that the error tolerance, r, should be reset after each coarse-graining operation, as coarse-graining alters the standard deviation of the signal. Costa and colleagues, the originators of this technique, believe this is inappropriate, as the entropy calculation accounts for these changes. Here we use their original approach and do not reset the error tolerance, r, after coarse graining but acknowledge that this is controversial. Problem 18 explores the difference in results that occurs when the error tolerance is reset after coarse graining.

FIGURE 10.15 The change in sample entropy with scale. As the resolution decreases, the signal becomes more predictable, as is reflected in the decreased entropy.

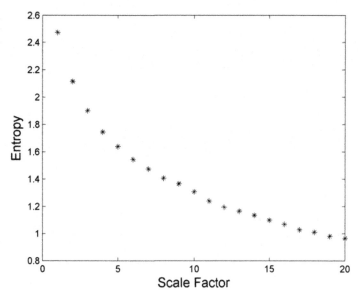

Sample entropy and MSE have been used clinically to evaluate heart rate variability. These measures are often made differentially: the change in sample entropy of MSE evaluated at different time periods is used as a diagnostic indicator. Heart rate is influenced by numerous feedback pathways, so it tends to be less predictable in healthy subjects. Several different diseases, some cardiovascular in nature but others more general such as infection (i.e., sepsis), can reduce the effectiveness of these pathways, so the heart rate becomes more predictable and the entropy decreases. Entropy measurements have shown promise in assessing patient health for both cardiac and systemic disjunctions. One of the problems applies MSE to a signal representing heart rate variability signal.[7]

10.3.3 Detrended Fluctuation Analysis

DFA has been introduced by Peng (1994) to determine if a signal has fractal properties and, if so, to determine the fractal scaling. In a fractal signal, small segments of the signal are similar, in some sense, to larger segments. This similarity could be an actual matching of data but more likely some property of the data such as the variance. The amplitude of the property scales proportionally to the timescaling, but the constant of proportionality is not a constant, but rather a power function. This amplitude and timescaling relationship is stated mathematically as follows:

$$x(st) = s^H x(t) \tag{10.11}$$

---

[7]Heart rate data are usually recorded as a sequence of interbeat intervals, so-called RR intervals, and need to be converted to evenly spaced data using the approach presented in Example 4.4.

where x is the signal, or more generally, some property of the signal such as variance, s is the length of a signal segment, and H is a constant known as the Hurst coefficient. The proportionality constant, s^H, specifies the relationship of the property's value to the timescale. Thus a property of signal segment s is the same as that for the entire signal but diminished by s^H. DFA searches the signal for this relationship and attempts to determine H or rather a factor related to H. The property that is usually evaluated is the signal RMS or variance (essentially the same for long signals). DFA has been shown to be particularly effective at finding long-term correlations in the signal.

DFA has been widely applied to diverse fields, including DNA sequencing, heart rate dynamics, neuron spiking, human gait, and economic time series and also to weather-related and earthquake signals. In heart rate analysis, scaling behavior has been used to identify several different disease states.

To determine if various portions of a signal are statistically similar, we can measure the RMS (or variance, or other similar measurements) of small segments and compare them over time. This is the approach taken in Example 10.1 to determine if a signal is ergodic. If we find variations in our metric segment to segment, as we would expect in nonlinear data, they could be due to influences from nearby processes that have short timescales or influences from distant portions of the signal due to processes with long timescales. If the segments are small, we would expect that most of the influence would be from nearby, short-term processes because long-term processes have less of an influence on short segments; their primary effect would be on longer segments. This is illustrated in Figure 10.16 (upper panel) where two sinusoidal components, one at a lower frequency (dotted line) and other at a higher frequency (dashed lines), have been added to a random signal. If we look at a small segment of that signal (Figure 10.16 (lower left panel)) the higher frequency (i.e., short-term) signal has a significant impact on that signal (in terms of its RMS or variance). The low frequency signal has little effect, although both sinusoids have the same amplitude. We could reduce the influence of the high frequency signal on the short data segment by subtracting a polynomial fitted to the data samples in that signal. In fact, we could just subtract a straight line or linear component as was done in Figure 10.16 (lower right panel). As described earlier in Section 10.2, the linear component is called the "trend," so removing it is called "detrending." If we apply detrending to a longer signal segment, for example, the whole record shown in the Figure 10.16 (upper panel), we will reduce the influence of the long-term component (but not the short-term component). Thus, as we increase the segment size, detrending removes the influence of ever longer-term processes. This is the basic concept behind DFA.

The scaling relationship between amplitude and timescale (Equation 10.11) is only valid for nonstationary signals.[8] To ensure that DFA works on both stationary and nonstationary signals, the signal is integrated to generate what is known as the "signal profile." Integration is a linear process and does not change the fractal characteristics of the signal, but it does convert a stationary signal to a nonstationary signal. If the signal is nonstationary, integration makes no difference, it is still nonstationary. The signal profile is then divided into short segments, and the best straight-line fit for each segments is found based on least-square error (Figure 10.17). This straight line is then subtracted from the data, detrending the data

---

[8]For stationary signals the variance will be the same at all timescales. That was one of the tests that is described in Section 10.2 to see if a signal was stationary.

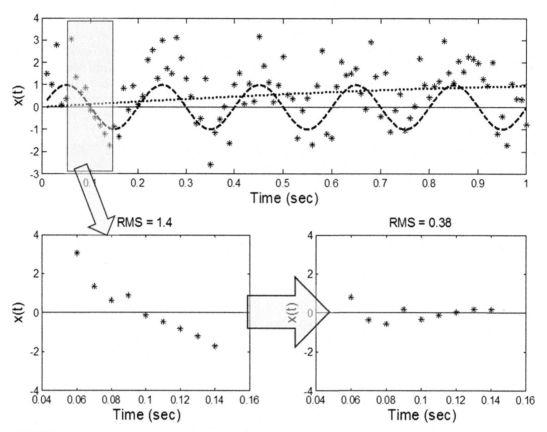

FIGURE 10.16 Illustration of the influence of short- and long-term processes on a short data segment. Upper panel: a random signal is influenced by both a short (dashed) and a long (dotted) sinusoidal process. Lower left panel: a small segment between 0.05 and 0.15 s is isolated from the overall signal and the influence of the short-term sinusoid is evident. Lower right panel: the influence of the short-term process has been reduced by subtracting the linear trend in the signal on the left. Removing the trend is called "detrending." Note the reduction in RMS after detrending.

segment. The RMS or variance of the detrended data segments is then taken. Rather than taking the average of RMS values from each section, it is simpler just to take the RMS of the overall detrended signal. To evaluate signal components having different timescales, the length of the detrended segments is varied. The RMS value represents the fluctuations of the detrended data, and they are plotted on a log–log scale against segment length.

Mathematically, the first step in the DFA, removal of the mean and digital integration, is represented as follows:

$$y_{int}[n] = \sum_{n=1}^{N} y[n] - \overline{y[n]} \tag{10.12}$$

where n is the time index and N is the number of samples in the signal.

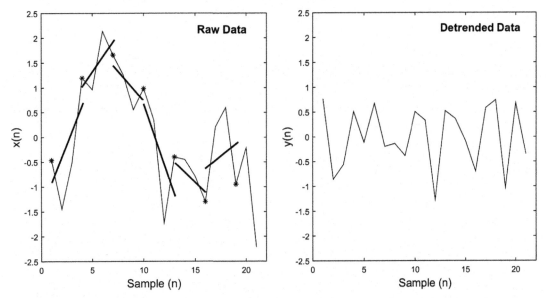

FIGURE 10.17   Detrending a signal. The raw data signal is divided into three-point segments. The * points indicated segment boundaries. Each segment is fitted by the best straight line based on least mean square error. These straight-line segments are subtracted from the raw data, leaving the detrended data shown in the right-hand plot. In DFA, the RMS of the detrended data is determined. These RMS values are determined over a range of segment lengths and plotted on a log–log scale against segment length.

The signal y_{int} is then divided into K segments of length m, and for each segment, a linear term, $y[n] - \overline{y[n]}$, is removed. (It is also possible to subtract out a higher-order polynomial.) After removing the linear trend from each segment, the remaining fluctuations are only those intrinsic signal variations minus the nearby (short-term) influences. The RMS value (or other property) is taken over the entire detrended signal length. These operations can be mathematically summarized as follows:

$$F(m) = \sqrt{\frac{1}{N} \sum_{k=1}^{K} [y_{int}(k) - y_{dt}(k)]^2} \tag{10.13}$$

where m is the segment length and $y_{int}(k)$ is the kth segment from which a linear fit, y_{dt}, has been subtracted. This is performed using different segment lengths, m, and results in a different number of segments, K. As segment length increases, the short-term processes are included in the segment fluctuations, as long-term processes are removed. Because we are looking for an exponential scaling coefficient, the natural log of the RMS value, $\ln(F(m))$, is plotted against the natural log of m. The scaling coefficient related to H is found by searching for a straight-line section in the resultant curve.

A linear region on the log–log plot identifies a range where $F(m) \propto m^\alpha$, where α is a generalization of the Hurst coefficient, H. The exponent α indicates the nature of signal correlations where

$\alpha < 0.5$ indicates anti-correlated data,
$\alpha \approx 0.5$ indicates uncorrelated data,

$\alpha > 0.5$ indicates correlated data, and

$\alpha > 1$ indicates a non-stationary process or an unbounded correlation such as a sinusoid.

The problems examine how different signal features influence the value of α. An application of DFA is shown in the next example.

EXAMPLE 10.9

Find the DFA of a 10,000-sample Gaussian random signal.

Solution: Construct the signal using randn. Integrate the signal using MATLAB's cumsum. Detrending can be achieved using MATLAB's detrend routine as follows:

```
y = detrend(x);     % Detrend signal x
```

where x is the input signal and y is the detrended output signal.[9]

Because the results will be plotted on a log–log scale, we make the segment lengths in terms of powers of 2. The range should be determined by trial and error to search for a scaling region. In DFA it is common to use a minimum segment length of four samples (i.e., 2^2) and, given the data length of 10,000 samples, the maximum segment length should be 2^{13} (i.e, 8192) samples. The MATLAB code then is as follows:

```
% Example 10.9 Example of DFA
%
N = 10000;                 % Data length
X = rand(1,N);             1% Construct signal
m = 2.^(2:13);             % Segment lengths: 14 powers of 2
x = cumsum(randn(1,N));    % Integrated Gaussian random data
for n = 1:length(m);       % Loop for different segment lengths
  seg = m(n);              % Segment length
  K = floor(N/seg)-1;      % Number of segments to detrend
  for k = 1:K
    ix = seg*(k-1) +1;
    y(ix:ix+seg) = detrend(x(ix:ix+seg));  % Detrend
  end
  F(n) = sqrt(mean(y.^2));                  % Take RMS
end
m = log(m);
F = log(F);
plot(m,F,'b');
    .......labels and grid.......
% Use ginput to find the most linear section
[x,y] = ginput(2);                        % Get 2 points on the curve
plot(x,y,'k','LineWidth',2);              % Plot line superimposed
slope = (y(2) - y(1))/(x(2) - x(1))       % Calculate linear slope
```

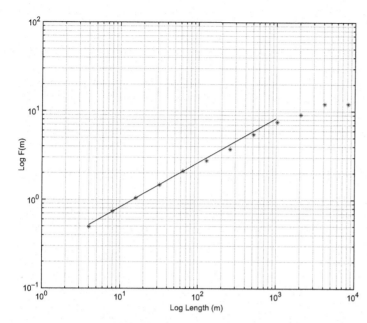

FIGURE 10.18 A plot of the log of detrended RMS signal value versus the log of segment length. The *solid reference line* has a slope of 0.5.

Result: The result is shown in Figure 10.18. Theoretically, Gaussian random data should have a scaling value, α, of 0.5 because it is uncorrelated. The plot of Figure 10.18 includes a reference line with a slope of 0.5. As it is seen here, the linear portion of the curve follows the reference line closely.

[9]There is a way to use detrending on segmented data with a single call, but it is much slower than using a loop to isolate the desired segments. In addition, it is not flexible, so it is not possible to use overlapping segments.

DFA is a technique that is relatively simple but powerful, if you believe that the data may contain signatures of long-term processes, but is dominated by unwanted or uninteresting events at the shorter terms.

10.4 SUMMARY

Stationarity is an important property of signals because all the signal analysis techniques we have studied thus far assume the signals are stationary. For a stationary signal, the basic signal properties of mean and variance do not change over time. For such a signal, the measurement of the mean or variance over only one segment is sufficient to estimate the signal's true mean. A more extreme case of stationarity is ergodicity, where not only are the mean and variance constant, but two higher moments, skew and kurtosis, are also constant over the length of signal. To determine whether a signal is stationary or ergodic requires multiple measurements over different signal segments and this usually entails a large amount of data.

Most biological signals have stationary means (termed ergodic in the mean) but are nonstationary in variance and other moments. We are, after all, creatures of change. In such cases, the most common remedy is to limit signal analyses to smaller segments where stationarity holds. Sometimes signal segmentation is arbitrary, but sometimes the underlying physiology suggests a segmentation strategy. With the ECG, a single heart beat is a natural segment, and an example of beat-to-beat segmentation and analysis of the ECG is presented in Example 10.3. This example introduces some basic concepts of pattern recognition, including feature detection, scattergrams, feature space boundaries, cluster analysis, and classification.

Signal nonlinearities can present analysis problems, as most methods we have studied thus far assume the signal is linear. However, nonlinearities also provide diagnostic opportunities. Current nonlinear analysis techniques attempt to identify fractal behavior and/or long-term correlations. Fractal behavior shows similarities within a signal when viewed at different timescales. This behavior is due to multiple physiological processes, each with a different timescale, acting on the signal. Analyses that seek to identify fractal behavior include correlation dimension, MSE, and DFA. The latter two are also useful for identifying long-term and short-term correlations.

Correlation dimension operates on a signal's phase space (or state space) trajectory to search for a fractal attractor, better known as a strange attractor. If only a one-dimensional signal is available, as is generally the case, a multidimensional phase trajectory can be approximated from delayed versions of the signal. The correlation sum is a measure of the space-filling characteristic of this phase trajectory. Fractal signals fill the space in which they are embedded and exhibit a fractal correlation dimension. Fractal correlation dimension is an indicator of fractal behavior. Unfortunately, the approach is sensitive to, and can be misled by, noise.

MSE grew out of information theory and attempts to estimate the entropy, or information content, of a signal at various resolutions. Information content is closely linked to signal unpredictability, so MSE looks for repeating patterns: the log ratio of two-sample patterns to three-point patterns is taken as a measure of unpredictability. Filtered downsampling is used to change signal scale, as each downsampling creates a signal at a lower scale (i.e., resolution).

DFA looks for fractal scaling and related long-term correlations. In fractal scaling, some signal property such as variance is found to be similar at different signal resolutions. However, the value or amplitude of this property is itself scaled in proportion to the timescaling, not linearly, but as an exponential. DFA measures this property over short signal segments, but first it removes any trend in the segment; that is, any linear change in signal baseline. Some versions even eliminate higher-order components in the segment. The idea is that such changes are due to nearby signal features, so if they are removed, the remaining signal contains only variation intrinsic to that segment. As the segments become longer, the detrending acts to reduce long-term interactions, but the larger segments now include the short-term fluctuations. Because we are looking for an exponential scale factor, the natural log of the RMS value is plotted against the natural log of the segment size. The slope of this curve is related to the exponential, H, known as the Hurst coefficient, and quantifies the scaling of the measured parameter. DFA is a simple technique, but not without controversy as some researchers suggest the approach has an inherent bias. Nonetheless, it is widely applied in biological signal analysis.

PROBLEMS

1. Repeat Example 10.2 using three different window sizes: 10, 60, and 240 s, all with 50% overlap. Plot the three results superimposed, and use a dashed line for the 10-s window and heavier linewidths for the longer data. Note how increasing window length smooths the data but with loss of detail.

2. Follow the approach of Example 10.2 and plot the alpha wave activity for two EEG signals in files `EEG_FP1_F4.mat` and `EEG_FP1_F3.mat`. Each signal is found as variable `val` in their respective files. These signals are taken at the same time from electrodes that were close together. Load both data files and analyze both signals simultaneously as in Example 10.2 using a 20-s window with a 50% overlap. Plot the resulting activities superimposed, but in different colors. Next take the cross-correlation between the two normalized RMS values. Is there a time delay between the two signals and if so what is its value? (Hint. You can use `crosscorr.m` to perform the cross-correlation, but remember that the correct lag value of each cross-correlation point is in the second output variable of this routine.)

3. Follow the approach of Example 10.2 to find the alpha wave and theta wave activity for the EEG signal given as variable `val` in the file `EEG_FP1_F3.mat`. The frequency range of the theta wave is 3.5 to 7.5 Hz. Plot the two RMS values superimposed. Use `corrcoef` to find the Pearson correlation between the two signals. (Hint. When using `corrcoef`, you may have to transpose the RMS data.)

4. Close examination of the clusters in Figure 10.3 suggests that the lower left cluster (cluster 1) may consist of a subcluster: a group of points that have a mean value less than 30. Modify the code of Example 10.4 to plot examples from two regions of cluster 1, above and below a mean of 30. Are the waveforms of these two subcultures different, i.e., do they represent two different classes? (Hint. When isolating the two subclusters, there are two criteria: mean $>$ or < -30 and RMS < 230. You must apply both or you could get samples from cluster 2 that are > -30.)

5. Modify Example 10.3 to change the second feature (i.e., `f2`). Divide each ECG segment in half and make the second feature the mean of the first half minus the mean of the second half. You should now get three different classes. Most beats cluster in the region where the first feature (RMS) is < 230 and the second feature is between 50 and 150. However, the beats having an RMS > 230 now fall into two different groups. Plot up to four beats from each of the two classes with RMS values > 230. Note the very different behaviors.

6. Apply cluster analysis to the EEG signal analyzed in Example 10.2 (variable `val` in the file `EEG_FP1_F7`). Use the same window size (10 s) and overlap (50%). For each isolated segment, extract two features: the normalized RMS of the alpha wave (as in Example 10.2) and the normalized RMS of the theta wave (3.5 to 7.5 Hz). You should be able to identify two possibly overlapping clusters. Use your best judgment on cluster boundaries and plot examples from each cluster. Because the signals are noisy, plot only three examples from each cluster and use different colors.

7. Modify Example 10.3 by adding an extra feature: the maximum value of the derivative. Use a derivative filter with a skip factor of 4. (Hint. If you have forgotten the

details of constructing a derivative filter (wonderfully presented in Section 8.5.1), you can borrow the code in `qrs_detect.m` given as part of Example 10.3.) Plot all three features on a 3D grid using `plot3` (use option `grid on` to improve visibility of the clusters). Note that there is now a third small cluster (only three points) that has a higher peak derivative than the rest of cluster 2. (You may want to rotate the 3D plot for a better view of the three clusters.) Plot two examples from these two "high-RMS" clusters. (Hint. You will have to select waveforms that have high RMS (> 230) and higher or lower peak velocities. The boundary is around 600 to 800.) Note that this isolates the same ectopic beats found in Problem 5.

8. In Chapter 7, Example 7.8 uses the transfer function of human wrist mechanics to solve for the response of the wrist to an impulse of torque. The time domain solution was as follows:

$$\theta(t) = Ae^{-3t}(\sin(17.3t) radians$$

    Plot the phase-plane trajectory of this response to three different values of A: $A = 0.1, 0.173$, and 0.25. Plot each trajectory in a different color and note that all trajectories end at the same fixed point attractor. Also plot the horizontal and vertical axes.

9. Use the simulation model constructed in Problem 16 of Chapter 9 to plot its phase-plane trajectory. This model is second order and based on the nonlinear van der Pol oscillator equation:

$$\frac{d^2x}{dt^2} = c(1 - x^2)\frac{dx}{dt} - x$$

    Plot the phase plane of the system with an initial condition for x of 0.1 and a value for c of 2.0. This is an example of a limit cycle attractor. Simulate a 100-s time frame and plot the 0.0 horizontal and vertical axes. (Hint. The Simulink solver `ode23s` gives a smoother output than the default.)

10. Following Example 10.5, plot the 3D embedded trajectory of the EEG signal given as variable x in the file `EEG_1min.mat`. To find the appropriate delay, take the autocorrelation function, but you only need ±100 or so points on either side of zero lag. Use the lag at which the first zero occurs to set the delay and make $m = 3$. Does the resultant trajectory appear to fill the 3D space?

11. Apply correlation dimension analysis to a 10,000-sample Gaussian noise signal. To create an evenly spaced $\ln(r)$ for plotting, set r as an exponential, specifically use the MATLAB command: $r = \exp(-3:.05:-1)$. This gives about the same range of r as used in Example 10.6. Embed the Gaussian noise in 3D space using the routine `delay_emb`. Because Gaussian noise decorrelates after only one sample, you can use a short embedding delay, around four or five samples. One of the properties of Gaussian noise is that it tends to completely fill the phase space in which it is embedded, so you would expect a correlation dimension of around 3.0, the dimension of the phase space. (This problem may take about 9 min to run.)

12. Apply correlation dimension analysis to the signal heart rate signal given as variable hr in the file hr_data.mat. This is the same signal whose phase trajectory was developed in Example 10.5 and shown in Figure 10.9. Use the approach outlined in Example 10.6 and make r range between 0.02 and 0.14 in steps of 0.005. Use an embedding delay of 40 as justified in Example 10.5. Estimate the scaling dimension from a plot of ln C_r versus ln r as in Example 10.6. Make sure to eliminate the signal mean. (This problem may take about 12 min to run.)

13. Test the influence of noise in sample entropy. In a loop, use sig_noise (from Chapters 3 and 4) to construct a sine wave with added noise. To get a good estimate of sample entropy, make $N = 10,000$. Recall sig_noise assumes $f_s = 1000$ and make the sine wave 1 Hz (i.e., 10 cycles). Make sure the signal mean is 0 and scale by the standard deviation. Vary the signal-to-noise ratio (SNR) between −12 dB to +20 dB in 1 dB increments. Plot sample entropy against noise in dB. Note the dependence of sample entropy on SNR, although (as noted in the text) neither a sine wave nor the noise carries any information. (Run time: approximately 2.5 min.)

14. The file rr_data.mat contains two heart rate signals as a function of time. These data were obtained by applying qrs_detect to a 30 min ECG signal and taking the difference to find the RR interval. These RR intervals were converted to seconds by dividing by f_s (100 Hz). The approach described in Example 4.4 was used to convert the RR intervals to evenly spaced data ($f_{resamp} = 4$ Hz). The resulting signal was divided into two segments and stored as variables rr1 and rr2.

   Apply MSE to the two signals using a maximum scale factor of 20 and a maximum error of 0.15 times the standard deviation of the signal. Follow the procedure in Example 10.8, but run the code twice for each signal and superimpose the two plots using different colors or symbols. Both recordings show short-term correlations corresponding to a scale factor of 5 or 6. Although both show long-term interactions, one record has stronger long-term interactions. (Suggestion: You can do the two runs using a loop structure, which can be modified for use in the next four problems.)

15. Compute the MSE of two pure sine waves, one with a frequency of 0.5 Hz and the other with a frequency of 1 Hz. In theory the sine waves should contain no entropy and that is roughly the case for the unscaled entropy (i.e., scale factor $= 1.0$). The small increase in entropy observed is an artifact of the scaling process. As the signals are downsampled through scaling, the effective sampling frequency decreases and the sine waves become poorly sampled and less predictable, thus having higher entropy. (Note the entropy increase of the 1 Hz sine wave is greater than that of the 0.5 Hz sine wave.)

16. Compute the MSE of two sine waves used in Problem 15 but with a small amount of added Gaussian noise. Make the noise 0.1 times the amplitude of the sine waves. Compare your results with the MSE from pure Gaussian noise shown in Figure 10.15. Note that the unscaled entropy is much lower due to the low-entropy sine waves. As scaling increases, the entropy decreases as in Figure 10.15, but the sine wave artifact observed in Problem 15 eventually causes an increase in entropy. The scale at which the entropy begins to increase depends on the sine wave frequency as expected based on the results of Problem 15.

17. Generate a signal consisting of integrated Gaussian. (You can use MATLAB's cumsum to perform the integration.) This signal has long-term correlations and should show an increasing MSE with scaling. Apply MSE using three values of maximum error: 0.10, 0.15, and 0.20. Use a loop to compute the MSE for the three values of maximum error and plot the results superimposed using different colors or symbols. Note that as the maximum error increases, the MSE decreases, but that the overall shape of the curve is independent of maximum error.

18. Modify Example 10.8 so that the maximum error, r, is modified after each scaling to be 0.15 times the standard deviation of the scaled signal. The result shows entropy no longer decreases significantly with scale but stays roughly the same. (The increased variability seen at the higher scaling is probably due to the shorter data segments at these scales.) This indicates that the decrease in entropy with scaling is due largely to the reduction of amplitude produced by downsampling. As discussed in footnote 6, whether or not to readjust the maximum error after scaling remains controversial.

19. Apply DFA to the two sine waves used in Problem 15 and find α. Because the sine waves have unbounded correlations, we would expect α to be >1.0.

20. Apply DFA to the two heart rate signals used in Problem 14 (variables rr1 and rr2 in the file rr_data.mat). Find α for both from the most linear section of the curve. Because these signals likely contain long-term correlations we would expect α to be between 0.5 and 1.0. (Suggestion: It is difficult to identify the most linear section, but the slopes do not vary all that much, so any reasonable region should give a similar answer. Just be consistent for the two curves.)

21. Apply DFA to the EEG signal found as variable val in the file EEG_FP1_F3. Estimate α from the most linear portion of the curve. Because EEG signals are likely to be nonstationary, we would expect α to be >1.0. (Run time about 1 min.)

22. Analyze high pass–filtered and low pass–filtered noise using both DFA and MSE. The high pass–filtered noise is given as variable xb(1,:) in the file prob10_22dat.mat and the low pass–filtered noise is given as xb(2,:) in the same file. High pass–filtered signals are anticorrelated and low pass–filtered signals are correlated, so for the DFA analysis we would expect α to be <0.5 for the high pass–filtered noise and >0.5 for the low pass–filtered noise. For MSE, the low pass–filtered noise should show long-term correlations, but the anticorrelated (high pass–filtered) noise should show only short-term correlations. (Suggestion. Because the two analyses are very different, it is easier to use a separate program for each.)

23. Analyze one component of the Lorenz attractor and low pass–filtered (i.e., correlated) noise using both DFA and MSE. The x component of the Lorenz attractor system is given as variable xa(1,:) in the file prob10_23dat.mat and the filtered noise is given as xa(2,:) in the same file.

    In the DFA analysis the Lorenz attractor shows a high α (> 1.0), whereas α for the filtered noise is near 0.5 because of the random noise. The MSE shows larger long-term correlations for the filtered noise, but not for the Lorenz system. However, because the Lorenz attractor is chaotic, it gives the same sample entropy across a wide range of scales. These results show that different nonlinear analyses provide different information about the signals. (Suggestion: Because the two results produce different graphs, it is easier to use a separate program for each analysis.)

24. Analyze low pass—filtered noise having two different bandwidths using both DFA and MSE. The filtered noise given as variable `xb(1,:)` in the file `prob22_24data.mat` has a bandwidth of 0.4 times $f_s/2$ whereas that in `xb(2,:)` of the same file has a bandwidth of 0.1 times $f_s/2$. Both signals were produced by filtering Gaussian noise with fourth-order Butterworth low-pass filters.

In the DFA analysis, the two signals produce similar curves with only slightly different slopes. In the MSE analysis, the signal with the larger bandwidth shows more short-term correlations that decrease with scale, whereas the lower bandwidth signal shows more long-term correlations. Again this shows that these two analyses provide different information.(Suggestion: Again because the two analyses are very different, it is easier to use a separate program for each.)

DETRENDED FLUCTUATION ANALYSIS

Ihlen, E. Introduction to multifractal detrended fluctuation analysis in Matlab. http://journal.frontiersin.org/article/10.3389/fphys.2012.00141/full.

Heartstone, R. et al. Detrended fluctuation analysis scale-free view on neuronal oscillations. Front. Physiol, November 30, 2012 https://doi.org/10.3389/fphys.2012.00450

11

Two-Dimensional Signals—Basic Image Analysis

11.1 GOALS OF THIS CHAPTER

The idea that an image is nothing more than a two-dimensional (2-D) signal was introduced in Chapter 1 (Section 1.2.5). In fact, many signal processing operations explained in previous chapters apply directly to image analysis: the Fourier transform, convolution, filtering, and some nonlinear transformations. The images we work with here are digital images; they are the result of sampling, just as digital signals are sampled. As they are 2-D signals, they are stored in arrays or matrices rather than vectors. As with one-dimensional (1-D) signals, the sampling frequency is important, and, as shown in the next section, undersampling an image also results in aliasing. With images, the independent variable is length rather than time, so the image sampling frequency, termed "spatial frequency," is in samples/cm or samples/in.[1]

For images, as noted in Chapter 1, each sample is termed a pixel, or if the sample represents a three-dimensional (3-D) space, a "voxel." Because images are represented as arrays, memory requirements can become quite large. An $8^1/_2$ by 11 in. image that is sampled at a frequency of 300 samples/in. would have 8,415,000 pixels. However, in MAT-LAB a variable requires 8 bytes, so this image would require 67 Mb of memory. The MAT-LAB Image Processing Toolbox can represent pixels using only one or two bytes, substantially reducing memory requirements. On the other hand, the type of computations that can be performed on these reduced formats are limited, whereas images represented as standard 8-byte variables are amenable to the full range of MATLAB operations.

For heavy duty image processing, you might want to use the Image Processing Toolbox, or other software specifically designed for images, but the basic and most important concepts

---

[1]Printer resolution usually is given in terms of sampling frequency but stated as dots/inch where a dot is the same as a sample.

of image processing and image analysis can be examined using standard MATLAB. Topics in this chapter include the following:

- Basic image format and display
- Sampling and the 2-D Fourier transform
- 2-D convolution
- Image filtering, including filter construction
- Identifying regions of interest (ROIs): basics of image segmentation, including edge detection and texture analysis.

11.2 IMAGE FORMAT AND DISPLAY

Images are stored in arrays, and when we need to refer to specific image elements we use the coordinate system that MATLAB terms the "pixel coordinate system" shown in Figure 11.1. In this format, the matrix element (1,1) represents the upper left-hand pixel in the image and element (m,n), the bottom right-hand pixel, where m is the number of rows in the matrix and n the number of columns.

The matrix values can represent image intensities or they might be pointers to another array known as a colormap. A colormap is a three-by-n matrix where the three columns contain values of red, blue, and green intensity values. In "colormapped" images, the pixel value indicates a row in the colormap that corresponds to the desired mixture of red, blue, and green. This mapping process allows color images to be stored in a single matrix plane. The appropriate colormap must be stored with the image. Neighborhood operations, operations that work on groups of adjoining pixels, may not be appropriate for colormapped images.

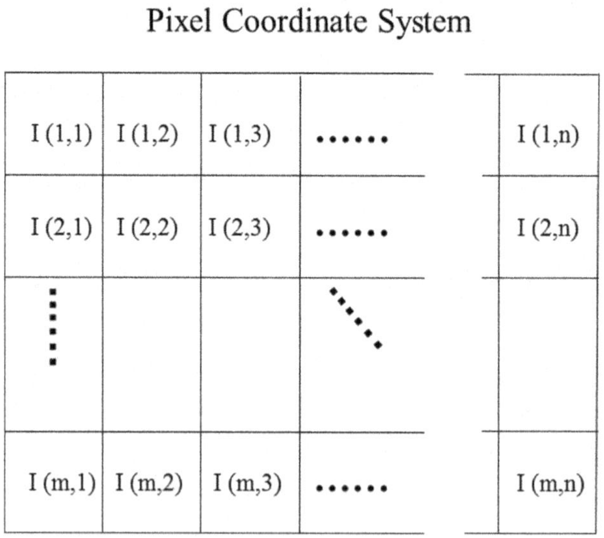

Pixel Coordinate System

I (1,1)	I (1,2)	I (1,3)	• • • • • •	I (1,n)
I (2,1)	I (2,2)	I (2,3)	• • • • • •	I (2,n)
⋮			⋱	
I (m,1)	I (m,2)	I (m,3)	• • • • • •	I (m,n)

FIGURE 11.1 Indexing format for MATLAB images using the pixel coordinate system. This indexing protocol follows the standard matrix notation.

For all the images used in this chapter, pixel values describe the image intensity at that location. We use two different types of images: black and white (BW) images where the image matrix contains only two intensity values, and grayscale images where the matrix has a range of values. For all of our images, 0.0 is assumed to represent black for both BW and grayscale images. For BW images, 1.0 is usually used to represent white and we follow that convention. For grayscale images, white is 255, but in some conventions, white is 1.0 and intensity values range between 0 and 1.0. The images used here are stored without a colormap.

All images used in this chapter are stored as tiff files. They are loaded into MATLAB workspace using the following routine:

```
I = importdata('filename.tif');  % Load tiff file image in matrix I
```

where I is the matrix that contains the image and `filename` is the name of the particular image file. It is common to name variables that contain grayscale images with words beginning with capital I. Similarly, it is the convention to name variables that contain BW images with words that begin with the capitals BW. Once loaded, the intensity values in matrix, I, are available for all suitable MATLAB operations.

In all the operations in this chapter, we assume that the image is stored as a MATLAB variable double format. This allows us to use all standard MATLAB operations. As many of our tiff file images are stored in `uint8` format,[2] we need to convert them to the double format using:

```
I = double(I);   % Convert to double format.
```

If the image is already in double format, the `double` command has no effect.

To display an image, we could use the approach used in Section 1.2.5 that is based on the MATLAB routine `pcolor`. In this chapter we use the image display routine, `image`. The `image` routine is more flexible in that it can display matrices that have three color planes as well as matrices that are in uint8 and uint16 formats (see Footnote 2). The routine displays images in either levels of gray or color by mapping image intensity values to a colormap. Because our images are stored without a colormap, we need to assign an appropriate colormap to our displayed image. A colormap is assigned to an image by the command `colormap(map_name)`. In textbook presentations, we limit images to either grayscale or BW, so we use `colormap(gray)` or `colormap(bone)` to assign a grayscale to our image. In the problems, you can try some of the more colorful colormaps that produce "pseudocolor" images.

Because the intensity values in our image arrays range between 0 and 255, we need to set the colormap range to these values. This entails using the `image` option `"CDataMapping,"` `"scaled"` in conjunction with the command `caxis([0 255])`, or `caxis([0 1])` for BW images.

---

[2]The `uint8` format uses a single byte (8 bits) to store a pixel, whereas the MATLAB double format uses 8 bytes. The `uint8` format requires one-eighth the storage and memory space, so it is popular for storing images that usually require large arrays. However, most MATLAB operations require that variables be stored in the double format. Occasionally images are stored in a 2-byte format known as `uint16`. Note that uint8 and uint16 are common terms for 1- and 2-byte variables in other computer languages.

The `caxis` command sets the range of the colormap. When using `image` for image display, two additional MATLAB visualization commands are helpful: (1) the `axis image` command ensures that each pixel is represented as having equal horizontal and vertical dimensions, and (2) the `axis off` command eliminates the numbers on the side of the image. Thus the commands needed to display an image are:

```
image(I,'CDataMapping','scaled'); caxis([0 255]); % Display image I
colormap(bone); axis image; axis off;             % Image options
```

The first example demonstrates these operations.

EXAMPLE 11.1

Load the image of blood cells in the file `blood1.tif` and display. Invert the image so that dark areas are light and vice versa and display.

Solution: Load the image using `importdata` and display using `image` with the appropriate options. To invert the image, subtract the matrix values from 255, the value that represents white.

```
% Example 11.1 Load and display an image.  Invert grayscale and display
%
I = importdata('blood1.tif');       % Load image into matrix I
I = double(I);                      % Convert to double format
subplot(1,2,1);
image(I,'CDataMapping','scaled'); caxis([0 255]);   % Display image
colormap(bone); axis image; axis off;  % Image options
title('Normal Image','FontSize',14);
subplot(1,2,2);
I_invert = 255 - I; % Invert image
image(I_invert,'CDataMapping','scaled')
......title.......
```

Results: The images produced in this example are shown in Figure 11.2.

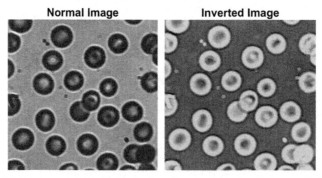

FIGURE 11.2   Image of blood cells displayed normally and with an inverted grayscale.

11.3 THE TWO-DIMENSIONAL FOURIER TRANSFORM

The Fourier transform and the efficient algorithm for computing it, the fast Fourier transform, extend in a straightforward manner to two dimensions. The 2-D version of the Fourier transform can be applied to images providing a spectral analysis of the image content. Of course, the resulting spectrum will be in two dimensions and is more difficult to interpret than a 1-D spectrum. Nonetheless, it can be a useful analysis tool, both for describing the contents of an image and to aid in the construction of imaging filters as described in the next section.

As mentioned, image sampling frequency is in terms of samples/length, such as samples/in. Undersampling an image will lead to aliasing, just as in 1-D signals. In an undersampled image, the spatial frequency content of the original image is greater than $f_s/2$, where f_s now is 1/(pixel size). Figure 11.3 shows an example of aliasing in the frequency domain. The upper left panel shows an image that varies sinusoidally, and the spatial frequency of this sinusoidal variation increases with horizontal position left to right. This is analogous to the 1-D chirp signal and, in fact, was generated by extending a horizontal chirp signal in the vertical direction. The higher-frequency elements on the right side of this image are adequately sampled in the left panel. The same pattern is shown in the upper right panel image except that the sampling frequency has been reduced by a factor of six. The low-frequency variations in intensity are unchanged, but the high-frequency variations have additional frequencies mixed in as aliasing folds in

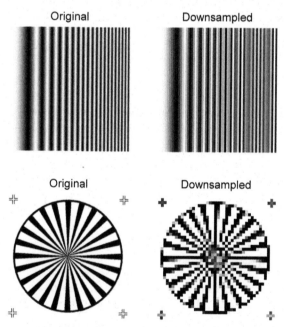

FIGURE 11.3 The influence of aliasing due to undersampling on two images with high spatial frequencies. Aliasing in these images manifests as additional sinusoidal frequencies in the upper right panel and jagged diagonals in the lower right panel.

the frequencies above $f_s/2$ (see Section 4.1.1). The lower panels show the influence of aliasing on a diagonal pattern. The jagged diagonals are characteristic of aliasing, as are the moiré patterns seen in other images. The problem of determining an appropriate sampling size is even more critical in image acquisition because oversampling can quickly lead to excessive memory storage requirements.

The 2-D Fourier transform in continuous form is a direct extension of the equation given in Chapter 3:

$$F(\omega_1 \omega_2) = \int_{m=-\infty}^{\infty} \int_{n=-\infty}^{\infty} f(m,n) e^{-j\omega_1 m} e^{-j\omega_2 n} dm\, dn \tag{11.1}$$

The variables ω_1 and ω_2 are still frequency variables, although their units are in radians per sample size. As with the time domain spectrum, the image spectrum, $F(\omega_1, \omega_2)$, is a complex-valued function that is infinitely periodic in both ω_1 and ω_2. Usually only a single period of the spectral function is displayed as with the time domain analog.

The inverse 2-D Fourier transform is defined as:

$$f(x,y) = \frac{1}{4p^2} \int_{\omega_1=-p}^{p} \int_{\omega_2=-p}^{p} F(\omega_1 \omega_2) e^{-j\omega_1 x} e^{-j\omega_2 y} d\omega_1 d\omega_2 \tag{11.2}$$

This statement is a 2-D extension of the 1-D equivalent; any image can be represented by a series (possibly infinite) of sinusoids, but now the sinusoids extend over two dimensions.

The discrete forms of Equations 11.1 and 11.2 are again similar to their time domain analogs. For an image size of M by N, the discrete Fourier transform becomes:

$$F(p,q) = \sum_{m=0}^{M-1} \sum_{n=0}^{N-1} f(m,n) e^{-j(2pm/M)} e^{-j(2qn/N)} \tag{11.3}$$

$$p = 0, 1, ..., M-1; \quad q = 0, 1, ..., N-1$$

The values $F(p,q)$ are the Fourier transform coefficients of $f(m,n)$. For images, $f(m,n)$ is just the image matrix. The discrete form of the inverse Fourier transform becomes:

$$F(m,n) = \frac{1}{MN} \sum_{p=0}^{M-1} \sum_{q=0}^{N-1} F(p,q) e^{-(2pm/M)} e^{-j(2qn/N)} \tag{11.4}$$

$$m = 0, 1, ..., M-1; \quad n = 0, 1, ..., N-1$$

11.3.1 MATLAB Implementation

Both the Fourier transform and the inverse Fourier transform are supported in two dimensions by MATLAB functions. The 2-D Fourier transform is invoked as:

```
F = fft2(x,M,N);    % Two dimensional Fourier transform
```

where F is the output matrix and x is the input matrix. M and N are optional arguments that specify padding for the vertical and horizontal dimensions, respectively, just as in the 1-D fast Fourier transform (fft). In the time domain, the frequency spectrum of simple waveforms can often be anticipated and the spectra of even relatively complicated waveforms can usually be understood. With two dimensions, it becomes more difficult to visualize the expected Fourier transform even for fairly simple images. In Example 11.2, a simple image is constructed consisting of a thin rectangular bar. The Fourier transform of the object is determined, and the resultant spatial frequency function is plotted as a 3-D function.

EXAMPLE 11.2

Determine and display the 2-D Fourier transform of a thin rectangular object. The object should be three-by-nine pixels in size and solid white against a black background. Display the Fourier transform as a 3-D function using MATLAB's mesh routine.

Solution: We construct the image by first setting up a black background that is a 22-by-30 matrix of zeros. We do not need a large image because we use the zero-padding option in the Fourier transform routine to extend the matrix to 128-by-128 samples. To complete the image, we insert a narrow vertical center strip of white by assigning the value 1.0 to matrix columns 10 through 12 and rows 11 through 19. The image is plotted using image but with the scaling changed to range between 0 and 1 as is common for BW images.

The 2-D Fourier transform is constructed using fft2 with padding as noted earlier. The fft2 routine places the zero frequency (DC) component in the upper left corner. This approach to plotting the 2-D Fourier transform is logical but even more difficult to interpret. Interpreting the spectrum can be a little easier if it is shifted so that the zero frequency component falls in the center of the plot. The MATLAB routine fftshift swaps the first and third quadrants and the second and fourth quadrants of the spectrum so that the DC component falls in the center position. This shifted spectrum is then plotted using routine mesh, which displays the spectrum as a 3-D surface.

Example 11.2 Fourier transform of a simple rectangular image

```
%
I = zeros(22,30);          % Original figure can be any size since it
I(11:19,10:12) = 1;        % will be padded
F = fft2(I,128,128);       % Take Fourier transform padded to 128
F = abs(fftshift(F));      % Shift center; get magnitude
%
image(I,'CDataMapping','scaled');              % Display image
colormap(bone); caxis([0 1]); axis image; axis off; % Image options
......label and new figure.......
mesh(F); colormap(bone);   % Plot Fourier transform as function
......labels, axis, and new figure.......
```

Result: The image and its spectrum are shown in Figure 11.4. The spectrum has been shifted and displayed as a 3-D surface. With a little thought, the spectrum can be interpreted as the combined spectra resulting from two pulse signals. The image is effectively a wide vertical pulse change in intensity combined with a narrow horizontal pulse change in intensity. Recall the inverse

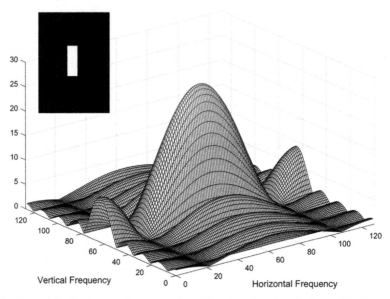

FIGURE 11.4 Upper left: the image of a rectangular object three-by-nine pixels used in Example 11.2. Lower main: the magnitude Fourier transform of this image. More energy is seen, particularly at the higher frequencies, along the horizontal axis because the image's vertical cross-section appears as a narrower pulse change in intensity. The broader vertical cross-section produces frequency characteristics that fall off more rapidly at higher spatial frequencies. This is analogous to what is found in 1-D pulse spectra (see Example 3.3).

relationship between pulse width and spectral width: the wider the pulse, the narrower its spectral peaks (see Example 3.3, and particularly Figure 3.12, which shows the spectra of a narrow and wide pulse). Extrapolating this to our 2-D image, we would expect our spectrum to have a narrower vertical peak (because the strip is wider in the vertical dimension) and a broader horizontal peak. For both directions, the spectrum should have the general shape of $|\sin(x)/x|$ as is seen in Figure 11.4.

Another easy-to-interpret 2-D spectrum is shown in Figure 11.5 and is generated from the image in the upper left panel of Figure 11.3. This image consists of a horizontal chirp signal that has been extended in the vertical direction. The fact that this image changes in only one direction, the horizontal direction, is reflected in the Fourier transform. The linear increase in spatial frequency in the horizontal direction produces an approximately constant spectral value over the horizontal spatial frequency. The constant image values in the vertical direction produce zero values over all vertical spatial frequencies other than the zero frequency value. Again, most images produce complex spectra that do not lend themselves to easy interpretation. However, the most useful application of the 2-D Fourier transform is not in image analysis but in the design and evaluation of linear imaging filters. Image filtering is the featured topic of the next section.

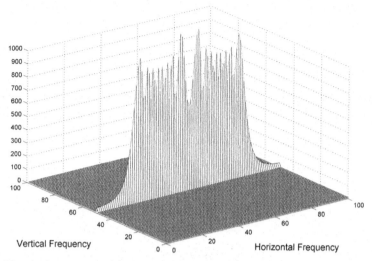

FIGURE 11.5 Magnitude spectrum of the horizontal chirp signal image shown in Figure 11.3, upper left panel. The spatial frequency characteristics of this image are zero for all vertical frequencies (except at zero frequency) because the image has constant intensities in the vertical direction. The linear increase in spatial frequency in the horizontal direction is reflected in the approximately constant amplitude of the spectrum over the horizontal spatial frequencies.

11.4 LINEAR FILTERING

The techniques of linear filtering described in Chapter 8 can be directly extended to two dimensions and applied to images. In image processing, finite impulse response (FIR) filters are used because of their linear phase characteristics. A sliding neighborhood operation is used to apply a 1-D FIR filter to a signal (see Section 8.5). A new sample of the filtered signal is calculated as the summation of a sequence of samples in the original signal scaled by the filter weights. This operation is often applied symmetrically so that the new sample lies at the center point of the filter coefficients, Figure 11.6. The filter coefficients then shift one position along the original signal and a new filtered sample is calculated. This operation is implemented through convolution.

Filtering an image is also a neighborhood operation, but now the neighborhood extends in two directions around a given pixel. In image filtering, the value of a filtered pixel is determined from the surrounding pixels, Figure 11.7. Pixels in a region of the original image are scaled by a matrix of filter coefficients. A summation of scaled pixel values is used as the new pixel and is placed in a position corresponding to the center of the filter matrix. The matrix of filter coefficients is then shifted across the original image both horizontally and then vertically. This operation is equivalent to 2-D convolution.

As described in Chapter 8, FIR filters are uniquely defined by their filter coefficients, $b[k]$. Filter design is a matter of finding those coefficients that produce the desired

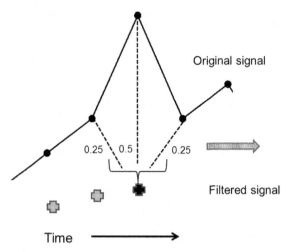

FIGURE 11.6   A finite impulse response filter is implemented as a weighted sum of filter coefficients applied to the original signal. This summation is often implemented symmetrically so that the new filter point is placed on the new signal at a point corresponding to the center of the filter coefficients. The coefficients slide over the original signal sample by sample to produce the filtered signal. This is illustrated for a three-weight filter.

modification of signal spectrum. An image filter is also uniquely defined by filter coefficients, but now these coefficients form a matrix in two dimensions, $b[k_1,k_2]$. These 2-D filter coefficients are applied to the image using 2-D convolution in an analogous approach to 1-D filtering described in Chapter 8.

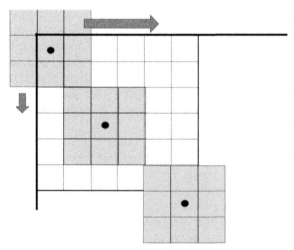

FIGURE 11.7   Implementation of a 2-D filter consisting of a three-by-three set of filter weights (*shaded boxes*). The filtered image is constructed by scaling the original image pixels by the filter coefficients. The weighted summation becomes a filtered image pixel that corresponds in position to the center of the filter coefficient matrix. The filter coefficient matrix slides over the image pixel by pixel in two dimensions. This is the 2-D extension of convolution.

11.4.1 Convolution in Two Dimensions

Using convolution to perform image filtering parallels its use in signal processing: the image array is convolved with a set of filter coefficients. However, in image analysis the filter coefficients are defined in two dimensions, $b[k_1,k_2]$ as a matrix. The equation for convolution in one dimension was given in Chapter 5 (Equation 5.3) and is repeated here:

$$y[n] = \sum_{k=-\infty}^{\infty} x[k]b[n-k] \tag{11.5}$$

where $x[k]$ is the input signal and $b[k]$ are the filter coefficients, and n slides the coefficients across the signal. Two-dimensional convolution is a straightforward extension of this equation:

$$y[m,n] = \sum_{k_1=-\infty}^{\infty} \sum_{k_2=-\infty}^{\infty} x[k_1,k_2]b[m-k_1,n-k_2] \tag{11.6}$$

where $x[k_1,k_2]$ is the image, $b[m,n]$ are the 2-D filter coefficients, and k_1 and k_2 slide these coefficients across the image in the horizontal (k_1) and vertical (k_2) directions. Although this equation would not be difficult to implement in MATLAB, there is, naturally, a MATLAB function that implements 2-D convolution directly, conv2:

```
I2 = conv2(I1,b,shape);    % Two-dimensional convolution
```

where I1 and b are image and filter matrices (or more generally, simply two matrices) to be convolved and shape is an optional argument that controls the size of the output image. If shape is "full," the default, then the size of the output matrix follows the same rules as in 1-D convolution: each dimension of the output is the sum of the two matrix lengths along that dimension minus one. Hence, if the two matrices have sizes I1(M1,N1) and b(M2,N2), the output size is M1+M2-1 by N1+N2-1. If shape is "valid," then every pixel evaluation that requires image padding is ignored, leading to an output image size of M1-M2+1 by N1-N2 +1. For our applications, the most useful option for image filtering is "same." Using this option results in an output matrix that is of the same size as the input image, I1, that is, M1-by-N1 and is similar to 1-D convolution using that option. Using the "same" option eliminates the need for dealing with the additional, or missing, pixels generated by convolution.

When convolution is used to apply a series of filter weights to either an image or a signal, the weights are applied to the data in reverse order as indicated by the negative sign in the 1-D and 2-D convolution equations (Equations 11.5 and 11.6). This can be a source of mild confusion in 1-D applications and becomes even more confusing in 2-D applications. It becomes difficult to conceptualize how a given filter matrix alters an image after it has been reversed in two directions. One way around this is to apply the filter matrix using correlation rather than convolution. The correlation equation is very similar to the convolution equation except that the negative signs are now positive:

$$y[m,n] = \sum_{k_1=-\infty}^{\infty} \sum_{k_2=-\infty}^{\infty} x[k_1,k_2]b[m+k_1,n+k_2] \tag{11.7}$$

When correlation is used, the set of filter coefficients is termed the "correlation kernel" to distinguish it from the standard filter coefficients. In fact, the operations of correlation and convolution both involve weighted sums of neighboring pixels, and the only difference between correlation kernels and convolution kernels is a 180-degree rotation of the coefficient matrix. The MATLAB routine `filter2` uses correlation to implement an image filtering. The call is very similar to that of `conv2`:

```
I2 = filter2(I1,b,shape);    % Image filter using correlation
```

where again `I1` is the original image, `b` is the filter matrix, and shape determines the size of the output in the same manner as `conv2`. For image filtering, we use the shape option `'same'`, which is the default option in `filter2`. This produces a filtered image that is of the same size as the original image. The filtering routine in the Image Processing Toolbox also uses correlation kernels, again because their application is easier to conceptualize.

11.4.2 Linear Image Filters

Image filters are designed differently than 1-D filters. Filters produced using 1-D design strategies, such as the spectral window approach (Section 8.5), do not work well on images. Many image filters were designed heuristically, that is to say, they stem from an idea about what ought to work and then were found that they did work well. In this section we describe five popular filters: one that performs spatial low-pass filtering for image smoothing, a high-pass filter for image sharpening, a high-pass filter that takes a spatial second derivative, and two similar filters that enhance edges for edge detection. Some filters have variably sized coefficients, but many linear filters, including most described here, have coefficients arranged as a three-by-three matrix.

One of the earliest linear filters was a filter designed for edge detection by Irwin Sobel. This "Sobel" filter performs a vertical spatial derivative operation for enhancement of horizontal edges. To detect vertical edges, the coefficient matrix can be rotated by 90 degrees using transposition.

$$b[m,n]_{\text{Sobel}} = \begin{bmatrix} 1 & 2 & 1 \\ 0 & 0 & 0 \\ -1 & -2 & -1 \end{bmatrix} \tag{11.8}$$

An examination of the Sobel matrix reveals its operation. The correlation between the image and filter coefficient is greatest when the underlying image has a dark (i.e., 0's) region overlapping the lowest row of the filter matrix and a white (i.e., 1's) region overlapping the upper row. If there is no horizontal boundary and the upper and lower regions are the same, the correlation will be 0 (because the lower negative rows cancel the upper positive rows). An application of a Sobel filter to an image is found in the next section.

The upper and lower rows of the Sobel filter have a larger number in the center of the outer rows, which creates a smoothing effect on the resulting image. A "Prewitt" edge detector filter is a modification of the Sobel filter that does not have this smoothing feature.

$$b[m,n]_{\text{Prewitt}} = \begin{bmatrix} 1 & 1 & 1 \\ 0 & 0 & 0 \\ -1 & -1 & -1 \end{bmatrix}$$

The Gaussian image filter is a common low-pass filter and, as with the 1-D low-pass filter, is useful in eliminating high-frequency noise. The equation for a Gaussian filter is similar to the equation for the Gaussian distribution:

$$b[m, n] = e^{-(d/2\sigma)^2} \quad \text{where} \quad d = \sqrt{(m^2 + n^2)} \tag{11.9}$$

where σ adjusts the attenuation slope of the low-pass filter. A common value for σ of 0.5 produces a modest slope, whereas larger values provide steeper slopes. The filter matrix is square but can be of any size. Changing the dimension of the coefficient matrix alters the number of pixels over which the filter operates. This filter has particularly desirable properties when applied to an image; it provides an optimal compromise between smoothness and filter sharpness. For this reason, it is popular as a general purpose image low-pass filter. The next example constructs a Gaussian filter and determines its spectral characteristics.

EXAMPLE 11.3

Construct a Gaussian filter having a five-by-five coefficients matrix. Determine the magnitude frequency spectrum of this filter for two values of σ: 0.5 and 1.9.

Solution: To create the Gaussian filter, we write a function that follows Equation 11.9. As in 1-D filtering, the filter coefficients are also the 2-D impulse response of the filter. So the filter's spectrum can be obtained from the 2-D Fourier transform of the filter coefficient matrix. The magnitude spectra are plotted for both values of σ using mesh as in the previous example.

To generate the filter coefficients, we construct a routine called gaussian. In this routine, the input variables are matrix dimension and σ, whereas the output variable is the filter coefficient matrix. The value of $2\sigma^2$ can be determined first because it is a constant in the equation. The variable d in Equation 11.9 represents the distance of a given matrix coefficient from the center of the matrix. To determine this value we need to calculate the indices of the center location: x_{center}, y_{center}. The distance then becomes $d = \sqrt{(x_{index} - x_{center})^2 + (y_{index} - y_{center})^2}$. In the gaussian routine, this distance variable is determined for every matrix position using a double loop. Because Equation 11.9 calls for d^2 in the exponent, the square root is not taken. The coefficients are then found by taking the exponential of d^2 divided by $2\sigma^2$. It is common to normalize the coefficients so they sum to 1.0, so the output image has the same overall intensities as the input image.

```
function b = gaussian(N,sigma)
% Function to construct a Gaussian lowpass filter
%
sig_sq = 2*(sigma2);        % Calculate 2 sigma squared
center = (N+1)/2;           % Find center point
for k1 = 1:N
  for k2 = 1:N
    d_sq(k1,k2) = (k1-center)2 + (k2-center)2; % d squared
  end
end
b = exp(-d_sq/sig_sq);      % Calculate filter coefficients
b = b/sum(b(:));            % Normalize filter coefficients
```

The main program simply calls the `gaussian` routine and plots the magnitude of Fourier transforms for both values of σ.

```
% Example 11.3 Program to plot the magnitude spectrum of a Gaussian filter
% for two values of sigma
%
N = 5;                          % Filter dimension
sigma = [0.5 1.0];              % Values of sigma
for k = 1: length(sigma)
   b = gaussian(N,sigma(k));    % Determine filter coefficients
   F = fft2(b,128,128);         % Calculate magnitude spectrum
   F = abs(fftshift(F));        % Shift spectrum
subplot(1,2,k);
mesh(F);                        % Plot spectrum
......labels, title, color axis, colormap......
end
```

Results: The two spectra are shown in Figure 11.8. As expected, the filter with the higher value of σ shows a sharper attenuation. The effect of the attenuation slope on a filtered image is evaluated in the next section.

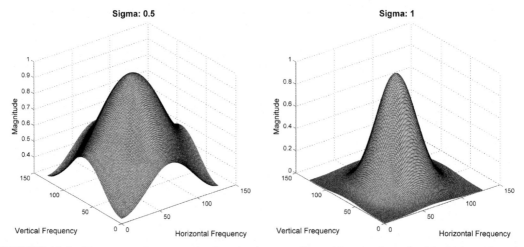

FIGURE 11.8   The magnitude spectra of a Gaussian low-pass filter with two values of σ. The higher value of σ produces a sharper cutoff. The effect of σ, as well as filter dimensions, is examined in the next section and in the problems.

The high-pass filter termed the "Laplacian of Gaussian" filter approximates a spatial second derivative operation: $\partial^2 I / \partial x^2$ and $\partial^2 I / \partial y^2$. A three-by-three version of this filter is shown in Equation 11.10.

$$b[m \cdot n]_{\text{Laplacian}} = \begin{bmatrix} \dfrac{\alpha}{\alpha + 1} & \dfrac{1 - \alpha}{\alpha + 1} & \dfrac{\alpha}{\alpha + 1} \\[2mm] \dfrac{1 - \alpha}{\alpha + 1} & \dfrac{4}{\alpha + 1} & \dfrac{1 - \alpha}{\alpha + 1} \\[2mm] \dfrac{\alpha}{\alpha + 1} & \dfrac{1 - \alpha}{\alpha + 1} & \dfrac{\alpha}{\alpha + 1} \end{bmatrix} \tag{11.10}$$

where the variable α ranges between zero and one and adjusts the steepness of the spectral curve. You might guess that because this filter functions as a high-pass filter, its shape would be roughly the inverse of the low-pass filters shown in Figure 11.8. In the next example, we show that this is indeed the case.

EXAMPLE 11.4

Plot the magnitude of the Laplacian of Gaussian filter for two values of α: 0.2 and 0.8.

Solution: Again, we write a special routine, lgauss, to construct the filter coefficient matrix based on Equation 11.10. We then use this routine as in Example 11.3 to determine and plot the magnitude spectra.

```
function h_l = lgauss(alpha)
% Function to design a 3-by-3 Laplacian of Gaussian filter
%
h1 = alpha/(alpha+1);        % Set up end filter coefficients
h2 = (1-alpha)/(alpha+1);    % Center end coefficients
h_lap = [h1 h2 h1; h2 -4/(alpha+1) h2; h1 h2 h1];   % Define coeff. matrix
```

The main routine is as follows:

```
Example 11.4 Plot the magnitude spectrum of Laplacian of Gaussian filter
% for two values of alpha: 0.2 and 0.8.
%
alpha = [0.2 0.8];           % Define values of alpha
for k = 1:length(alpha)
  b = lgauss(alpha(k));      % Get filter coefficients
  F = fft2(b,128,128);       % Calculate magnitude spectrum
  F = abs(fftshift(F));      % Shift spectrum
  subplot(1,2,k);
  mesh(F);
  ......labels, titles, z axis scaling, caxis scaling, colormap
end
```

Results: The spectra produced in this example are shown in Figure 11.9, and the general form is inverted with respect to the low-pass filter spectra shown in Figure 11.8. Again the effect of α on the slope of the spectra is apparent. For this filter, increasing α decreases the spectrum's slope.

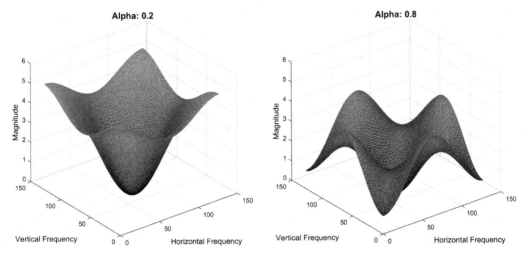

FIGURE 11.9   The magnitude spectra of the Laplacian of Gaussian filter for two different values of the constant α. This filter approximates a second derivative operation on an image. The magnitude spectrum shows that increasing α decreases the upward slope of the spectrum.

The other high-pass filter is the "unsharp" filter, which produces contrast enhancement. The unsharp filter gets its name from an abbreviation of the term "unsharp masking," a double negative that indicates that the unsharp or low spatial frequencies are suppressed (i.e., masked). Suppressing low frequencies is the same as enhancing high frequencies, so this is a high-pass filter. In fact, as shown in one of the problems, the magnitude spectrum of the unsharp filter is almost identical to that of the Laplacian of Gaussian filter, but the phase characteristics are very different. Except for the center coefficient, the unsharp filter coefficients are the negative of the Laplacian of Gaussian filter coefficients. The center coefficient is one minus the center term of the Laplacian of Gaussian filter:

$$b[m \cdot n]_{\text{Laplacian}} = \begin{bmatrix} \dfrac{-\alpha}{\alpha+1} & -\left(\dfrac{1-\alpha}{\alpha+1}\right) & \dfrac{-\alpha}{\alpha+1} \\[2ex] -\left(\dfrac{1-\alpha}{\alpha+1}\right) & 1 - \left(\dfrac{4}{\alpha+1}\right) & -\left(\dfrac{1-\alpha}{\alpha+1}\right) \\[2ex] \dfrac{-\alpha}{\alpha+1} & -\left(\dfrac{1-\alpha}{\alpha+1}\right) & \dfrac{-\alpha}{\alpha+1} \end{bmatrix} \tag{11.11}$$

It is easy to generate this filter in MATLAB from the Laplacian of Gaussian filter:

```
h_unsharp = [0 0 0;0 1 0;0 0 0] - h_laplacian;    % Generate the unsharp filter
```

This code has been added to routine lgauss, and this routine outputs the unsharp coefficients as a second argument. The influence of these two filters on images is explored in the next section and in the problems.

11.4.3 Linear Filter Application

Now it is time to apply some of these filters to images, which we do in the next two examples.

EXAMPLE 11.5

Load the magnetic resonance (MR) image of the brain found in brain1.tif. Sharpen the image with the unsharp masking filter described earlier. Use an alpha of 0.2.

Solution: Load the image using importdata as in Example 11.1. The unsharp masking filter coefficients are determined from the Laplacian of Gaussian filter as described earlier and are found as the second output argument of lgauss. Apply these coefficients to the image using filter2. Display the image using image with the necessary options as in Example 11.1.

```
% Example 11.5 Sharpen an MR image with the unsharp masking filter
%
alpha = 0.2;                          % Unsharp filter alpha
I = importdata('brain1.tif');         % Load image
I = double(I);                        % Convert to double format
[~,b_unsharp] = lgauss(alpha);        % Unsharp filter coefficients (second output)
%
I_unsharp = filter2(b_unsharp,I);  % Filter image
subplot(1,2,1);
  image(I,'CDataMapping','scaled'); caxis([0 255]);          % Display
  colormap(gray); axis off; axis image;                      % Necessary options
  ......title......
subplot(1,2,2);
  image(I_unsharp,'CDataMapping','scaled'); caxis([0 255]);  % Display
  colormap(gray); axis off; axis image;                      % Necessary options
  title('Unsharp Filter','FontSize',14);
```

Results: The filtered and unfiltered images are displayed in Figure 11.10. The filtered image shows more detail and has sharper, better defined boundaries, but the background neural tissue shows some noise. The unsharp masking filter acts as a high-pass filter, hence it performs a kind of spatial differentiation. We know that 1-D differentiation enhances noise in a signal (see Section 8.5.1), so

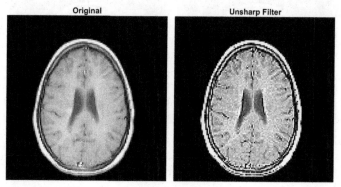

FIGURE 11.10 Left: MR image of the brain. Right: the original image is now sharper after application of the unsharp masking filter in Example 11.5.

it is not surprising that making an image sharper using spatial differentiation enhances noise in the image. This is a traditional engineering compromise. If we wanted to reduce noise, we could low-pass filter the image, but this would lead to some blurring. This is demonstrated in the next example.

EXAMPLE 11.6

Load the image stored as variable I_noise in the file brain_noise.mat. (Note, I_noise is already in double format.) This image contains what is known as "salt and pepper" noise. This is the type of noise generated when individual pixels drop out and become either white or black speckles. Apply a low-pass Gaussian filter to reduce the noise. Use a three-by-three coefficient matrix and a sigma of 2.0 (just because it seems to work well). Display the noisy and filtered images side by side.

Solution: Load the image as in Example 11.5 and use routine gaussian to get the low-pass filter coefficients. Apply the coefficient to the noisy image and display as in Example 11.5.

```
% Example 11.6 Filter noisy MR image of the brain using a lowpass Gaussian filter
%
N = 3;                            % Filter dimension
sigma = 0.7;                      % Gaussian sigma
load('brain_noise.tif');          % Load image
h_gauss = gaussian(N,sigma);      % Construct Gaussian lowpass filter.
I_lowpass = filter2(h_gauss,I);   % Apply filter

......display images following same code as in  Example 11.5.......
```

Results: The results are shown in Figure 11.11. The salt and pepper noise in the original figure appear as BW speckles. Although some dark speckles can still be seen in the filtered image, the noise is substantially reduced.

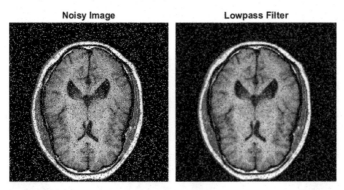

FIGURE 11.11   Left: an MR image of the brain with salt and pepper noise, noise that occurs when individual pixels drop out. Right: the noisy image after filtering with a low-pass Gaussian filter.

Linear filters are also used to aid in identifying physiological structures in medical images. Identifying and isolating regions of particular medical interest, such as bone, organs, or soft tissue structures, is perhaps the greatest challenge in biomedical image analysis. Identifying such ROIs is known as image "segmentation." It is a rich and deep area of biomedical image processing with much ongoing research. In the next section, we explore some basics of image segmentation.

11.5 IMAGE SEGMENTATION

The problems associated with segmentation have been well studied and a large number of approaches have been developed, many specific to particular image features. General approaches to segmentation can be grouped into four classes: pixel-based, edge-based, regional, and morphological methods. Pixel-based methods are the easiest to understand and to implement but are also the least powerful. Because they operate on one element at time, they are also susceptible to noise. Continuity-based and edge-based methods approach the segmentation problem from opposing sides: edge-based methods search for differences, whereas continuity-based methods search for similarities. Morphological methods use information about the shape, properties, and/or mechanical properties of a particular organ or tissue to aid the image segmentation. As these methods tend to be specific to a specific organ, they are not covered in this brief overview.

11.5.1 Pixel-Based Methods

The most straightforward and common of the pixel-based methods is "thresholding" in which all pixels having intensity values above or below some level are classified as part of the segment. Thresholding is also used to convert a grayscale image to a BW (or binary) image. Thresholds can be used in conjunction with other methods to ultimately produce a segmentation mask, a BW template that identifies the region of interest.

Thresholding is usually quite fast and can be done in real time, allowing for interactive adjustment of the threshold. The basic concept of thresholding can be extended to include both upper and lower boundaries, an operation termed "slicing," because it isolates a specific range of pixel intensities. Slicing can be generalized to include any number of different upper and lower boundaries having different intensities.

A technique that can aid in all image analyses, but is particularly useful in pixel-based methods, is intensity remapping. In this global procedure, pixel values are rescaled so as to extend over different maximum and minimum values. Usually the rescaling is linear, so each point is adjusted proportionally with a possible offset. Rescaling operations are easily done in MATLAB using basic arithmetic.

11.5.1.1 Threshold Level Adjustment

The essential problem in pixel-based methods is setting the threshold(s) or slicing level(s) appropriately. Usually these levels are set by the program, although they can be set interactively by the user. Finding an appropriate threshold level can be aided by a plot of distributions of pixel intensity over the image. Such a plot is termed the "intensity histogram." A histogram was used in Example 1.4 and the MATLAB routine `hist` is described in that example.

Although intensity histograms contain no information on position, they can still be useful for segmentation, particularly for estimating threshold(s) from the histogram. If the intensity histogram is, or is assumed to be, bimodal (or multimodal), a common strategy is to search for low points or minima in the histogram. This histogram-based strategy is used in the next example to aid in setting a threshold.

EXAMPLE 11.7

Load the x-ray image of the spine in the file `spine.tiff`. Plot the intensity histogram and determine a threshold that best separates the spine from the black background. Superimpose the threshold value on the histogram and then display the original image and the thresholded image.

Solution: First we need a routine to threshold the image. Because the thresholded image is a BW image, its intensity values are either 0 or 1.0 as is the convention for BW images. Our `thresh_image` routine takes the image and threshold as input arguments and outputs a BW thresholded image. To accomplish this, the routine first sets up a matrix of zeros the same size as the input image. MATLAB's `find` routine is used to locate input image pixels > `threshold` and set these values to 1.0 in the output matrix.

```
function BW = thresh_image(I,thresh)
% Function to threshold an input image and create a new BW output image
%
BW = zeros(size(I));        % Set up output matrix
BW(find(I > thresh)) = 1;   % Set appropriate locations to 1.0
```

In the main routine, we load the image and use histogram[3] to get the intensity histogram. From the histogram, we select and input (manually) a value that seems to divide the black background from the image of the spine. We display the selected value as a line superimposed on the histogram plot. We threshold the image using `thresh_image` and then display the original and thresholded images as in previous examples.

```
% Example 11.7 Example of segmentation using thresholding.
%
I = importdata('spine.tif');           % Load image
I = double(I);                         % Convert to double format
histogram(I);³ xlim([0 255]); hold on; % Plot histogram
threshold = input('Input threshold: '); % Input threshold manually
plot([threshold threshold],[0 40000],'k','LineWidth',2); % Display threshold
BW = thresh_image(I,threshold);        % Threshold image
......display images as in previous examples
```

Results: The histogram is shown in Figure 11.12, upper plot. As seen the distribution is bimodal with a large number of 0.0 values originating from the black background. The lowest point between the two distributions occurs around an intensity value of 30—40. A value of 40 was selected and is plotted as a thick line superimposed on the histogram, Figure 11.12, upper plot. The original spine x-ray is shown in the lower left image of Figure 11.12, and the threshold image is shown as the

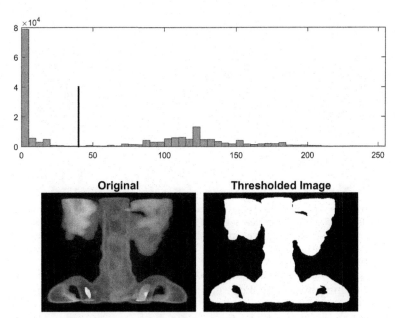

FIGURE 11.12 Upper plot: intensity histogram of the x-ray image of a spine showing a clear bimodal distribution. The lowest point between the two distributions occurs around an intensity value of 30–40, and a value of 40 was selected manually as a threshold value. Lower left: original x-ray image of the spine. Lower right: black and white image after thresholding. *Original image courtesy of MATLAB.*

lower right image. The BW threshold image isolates the spine image and can be used as a "mask" to distinguish this region from the black background.

[3]Older versions of MATLAB may not have the histogram function. If histogram is not available to you, use histogrm(I) found in the associated files instead. This routine first aligns the image into a single row and then calls MATLAB's hist routine. As with histogram, the second argument is optional and specifies the number of bins. The default value of 60 seems appropriate for most images.

Pixel-based approaches can lead to serious errors, even when the average intensities of the various segments are clearly different, because of noise-induced intensity variations within the structure. Such variations could be acquired during image acquisition but could also be inherent in the structure itself. For example, the left image in Figure 11.13 shows two regions with quite different average intensities. Even with optimal threshold selection, many inappropriate pixels are found in both segments because of intensity variations within the segments, Figure 11.13, right image. Techniques for improving separation in such images are explored in the sections on continuity-based approaches.

11.5.2 Edge Detection

Edge detection methods can be more powerful than pixel-based methods because they can involve multiple criteria. For example, edges might be required to be continuous so small

Original Image   **Thresholded**

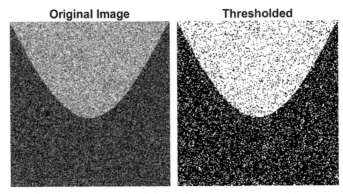

FIGURE 11.13   An image with two regions that have different average gray levels. The two regions are clearly distinguishable, but it is not possible to accurately segment the two regions using thresholding alone because of noise. In the section on continuity-based methods, we separate these two regions.

gaps between edges can be identified and filled. Shape criteria can also be imposed on edges based on the underlying physiology. Edges might be forced to be convex or, in other circumstances, concave. Edges might be constrained to form closed boundaries. For example, the boundary of a red blood cell should be both convex and closed.

Often advanced edge detection routines begin with a set of candidate edges found using more fundamental methods, such as taking the spatial derivatives. Spatial derivatives enhance changes in intensity usually associated with boundaries. In this section, we use the Sobel derivative filter (Equation 11.8) to take the spatial derivative followed by thresholding to isolate the boundary. Figure 11.14, from the next example, shows the Sobel filter applied to an image of blood cells.

The Sobel filter takes the spatial derivative in only one orientation, so it enhances boundaries going from light to dark but only in the vertical direction and only from top to bottom. To enhance all sides of the cell boundaries would require rotating the filter to the three other

Blood Cells   Sobel Filter

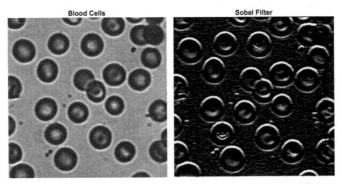

FIGURE 11.14   The right-hand image was generated by applying the Sobel filter defined in Equation 11.8 to the image of blood cells on the left. The boundaries going from light to dark and from top to bottom are enhanced by this filter. To enhance boundaries in other directions would require applying rotated versions of this filter to the image. This is done in Example 11.8. *Image courtesy of MATLAB.*

possible alignments and applying the rotated coefficients to the filter. After thresholding, the result would be four BW images highlighting the boundaries in the four directions. Because BW images are essentially binary images (they consist only of 0's and 1's) we could use the 'or' operation to combine them to get a binary image of the total cell boundary. This is the goal of the next example.

EXAMPLE 11.8

Load the image of blood cells in the file blood.tiff. Apply the Sobel filter to enhance the cell boundaries and determine a threshold that extracts the boundary as a BW, or binary, image. Rotate the Sobel coefficient matrix so that the boundaries in all four directions are highlighted and thresholded. Combine the thresholded images using the 'or' operation to display the combined image.

Solution: Load the blood cell image and display as in previous examples. Generate the Sobel and apply it to the image using filter2. The intensity histogram is not so useful in this application, so find the threshold empirically. Examining the right-hand image in Figure 11.14 indicates that the enhanced boundaries are quite light suggesting a high value for the threshold. In fact a threshold of 240, near the maximum value of 255, works well.

To enhance the boundaries in the four directions, it is necessary to rotate the Sobel matrix, and then reapply it to the image and take the threshold of that new filtered image. Rotation can be done by transposing the matrix, flipping the matrix, or a combination of the two as we do here. The four binary boundary images are combined through the 'or' operation and the resultant image displayed.

```
% Example 11.8 Blood cell boundary identification
%
I = importdata('blood1.tif');           % Load image
I = double(I);                          % Convert data
h_sobel = [1 2 1; 0 0 0; -1 -2 -1];     % Sobel filter coefficient
I_sobel = filter2(h_sobel,I);           % Apply Sobel filter
.......display original and filter images using image.m.......
%
I_sobel1 = filter2(h_sobel',I);         % Rotate 90 deg counterclockwise
I_sobel2 = filter2(flipud(h_sobel),I);  % Rotate 180
I_sobel3 = filter2(flipud(h_sobel)',I); % Rotate 90 clockwise
%
% Threshold all images
threshold = 240;                        % Threshold. Determined empirically
I_sobel = thresh_image(I_sobel,threshold);   % Threshold the filtered images
I_sobel1 = thresh_image(I_sobel1,threshold);
I_sobel2 = thresh_image(I_sobel2,threshold);
I_sobel3 = thresh_image(I_sobel3,threshold);
......display as in previous examples but make caxis([0 1].......
%
BW = I_sobel | I_sobel1 | I_sobel2 | I_sobel3; % Combine the images
......display binary image........
```

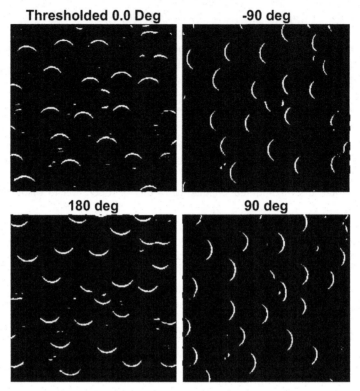

FIGURE 11.15 Four Sobel filtered images were thresholded to produce the BW (binary) images shown here. The filtered images were generated by applying Sobel filters to the original filters in four different orientations. Each orientation enhances the cell boundaries in a specific direction.

Results: The original and one filtered image have already been shown in Figure 11.14. The four thresholded images in Figure 11.15 highlight the cell boundaries in their respective directions. Combining these four images using the 'or' operation produces the image shown in Figure 11.16. The cell boundaries are well delineated. To segment (i.e., isolate) these cells, we need an algorithm

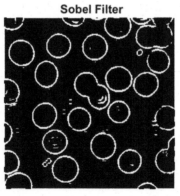

FIGURE 11.16 Edges of the blood cells shown in Figure 11.14, left side. This BW image was made by combining the four images in Figure 11.15 using the logical 'or.'

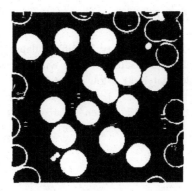

FIGURE 11.17 A binary image constructed by filling the closed blood cell boundaries shown in Figure 11.16. A routine from MATLAB's Image Processing Toolbox, imfill, was used to generate this binary image from the boundary outlines. Only boundaries that are complete and unbroken are identified and filled by this routine.

that detects, and marks, the interiors of closed boundaries. MATLAB's Image Processing Toolbox has such a routine called imfill. When it is applied to the image in Figure 11.16, it produces the binary image in Figure 11.17 that fills the interior of cells, producing a mask that can be used to isolate the cells. Some cells are not identified by this routine because they have broken or incomplete borders and, thus, are not closed boundaries.

The next section looks at methods based on image similarities. It is possible, and common, to combine different methods to isolate medical features.

11.5.3 Continuity-Based Methods

Continuity-based approaches look for similarity or consistency in the search for feature elements. These approaches can be effective in segmentation tasks, but they all tend to reduce edge definition. This is because they are based on neighborhood operations that operate on a local area and blur the distinction between edge and feature regions. The larger the neighborhood used, the more poorly the edges will be defined. Increasing neighborhood size usually improves the power of any given continuity-based operation, setting up a compromise between identification ability and edge definition.

One simple continuity-based technique is low-pass filtering. Because a low-pass filter is a sliding neighborhood operation that takes a weighted average over a region, it enhances similarities and consistent features. In the next example we use a Gaussian low-pass filter to segment the image in Figure 11.13.

EXAMPLE 11.9

Use a low-pass filter to enhance the similarities of the two regions in Figure 11.13. The pattern is found in the file texture1.tif. Plot the histograms of the original and filtered signal, and select a boundary from the histogram of the filtered signal. Threshold the signal, and display the filtered and thresholded images.

Solution: The image is loaded in the usual way. To differentiate the two regions, we need to filter over a large area, so we use a 10-by-10 Gaussian filter with a σ of 1.5. From the filtered image histogram, we select a threshold that is at a low point between the two distributions. A BW thresholded image is generated using thresh_image and displayed.

```
% Example 11.9 Demonstrates the effectiveness of simple linear filtering
% in separating two segments containing noise.
%
I = importdata('texture1.tif');          % Load image
I = double(I);                           % Convert data
b = gaussian(10, 1.5);                   % Gaussian lowpass filter
I_lowpass = filter2(b,I);                % Apply Gaussian filter
% Plot histograms
subplot(2,1,1);
  histogram(I);
  ......title and labels.......
subplot(2,1,2);
  histogram(I_lowpass);
  ......title and labels.......
threshold = input('Input threshold: ');  % Input threshold (approx.. 110)
BW = thresh_image(I_lowpass,threshold);  % Apply threshold
......new figure; display images; add titles.......
```

Results: The histograms are shown in Figure 11.18. The effect of low-pass filtering is evident from the histogram. The histogram of the original signal has a unimodal distribution, whereas the histogram of the filtered signal is bimodal, reflecting the two regions. The low point between the two distributions occurs at an intensity of around 110. Thresholding the filtered image at this value results in perfect separation of the two regions. The filtered image and the BW image separating the two regions are shown in Figure 11.19. This example shows the power of simple low-pass filtering to identify ROIs based on similarities.

11.5.3.1 *Texture Analysis and Nonlinear Filtering*

Image features related to "texture" can be particularly useful in segmentation. Figure 11.20 shows three regions that have approximately the same average intensity values but are distinguished visually because of differences in texture. Several neighborhood-based operations can be used to distinguish textures: the small segment Fourier transform, local variance (or standard deviation), the "Laplacian" operator, the "range" operator (the difference between maximum and minimum pixel values in the neighborhood), the "Hurst" operator (maximum difference as a function of pixel separation), and the "Haralick" operator (a measure of distance moment). These operations are neighborhood operations, but, unlike linear filtering, they involve nonlinear processes. The Image Processing Toolbox has a routine, nlfilter, that can implement some of these operations, but it is not too difficult to code some of these features in standard MATLAB.

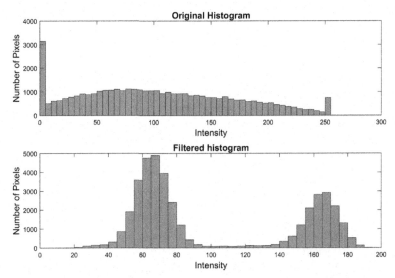

FIGURE 11.18 Histogram of the image shown in Figure 11.13 before (upper) and after (lower) low-pass filtering. Before filtering, the two regions overlap to such an extent that they cannot be identified in the histogram. After low-pass filtering, the two regions are evident in the bimodal distribution. An intensity value around 110 is a low point between the two distributions. Thresholding the filtered image at this value results in perfect separation as shown in Figure 11.19.

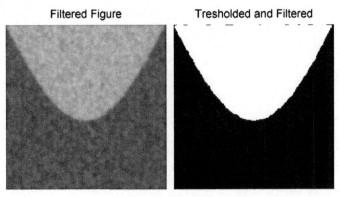

FIGURE 11.19 Left: the same image as in Figure 11.13 after low-pass filtering. Right: the two regions can now be separated perfectly by thresholding.

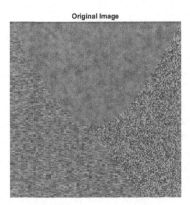

FIGURE 11.20 An image containing three regions having approximately the same intensity but different textures. Although these areas can be distinguished visually, separation based on intensity or edges will surely fail.

EXAMPLE 11.10

Separate out the three segments in Figure 11.20 that differ only in texture. The image is found in the file texture3.tif. Use one of the texture operators described above, the "range" operator, and demonstrate the improvement in separability through histogram plots. Determine appropriate threshold levels for the three segments from the histogram plot, and plot the segments.

Solution: First we need a program to code the nonlinear filter. The function nonlfilter takes in an image and produces a nonlinearly filtered output image of the same size. The routine first extends the horizontal and vertical extent of the original image with zero padding to take care of edge effects. Then a sliding neighborhood operation is applied to compute the range operation. The program defines an ROI (variable roi) in the image corresponding to the filter's dimensions. The range operator value is determined as the difference in maximum and minimum values within this ROI. The ROI is then shifted one pixel position, first horizontally then vertically. At each position, a new range value is computed into the output image matrix.

Because we have done all this work to isolate the ROIs, we might as well throw in some possible other nonlinear computations. Adding the input argument type, we can specify a standard deviation operation ('std') and a variance operation ('var') in addition to the range operation ('range'). Because the ROI is 2-D, we need to use the MATLAB commands std2 and var2 to compute the regional standard deviation and variance.

```
function I1 = nonlfilter(I,type,m,n)
% Nonlinear filtering. m and m specify filter size.
%
[M,N] = size(I);                    % Original image size
h_pad =ceil( (n-1)/2);              % Padding dimensions
v_pad = ceil((m-1)/2);              % m and n must be odd
I = [zeros(v_pad,N); I; zeros(v_pad,N)];   % Pad top and bottom
[M1,N1] = size(I);
I = [zeros(M1,h_pad), I, zeros(M1,h_pad)]; % Pad left and right
%
for n1 = 1:N                        % Isolate a region of interest
    for m1 = 1:M
        roi = I(m1:m1+m-1,n1:n1+n-1);       % Define region of interest
        if type(1) == 's'
           I1(m1,n1) = std2(roi);           % Apply operations standard deviation
        elseif type(1) == 'v'
           I1(m1,n1) = var(roi);            % Variance
        elseif type(1) == 'r'
           I1(m1,n1) = max(max(roi)) - min(min(roi)); % Range operator
        end
    end
end
```

Next we use the nonlinear range operator to convert the textural patterns into differences in intensity. The range operator is a sliding neighborhood procedure that sets the center pixel to the difference between the maximum and minimum pixel value found within the neighborhood. We implement this operation in the main program over a seven-by-seven neighborhood. This neighborhood size was found empirically to produce good results.

The three regions are then thresholded using thresholds determined from the histograms. The right-side texture is segmented by an upper threshold. The upper texture is segmented by inverting the BW image and applying the lower threshold to this inverted image. Because these images are binary, inverting can be accomplished using the logical 'not' operation represented in MATLAB by the symbol $\sim$. The remaining texture can be found using a logical combination (the 'and' operation) applied to the two isolated images after inverting.

```
% Example 11.10 Texture analysis using nonlinear filtering
%
M = 7;          % Define filter neighborhood size, horizontal
N = 7;          % and vertical
I = importdata('texture3.tif');      % Load image
I = double(I);                        % Convert data
I_nl = nonlfilter(I,'range',M,N);    % Apply nonlinear filter
% Plot histograms
subplot(2,1,1);
  histogram(I);
......title and labels1........
  histogram(I_nl);
  ......title and labels1........
thresh1 = input('Input lower threshold: ');   % Get upper and lower threshold
thresh2 = input('Input upper threshold: ');
BW_upper = ~thresh_image(I_nl,thresh1);   % Isolate upper image (note invert)
BW_right = thresh_image(I_nl,thresh2);    % Isolate right image
BW_left = ~I_upper & ~I_right;            % Isolate left from other two
```

Results: The image produced by the range filter is shown in Figure 11.21, and a clear distinction in intensity level can now be seen between the three regions. This is also demonstrated in the histogram plots of Figure 11.22. The histogram of the original figure (upper plot) shows a single Gaussian-like distribution with no evidence of the three patterns.[4] After nonlinear filtering, the three patterns emerge as three distinct distributions. Using this distribution, two thresholds are chosen where the minimum value occurs between the distributions: 50 and 110.

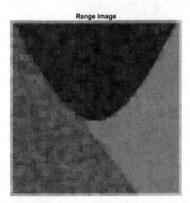

FIGURE 11.21 The texture pattern shown in Figure 11.20 after application of the nonlinear range operation. This operator converts the textural properties in the original figure into a difference in intensities. The three regions are now clearly visible as intensity differences and can be isolated using thresholding.

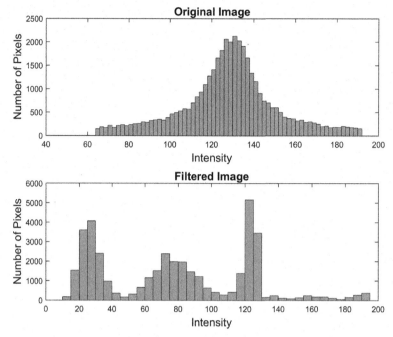

FIGURE 11.22   Histogram of original texture pattern before (upper) and after nonlinear filtering using the *range* operator (lower). After filtering, the three intensity regions are seen as three different distributions. The thresholds used to isolate the three segments are 50 and 110.

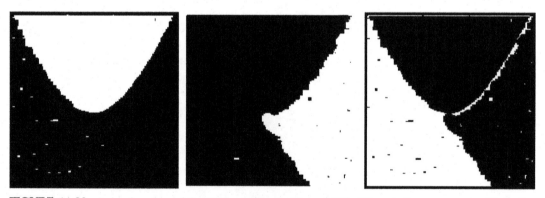

FIGURE 11.23   Isolated regions of the texture pattern in Figure 11.20. Although there are some artifacts, the segmentation is quite good considering the original image. Low-pass filtering and other techniques can be used to improve segmentation as shown in the problems.

The three segments are isolated based on these thresholds in conjunction with logical operations. The three fairly well-separated regions are shown in Figure 11.23. A few artifacts remain in the isolated images. The separation can also be improved by applying low-pass filtering to the range image and other minor modifications as demonstrated in the problems.

---

[4]In fact, the distribution is Gaussian because the image patterns were generated by filtering an array filled with Gaussainly distributed numbers generated by `randn`.

FIGURE 11.24 Textural pattern used in Problem 18. The upper and left diagonal features have similar statistical properties and would not be separable with a standard range operator. However, they do have different orientations. As in Example 11.10, all three features have the same average intensity.

Occasionally, segments have similar intensities and textural properties, except that the texture differs in orientation. Such patterns can be distinguished using a variety of nonlinear operators that have orientation-specific properties. The local Fourier transform can also be used to distinguish orientation. Figure 11.24 shows a pattern with texture regions that are different only in terms of their orientation. This figure is segmented into the three texture-specific features in one of the problems. When textures differ in orientation, a direction-specific operator followed by a low-pass filter is used.

11.6 SUMMARY

An image can be considered as a 2-D signal and many signal processing tools can be extended to images. MATLAB has routines that implement many signal processing operations on 2-D data, including the Fourier transform, convolution, and correlation (filtering). A single image sample, or pixel, can be considered one element of a 2-D array and is referenced by its row and column location in the array. When converted to the frequency domain, an image spectrum is a function of two spatial frequencies. Images usually contain a large number of pixels, so they are often stored using only one or two bytes per pixel, but they can be converted to MATLAB's `double` format and treated as standard MAT-LAB variables.

Linear image filtering is done using FIR filters where a filter coefficients matrix, $b[k_1,k_2]$, is applied to the image using a modified form of 2-D convolution. Although the filter matrix can be of any size, many image filters use a three-by-three coefficients matrix. Popular filter matrices include the Sobel and Prewitt filters for edge detection, the Gaussian filter for low-pass filtering, the Laplacian of Gaussian filter that approximates a spatial second derivative, and the unsharp (or unsharp masking) filter for high-pass filtering.

Separating out important sections of an image is known as image segmentation and is often used to identify ROIs such as particular organs in a medical image. Image segmentation approaches can be grouped into four classes: pixel-based, edge-based, regional, and morphological methods. Pixel-based methods operate on one pixel at a time and are easy to implement but are not as powerful as the other methods. A common pixel-based method is thresholding, which works well if the ROI is brighter or darker than the surrounding tissue. Multiple thresholds can be applied if the ROI has a specific range of intensity values. An intensity histogram that shows the distribution of intensity values in an image can be helpful in setting the threshold.

Edge detection methods search for intensity differences, i.e., edges, and usually use multiple criteria to outline the ROI. For example, edges can be forced to be continuous and/or conform to a specific shape or be constrained to form closed loops. Continuity-based methods take the opposite tack: they search for similarities with the ROI. One of the most common continuity-based methods is low-pass filtering. Continuity-based methods also use nonlinear filters that enhance specific textural features in the ROI. Finally, morphological methods use information about the shape, properties, and/or mechanical properties of a particular organ or tissue to aid the image segmentation.

PROBLEMS

1. Load the image of the brain in the file `brain2.tif`. Brighten the image by multiplying it by 1.4. Construct a new image that is the same as the original except that any point between 10 and 120 is white. Finally, construct a new image that is the same as the original except that any point that is less than 120 or greater than 215 is black. This operation tends to highlight the gray matter in the brain. Plot the original and three new images together using `subplot`.

2. Load the MATLAB test pattern image found in `testpat1.png`. Plot the image along with a plot of the magnitude Fourier transform of this image. Use `mesh` to plot the Fourier transform and shift the transform before plotting using `fftshift`. (Note `fftshift` shifts only valid points, i.e., $f_s/2$, so you do not need to restrict the size of the Fourier transform when shifting.) Then eliminate the mean in the Fourier transform by shifting only the second data point to the end, i.e., shift `F(2:end,2:end)`. (Recall, the first point of the Fourier transform contains the mean or average value.) Note the greater detail visible in the spectrum when you eliminate the DC (i.e., zero frequency) component. Nonetheless, the spectrum of this fairly simple image is still difficult to interpret.

3. Load the image in `double_chirp.tif`, which has a chirp going both horizontally and vertically. Plot this image and take the magnitude Fourier transform. As in Problem 2, use `fftshift` to shift the spectrum and `mesh` to plot. The magnitude spectrum is dominated by the DC component and is difficult to interpret. Plot the shifted Fourier transform after removing the DC component from the magnitude spectrum (see Problem 2). Because this is a chirp both horizontally and vertically, we would expect to see a relatively constant spectrum in both spatial frequency directions producing a plateau-like structure, as indeed we do.

4. Plot the magnitude spectrum of the Sobel and Prewitt filters. The spectrum of each filter shows two positive peaks because you are plotting the magnitude spectra. The phase spectra would show that one of the peaks is actually inverted. Although the two spectra are very similar, there are differences at the higher spatial frequencies.

5. Plot the magnitude and phase spectrum of the Laplacian of Gaussian and the unsharp masking filters. You can use routine `lgauss` to get the coefficients of both of these filters. (The first output argument is Laplacian of Gaussian coefficients and the second the unsharp masking coefficients.) Make $\alpha = 0.5$. Note that the two magnitude spectra are almost identical; however, the phase characteristics are very different. Use `subplot` to plot the magnitude and phase together. (The phase plots are easier to read if you change to view to Az $= 30$, El $= 35$, i.e., `view([30,35])`. Also note the phase wrapping in the phase plots;.)

6. Plot the magnitude and phase spectrum of a three-by-three Gaussian low-pass filter and a three-by-three averaging filter, i.e., nine equal weights of 1/9. Use routine `gaussian` to get the Gaussian filter coefficients, but construct the averaging filter yourself. (Suggestion: Use MATLAB's `ones` operator.) Use a value of sigma of 0.5. Observe the complicated phase spectrum of the averaging filter. Also observe that, although the averaging filter has a sharper cutoff, it has what are known as sidelobes. The next problem compares the operation of the two filters. Use subplot to plot the magnitude and phase together. (The phase plot is easier to read if you change to view to Az $= 30$, El $= 35$; i.e., `view([30,35])`; Again note the phase wrapping in the phase plots, particularly for the averaging filter.)

7. Load the image `blood2.tif`, which is an image of blood cells that contains salt and pepper noise. Apply a five-by-five Gaussian low-pass filter and a five-by-five averaging filter. Adjust the Gaussian filter's σ to maximally reduce the noise without making the image of the cell too blurry. Display the original and both filtered images together. Note that, although the ability of both filters to reduce the salt and pepper noise is similar, the Gaussian filter produces a slightly sharper image. (Hint: Use MATLAB's `ones` operator to construct the averaging filter.)

8. Load the image of blood cells found in `blood2.tif`. This image has added salt and pepper noise. Filter this image with two Gaussian low-pass filters of different strengths. A three-by-three filter with $\sigma = 0.5$ and a five-by-five filter with $\sigma = 3.0$. Also plot the magnitude spectra of the two filters. Display the original and both filtered images together.

9. Compare the Laplacian of Gaussian filter with the related unsharp masking filter. Apply both to the MR image of the brain in `brain1.tif`, the image used in Example 11.5. Compare the images from the two filters side by side. The Laplacian of Gaussian filter takes the second derivative so that the output image has both positive and negative values. Hence, when displaying this image set the colormap to range between ±255 using `caxis([-255 255]);`. The background will be gray, not black. The only features in this image are found where the most severe boundaries occur in the original image. Display the original and two filtered images.

10. Load the image `noise_brain2.tif`, which is an MR image of the brain that contains Gaussian noise. Apply a Gaussian low-pass filter that reduces the noise. Adjust filter dimension and σ to maximally reduce the noise without making the image of the cell too blurry. You should be able to get a clean image. Display the original and filtered images.

11. Modify Example 11.8 to apply the four versions of the Sobel filter to the image of bacteria in the file `bacteria.tif`. Plot the histogram of one of the Sobel filtered images and determine the threshold. (Hint: Look for a notch somewhere below the 50%

maximum intensity level.) Apply `thresh_image` to get an outline of the bacteria cells. Plot the original image and its histogram together, the four Sobel filtered images together, and the final combined image. You should get a good outline of the cells. (Recall that if you do not have MATLAB's `histogram` you can use `histogrm` found in the associated files.)

12. Isolate the bacterial cells in `bacteria.tiff` using a Laplacian of Gaussian filter. Use `lgauss` to generate the filter coefficients and then filter the image with `filter2`. The resulting image has very low spatial contrast, and because this filter calculates the second spatial derivative of the image the mean value of the filtered image is near 0.0. Display this image using a scaled-down colormap to enhance the contrast (suggestion: `caxis([ -4 , 40])`. The histogram is not so useful in determining the threshold from this filtered image. However, you know the threshold will be near zero, so try a small positive threshold (between 4 and 8) to isolate the features in the cell. Display the original and filtered images as well as thresholded BW image.

13. Isolate the cell in `cell.tif` using multiple thresholding. This problem uses multiple thresholds to partially isolate a changeling image. The next problem completes the segmentation task. Again the histogram is not useful in thresholding this image, but from the image you can see that the cells appear to be slightly lighter and darker than the gray background. This suggests two thresholding operations. First, threshold the image slightly above 128 (i.e., 50% of the maximum value) to isolate the lighter sections of the cell. Then threshold a grayscale inverted version of this image somewhat below 128 to isolate the darker sections of the cell. You can combine the two BW images, but you need to invert the second image after thresholding so the areas identified as darker are now white. (You can invert the BW image using the not (~) operation and combine using the or (|) because these can be treated as logical variables, but to invert the original grayscale image you need to subtract it from 255 as in Example 11.1.) The resulting image will have some dark areas remaining inside the cell, but these are removed in the next problem. Plot the original images, the two thresholded images, and the combined images. You should be able to generate an image similar to that stored in `Prob13_data.mat`.

14. The variable BW3 in the file `Prob13_data.mat` is a BW image that was obtained by double thresholding an image of bacterial cells in Problem 13. To remove the remaining material inside the cells, apply a Gaussian low-pass filter and then rethreshold the filtered image. Note that the filtered image is between 0 and 1, so the threshold should be very small (<0.1). Use the weakest filter that completely whitens the cell, i.e., the lowest dimension and smallest σ. Display the original and segmented images.

15. Load the blood cell image in `blood1.tif`. Filter the image with two Gaussian low-pass filters having a dimension of 20. One filter should have a gentle spectral slope (for example, $\sigma = 0.5$) and the other should have a sharper slope ($\sigma = 4.5$). Threshold the two filtered images based on the histogram, but adjust for optimum images, i.e., minimum background spots with relatively unbroken cell walls. (Suggestion: The threshold of the image produced by the weaker filter can be about 10%—30% lower than that used for the image from the stronger filter.) Display the filtered images along with their histograms. Also display the thresholded images side by side. (Hint: Good thresholds can be found around 50% of maximum intensity (128).)

16. Repeat Problem 11.15 for the original image shown in Figure 11.13 and found in the file `Prob11_16data.tif`. The histogram provides guidance on thresholds for both filtered images. Note the difference between the two histograms. Also note that the histogram of the weaker filter indicates that the segments cannot be perfectly separated but that complete segmentation is possible with the image from the stronger filter. The downside is that the border is less precisely defined in the heavily filtered image.

17. Recall the Laplacian of Gaussian filter that calculates the second derivative. It can be useful to highlight texture created by small edges in the image. Load the image of the spine found in `spine.tif`. Filter this image using the Laplacian of Gaussian filter obtained from `lgauss` with $\alpha = 0.5$. Then threshold this image. In this image the edges are located where the second derivative is near zero, so you want to take a low threshold (around 0). Note the extensive tracing of subtle boundaries that this filter produces.

18. In this problem we segment the complicated texture image shown as a teaser in Figure 11.24 and found in file `texture4.tif`. We again apply the range nonlinearity but with a nonsquare region of interest to isolate areas that have different orientations. In particular, we make $m = 9$ and $n = 1$ in our call to `nonlfilter: nonlfilter(I,'range', 9,1)`. Also apply an additional Gaussain low-pass filter with a dimension of 9 and $\sigma = 3$. Plot the two filtered images and the histogram for each. The histograms show the improvement in the ability to separate the textures because of low-pass filtering. Although the range and low-pass filtered histogram provides guidance on the placement of thresholds, you may be able to improve separation by adjusting the thresholds. You should be able to achieve almost the same level of separation as in Figure 11.23. Submit the original and segmented images.

19. The last three problems explore methods to improve the separation of the complex textures shown in Figures 11.20 and 11.24. Load the texture orientation image used in Problem 18 and shown in Figure 11.24 that is found in `texture4.tif`. Separate the segments by first preprocessing the image with a Sobel filter that enhances horizontal edges that are brighter on top. Then apply a nonlinear filter using a standard deviation operation determined over a two-by-nine pixel grid. Finally postprocess the image from the nonlinear filter using a strong Gaussian low-pass filter. (Note that you should multiply the output of the nonlinear filter by around 3.5 before applying the Gaussian filter to get it into an appropriate range.) Plot the output image of the nonlinear filter and that of the low-pass-filtered image and the two corresponding histograms. Separate into three segments as in Example 11.10. Use the histogram of the low-pass-filtered image to determine the best boundaries for separating the three segments, although you may need to adjust the thresholds for best separation. Display the three segments as white objects. You should be able to get clean separation for two of the three textures. Submit the original and segmented images.

20. Now we need to improve the separation of the texture image used in Example 11.10 with a low-pass filter. Follow the same procedure used in Example 11.10. Load the image in the file `texture3.tif` and apply the range nonlinearity with a seven-by-seven ROI. Then filter the nonlinear image with a Gaussian low-pass filter. This is the same strategy used in Problem 19 with the textures found in `texture4.tif`. As in Problem 19 you should be able to get perfect isolation of the left and center images.

Set the low-pass filter's dimensions and σ as low as possible yet still with good separation. Submit the original and segmented images.

21. Next we improve the separation of the texture image used in Example 11.10 and Problem 20. Follow the same procedure as in Problem 20 using the same thresholds. Before displaying the three BW images, low-pass filter BW3 using the same Gaussian filter that was applied to the image from the nonlinear filter. Again threshold this low-pass-filtered image to produce an image that affords perfect separation. (Hint: For this last image you need a high threshold, but remember this filtered image ranges between 0 and 1, so >0.85.)

A caveat on this and other segmentation problems in this chapter: Although the relatively simple methods we use here achieve good image separations, they may not be good examples of segmentation. The problem is that when we use low-pass and other filters, the borders we obtain do not precisely align with the edges of the tissue we want to isolate. That is, the segments we have identified in these problems do not exactly overlap the tissue (or texture) of interest. Greater care in identifying boundaries is required and more advanced approaches exist for boundary detection of tissue segment. Image segmentation is a very challenging area of biomedical engineering and, as mentioned previously, an active research area. Nonetheless, these problems do give a good introduction to the tasks and techniques of biomedical image analysis.

PART III

CIRCUITS

12

Circuit Elements and Circuit Variables

12.1 GOALS OF THIS CHAPTER

In this section of the book we turn our attention to two analog systems: electrical circuits (or networks) and mechanical systems.[1] The overreaching goal of the next four chapters is to find a mathematical representation of signals (or variables) in a circuit or mechanical system and, if appropriate, find the transfer function. There are significant differences between the analog systems described in these chapters and the general systems described in Chapters 6 and 7. The elements appear to be different, but in fact they are described by the same simple calculus equations used for general systems elements. This means we can use the same simplifying approaches developed in Chapters 6 and 7 analyzing these systems in either the frequency domain (i.e., phasor) or Laplace domain. What is different is the way in which the elements interact, so we need to develop a different set of rules to find the transfer function of an analog system. Once the transfer function is determined, the behavior of a mechanical or electrical system can be evaluated by direct application of our time domain, frequency domain, or Laplace domain toolkits (convolution, Bode plots, inverse Fourier transform, and inverse Laplace transform).

This introductory chapter is devoted largely to definitions employed in mechanical and electrical systems: defining the signal variables and the basic elements. The main signal variable in electrical systems is voltage, but another electrical variable, current, can also be important. The analogous mechanical variables are force and velocity. Mechanical and electrical elements are defined using calculus operations and readily slip into Laplace or frequency domain (i.e., phasor) representation.

Choosing between phasor and Laplace analysis for a specific circuit or mechanical system follows the same rules used for the more general systems discussed in Chapters 6 and 7. The Laplace transform is more general as it applies to systems exposed to a wider variety of

---

[1]The terms "network" and "circuit" are completely interchangeable and I use them more or less randomly. "Analog system" is a general term that includes mechanical and electrical systems as well as most biological systems.

signals, including transient or step-like signals, or to circuits with nonzero initial conditions. Phasor analysis is easier to use, but applies only to circuits and mechanical systems in steady state with signals that can be decomposed into sinusoids. The basic rules for analyzing a circuit or mechanical system are independent of which technique, Laplace or phasor, is used to handle the resulting differential equations. Of course, these approaches still require that the underlying processes be linear, but electrical elements are surprisingly linear and electrical systems are as close to LTI systems as is found in the real world.

Specific topics of this chapter include:

- Define the signal variables of mechanical and electrical systems.
- Define variables associated with specific elements such as power and energy.
- Describe the general features of mechanical and electrical systems (or models based on mechanical or electrical elements) and show how they differ from systems models described in Chapters 6 and 7.
- Define the basic active and passive elements in electrical systems in the time, frequency (phasor), and Laplace domain.
- Define the basic active and passive elements of mechanical systems in the same three domains.

12.2 SYSTEM VARIABLES: THE SIGNALS OF ELECTRICAL AND MECHANICAL SYSTEMS

Electric and electronic circuits consist of arrangements of basic elements that define relationships between voltages and currents. The mechanical systems covered here are arrangements of mechanical elements that define relationships between force and velocity. In both electric circuits and mechanical systems only two variables, voltage/current and force/velocity, define the behavior of an element.[2] From another perspective, an element, electrical or mechanical, can be viewed as forcing a specific relationship between the two variables. Although voltage and current seem quite different from force and velocity, they have much in common: one variable, voltage or force, is associated with potential energy; the other variable, current or velocity is related to kinetic energy. The potential energy variable may be viewed as the cause of an action, whereas the kinetic energy variable is the effect: voltage across an element causes current to flow and force applied to an element causes it to move. To emphasize the strong relationship between mechanical and electrical systems, both these variables are defined in the next section.

12.2.1 Electrical and Mechanical Variables

For electric circuits, the major variables are voltage and current. Voltage, the potential energy variable, is sometimes called "potential." When voltage is applied to a circuit element

---

[2]There are also energy considerations—how much energy an element dissipates or supplies—but these variables are associated with a specific element and do not interact with, or influence, other elements in the system.

it causes current to flow through that element. Voltage is the push behind current. It is defined as a potential energy: the energy with respect to charge:

$$v = \frac{dE}{dq} \left(\frac{J}{C}\right) V \tag{12.1}$$

where v is voltage, E is the energy of an electric field, and q is charge. The basic units of V (volts) is J (joules) per C (coulomb). (Slightly different typeface will be used in this text to represent voltage, v, and velocity, v, to minimize confusion.)

The kinetic energy or flow variable that results from voltage is current. Current is the flow of charge: the differential change in charge with time:

$$i = \frac{dq}{dt} \left(\frac{C}{s}\right) A \tag{12.2}$$

where i is current and q is charge. The basic units of A (amps) is C (coulombs) per s (seconds).

A variable that is uniquely associated with a specific element or an entire system defines the energy used (or supplied) over a given time period. Energy per unit time is termed "power" and for electrical elements is given as:

$$P = \frac{dE}{dt} \left(\frac{J}{s}\right) W \tag{12.3}$$

where W (watts) are defined as J/s (Joules per s), To relate power to the variables v and i, note that from Equation 12.1:

$$dE = vdq \tag{12.4}$$

Substituting this into the definition of energy for electrical elements (Equation 12.3) gives:

$$P = \frac{dE}{dt} = v\left(\frac{dq}{dt}\right) = vi \ W \tag{12.5}$$

In mechanical systems the two variables are force, which is related to potential energy and velocity, which is related to kinetic energy. The relationship between force and energy can be derived from the basic definition of energy in mechanics:

$$E = \text{Work} = \int F \, dx = Fx \ J \quad \text{(if } F \text{ constant over } x) \tag{12.6}$$

Solving Equation 12.6 for F by differentiating both sides:

$$F = dE/dx \tag{12.7}$$

where F is force in dynes (dyn) and x is distance in centimeters (cm).

TABLE 12.1 Major Variables in Mechanical and Electrical Systems

Domain	Potential Energy Variable (Units)	Kinetic Energy Variable (Units)
Mechanical	Force, F $F = \frac{dE}{dx}$ (J/cm = dyn)	Velocity, v $v = \frac{dx}{dt}$ (cm/s)
Electrical	Voltage, v $v = \frac{dE}{dq}$ (J/C = V)	Current, i $i = \frac{dq}{dt}$ (C/s = A)

The kinetic energy variable, the variable caused by force, is velocity:

$$v = \frac{dx}{dt} \ \text{cm/s} \tag{12.8}$$

Power in mechanical systems can be shown to again be the product of the potential and kinetic energy variables (Equation 12.3). Starting with the basic definition of power (Equation 12.3):

$$P = \frac{dE}{dt}$$

From the definition of force (Equation 12.7):

$$P = \left(\frac{dE}{dx}\right)\left(\frac{dx}{dt}\right) = F\left(\frac{dx}{dt}\right) = Fv \tag{12.9}$$

where P is power, F is force, x is distance, and v is velocity.

Table 12.1 summarizes the variables used to describe the behavior of mechanical and electrical systems. The tools developed in this chapter are first introduced in terms of electrical circuits, and later in this chapter are applied to certain mechanical systems. The mechanical analysis described later in this chapter could be applied to mechanical systems, but for biomedical engineers the most likely application is to analog models of physiological processes that use mechanical elements.

12.2.2 Voltage and Current Definitions

Analyzing a circuit usually means finding the voltages (or currents) in the circuit or finding its transfer function. In electrical systems, the transfer function is almost always a ratio of voltages, specifically, the voltages considered as input and output, $TF = V_{out}/V_{in}$.

Voltage is a relative variable: it is the difference between the voltages at two points. In fact the proper term for voltage is "potential difference" (abbreviated p.d.), but this term is rarely used by engineers. Subscripts are sometimes used to indicate the points from which the potential difference is measured. For example, in Equation 12.10 the notation v_{ba} means "the voltage at point b with respect to point a":

$$v_{ba} = v_b - v_a \tag{12.10}$$

The positive voltage point, point "b" in Equation 12.10, is indicated by a plus sign when drawn on a circuit diagram, as shown in Figure 12.1. Given the position of the plus sign in Figure 12.1, it is common to say that there is a voltage drop from point b to point a (i.e., left to right), or a voltage rise from a to b. By this convention, it is logical that v_{ab} should be the negative of v_{ba}: $v_{ab} = -v_{ba}$. Voltage always has a direction, or "polarity," that is usually indicated by a "+" sign to show the side assumed to have a greater voltage, Figure 12.1.

A source of considerable confusion is that the "+" sign indicates only the point that is assumed to have a more positive value for the purpose of analysis or discussion. Sometimes the plus position is selected arbitrarily. It could be that the voltage polarity is actually the opposite of what was originally assigned. As an example, suppose that b was, in fact, more negative than a, i.e., there is actually a rise in voltage from b to a for the element in Figure 12.1. Even though we have already made the assumption that b was the more positive as indicated by the "+" sign, we do not change our original polarity assignment. We merely state that v_{ba} has a negative value. So a negative voltage does not imply negative potential energy, it is just that the actual polarity is the reverse of that assumed.

By convention, but with some justification as is shown later, the voltage of the earth is assumed to be at 0.0 V, so voltages are often measured with respect to the voltage of the earth, or some common point referred to as "ground." A common ground point is indicated by either of the two symbols shown at the bottom of the simple circuit shown in Figure 12.4. Some conventions use the symbol on the right side to mean a ground that is actually connected to earth, whereas the symbol on the left side indicates a common reference point not necessarily connected to the earth, but still assumed to be 0.0 V. However, this usage is not standardized and the only assumption that can be made with certainty is that both symbols represent a common grounding point that may or may not be connected to earth, but is assumed to be 0.0 V with respect to the rest of the circuit. Hence when a voltage is given with only one subscript, v_a, it is understood that this is voltage at a with respect to ground or a common reference point.

Current is a flow so it must have a direction. This direction is indicated by an arrow as in Figures 12.1 and 12.2, but it is an assumed direction as we often do not yet know the actual direction when we assign the direction. By convention, the direction of the arrow indicates the direction of assumed positive charge flow. In electric or electronic circuits, charge is

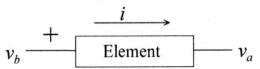

FIGURE 12.1 A generic electric circuit element demonstrating how voltage and current directions are specified. The *straight lines* on either side indicate wires connected to the element.

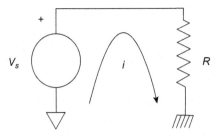

FIGURE 12.2 A simple circuit consisting of a voltage source, V_s, and a resistor R. Two different symbols for grounding points are shown. Sometimes the symbol on the right is taken to mean an actual connection to earth, but this is not standardized.

usually carried by electrons, which have a negative value for charge, so the particles that are actually flowing, the electrons, are flowing in the opposite direction of (assumed) positive charge flow. Nonetheless, the convention of defining positive charge flow was established by Benjamin Franklin before the existence of electrons was known and has not been modified because it really does not matter which direction is taken as positive as long as we are consistent.

As with voltage, it may turn out that positive charge flow is actually in the direction opposite to that indicated by the arrow. Again, we do not change the direction of the arrow, rather we assign the current a negative value. So a negative value of current flow does not mean that some strange positrons or anti-particles are flowing but only that the actual current direction is opposite to our assumed direction.

This approach to voltage polarity and current direction may seem confusing, but it is actually quite liberating. It means that we can make our polarity and direction assignments without worrying about reality, that is, without concern for the voltage polarity or current direction that actually exists in a given circuit. We can make these assignments more or less arbitrarily (there are some rules that must be followed as described later) and, after the circuit is analyzed, allow the positive or negative values to indicate the actual direction or polarity.

12.3 ANALOG SYSTEM VERSUS GENERAL SYSTEMS

Both analog systems and the general systems studied in Chapters 6 and 7 consist of collections of elements. Moreover, we find in the next section that analog elements can be defined by the same basic equations used for general system elements. But the way in which the elements interact is different.

Figure 12.3 shows a generic two-element system and a two-element electric circuit. In the system model, Figure 12.3A, Element 1 acts on Element 2, but Element 2 has no influence on Element 1. If there was such an interaction, it would be explicitly shown as a feedback

(A)   **(B)**

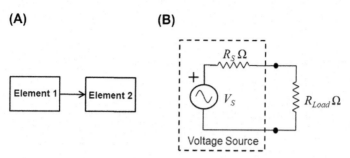

FIGURE 12.3 (A) A generic two-element system. Element 1 influences Element 2, but Element 2 has no influence on Element 1. (B) A typical two-element circuit. The realistic voltage source (*dashed line*) influences the load resistor, R_{Load}, but this resistor also influences the voltage source.

pathway. In the electrical circuit, Figure 12.3B, Element 1 is a realistic battery consisting of an ideal source and a resistor, and the second element, R_{Load}, is a load resistor. When the voltage source is connected to the load resistor it creates a voltage across the resistor, which causes current to flow through this element. But the resistor also has an effect on the voltage source because it draws current, which decreases its output voltage because of its internal resistance, R_S. So there is a hidden feedback pathway between the voltage source and the load resistor.

In systems representations all influences between elements are explicitly shown, but in analog models the interactions between elements are usually implicit. This means that the techniques used to develop equations are different for the two types of systems. In systems models, we simply multiply the individual transfer functions together to find the transfer function for any number of series elements. If feedback is involved, we apply the feedback equation (Equation 6.7) to the combined transfer function of feedforward and feedback pathways. In analog models, the approach to developing and overall transfer function is more complicated, but still consists of a set of rules that, if correctly applied, produce the desired equation. These rules are detailed in the next chapter.

12.4 ELECTRICAL ELEMENTS

The elements as described later are idealizations: true elements only approximate the characteristics described. However, actual electrical elements come quite close to these idealizations, so their shortcomings can usually be ignored, at least with respect to that famous engineering phrase "for all practical purposes."

Electrical elements are defined by the mathematical relationship they impose between the voltage and current. They are divided into two categories based on their energy use: active elements can supply energy to a circuit; passive elements cannot. Active elements do not always supply energy; in some situations they actually absorb energy (imagine charging a battery), but they have the ability to supply energy.

12.4.1 Passive Electrical Elements

Passive elements are divided into two subcategories: those that use up, or dissipate, energy and those that store energy. Only one passive element falls into the first category, the resistor. This element is described first.

12.4.1.1 Energy Users: Resistors

The resistor is the only element in the first group of passive elements: elements that use up energy. Resistors dissipate energy as heat. The defining equation for a resistor, the basic voltage−current relationship, is the classic Ohm's law:

$$v_R = R\,i_R \ \text{V} \tag{12.11}$$

where R is the resistance in volts/amp, better known as ohms (Ω), i is the current in amps (A), and v is the voltage in volts (V). The resistance value of a resistor is a consequence of a property of the material from which it is made, known as resistivity, ρ. The resistance value is determined by this resistivity and the geometry and is given by:

$$R = \rho\frac{l}{A} \ \Omega \tag{12.12}$$

where ρ is the resistivity of the resistor material, l is the length of the material, and A is the area of the material. Table 12.2 shows the resistivity, ρ, of several materials commonly used in electric components.

The power that is dissipated by a resistor can be determined by combining Equation 12.5, the basic power equation for electrical elements, and the resistor-defining equation, Equation 12.11:

$$P = vi = v\left(\frac{v}{R}\right) = \frac{v^2}{R} \ \text{W}$$

or:

$$P = vi = (iR)i = i^2R \ \text{W} \tag{12.13}$$

TABLE 12.2 Resistivity of Common Conductors and Insulators

Conductors	ρ (Ohm-m)	Insulators	ρ (Ohm-m)
Aluminum	2.74×10^{-8}	Glass	$10^{10}-10^{14}$
Nickel	7.04×10^{-8}	Lucite	$>10^{13}$
Copper	1.70×10^{-8}	Mica	$10^{11}-10^{15}$
Silver	1.61×10^{-8}	Quartz	75×10^{16}
Tungsten	5.33×10^{-8}	Teflon	$>10^{13}$

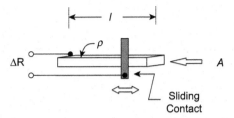

FIGURE 12.4 A variable resistor made by changing the effective length, Δl, of the resistive material.

The voltage–current relationship expressed by Equation 12.11 can also be stated in terms of the current:

$$i = \frac{1}{R}v = Gv \text{ A} \tag{12.14}$$

The inverse of resistance, R, is termed the "conductance," G, and has the units of mhos (ohms spelled backward, a rare example of engineering humor) and is symbolized by the upside down omega, $\mho$. Equation 12.12 can be exploited to make a device that varies in resistance, usually by varying the length l, as shown in Figure 12.4. Such a device is termed a "potentiometer" or "pot" for short. The two symbols that are used to indicate a variable resistor are shown in Figure 12.5B.

By convention, power is positive when it is being lost or dissipated. Hence resistors must always have a positive value for power. In fact, one way to define a resistor is to say that it is a device for which $P > 0$ for all t. For P to be positive, the voltage and current must point in the same direction, that is, they must have the same orientation. In other words the current direction must point in the direction of the voltage drop. This polarity restriction is indicated in Figure 12.5A along with the symbol that is used for a resistor in electric circuit diagrams or "schematics." To meet the positive power criterion, the voltage and current polarities must be set so that current flows into the positive side of the resistor as shown in Figure 12.5A. Either the voltage direction (+ side) or the current direction can be chosen arbitrarily, but not both. Once either the voltage polarity or the current direction is selected, the other is fixed by power considerations. The same will be true for other passive elements, but not for source elements. This is because source elements can, and usually do, supply energy, so their associated power usage is usually negative.

(A) I_R $R\,\Omega$

   + V_R

(B) $R\,\Omega$

$R\,\Omega$

FIGURE 12.5 (A) The symbol for a resistor along with its polarity conventions. For a resistor, as with all passive elements, the current direction must be such that it flows from positive to negative. In other words, current flows into the positive side of the element. (B) Two symbols that denote a variable resistor or potentiometer.

Figure 12.5B shows two symbols used to denote a variable resistor such as shown in Figure 12.4.

EXAMPLE 12.1

Determine the resistance of 100 ft of #14 AWG copper wire.

Solution: A wire of size #14 AWG (American Wire Gauge or B. & S. gauge) has a diameter of 0.064 in. (see Appendix D). The value of ρ for copper is 1.70×10^{-8} Ω-m (Table 12.2). Convert all units to the cgs system and then apply Equation 12.12. To ensure proper unit conversion, we will carry the dimensions into the equation and make sure they cancel to give us the desired dimension. This approach is sometimes known as dimensional analysis and can be very helpful where there are a lot of unit conversions.

$$l = 100\mathrm{ft}\left(\frac{12\,in}{1\,\mathrm{ft}}\right)\left(\frac{2.54\mathrm{cm}}{in}\right) = 3048\ \mathrm{cm}$$

$$A = \pi r^2 = \pi\left(\frac{d\ \mathrm{cm}}{2}\right)^2 = \pi\left(\frac{2.54\ \mathrm{cm}}{in}\frac{0.064\ in}{2}\right)^2 = 0.0208\ \mathrm{cm}^2$$

$$R = \rho\left(\frac{l}{A}\right) = 1.7x10^{-8}\Omega\mathrm{m}\left(\frac{100\mathrm{cm}}{1\mathrm{m}}\right)\left(\frac{3048\mathrm{cm}}{0.0208\,cm^2}\right) = 0.2491\Omega$$

12.4.1.2 *Energy Storage Devises: Inductors and Capacitors*

Energy storage devices can be divided into two classes: "inertial" elements and "capacitive" elements. The corresponding electrical elements are the inductor and capacitor, respectively, and the voltage–current equations for these elements involve differential or integral equations.

12.4.1.2.1 INDUCTOR

Current flowing into an inductor carries energy that is stored in a magnetic field. The voltage across an inductor is the result of a self-induced electromotive force that opposes that voltage and is proportional to the time derivative of the current:

$$v = L\frac{di}{dt} \tag{12.15}$$

where L is the constant of proportionality termed the "inductance" measured in henrys (h). (The henry is actually Weber-turns per amp, or volts per amp/s, and is named for the American physicist Joseph Henry, 1797–1878.) An inductor is simply a coil of wire that utilizes mutual flux coupling (i.e., mutual inductance) between the wire coils. The inductance is related to the magnetic flux, Φ, carried by the inductor and by the geometry of the coil and the number of loops, or "turns," N:

$$L = \frac{N\Phi}{i}\ \mathrm{h} \tag{12.16}$$

This equation is not as useful as the corresponding resistor equation (Equation 12.11) for determining the value of a coil's inductance because the actual flux in the coil depends on the shape of the coil. In practice, coil inductance is determined empirically.

The energy stored can be determined from the equation for power (Equation 12.5) and the voltage–current relationship of the inductor (Equation 12.15):

$$P = vi = Li\left(\frac{di}{dt}\right) = \frac{dE}{dt} \quad \text{solving for } i$$

$$dE = Pdt = Li\left(\frac{di}{dt}\right)dt = Li\ di$$

The total energy stored as current increases from 0 to a value I given by:

$$E = \int dE = L\int_0^I idi = \frac{1}{2}LI^2 \tag{12.17}$$

Later in this chapter we find there is a similarity between the equation for kinetic energy of a mass (Equation 12.55) and the energy in an inductor. Equation 12.17 explains why an inductor is considered an inertial element. It behaves as if the energy is stored as kinetic energy associated with a flow of moving electrons and that is a good way to conceptualize the behavior of an inductor, even though the energy is actually stored in an electromagnetic field. Inductors follow the same polarity rules as resistors. Figure 12.6 shows the symbol for an inductor, a schematic representation of a coil, with the appropriate current–voltage directions.

If the current through an inductor is constant (i.e., "direct current" or "DC"), then there will be no energy stored in the inductor and the voltage across the inductor will be zero, regardless of the amount of steady current flowing through the inductor.[3] The condition when voltage across an element is zero irrespective of the current through the element is known as a "short circuit." Hence an inductor appears as a short circuit to a DC current, a feature that can be used to solve certain electrical circuit problems encountered later.

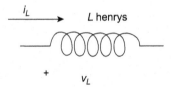

FIGURE 12.6 Symbol for an inductor showing the polarity conventions for this passive element.

[3]The term "DC" originally stood for direct current, but it has been modified to mean "constant value" so it can be applied to either current or voltage: as in "DC current" or "DC voltage."

The *v-i* relationship of an inductor can also be expressed in terms of current. Solving Equation 12.15 for i:

$$v_L = L\frac{di_L}{dt}; \quad di_L = \frac{1}{L}v_L dt; \quad \int di_L = \int \frac{1}{L}v_L dt$$

$$i_L = \frac{1}{L}\int v_L dt \tag{12.18}$$

The integral of any function will be continuous, even if that function contains a discontinuity as long as that discontinuity is finite. A continuous function is one that does not change instantaneously, i.e., for a continuous function:

$$f(t-) = f(t+) \quad \text{for any } t \tag{12.19}$$

Since the current through an inductor is the integral of the voltage across the inductor (Equation 12.18), the current will be continuous in real situations because any voltage discontinuities that occur will surely be finite. Thus the current through an inductor can change slowly or rapidly (depending on the current), but it can never change in a discontinuous or step-like manner. In mathematical terms, for an inductor:

$$i_L(t-) = i_L(t+) \tag{12.20}$$

Since the current passing through an inductor is always continuous, one of the popular applications of an inductor is to reduce current spikes (i.e., discontinuities). Current that is passed through an inductor will have any spikes "choked off," so an inductor used in the purpose of spike reductions is sometimes called a "choke."

12.4.1.2.2 CAPACITOR

A capacitor also stores energy, in this case in an electromagnetic field created by oppositely charged plates.[4] In the case of a capacitor, the energy stored is proportional to the charge on the capacitor and charge is related to the time integral of current flowing through the capacitor. This gives rise to voltage–current relationships that are the reverse of the relationships for an inductor:

$$v = \frac{1}{C}\int i \, dt \tag{12.21}$$

or solving for i_C:

$$i_C = C\frac{dv}{dt} \tag{12.22}$$

---

[4]Capacitors are nicknamed "caps" and engineers frequently use that term. Curiously, no such nicknames exist for resistors or inductors, except for the occasional use of "choke" for an inductor as noted earlier.

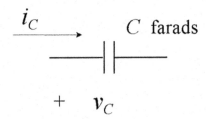

FIGURE 12.7 The symbol for a capacitor showing the polarity conventions.

where C, the "capacitance," is the constant of proportionality and is given in units of "farads," which are coulombs/volt. (The farad is named after Michael Faraday, an English chemist and physicist who, in 1831, discovered electromagnetic induction.) The inverse relationship between the voltage–current equations of inductors and capacitors is an example of "duality," a property that occurs often in electric circuits and in physics in general. The symbol for a capacitor is two short parallel lines reflecting the parallel plates of a typical capacitor, Figure 12.7.

The capacitance is well named because it describes the ability (or capacity) of a capacitor to store (or release) charge without much change in voltage. Stated mathematically, this relationship between charge and voltage is:

$$C = \frac{q}{v} \text{ F} \tag{12.23}$$

where q is charge in coulombs and v is volts. A large capacitor can take on or release charge, q, with little change in voltage, whereas a small capacitor shows a greater voltage change for a given charge flow. The largest capacitor readily available to us, the earth, is considered to be a near-infinite capacitor: its voltage remains constant (for all practical purposes) no matter how much current flows into or out of it. This is why the earth is a popular ground point or reference voltage; it is always at the same voltage and we all agree that this voltage is 0.0 V.

Most capacitors are constructed from two approximately parallel plates. Sometimes the plates are rolled into a circular tube to reduce volume. The capacitance for such a parallel plate capacitor is given as:

$$C = \frac{q}{v} = \varepsilon\frac{A}{d} \tag{12.24}$$

where A is the area of the plates, d is the distance separating the plates, and ε is a property of the material separating the plates termed the "dielectric constant." Although Equation 12.24 is only an approximation for a real capacitor, it does correctly indicate that capacitance can be increased either by increasing the plate area, A, or by decreasing the plate separation, d. However, if the plates are closely spaced and a high voltage is applied, the material between the plates, the "dielectric material," may break down, allowing current to flow directly between the plates. Sometimes this leads to dramatic failure involving smoke or even a tiny explosion. The device, if it survives, is now a resistor (or short circuit) not a capacitor. So capacitors that must work at high voltages need more material, hence more distance, between their plates. For a given capacitance, increasing the distance, d, means the area, A, must be increased. This leads to a larger physical volume so capacitors that can handle large voltages are physically large, much larger than low-voltage capacitors. Alternatively, there are special dielectrics that can sustain higher voltages with smaller distances between plates. Such capacitors are more expensive, a classic engineering trade-off. Referring again to Equation 12.24, capacitors having larger capacitance values also require more plate area again leading to larger physical volume. For a given dielectric material, the size of a capacitor is related to the product of its capacitance and voltage rating, at least for larger capacitance/voltage sizes.

EXAMPLE 12.2

Calculate the dimensions of a 1-F capacitor. Assume a plate separation of 1.0 mm. with air between the plates.

Solution: Use Equation 12.24 and the dielectric constant for a vacuum. The dielectric constant for a vacuum is $\varepsilon_0 = 8.85 \times 10^{-12}\,\mathrm{C^2/N\,m^2}$ and is also used for air.

$$A = \frac{Cd}{\varepsilon_0} = \frac{1\,\mathrm{C}/v(10^{-3}\mathrm{m})}{8.85 \times 10^{-12}\mathrm{C^2/N\,m^2}} = \frac{\dfrac{10^{-3}\,\mathrm{C}}{\mathrm{N\,m/C}}}{8.85 \times 10^{-12}\mathrm{C^2/N\,m^2}} = 1.13 \times 10^8\,\mathrm{m^2}$$

This is an area of about 6.5 miles on a side! This large size is related to the units of farads, which are very large for practical capacitors. Typical capacitors are in the microfarads ($1\,\mu\mathrm{F} = 10^{-6}\mathrm{F}$) or picofarads ($1\,\mathrm{pF} = 10^{-12}\,\mathrm{F}$), giving rise to much smaller plate sizes. An example calculating the dimensions of a practical capacitor is given in the problems.

The energy stored in a capacitor can be determined using modified versions of Equations 12.4 and 12.23:

$$v = q/C \quad \text{and from Equation 12.4:} \quad dE = v\,dq = \frac{q}{C}dq$$

Hence, for a capacitor holding a specific charge, Q:

$$E = \int dE = \frac{1}{C}\int_0^Q q\,dq = \frac{1}{2}\frac{Q^2}{C} \quad \text{Substituting } V = \frac{Q}{C}\;\mathrm{J}$$

$$E = \frac{1}{2}CV^2\;\mathrm{J}$$

(12.25)

Capacitors in parallel essentially increase the effective size of the capacitor plates, so when two or more capacitors are connected in parallel, their values add. If they are connected in series, their values add as reciprocals. Such series and/or parallel combinations are discussed at length in Chapter 14.

Inductors do not allow an instantaneous change in current; capacitors do not allow an instantaneous change in voltage. Since the voltage across a capacitor is the integral of the

TABLE 12.3 Energy Storage and Response to Discontinuous and Direct Current (DC) Variables in Inductors and Capacitors

Element	Energy Stored	Continuity Property	DC Property
Inductor	$E = \frac{1}{2}LI^2$	Current continuous $i_L(0-) = i_L(0+)$	If i_L = constant (DC current) $v_L = 0$ (short circuit)
Capacitor	$E = \frac{1}{2}CV^2$	Voltage continuous $v_C(0-) = v_C(0+)$	If v_C = constant (DC voltage) $i_C = 0$ (open circuit)

current, capacitor voltage will be continuous based on the same arguments used for inductor current. Thus, for a capacitor:

$$v_c(t-) = v_c(t+) \tag{12.26}$$

It is possible to change the voltage across a capacitor either slowly or rapidly depending on the current, but never instantaneously. For this reason, capacitors are frequently used to reduce voltage spikes just as inductors are sometimes used to reduce current spikes. The fact that the behavior of voltage across a capacitor is similar to the behavior of current through an inductor is another example of duality. Again, this behavior is useful in the solution of certain types of problems encountered later in this text.

Capacitors and inductors have reciprocal responses to situations where voltages and currents are constant; i.e., DC conditions. Since the current through a capacitor is proportional to the derivative of voltage (Equation 12.22), if the voltage across a capacitor is constant the capacitor current will be zero irrespective of the value of the voltage (the derivative of a constant is zero). An "open circuit" is defined as an element having zero current for any voltage; hence, capacitors appear as open circuits to DC current. For this reason, capacitors are said to "block DC" and are sometimes used for exactly that purpose.

The continuity and DC properties of inductors and capacitors are summarized in Table 12.3. A general summary of passive and active electrical elements is presented later in Table 12.4.

12.4.1.3 Electrical Elements: Reality Check

The equations given earlier for passive electrical elements are idealizations of the actual elements. In fact, real electrical elements do have fairly linear voltage–current characteristics. However, all real electric elements will contain a combination of resistance, inductance, and capacitance irrespective of their primary function. The undesired properties are termed "parasitic" elements. For example, a real resistor will have some inductor- and capacitor-like characteristics, although these will generally be small and can be ignored except at very high frequencies. (Resistors made by winding resistance wire around a core, so-called wire-wound resistors, have a large inductance as might be expected for this coil-like configuration. However, these are rarely used nowadays.) Similarly, real capacitors also closely approximate ideal capacitors except for some parasitic resistance. This parasitic element appears as a large resistance in parallel with the capacitor, Figure 12.8, leading to a small

TABLE 12.4 Electrical Elements—Basic Properties

Element	Units	Equation $v(t) = f[i(t)]$	Symbol
Resistor (R)	$\Omega \sim$ ohms (volts/amp)	$v(t) = R\, i(t)$	
Inductor (L)	h $\sim$ henry (weber turns/amp)	$v(t) = L\frac{di}{dt}$	
Capacitor (C)	f $\sim$ farad (coulombs/volt)	$v(t) = \frac{1}{C}\int i\,dt$	
Voltage source (V_S)	v $\sim$ volts (joules/coulomb)	$v(t) = V_S(t)$	
Current source (I_S)	a $\sim$ amperes (coulombs/sec)	$i(t) = I_S(t)$	

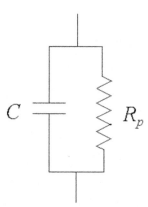

FIGURE 12.8 The schematic of a real capacitor showing the parasitic resistance that would lead to a leakage of current through the capacitor.

"leakage" current through the capacitor. Low-leakage capacitors can be obtained at additional cost with parallel resistances exceeding 12^{12}–$12^{14}\,\Omega$, resulting in very small leakage currents. Inductors are constructed by winding a wire into coil configuration. Since all wire contains some resistance, and a considerable amount of wire may be used in an inductor, real inductors generally include a fair amount of series resistance. This resistance can be reduced by using wire with a larger diameter (as suggested by Equation 12.11), but at the expense of increased physical size.

In most electrical applications, the errors introduced by real elements can be ignored. It is only under extreme conditions, involving high-frequency signals or the need for very high resistances, that these parasitic contributions need be taken into account. The inductor is the least ideal of the three passive elements; it is also the least used in conventional electronic circuitry, so its shortcomings are not that consequential.

12.4.2 Electrical Elements: Active Elements or Sources

Active elements can supply energy to a system, and in the electrical domain they come in two flavors: voltage sources and current sources. These two devices are somewhat self-explanatory. Voltage sources supply a specific voltage, which may be constant or time varying but is always defined by the element. In the ideal case, the voltage is independent of the current going through the source: a voltage source is concerned only about maintaining its specified voltage; it does not care about what the current is doing.

Voltage polarity is part of the voltage source definition and must be included with the symbol for a voltage source as shown in Figure 12.9. The current through the source can be in either direction (again, the source does not care). If the current is flowing into the positive end of the source, the source is being "charged" and is removing energy from the circuit. If current flows out of the positive side, then the source is supplying energy.

The voltage–current equation for a voltage source is just $v = V_{Source}$. Not only are voltages unconcerned with their currents, they really have no control over those currents, at least in theory. The current through a voltage source depends on the circuit elements that are connected to the source.

The energy supplied or taken up by the source is still given by Equation 12.5: $P = vi$. The voltage source in Figure 12.9 is shown as "grounded," that is, one side is connected to

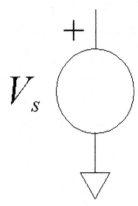

FIGURE 12.9 Schematic representation of a voltage source, V_s. This element specifies only the voltage, including the direction or polarity. The current value and direction are unspecified and depend on the rest of the circuit. Voltage sources are often used with one side connected to ground as shown.

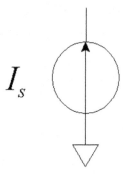

FIGURE 12.10 Schematic representation of a current source, I_s. This element specifies only the current: the voltage value and polarity are unspecified and will be whatever needed to produce the specified current.

ground. Voltage sources are commonly used in this manner and many commercial power supplies have this grounding built in. Voltage sources that are not grounded are sometimes referred to as "floating" sources. A battery could be used as a floating voltage source since neither its positive or negative terminal is necessarily connected to ground.

An ideal current source supplies a specified current, which again can be fixed or time-varying. It cares only about the current running through it. Current sources are less intuitive since current is thought of as an effect (of voltage and circuit elements), not as a cause. One way to think about a current source is that it is really a voltage source whose output voltage is somehow automatically adjusted to produce the desired current. A current source manipulates the cause, voltage, to produce a desired effect, current. The current source does not directly control the voltage across it: it will be whatever it has to be to produce the desired current. Figure 12.10 shows the symbol used to represent an ideal current source.

Current direction is part of the current source specification and is indicated by an arrow, Figure 12.10. Since a current source does care about voltage (except indirectly) it does not specify a voltage polarity. The actual voltage and voltage polarity will be whatever it has to be to produce the desired current.

Again, these are idealizations and real current and voltage sources usually fall short. Real voltage sources do care about the current they have to produce, at least if it gets too large, and their voltages will drop off if the current requirement becomes too high. Similarly, real current sources do care about the voltage across them, and their current output will decrease if the voltage required to produce the desired current gets too large. More realistic representations for voltage and current sources are given in Chapter 14 under the topics of Thévenin and Norton equivalent circuits.

Table 12.4 summarizes the various electrical elements giving the associated units, the defining equation, and the symbol used to represent that element in a circuit diagram.

12.4.3 The Fluid Analogy

One of the reasons analog modeling is popular is that it parallels human intuitive reasoning. To understand a complex notion, we often use metaphors that describe something similar that is easier to comprehend. Some intuitive insight into the characteristics of electrical elements can be made using an analogy based on the flow of a fluid such as water. In this analogy the flow volume of the water would be analogous to the flow of charge in an electric circuit (i.e., current), and the pressure behind that flow would be analogous to voltage. In this analogy, a resistor would be a constriction, or pipe, inserted into the path of water flow. As with a resistor, the flow through this pipe would be linearly dependent on the pressure (voltage) and inversely related to the resistance offered by the constructing pipe. The equivalent of Ohm's law (i.e., $v = i/R$) would be: pressure $=$ flow/resistance. Also as with a resistor, the resistance to flow generated by the pipe would increase linearly with its length and decrease with its cross-sectional area, so the analogy to Equation 12.11 $\left(R = \rho\, l/A \right)$ would be: pipe resistance $=$ constant (length/area).

The fluid analogy to a capacitor would be that of a container with a given cross-sectional area. The pressure at the bottom of the container would be analogous to the voltage across the capacitor, and the water flowing into or out of the container would be analogous to current. As with a capacitor, the pressure at the bottom would be proportional to the height of the water. This pressure or water height would be linearly related to the integral of water flow and inversely related to the area of the container, Figure 12.11.

A container with a large area (i.e., a larger capacity) would be analogous to a large capacitor; it would have the ability to accept larger amounts of water (charge) with little change in bottom pressure (voltage). Conversely, a vessel with a small area would fill quickly so the change in pressure at the bottom would change dramatically with only minor changes in the amount of water in the vessel. Just as in a capacitor, it would be impossible to change the height of the water, and therefore the pressure at the bottom, instantly unless you had an infinite flow of water. With a high flow, you could change the height and related bottom pressure quickly, but not instantaneously.

Water flowing out of the bottom of the vessel would continue to flow until the vessel is empty. This is analogous to fully discharging a capacitor. In fact, even the time course of the outward flow (an ever decreasing exponential) would parallel that of a discharging capacitor. Also, for the pressure at the bottom of the vessel to remain constant, the flow into or out of the vessel would have to be zero, just as the current must be zero for constant capacitor voltage.

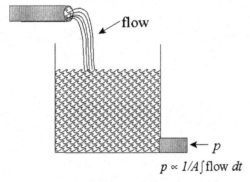

$$p \propto 1/A \int \text{flow}\ dt$$

FIGURE 12.11 Water analogy of a capacitor. Water pressure at the bottom is analogous to voltage across a capacitor, and water flow is analogous to the flow of charge, current. The amount of water contained in the vessel is analogous to the charge on a capacitor.

A water container, like a capacitor, stores energy. In a container, the energy is stored as the potential energy of the contained water. Using a dam as an example, the amount of energy stored is proportional to the amount of water contained behind the dam and the pressure squared. A dam, or any other real container, would have a limited height, and if the inflow of water continued for too long it would overflow. This is analogous to exceeding the voltage rating of the capacitor where the inflow of charge would cause the voltage to rise until some type of failure occurred. It is possible to increase the overflow value of a container by increasing its height, but this would lead to an increase in physical size just as in the capacitor. A real container might also leak, in which case water stored in the container would be lost, either rapidly or slowly depending on the size of the leak. This is analogous to the leakage current that exists in all real capacitors. Even if there was no explicit current outflow, eventually all charge on the capacitor will be lost due to leakage and a capacitor's voltage reduced to zero.

In the fluid analogy, the element analogous to an inductor would be a large pipe with negligible resistance to flow, but in which any change in flow would require some pressure just to overcome the inertia of the fluid. This parallel with inertial properties of a fluid demonstrates why an inductor is sometimes referred to as an "inertial element." For water traveling in this large pipe, the change in flow velocity (d (flow)$/dt$) would be proportional to the pressure applied. The proportionality constant would be related to the mass of the water. Hence the relationship between pressure and flow in such an element would be:

$$p = k \text{ flow velocity} = \frac{k\,d(\text{flow})}{dt} \tag{12.27}$$

which is analogous to Equation 12.15, the defining equation for an inductor. Energy would be stored in this pipe as kinetic energy of the moving water.

The greater the applied pressure, the faster the velocity of the water would change, but just as with an inductor, it would not be possible to change the flow of a mass of water instantaneously using finite pressures. Also as with an inductor, it would be difficult to construct a pipe holding a substantial mass of water without some associated resistance to flow analogous to the parasitic resistance found in an inductor.

In the fluid analogy a current source would be an ideal, constant-flow pump. It would generate whatever pressure was required to maintain its specified flow. A voltage source would be similar to a very-large-capacity vessel, such as a dam. It would supply the same pressure stream, no matter how much water was flowing out of it, or even if water was flowing into it, or if there was no flow at all.

12.5 PHASOR ANALYSIS

If a system or element is being driven by sinusoidal signals or signals that can be converted to sinusoids via the Fourier series or Fourier transform, then phasor analysis techniques can be used to analyze the system's response (see Chapter 6, Section 6.3). The defining equations for an inductor (Equation 12.15) and capacitor (Equation 12.21) contain the calculus operations of differentiation and integration, respectively. In Chapter 6, we found that when the system variables were in phasor notation, differentiation was replaced by multiplying by $j\omega$ (Equation 6.15) and integration was replaced by dividing by $j\omega$ (Equation 6.17). If we use phasor equations to define an inductor and capacitor, the associated equations become algebraic like the equation for a resistor (Equation 12.11).

Since phasor analysis will be an important component of circuit analysis, a quick review of Section 6.3 might be worthwhile. In those cases where the variables are not sinusoidal, or cannot be reduced to sinusoids, we again turn to Laplace transforms to convert calculus operations into algebraic operations. The Laplace domain representation of electrical elements is covered in Section 12.6.

12.5.1 Phasor Representation—Electrical Elements

Converting the voltage–current equation for a resistor from time to phasor domain is not difficult, nor is it particularly consequential since the time domain equation (Equation 12.10) is already an algebraic relationship. Accordingly, the conversion is only a matter of restating the voltage and current variables in phasor notation:

$$V(\omega) = RI(\omega) \tag{12.28}$$

Rearranged as a voltage–current ratio:

$$R = \frac{V(\omega)}{I(\omega)} \tag{12.29}$$

Converting the voltage–current equation of an inductor or capacitor to phasor notation is more consequential as the differential or integral relationships then become algebraic. For an inductor the voltage–current equation in the time domain is given in Equation 12.15 and repeated here:

$$v_L = \frac{L\frac{di(t)}{dt}}{dt} \tag{12.30}$$

But in phasor notation, the derivative operation becomes multiplication by $j\omega$: $L\frac{di(t)}{dt} \Leftrightarrow j\omega\, I(\omega)$ so the voltage–current operation for an inductor becomes:

$$V_L(\omega) = Lj\omega\, I(\omega) = j\omega\, LI(\omega) \tag{12.31}$$

Since the calculus operation has been removed it is now possible to rearrange Equation 12.31 to obtain a voltage-to-current ratio similar to that for the resistor (Equation 12.29):

$$\frac{V(\omega)}{I(\omega)} = j\omega L\ \Omega \tag{12.32}$$

The ability to express the voltage–current relationship as a ratio is part of the power of the phasor domain method. Thus the term $j\omega L$ becomes something like the equivalent resistance of an inductor. This resistor-like ratio is termed "impedance," represented by the letter "Z," and has the units of ohms (volts/amp), the same as for a resistor:

$$Z_L(\omega) = \frac{V_L(\omega)}{I_L(\omega)} = j\omega L \ \Omega \qquad (12.33)$$

Impedance, the ratio of voltage to current, is not defined for inductors or capacitors in the time domain since the voltage–current relationships for these elements contain integrals or differentials and it is not possible to determine a V/I ratio. Impedance is a function of frequency except for resistors. Often impedance is written simply as Z with the frequency term understood. Since impedance is a generalization of the concept of resistance (it is the V/I ratio for any passive element), the term is often used in discussion of any V/I relationships, even if only resistances are involved. The concept of impedance extends to mechanical and thermal systems as well. For example, in mechanical systems, the impedance of an element would be the ratio of force to velocity defined by the element: $Z(\omega) = \frac{F(\omega)}{v(\omega)}$.

To apply phasor analysis and the concept of impedance to a capacitor, we start with the basic voltage–current equation for a capacitor (Equation 12.21), repeated here:

$$v_C(t) = \frac{1}{C} \int i_c(t)dt \qquad (12.34)$$

Noting that integration becomes the operation of dividing by $j\omega$ in the phasor domain:

$$\int i(t)dt \Leftrightarrow \frac{I(\omega)}{j\omega}$$

so the phasor voltage–current equation for a capacitor becomes:

$$V(\omega) = \frac{I(\omega)}{j\omega C} \qquad (12.35)$$

The capacitor impedance then becomes:

$$Z_C(\omega) = \frac{V_C(\omega)}{I_C(\omega)} = \frac{1}{j\omega C} = \frac{-j}{\omega C} \ \Omega \qquad (12.36)$$

Active elements producing sinusoidal voltages or currents can also be represented in the phasor domain by returning to the original phasor description of sinusoid given as Equation 6.13 in Chapter 6:

$$A \cos(\omega t + \theta) \Leftrightarrow Ae^{j\theta} \qquad (12.37)$$

Using this equation, the phasor representation for a voltage source becomes:

$$V_s(t) = V_s \cos(\omega t + \theta) \Leftrightarrow V_s e^{j\theta} \equiv V_s \angle \theta \qquad (12.38)$$

and for a current source:

$$I_s(t) = I_s \cos(\omega t + \theta) \Leftrightarrow I_s e^{j\theta} \equiv I_s \angle \theta \qquad (12.39)$$

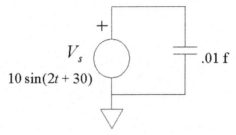

FIGURE 12.12 A simple circuit consisting of a voltage source and a capacitor used in Example 12.3. The two elements are represented by their voltage/current relationships in the phasor domain.

(Recall, capital letters are used for phasor variables.) Usually the frequency, ω, is not explicitly included in the phasor expression of a current or voltage, but it is often shown in its symbolic representation on the circuit diagram as shown in Figure 12.12. Here the frequency is defined in the definition of the voltage source as $\omega = 2$. An elementary use of phasors is demonstrated in the following example.

EXAMPLE 12.3

Find the current through the capacitor in the circuit of Figure 12.12.

Solution: Since the voltage across the capacitor is known (it is the just voltage of the voltage source, V_s), the current through the capacitor can be determined by the phasor extension of Ohm's law: $V(\omega) = I(\omega)\, Z(\omega)$:

Solving Ohm's law for $I_C(\omega)$: $I_C(\omega) = \frac{V_C(\omega)}{Z_C(\omega)}$.

The voltage across the capacitor is the same as the source voltage, $V_C = V_S$:

$$V_c = V_s = 10\sin(2t + 30) = 10\sin(2t - 60) \Rightarrow V_c(\omega) = V_s(\omega) = 10\angle -60$$

Recall, phasor representation is based on the cosine so the sine is converted to a cosine before converting to phasor representation. Next, we find the phasor notation of the capacitor:

$$Z_c = \frac{1}{j\omega C} = \frac{1}{j2(.01)} = -j50\ \Omega = 50\angle -90\ \Omega$$

Then solving for I_C using Ohm's law:

$$I_C(\omega) = \frac{V_c(\omega)}{Z_c(\omega)} = \frac{10\angle -60}{50\angle -90}$$

Recall, the rule for dividing two complex numbers is to convert them to polar notation (these already are in polar notation) and divide the magnitudes and subtract the denominator angle from the numerator angle (see Appendix E for details.).

$$I_C(\omega) = \frac{10\angle -60}{50\angle -90} = 0.2\angle 30\ \text{A}$$

Converting from the phasor domain to the time domain:

$$i_C(t) = 0.2\cos(2t + 30)\ \text{A}$$

The solution to the problem requires only algebra, although it is complex algebra. This is true for all electrical network problems as long as we operate in the phasor (i.e., frequency) domain.

III. CIRCUITS

The phasor approach can be used whenever the voltages and currents are sinusoids or can be decomposed to sinusoids. In the latter case, the circuit needs to be solved separately for each sinusoidal frequency applying the law of superposition. Like Fourier series analysis, this can become quite tedious if manual calculations are involved, but such analyses are not that difficult using MATLAB. An example of solving a simple circuit over a large number of frequencies using MATLAB is shown in the following discussion.

EXAMPLE 12.4

The circuit shown in Figure 12.13 was one of the earliest models of the cardiovascular system. This model, the most basic of the "Windkessel" models described in Chapter 1 (Section 1.4.5.1), represents the cardiovascular load on the left heart. The voltage source, $v(t)$, represents the pressure in the aorta and the current, $i(t)$, the flow of blood into the periphery. The resistor represents the resistance to flow of the arterial system and the capacitor represents the compliance of the arteries (i.e., stretchiness of the arteries). Of course the heart output is pulsatile not sinusoidal, so the analysis of this model is better suited to Laplace methods, but a sinusoidal analysis can be used to find the frequency characteristics of the load, that is, the effective impedance (resistance to blood flow), of the vascular system.

Assume that the voltage source produces a sinusoidal series consisting of frequencies between 0.01 and 10 Hz in 0.01-Hz intervals. This will allow us to generate a Bode plot of the impedance characteristics of the peripheral vessels in the Windkessel model. Plot the log of the magnitude impedance against log frequency as used in constructing Bode plots.

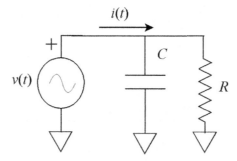

FIGURE 12.13   The least complicated Windkessel representation of the cardiovascular load on the heart. Electrical elements and variables are used to represent analogous cardiovascular elements. The voltage, $v(t)$, represents the pressure generated in the aorta in mmHg; the current, $i(t)$, the blood flowing into the periphery in mL/s; R the resistance in units of mmHg s/mL; and C the effective compliance (stretchiness) of the peripheral vessels in mL/mmHg. In Example 12.4, we determine the frequency characteristics (i.e., Bode plot) for the arteries loading the heart between 0.01 and 10 Hz.

Solution: The peripheral impedance is just $Z(f) = V(f)/I(f)$ where $V(f)$ is the phasor representation of the aortic pressure and $I(f)$ is the phasor notation for the blood flow, $i(t)$. Since the problem gives frequency ranges in hertz instead of radians, we will use f instead of ω. One approach is to apply a $V(f)$ having a known value and solving for the current, $I(f)$. In this case, $V(f)$ will be a sinusoidal series between 0.01 and 10 Hz. The impedance as a function of frequency is the ratio of $V(f)/I(f)$. Since the actual input pressure is arbitrary, we make it a series of cosine waves with amplitudes of 1.0; i.e., $v(t) = \cos(2\pi ft)$ for $f = 0.01-10$ Hz. This leads to a phasor representation of $V(f) = 1.0$.

First convert the elements into phasor notation. The resistor does not change, the voltage source becomes 1.0, and the capacitor would be $1/j\omega C = 1/j2\pi fC$. Note that since the voltage represents pressure it is in mmHg and the current will be in mL/s. Typical values for R and C are: $R = 1.05$ mmHg s/mL (pressure/volume flow) and $C = 1.1$ mL/mmHg (volume/pressure).

In the next chapter, we develop an algorithmic approach to solving any circuit problem, but for now note that the total current flow is just the current through the resistor plus the current through the capacitor. By Ohm's law for the resistor and capacitor (Equation 12.35):

$$V(f) = I_R(f)R; \quad I_R(f) = \frac{V(f)}{R} \quad V(f) = \frac{I_C(f)}{j2\pi fC}; \quad I_C(f) = V(f)(j2\pi fC)$$

So the total current (flow) becomes:

$$I(f) = I_R(f) + I_C(f) = \frac{V(f)}{R} + V(f)(2\pi fC) = V(f)\left(\frac{1}{R} + j2\pi fC\right)$$

Substituting in appropriate values for R and C:

$$I(f) = V(f)\left(\frac{1}{1.05} + j2\pi f1.1\right) = V(f)(0.95 + j6.9f)$$

Solving for the impedance of the arteries:

$$Z(f) = \frac{V(f)}{I(f)} = \frac{1.0}{0.95 + j6.9f} = \frac{1.05}{1 + j6.55f}$$

Results: From our knowledge of Bode plot primitives (recall Section 6.5.1), this equation has the form of a first-order system with a cutoff frequency of $1/6.55 = 0.153$ Hz.[5] The Bode plot for the magnitude and phase of $Z(f)$ can be plotted using Bode plot primitives (see the generic magnitude plot, Figure 6.13, and phase plot in Figure 6.14). The plots (actually generated using MATLAB) are shown in Figure 12.14. These show that the arterial impedance decreases above 0.152 Hz due to the compliance of the arteries (capacitance in the model). This means that at higher frequencies flow increases for a given pressure, but remember this is a much simplified model of the arterial system.

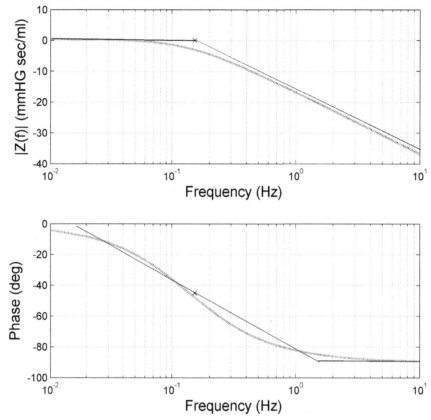

FIGURE 12.14 The impedance of the cardiovascular system as a function of frequency as determined from a simple two-element Windkessel model. The impedance decreases above 0.152 Hz due to the compliance of the arteries (capacitance in the model).

A more complicated version of the Windkessel model with a more realistic pulse-like pressure signal is analyzed in the next chapter.

[5]This is the same as Equation 6.31 except that frequency in now in Hz instead of rad/s. As biomedical engineers, we should be comfortable expressing frequency in either unit.

12.6 LAPLACE DOMAIN—ELECTRICAL ELEMENTS

If the voltages and currents are not sinusoids, periodic, or aperiodic signals, then they must be represented in the Laplace domain. The Laplace transform cannot be used in situations where $t < 0$ since the transform equation (Equation 7.4) diverges for negative time. In the Laplace domain, signals are valid only for $t \geq 0$. With this restriction, analysis in the Laplace domain is a straightforward extension of phasor analysis. Recall that activity before $t = 0$ can be summarized in terms of nonzero initial conditions as shown in Section 12.6.2, but first we define the Laplace domain representation of electrical elements (R, L, and C) assuming zero initial conditions.

12.6.1 Electrical Elements With Zero Initial Conditions

The impedance of a resistor, R, is the same in both the time and Laplace domains since resistors have a constant relationship between voltage and current:

$$V(s) = R\,I(s) \tag{12.40}$$

The impedance of a resistor is $V(s)/I(s)$:

$$Z_R(s) = \frac{V_s}{I_s} = R \tag{12.41}$$

For inductors and capacitors with zero initial conditions the Laplace representation is straightforward: differentiation and integration are represented by the Laplace operators s and $1/s$, respectively. For an inductor, the time derivative in Equation 12.16 is replaced by s and the defining equation becomes:

$$V(s) = sLI(s) \tag{12.42}$$

The Laplace impedance for an inductor is:

$$Z_L(s) = \frac{V(s)}{I(s)} = sL \ \Omega \tag{12.43}$$

Impedance in the Laplace domain, like impedance in the phasor domain, is given in ohms. For a capacitor, the time integral is represented by $1/s$ and the defining equation is:

$$V(s) = \frac{1}{Cs}I(s) \tag{12.44}$$

The impedance of a capacitor is:

$$Z_C(S) = \frac{v(S)}{i(S)} = \frac{1}{Cs} \ \Omega \tag{12.45}$$

The defining equations and the impedance relationships are summarized for electrical elements in Table 12.5 for the time, phasor, and Laplace domain representations.

12.6.2 Nonzero Initial Conditions

Circuit activity that occurs for $t < 0$ can be included in a Laplace analysis as initial voltages and/or currents. Essentially we are taking all of the past history and boiling it down to a

TABLE 12.5 *V-I* Relationships and Impedances for Electrical Elements

Element	*v/i* **Time Domain**	Phasor Domain		Laplace Domain	
		V-I Equation	**Z(ω)**	**V-I Equation**	**Z(s)**
Resistor	$v = R\,i$	$V(\omega) = R\,I(\omega)$	$R\ \Omega$	$V(s) = R\,I(s)$	$R\ \Omega$
Inductor	$v = L\frac{di}{dt}$	$V(\omega) = j\omega L\,I(\omega)$	$j\omega L\ \Omega$	$V(s) = sL\,I(s)$	$sL\ \Omega$
Capacitor	$v = \frac{1}{C}\int i\,dt$	$V(\omega) = \frac{I(\omega)}{j\omega C}$	$\frac{1}{j\omega C}\ \Omega$	$V(s) = \frac{1}{sC}I(s)$	$\frac{1}{sC}\ \Omega$

voltage or current at $t = 0$. Only the passive energy storage elements, *L* or *C*, have nonzero initial conditions; specifically, capacitors can have nonzero initial voltages and inductors can have nonzero initial currents.

In an inductor, energy is stored in the form of current flow, so the salient initial condition for an inductor is the current at $t = 0$. This can be obtained based on the past history of the inductor's voltage:

$$I_L(0) = \frac{1}{L}\int_{-\infty}^{0} v_L\,dt \tag{12.46}$$

An inductor treats the current that flows through it the same way a capacitor treats voltage: the current flowing through an inductor is continuous and cannot change instantaneously. So no matter what change occurs at $t = 0$, the current just before $t = 0$ will be the same as the current immediately after $t = 0$. Stated mathematically: $I_L(0-) = I_L(0+)$.

For an inductor with nonzero initial conditions, consider the equation that defines the derivative operation in the Laplace domain, Equation 7.5, repeated here:

$$\mathcal{L}\frac{dx(t)}{dt} = sX(s) - x(0-) \tag{12.47}$$

To get the equation defining an inductor with an initial current, we start with the basic time domain equation, $v(t) = L\frac{di(t)}{dt}$, and apply Equation 12.47 for the derivative operation. Note that since the inductance *L* is a constant, its Laplace transform is just *L* and the Laplace domain equation for an inductor is:

$$\mathcal{L}v(t) = \mathcal{L}\left[L\frac{di(t)}{dt}\right] = L\,\mathcal{L}\left[\frac{di(t)}{dt}\right]$$

Applying Equation 12.47 substituting *i(t)* for *x(t)*, the Laplace domain equation for the voltage across an inductor becomes:

$$V_L(s) = L(sI(s) - i(0)) = sLI(s) - Li(0) \tag{12.48}$$

So the initial current $i(0)$ adds a second term to the voltage–current relationship of the inductor. Often these values are known, but in some cases they must be calculated from the past history of the system.

Regarding Equation 12.48 it seems impossible to get a simple impedance term, i.e., a single term for the ratio $V(s)/I(s)$. However, there is a clever way to deal with the initial condition term, the $Li(0)$ term, and still retain the concept of impedance. Dissecting the two terms in Equation 12.48, the first term is the impedance, sL, the same as Equation 12.43, and the second term is a constant, $Li(0)$. The second term can be viewed as a constant voltage source with a value of $Li(0)$. It is a strange voltage source, its value being dependent on both the initial current and the inductance, but it does look like a constant voltage source in the Laplace domain.[6] So the symbol for an inductor with a nonzero initial condition in the Laplace domain would actually be a combination of two elements: a Laplace impedance representing the inductor in series with the voltage source representing the initial current condition, Figure 12.15.

For a capacitor, the entire $t < 0$ history can be summarized as a single voltage at $t = 0$ using the basic equation that defines the voltage–current relationship of a capacitor:

$$V_C(t = 0) = \frac{1}{C} \int_{-\infty}^{0} i_C(t)dt \qquad (12.48)$$

This equation has the same form as the right-hand term of the Laplace transform of the integration operation, Equation 7.7 repeated here:

$$\mathcal{L}\left[\int_{0}^{T} x(t)dt\right] = \frac{1}{s}X(s) + \frac{1}{s}\int_{-\infty}^{0} x(t)dt \qquad (12.49)$$

Substituting in $i(t)$ for $x(t)$ in Equation 12.49 and noting that $1/C$ is a constant, the Laplace equation for the voltage across a capacitor becomes:

$$\mathcal{L}v(t) = \mathcal{L}\left[\frac{1}{C}\int i(t)dt\right] = \frac{1}{sC}I(s) + \frac{1}{sC}\int_{-\infty}^{0} i(t)dt \qquad (12.50)$$

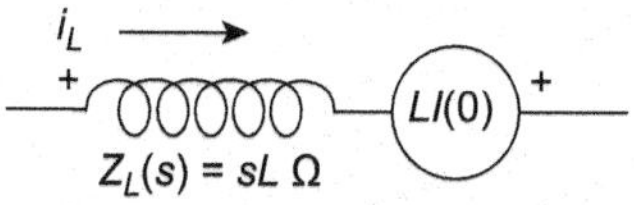

FIGURE 12.15 The Laplace domain representation of an inductor with a nonzero initial current. The inductor becomes two elements: a Laplace domain inductor having an impedance of sL, and a voltage source with a value of $Li(0)$ where $i(0)$ is the initial current. Note the polarity of the voltage source, which is based on the negative sign in Equation 12.48.

[6]Look at it this way: if you set the first term to zero, you are left with $V(s) = Li(0) = V_L$, a constant since both L and $i(0)$ are constants.

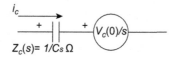

FIGURE 12.16 Laplace domain representation for a capacitor with an initial voltage. The element consists of two components: an impedance element, $1/Cs$, and a voltage source element representing the initial condition $V_C(0)/s$. The polarity of the voltage element is in the same direction as polarity of the impedance element as given by Equation (12.51).

Again the nonzero initial voltage adds a second term to the voltage current relationship. The first term is just the standard impedance of a capacitor, Equation 12.45. The second term is actually a voltage. From the basic time domain definition of a capacitor:

$$\frac{1}{C} \int_{-\infty}^{-0} i(t)dt = V_c(-0)$$

So the second term is just the voltage on a capacitor at $t = 0$ divided by s (i.e., $V_C(0-)/s$). The voltage across a capacitor cannot change instantaneously so $V_C(0-) = V_C(0+) = V_C(0)$ and Equation 12.50 becomes:

$$V(s) = \frac{1}{sC}I(s) + \frac{V_C(0)}{s} \tag{12.51}$$

The Laplace domain representation of a capacitor having an initial voltage, Equation 12.51, can be interpreted as capacitance impedance, $1/sC$, in series with a voltage source. In this case, the voltage source is $V_C(0)/s$. This leads to the combined Laplace elements shown in Figure 12.16. The polarity of the voltage source in this case has the same polarity as the initial voltage on the capacitor.

EXAMPLE 12.5

A 0.1-F capacitor with an initial voltage of 10 V is connected to a 100-Ω resistor at $t = 0$. Find the value of the resistor voltage for $t \geq 0$.

Solution: Use Figure 12.16 to configure a Laplace domain capacitor with nonzero initial conditions. Adding a resistor results in the circuit shown in Figure 12.17.

From Figure 12.17 we see that the voltage source representing the initial condition, $V_C(0)$, equals the voltage across the series resistor and capacitor. By Ohm's law, this voltage must equal the current in the circuit times the combined impedance of resistor and capacitor:

$$V_C = Z_C I(s) + RI(s) = \frac{1}{sC}I(s) + RI(s)$$

$$\frac{10}{s} = \frac{1}{0.1s}I(s) + 100I(S) = I(s)\left(\frac{10}{s} + 100\right)$$

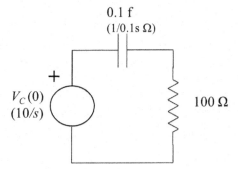

FIGURE 12.17   The network used in Example 12.5 shown in Laplace notation. Note that the 0.1 F capacitor is also shown with its impedance value of $\frac{1}{sC} = \frac{1}{0.1s}$ Ω. We use the ohm symbol to indicate that the capacitance has been converted to impedance.

Solving for current, $I(s)$:

$$I(s) = \frac{10}{s\left(\dfrac{10}{s} + 100\right)} = \frac{10}{100s + 10} = \frac{0.1}{s + 0.1}$$

To find the voltage across the resistor, again apply Ohm's law:

$$V_R(s) = RI(s) = 100\frac{.1}{s + 0.1} = \frac{10}{s + 0.1}$$

Converting to the time domain using entry #3 of the Laplace transform table in Appendix B gives the required answer:

$$v_R(t) = 10e^{-0.1t} \text{ V}$$

12.7 SUMMARY: ELECTRICAL ELEMENTS

In the time domain, impedance, the ratio of voltage to current for a component, can be defined only for a resistor (i.e., Ohm's law). In the phasor and Laplace domains, impedance can be defined for inductors and capacitors as well. Using phasors or Laplace notation it is possible to treat these "reactive elements" (inductors and capacitors) as if they were resistors, at least from a mathematical standpoint. This allows us to generalize Ohm's law, $V = I\,Z$, to include inductors and capacitors and to treat them mathematically using only algebra. Applications using this extension of Ohm's law are given in Examples 12.3–12.5.

In the next chapter, rules will be introduced that capitalize on the generalized form of Ohm's law. These rules will lead to a step-by-step process to analyze any network of sources and passive elements no matter how complicated. However, the first step in circuit analysis is always the same as shown in the examples here: convert the electrical elements into their phasor or Laplace representations. This could include the generation of additional elements if the energy storage elements have initial conditions. In the last chapter of this text, the analysis of circuits containing electronic elements will be presented.

12.8 MECHANICAL ELEMENTS

The mechanical properties of many materials often vary across and through the material so that analysis requires an involved mathematical approach known as "continuum mechanics." However, if only the overall behavior of an element or collection of elements is needed, then the properties of each element can be grouped together and a "lumped-parameter" analysis can be performed. An intermediate approach facilitated by high-speed computers is to apply lump-parameter analysis to small segments of the material, and then compute how each of these segments interacts with its neighbors. This approach is often used in biomechanics and is known as "finite element analysis."

Lumped-parameter mechanical analysis is similar to that used for electrical elements and most of the mathematical techniques described earlier and elsewhere in this text can be applied to this type of mechanical analyses. In lumped-parameter mechanical analysis, the major variables are force and velocity. Mechanical elements have well-defined relationships between these variables, a relationship very similar to the voltage–current relationship defined by electrical elements. In mechanical systems, the flow-like variable analogous to current is velocity while the potential energy variable analogous to voltage is force. Mechanical elements can be either active or passive and, as with electrical elements, passive elements can either dissipate or store energy.

12.8.1 Passive Mechanical Elements

Dynamic friction is the only mechanical element that dissipates energy and, as with the resistor, that energy is converted to heat. The force-velocity relationship for a friction element is also similar to a resistor: the force generated by the friction element is proportional to its velocity:

$$F = k_f v \tag{12.52}$$

where k_f is the constant proportionality and is termed "friction," F is force, and v is velocity.

In the "cgs" (centimeters, grams, dynes) metric system used in this text, the unit of force is dynes and the unit of velocity is cm/s, so the units for friction are dyn/cm/s (force in dynes divided by velocity in cm/s) or dyne-s/cm. Another commonly used measurement system is the "mks" (meters, kilograms, seconds) system preferable for systems having larger forces and velocities than those generally found in biological systems. Conversion between the two is straightforward (see Appendix D).

The equation for the power lost as heat in a friction element is analogous to that of a resistor:

$$P = Fv \tag{12.53}$$

The symbol for such a friction element is termed a "dashpot," and is shown in Figure 12.18. Friction is often a parasitic element, but can also arise from a device specifically constructed to

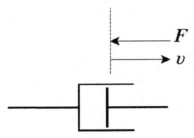

FIGURE 12.18 The schematic representation of a friction element showing the convention for the direction of force and velocity. This element is also referred to as a dashpot.

produce it. Devices designed specifically to produce friction are sometimes made using a piston that moves through a fluid (or air), for example, shock absorbers on a car or some door-closing mechanisms. This construction approach, a moving piston, forms the basis for the schematic representation of the friction element shown in Figure 12.18.

As with passive electrical elements, passive mechanical elements have a specified directional relationship between force and velocity: specifically, the direction of positive force is opposite to that of positive velocity. The direction of one of the variables can be chosen arbitrarily after which the direction of the other variable is determined. These conventions are illustrated in Figure 12.18.

In addition to elements specifically designed to produce friction (such as shock absorbers), friction occurs in association with other elements, just as resistance is unavoidable in other electrical elements (particularly inductors). For example, a mass sliding on a surface would exhibit some friction no matter how smooth the surface. Irrespective of whether friction arises from a dashpot element specifically designed to create friction or is associated with another element, it is usually represented by the dashpot schematic shown in Figure 12.18.

There are two mechanical elements that store energy just as there are two energy-storing electrical elements. The "inertial type" element corresponding to inductance is, not surprisingly, inertia associated with mass. It is termed simply "mass," and is represented by the letter m. The force–velocity relationship associated with mass is a version of Newton's law:

$$F = ma = m\frac{dv}{dt} \tag{12.54}$$

The mass element is schematically represented as a rectangle, again with force and velocity in opposite directions, Figure 12.19.

A mass element stores energy as kinetic energy following the well-known equation for kinetic energy.

$$E = \frac{1}{2}mv^2 \tag{12.55}$$

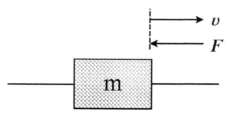

FIGURE 12.19   The schematic representation of a mass element that features the same direction conventions as the friction element.

The parallel between the inertial electrical element and its analogous mechanical element, mass, includes the continuity relationship imposed on the flow variable. Just as current moving through an inductor must be continuous and cannot be changed instantaneously, moving objects tend to continue moving (to paraphrase Newton) so the velocity of a mass cannot be changed instantaneously without applying infinite force. Hence, the velocity of a mass is continuous, so that $v_m(0-) = v_m(0+)$. It is possible to change the force on a mass instantaneously, just as it is possible to change the voltage applied to an inductor instantaneously, but not the velocity.

The mechanical energy storage element analogous to a capacitor is a spring and it has a force–velocity equation that is similar to that of a capacitor:

$$F = k_e x(t) = k_e \int v dt \tag{12.56}$$

where k_e is the "spring constant" in dyn/cm. A related term frequently used is the compliance, C_k, which is just the inverse of the spring constant ($C_k = 1/k_e$), and its use makes the equation of spring and capacitor even more similar to that of a capacitor:

$$F = \frac{1}{C_k} x(t) = \frac{1}{C_k} \int v dt \tag{12.57}$$

Although a spring is analogous to a capacitor, the symbol used for a spring is similar to that used for an inductor as shown in Figure 12.20. Springs look schematically like inductors, but they act like capacitors; no analogy is perfect.

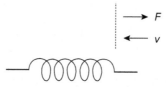

FIGURE 12.20   The symbol for a spring showing the direction conventions for this passive element. Schematically, the spring looks like an inductor: an unfortunate coincidence because it acts like a capacitor.

TABLE 12.6 Energy Storage and Response to Discontinuous and DC Variables in Mass and Elasticity

Element	Energy Stored	Continuity Property	DC Property
Mass	$E = \frac{1}{2}LI^2$	Velocity continuous $v_m(0-) = v_m(0+)$	If v_m = constant (DC velocity) $F = 0$
Elastic element	$E = \frac{1}{2}k_e x^2$	Force continuous $F_e(0-) = F_e(0+)$	If F_e = constant (DC force) $v_m = 0$

As with a capacitor, a spring stores energy as potential energy. A spring that is stretched or compressed generates a force that can do work if allowed to move through a distance. The work or energy stored in a spring is:

$$E = \int F dx = \int k_e x dx = \frac{1}{2}k_e x^2 \tag{12.58}$$

Displacement, x, is analogous to charge, q, in the electrical domain, so the equation for energy stored in a spring is analogous to the equation for energy stored in a capacitor found in the derivation of Equation 12.25 and repeated here:

$$E = \frac{Q^2}{2C} \tag{12.59}$$

As with a capacitor, it is impossible to change the force on a spring instantaneously using finite velocities. This is because force is proportional to length ($F_s = k_e x$) and the length of a spring cannot change instantaneously. Using high velocities, it is possible to change spring force quickly, but not instantaneously; hence a spring force is continuous: $F_S(0-) = F_S(0+)$.

Since passive mechanical elements have defining equations similar to those of electrical elements, the same analysis techniques, such as phasor and Laplace analyses, can be applied. Moreover, the rules for analytically describing combinations of elements (i.e., mechanical systems) are similar to those for describing electrical circuits. Table 12.6 is analogous to Table 12.3 and shows the energy, continuity, and DC properties of mass and elasticity.

12.8.2 Elasticity

Elasticity relates to the spring constant of a spring, but because it is such an important component in biomechanics, a few additional definitions related to the spring constant and compliance are presented here. Elasticity is most often distributed through or within a material and is defined by the relationship between "stress" and "strain." Stress is a *normalized force*, one that is normalized by the cross-sectional area:

$$\text{Stress} = \frac{\Delta F}{A} \tag{12.60}$$

Strain is a *normalized stretching* or elongation. Strain is the change in length with respect to the rest length, which is the length the material would assume if no force were applied:

$$\text{Strain} = \frac{\Delta\ell}{\ell} \tag{12.61}$$

The ratio of stress to strain is a normalized measure of the ability of a material to stretch and is given as an elastic coefficient termed Young's modulus:

$$Y_M = \frac{\text{Stress}}{\text{Strain}} = \frac{\dfrac{\Delta F}{A}}{\dfrac{\Delta\ell}{\ell}} \tag{12.62}$$

If a material is stretched by a load or weight produced by a mass, m, then the equation for Young's modulus can be written as:

$$Y_M = \frac{\dfrac{mg}{\pi r^2}}{\dfrac{\Delta\ell}{\ell}} \tag{12.63}$$

where g is the earth's gravitational constant, 980.665 cm/s^2. Values for Young's modulus for a wide range of materials can be found in traditional references such as the *Handbook of Physics and Chemistry* (CRC Press). Some values for typical materials are shown in Table 12.7. The following examples illustrate applications of Young's modulus and the related equations given earlier.

TABLE 12.7   Young's Modulus of Selected Materials

Material	Y_M (dyn/cm^2)
Steel (drawn)	19.22×10^{10}
Copper (wire)	10.12×10^{10}
Aluminum (rolled)	$6.8–7.0 \times 10^{10}$
Nickel	$20.01–21.38 \times 10^{10}$
Constantan	$14.51–14.89 \times 10^{-10}$
Silver (drawn)	7.75×10^{10}
Tungsten (drawn)	35.5×10^{10}

EXAMPLE 12.6

A 10 lb. weight is suspended by a #12 (AWG) wire 12 in. long. How much does the wire stretch?

Solution: To find the new length of the wire use Equation 12.63 and solve for $\Delta\ell$. First, convert all constants to cgs units:

$$m = 10\,\text{lb} = 10\,\text{lb}\left(\frac{1\,\text{kg}}{20.4\,\text{lb}}\right)\left(\frac{1000\,\text{gm}}{1\,\text{kg}}\right) = 490.2\,\text{gm}$$

$$\ell = 10\,\text{in}\left(\frac{2.54\,\text{cm}}{1\,\text{in}}\right) = 25.4\,\text{cm}$$

To find the diameter of the 12-gauge (AWG) wire use Table 5 in Appendix D:

$$d = 0.081\,\text{in} \quad \text{From Table 5 in Appendex D}$$

$$r = \frac{d}{2} = \left(\frac{0.081\,\text{in}}{2}\right)\left(\frac{2.54\,\text{cm}}{1\,\text{in}}\right) = 0.103\,\text{cm}$$

Then solve for $\Delta\ell$; use the value of Y_M for copper from Table 12.7:

$$Y_M = \frac{\frac{mg}{A}}{\frac{\Delta\ell}{\ell}} = \frac{\frac{mg}{\pi r^2}}{\frac{\Delta\ell}{\ell}} \quad \Delta\ell = \frac{mg\ell}{\pi r^2 Y_M} = \frac{(490.2)(980.6)(25.4)}{\pi(0.103)^2(10.12 \times 10^{10})} = 0.0036\,\text{cm}$$

EXAMPLE 12.7

Find the elastic coefficient of a steel bar with a diameter of 0.5 mm and length of 0.5 m.

Solution: From Equation 12.56: $F = k_e x$, so $k_e = F/x$ where in this case $x = \Delta\ell$. After rearranging Equation 12.62, k_e is found in terms of Young's modulus:

$$k_e = \frac{F}{x} = \frac{F}{\Delta\ell} = \frac{Y_M A}{\ell}$$

Use the dimensions given and the material in Equation 12.62 to find Young's modulus:

$$Y_M = \frac{\frac{\Delta F}{A}}{\frac{\Delta\ell}{\ell}} = \frac{F\ell}{\Delta\ell A}; \quad F = \frac{Y_M A \Delta\ell}{\ell}; \quad k_e = \frac{F}{\Delta\ell} = \frac{Y_M A}{\ell} = \frac{Y_M \pi\left(\frac{d}{2}\right)^2}{\ell}$$

$$k_e = \frac{19.22 \times 10^{10}\pi\left(\frac{.05}{2}\right)^2}{50} = \frac{3.77 \times 10^8}{50} = 7.55 \times 10^6\,\text{dyn/cm}^2$$

12.8.3 Mechanical Sources

Sources supply mechanical energy and can be sources of force, velocity, or displacement. Displacement is another word for "a change in position," and is just the integral of velocity: $x = \int v\,dt$. As mentioned earlier, displacement is analogous to charge in electrical circuits since $q = \int i\,dt$, and current is analogous to velocity. Although sources of constant force or constant velocity do occasionally occur in mechanical systems, most sources of mechanical energy are much less ideal than their electrical counterparts. Sometimes a mechanical source can look like either a velocity (or displacement) generator or a force generator depending on the characteristics of the load, that is, the mechanical properties of the elements connected to the source. For example, a muscle contracting under a light, constant load, a so-called isotonic contraction because the force (i.e., "tonus") opposing the contraction is constant (i.e., "iso"), would appear to be a velocity generator, although the velocity would not be constant throughout the contraction. However, if the muscle's end points were not allowed to move, a so-called isometric contraction because the muscle's length (i.e., "metric") is constant (again "iso"), then the muscle would look like a force generator.

In fact, a muscle is neither an ideal force generator nor an ideal velocity generator. An ideal force generator would put out the same force no matter what the conditions; however, the maximum force developed by a muscle depends strongly on its initial length. Figure 12.21 shows the classic "length–tension curve" for skeletal muscle and shows how maximum force

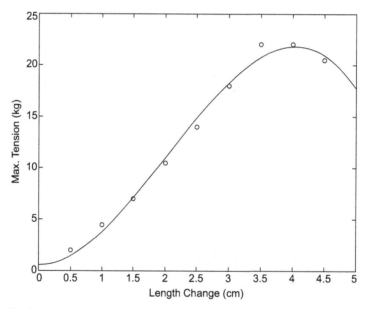

FIGURE 12.21   The length–tension relationship of skeletal muscle: the relationship between the maximum force a muscle can produce depends strongly on its length. An ideal force generator would produce the same force irrespective of its length, or its velocity for that matter. (This curve is based on historical measurements made on the human triceps muscle.)

depends on position with respect to rest length. (Again, the rest length is the position the muscle assumes where there is no force applied to the muscle.) When operating as a velocity generator under constant load, muscle is far from ideal since the velocity generated is highly dependent on the load. As shown by another classic, the "force–velocity curve," as the force resisting the contraction of a muscle is increased its velocity decreases and can even reverse if the opposing force becomes great enough, Figure 12.22. Of course, electrical sources are not ideal either, but they are generally more nearly ideal than mechanical sources. The characteristics of real sources, mechanical and electrical, are explored in Chapter 14.

With these practical considerations in mind, a force generator is usually represented by a circle or simply an F with a directional arrow, Figure 12.23.

A mass placed in a gravitational field looks like a force generator. The force developed by gravitation is in addition to its inertial force described by Equation 12.54. The forces devel-

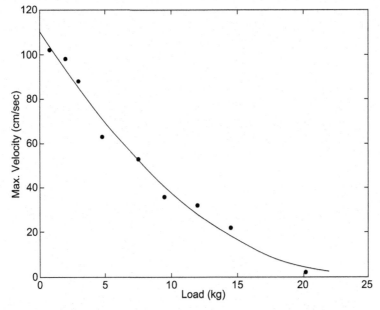

FIGURE 12.22 As a velocity generator, muscle is hardly ideal. As the load increases the maximum velocity does not stay constant as would be expected of an ideal source, but decreases with increasing force and can even reverse direction if the force becomes too high. This is known as the force–velocity characteristics of muscle. (This curve is based on historical measurements made on the human pectoralis major muscle.)

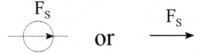

FIGURE 12.23 Two schematic representations of an ideal force generator showing direction of force.

oped by the inertial properties of a mass, or "inertial mass," and its gravitational properties, "gravitational mass," need not necessarily be coupled if they are the result of separate physical mechanisms. However, very careful experiments have shown them to be linked down to very high resolutions, indicating that they are related to the same underlying physics. The force is proportional to the value of the mass and the earth's gravitational constant:

$$F = mg \qquad (12.64)$$

where m is the mass in grams and g is the gravitational constant in cm/sec^2 Note that a g-cm/s^2 equals a dyne of force. The average value of g at sea level is 980.665 cm/s^2 Provided the mass does not change significantly in altitude, the force produced by a mass due to gravity is nearly ideal: the force produced is even independent of velocity, although, if it is not moving in a vacuum, a frictional force due to wind resistance would be present.

In some mechanical systems that include mass, the force due to gravity must be considered, whereas in others it is cancelled by some sort of support structure. In Figure 12.24, the system on the left side has a mass supported by a surface (either a frictionless surface or with the friction incorporated in k_f) and only the inertial force defined in Equation 12.54 is considered. In the system on the right-hand side, the mass is under the influence of gravity and produces both an inertial force that is a function of velocity (Equation 12.54) and a gravitational force that is constant and defined by Equation 12.64. This additional force would be represented as a force generator acting in the downward direction with a force of mg.

A velocity or displacement generator would be represented as in Figure 12.23, but the letters used would be either V_S if it were a velocity generator or X_S for a displacement generator. Motors, can be viewed either as sources of rotational velocity or of rotational force (i.e., torque). The schematic representation of a motor is shown in Figure 12.25.

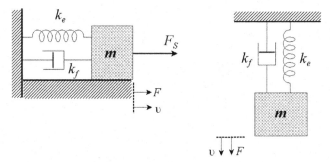

FIGURE 12.24   Two mechanical systems containing mass, m. In the left-hand system, the mass is supported by a surface so the only force involved with this element is the inertial force. In the right-hand system, gravity is acting on the mass so that it produces two forces: a constant force due to gravity (mg) and its inertial force.

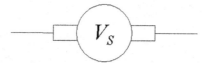

FIGURE 12.25 Symbol used to represent a motor. A motor can be either a force (torque) generator or a velocity (rpm).

12.8.4 Phasor Analysis of Mechanical Systems: Mechanical Impedance

Phasor analysis of mechanical systems is the same as for electrical systems, only the names, and associated letters, change. The mechanical elements, their differential and integral equations, and their phasor representations are summarized in Table 12.8 just as the electrical elements are summarized in Table 12.4. Impedance is defined for mechanical elements as:

$$Z(\omega) = \frac{F(\omega)}{v(\omega)} \tag{12.65}$$

Mechanical impedance has the units of dyn-cm/s.

The application of phasors to mechanical systems is given in the next example. More complicated systems are presented in the next chapter.

TABLE 12.8 Mechanical Elements

Element (Units)	Equation $F(t) = f[v(t)]$	Phasor Equation Laplace Equation	Impedance $Z(\omega)\ Z(s)$	Symbol
Friction (k_f) (dyn-s/cm)	$F(t) = k_f v(t)$	$F(\omega) = k_f\, v(\omega)$	k_f	
		$F(s) = k_f\, v(s)$	k_f	
Mass (m) (g)	$F(t) = m\frac{dv}{dt}$	$F(\omega) = j\omega m\, v(\omega)$	$j\omega m$	m
		$F(s) = sm\, v(s)$	sm	
Elasticity (k_r) (spring) (dyn/cm)	$F(t) = k_e \int v\, dt$	$F(\omega) = \frac{k_e}{j\omega} v(\omega)$	$\frac{k_e}{j\omega}$	
		$F(\omega) = \frac{k_e}{s} v(\omega)$	$\frac{k_e}{s}$	
Force generator (F_S)	$F(t) = FS(t)$	$F(\omega) = F_S(\omega)$	–	F_S
		$F(s) = F_S(s)$		
Velocity or displacement generator (V_S or X_S)	$v(t) = VS(t)$	$v(\omega) = V_S(\omega)$	–	V_S
	$x(t) = XS(t)$	$v(s) = V_S(s)$		

EXAMPLE 12.8

Find the velocity of the mass in the mechanical system shown in Figure 12.26. The force, F_S, is 5cos(12t) dynes and the mass is 5 g. The mass is supported by a frictionless surface. (Note that force and velocity are defined in the same, arbitrary, direction. Since the force produced by the mass is opposite to the defined velocity direction, it will appear on the opposite side of the equation from F_S.)

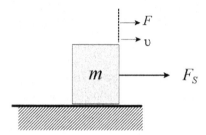

FIGURE 12.26   Mechanical system consisting of a mass with a force applied used in Example 12.8.

Solution: Convert the force to a phasor and apply the appropriate phasor equation from Table 12.8. Solve for $v(\omega)$. Converting the force to phasor notation:

$$5 \cos(12t) \qquad \Leftrightarrow \qquad 5\angle 0 \text{ dyn}$$

$$F(\omega) = j\omega m v(\omega); \quad v(\omega) = \frac{F(\omega)}{j\omega m}$$

$$v(\omega) = \frac{5\angle 0}{j9(5)} = \frac{5\angle 0}{45\angle 90} = 0.11\angle - 90 \text{ cm/s}$$

Converting back to the time domain (if desired):

$$v(t) = 0.11 \cos(12t - 90) = 0.11 \sin(12t)\text{cm/s}.$$

12.8.5 Laplace Domain Representations of Mechanical Elements With Nonzero Initial Conditions

The analogy between electrical and mechanical elements holds for initial conditions as well. The two energy storage mechanical elements can have initial conditions that need to be taken into account in the analysis. A mass can have an initial velocity, which will clearly produce a force, and a spring can have a nonzero rest length, which also produces a force.

For a mass, an initial velocity produces a force that is equal to the mass times the initial velocity. Taking the Laplace transform of the equation defining the force–velocity relationship of a mass (Newton's law; Equation 12.54):

$$\mathscr{L}F(t) = \mathscr{L}\left[m\frac{dv(t)}{dt}\right] = m\mathscr{L}\left[\frac{dv(t)}{dt}\right]$$

Applying the Laplace transform equation for the derivative operation, Equation 12.46, the force–velocity relationship of a mass with initial conditions becomes:

$$F(s) = m(sV(s) - v(0)) = smV(s) - mv(0) \tag{12.66}$$

The Laplace representation of a mass with an initial velocity consists of two elements: an impedance term related to the Laplace velocity and a force generator, Figure 12.27A. When solving mechanical systems that contain mass with an initial velocity, the Laplace elements in Figure 12.27A should be used to represent the mass.

The same approach can be used to determine the Laplace representation of a spring with an initial nonzero rest length. Applying the Laplace transform to both sides of the equation defining the force–velocity relationship of a spring (Equation 12.57):

$$\mathscr{L}F(t) = \mathscr{L}\left[k_e \int v \, dt \right] = k_e \mathscr{L}\left[\int v \, dt \right]$$

Applying the Laplace transform for an integral operation:

$$F(s) = \frac{k_e}{s}\left(V(s) + \int_{-\infty}^{0} v(t)dt \right) = \frac{k_e}{s} V(s) + \frac{k_e}{s} \int_{-\infty}^{0} v(t)dt$$

The second term is just the initial displacement $\left(x(0) = \int_{-\infty}^{0} v \, dt \right)$ times the spring constant divided by s, so the force–velocity equation becomes:

$$F(s) = \frac{k_e}{s} V(s) + \frac{k_e x(0)}{s}$$

Note that $k_e x(0)$ is also the initial force on the spring. Thus the Laplace representation of a spring having a nonzero rest length again includes two elements: the impedance of the spring in parallel with a force generator having a value of $k_e x(0)/s$, Figure 12.27B.

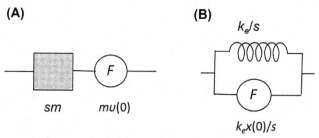

(A)   **(B)** k_e/s

sm $mv(0)$

F

$k_e x(0)/s$

FIGURE 12.27   The Laplace representation of mechanical energy storage elements with nonzero initial conditions. (A) A mass with a nonzero initial velocity is represented in the Laplace domain as a mass impedance plus a series force generator having a value of $mv(0)$. (B) A spring having a nonzero rest length is represented in the Laplace domain as a spring impedance with a parallel force generator of $k_e x(0)/s$, which is also the initial force divided by s.

An example of a mechanical system having energy storage elements with nonzero initial conditions is given next with more examples in the next chapter.

EXAMPLE 12.9

Find the velocity of the mass in Figure 12.28A for $t \geq 0$. Assume the mass is 5 g, the spring constant is 0.4 dyn/cm, and the spring is initially stretched 2 cm beyond its rest length.

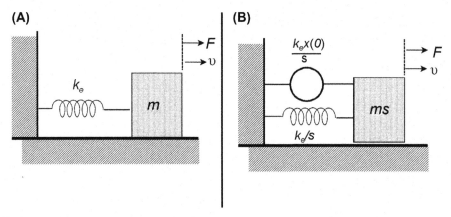

FIGURE 12.28   A mechanical system composed of a mass and a spring used in Example 12.9. The mass is 5 g, the spring constant is 0.4 dyn/cm and the spring is stretched 2 cm beyond its rest length.

Solution: Replace the time variables by their Laplace equivalent and add a force generator to account for the initial condition on the spring, Figure 12.28B. The force generator will have a value of:

$$F_s(0) = \frac{k_e x(0)}{s} = \frac{0.4(2)}{s} = \frac{0.8}{s} \text{ dyn}$$

This initial force is directly applied to both the mass and the spring, so

$$Fs(0) = Fm(s) + Fk(s); \quad \frac{0.8}{s} = msV(s) + \frac{k_e}{s}V(s) = V(s)\left(5s + \frac{0.4}{s}\right)$$

Solving for $V(s)$:

$$V(s) = \frac{\dfrac{0.8}{s}}{5s + \dfrac{0.4}{s}} = \frac{0.8}{5s^2 + 0.4} = \frac{0.16}{s^2 + 0.08}$$

This matches entry #6 in the Laplace transform table of Appendix B where $\beta = \sqrt{0.08} = 0.283$.

$$V(s) = 0.565\frac{0.283}{s^2 + 0.283^2} \Leftrightarrow v(t) = 0.565\sin(0.283t) \text{ dyn/s}$$

The next chapter will explore solutions to more complicated mechanical systems using an algorithmic-like approach that parallels an approach developed for electric circuits.

12.9 SUMMARY

The most complicated electrical and mechanical systems are constructed from a small set of basic elements. These elements fall into two general categories: active elements, which usually supply energy to the system, and passive elements, which either dissipate or store energy. In electrical systems, the passive elements are described and defined by the relationship they enforce between voltage and current. In mechanical systems, the defining relationships are between force and velocity. These relationships are linear and involve only scaling, differentiation, or integration. Active electrical elements supply either a specific voltage and are logically termed voltage sources or a well-defined current and are called current sources. Active mechanical elements are categorized as either sources of force or sources of velocity (or displacement), although mechanical sources are generally far from ideal. All of these elements are defined as idealizations. Many practical elements, particularly electrical elements, approach these idealizations and some of the major deviations have been described.

These basic elements are combined to construct electrical and mechanical systems. Since some of the passive elements involve calculus operations, differential equations are required to describe most electrical and mechanical systems. Using phasor or Laplace analysis, it is possible to represent these elements so that only algebra is needed for solution.

PROBLEMS

1. A resistor is constructed of thin copper wire wound into a coil (a "wire-wound" resistor). The wire has a diameter of 1 mm.
   a. How long is the wire required to be to make a resistor of 12 Ω?
   b. If this resistor is connected to a 5-V source, how much power will it dissipate as heat?
2. a. A length of size #12 copper wire has a resistance of 0.05 Ω. It is replaced by #16 (AWG) wire. What is the resistance of this new wire?
   b. Assuming both wires carry 2 A of current, what is the power lost in the two wires?
3. The following figure shows the current passing through a 2-h inductor.
   a. What is the voltage drop across the inductor?
   b. What is the energy stored in the inductor after 2 s?

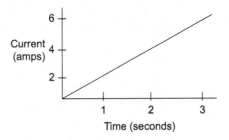

4. The voltage drop across a 10-h inductor is measured as 10 cos (20t) V. What is the current through the inductor?

5. A parallel plate capacitor has a value of 1 μF (10^{-6} F). The separation between the two plates is 0.2 mm. What is the area of the plates?

6. The current waveform shown in Problem 3 passes through a 0.1-F capacitor.
   a. What is the equation for the voltage across the capacitor?
   b. What is the charge, q, contained in the capacitor after 2 s (assuming it was unchanged or $t = 0$)?

7. A current of 1 A has been flowing through a 1-F capacitor for 1 s.
   a. What is the voltage across the capacitor, and what is the total energy stored in the capacitor?
   b. Repeat for a 120-F capacitor.

8. In the following circuit, find the value of the current, $i(t)$, through the inductor using the phasor extension of Ohm's law.

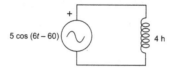

5 cos (6t – 60)   4 h

9. The sinusoidal source in Problem 8 is replaced by source that generates a step function at $t = 0$. Find the current, $i(t)$, through the inductor for $t \geq 0$. (Hint: Use the same approach as in Problem 8, but replace the source and inductor by their Laplace representations.)

10. In Problem 8, assume the inductor has an initial current of 2 A going toward ground (i.e., from top to bottom in the schematic). Find the current, $i(t)$, through the inductor for $t \geq 0$. [Hint: The problem is similar except you now have two voltage sources in series. These two sources can be combined into one by algebraic summation (be sure to get the signs right).]

11. The following 5-h inductor has an initial current of 0.5 A in the direction shown. Find the voltage across the 50-Ω resistor. (Hint: This is the same as Example 12.5 except an inductor is used instead of the capacitor. The approach will be the same.)

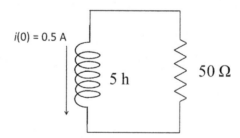

$i(0) = 0.5$ A   5 h   50 Ω

12. In the following circuit, the voltage source is a sinusoid that varies in frequency between 10 and 10,000 Hz in 10-Hz intervals. Use MATLAB to find the voltage across the resistor for each of the frequencies and plot this voltage as a function of frequency. The plot should be in dB versus log frequency. The resulting curve should look familiar. (Hint: The same current flows thought all three elements. Convert the elements to the phasor representation, write the extended Ohm's law equation, solve for $I(f)$, and solve for $V_R(f)$. In MATLAB, define a frequency vector, f, that ranges between 10 and 10,000 in steps of 10 Hz and use f to find $V_R(f)$. Since the code that solves for $V_R(f)$ will contain a vector, you need to use the ./divide operator to perform the division.)

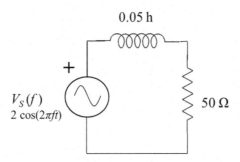

0.05 h

$V_S(f)$
$2\cos(2\pi f t)$

50 Ω

13. In a physiological preparation, the left heart of a frog is replaced by a sinusoidal pump that has a pressure output of $v(t) = \cos(2\pi t)$ mmHg ($f = 1$ Hz). Assuming that the Windkessel model of Figure 12.14 accurately represents the aorta and vasculature system of the frog, what is the resulting blood flow? If the pump frequency is increased to 4 Hz, what is the blood flow?

14. Use MATLAB to find cardiac pressure ($v(t)$ in Figure 12.13) of the Windkessel model used in Example 12.4 to a more realistic waveform of blood flow. In particular, the aortic flow ($i(t)$ in Figure 12.14) should be a periodic function having a period T and defined by:

$$i(t) = \begin{cases} I_o \sin^2\left(\dfrac{\pi t}{T_1}\right) & 0 \le t < \dfrac{T_1}{2} \\[2mm] 0 & \dfrac{T_1}{2} \le t < T \end{cases}$$

This function will have the appearance of a sharp half-rectified sine wave. Assume $T_1 = 0.3$ s, the period, T, is 1 s and $I_o = 500$ mL/s. (Hint: Since $i(t)$ is periodic it can be decomposed into a series of sinusoids using the Fourier transform, multiplied by $V(f)/I(f)$, the inverse Fourier transform taken to get the pressure wave $v(t)$. Use a sampling frequency define a 1-s time vector and use it to define $i(t)$. Set values of $i(t)$ above 0.3 s to zero. Because of round off errors, you need to take the real part of $v(t)$ before plotting.)

III. CIRCUITS

15. A constant force of 12 dyn is applied to a 5- g mass. The force is initially applied at $t = 0$ when the mass is at rest.

  a. At what value of t does the speed of the mass equal 6 dyn/s?

  b. What is the energy stored in the mass after 2 s?

16. A force of 12 cos $(6t + 30)$ dyn is applied to a spring having a spring constant of 20 dyn/cm.

  a. What is the equation for the velocity of the spring?

  b. What is the instantaneous energy stored in the spring at $t = 2.0$ s?

17. A 120-foot length of silver wire having a diameter of 0.02 in. is stretched by 0.5 in. What is the tension (stretching force) on the wire?

18. Use MATLAB to find the velocity of the mass in Example 12.8 for the 5-dyn cosine source where frequency varies from 1 to 40 rad/s in increments of 1 rad/s. Plot the velocity in dB as a function of log frequency in radians. (Hint: The approach is analogous to that of Problem 12. Generate a frequency vector between 1 and 40 using the MATLAB command w = 1:40, and then solve for a velocity vector, vel, by dividing the source value, 5, by j*5*w. Since w is a vector you will need to do point-by-point division using the ./ command. Plot the magnitude of the velocity vector.)

13

Analysis of Analog Circuits and Models

13.1 GOALS OF THIS CHAPTER

In Chapter 12 we defined the basic players that are featured in the next three chapters: electrical and mechanical analog elements. We found that they can be distinguished by the way they treat energy: they supply it, store it, or use (dissipate) it. They also establish relationships between the potential energy variables and the kinetic energy variables, that is, between the force and movement variables. In electrical circuits, voltage is the potential energy variable, whereas current is the kinetic energy variable; in mechanical systems, force is the potential energy variable and velocity is the kinetic energy variable. In this chapter, we learn how to analyze collections of these elements. Collections of electrical elements are called "networks" or "circuits,"[1] whereas collections of mechanical elements are simply "mechanical systems." Often the circuit or mechanical system we are analyzing is meant to represent a real circuit or mechanical system, but occasionally it may be the analog representation of a physiological mechanism. For example, the Windkessel models introduced in Chapter 1 (Section 1.4.5.1) are circuits meant to represent the mechanical load on the heart.

In circuit or mechanical system analysis, we find representations for one or more of the system's variables. For example, in Example 12.5 we found the voltage across a simple circuit and in Example 12.9 we found the velocity of a mass in a mechanical system. We may only want to find one voltage or current in a circuit, or one force of velocity, but the tools we develop here enable us to find any variable in the system and/or to construct a transfer function for a circuit or mechanical system. Once we have the transfer function, we can use any of the techniques developed in previous chapters, such as Bode plots to get the frequency characteristic or the inverse Laplace transform to find the time domain behavior of the system.

Having defined the players in the last chapter, we now need to find the rules of the game: the rules that describe the interactions between elements. For both mechanical and electrical elements, the rules are based on two fundamental conservation laws: (1) conservation of energy and (2) conservation of mass or charge. For electrical elements, these rules are termed

---

[1]The terms "circuits" and "networks" are used interchangeably.

"Kirchhoff's voltage law" (KVL) for conservation of energy and "Kirchhoff's current law" (KCL) for conservation of charge. For mechanical systems, we only use conservation of energy which manifests as an extension of Newton's famous force law, $F = Ma$ and as illustrated in Example 12.9. With these rules and some basic matrix techniques, we can analyze any linear network or mechanical system.

A unique behavior of systems covered in this chapter is the phenomenon of "resonance." Certain circuits or mechanical systems respond most favorably to a small set of frequencies. Such systems are called "resonant circuits" or "resonant systems" and can be useful in selecting out a specific frequency or range of frequency. Resonance is a common, and usually undesirable, feature in mechanical systems.

Specific goals of this chapter include:

- Provide definitions of circuit conversation laws: Kirchhoff's law,
- Show how to analyze single-loop circuits: mesh analysis,
- Show how to analyze multiple loop circuits using matrices: mesh analysis,
- Show how to analyze one node and multiple node circuits: nodal analysis,
- Define the mechanical system conservation law, Newton's force law, and demonstrate its application to more complex mechanical systems,
- Describe resonance circuits and resonant mechanical systems.

13.2 CONSERVATION LAWS: KIRCHHOFF'S VOLTAGE LAW

KVL is based on the conservation of potential energy: the total potential energy in a closed loop must be zero. Since voltage is a measure of potential energy, the law implies that all voltage increases or decreases around a closed loop must sum to zero. Simply stated, what goes up must come down (in voltage):

$$\sum_{\text{Loop}} v = 0 \tag{13.1}$$

To do anything useful, electrical elements must have current flowing through them. Otherwise, the voltage across the element, any element, is zero, and for all practical purposes the element does nothing and can be ignored. But current can only flow in a loop; it cannot simply fall out of the end of an unconnected element. So, by reverse logic, all elements that do anything must be connected in a closed path, a loop. KVL applies to all elements in the circuit and allows us to write an equation for all electrical elements connected in a loop. Figure 13.1 illustrates KVL. It also gives an example of a "useless" element that is connected to the loop but is not itself in a loop. Since no current can flow through this element, the voltage across the element, V_4, must be zero. The voltage on one side of this element is the same as the voltage on the other side, so it might as well not be there. (The story changes if something is connected to that element that creates a closed loop for that element. Then current flows through that element and voltage V_4 is no longer zero.)

More complicated circuits may contain a number of loops, and some elements may be involved in more than one loop, but KVL still applies. The analysis of a circuit with any number of loops is a straightforward extension of the analysis for a single loop.

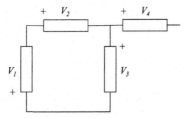

FIGURE 13.1 Illustration of Kirchhoff's voltage law. Three generic elements form a loop. The three voltages must sum to zero: $V_1 + V_2 + V_3 = 0$. This circuit also includes an element that is not part of a loop. Since this element is not in a loop there is no current through it and no voltage across it. So V_4 is zero and the voltage on either side of this element is the same. The element acts like a wire: it does nothing, and the circuit behaves as if it were not there.

Although all circuits can be analyzed using only KVL, in some situations the analysis is simplified by using the other conservation law that is based on the conservation of kinetic energy, that is, conservation of charge. This law is known as KCL and states that the sum of currents into a connection point (otherwise known as a "node") must be zero:

$$\sum_{\text{Node}} i = 0 \tag{13.2}$$

In other words, what goes in must come out (with respect to charge at a connection point). For example, consider the three currents going into the node in Figure 13.2. According to KCL the three currents must sum to zero: $i_1 + i_2 + i_3 = 0$. Of course we know that one, or maybe two, of the currents is actually flowing out of the node, but this just means that one (or two) of the current values will be negative.

In the analysis of some electronic circuits, both KVL and KCL are applied, but to analyze the networks covered in this chapter only one of the two rules is required at any given time. As mentioned previously, network analysis involves the determination of the network's variables or transfer function. In the case of networks, the transfer function is the ratio of a

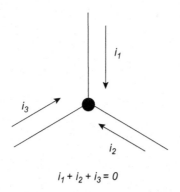

$$i_1 + i_2 + i_3 = 0$$

FIGURE 13.2 Illustration of Kirchhoff's current law. The sum of the three currents flowing into the connection point, or node, must be zero. In reality, one (or two) of these currents is flowing out of the node, which means that one (or two) current(s) has negative values.

specific output variable to a specific input variable, usually both voltages. Although either rule can be used to analyze the networks in this chapter, the rule of choice is the one that results in the least number of equations.

"Mesh analysis" is the term given to the analysis that uses KVL. This terminology makes more sense if you know that the word "mesh" is technical jargon for a circuit loop. Mesh analysis using KVL generates one equation for each loop in the circuit. "Nodal analysis" is the term used for the circuit analysis approach that uses KCL, and it generates one equation for each node in the circuit minus one node. (In all circuits one node provides a reference and is assumed to be at 0.0 V. A node that is assumed to be at 0.0 V is said to be "grounded.") If multiple equations are generated by either method they must be solved simultaneously. The method that leads to the fewest number of simultaneous equations is an obvious choice, particularly if we must find the solutions manually. However, if we use MATLAB to solve the circuit equations, it matters little how many equations we need to solve, so we could stick with just one approach, usually KVL. Nonetheless, some electronic networks require the use of both rules, so we learn to use them both.

13.2.1 Mesh Analysis: Single Loops

Mesh analysis employs KVL (Equation 13.1) to generate the circuit equations. In mesh analysis you write equations based on voltages but solve for currents. Once you find the loop currents, you can go back and find any of the voltages in the loop by applying the basic voltage/current definitions given in Chapter 12. (As you might guess, in nodal analysis it is the opposite: you write an equation(s) based on currents, but end up solving for node voltages.) Mesh analysis can be done using an algorithm that provides a step-by-step procedure that can be applied to almost any circuit.[2] In our now classic learn-by-doing approach, an example of this mesh analysis algorithm is given in Example 13.1.

EXAMPLE 13.1

Find the voltage across the capacitor in the network of Figure 13.3. Note that the source is sinusoidal, so phasor analysis is used.

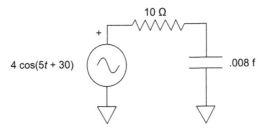

FIGURE 13.3   Single mesh (loop) circuit used in Example 13.1.

[2]Circuits containing electronic elements such as transistors, or integrated circuits, require a more complicated analysis procedure, but the basic ideas are the same.

Solution: We have already solved a similar network in Chapter 12 (Example 12.5) using an extended version of Ohm's law. However, the algorithmic approach used here is more general and can be applied to more complicated networks.

The circuit has one mesh (i.e., loop) and three nodes. Nodal analysis requires the simultaneous solution of two equations, whereas mesh analysis requires the solution of only one equation, making it the analysis of choice. Remember that by KVL what goes up must come down, but the trick is to keep accurate tabs on the direction, or "polarity," of the voltage changes: up or down corresponding to voltage increases or decreases. This bookkeeping problem is simplified by the algorithmic approach described here. The steps are presented in considerable detail for this example, but are easily applied to a wide range of network problems.

Step 1. Apply a transformation to the network so that all elements are represented in either the phasor or Laplace domain. Table 12.5 can be used to get the phasor or Laplace representations of the various elements. Since we are dealing with a sinusoidal stimulus, we will use the former. The sources are represented as phasor constants, $V_s \angle \theta$, whereas passive elements are given their respective phasor impedances: R Ω, $j\omega L$ Ω, or $1/j\omega C$ Ω. Putting the Ω symbol on the circuit diagram is a good way to show that the elements have been transformed into phasor (or Laplace) notation. Sometimes voltage sources use root mean square, but peak values will be used in this text as is more common. It really does not matter as long as you are consistent and know which units are being used.

Step 2. This step consists of defining the mesh current (or currents if more than one loop is involved). The loop in this example is closed by the two grounds since the two grounds are at the same voltage and are actually connected. The mesh current goes completely around the loop in either a clockwise or counterclockwise direction, theoretically your choice. To be consistent, in this book we always assume that the mesh current travels clockwise around the mesh. Maybe the current is traveling in the opposite direction, but that is not a problem because then the value we find for this current will be negative. Defining the current direction automatically defines the voltage polarities for the passive elements, since by definition current must flow into the positive side of a passive element. Recall, the polarity of a voltage source is defined by the element itself. After completion of these two steps, the circuit looks as shown in Figure 13.4.

Step 3. Apply KVL. We mentally go around the mesh (again clockwise) summing the voltages, but it is an algebraic summation. We assign positive values if there is an increase in voltage and negative values if there is a decrease in voltage. Start at the lower left corner (below the source) and

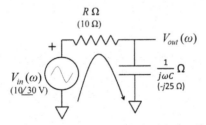

FIGURE 13.4 The circuit in Example 13.1 after Steps 1 and 2. The passive elements and source have been converted to their phasor representation (indicated by the Ω symbol). The direction of mesh current, $I(\omega)$, is arbitrary, but has been assigned as clockwise, a convention used throughout this book. If actual current flow is counterclockwise, $I(\omega)$ will turn out to have a negative value.

proceed around the loop in a clockwise direction. Traversing the source leads to a voltage rise, so this entry is positive; the next two components have a voltage drop (from + to −), so their entries are negative:

$$V_S - V_R - V_C = 0$$

Substituting in the element impedances, $V_R = R$, $V_C = \frac{1}{j\omega C}I(\omega)$, this equation becomes:

$$V_S - RI(\omega) - \frac{1}{j\omega C}I(\omega) = 0$$

$$V_S - I(\omega)\left(R + \frac{1}{j\omega C}\right) = 0$$

In fact, to develop this mesh equation we could have started anywhere in the loop and gone in either the clockwise or counterclockwise direction, but again for consistency, and since it really does not matter, we will always go clockwise and always begin at the lower left corner of the circuit.

Step 4. Solve for the current. Put the source (or sources, if more than one) on one side of the equation and the terms for the passive elements on the other. Then solve for $I(\omega)$.

$$V_S(\omega) = (R + Z_C)I(\omega)$$

$$I(\omega) = \frac{V_S}{R + Z_C}$$

Step 5. Solve for any voltages of interest. In this problem, we want the voltage across the capacitor. The voltage–current relationship for a capacitor is $V_C(\omega) = Z_C(\omega)I(\omega)$. Substituting our solution for $I(\omega)$ into this equation:

$$V_C = Z_C(\omega)I(\omega) = \frac{Z_C(\omega)V_S}{R + Z_C(\omega)}$$

Sometimes you can leave the solution in this form, with variables for element values, for example, when you do not know the specific values, when several different values may be used in the circuit, or when the problem will be solved on the computer. In this case we have specific values for our elements, so we can substitute in the values for R, Z_C, and V_S and solve:

$$V_C = \frac{(-j25)4\angle 30}{10 - j25} = \frac{(25\angle -90)4\angle 30}{27\angle -68} = \frac{100\angle -60}{27\angle -68} = 3.7\angle 8 \text{ V}$$

The next example applies this five-step algorithm to a slightly more complicated single-loop circuit.

EXAMPLE 13.2

Find the general solution for V_{out} for the circuit in Figure 13.5. Since it contains a resistor, inductor, and capacitor, it is sometimes called an "RLC" circuit. The arrows on either side of V_{out} indicate that this output voltage is the voltage across the capacitor. Again the two ground points are effectively connected.

Steps 1 and 2 lead to the circuit shown in Figure 13.6 below. Passive elements are shown with units in ohms to help remind us that we are now in the phasor domain.

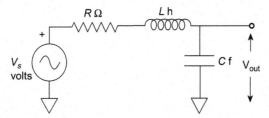

FIGURE 13.5   The "RLC" network used in Example 13.2.

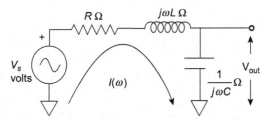

FIGURE 13.6   The network in Figure 13.5 after replacing the passive elements by their impedances, representing the source as a phasor voltage, and assigning the mesh current $I(\omega)$ as clockwise.

Step 3. Now write the basic equation going around the loop:

$$V_S(\omega) - RI(\omega) - j\omega LI(\omega) - \frac{1}{j\omega C}I(\omega) = 0$$

$$V_S(\omega) - I(\omega)\left(R + j\omega L + \frac{1}{j\omega C}\right) = 0$$

Step 4. Solve for $I(\omega)$:

$$V_S(\omega) - I(\omega)\left(R + j\omega L + \frac{1}{j\omega C}\right)$$

To clean things up a bit, clear the fraction in the denominator and rearrange the right-hand side into real and imaginary parts. (Recall $j^2 = -1$.)

$$I(\omega) = \frac{V_S(\omega)j\omega C}{R(j\omega C) + j\omega L(j\omega C) + 1} = \frac{V_S(\omega)j\omega C}{1 - \omega^2 LC + j\omega RC} \quad A$$

Step 5. Now find the desired voltage, the output voltage, V_{out}. Use the same strategy used in the last example: multiply $I(\omega)$ by the capacitance impedance, $1/j\omega C$:

$$V_{out}(\omega) = \frac{V_S(\omega)j\omega C}{1 - \omega^2 LC + j\omega RC}\left(\frac{1}{j\omega C}\right) = \frac{V_S(\omega)}{1 - \omega^2 LC + j\omega RC} \quad V \qquad (13.3)$$

To find a specific value for V_{out} it is necessary to put in specific values for R, L, and C as well as for V_S and ω. The source defines ω. However, by using some of the tools developed in Chapter 6, much can be learned from just the form of the equation. For example, Equation 13.3 is a second-order equation in phasor notation. By equating coefficients, we could even determine values for ω_n and δ in terms of R, L, and C without actually solving the equation. This is done in a later example.

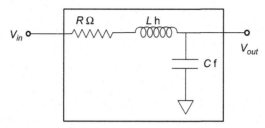

FIGURE 13.7 The network of Figure 13.6 viewed as an input–output system with V_S now represented as an input, V_{in}, and V_{out} as the output. The transfer function for this system is $TF(\omega) = V_{out}(\omega)/V_{in}(\omega)$.

The network in Figure 13.6 can also be viewed as an input–output system having a transfer function where the voltage source, V_S, becomes the input voltage and V_{out} is the output. When thought of in these terms, the network might be drawn as in Figure 13.7.

If the network is thought of as an input–output system, then the transfer function for this network is $TF(\omega) = V_{out}(\omega)/V_{in}(\omega)$. To repeat the caveat stated in the Laplace transform chapter, Chapter 7, the term "transfer function" should technically only be used for a function that is written in terms of the Laplace variables. However, the concept is so powerful that it is used to describe almost any input–output relationship, even qualitative relationships. To find the transfer function for the system shown in Figure 13.7, simply divide both sides of Equation 13.3 by $V_S(\omega)$.

$$\frac{V_{out}(\omega)}{V_{in}(\omega)} = \frac{1}{1 - \omega^2 LC + j\omega RC} \qquad (13.4)$$

This is clearly the transfer function of a second-order system (compare Equations 6.42 and 6.43). Since this transfer function is in the phasor domain, it is limited to sinusoidal functions or general periodic functions if we bring in the Fourier transform. However, this circuit could just as easily be analyzed using Laplace domain variables as illustrated in the next example.

EXAMPLE 13.3

Find the Laplace domain transfer function for the system/network shown in Figure 13.8.

Solution: The analysis of this network in the Laplace domain follows the same steps for the phasor domain analysis, except that in Step 1 the elements are represented by their Laplace representations: R, sL, and $1/sC$.

Step 1. The elements' values are replaced by the Laplace domain equivalents. These modified element values carry the unit of ohms to show they are impedances.

Step 2. The mesh current is assigned, again clockwise, but as a Laplace variable: $I(s)$. After the application of these two steps, the circuit appears as shown in Figure 13.9.

Step 3. Writing the loop equation:

$$V_{in}(s) - RI(s) - sLI(s) - I(s)/sC = 0$$
$$V_{in}(s) - I(s)(R + sL + 1/sC) = 0; \quad V_{in}(s) = I(s)(R + sL + 1/sC)$$

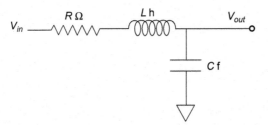

FIGURE 13.8   In Example 13.8, we solve for the Laplace transfer function of this network, $V_{out}(s)/V_{in}(s)$.

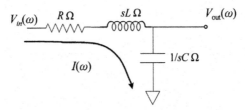

FIGURE 13.9   The circuit of Figure 13.8 with elements represented in Laplace notation and the mesh current assigned.

Step 4. Solving for $I(s)$:

$$I(s) = \frac{V_{in}(s)}{R + sL + 1/sC}$$

Step 5. Solving for $V_{out}(s) = I(s)/sC$:

$$V_{out}(s) = \frac{V_{in}(s)sC}{R + sL + 1/sC}\left(\frac{1}{sC}\right) = \frac{V_{in}(s)}{sRC + s^2CL + 1}$$

Rearranging into standard Laplace format where the highest power of s (in this case the s^2) has a coefficient of 1.0:

$$V_{out}(s) = \frac{V_{in}(s)/CL}{s^2 + R/L^s + 1/LC}$$

To find the transfer function, just divide by $V_S(s)$:

$$TF(s) = \frac{V_{out}(s)}{V_{in}(s)} = \frac{1/CL}{s^2 + R/L^s + 1/LC} \tag{13.5}$$

This is the Laplace transform transfer function of a second-order system having the same format as Equation 7.31, repeated here.

$$TF(s) = \frac{\omega_n^2}{s^2 + 2\delta\omega_n s + \omega_n^2} \tag{13.6}$$

Equating coefficients to solve for ω_n in terms of R, L, and C:

$$\omega_n^2 = 1/LC; \quad \omega_n = \sqrt{\frac{1}{LC}} \tag{13.7}$$

Solving for δ:

$2\delta\omega_n = \frac{R}{L}$; $\delta = \frac{R}{2L\omega_n}$ Substituting in for ω_n

$$\delta = \frac{R}{2L\left(1/\sqrt{LC}\right)} = \frac{R\sqrt{LC}}{2L} = \frac{R}{2}\sqrt{\frac{C}{L}} \tag{13.8}$$

The values of ω_n and δ are of particular importance in resonant systems as described in the section on resonance.

There are two implicit assumptions in the development of the transfer function of Equation 13.5: (1) the input, $V_{in}(s)$, is an ideal voltage source, that is, $V_{in}(s)$ will produce whatever current is necessary to maintain its prescribed voltage and (2) nothing is connected to the output, nothing meaning that no current flows out of the output terminals. Another way to state the second assumption is that $V_{out}(s)$ is connected to an "ideal load." (An ideal load is one where the load, Z_L, is infinite, which means there is really no load at all.) Although these assumptions may not always be true, in many real circuits input and output conditions are surprisingly close to these idealizations.

13.2.2 Mesh Analysis: Multiple Loops

Any single-loop circuit can be solved using the five-step process and a large number of useful circuits consist of only a single loop. Nonetheless, it is not difficult to extend the approach to contend with two or more loops, although the complex arithmetic can become tedious for three loops or more. Again, this is not really a problem since MATLAB can handle the necessary math for a large number of loops without breaking a sweat. The following example uses the five-step approach to solve a two-loop network and indicates how larger networks can be solved.

EXAMPLE 13.4

An extension of the Windkessel model used in Example 12.4 is shown in Figure 13.10. In this version, an additional resistor has been added to account for the resistance of the aorta, so the model is now a two-mesh circuit. In this example, we find the relationship between $v(t)$, which represents the pressure in the left heart, and the current $i(t)$, which represents the blood flow in the aorta. This is done using phasor notation in conjunction with an extension of the approach developed for single-loop circuits. For this circuit $R_A = 0.79$ mmHg s/mL, $R_p = 0.0033$ mmHg s/mL, and $C = 1.75$ mL/mmHg.

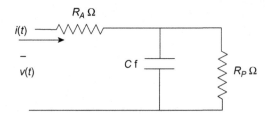

FIGURE 13.10 A two-mesh (two-loop) representation of the Windkessel model that is analyzed in Example 13.4.

Solution: To find the phasor relationship between v_{in} and i_{in}, follow the same five-step plan, but with modifications to Step 2. Since this problem is solved manually, we substitute in the actual component values to simplify the complex algebra.

Step 1. Represent all elements by their equivalent phasor or Laplace representations. This step is always the same in any analysis. Note that the capacitor impedance in phasor notation becomes:

$$\frac{1}{j\omega C} = \frac{1}{j1.75\omega} = \frac{0.57}{j\omega}$$

Step 2. Define the mesh currents. This step is essentially the same as for single-loop circuits. The only trick is that the mesh currents go around each loop and are defined as limited to their respective loop. These currents also are defined as going in the same direction, clockwise or counterclockwise. Again, for consistency, we always define mesh current in the clockwise direction. Of course, real currents are not limited to an individual loop in such an organized fashion. Mesh currents are artificial constructs that aid in solving multiloop problems. Nonetheless, the two mesh currents do account for all of the currents in the circuit. For example, the current through the capacitor would be the difference between the two mesh currents, $I_1(\omega) - I_2(\omega)$. This is not a problem as long as this difference in current is used when solving for the voltage across that capacitor.

Steps 1 and 2 lead to the circuit in Figure 13.11.

Step 3. Apply KVL around each loop, keeping in mind that the voltage drop (or rise) across the resistor shared by both meshes will be due to two currents, and since the currents are flowing in opposite directions their voltage contributions will have opposite signs: $I_1(\omega)$ produces the usual voltage drop, but $I_2(\omega)$ produces a voltage gain since it flows into the bottom of the resistor (again, mentally going clockwise around the loop). Each loop produces a separate equation.

Mesh 1: KVL following standard procedure, beginning in the lower left-hand corner.

$$V(\omega) - 0.79I_1(\omega) - \frac{0.57}{j\omega}I_1(\omega) + \frac{0.57}{j\omega}I_2(\omega) = 0$$

The capacitor produces a voltage term that is due to both I_1 and I_2, and the term produced by I_2 is positive because mesh current I_2 produces a voltage rise when going around the loop clockwise.

Mesh 2: KVL using the same procedure:

$$\frac{0.57}{j\omega}I_1(\omega) - \frac{0.57}{j\omega}I_2(\omega) - 0.033I_2(\omega) = 0$$

Now the capacitor contributes a voltage rise from current I_1 as we mentally go around the second loop clockwise, in addition to a voltage drop from I_2.

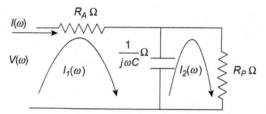

FIGURE 13.11 Windkessel model components have been converted to impedances and the mesh currents defined.

Step 4. Solve for the current(s). Rearranging the two equations, placing current on the right side and sources on the left, and separating the coefficients of the two current variables gives us two equations to be solved simultaneously. Pay particular attention to keeping the signs straight.

$$V(\omega) = \left(0.033 + \frac{0.57}{j\omega}\right)I_1(\omega) - \frac{0.57}{j\omega}I_2(\omega)$$

$$0 = -\frac{0.57}{j\omega}I_1(\omega) + \left(0.79 + \frac{0.57}{j\omega}\right)I_2(\omega)$$

With only two equations, it is possible to solve for the currents using substitution, avoiding matrix methods. However, with more than two meshes, matrix methods are easier and lend themselves to computer solutions. (The solution of a three-mesh circuit using MATLAB is given in the following discussion.) The spacing used in the above-mentioned equations facilitates transforming them into matrix notation:

$$\begin{vmatrix} V(\omega) \\ 0 \end{vmatrix} = \begin{vmatrix} 0.033 + \dfrac{0.57}{j\omega} & -\dfrac{0.57}{j\omega} \\ -\dfrac{0.57}{j\omega} & 0.79 + \dfrac{0.57}{j\omega} \end{vmatrix}\begin{vmatrix} I_1(\omega) \\ I_2(\omega) \end{vmatrix}$$

(13.9)

Solve for $I(\omega) \equiv I_1(\omega)$, using the method of determinants (Appendix G):

$$I_1(\omega) = \frac{\begin{vmatrix} V(\omega) & -\dfrac{0.57}{j\omega} \\ 0 & 0.79 + \dfrac{0.57}{j\omega} \end{vmatrix}}{\begin{vmatrix} 0.033 + \dfrac{0.57}{j\omega} & -\dfrac{0.57}{j\omega} \\ -\dfrac{0.57}{j\omega} & 0.79 + \dfrac{0.57}{j\omega} \end{vmatrix}} = \frac{V(\omega)\left(0.79 + \dfrac{0.57}{j\omega}\right)}{0.026 + \dfrac{0.02}{j\omega} + \dfrac{0.45}{j\omega} + \left(\dfrac{0.57}{j\omega}\right)^2 - \left(\dfrac{0.57}{j\omega}\right)^2}$$

$$I_1(\omega) = \frac{V(\omega)\left(0.79 + \dfrac{0.57}{j\omega}\right)}{0.026 + \dfrac{0.47}{j\omega}} = \frac{V(\omega)(0.57 + j0.79\omega)}{0.47 + j0.026\omega}$$

$$I_1(\omega) = \frac{0.47}{0.57}\frac{V(\omega)(1 + j1.39\omega)}{1 + j0.055\omega} = \frac{0.82V(\omega)(1 + j1.39\omega)}{1 + j0.055\omega}$$

Even this relatively simple two-mesh circuit involves considerable complex arithmetic, but it is easy to solve these problems using MATLAB.

Step 5. In this case we want the cardiac pressure as a function of blood flow. In this model, pressure is analogous to the input voltage, $V(\omega)$, and flow to the current, $I(\omega)$. To find $V(\omega)$ as a function of $I(\omega)$, we just factor out $V(\omega)$ and then invert the equation:

$$\frac{V(\omega)}{I(\omega)} = \frac{1.22(1 + j0.055\omega)}{1 + j1.39\omega}$$

(13.10)

The relationship between cardiac blood flow, $i(t)$ in this model, and cardiac pressure, $v(t)$ here, can be used to determine the pressure given the flow or vice versa. It is only necessary to have a quantitative description of the flow or the pressure. The pressure or flow need not be sinusoidal (which would not be very realistic); as long as it is periodic it can be decomposed into sinusoids using the Fourier transform. This approach is illustrated in the next example.

EXAMPLE 13.5

Use the three-element Windkessel model as described by Equation 13.10 to find the cardiac pressure given the blood flow. Use a realistic pulsatile waveform for blood flow as given by Equation 13.11. (This waveform is also used in Problem 14 of Chapter 12.) It might seem more logical to have cardiac pressure as the input and blood flow as the output, following the causal relationship, but in certain experimental situations we might measure blood flow and want to use the model to determine the cardiac pressure that produced that flow.

$$i(t) = \begin{cases} I_o \sin^2\left(\dfrac{\pi t}{T_1}\right) & 0 \le t < T_1 \\ 0 & T_1 \le t < T \end{cases} \tag{13.11}$$

Solution: To find the pressure we use the standard frequency domain approach: convert $i(t)$ to the frequency domain using the Fourier transform, multiply $I(\omega)$ by Equation 13.10 to get $V(\omega)$, and then take the inverse Fourier transform of $V(\omega)$ to get $v(t)$, the pressure. It is good to remember that in this approach you are solving for $V(\omega)$ a large number of times, each at a different frequency, i.e., the frequencies determined by the Fourier transform of $i(t)$. The fact that the overall answer can be obtained as just the sum of all the individual solutions is possible because superposition holds for this linear system.

Solutions are obtained for the pressure waveform over one cardiac cycle, which we assume takes 1.0 s. We make $f_s = 1.0$ kHz, so we need a 1000-point time vector to construct a 1.0-s blood flow waveform given in Equation 13.11. We assume that $I_o = 500$, the period of the heartbeat, T, is 1 s, and T_1 in Equation 13.11 is 0.3 s; in other words, blood flow occurs only during approximately one-third of the cardiac cycle (the "systolic" time period).

After we take the Fourier transform of $i(t)$ to get $I(\omega)$, we multiply it by the transfer function, $TF(\omega)$, given in Equation 13.10. To implement this transfer function, we construct a frequency vector, ω, that is of the same length as $i(t)$ (and, hence, the same length as $I(\omega)$) and has the same range as $I(\omega)$, from 1 to f_s Hz. After multiplying $I(\omega)$ by $TF(\omega)$ (point by point), we take the inverse Fourier transform. Although the output of the inverse Fourier transform should be real, computational errors can produce nonzero imaginary values, so we take the real part when plotting.

```
% Example 13.5 Plot the cardiac pressure from the Windkessel model.
%
fs = 1000;              % Sampling frequency
t = (1:1000)/fs;        % Time vector. One sec long
T1 = .3;                % Period of blood flow in sec
it = round(T1*fs);      % Index of blood flow
```

```
w = 2*pi*(1:fs);                          % Frequency vector in radians
i = 500*(sin(pi*t/T1)).2;                 % Define blood flow waveform, i(t).
                                            Equation 10.11
i(it:1000) = 0;                           % Zero period when no blood flow
subplot(2,1,1);
plot(t,i,'k');                            % Plot blood flow waveform
   .......labels and title.......
I = fft(i);                               % Take Fourier transform
TF = 1.22*(1+j*0.055*w)./(1+ j*1.39*w);   % Pressure/flow relationship
V = TF.*I;                                % Find V (in phasor)
v = ifft(V);                              % Find v(t). Inverse FT
subplot(2,1,2);
plot(t,real(v),'k');                      % Plot pressure waveform
....... Labels.......
```

Results: The plots produced by this program are shown in Figure 13.12. Cardiac pressure follows flow during the systolic period, but when flow stops (when the aortic valve closes) pressure falls slowly but does not become zero. A more complicated four-element Windkessel model is analyzed in one of the problems.

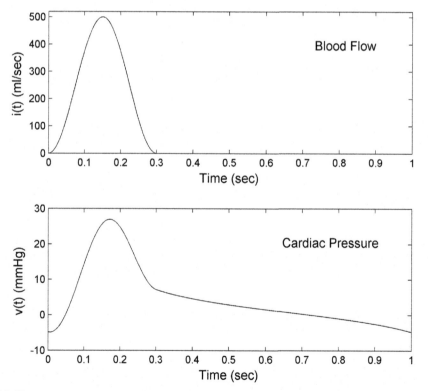

FIGURE 13.12   Pressure in the aorta left heart (lower curve) as a function of time as determined from the three-element Windkessel model. The upper curve shows the assumed blood flow pattern that produces this pressure.

13.2.2.1 *Shortcut Method for Multimesh Circuits*

There is a shortcut that makes it possible to write the matrix equation directly from inspection of the circuit. Regard the matrix equation, Equation 13.9, derived for the circuit in Figure 13.10 and repeated here:

$$\underbrace{\begin{vmatrix} V(\omega) \\ 0 \end{vmatrix}}_{V} = \underbrace{\begin{vmatrix} 0.033 + \dfrac{0.57}{j\omega} & -\dfrac{0.57}{j\omega} \\ -\dfrac{0.57}{j\omega} & 0.79 + \dfrac{0.57}{j\omega} \end{vmatrix}}_{Z} \underbrace{\begin{vmatrix} I_1(\omega) \\ I_2(\omega) \end{vmatrix}}_{I}$$

This equation has the general form of $V = Z\, I$ as indicated by the brackets. The left-hand side contains only the sources; the right-hand side contains a matrix of impedances multiplied by a vector containing the mesh currents. The source column vector contains the source in Mesh 1 in the upper position and the source in Mesh 2 in the lower position, which, in Example 13.4, is 0.0 because there are no sources in Mesh 2. The current column vector consists of just the two mesh currents. The impedance matrix also relates topographically to the circuit: the upper left entry is the sum of impedances in Mesh 1, the lower right is the sum of impedances in Mesh 2, and the off-diagonals (upper right and lower left) contain the negative of the sum of impedances common to both loops. In this circuit there is only one element common to both elements, the capacitor, so the off-diagonals contain the negative of this element, but other circuits could have multiple elements common to the two meshes. Putting this verbal description into mathematical form:

$$\begin{vmatrix} \Sigma V_S \text{ Mesh 1} \\ \Sigma V_S \text{ Mesh 2} \end{vmatrix} = \begin{vmatrix} \Sigma Z \text{ Mesh 1} & -\Sigma Z \text{ Mesh 1\&2} \\ -\Sigma Z \text{ Mesh 1\&2} & \Sigma Z \text{ Mesh 2} \end{vmatrix} \begin{vmatrix} I_1 \\ I_2 \end{vmatrix} \qquad (13.12)$$

In this equation the impedance sums, ΣZ, are additions; there is no need to worry about signs except if the impedance itself is negative, as for a capacitor. However, the summation of voltage sources, the ΣV_S, still requires some care, since the summations must take the voltage source signs into consideration. For example, the source in Mesh 1 (ΣV_S Mesh 1) in Example 10.4 has a positive sign because it represents a voltage rise when going around the loop clockwise. If more than one source appears in a mesh, these sources are algebraically summed using the rule for keeping track of signs.

This shortcut rule also applies to circuits that use Laplace representations. The approach can easily be extended to circuits having any number of meshes, although the subsequent calculations become tedious for three or more meshes unless computer assistance is used. The extension to three meshes is given in the following example, but the solution is determined using MATLAB.

13.2.3 Mesh Analysis: MATLAB Implementation

EXAMPLE 13.6

Solve for V_{out} in the three-mesh network in Figure 13.13. This circuit uses realistic values for R, L, and C.

Solution: Follow the steps used in the previous examples, but use the shortcut method in Step 3. In Step 4, solve for the currents using MATLAB, and in Step 5 solve for V_{out} using MATLAB.

Steps 1 and 2. Figure 13.14 shows both the original circuit and the circuit after Step 1 and 2 with the elements represented in phasor representation and the phasor currents defined.

The impedances for L and C are determined as:

$$Z_L = j\omega L = j2\pi f L = j2\pi 10^4 (10 \times 10^{-3}) = j628 \ \Omega$$

$$Z_{C2} = \frac{1}{j\omega C_2} = \frac{1}{j2\pi f C_2} = \frac{1}{j2\pi 10^4 (0.022 \times 10^{-6})} = \frac{1}{j1.38 \times 10^{-3}} = -j723 \ \Omega$$

$$\text{Similarly}: Z_{C2} = \frac{1}{j2\pi f C_2} = \frac{1}{j2\pi 10^4 (0.01 \times 10^{-6})} = j1592 \ \Omega$$

Step 3. The matrix equation for Step 3 is only a modification of Equation 13.11 where the voltage and current vectors have three elements each and the impedance matrix is extended to a 3×3 matrix written as:

$$\begin{vmatrix} \Sigma V_S \text{ Mesh 1} \\ \Sigma V_S \text{ Mesh 2} \\ \Sigma V_S \text{ Mesh 3} \end{vmatrix} = \begin{vmatrix} \Sigma Z \text{ Mesh 1} & -\Sigma Z \text{ Mesh 1\&2} & -\Sigma Z \text{ Mesh 1\&3} \\ -\Sigma Z \text{ Mesh 1\&2} & \Sigma Z \text{ Mesh 2} & -\Sigma Z \text{ Mesh 2\&3} \\ -\Sigma Z \text{ Mesh 1\&3} & -\Sigma Z \text{ Mesh 2\&3} & \Sigma Z \text{ Mesh 3} \end{vmatrix} \begin{vmatrix} I_1 \\ I_2 \\ I_3 \end{vmatrix} \qquad (13.13)$$

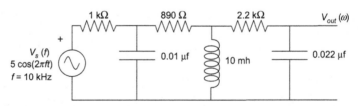

FIGURE 13.13   A three-mesh circuit analyzed in Example 13.6.

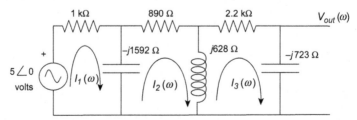

FIGURE 13.14   The circuit shown in Figure 13.13 after the mesh currents have been assigned, the voltage source has been converted to phasor notation, and the elements have been converted to their equivalent impedances.

where $\Sigma\, Z$ Mesh 1, $\Sigma\, Z$ Mesh 2, and $\Sigma\, Z$ Mesh 3 are the sums of impedances in the three meshes; $\Sigma\, Z$ Mesh 1 & 2 is the sum of impedances common to Meshes 1 and 2; $\Sigma\, Z$ Mesh 1 & 3 is the sum of impedances common to Meshes 1 and 3; and $\Sigma\, Z$ Mesh 2 & 3 is the sum of impedances common to Meshes 2 and 3. Note that all the off-diagonals are negative sums and the impedance matrix has symmetry about the diagonal. Such symmetry is often found in matrix algebra, and a matrix with this symmetry is termed a "Toplitz matrix." In this particular network there are no impedances common to Meshes 1 and 3, but one of the problems has a three-mesh circuit in which all three meshes have at least one element in common.

Applying Equation 13.13 to the network in Figure 13.14 gives rise to the matrix equation for this network:

$$\begin{vmatrix} 5 \\ 0 \\ 0 \end{vmatrix} = \begin{vmatrix} 1000 - j1592 & j1592 & 0 \\ j1592 & 890 - j964 & -j628 \\ 0 & -j628 & 2200 - j95 \end{vmatrix} \begin{vmatrix} I_1(\omega) \\ I_2(\omega) \\ I_3(\omega) \end{vmatrix}$$

Note that the $-j964$ in the middle term is the algebraic sum of $j628 - j1592$ and the $-j95$ in the lower right is the sum of $j628 - j723$.

Steps 4 and 5. To find V_{out} we need only the current through the final 0.022-μF capacitor, $I_3(\omega)$. Solve for this current, then the desired voltage is $-j723\, I_3(\omega)$. The MATLAB program does both. First it defines the voltage vector and impedance matrix and then solves the matrix equation for the current vector. The current, $I(3)$, is then used to find the output voltage.

```
% Example 13.6 Solution of a three-mesh network
%
V = [10 0 0]';          % Note the use of the transpose symbol, '
% Define the impedance matrix
Z= [1000 - 1i*1592, 1i*1592, 0;...
   1i*1592, 890-1i*964, -1i*628;...
   0, -1i*628, 2200-1i*95];
I = Z\V                 % Solve for the currents
Vout = I(3)*(-723*1i)   % Find and output the requested voltage
Vmag = abs(Vout)        % also as magnitude and phase
Vphase = angle(Vout)*306/(2*pi)
```

Results: The output of this program for the three-mesh currents is:

```
I =
 0.0042 + 0.0007i
 0.0037 - 0.0029i
 0.0008 + 0.0011i
```

The output voltages are:

```
Vout =  0.7908 - 0.5659i
Vmag =  0.9724 V phase =  -30.2476 deg
```

III. CIRCUITS

Again, MATLAB accepts either i, $1i$, or j to represent a complex number, but outputs only using i. (In newer versions of MATLAB, the symbol $1i$ is recommended to represent an imaginary number as this gives improved speed and stability.)

The time domain output is determined directly from the phasor output given earlier. Recall that $\omega = 10^4$ rad/s:

$$V_{out}(t) = 0.97 \cos(2\pi10^4 t - 30.2) \text{ V}.$$

Analysis: In the MATLAB program, the voltage vector is written as a row vector, but it should be a column vector as in Equation 13.13, so the MATLAB transpose operator (single quote) is used.[3] The second line defines the impedance matrix using standard MATLAB notation. The third line solves for the three currents by matrix inversion, implementing the equation $I = Z^{-1} V$ using the backslash (\) operator to invert the impedance matrix. The fourth line multiplies the third mesh current, $I_3(\omega)$, by the capacitor impedance to get V_{out}. The next two lines convert V_{out} from real and imaginary components into the more conventional polar form.

---

[3]It could just as easily been entered as a column vector directly ($V = [10; 0; 0]$); the transposed row vector was just the whim of the programmer.

If MATLAB can solve one matrix equation, it can solve it again and again. By making the inputs sinusoids of varying frequency, we can generate the Bode plot of any network. The next example treats a network as an input–output system where the source voltage is the input and the voltage across one of the elements is the output. MATLAB is called upon to solve the problem many times over at different sinusoidal frequencies and for three component values.

EXAMPLE 13.7

Plot the magnitude Bode plot in hertz of the transfer function of the network in Figure 13.15 with inputs and outputs as shown. Plot the Bode plot for three values of the capacitance: 0.01, 0.1, and 1.0 μf. The three capacitors all have the same values and the three resistors are all 2.0 kΩ.

Solution: We could first manually determine the frequency domain transfer function and then plot that function for the three capacitance values. However, that would require some tedious

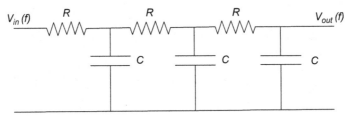

FIGURE 13.15 A network that can be viewed as an input–output system where the transfer function is $TF = V_{out}/V_{in}$. The Bode plot (i.e., system frequency spectrum) of this system is found in Example 13.7. The three resistors are all 2.0 kΩ. The three capacitors also have the same value, but the Bode plot is determined for three different capacitance values: 0.01, 0.1, and 1.0 μf.

algebra, particularly as we have to leave the capacitance values as variables. It is easier for us to rely more heavily on MATLAB. Set up the KVL matrix equations and the output equation, solve these equations in MATLAB over a range of sinusoidal frequencies, and plot the output voltages for these frequencies.

Follow the five-step procedure to generate the KVL matrix problem, then find the output in terms of the third mesh current. Use MATLAB to solve for this output current from the KVL matrix, then solve for V_{out}, the voltage across the output capacitor. Assume V_{in} is an ideal sinusoidal source that varies in frequency. Select a frequency range that includes any interesting changes in the Bode plot. This might require trial and error, so, to begin, let us try a large frequency range of between 10 Hz and 80 kHz. To keep the computational time down, use 20-Hz intervals. Use one loop to vary the frequency and another outside loop to change the capacitor values.

Steps 1 and 2. At this point you can implement these steps by inspection. The capacitors become $1/j\omega C$ and the mesh currents are $I_1(\omega)$, $I_2(\omega)$, and $I_3(\omega)$. No need to redraw the network.

Step 3. Again the KVL equation for this circuit can be done by inspection.

$$\begin{vmatrix} V_{in}(s) \\ 0 \\ 0 \end{vmatrix} = \begin{vmatrix} R + 1/j\omega C & -1/j\omega C & 0 \\ -1/j\omega C & R + 2/j\omega C & -1/j\omega C \\ 0 & -1/j\omega C & R + 2/j\omega C \end{vmatrix} \begin{vmatrix} I_1(\omega) \\ I_2(\omega) \\ I_3(\omega) \end{vmatrix}$$

Steps 4 and 5. These steps are implemented in the following MATLAB program noting that:

$$V_{out}(\omega) = 1/j\omega C^{I_3(\omega)}$$

The MATLAB implementation of the KVL and output equations is shown in the following discussion. The matrix equation and its solutions are placed in a loop that solves the equation for 4000 values of frequency ranging from 10 to 80 kHz. In this example, frequency is requested in hertz. The desired voltage V_{out} is determined from mesh current I_3. A Bode plot is constructed by plotting 20 log $|V_{out}|$ against log frequency since $V_{in} = 1$ at all frequencies. This loop is placed within another loop that repeats the 4000 solutions for the three values of capacitance.

```
% Example 13.7 Solution of a 3-mesh network
R = 2000;                        % Resistor values
C = [1, 0.1, 0.01]*1e-6;         % Capacitance values
V = [ 1; 0; 0];                  % Mesh voltages: Vin = 1
for i = 1:3                      % Capacitance value loop
  for k = 1:4000                 % Frequency loop
    f(k) = 10 + (k-1)*20;        % Determine frequency: 10- 80 kHz
    w = 2*pi*f(k);               % Calculate frequency in radians
    Xc = 1/(j*w*C(i));           % Calculate capacitance impedance
    Z = [R+Xc, -Xc, 0; -Xc, R+2*Xc, -Xc; 0, -Xc, R+2*Xc]; % Mesh equation
    I = Z\V;                     % Solve for currents
    Vout(k) = Xc*I(3);           % Solve for Vout
  .....label capacitor value.....
  end
  semilogx(f,20*log10(abs(Vout))); hold on;  % Bode plot
end
```

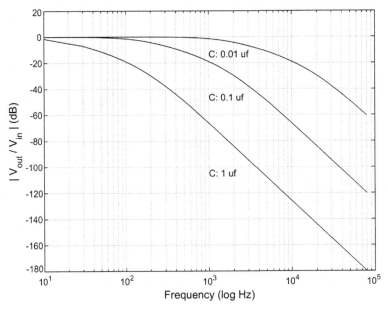

FIGURE 13.16 The Bode plot of the network shown in Figure 13.16 where V_{out} is the output and V_{in} the input. The frequency characteristics are shown for three capacitor values as indicated. These plots show that the network is a low-pass filter with a slope of 60 dB/decade. Increasing the capacitor values decreases the cutoff frequency.

The Bode plots generated by this program are shown in Figure 13.16. The plots of the three curves look like low-pass filters with downward slopes of 60 dB/decade, each having a different cutoff frequency (i.e., −3 dB attenuation points). Based on the Bode plot slope, this network is a third-order low-pass filter with a cutoff frequency that varies with the value of C. In fact, the cutoff frequency is approximately $1/2\pi RC$. Usually filters above first-order are constructed using active elements (see Chapter 15), but the circuit of Figure 13.15 is easy to construct and is occasionally used as a convenient third-order filter in real-world electronics.

13.3 CONSERVATION LAWS: KIRCHHOFF'S CURRENT LAW—NODAL ANALYSIS

KCL can also be used to analyze circuits. This law, based on the conservation of charge, was given in Equation 13.2 and is repeated here:

$$\sum_{Node} i = 0 \qquad (13.14)$$

KCL is best suited for analyzing circuits with many loops but only a few connection points. Figure 13.17 shows the Hodgkin—Huxley model for the nerve membrane. The three voltage—resistor combinations represent the potassium membrane channel, the sodium membrane channel, and the chloride membrane channel, whereas C is the membrane capacitance. Analyzing this circuit requires four mesh equations (V is actually a voltage source) but only

Axoplasm

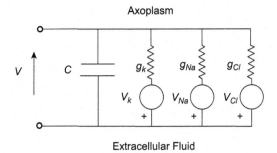

Extracellular Fluid

FIGURE 13.17   The Hodgkin–Huxley model for nerve membrane. The three voltage–resistor combinations represent the potassium membrane channel (K), the sodium membrane channel (Na), and the chloride membrane channel (Cl), whereas C is the membrane capacitance. This circuit contains four meshes, but only one independent node (assuming the bottom node is ground).

one nodal equation. In this model, most of the components are nonlinear, at least during an action potential, so the model cannot be solved analytically. Nonetheless, the defining equation(s) would be generated using nodal analysis and can be solved using computer simulation.

Another example of a circuit appropriate for nodal analysis is shown in Figure 13.18. This circuit has four meshes and mesh analysis would give rise to four simultaneous equations. This same circuit has only two nodes (marked A and B; again ground points do not count) and requires solving only two nodal equations. If MATLAB is used, then solving a four-equation problem is really not all that much harder than solving a two-equation problem; it is just a matter of adding a few more entries into the voltage vector and impedance matrix. However, when circuits are used as models representing physiological processes as in Figure 13.17, the more concise description given by nodal equations is of great value.

The circuit in Figure 13.18 contains a current source, not the familiar voltage source. This is because nodal analysis is an application of a current law, so it is easier to implement if the sources are current sources. A similar statement could be made about mesh analysis: mesh analysis involves voltage summation and it is easier to implement if all sources are voltage sources. The need to have only current sources may seem like a drawback to the application of nodal analysis, but we see in Chapter 14 that it is easy to convert voltage sources to equivalent current sources and vice versa, so this requirement is not really a handicap. In this

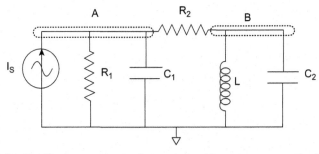

FIGURE 13.18   A circuit with four meshes and two nodes. The power source in this circuit is a current generator, I_S.

chapter, all nodal analysis examples use current sources, but the technique can be applied equally well to voltage sources after the simple conversion described in the next chapter.

Analyzing circuits using nodal analysis follows the same five-step procedure as that used in mesh analysis. In fact, Steps 4 and 5 are the same. Step 1 could also be the same, but often elements are converted to $1/Z$, instead of simply Z. Inverse impedance, $Y = 1/Z$, is termed "admittance" since it describes how current is admitted as opposed to impeded. In Step 2 the node voltages are assigned rather than the mesh currents and in Step 3 the equations are generated using KCL.

The equations developed from KCL have a sort of inverse symmetry with those of mesh analysis. In mesh analysis, we are writing matrix equations of the form:

$$v = Zi \tag{13.15}$$

where v is a voltage vector, i is a current vector, and Z is an impedance matrix (Equations 13.12 and 13.13). In nodal analysis we are writing matrix equations in the form:

$$i = Yv \tag{13.16}$$

where Y is a matrix, termed the "admittance matrix" containing the inverse impedances. The terms v and i are vectors as in Equation 13.15, although they occupy different positions in the equation.

EXAMPLE 13.8

Find the voltage V_A in the circuit in Figure 13.19.

Solution: This circuit requires two mesh equations (after conversion of the current source to an equivalent voltage source as explained in Chapter 14), but only one nodal equation. Moreover, it conveniently contains a current source, making nodal analysis even easier. There are four currents flowing into or out of the single node at the top of the circuit labeled A. The current in the current source branch is $0.1 \cos (2\pi 10t)$, and the current in the other three branches is equal to the voltage, V_A, divided by the impedance of the branch, i.e., $I(\omega) = V_A(\omega)/Z_{Branch}(\omega)$. By KCL, these four currents sum to zero.

After applying Steps 1 and 2, the network becomes as shown in Figure 13.20. If we define $V_A(\omega)$ as a positive voltage, then the current through the passive elements flows downward as shown owing to the voltage–current polarity rule for passive elements. The frequency in radians is $\omega = 2\pi f = 2\pi 10 = 62.8$ rad/s.

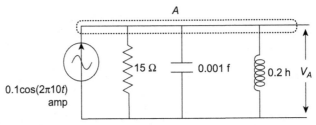

FIGURE 13.19   The network analyzed in Example 13.8. This network contains only one node and can be analyzed with a single KCL equation.

$V_A(\omega)$

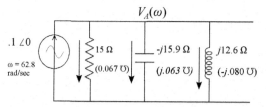

FIGURE 13.20 The circuit of Figure 13.19 after assigning nodal currents and converting the components values to impedances and admittances. (The units for admittance are called mhos; that is, ohms spelled backward. The symbol for admittance is the inverted symbol for ohms, ℧.)

Step 3 applies KCL: the four currents sum to zero. As with mesh analysis, care has to be taken that the signs are correct. The current source flows into node A so it is positive, but the other three currents flow out so they are negative.

$$i_S(\omega) - i_R(\omega) - i_C(\omega) - i_L(\omega) = 0 \quad \text{(KCL)}$$

$$I_S - \frac{V_A(\omega)}{R} - \frac{V_A(\omega)}{1/j\omega C} - \frac{V_A(\omega)}{j\omega L} = 0$$

$$0.1 + \frac{V_A(\omega)}{15} + \frac{V_A(\omega)}{-j15.4} + \frac{V_A(\omega)}{j13} = 0$$

Now we can solve this single equation for $V_A(\omega)$. The equation is easier written in terms of admittances: $Y = 1/Z$. The values of the admittances are shown in parentheses in the circuit given previously. Using admittances:

$$I_S + Y_R V_A(\omega) + Y_C V_A(\omega) + Y_L V_A(\omega) = 0$$

$$I_S + V_A(\omega)(1/R + j\omega C + 1/j\omega L) = 0$$

$$0.1 + V_A(\omega)(0.067 + j0.065 - j0.077) = 0$$

$$V_A(\omega) = \frac{0.1}{0.067 + j0.065 - j0.077} = \frac{0.1}{0.067 - j0.012}$$

$$V_A(\omega) = \frac{0.1}{0.068 \angle - 10} = 1.47 \angle 10 \text{ V}$$

Moving to multinodal systems, we go directly to the shortcut, matrix equation. If we apply KCL to circuits with multiple nodes, we find that the equations fall into a pattern similar to that of mesh analysis, except that they have the form of Equation 13.16: $i = Yv$. The admittance matrix consists of the summed admittances (i.e., $1/Z$'s) that are common to each node along the diagonal and the summed admittances between nodes on the off-diagonals. This general format is shown here for a three-node circuit:

$$\begin{vmatrix} \Sigma I_1 \\ \Sigma I_2 \\ \Sigma I_3 \end{vmatrix} = \begin{vmatrix} \Sigma Y \text{ Node 1} & -\Sigma Y \text{ Node 1\&2} & -\Sigma Y \text{ Node \&3} \\ -\Sigma Z \text{ Mesh 1\&2} & \Sigma Y \text{ Node 2} & -\Sigma Y \text{ Node 2\&3} \\ -\Sigma Y \text{ Node 1\&3} & -\Sigma Y \text{ Node 2\&3} & \Sigma Y \text{ Node 3} \end{vmatrix} \begin{vmatrix} V_1 \\ V_2 \\ V_3 \end{vmatrix} \qquad (13.17)$$

The application of Equation 13.17 is straightforward and follows the same pattern as in mesh analysis. An example of nodal analysis to the two-node circuit is given in Example 13.9.

EXAMPLE 13.9

Find the voltage, v_2, in the circuit shown in Figure 13.21. This circuit, like the one in Figure 13.18, has two nodes having voltages v_1 and v_2.

Solution: Apply nodal analysis to this two-node circuit. Follow the step-by-step procedure outlined earlier, but in Step 3 write the matrix equation directly, modifying Equation 13.17 for two nodes. Implement Step 4 to solve for V_B using MATLAB.

Step 1. Convert all the elements to phasor admittances. Note that $\omega = 20$ rad/s.

Step 2. Assign nodal voltages. This has already been done in the circuit. The circuit after modification by these two steps is shown in Figure 13.22.

Step 3. Generate the matrix equations directly following a reduced version of Equation 13.15. Note that inductors now have $-j$ values, whereas conductors have $+j$ values. Also note that the two nodes share two components, so the shared admittance will be the sum of the admittances from each component:

$$\sum Y_{nodel\ 1,2} = .004 - j.007$$

Hence the KCL circuit equation becomes:

$$\begin{vmatrix} 0.5 \\ 0 \end{vmatrix} = \begin{vmatrix} 0.01 + 0.004 + j0.01 - j0.007 & -0.004 + j0.007 \\ -0.004 + j0.007 & 0.004 - j0.005 - j0.007 + j0.04 \end{vmatrix} \begin{vmatrix} V_1 \\ V_2 \end{vmatrix}$$

$$\begin{vmatrix} 0.5 \\ 0 \end{vmatrix} = \begin{vmatrix} .014 + j.003 & -0.004 + j0.007 \\ -.004 + j.007 & 0.004 - j0.028 \end{vmatrix} \begin{vmatrix} V_1 \\ V_2 \end{vmatrix}$$

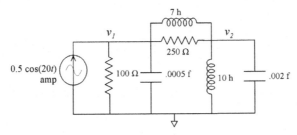

FIGURE 13.21 Two-node circuit analyzed using KCL matrix equations in Example 13.9.

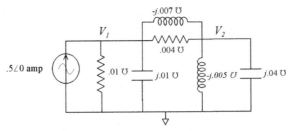

FIGURE 13.22 The circuit shown in Figure 13.21 after assignment of nodal currents and converting element values to admittances. Note that the component values are given as admittances in mhos (℧).

Step 4. This matrix equation can easily be solved using MATLAB as illustrated by the following code.

```
% Example 13.9  Solution of two-node matrix equation
%
I = [.5; 0];                          % Assign current vector
Y11 = 0.01 +.004 + 1i*.01 -1i*.007;   % Assign admittances
Y12 = (.004-1i*.007);
Y22 = 0.004 - 1i*.005 - 1i*.007 + 1i*.04;
Y1 = [Y11 -Y12; -Y12 Y22];            % Admittance matrix
V = Y\I;                              % Solve for voltages
Magnitude = abs(V(2))                 % Magnitude and phase of V2
Phase = angle(V(2))*360/(2*pi)
```

The output gives the magnitude and phase of V_2 as:

```
Mag = 8.9234;  Phase  -149.3235
```

Converting to the time domain:

$$v_2(t) = 8.92 \cos(20t - 149) \text{ V}$$

The approach used in Example 13.9 can be extended to three-node or even higher-node circuits without great difficulty. A three-node problem is given at the end of the chapter. Nodal analysis applies equally well to networks represented in Laplace notation. The basic five-step approach can also be applied to the analysis of lumped-parameter mechanical systems as described in the next section.

13.4 CONSERVATION LAWS: NEWTON'S LAW—MECHANICAL SYSTEMS

The analysis of lumped-parameter mechanical systems also uses a conservation law, one based on the conservation of energy. In mechanical systems, force is potential energy, since force acting through a distance produces work: $W = \int F dx$. The mechanical version of KVL (the electrical law based on the conservation of energy) states that the sum of the forces around any one connection point must be zero. This is a form of the classic law associated with Newton:

$$\sum_{\text{Point}} F = 0 \tag{13.18}$$

In this application, a connection point includes all connections between the mechanical elements that are at the same velocity (just as a node is all the points at the same voltage).

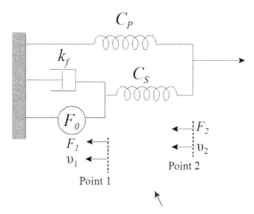

FIGURE 13.23 Linear mechanical model of skeletal muscle. F_0 is the force produced by the active contractile element, C_P and C_S are the parallel and series elasticities, and k_f is the viscous damping associated with the tissue.

Figure 13.23 shows one version of the linear model for skeletal muscle. Skeletal muscle has two different elastic elements: a parallel elastic element, C_p, and a series elastic element, C_s. (In mechanical systems, the symbol C is used to indicate a component's compliance, the inverse of elasticity, $1/k_e$.) The force generator, F_0, represents the active contractile element, and k_f represents viscous damping inherent in the muscle tissue. The muscle model has two connection points that may have different velocities, labeled Point 1 and Point 2. The positive force is defined inward, reflecting the fact that muscles can generate only contractile force. This is the reason they are so often found in agonist—antagonist pairs.

Since this system has two different velocities, its analysis would require the simultaneous solution of equations. The system is the mechanical equivalent of a two-mesh electrical circuit. The two equations would be written around Points 1 and 2: the sum of forces around each point must be zero. The graphic on the left represents a zero-velocity point or a solid wall, the analog of a ground point in an electrical system.

The muscle model will be analyzed in Example 13.11, but for a first example of the application of Newton's law (Equation 13.18) we turn to a less complicated, single equation system with a single velocity point as shown in Figure 13.24.

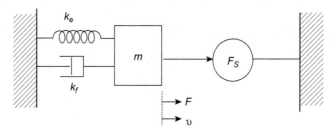

FIGURE 13.24 A simple mechanical system consisting of a force generator, F_S, mass, m, friction element, k_f, and elasticity, k_e. The velocity of the mass, v, is found using Equation 13.18 in Example 13.10.

EXAMPLE 13.10

Find the velocity and displacement of the mass in the mechanical system shown in Figure 13.24. The force is $F_S(t) = 5 \cos(2t + 30)$ dynes. The following parameters also apply:

$$k_f = 5\,\text{s/cm}; \quad k_e = 8\,\text{dyn/cm}; \quad m = 3\,\text{g.}$$

Solution: In this example, all units are in the cgs metric system and therefore are comparable. This is not always the case in mechanical systems; in Example 13.11, conversion of units is required. To analyze this system, we follow the same five-step plan developed for electric circuits.

Step 1. Convert the variables to phasor notation and represent the passive elements by their phasor impedances. Since $\omega = 2$:

$$Z_f = k_f = 5; \quad Z_e = \frac{k_e}{j\omega} = \frac{8}{j2} = -j4; \quad Z_m = j\omega m = j2(3) = j6$$

Step 2. Assign variable directions. In mechanical systems, we use the convention of assigning the force and velocity in the same direction, but the direction (right or left) is arbitrary (in this example it is to the right). This is analogous to assigning currents as counterclockwise and keeping track of voltage polarities by going in the same direction. By assigning force and velocity in the same direction, the polarity of passive elements is always negative just as in electric circuits, so the equations look similar.

Step 3. Apply Newton's law about the force–velocity point (next to the mass):

$$\sum F = 0; \quad F_S(\omega) - k_f v(\omega) - \frac{k_e}{j\omega} v(\omega) - j\omega m v(\omega) = 0$$

$$5 \angle 30 - v(\omega)(5 - j4 + j6) = 0$$

The first three steps follow a path parallel to that followed in the KVL analysis, whereas the last two steps are essentially identical: solve for the velocity (analogous to current), then any other variable of interest such as a force or, in this case, a displacement.

Step 4. Solve for the phasor velocity:

$$v(\omega) = \frac{5 \angle 30}{5 - j4 + j6} = \frac{5 \angle 30}{5 + j2} = \frac{5 \angle 30}{5.39 \angle 21.8} = 0.93 \angle 8.2 \ \text{cm/s}$$

Step 5. Solve for displacement. Since $x(t) = \int v\,dt$, integration in the phasor domain is division by $j\omega$; so $x(\omega) = v(\omega)/j\omega$:

$$x(\omega) = \frac{v(\omega)}{j\omega} = \frac{0.93 \angle 8.2}{j2} = \frac{0.93 \angle 8.2}{2 \angle 90} = 0.465 \angle -81.8 \ \text{cm}$$

Both $v(\omega)$ and $x(\omega)$ can be converted to the time domain if desired:

$$v(t) = 0.93 \ \cos(2t + 8.2) \ \text{cm/s}$$
$$x(t) = 0.4610 \ \cos(2t - 81.8) \ \text{cm}$$

The next example is more complicated in two ways: (1) the component values are not given and must be carried as variables through the algebra, and (2) there are two summation points in the problem, so two equations are required that must be solved simultaneously.

EXAMPLE 13.11

Find the force out of the skeletal muscle model in Figure 13.23.

Solution: After converting to the phasor domain, write Newton's law (Equation 13.18) around points 1 and 2. Solve for $v_2(\omega)$, then $F_2(\omega) = v_2(\omega)Z_{C_p}(\omega) = v_2(\omega)1/j\omega C_p$. In this solution the algebra is a bit tedious because the various parameters remain as variables, but the procedure is otherwise straightforward.

Steps 1 and 2. The force and velocity assignments are given in Figure 13.22. The various components become:

$$F_o \rightarrow F_o(\omega); \quad k_f \rightarrow k_f; \quad C_p \rightarrow \frac{1}{j\omega C_p} \quad C_S \rightarrow \frac{1}{j\omega C_S}$$

The equation around Point 1 is:

$$F_o - k_f v_1(\omega) - \frac{1}{j\omega C_s}(v_1(\omega) - v_2(\omega)) = 0$$

Note that the force generated by the elastic element C_s depends on the difference in velocities between Points 1 and 2. The force that is generated by $v_1(\omega) - v_2(\omega)$ is in the opposite direction of F_1, which accounts for the negative sign in front of this term.

Step 3. The equation around Point 2 is:

$$0 - \frac{1}{j\omega C_P}v_2(\omega) - \frac{1}{j\omega C_s}(v_2(\omega) - v_1(\omega)) = 0$$

Rearranging to separate coefficients of $v_1(\omega)$ and $v_2(\omega)$:

$$F_o = \left(k_f + \frac{1}{j\omega C_s}\right)v_1(\omega) - \frac{1}{j\omega C_s}v_2(\omega)$$

$$0 = \frac{1}{j\omega C_s}v_1(\omega) + \left(\frac{1}{j\omega C_s} + \frac{1}{j\omega C_p}\right)v_2(\omega)$$

Step 4. Solving $v_2(\omega)$:

$$v2(\omega) = \cfrac{\begin{vmatrix} F_o & -\dfrac{1}{j\omega C_s} \\[3ex] 0 & \dfrac{1}{j\omega C_s} + \dfrac{1}{j\omega C_p} \end{vmatrix}}{\begin{vmatrix} k_f + \dfrac{1}{j\omega C_s} & -\dfrac{1}{j\omega C_s} \\[3ex] -\dfrac{1}{j\omega C_s} & \dfrac{1}{j\omega C_s} + \dfrac{1}{j\omega C_p} \end{vmatrix}}$$

$$= \cfrac{F_o\left(\dfrac{1}{j\omega C_s} + \dfrac{1}{j\omega C_p}\right)}{\dfrac{k_f}{j\omega C_s} + \dfrac{k_f}{j\omega C_p} + \left(\dfrac{1}{j\omega C_s}\right)^2 + \dfrac{1}{(j\omega)^2 C_p C_s} - \left(\dfrac{1}{j\omega C_s}\right)^2}$$

Canceling the two terms, noting that $j^2 = -1$, and multiplying through by $(j\omega)^2$:

$$v_2(\omega) = \frac{j\omega F_o\left(\dfrac{1}{C_s} + \dfrac{1}{C_p}\right)}{\dfrac{1}{C_pC_s} + j\omega\left(\dfrac{k_f}{C_s} + \dfrac{k_f}{C_p}\right)} = \frac{j\omega F_o(C_s + C_p)}{1 + j\omega(k_f C_s + k_f C_p)}$$

Step 5. Now to find the force at the output, multiply $v_2(\omega)$ by the impedance of the parallel elastic element, C_p.

$$F(\omega) = v_2(\omega)\left(\frac{1}{j\omega C_p}\right) = \frac{j\omega F_o(C_s + C_p)}{1 + j\omega(k_f C_s + k_f C_p)}\left(\frac{1}{j\omega C_p}\right) = \frac{F_o(C_s + C_p)/C_p}{1 + j\omega(k_f C_s + k_f C_p)}$$

In the next chapter we find that this equation, if viewed as a transfer function $F(\omega)/v_2(\omega)$, has the same general properties as those of a low-pass filter. Although real muscle can be fairly well described by the elements in Figure 13.23, the equation does not take into account the component nonlinearities. Nonetheless, this linear analysis provides a starting point for more complicated models and analyses.

The final example solves a two-equation system that includes a mass. This somewhat involved example also demonstrates how to convert various measurement units to the cgs system.

EXAMPLE 13.12

Find the velocity of the mass in the system shown in Figure 13.25. Assume that $F_S(t) = 0.001 \cos(20t)$ N and that the following parameter assignments apply:

$$m = 1.0 \text{ oz}; \quad k_{f1} = 200 \text{ dyn} - \text{s/cm};$$
$$k_{e1} = 6000 \text{ dyn/s}; \quad k_{e2} = 0.05 \text{ lbs/in}.$$

Solution: We first need to convert F_S, m, and k_{e2} to cgs units. Converting F_S is relatively straightforward since it is already in MKS metric units. Use the conversion factors in Appendix D:

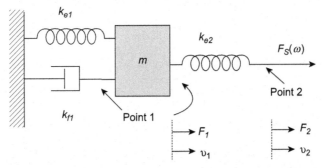

FIGURE 13.25 The mechanical system used in Example 13.12. This system has two independent velocities at points 1 and 2, so as with a two-mesh or two-node circuit, it requires two equations for analysis.

$1\,\text{N} = 10^5$ dyn. Hence $F_S = 100\sin(20t)$ dyn. To convert m from English units (oz) to cgs metric units, use the conversion factors in Appendix D in conjunction with dimensional analysis:

$$m = 1.0\,\text{oz}\,\frac{1\,\text{lb}}{16\,\text{oz}}\,\frac{1\,\text{kg}}{2.2046\,\text{lb}}\,\frac{1000\,\text{gm}}{1\,\text{kg}} = 28.4\,\text{gm}$$

In the English system the pound is actually a measure of mass but is often used as a measure of force as is the case here. The assumption is that a pound weight is equal to the force produced by a pound accelerated by gravity (i.e., $F = mg$). So it is not surprising that the cgs equivalent of 1 lb is $4103.109\,\text{gm}^4$, a measure of mass; however, the cgs equivalent of a 1-lb weight is 4.448 N, a measure of force. The cgs system is not without some ambiguity: grams are sometimes used to represent a weight as well as a mass. In cgs, the earth's gravitational constant, g, is equal to $980.6610\,\text{cm/s}^2$, so a gram weight is 980.6610 dyn. (Similarly a kilogram weight is 9.807 N). Applying the conversion factors from pounds weight to dynes and inches to centimeters:

$$k_{e2} = 0.05\,\frac{\text{lb}}{\text{in}}\,\frac{4.448\,\text{N}}{\text{lb. wt}}\,\frac{10^5\,\text{dyn}}{1\,\text{N}}\,\frac{1\,\text{in}}{2.54\,\text{cm}} = 8756\,\text{dyn/cm}$$

This is a lot of work to go through before we even get to Step 1, but real-world problems often involve messy situations like inconsistent units.

Step 1. Following the conversion to cgs units, the determination of the phasor impedances is straightforward. With $\omega = 20\,\text{rad/s}$:

$$Z_{f1} = k_{f1} = 200\,\text{dyn/cm}; \quad Z_m = j\omega m = j20(28.4) = 568\,\text{dyn/cm}$$

$$Z_{e1} = \frac{k_{e1}}{j\omega} = \frac{6000}{j20} = -j300\,\text{dyn/cm}; \quad Z_{e2} = \frac{k_{e2}}{j\omega} = \frac{8756}{j20} = -j438\,\text{dyn/cm}$$

Step 2. The force and velocity directions have already been assigned with F_1 and F_2 as positive to the right. After the first two steps, the phasor representation of the system is as shown in Figure 13.26.

In writing the equations about the two points, we must take into account the fact that the spring and friction on the right side of the mass have nonzero velocities on both sides. Therefore, the net

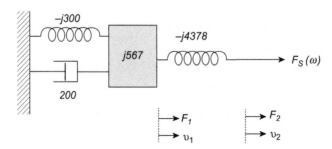

FIGURE 13.26   The mechanical system in Figure 13.25 after the component values have been converted to mechanical impedances. Before converting to impedances, it was first necessary to convert all the units to the cgs metric system.

velocity across these two elements is $v_2 - v_1$. Thus the force across the right-hand spring is $k_{e2}/j\omega$ $(v_2 - v_1)$. With this in mind, the equation around Point 2 becomes:

$$F_S - \frac{k_{e2}}{j\omega}(v_2 - v_1) = F_S - (-j438)(v_2 - v_1) = 0$$

For Point 2, the force exerted by this spring has the same magnitude, but is opposite in direction (i.e., positive with respect to F_1). Thus the equation for the force around Point 1 is:

$$\frac{k_{e2}}{j\omega}(v_2 - v_1) - \left(j\omega m - k_f - \frac{k_{e1}}{j\omega}\right)v_1 = 0$$

$$-j438(v_2 - v_1) - j568v_1 - 200v_1 - (-j300)v_1 = 0$$

Multiplying through by -1 and rearranging the two equations as coefficients of v_1 and v_2:

$$0 = (200 + j568 - j300 - j438)v_1(\omega) - (-j438)v_2(\omega)$$
$$100 = -(-j438)v_1(\omega) - j438v_2(\omega)$$

and in matrix notation:

$$\begin{vmatrix} 0 \\ 100 \end{vmatrix} = \begin{vmatrix} 200 - j170 & j438 \\ j438 & -j438 \end{vmatrix} \begin{vmatrix} v_1(\omega) \\ v_2(\omega) \end{vmatrix}$$

Solving for v_2, the velocity, this time manually:

$$v_2(\omega) = \frac{\begin{vmatrix} 200 - j170 & 0 \\ j438 & 100 \end{vmatrix}}{\begin{vmatrix} 200 - j170 & j438 \\ j438 & -j438 \end{vmatrix}} = \frac{20000 - j17000}{-j85600 - 74460 + 191844}$$

$$v_2(\omega) = \frac{20000 - j17000}{117384 - j87600} = \frac{26249 \angle -40}{146470 \angle -36.7} = 0.18 \angle -3.3 \text{ cm/s}$$

Of course, this solution could have been more easily obtained using MATLAB.

[4]To make matters even more confusing there are two units of mass in the English system termed pounds and abbreviated lbs. The most commonly used pound is termed the commercial or "avoirdupois pound," whereas a less commonly used measure is the "troy" or "apothecary pound." To convert: 1 troy lb. = 0.822857 avoirdupois lb. In this text only avoirdupois pounds are used, but conversions for both can be found in Appendix D.

13.5 RESONANCE

Resonance is a frequency-dependent behavior characterized by a sharp increase (or decrease) in some system variable(s) over a limited range of frequencies. It can occur in almost any feedback system, including mechanical and electrical systems as well as chemical and molecular systems. Often it is beneficial and exploited as in proton resonance used in magnetic resonance imaging, optical resonance used in spectroscopy for identifying molecular systems, or electrical resonance used to isolate frequencies or generate sinusoidal signals. However, it can also be undesirable, particularly in mechanical systems. For example, shock

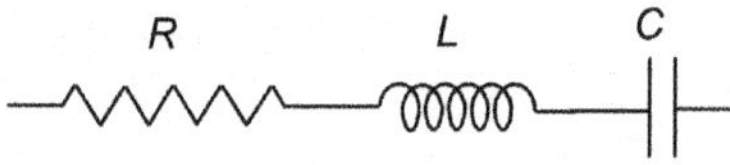

FIGURE 13.27 A series RLC circuit that can exhibit the properties of resonance at a select frequency.

absorbers are friction elements placed on cars to increase damping and reduce the resonant properties of automotive suspension systems. Resonance is discussed in terms of mechanical and electrical systems since these are the systems we have been studying, but the basic concepts are applicable to other systems.

13.5.1 Resonant Frequency

In electrical and mechanical systems resonance occurs when the impedance of an inertial-type element (mass or inductor) equals and cancels the impedance of a capacitive-type element. Consider the impedance of a series RLC circuit shown in Figure 13.27.

In the frequency (phasor) domain, the impedance of this series combination is:

$$Z(\omega) = R + j\omega L + \frac{1}{j\omega C} = R + j\left(\omega L - \frac{1}{\omega C}\right) \Omega \qquad (13.19)$$

At some value of ω, the capacitor's impedance will be equal to the inductor's impedance and the two impedances will cancel. This will leave only the resistor to contribute to the total impedance. To determine the frequency at which this cancellation takes place, simply set the impedances equal and solve for frequency:

$$\omega_o L = \frac{1}{\omega_o C}; \quad \omega_o L(\omega_o C) = 1; \quad \omega_o^2 = \frac{1}{LC}$$

$$\omega_o = \frac{1}{\sqrt{LC}} \text{ rad/s} \qquad (13.20)$$

where ω_o is the "resonant frequency." Note that this is the same equation as for the undamped natural frequency, ω_n, in a second-order representation of the RLC circuit (see Equation 13.7). If we plot the magnitude of the impedance in Equation 13.19, we get a curve that reaches a minimum value of R at $\omega = \omega_o$ and increases on either side, Figure 13.28. The sharpness of the curve relates to the bandwidth of the resonant system as discussed in the next section.

13.5.2 Resonant Bandwidth, Q

When a system approaches the resonant frequency, the system variables (voltage–current or force–velocity) will increase (or decrease) to a maximum (or minimum). The sharpness of that curve depends on the energy dissipation element (resistance or friction). Figure 13.29 shows an RLC circuit configured as an input–output system. In the next example, we find and plot the Bode plot of the transfer function of this system for different values of R and show how the sharpness of the resonant peak depends on R: as R decreases the sharpness increases.

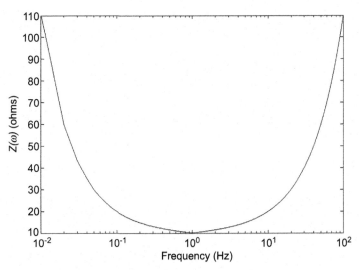

FIGURE 13.28 A plot of impedance versus frequency for a series RLC circuit as given in Equation 13.19. The impedance reaches a minimum at $\omega = \omega_0$.

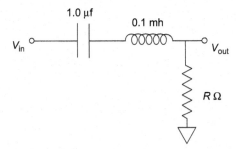

FIGURE 13.29 An RLC circuit configured as an input–output system. This Bode plot (i.e., system frequency spectrum) of this system is determined for different values of R in Example 13.13.

EXAMPLE 13.13

Plot the Bode plot of the transfer function of the network shown in Figure 13.28 for four values of resistance: 0.1, 1.0, 10, and 100 Ω. Plot the Bode plot in radians per seconds (for variety).

Combining **Step 1** (conversion), **Step 2** (current assignment), and **Step 3** (KVL), the equation for the circuit current becomes:

$$V_{in}(\omega) = I(\omega)\left(R + j\omega L + \frac{1}{j\omega C}\right)$$

Steps 4 and 5. Since we will be using MATLAB to plot the Bode plot, we can leave the L and C as variables. Solving for $I(\omega)$, then finding $V_{out}(\omega)$:

$$I(\omega) = \frac{V_{in}(\omega)}{R + j\omega L + 1/j\omega C}$$

$$V_{out}(\omega) = I(\omega)R = \frac{RV_{in}(\omega)}{R + j\omega L + 1/j\omega C}$$

The transfer function becomes:

$$TF(\omega) = \frac{V_{out}(\omega)}{V_{in}(\omega)} = \frac{R}{R + j\omega L + 1/j\omega C} \qquad (13.21)$$

Rearranging into the standard Bode plot format with the lowest coefficient of ω equal to 1 (in this case the constant):

$$TF(\omega) = \frac{V_{out}(\omega)}{V_{in}(\omega)} = \frac{j\omega RC}{j\omega RC + j^2\omega^2 LC + 1} = \frac{j\omega RC}{j\omega RC - \omega^2 LC + 1}$$

$$TF(\omega) = \frac{V_{out}(\omega)}{V_{in}(\omega)} = \frac{j\omega RC}{1 - \omega^2 LC + j\omega RC} \qquad (13.22)$$

Note that since we will be solving the transfer function equation on the computer it would have been just as easy to use Equation 13.21 directly; the extra work of putting it into standard Bode plot form was just an exercise.

The resonant frequency of the transfer function is $\omega_o = 1/\sqrt{LC} = 1 \times 10^5$ rad/s and the Bode plot should include a couple of orders frequency above and below this frequency, say, from 10^3 to 10^7 rad/s. Solving Equation 13.22 over this frequency range leads to the following MATLAB program.

```
% Example 13.13 Bode plot of the TF of an RLC circuit with 4 different values of R.
%
R= [0.1 1 10 100];          % Resistance values
L = 10-4;                   % Inductance value
C= 10-6;                    % Capacitance value
w = (1000:1000:10000000);   % Frequency range
for k = 1:length(R)
  TF = (j*w*R(k)*C)./(1 - w.^2* L*C + j*w*R(k)*C); % TF equation
  TF = 20*log10(abs(TF));   % Convert to dB
  semilogx(w,TF,'k');       % Plot as Bode plot
  ....... text labels......
end
  ....... labels and axis.......
```

Results: This program produces the graph shown in Figure 13.30.

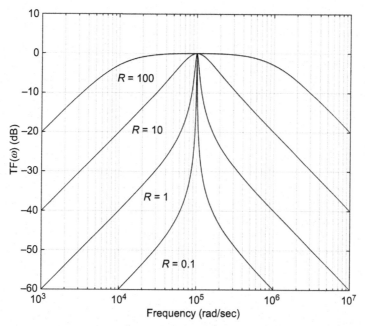

FIGURE 13.30 The Bode plot (system frequency spectrum) of the RLC system shown in Figure 13.29 for four different values of R. This input–output system shows a resonance at 10,000 rad/s for all values of R, but the resonance peak increases dramatically with decreasing values of R.

Figure 13.30 shows that the resonance peak of the transfer function occurs at the same frequency for all values of R. This is expected since the resonant frequency is a function of only L and C, Equation 13.18. Specifically, the peak occurs at:

$$\omega_o = 1/\sqrt{LC} = 1/\sqrt{10^{-4}10^{-6}} = 100,000 \text{ rad/s.}$$

Peak sharpness increases as R decreases. In the configuration shown in Figure 13.29, the RLC network is a band-pass filter; however, if the resistor and capacitor are interchanged, the circuit becomes a low-pass filter as used in previous examples. Nonetheless, this arrangement of an RLC network still has a resonant peak at a frequency determined by L and C (Equation 13.18), and the sharpness of the peak varies with R in a similar manner. The behavior of a low-pass RLC network is examined in a problem.

The sharpness of the frequency curve around the resonant frequency is an important property of resonant systems. This characteristic is often described by a number known as "Q," which is defined as the resonant frequency divided by the bandwidth:

$$Q = \frac{\omega_o}{BW} \tag{13.23}$$

where ω_o is the resonant frequency and BW is the bandwidth using the standard definition: the difference between the high-frequency cutoff and the low-frequency cutoff (−3 dB points). Occasionally, Equation 13.23 is referred to as "selectivity." To confuse things further, Q also has an alternate, more fundamental definition.

Classically, Q is defined in terms of energy storage and energy loss to describe how close energy storage elements such as inductors and capacitors approach the ideal. Ideally such elements should only store energy, but real elements also dissipate energy owing to parasitic resistance. In this context, Q is defined as the energy stored over the energy lost in one cycle:

$$Q = 2\pi \frac{\text{Energy stored}}{\text{Energy dissipated}} \tag{13.24}$$

Using Equation 13.24 we can calculate Q for a single element such as an inductor. Assume an inductor has not only inductance L h, but also a parasitic series resistance of R Ω. Since inductors are constructed as wire coils, often with quite a bit of wire, inductors inevitably have some resistance. (Inductors are the least ideal of the passive electrical elements.) The energy lost in the wire's resistance over one sinusoidal cycle is equal to the power integrated over the cycle, and the power is just vi, or for a resistor, $R\,i^2$. Assuming a sinusoidal current through the resistor, $i_R(t) = I \sin(\omega t)$, the energy lost over one cycle becomes:

$$E_{lost} = \int_{Cyc} vi\,dt = \int_0^{2\pi} R(I \sin(\omega t))^2\,dt = \frac{2\pi R I^2}{2\omega}\ \text{J}$$

The energy stored in an inductor is also the integral of vi over one cycle. Since the parasitic R is effectively in series with the inductor, the current through the inductor is the same as through the resistor $i_L(t) = I \sin(\omega t)$ and, from the definition of an inductor (Equation 12.15), the voltage is L times the derivative of the current:

$$v_L(t) = L\frac{di}{dt} = \frac{d(I \sin(\omega t))}{dt} = \omega I \cos(\omega t)$$

$$E_{Stored} = \int_{Cyc} vi\,dt = \int_0^{2\pi} \omega I^2 \cos(\omega t)\sin(\omega t)dt = \frac{\omega L I^2}{2\omega}$$

Plugging in the two energies into Equation 13.24, the Q of an inductor becomes:

$$Q_L = \frac{2\pi\left(\dfrac{\omega L I^2}{2\omega}\right)}{\left(\dfrac{2\pi R I^2}{2\omega}\right)} = \frac{\omega L}{R} \tag{13.25}$$

where L is the inductance, ω the frequency in rad/s, and R the parasitic resistance.

Similarly, it is possible to derive the Q of a capacitor having a capacitance value C and a parasitic resistance R as:

$$Q_C = \frac{1}{\omega RC} \tag{13.26}$$

However, in a circuit that includes both an inductor and a capacitor, most of the parasitic resistance is from the inductor. Any contribution from the capacitor is usually ignored and Equation 13.25 can be used to find the Q.

Based on these definitions, it is possible to derive Equation 13.23 from the definition of bandwidth. Returning to the RLC circuit in Figure 13.29, the transfer function of this circuit is given in Equation 13.21 is:

$$TF(\omega) = \frac{R}{R + j\omega L + 1/J\omega C} = \frac{R}{Z(\omega)} = \frac{1}{Z(\omega)/R}$$

where $Z(\omega)$ is the series R, L, C impedance.

$Z(\omega)/R$ can also be written as:

$$\frac{Z}{R} = \frac{R + j\omega L - \dfrac{j}{j\omega C}}{R} = 1 + j\left(\frac{\omega L}{R} - \frac{1}{\omega C}\right); \quad \text{multiplying both imaginary terms by } \frac{\omega_o}{\omega_o}$$

$$\frac{Z}{R} = 1 + j\left(\frac{\omega}{\omega_o}\left(\frac{\omega_o L}{R}\right) - \frac{\omega_o}{\omega}\left(\frac{1}{\omega_o C}\right)\right)$$

This allows us to substitute the definition of Q into the equation for Z/R:

$$\frac{Z}{R} = 1 + j\left(\frac{\omega Q}{\omega_o} - \frac{\omega_o Q}{\omega}\right) = 1 + jQ\left(\frac{\omega}{\omega_o} - \frac{\omega_o}{\omega}\right) \tag{13.27}$$

At the cutoff frequencies, $|TF(\omega_{H,L})| = 0.707 \ |TF(\omega = \omega_o)|$. However, at the resonant frequency, $Z(\omega = \omega_o) = R$ and $TF(\omega = \omega_o)$. Hence, at the cutoff frequencies $|TF(\omega_{H,L})| = 0.707$.

So to find the bandwidth (which is just the difference between the cutoff frequencies), set $\left|TF(\omega)\right| = 0.707\left|TF(\omega)\right|$ so $\left|\dfrac{Z}{R}\right| = \left|\dfrac{1}{TF(\omega)}\right| = \dfrac{1}{.707} = 1.414$ and solve for ω_H and ω_L.

Setting the magnitude of Equation 13.27 to $1.414 = \sqrt{2}$, we get:

$$\left|1 + jQ\left(\frac{\omega}{\omega_o} - \frac{\omega_o}{\omega}\right)\right| = \sqrt{2}; \quad \left(\text{For}|1 + jB| = \sqrt{2}, B = \pm 1\right)$$

$$Q\left(\frac{\omega}{\omega_o} - \frac{\omega_o}{\omega}\right) = \pm 1$$

There are two solutions to the equation for $+1$ and -1:

$$\omega_L = \omega_o\left(1 - \frac{1}{2Q}\right); \quad \omega_H = \omega_o\left(1 + \frac{1}{2Q}\right)$$

$$BW = \omega_H - \omega_L = \frac{\omega_0}{Q}; \quad Q = \frac{\omega_0}{BW}$$

We can also relate Q to the standard coefficients of a second-order underdamped equation. Again referring to the RLC circuit in Figure 13.29, the transfer function can be easily determined in terms of the Laplace variable, s. To make the solution more general, we continue using variables L and C to represent the inductance and capacitance:

Steps 1, 2, and 3:

$$V_{in}(s) = I(s)(R + sL + 1/sC)$$

Steps 4 and 5:

$$I(s) = \frac{V_{in}(s)}{R + sL + 1/sC}; \quad V_{out}(s) = I(s)R = \frac{RV_{in}(s)}{R + sL + 1/sC}$$

Solving for the transfer function, $V_{out}(s)/V_{in}(s)$:

$$TF(s) = \frac{V_{out}(s)}{V_{in}(s)} = \frac{R}{R + sL + 1/sC}$$

Putting the transfer function equation into standard Laplace domain format where the highest power of s has a coefficient of 1:

$$TF(s) = \frac{V_{out}(s)}{V_{in}(s)} = \frac{sR}{sR + s^2L + 1/C} = \frac{sR/L}{sR/L + s^2 + 1/LC}$$

Rearranging and comparing with the standard form of a second-order equation analyzed in Chapter 7 (Equation 7.31):

$$TF(s) = \frac{R/L^s}{s^2 + R/Ls + 1/LC} = (RCs)\frac{1/LC}{s^2 + R/Ls + 1/LC} = (RCs)\frac{\omega_n^2}{s^2 + 2\delta\omega_n s + \omega_n^2}$$

where $\omega_n = \omega$. Equating coefficients:

$$2\delta\omega = \frac{R}{L}; \quad \delta = \frac{R}{2\omega L}; \quad \text{and } Q = \frac{\omega L}{R}$$

$$\delta = \frac{1}{2Q}; \quad Q = \frac{1}{2\delta} \tag{13.28}$$

The relationship between Q and δ is amazingly simple.

The characteristics of resonance and the various definitions and relationships described apply equally to mechanical systems. In fact, most mechanical systems exhibit some resonant behavior and often that behavior is detrimental to the system's performance (consider a car with bad shock absorbers). An investigation of the resonant properties of a mechanical system is given in the next example.

EXAMPLE 13.14

Find the Q of the mechanical system shown in Figure 13.31. The system coefficients are $k_f = 6$ dyn s/cm; $m = 8$ g; $k_e = 10$ dyn/cm.

Solution: Q can be determined directly from δ in transfer function using Equation 13.28. So we need to find the transfer function $v(s)/F_s(s)$. Applying the standard analysis based Equation 13.18:

$$F_s(s) - v(s)(k_f + ms + k_e/s) = 0; \quad v(s) = \frac{F_s(s)}{k_f + ms + k_e/s}$$

$$\frac{v(s)}{F_s(s)} = \frac{s/m}{s^2 + \frac{k_f}{m}s + \frac{k_e}{m}} = \frac{s/m}{s^2 + 2\delta\omega_n s + \omega_n^2}$$

Equating coefficients:

$$\omega_n = \sqrt{\frac{k_e}{m}} = \sqrt{\frac{10}{8}} = 1.1 \text{ rad/sec}; \quad 2\delta\omega_n = \frac{k_f}{m}; \quad \delta = \frac{k_f}{2m\omega_n} = \frac{k_f}{2\sqrt{k_e m}} = \frac{6}{2\sqrt{10(8)}} = 0.335$$

$$Q = \frac{1}{2\delta} = 1.49$$

The next example uses MATLAB to explore the ways in which the Bode plot and impulse response of a generic second-order system vary for different values of Q.

EXAMPLE 13.15

Plot the frequency characteristics and impulse response of a second-order system in which $Q = 1$, 10, and 100. Use MATLAB and assume a resonant frequency, ω_n^2 of 1000 rad/s. Note that it could be either a mechanical or electrical system; the equations would be the same.

Solution: Begin with the standard second-order Laplace transfer function equation, but substitute Q for δ. Convert this equation to the time domain by taking the inverse Laplace transform to get the impulse response. (Recall, the inverse Laplace transform of the transfer function is the impulse response.) Then convert the transfer function equation to the phasor domain to get the Bode plot equation. Use MATLAB to plot both the Bode plot and the time response.

The standard Laplace second-order equation is:

$$TF(s) = \frac{\omega_n^2}{s^2 + 2\delta\omega_n s + \omega_n^2};$$

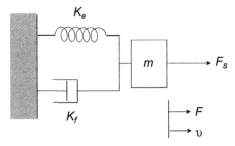

FIGURE 13.31 A mechanical system consisting of a mass, friction, and elasticity used in Example 13.14.

Substituting in $\delta = \frac{1}{2Q}$ and $\omega_n = 1000$:

$$TF(s) = \frac{\omega_n^2}{s^2 + \frac{\omega_n}{Q}s + \omega_n^2} = \frac{10^6}{s^2 + \frac{1000}{Q}s + 10^6}$$

To find the impulse response, take the inverse Laplace transform of the transfer function. Since $Q \geq 1$, δ will be ≤ 0.5, so the system is underdamped for all values of Q and entry 15 in the Laplace transform table can be used:

$$x(t) = \frac{\omega_n}{\sqrt{1-\delta^2}}\left[e^{-\delta\omega_n t}\sin\left(\omega_n\sqrt{1-\delta^2}t\right)\right] = \frac{10^3}{\sqrt{1-\frac{1}{4Q^2}}}\left[e^{-1000/2Q}\sin\left(1000\sqrt{1-\frac{1}{4Q^2}}t\right)\right]$$

To find the frequency response, convert to phasor and plot for the requested values of Q:

$$TF(\omega) = \frac{10^6}{(j\omega)^2 + j\omega\frac{1000}{Q} + 10^6} = \frac{1}{1 - \left(\frac{\omega}{10^3}\right)^2 + \frac{j\omega}{Q10^3}}$$

Plotting of the frequency and time domain equations is done in the following program.

```
% Example 13.15 Frequency response and impulse response of a
%   second-order system.
%
wn = 1000;                      % Define resonant frequency
w = (100:10:10000);             % Define a frequency vector for Bode plot
t = (10^-5:10^-5:.2);           % Define a time vector
Q = [1 10 100];                 % Define Q's
for k = 1:length(Q)             % Calculate and plot the frequency plots
   TF = 1./(1-(w/1000).^2 + j*w/(Q(k)*1000));  % Frequency equation
   TF = 20*log10(abs(TF));      % Convert to dB
   semilogx(w,TF,'k');          % Bode plot
end

   .......labels and axis.......

%
% Now construct the impulse responses
figure;
for k = 1:3                     % Cal. and plot the time response
   d =sqrt(1-1/(4*Q(k)^2));     % Define square root of 1-δ2
   x = (wn/d)*(exp(-wn*t/(2*Q(k)))).*sin(wn*d*t);
   subplot(3,1,k);              % Plot separately for clarity
     plot(t,x,'k');
     ylabel('x(t)');
end
   .......labels.......
```

This program generates the following plots. Figure 13.32 reiterates the message in Figure 13.30, that high-Q systems can have very sharp resonance peaks.

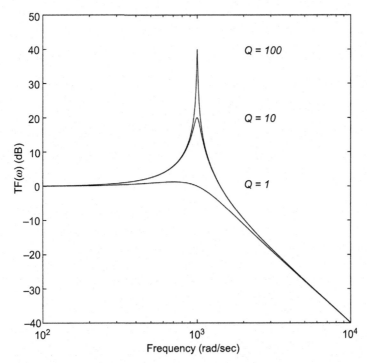

FIGURE 13.32 The frequency response of the transfer function of a second-order system to three different values of Q. The impulse response from this system is shown in Figure 13.33 for the same values of Q.

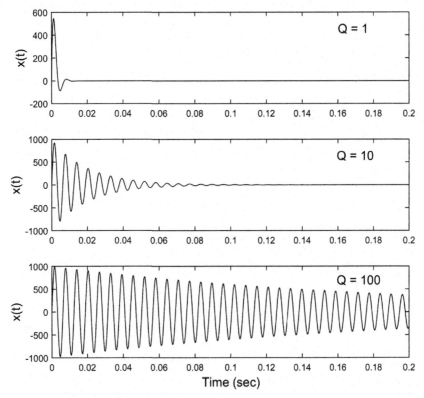

FIGURE 13.33 The impulse response of a second-order system with three values of Q. The frequency response of the transfer function of this system with the three values of Q is shown in Figure 13.32.

In the time domain, sharp resonance peaks correspond to sustained oscillations at the resonance frequency that diminishe slowly, Figure 13.33.

A common example of a high-Q mechanical system is a large bell. Striking a bell is like applying a mechanical impulse and the tone continues to sound long after it is struck. In fact, sustained oscillation is a characteristic of high-Q systems and is sometimes referred to as "ringing" even in electrical systems.

If a high-Q circuit is used as a filter, it can be quite selective with regard to the frequencies it allows to pass through. This is demonstrated in the final example.

EXAMPLE 13.16

Demonstrate the filtering characteristic of a high-Q system. Generate a 1.0-s chirp waveform that continually increases in frequency from 1 Hz to 1 kHz. (Recall, a chirp waveform increases in frequency with time usually linearly.) Assume a sampling frequency of 5 kHz.

Modify the RLC circuit of Figure 13.29 so that the resonant frequency, f_o, is 100 Hz and the circuit has a Q of 10. Assume the capacitor value is the same as that shown in Figure 13.29 (i.e., 1.0 μf), and adjust L to get the desired resonant frequency. Simulate the operation of passing the chirp waveform through the filter and plot the output. Repeat for two other values of resonant frequency: $f_0 = 250$ and 500 Hz.

To simulate the operation the circuit performs on the signal, find the circuit's impulse response and convolve that response with the input to get the simulated output.

Solution: First we are requested to design the network to have the desired resonant frequency and Q. The resonant frequency is entirely determined by L and C. Since we already have a value for C, we can determine the value of L from Equation 13.21.

$$\omega_0 = 2\pi f_0 = \frac{1}{\sqrt{LC}}; \quad f = \frac{1}{2\pi\sqrt{LC}}$$

Squaring both sides: $f_0^2 = \dfrac{1}{4\pi^2 LC}$

Solving for L: $L = \dfrac{1}{4\pi^2 f_0^2 C} = \dfrac{1}{4\pi^2 (10^4)(10^{-6})} = 2.53\,\text{h}$ when $f_0 = 100\,\text{Hz}$

For the other two resonant frequencies:

$$L = \frac{1}{4\pi^2 f_0^2 C} = \frac{1}{4\pi^2 (250^2)(10^{-6})} = 0.41\,\text{h} \quad f_0 = 250\,\text{Hz}$$

$$L = \frac{1}{4\pi^2 (500^2)(10^{-6})} = 0.1\,\text{h} \quad f_0 = 500\,\text{Hz}$$

To find the resistor values required to give the desired Q, we can use Equation 13.28:

$$Q = \frac{\omega_0 L}{R} = \frac{2\pi f_0 L}{R}; \quad R = \frac{2\pi f_0 L}{Q}$$

$$R = \frac{2\pi(100)(2.53)}{10} = 158.9\,\Omega \quad f_0 = 100\,\text{Hz}$$

TABLE 13.1 Resonant Frequencies and Respective Component Values of the Circuit Given in Figure 13.28

Resonant Frequency (Hz)	R (Ω)	L (h)	C (μf)
100	158.9	2.53	1.0
250	64.4	0.41	1.0
500	31.4	0.10	1.0

Solving for the other two values of L and R, the component values for the three resonant frequencies are summarized in Table 13.1.

Now that we have the circuit designed we need to find the impulse response. We can get this from the inverse of the Laplace transfer function. In Example 13.11 we find the transfer function of the network in Figure 13.29. Substituting in s for $j\omega$ in Equation 13.21 we get:

$$TF(s) = \frac{R}{R + Ls + 1/CS} = \frac{RCs}{RCs + CLs^2 + 1} = \frac{R/Ls}{s2 + R/Ls + 1/CL} \tag{13.29}$$

To find the inverse Laplace transform, we note that $Q = 10$ in all cases, thus $\delta = 1/2\,Q = 0.05$, so the system is underdamped. We can use entry 13 of the Laplace transform table (Appendix B) where:

$$b = R/L; \quad \alpha = R/2L; \quad c = 0$$
$$\alpha^2 + \beta^2 = 1/LC; \quad \beta^2 = 1/LC - \alpha^2 = 1/LC - \alpha^2; \quad \beta = \sqrt{1/LC} - \alpha$$

Taking the inverse Laplace transform, the impulse response becomes:

$$\delta(t) = e^{-\alpha t}\left[\left(\frac{-(R/L)\alpha}{\beta}\right)\sin \beta t + R/L \cos \beta t\right] \tag{13.30}$$

where α and β were defined earlier (it is easier to program if we leave these as variables and substitute in their specific values in the MATLAB routine). We could combine this equation into a single sinusoid with phase, but since we are going to program it in MATLAB further reduction is not necessary.

To construct the chirp signal, we define a time vector, t, that goes from 0 to 1 in steps of T_s where $T_s = 1/f_s = 1/5000$. This time vector is then used to construct a frequency vector of the same length that ranges from 1 to 1000. The chirp signal is then constructed as the sine of the product of f and t (i.e., `sin(pi*f.*t)`) so that the frequency increases linearly as t increases. (Note: use π, not 2π, to make the final frequency 1000 Hz).

We then program the impulse response using the above-mentioned equation and the various values of R, L, and C. Convolving the input signal with the impulse response produces the output signal. The solution of Equation 13.30 is performed three times in a loop for the various values of R, L, and C and plotted.

```
% Example 13.16 Demonstrate the filter characteristics of a high-Q system.
%
fs = 5000;                            % Sample frequency
Ts = 1/fs;                            % Sample interval
C = 0.000001;                         % Capacitor value (fixed)
R = [158.9, 64.4, 31.4];              % Resistor values
L = [2.53, 0.41, 0.1];                % Inductor values
f_0 = [100, 250, 500];                % Plot labels
t = (0:Ts:1);                         % Define time vector (0 - 5 sec; Ts = .0002)
f = t*1000;                           % Frequency goes from 1 to 1000 Hz
xin = sin(pi*f.*t);                   % Construct "chirp"
%
for k = 1:3                           % Calculate impulse response
  alpha = R(k)/(2*L(k));              % Define alpha and beta, then the
                                         impulse response
  beta = sqrt(1/(L(k)*C)) - alpha;
  h   =   exp(-alpha*t).*((-R(k)*alpha)/(L(k)*beta)*sin(beta*t)   +   (R(k)/L(k))
*cos(beta*t));
  xout = conv(h,xin);                 % Filter vin
  xout = xout(1:length(t));           % Remove extra points
  subplot(3,1,k);                     % Plot separately
    plot(t,xout,'k');
  ....... Labels and title.......
end
%
figure;                     % Plot spectrum of chirp
XIN= abs(fft(xin));                   % FFT
f1 = (1:length(t))*fs/length(t);     % Construct frequency vector for
                                        plotting
plot(f1(1:2500),XIN(1:2500),'k');    % Plot only valid points
    .......labels......
```

Analysis: The initial portion of the program defines the sampling characteristics, the component values, and the chirp. (If you have a sound system on your computer, try typing sound(xin) in MATLAB after executing this code. The resulting chirp sound is both startling and amusing.)

A loop is used to evaluate the impulse response for the three component configurations and find the circuit's output to the chirp input using convolution. This output is then plotted and is shown in Figure 13.34. The RLC circuit functions as a band-pass filter with a center frequency that is equal to the resonant frequency. The chirp signal's frequency depends on time and is approximately equal to the time in milliseconds. The signal appears at the output only when the chirp frequency corresponds to the circuit's resonant frequency. So when the resonant frequency is 100 Hz, the chirp signal passes through the filter at approximately 100 ms, Figure 13.34 (upper curve), and when it is

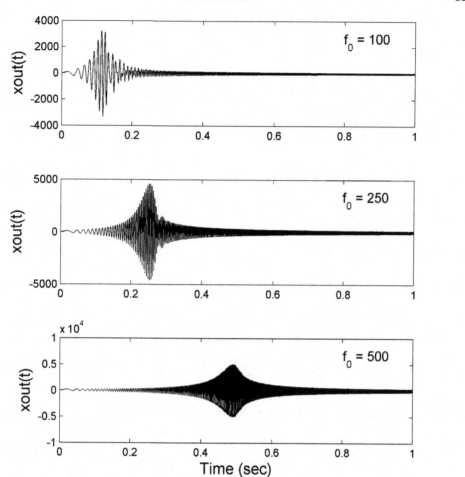

FIGURE 13.34 The output of a chirp signal after passing through a high-Q second-order system. The chirp ranges in frequency between 10 and 1000 Hz over a 1.0-s time period, so the instantaneous frequency is approximately equal to the time in milliseconds. The chirp signal passes through the system at a time when its frequency corresponds with the resonant frequency of the system.

500 Hz, it passes through the filter at approximately 0.5 s, Figure 13.34 (lower curve). The selective ability of a high-Q filter is well demonstrated in this example. If Q increases, the selectivity also increases as is demonstrated in one of the problems.

The last section of the program plots the frequency spectrum of the chirp signal using the standard fft command. The spectrum is seen in Figure 13.35 to be relatively flat up to 1000 Hz as expected.

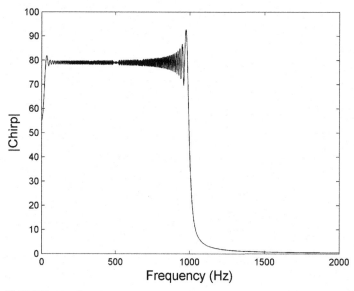

FIGURE 13.35 The spectrum of the chirp signal used in Example 13.16.

13.6 SUMMARY

Conservation laws are invoked to generate an orderly set of descriptive equations for any collection of mechanical or electrical elements. In electric circuits, the law of conservation of energy leads directly to KVL, which states that the voltages around the loop must sum to 0.0. Combining this rule with the phasor representation of network elements leads to an analysis technique known as mesh analysis. In mesh analysis, matrix equations are constructed that have the form of Ohm's law: $v = Zi$, where v is a voltage vector, i is a current vector, and Z an impedance matrix. These equations are solved for the mesh currents, i, and can be used to determine any voltage in the system.

The law of conservation of charge leads to KCL, which states that the sum of voltages into a node must be 0.0. As with KVL, it can be used to find the voltages and currents in any network. The application of KCL leads to a matrix equation of the form $i = v/Z$, or $i = vY$, where Y is the admittance matrix consisting of inverse impedances. This equation can be solved for the node voltages, v. From the node voltages any current in the circuit can be found. This analysis, termed nodal analysis, leads to fewer equations in networks that contain many loops but only a few nodes.

The conservation law that applies to mechanical systems is Newton's law, which, keeping with the sum to 0.0 idea, states that the forces on any element must sum to 0.0. Again using the phasor representation of mechanical elements, this law can be applied to generate equations of the form $F = Zv$, where F is a force vector, v is a velocity vector, and Z is a matrix of mechanical impedances. These equations are solved for velocities and these velocities can be used to determine all of the forces in the system.

These conservation laws, and the analysis procedures they lead to, allow us to develop equations for even very complex electrical or mechanical systems. Large systems may generate equations that are difficult to solve analytically, but MATLAB makes short work out of even the most challenging equations.

Resonance is a phenomenon commonly found in nature. In electrical and mechanical systems, it occurs when two different energy storage devices have equal, but oppositely signed, impedances. During resonance, energy is passed back and forth between the two energy storage devices. For example, in an oscillating mechanical system, the moving mass stretches the spring transferring the kinetic energy of the mass to potential energy in the spring. Once the spring is fully compressed (or extended), the energy is passed back to the mass in terms of momentum as the spring recoils. Without friction this process would continue forever, but friction removes energy from the system, so the oscillation gradually decays. Electrical RLC circuits behave in exactly the same way, passing energy between the inductor and capacitor while the resistor removes energy from the system. (Recall, both friction and resistance remove energy in the form of heat.)

The quantity Q is a measure of the ratio of energy storage to energy dissipation of a system. The Q of a system is inversely proportional to the damping factor, δ. In the frequency domain, a higher Q corresponds to a sharper resonance peak. In the time domain, higher Q systems have longer impulse responses: they continue "ringing" long after the impulse because energy is only slowly removed from the system. Such systems are useful in frequency selection: with their sharp resonant peaks they are able to select out a narrow range of frequencies from a broadband signal. Standard systems analysis techniques such as Bode plots and the transfer function can be used to describe the behaviors of these systems.

PROBLEMS

1. In the following circuit, the voltage across element 1 is $2 \cos(2t + 60)$. What is the voltage, V_2, across element 2? Assume the plus side of element 1 is to the right and define positive V_2 as upward.

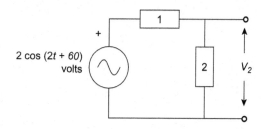

2. Find the voltage across the 50-Ω resistor.

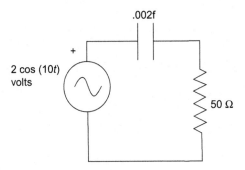

3. The loop current (clockwise) in the following circuit is 0.2 cos (10t − 103). What is element 2 (i.e., is it an R, L, or C) and what is its value?

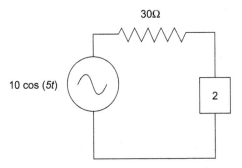

4. Find the voltage across the inductor, v_L, for $\omega = 10$ and 20 rad/s.

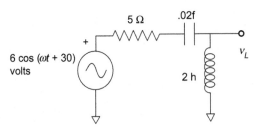

III. CIRCUITS

5. Find the voltage across the 5-h inductor in the following network.

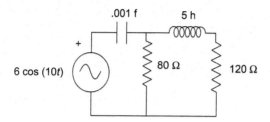

6. What is the voltage in the center 80-Ω resistor in the network given in Problem 5? (Note the total current through the resistor is $i_R = i_1 - i_2$.)

7. Find the current through the 2-h inductor in the following circuit. Is the voltage source, V_2, supplying energy or storing energy?

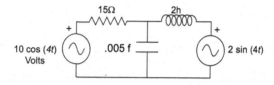

8. Find V_1.

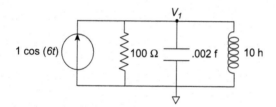

9. Find the voltage across the 10-Ω resistor.

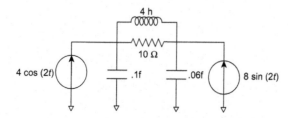

III. CIRCUITS

10. The following circuit is a four-element Windkessel model. The element values in this model of the cardiovascular system are $R_A = 0.76$ mmHg/mL/s, $C = 1.75$ mL/mmHg, $R_p = 0.33$ mmHg/mL/s, and $L = 0.005$ mmHg/mL. Write the three-mesh equation and use the frequency domain approach of Example 13.5 to find the pressure in the aorta (i.e., $v(t)$) assuming the blood flow as given in Equation 13.11. As in Example 13.5, this problem requires both manual calculations and MATLAB.

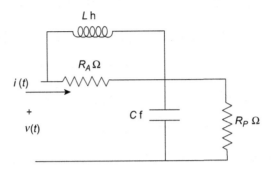

11. The following mechanical system has $k_f = 8$ dyn s/cm; $k_e = 12$ dyn/cm; $m = 2$ g; $F_s(t) = 10 \cos (2t)$ dyn.

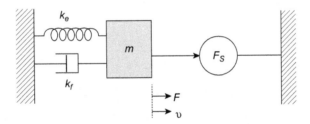

Find the length of the spring when $t = 0.10$ s. (Hint: solve for $x(\omega)$ where $x(\omega) = v(\omega/j\omega)$. Then convert to time domain $x(t)$ and solve for $t = .10$ s.)

12. In the mechanical system of Problem 11, what must be the value of k_f to limit the maximum velocity to ± 1 cm/s?

13. Find V_{out} using MATLAB. (Hint: Example 13.7 provides a framework for the code required to solve this three-mesh circuit, but you should let MATLAB do most of the work.)

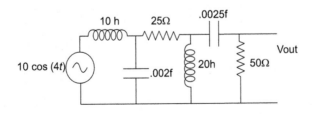

14. Find the voltage across the right-hand current generator, using MATLAB and nodal analysis. Remember, $10 \sin(10t) = 10 \cos(10t - 90)$. This is a three-node circuit, so the admittance matrix has the same format as the impedance matrix in Problem 13. (Hint: You can solve this entirely in MATLAB. First define the current sources directly in MATLAB, then directly define the admittance matrix. Solve the resultant equation for V. The voltage across the right-hand current generator should be V(3) in the voltage vector.)

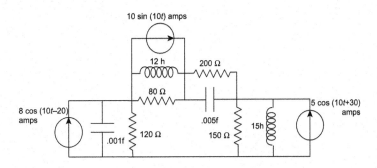

15. Find v_{out} for the following circuit for frequencies ranging between 1 and 1000 rad/s. Plot the magnitude of v_{out} as a function of frequency. (Hint: Set up ω as a vector and then directly write the impedance matrix in MATLAB keeping frequency as a variable. Solve for v_{out}. The vector ω should be a vector range between 1 and 1000 rad/s in increments of 1 rad/s. Then plot $\text{abs}(v_{out}(\omega))$.)

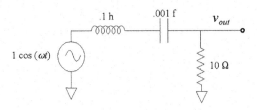

16. Repeat problem 15 for the mechanical system shown in Problem 11. That is, find v_{out} (which is now a velocity) for that mechanical system for frequencies ranging between 1 and 1000 rad/s. Plot the magnitude of v_{out} as a function of frequency. Use the parameter values: $k_f = 10$ dyn s/cm; $k_e = 1100$ dyn/cm; $m = 1.10$ g; $F_S(t) = \cos(\omega t)$.

17. Find and plot $V_{out}(\omega)$ over a range of frequencies from 11 to 10,000 rad/s. Plot V_{out} in dB as a function of $\log \omega$ (i.e., on a semi-log axis).

18. Use the graphical technique (not MATLAB) developed in Chapter 6 to plot the magnitude Bode Plot for the following RLC circuit. Set the value of the components: $L = 1$ h, $C = 0.0001$ f, and $R = 10 \, \Omega$.

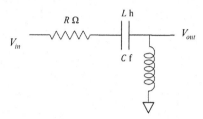

19. Use MATLAB to plot the magnitude transfer function of the circuit in Problem 18 for values of $R = 1, 10, 70$, and $500 \, \Omega$. Assume that $L = 1$ h and $C = .0001$ f. Frequency, ω, should range between 10 and 1000 rad/s. Use a small enough frequency increment to show the details of the frequency plot. If you are a reasonably skilled MATLAB programmer, you can use different colors for the four plots. How does the resistance affect the shape of the frequency plot?

20. Use MATLAB to plot the Bode Plot (magnitude and phase) for the transfer function for the mechanical system in Problem 10 for values of $k_e = 2$ dyn/cm, m = 3 g, and $k_f = 1, 10, 70$, and 500 dyn s/cm. How does the friction element affect the shape of the frequency plot?

21. Use MATLAB to plot the magnitude Bode plot of the transfer function of the following circuit . Note the realistic units used for the elements. What type of filter is it? Plot between 100 and 10^6 rad/s, but plot in hertz.

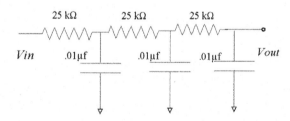

22. Reverse the resistors and capacitors for the network of Problem 21 and plot the magnitude Bode plot. How does it compare with that of the original network?

23. Plot the Bode Plot (magnitude and phase) for the transfer function of the following RLC circuit. Reverse the position of the resistor and capacitor and replot. That is, put the resistors across the top and connect the capacitor to ground. How does this change the frequency characteristics of the filter?

24. Plot the Bode plot of the network used in Example 13.16 from the transfer function. Note that the transfer function equation was derived in Example 13.13 and is given by Equation 13.21. Plot the Bode plot for the three configurations of component values given in Table 13.1 over a frequency range of 1 to 10,000 Hz.

25. Plot the Bode plot of the network used in Example 13.16 as in Problem 24, but use the time domain approach and the impulse response. Plot the Bode plot for the three configurations of component values over a frequency range of 1 to 1000 Hz. Plot in decibel versus log frequency and take care to construct the correct frequency vector. (Hint: The code in Example 13.15 already determines the impulse response; you just need to add the code that determines the Bode plot and plots frequency response. Recall, the spectrum is the Fourier transform of the impulse response.)

26. Modify the code of Example 13.16 so that the simulated circuit has a resonant frequency of 400 Hz and plot the circuit's response to the chirp for three different values of Q: 2, 10, and 20. Note the effect of increasing and decreasing the filter's selectivity.

CHAPTER

14

Circuit Reduction: Simplifications

14.1 GOALS OF THE CHAPTER

Sometimes it is desirable to simplify a circuit or system by combining elements. Many quite complex circuits can be reduced to just a few elements using an approach known as "network reduction." Such simplifications make it easier to analyze and understand the network by providing a summary-like representation of a complicated system. New insights can be had about the properties of a network after it has been simplified. Network reduction can be particularly useful when two networks or systems are to be connected together. We only need the reduced networks to figure out how connecting them together affects the passage of information between them. Finally, the principles of network reduction are useful in understanding the behavior of real sources.

Network reduction principles also help us understand the problems that arise when making medical measurements. Making a measurement on a biological system is essentially connecting the measurement system to the biological system. All measurements require drawing some energy from the system being measured. The energy required depends on the match between the biological and measurement systems. This match can be quantified in terms of a generalized concept of impedance: the difference between the impedance of the biological system and the input impedance of the measurement system.

In this chapter we will learn how to:

- Combine series and parallel elements and apply this approach to complex configurations of passive elements.
- Reduce complex networks of circuits to a single source and impedance by successively combining elements.
- Reduce complex circuits to a single source and impedance using an applied voltage source (either real or theoretical) and calculating (or measuring voltage and current).
- Use passive elements to construct circuits that have sharp resonance characteristics.
- How to represent real current and voltage sources using ideal elements.
- Determine if, when one circuit is connected to another, the influence of the second circuit on the first can be ignored.
- Determine how, when one circuit is connected to another, maximum power from the first circuit to the second is transferred.

- Determine the equivalent impedance of a mechanical system and apply the concepts of source and load impedance to mechanical systems.
- Determine the influence, when one mechanical system is connected to another, of the loading mechanical system on the source mechanical system, particularly in measurement situations.

14.2 SYSTEM SIMPLIFICATIONS—PASSIVE NETWORK REDUCTION

Before we can reduce complex networks or systems, we must first learn to reduce simple configurations of elements such as series and parallel combinations. Network reduction is based on a few simple rules for combining series and/or parallel elements. The approach is straightforward, although implementation can become tedious for large networks (that is when we turn to MATLAB). After we introduce the reduction rules for networks consisting only of passive elements, we will expand our guidelines to include networks with sources. These reduction tools can be applied whenever two systems are interconnected.

14.2.1 Series Electrical Elements

Electrical elements are said to be in "series" when they are connected to each other and no other elements share that common connection, Figure 14.1A. Although series elements are often drawn in line with one another, they can be drawn in any configuration and still be in series as long as they follow the "no other connection" rule. The three elements in Figure 14.1B are also in series as long as no other elements are connected between the elements. A simple application of Kirchoff's voltage law (KVL) demonstrates that when

(A)

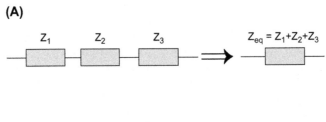

(B)

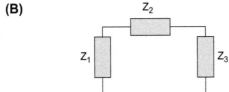

FIGURE 14.1   (A) Three elements in series, Z_1, Z_2, and Z_3, can be converted into a single equivalent element, Z_{eq}, that is the sum of the three individual elements. Elements are in series when they share one node and no other elements share this node. (B) Series elements are often drawn in *line*, but these elements are also in series as long as nothing else is connected between the elements.

elements are in series their impedances add. The voltage across three series elements in Figure 14.1B is:

$$v_{total} = v_1 + v_2 + v_3 = (Z_1 + Z_2 + Z_3)i$$

The total voltage can also be written as:

$$v_{total} = Z_{eq}\, i;$$

where $Z_{eq} = Z_1 + Z_2 + Z_3$.

So series elements can be represented by a single element that is the sum of the individual elements, Equation 14.1.

$$Z_{eq} = Z_1 + Z_2 + Z_3 + \ldots + Z_N \tag{14.1}$$

If the series elements are all resistors or all inductors, they can be represented by a single resistor or inductor that is the sum of the individual elements:

$$R_{eq} = R_1 + R_2 + R_3 + \ldots \tag{14.2}$$

$$L_{eq} = L_1 + L_2 + L_3 + \ldots \tag{14.3}$$

If the elements are all capacitors, their reciprocals add, since the impedance of a capacitor is a function of $1/C$, i.e., $1/j\omega C$ or $1/sC$:

$$1/C_{eq} = 1/C_1 + 1/C_2 + 1/C_3 + \ldots \tag{14.4}$$

If the string of elements includes different element types, the individual impedances can be added using complex arithmetic to determine a single equivalent impedance. In general, this single impedance is complex, as shown in the following example.

EXAMPLE 14.1

Series element combination. Find the equivalent single impedance, Z_{eq}, of the series combination in Figure 14.2.

Solution: First combine the two resistors into a single 25-Ω resistor, then combine the two inductors into a single 11-h inductor, and then add the three impedances (R_{eq}, $j\omega L_{eq}$, and $1/j\omega C$). Alternatively, convert each element to its equivalent phasor impedance, and then add these impedances.

$$Z_{eq} = 10 + j\omega 5 + \frac{1}{j\omega.01} + 15 + j\omega 6 = 25 + j\omega 11 + \frac{100}{j\omega}$$

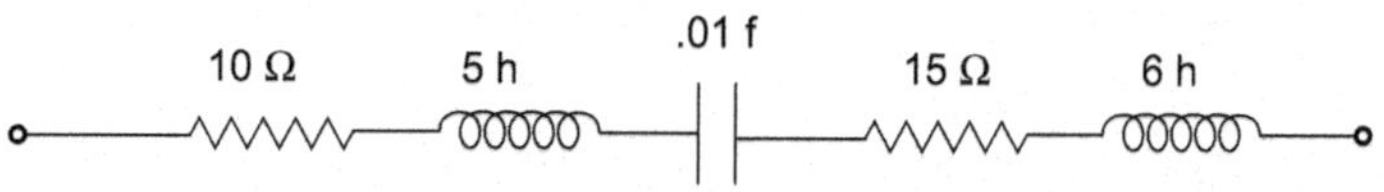

| 10 Ω | 5 h | .01 f | 15 Ω | 6 h |

FIGURE 14.2   A combination of different series elements that can be mathematically combined into a single element. Usually the resulting element is complex (i.e., contains a real and an imaginary part).

If a specific frequency is given, for example, $\omega = 2.0$ rad/s, then Z_{eq} can be evaluated as a single complex number.

$$Z(\omega = 2) = 25 + j(11(2) - 100/2) = 25 - j28 \ \Omega = 37.5\angle -48 \ \Omega$$

EXAMPLE 14.2

Find the equivalent capacitance for the three capacitors in series in Figure 14.3.
Solution: Since all the elements are capacitors, they add as reciprocals:

$$1/C_{eq} = 1/C_1 + 1/C_2 + 1/C_3 = 1/0.1 + 1/0.5 + 1/0.2 = 10 + 2 + 5 = 17$$

$$C_{eq} = 1/17 = 0.059 \ f$$

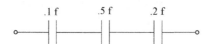

FIGURE 14.3   Three series capacitors. They can be combined into a single capacitor by adding reciprocals as shown in Example 14.2.

14.2.2 Parallel Elements

Elements are in parallel when they share both connection points as shown in Figure 14.4. For parallel electrical elements, it does not matter if other elements share these mutual connection points, as long as both ends of the elements are connected to each other.

When you are looking at electrical schematics, it is important to keep in mind the definition of parallel and series elements because series elements may not be drawn in line (Figure 14.1B) and parallel elements may not be drawn as geometrically parallel. For example, the two elements, Z_1 and Z_2, on the left side of Figure 14.5 are in parallel because they connect at both ends, even though they are not drawn in parallel geometrically. Conversely, elements Z_1 and Z_2 are drawn parallel, but they are not in parallel electrically because they are not connected.

As can be shown by KCL, parallel elements combine as the reciprocal of the sum of the reciprocals of each impedance. With application of KCL to the upper node of the three parallel elements in Figure 14.4, the total current flowing through the three impedances is:

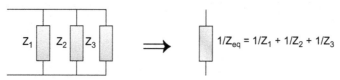

FIGURE 14.4   Parallel elements share connection points and both ends. Three elements in parallel, Z_1, Z_2, and Z_3, can be converted into a single equivalent impedance, Z_{eq}, that is the reciprocal of the sum of the reciprocals of the three individual impedances.

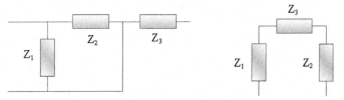

FIGURE 14.5 Left circuit: the elements Z_1 and Z_2 are connected at both ends and are therefore electrically in parallel, even though they are not drawn parallel. Right circuit: although they are drawn parallel, elements Z_1 and Z_2 are not electrically parallel because they are not connected at both ends (in fact they are not connected at either end).

$i_{total} = i_1 + i_2 + i_3$. Substituting in v/Z for the currents through the impedances, the total current is:

$$i_{total} = v/Z_1 + v/Z_2 + v/Z_3 = v(1/Z_1 + 1/Z_2 + 1/Z_3).$$

This equation, restated in terms of an equivalent impedance, becomes:

$$i_{total} = v/Z_{eq},$$

where $1/Z_{eq} = 1/Z_1 + 1/Z_2 + 1/Z_3$.
Hence:

$$1/Z_{eq} = 1/Z_1 + 1/Z_2 + 1/Z_3 + \dots \tag{14.5}$$

Equation 14.5 also holds for the value of parallel resistors and inductors:

$$1/R_{eq} = 1/R_1 + 1/R_2 + 1/R_3 + \dots \tag{14.6}$$

$$1/L_{eq} = 1/L_1 + 1/L_2 + 1/L_3 + \dots \tag{14.7}$$

Parallel capacitors, however, simply add.

$$C_{eq} = C_1 + C_2 + C_3 + \dots \tag{14.8}$$

If the three capacitors in Example 14.2 were in parallel, their equivalent capacitance would be the addition of the three capacitance values:

$$C_{eq} = 0.1 + 0.5 + 0.2 = 0.8 \, \text{f}$$

EXAMPLE 14.3

Find the equivalent single impedance for the parallel resistor, inductor, capacitor (RLC) combination in Figure 14.6.

Solution: First take the reciprocals of the impedances:

$$1/R = 1/10 = 0.1 \, \Omega; \quad 1/j\omega L = 1/j5\omega \, \Omega; \quad j\omega C = j.01\omega \, \Omega$$

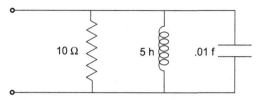

FIGURE 14.6   Three parallel elements can be combined into a single element as shown in Example 14.3.

Then add, and invert:

$$\frac{1}{Z_{eq}} = Y_{eq} = 0.1 + \frac{1}{j\omega 5} + j\omega .01 = 0.1 + j\left(.01\omega - \frac{1}{5\omega}\right)$$

$$Z_{eq} = \frac{1}{Y_{eq}} = \frac{1}{0.1 + j\left(.01\omega - \frac{1}{5\omega}\right)} \, \Omega$$

Once a value of frequency, ω, is given, this equation can be solved for a specific impedance value. Alternatively, we can solve for Z_{eq} over a range of frequencies using MATLAB. This is given as a problem at the end of this chapter.

EXAMPLE 14.4

Find the equivalent resistance of the parallel combination of three resistors: 10, 15, and 20 Ω. Solution/Result: Calculate reciprocals, add them and invert:

$$1/R_{eq} = 1/R_1 + 1/R_2 + 1/R_3 = 1/10 + 1/15 + 1/20 = 0.1 + 0.0667 + .05 = 0.217 \, \Omega$$

$$R_{eq} = 1/0.217 = 4.61 \, \Omega$$

Note that the equivalent resistance of a parallel combination of resistors is always less than the smallest resistor in the group. The same is true of inductors, but the opposite is true of parallel capacitors—the parallel combination of capacitors is always larger than the largest capacitor in the group.

14.2.2.1 *Combining Two Parallel Impedances*

Combining parallel elements via Equation 14.5 is mildly irritating, with its double inversions. Most parallel element combinations involve only two elements, so it is useful to have an equation that directly states the equivalent impedance of two parallel elements without the inversion. Starting with Equation 14.5 for two elements:

$$\frac{1}{Z_{eq}} = \frac{1}{Z_1} + \frac{1}{Z_2} = \frac{Z_2}{Z_2 Z_1} + \frac{Z_1}{Z_1 Z_2} = \frac{Z_1 + Z_2}{Z_1 Z_2}$$

$$\text{Inverting:} \quad Z_{eq} = \frac{Z_1 Z_2}{Z_1 + Z_2}$$

(14.9)

Hence the equivalent impedance, Z_{eq}, of two parallel elements equals the product of the two impedances divided by their sum.

14.3 NETWORK REDUCTION—PASSIVE NETWORKS

The rules for combining series and parallel elements can be applied to networks that include a number of elements. Even very involved configurations of passive elements can usually be reduced to a single element. Obviously, it is easier to grasp the significance of a single element than a confusing combination of many elements.

14.3.1 Network Reduction—Successive Series—Parallel Combination

In the last section of this chapter, we see that it is possible to combine a number of series or parallel combinations. Even when most of the elements are not in either series or parallel configurations, it is possible to combine them into a single impedance using the techniques of network reduction. In this section, the networks being reduced consist only of passive elements, but in Section 14.4 we learn how to expand network reduction to include networks with sources.

In the network in Figure 14.7, most of the elements are neither in series nor in parallel. It is important to realize that the elements across the top of this network, inductor—resistor—inductor, are not in series because their connection points are shared by other elements, capacitors in this case. To be in series, the elements must share one connection point and must be the only elements to share that point. If we could somehow eliminate the two capacitors (we cannot), then these three elements would be in series. Nor are any of the elements in parallel, since no elements share both connection points. If they did, they would be in parallel even if other elements share these connection points. However, there are two elements in series: the 4-h inductor and the 20-Ω resistor on the right-hand side of the network. We could combine these two elements using Equation 14.1. After combining these two elements, we now find that the new element is in parallel with the 0.02-f capacitor. We can then combine these two parallel elements using the parallel rule given in Equation 14.9. Although the

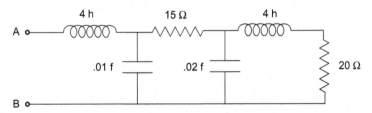

FIGURE 14.7 A network containing R, L, and C's where most of the elements are neither in series nor in parallel. Nonetheless, this network can be reduced to a single equivalent impedance (with respect to nodes A and B) as is shown in Example 14.5.

argument becomes difficult to follow at this point without actually going through it, the single element newly combined from the parallel elements is now in series with the 15-Ω resistor. Most reductions of passive networks proceed in this fashion: find a series or parallel combination to start with, combine them into a single element, and then look to see if the new combination produces a new series or parallel combination. Then just continue down the line:

The following example uses the approach based on sequential series–parallel combinations to reduce the network in Figure 14.7 to a single equivalent impedance.

EXAMPLE 14.5

Network reduction using sequential series–parallel combinations. Find the equivalent impedance between the nodes A and B in Figure 14.7. Find the impedance at only one frequency, $\omega = 5.0$ rad/s. Using network reduction, we can find the equivalent impedance leaving frequency as a variable (i.e., $Z(\omega)$), but this will make the algebra more difficult. By using a specific frequency, we are able to use complex arithmetic instead of complex algebra.

Solution: Convert all elements to their equivalent impedances at $\omega = 5.0$ rad/s (to simplify subsequent calculations). Then begin the reduction by combining the two series elements on the right-hand side. As a first step, first convert the elements to their phasor impedances (at $\omega = 5$ rad/s). Then combine the two series elements using Equation 14.1, leading to the network shown in Figure 14.8 on the right-hand side.

This combination puts two elements in parallel: the newly formed impedance and the $-j10$-Ω capacitor. These two parallel elements can be combined using Equation 14.9:

$$Z_{eq} = \frac{Z_1 Z_2}{Z_1 + Z_2} = \frac{(20 + j20)(-j10)}{20 + j20 - j10} = \frac{200 - j200}{20 + j10} = \frac{282.8\angle - 45}{22.36\angle 26.6}$$

$$Z_{eq} = 12.65\angle - 71.6\ \Omega = 4 - j12\ \Omega$$

This leaves a new series combination that can be combined as follows. In the third step of network reduction, the newly formed element from the parallel combination ($4 - j12\ \Omega$) is now in series with the 15-Ω resistor, and this equivalent series element will be in parallel with the 0.01-f capacitor, Figure 14.9.

Combining these two parallel elements:

$$Z_{eq} = \frac{-j20(19 - j12)}{-j20 + 19 - j12} = \frac{-240 - j380}{19 - j32} = \frac{449.4\angle 238}{37.2\angle - 59}$$

$$Z_{eq} = 12.07\angle 297\ \Omega = 5.49 - j10.76\ \Omega$$

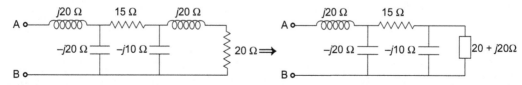

FIGURE 14.8 The network on the left is a phasor representation of the network in Figure 14.7; a partial reduction of this network is shown on the right.

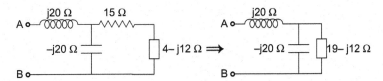

FIGURE 14.9 The next step in the reduction of the network is shown in Figure 14.7. The newly combined elements are now in series with the 15-Ω resistor, and the element formed from that combination will be in parallel with the 0.01-f capacitor.

Result: As shown in Figure 14.10, combining the parallel elements leads to the final series combination and, after combination, a single equivalent impedance with a value of:

$$Z_{eq} = 5.49 + j9.24 \ \Omega = 10.75 \angle 59 \ \Omega$$

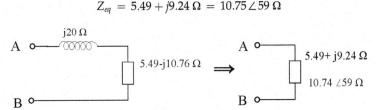

FIGURE 14.10 A single element represents the effective impedance between Nodes A and B in the network of Figure 14.7.

Network reduction is always done from the point of view of two nodes, such as nodes A and B in this example. In principle, any two nodes can be selected for analysis and the equivalent impedance can be determined between these nodes. Generally the nodes selected have some special significance, e.g., the nodes that make up the input or output of the circuit. Network reduction usually follows the format of this example: sequential combinations of series elements, then parallel elements, then series elements, and so on. In a few networks, there are no elements either in series or in parallel to start with, and an alternative method described in the next section must be used. This method works for all networks and any combination of two nodes, but it is usually more computationally intensive. On the other hand, it lends itself to a computer solution using MATLAB, which ends up being less computationally intensive.

14.3.1.1 *Resonance Revisited*

In Chapter 13 we observe the resonance that occurs in electrical and mechanical systems when the impedance of the two types of energy storage devices, inertial and capacitive, are equal and cancel. This leads to a minimum in total impedance: zero if there are no dissipative elements present. For example, in a series resistor, inductor, capacitor (RLC) circuit the total impedance is $R + j\omega L + 1/j\omega C$ and resonance occurs when the impedances of the inductor and capacitor are the same. The total impedance is just that of the resistor. If the two energy storage elements are in parallel, then an antiresonance occurs where the net impedance goes to infinity, since by Equation 14.9:

$$Z = \frac{Z_1 Z_2}{Z_1 + Z_2} = \frac{j\omega L (1/j\omega C)}{j\omega L + 1/j\omega C} \rightarrow \infty \quad \text{when } j\omega L = 1/j\omega C$$

An example of the impedance of a 5.0-μf capacitor in series with a 50-mh inductor is shown as a function of log frequency in Figure 14.11A. The sharp resonance peak at 2000 rad/s is evident.

A network can exhibit both resonance and antiresonance, at different frequencies of course. In the network shown in Figure 14.12, the parallel inductor and capacitor produce an antiresonance at 2000 rad/s, but at a higher frequency the combination becomes capacitive and that parallel combination will resonate with the series inductor. This is shown in the next example.

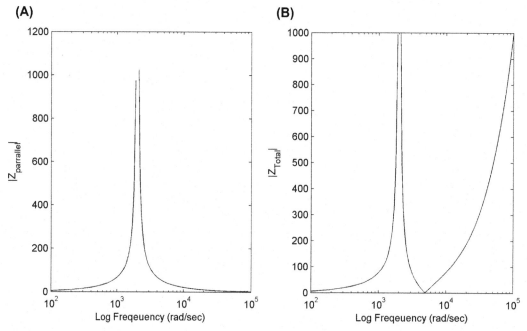

(A)

(B)

FIGURE 14.11   (A) The impedance of a parallel inductor-capacitor (LC) combination. An antiresonance peak (i.e., a very large impedance) is seen at 2000 rad/s when the impedances of the two elements are equal. (B) An example of both resonance and antiresonance phenomena produced by the network shown in Figure 14.12. This graph is constructed in Example 14.6 using MATLAB.

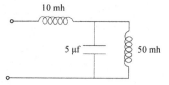

FIGURE 14.12   A circuit that exhibits both resonance and antiresonance. This circuit is analyzed in Example 14.6.

EXAMPLE 14.6

Use MATLAB to find the net impedance of the circuit in Figure 14.12 and plot the impedance as a function of log frequency. Also plot the impedance characteristic of the parallel inductor-capacitor (LC) separately.

Solution: After defining the component values, calculate the impedance of each component. Use the parallel element equation, Equation 14.9, to find the impedance of the parallel LC, and then add in the impedance of the series inductor to find the total impedance. The program is as follows.

```
% Example 14.6 Example to show resonance and antiresonance
%
w = (100:100:1000000);             % Define frequency vector
C = 05e-6;                         % Assign component values
L1 = 0.05;
L2 = .01;
ZL1 = j*w*L1;                      % Calculate component impedances
ZL2 = j*w*L2;
ZC = 1./(j*w*C);
Zp = (ZL1 .* ZC)./(ZL1 + ZC);      % Calculate parallel impedance
Z = ZL2 + Zp;                      % Calculate total impedance
    ......plot using semilogx and label.......
```

Result: The resulting plots from this program are shown in an earlier figure, Figure 14.11. Figure 14.11 B plots the total impedance and shows both the antiresonance peak, when $Z_{Total} \rightarrow \infty$ Ω at 2000 rad/s, and the resonance peak, when $Z_{Total} \rightarrow 0$ Ω at 5000 rad/s. Although the impedance of the network in Figure 14.12 was found for a fixed set of component values, it would be easy to rerun the program using different component values or even a range of component values.

14.3.2 Network Reduction—Voltage—Current Method

The other way to find the equivalent impedance of a network follows the approach that you would use in the real world on an existing physical system. Suppose you are asked to determine the impedance between two nodes of some physical network. This is called a "two-terminal" problem. Perhaps the actual network is inside a difficult-to-open box and all you have available are two nodes as shown in Figure 14.13.

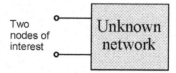

Two
nodes of
interest

Unknown
network

FIGURE 14.13 An unknown network that has only two terminals available for measurement, or for possible use. This is a classic "two-terminal" device.

The actual network could contain any number of nodes, but in this scenario only two are accessible for measurement. Without opening the box it is possible to determine the effective impedance between these two terminals. What you do in this situation (or, at least, what I would do) is to apply a known voltage to the two terminals, measure the resulting current, and calculate the equivalent impedance, Z_{eq}, using Ohm's law:

$$Z_{eq} = \frac{V_{known}}{I_{measured}} \qquad\qquad (14.10)$$

Of course, V_{known} has to be a sinusoidal source of known amplitude, phase, and frequency unless you know, *a priori*, that the network is purely resistive, in which case a direct current (DC) source suffices. Moreover, you are limited to determining Z_{eq} at only one frequency, the frequency of the voltage source, but most laboratory generators can produce a sine wave over a wide range of frequencies, so the impedance can determined for a range of frequencies. Before you apply this method, you should first see if there is a voltage between the two terminals. If so, the unknown network is not a collection of passive elements, but rather contains voltage or current sources. In this case, a slightly different strategy developed in the next section can be used.

This same approach can be applied to a network that exists only on paper, such as the network in the last example. Using the tools that we have acquired thus far, we simply connect a source of our choosing to the network and solve for the current into the network. The source can be anything. Moreover, this approach can be applied to simplify any network, even one that does not have any series or parallel elements. In the next example, this method is applied to the network in Example 14.5 and, in a subsequent example, to an even more challenging network.

EXAMPLE 14.7

Passive network reduction using the source–current method. Find the equivalent impedance between nodes A and B in the network of Figure 14.7 for a frequency of $\omega = 5$ rad/s.

Solution: To the given network, apply a known (if theoretical) source across nodes A and B and solve for the resulting current using mesh analysis. Theoretically we can choose any sinusoidal source as long as it has a frequency of 5 rad/s, but why not choose something simple like 1 V RMS at 0.0 degree phase: $v(t) = \cos(5t) \rightarrow V(\omega) = 1\angle 0$. The desired impedance Z_{ab} will then be: $Z_{ab} = 1/I_1(\omega)$. Mesh analysis can be used to solve for $I_1(\omega)$. The three-mesh network is shown using phasor notation in Figure 14.14.

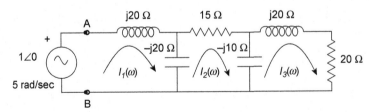

FIGURE 14.14 The network shown in Figure 14.7 with a voltage source attached to nodes A and B and represented in the phasor domain.

To find the input current, we analyze this as a straightforward three-mesh problem solving for I_1. However, several alternatives are also possible: using network reduction to convert the last three elements to a single element and solving as a two-mesh problem; converting the voltage source to an equivalent current source (as shown later in this chapter) and solving as a two-node problem; or combining these approaches and solving as a one-node problem. Here we solve the three-mesh problem directly, but use MATLAB to reduce the computational burden. Applying standard mesh analysis to the above-mentioned network, the basic matrix equation can be written as:

$$
\begin{vmatrix} 1.0 \\ 0 \\ 0 \end{vmatrix} = \begin{vmatrix} j20 - j20 & +j20 & 0 \\ +j20 & 15 - j20 - j10 & +j10 \\ 0 & +j10 & 20 + j20 - j10 \end{vmatrix} \begin{vmatrix} i_1 \\ i_2 \\ i_3 \end{vmatrix}
$$

Simplifying by complex addition:

$$
\begin{vmatrix} 1.0 \\ 0 \\ 0 \end{vmatrix} = \begin{vmatrix} 0 & j20 & 0 \\ j20 & 15 - j30 & j10 \\ 0 & j10 & 20 + j10 \end{vmatrix} \begin{vmatrix} i_1 \\ i_2 \\ i_3 \end{vmatrix}
$$

Solving for I_1, then Z_{eq}, using MATLAB:

```
% Example 14.7 Find the equivalent impedance of a network by applying a source and
solving for the resultant current.
%
% First assign values for v and Z
v = [1 0 0];
Z = [0 20j 0; 20j 15-30j 10j; 0 10j 20+ 10j];
%
i = Z\v                          % Solve for currents

Zeq = 1/i(1)                     % Solve for Zeq. Output as real and imaginary

Zeq = [abs(Zeq) angle(Zeq)*360/(2*pi)]   % Output in polar form
```

Result: The output from this program is: $Zeq = 5.49 + 9.24i$ Ω or $10.75 \angle 59$ Ω, which is the same as that found in the previous example using network reduction. Again, this approach could be used to reduce any passive network of any complexity to an equivalent impedance between any two terminals.

The latter portion of this chapter shows how to reduce, and think about, networks that also contain sources. Before proceeding to the next section, we look at another example that shows how to reduce a passive network when the two terminals of interest are in more complicated positions (for example, when separated by more than a single element). We will also solve for the impedance over a range of frequencies.

EXAMPLE 14.8

Find the equivalent impedance of the circuit shown in Figure 14.15 between terminals A and B. Use MATLAB to find Z_{eq} over frequencies and plot Z_{eq} as a function of log frequency. Use a frequency range of 0.01–100 rad/s. (This frequency range was established by trial and error in MATLAB.)

Solution: Apply a source between terminals A and B and solve for the current flowing out of the source. In this problem, the source has a fixed amplitude of 1.0 V with a phase of 0 degree, but the frequency varies so that Z_{eq} can be determined over the specified range of frequencies. Before you apply the source, it is helpful to rearrange the network so that the meshes can be more readily identified. Sometimes this topographical reconfiguration can be the most challenging part of the problem, especially to individuals who are spatially challenged.

In this case, simply rotating the network by 90 degrees makes the mesh arrangement evident as shown in Figure 14.16.

This network cannot be reduced by a simple series–parallel combination strategy as used previously. Although more complicated transformations can be used,[1] the voltage–current method coupled with computer evaluation is far easier. Again the problem is solved using standard network analysis. Apply a voltage source and define the mesh currents as shown in Figure 14.16. Note that the total current out of the source flows into both the 15-Ω resistor and the 0.01-f capacitor. Hence the current into node B, the current we are looking for, is the sum of $i_1 + i_2$.

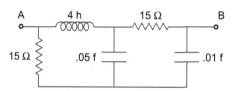

FIGURE 14.15   Network used in Example 14.8. The goal is to find the equivalent impedance between nodes A and B over a frequency range of 0.01 to 100 rad/s.

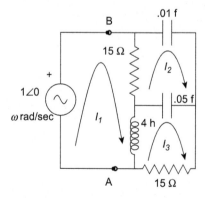

FIGURE 14.16   The network shown in Figure 14.15 after rotating by 90 degrees to make the three-mesh topology more evident.

After you convert all elements to phasor notation, write the matrix equation as:

$$\begin{vmatrix} 1 \\ 0 \\ 0 \end{vmatrix} = \begin{vmatrix} 15+j4\omega & -15 & -j4\omega \\ -15 & 15-j100/\omega-j20/\omega & j20/\omega \\ -j4\omega & j20/\omega & 15+j4\omega-j20/\omega \end{vmatrix} \begin{vmatrix} I_1 \\ I_2 \\ I_3 \end{vmatrix}$$

Note that the frequency, ω, remains a variable in this equation because we want to find the value of Z_{eq} over a range of frequencies. MATLAB is used to find the values of Z_{eq} over the desired range of frequencies and also plots these impedance values as a function of frequency.

```
% Example 14.8 Find and plot the values of an equivalent impedance
%  between .01 and 100 rad/sec
%
% Define frequency range, use .01 rad/sec increments
w = .01:.01:100;              % Define frequency vector
v = [1; 0; 0];                % Define voltage vector
%
% Loop over all frequencies, solving for Zeq²
for k = 1:length(w)
  % Define impedance vector (Use continuation statement)
  Z = [15+4j*w(k), -15, -4j*w(k); -15, 15-120j/w(k), 20j/w(k);
    -4j*w(k), 20j/w(k), 15+4j*w(k)-20j/w(k)];
  i = Z\v;                    % Solve for current
  Zeq(k) = 1/(i(1) + i(2));   % Solve for Zeq
end
  .......plot and label magnitude and phase........
```

The graph produced by this program is shown in Figure 14.17. Both the magnitude and phase of Z_{eq} are functions of frequency and both reach a maximum value at between 3 and 5 rad/s.

It is easy to find the maximum and minimum of the magnitude and phase impedances using the MATLAB max and min functions. Applying these functions to the Zeq:

```
[max_Zeq_abs, abs_freq] = max(abs(Zeq));
[max_Zeq_phase, phase_freq] = min(angle(Zeq)*360/(2*pi));
%
% Now display the maximum and minimum values including frequency
disp([max_Zeq_abs, w(abs_freq); max_Zeq_phase, w(phase_freq)])
```

This produces the values:

$$|Z_{eq}|_{max} = 15.43 \, \Omega \qquad f_{max} = 1.6 \, \text{rad/s}$$
$$\angle Z_{eq \ max} = -21.48 \, \text{degrees} \quad f_{max} = 8.15 \, \text{rad/s}$$

The max and min functions give the maximum or minimum value of their arguments along with an index indicating where these values occur. To convert the index to the appropriate frequency, use w(index) (where index is the second output of the max and min functions) as in the earlier code.

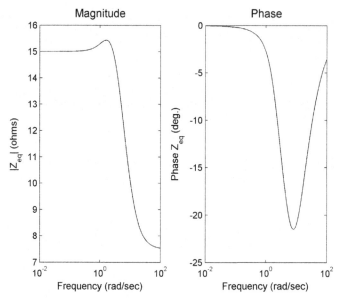

FIGURE 14.17   The value of Z_{eq} as a function of frequency for the network in Example 14.8. The impedance is plotted in terms of magnitude and phase. These plots are obtained from the MATLAB code used in Example 14.8.

[1]A slightly more complicated transformation exists that allows configurations such as this to be reduced by the reduction method, specifically a transformation known as the "Π (pi) to T" transformation. But the voltage–current approach used in this example applies to any network and lends itself well to computer analysis.
[2]You could avoid the loop and use MATLAB's vectorization capabilities. You would need to use ./ and .* operators. A loop to modify frequency is used here because it is easy and more understandable for the reader.

The remainder of this chapter examines the characteristics of sources, both real and ideal, and develops methods for reducing networks that contain sources. Many of the principles used here in passive networks are also used with these more general networks.

14.4 IDEAL AND REAL SOURCES

Before we develop methods to reduce networks that contain sources, it is helpful to revisit the properties of ideal sources described in Chapter 12 (Section 12.4.2) and examine how ideal and real sources differ. In this discussion, only constant output sources (i.e., DC sources) are considered, but the arguments presented generalize with only minor modifications to time-varying sources as shown in the next section.

14.4.1 The Voltage–Current or v–i Plot

Essentially an ideal source can supply any amount of energy that is required by whatever is connected to that source. Discussions of real and ideal sources often use plots of voltage against current (or force against velocity), which provide a visual representation of the source

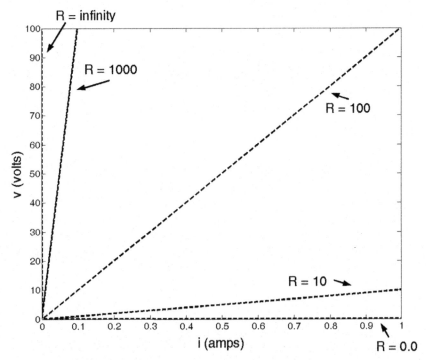

FIGURE 14.18 A $v–i$ plot, showing voltage against current, for resistors having five different resistance values between 0.0 Ω and infinity Ω. This shows that the $v–i$ plot of a resistor is a straight line passing through the origin and having a slope equal to the resistance.

characteristics. Such "$v–i$" plots are particularly effective at demonstrating the equivalent resistance of an element. Consider the $v–i$ plot of pure resistors as shown in Figure 14.18. The plots of five different resistors are shown: 0, 10, 100, 1000, and ∞ Ω. The voltage–current relationship for a resistor is given by Ohm's law, $v = Ri$. A comparison of Ohm's law with the equation for a straight line, $y = mx + b$, where m is the slope of the line and b is the intercept, shows that the voltage–current relationship for a resistor plot is a straight line with a slope equal to the value of the resistance and an intercept of 0.0.

The reverse argument says that an element that plots as a straight line on a $v–i$ plot is either a resistor or contains a resistance, and the slope of the line indicates the value of the resistance. The steeper the slope, the greater the resistance: a vertical line with a slope of infinity indicates the presence of an infinite resistance, whereas a horizontal line with a zero slope indicates the presence of a 0.0 Ω resistance.

The $v–i$ plot of an ideal DC voltage source follows directly from the definition: a source of voltage that is constant irrespective of the current through it. For a time-varying source, such as a sinusoidal source, the voltage varies as a function of time, but not as a function of the current. An ideal voltage source cares naught about the current through it. The current will be whatever it has to be to maintain the specified voltage. Hence the $v–i$ plot of an ideal DC voltage source, V_S, is a horizontal line intersecting the vertical (voltage) axis at $v = V_S$,

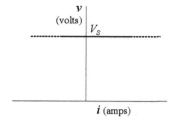

FIGURE 14.19 A v–i plot of an ideal voltage source. This plot shows that the resistor-like properties of a voltage source have a zero value.

Figure 14.19. If the voltage source were time varying, the v–i plot would look essentially the same except that the height of the horizontal line would vary as with time.

The v–i plot of a voltage source with its horizontal line demonstrates that the resistive component of an ideal voltage source is zero. In other words, the equivalent resistance of an ideal voltage source is 0.0 Ω. With regard to the v–i plot, an ideal source looks like a resistor of 0.0 Ω with an offset of V_S. It may seem strange to talk about the equivalent resistance of a voltage source, especially since it has a resistance of 0.0 Ω, but it turns out to be a very useful concept. The equivalent resistance of a source is its resistance ignoring other electrical properties. It is especially useful in describing real sources where the equivalent resistance is no longer zero. The concept of equivalent resistance or, more generally, equivalent impedance, is an important concept in network reduction and has a strong impact on biotransducer analysis and design.

Resistive elements having zero or infinite resistance have special significance and have their own terminology. Resistances of zero will produce no voltage drop regardless of the current running through them. As mentioned in Chapter 12, an element that has zero voltage for any current is termed a "short circuit" since current flows freely through such elements. Although it may seem contradictory, voltage sources are actually short-circuit devices, but with a voltage offset.

Resistors with infinite resistance have the opposite voltage–current relationship: they allow no current irrespective of the voltage (assuming that it is finite). Devices that have zero current irrespective of the voltage are termed "open circuits" since they do not provide a path for current. These elements have infinite resistance. As shown later, current sources also have nonintuitive equivalent resistance: they are open-circuit devices but with a current offset.

The v–i plot of an ideal current source is evident from its definition: an element that produces a specified current irrespective of the voltage across it. This leads to the v–i plot shown in Figure 14.20 of a vertical line that intersects the current axis at $i = I_S$. By the earlier arguments, the equivalent resistance of such an ideal current source is infinite.

The concepts of ideal voltage and ideal current sources are somewhat counterintuitive. An ideal voltage source has the resistive properties of a short circuit, but it also somehow maintains a nonzero voltage. The trick is to understand that an ideal voltage source is a short circuit with respect to current, but not with respect to voltage. A similar apparent contradiction applies to current sources: they are open circuits with respect to voltage, yet produce a specified current. Understanding these apparent contradictions is critical to understanding ideal sources.

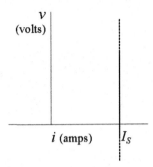

FIGURE 14.20 A v–i plot of an ideal current source. This plot shows that the resistor-like properties of a current source have an infinite value.

In this section, voltage and current sources have been described in terms of fixed values (i.e. DC sources), but the basic arguments do not change if V_S or I_S are time varying. This generalization also holds for the concepts presented next.

14.4.2 Real Voltage Sources—The Thévenin Source

Unlike ideal sources, real voltage sources are not immune to the current flowing through them, nor are real current sources immune to the voltage falling across them. In real voltage sources, the source voltage drops as more current is drawn from the source. This gives rise to a v–i plot such as shown in Figure 14.21, where the line is no longer horizontal but decreases with increasing current. The decrease indicates the presence of an internal, nonzero, resistance having a value equal to the negative slope of the line. The slope is negative because the voltage drop across the internal resistance is opposite in sign to V_s and hence subtracts from the values of V_s.

Real sources, then, are simply ideal sources with some nonzero resistance, so they can be represented as an ideal source in series with a resistor as shown in Figure 14.22. This configuration is also known as a Thévenin source, named after an engineer who developed a network reduction theory described later. Finding values for V_T and R_T given a physical (i.e., a real) source is fairly straightforward. The value of the internal ideal source, V_T, is

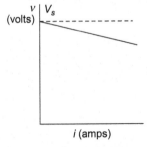

FIGURE 14.21 The v–i plot of a real voltage source (*solid line*). The nonzero slope of this line shows that the source contains an internal resistor. The slope indicates the value of this internal resistance.

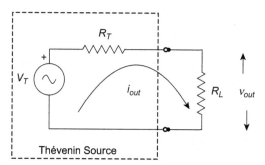

FIGURE 14.22 Representation of a real source using a Thévenin circuit. The Thévenin circuit is inside the *dashed rectangle.* The load resistor, R_L, draws current from the source. In a high-quality voltage source, the internal resistance, R_T, will be small.

the voltage that would be measured at the output, v_{out}, if no current was flowing through the circuit, that is, if the source is connected to an open circuit. For this reason, V_T is equivalent to the "open-circuit voltage," denoted v_{oc}. To find the value of the internal resistance, R_T, we need to draw current from the circuit and measure how much the output voltage decreases. A resistor placed across the output of the Thévenin source will do the trick. This resister, R_L, is often referred to as a "load resistor" or just the "load" because it makes the source do work by drawing current from the source $(P = v_{out}^2/R_L)$. The smaller the load resistor, the more current that will be drawn for the source, and the more power the source must supply.

Assuming a current i_{out} is being drawn from the source and the voltage measured is v_{out}, then the difference in voltage between the no-current voltage and current conditions is: $v_D = V_T - v_{out}$. Since the voltage difference v_D is due entirely to R_T, the value of R_T can be determined as:

$$R_T = \frac{v_D}{i_{out}} = \frac{(V_T - v_{out})}{i_{out}} \tag{14.11}$$

The maximum current the source is capable of producing occurs when the source is connected to a short circuit (i.e., $R_L \to 0\,\Omega$), and the current out of the source, the "short-circuit current," i_{sc}, is:

$$i_{sc} = \frac{V_T}{R_T} \tag{14.12}$$

Remembering that V_T is the voltage that would be measured under open-circuit conditions and defining the open-circuit voltage as v_{oc}, we write:

$$V_T = v_{oc} = R_T i_{sc}$$
$$R_T = \frac{v_{oc}}{i_{sc}} \tag{14.13}$$

Putting Equation 14.13 in words, the internal resistance is equal to the open-circuit voltage (v_{oc}) divided by the short-circuit current (i_{sc}). Using Equation 14.13 is a viable method for determining R_T in theoretical problems, but is not practical in real situations with real sources because shorting a real source may draw excessive current and damage the source.[3] It is safer to draw only a small amount of current out of a real source by placing a large resistor across the source, not a short circuit. In this case, Equation 14.11 is used to find R_T since v_{out} and i_{out} can be measured and V_T can be determined by a measurement of open-circuit voltage. Example 14.9 takes this approach to determine the internal resistance of a voltage source.

EXAMPLE 14.9

In the laboratory, the voltage of a real source is measured using a "voltmeter" that draws negligible current from the source; hence the voltage recorded can be taken as the open-circuit voltage. (Most high-quality voltmeters require very little current to measure a voltage.) The voltage measured is 9.0 V. Figure 14.22 shows a resistor, R_L, placed across the output terminals of the source. In this example, this load resistor results in a current flow out of the source of: $i_{out} = 5$ mA (5×10^{-3} A). Assume this current is measured using an ideal current measurement device, although it could also be calculated from v_{out} if the value of R_L is known, i.e., $i_{out} = v_{out}/R_L$. Under this load condition, the output voltage, v_{out}, falls to 8.6 V. What is the internal resistance of the source, R_T? What is the load resistance, R_L, that produced this current?

Solution/Result: When there is no load resistor (i.e., $R_L = \infty$ Ω), then $i_{out} = 0$ and $v_{out} = V_T$. When the load resistor is attached to the output, $i_{out} = 5$ mA, and $v_{out} = 8.6$ V. Applying Equation 14.11:

$$R_T = \frac{v_D}{i_{out}} = \frac{(V_T - v_{out})}{i_{out}} = \frac{(9.0 - 8.6)}{0.005} = \frac{0.4}{0.005} = 80 \ \Omega$$

To find the load resistor, R_L, use Ohm's law:

$$R_L = \frac{v_{out}}{i_{out}} = \frac{8.6}{0.005} = 1720 \ \Omega$$

In summary, a real voltage source can be represented by an ideal source with a series resistance. In the examples shown here the sources are DC and the series element a pure resistor. In the more general case, the Thévenin source could generate sinusoids or other waveforms (but would still be ideal), and the series element might be a Thévenin impedance, Z_T. In the case where the source is sinusoidal, the equations mentioned previously still hold, but require phasor analysis for their solution.

14.4.3 Real Current Sources

Current sources are difficult to grasp because we intuitively think of current as an effect of voltage: voltage pushes current through a circuit. Current sources can be viewed as sources

---

[3]Many real sources such as laboratory "power supplies" have short-circuit protection where the output voltage drops to zero if the source is short circuited. Alternatively, some laboratory sources have current limiting where the voltage is automatically lowered to maintain a given maximum current.

that somehow adjust their voltage to produce the desired current. For an ideal current source, the larger the load resistor, the more work it has to do since it must generate a larger voltage to produce the desired current. Current sources prefer small load resistors, the opposite of voltage sources. For a real current source, as the load resistor increases the voltage requirement increases and will eventually exceed the voltage capabilities of the current source so the output current will fall off. This is reflected by the v–i plot circuit in Figure 14.23, and can be represented by an internal resistor.

For current sources, the negative slope of the line in the v–i plot is equal to the internal resistance just as for voltage sources. The circuit diagram of a real current source is shown in Figure 14.24 to be an ideal current source in parallel with an internal resistance. This configuration is often referred to as a "Norton equivalent" circuit. Inspection of this circuit shows how it represents the falloff in output current that occurs when higher voltages are needed to push the current though a large resistor. As the voltage needed at the output increases, more current flows though the internal resistor, R_N, and less is available for the

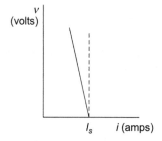

FIGURE 14.23   The v–i plot of a real current source (*solid line*). The less-than-infinite slope of this *line* shows that the source cannot produce the voltage required when the load resistance increases. The source is not capable of producing the higher voltages that are needed to attain the desired current. The resultant reduction in current output with increased voltage requirements can be represented by an internal resistor, and the slope is indicative of its value.

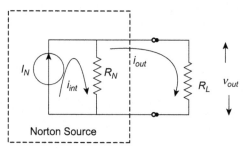

FIGURE 14.24   A circuit diagram of a real current source inside the *dashed rectangle* that is connected to a resistor load. The current source circuit is referred to as a Norton circuit. When a large external load is presented to this circuit, some of the current flows through the internal resistor, R_N. As R_L increases, a higher percentage of current flows through R_N and less through the output. In a high-quality current source R_N will be large.

output. If there were no internal resistor, all of the current would have to flow out of the source irrespective of the output voltage as expected from an ideal current source.

As shown in Figure 14.24, when the output is a short circuit (i.e., $R_L = 0.0\ \Omega$), the current that flows through R_N is zero, so all the current flows through the output. (Note that analogous to the open-circuit condition for a voltage source, the short-circuit condition produces the least work for a current source, in fact no work at all.) By KCL:

$$I_N - i_{R_N} - i_{out} = 0;$$
$$i_{out} = i_{sc} = I_N \tag{14.14}$$

Hence I_N equals the short-circuit current, i_{sc}. When R_L is not a short circuit, some of the current flows through R_N and i_{out} decreases. Essentially, the internal resistor steals current from the current source when the load resister is anything other than zero. Applying KCL to the upper node of the Norton circuit (Figure 14.24), paying attention to the current directions:

$$I_N - i_{R_N} - i_{out} = 0$$
$$I_N - \frac{v_{out}}{R_N} - i_{out} = 0 \tag{14.15}$$

Solving for R_N:

$$\frac{v_{out}}{R_N} = I_N - i_{out}; \quad R_N = \frac{v_{out}}{(I_N - i_{out})}$$
$$\text{Since } I_N = isc: \quad R_N = \frac{v_{out}}{(i_{sc} - i_{out})} \tag{14.16}$$

When the output of the Norton circuit is an open circuit (i.e., $R_L\ 6\ 4$), all the current flows through the internal resistor, R_N. Hence:

$$v_{oc} = I_N R_N \tag{14.17}$$

Combining this equation with Equation 14.14, we can solve for R_N in terms of the open-circuit voltage, v_{oc}, and the short-circuit current, i_{sc}:

$$v_{oc} = I_N R_N = i_{sc} R_N$$
$$R_N = \frac{v_{oc}}{i_{sc}} \tag{14.18}$$

This relationship is the same as for the Thévenin circuit as given in Equation 14.13 if we make $R_T = R_N$.

EXAMPLE 14.10

A real current source produces a current of 500 mA under short-circuit conditions and a current of 490 mA when the short is replaced by a 20-Ω resistor. Find the internal resistance.

Solution/Result: Find v_{out} when the load is 20 Ω, then apply Equation 14.16 to find R_N.

$$v_{out} = R_L i_{out} = 20(.49) = 9.8 \text{ V}$$

$$R_N = \frac{v_{out}}{i_{sc} - i_{out}} = \frac{9.8}{0.5 - 0.49} = \frac{9.8}{0.01} = 980 \ \Omega$$

The Thévenin and Norton circuits have been presented in terms of sources, but they can also be used to represent entire networks as well as mechanical and other nonelectrical systems. These representations can be especially helpful when two systems are being connected. Imagine you are connecting two systems and you want to know how the interconnection will affect the behavior of the overall system. If you represent the source system as a Thévenin or Norton equivalent and then determine the effective input impedance of the loading system, you are able to calculate the loss of signal due to the interconnection. The same could be stated for a biological measurement where the biological system is the source and the measurement system is the load. You may not have much control over the nature of the source, but, as a biomedical engineer, you have some control over the effective impedance of the load. These concepts are explored further in a later section.

14.4.4 Thévenin and Norton Circuit Conversion

It is easy to convert between the Thévenin and Norton equivalent circuits, that is, to determine a Norton circuit that has the same voltage–current relationship as a given Thévenin circuit and vice versa. Such conversions allow you to apply KVL to systems with current sources (by converting them to an equivalent voltage source) or to use Kirchoff's current law (KCL) in systems with voltage sources (by converting them to an equivalent current). Consider the voltage–current relationship shown in the $v–i$ plot of Figure 14.25. Since the curve is a straight line, it is uniquely determined by any two points. The horizontal and vertical intercepts, v_{oc} and i_{sc}, are particularly convenient points to use.

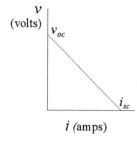

FIGURE 14.25   The $v–i$ plot of the output of a device that can be represented as either a Thévenin or Norton circuit. The voltage–current relationship plots as a *straight line* and can be uniquely represented by two points such as v_{oc} and i_{sc}.

The equations for equivalence are easy to derive based on previous definitions:

For a Thévenin circuit: $v_{ocT} = V_T$;   and since: $R_T = \dfrac{v_{ocT}}{i_{scT}}$; $i_{scT} = \dfrac{v_{ocT}}{R_T}$

For a Norton circuit: $I_N = i_{scN}$;   and since $R_N = \dfrac{v_{ocN}}{i_{scN}}$; $v_{ocN} = R_N i_{scN}$

For a Norton circuit to have the same $v-i$ relationship as a Thévenin, $i_{scN} = i_{scT}$ and $v_{ocN} = v_{ocT}$. Equating these terms in the earlier equations:

$$I_N = i_{scN} = i_{scT} = \frac{v_{ocT}}{R_T}; \quad \text{but } v_{ocT} = V_T$$

$$I_N = \frac{V_T}{R_T} \tag{14.19}$$

$$R_N = \frac{v_{ocN}}{i_{scN}} = \frac{v_{ocT}}{i_{scT}}; \quad \text{but } \frac{v_{ocT}}{i_{scT}} = R_T$$

$$R_N = R_T \tag{14.20}$$

To go the other way and convert from a Norton to an equivalent Thévenin:

$$V_T = v_{ocT} = v_{ocN}; \quad \text{but } v_{ocN} = I_N R_N$$

$$V_T = I_N R_N \tag{14.21}$$

$$R_T = \frac{v_{ocT}}{i_{scT}} = \frac{v_{ocN}}{i_{scN}}; \quad \text{but } \frac{v_{ocN}}{i_{scN}} = R_N$$

$$R_T = R_N \tag{14.22}$$

These four equations allow for easy conversion between Thévenin and Norton circuits. Note that the internal resistance, R_N or R_T, is the same for either configuration. This is reasonable, since the internal resistance defines the slope of the $v-i$ curve, so to achieve the same $v-i$ relationship you need the same-sloped curve and hence the same resistor.

The ability to represent any linear $v-i$ relationship by either a Thévenin or a Norton circuit implies that it is impossible to determine whether a real source is in fact a current or a voltage source based solely on external measurement of voltage and current. If the $v-i$ relationship of a source is more or less a vertical line as in Figure 14.24, indicating a large internal resistance, we might guess that the source is probably a current source. In fact, one simple technique for constructing a crude current source in practice is to place a voltage source in series with a large resistor. Alternatively, if the $v-i$ relationship is approximately horizontal as in Figure 14.21, a nonideal voltage source would be a better guess. However, if the $v-i$ curve is neither particularly vertical nor horizontal, as in Figure 14.25, it is anyone's guess as to whether it is a current or voltage source and either would be an equally appropriate representation unless other information was available.

Conversion between Thévenin and Norton circuits can be used for nodal analysis in circuits that contain voltage sources or for mesh analysis in circuits that contain current sources. This application of Thévenin–Norton conversion is shown in the following example.

EXAMPLE 14.11

Find the voltage, V_A, in the circuit shown in Figure 14.26 using nodal analysis.

Solution: This circuit can be viewed as containing two Thévenin circuits: a 5-V source and 10-Ω resistor, and a 10-V source and 40-Ω resistor. After converting these two Thévenin circuits to equivalent Norton circuits using Equations 14.19 and 14.20, we apply standard nodal analysis. Figure 14.27 shows the circuit after Thévenin–Norton conversion.

Result: Writing the KCL equation around node A.

$$i_{S1} + i_{S2} + i_{R1} + i_{R2} + i_{R3} = 0$$

$$0.5 + 0.25 - \frac{V_A}{R_1} - \frac{V_A}{R_2} - \frac{V_A}{R_3} = 0$$

$$0.5 + 0.25 - V_A \left(\frac{1}{10} + \frac{1}{20} + \frac{1}{40} \right) = 0$$

$$V_A = \frac{0.5 + 0.25}{1/10 + 1/20 + 1/40} = \frac{.75}{0.1 + 0.05 + 0.025} = \frac{0.75}{0.175} = 4.29 \text{ V}$$

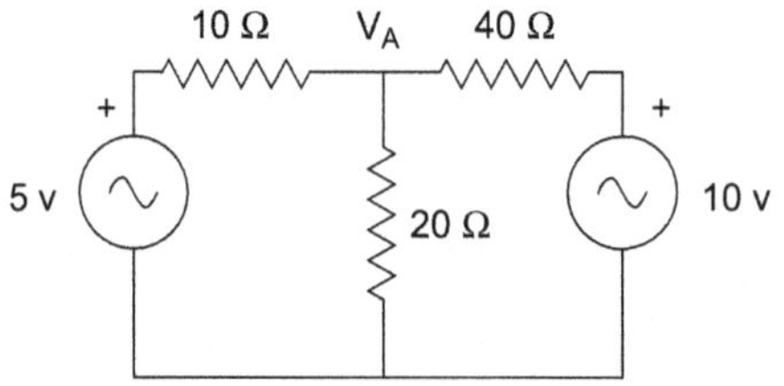

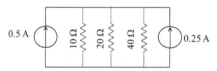

FIGURE 14.26　A two-mesh network containing voltage sources. If the two Thévenin circuits on either side are converted to their Norton equivalents, this circuit becomes a single-node circuit and can be evaluated using a single nodal equation. (The sources are shown as sinusoidal, but since the passive elements are all resistors, the equations will be algebraic.)

FIGURE 14.27　The network in Figure 14.26 after the two Thévenin circuits have been converted to their Norton equivalents. This is now a single-node network.

The Thévenin and Norton circuits and their interconversions are useful for network reduction of circuits that contain sources as shown in the next section. These concepts also apply to mechanical systems, with appropriate modifications, as illustrated at the end of this chapter.

14.5 THÉVENIN AND NORTON THEOREMS: NETWORK REDUCTION WITH SOURCES

The "Thévenin Theorem" states that any network of passive elements and sources can be reduced to a single voltage source and series impedance. Such a reduced network would look like a Thévenin circuit such as that shown in Figure 14.22, except that the internal resistance, R_T, would be replaced by a generalized impedance, $Z_T(\omega)$. The Norton theorem makes the same claim for Norton circuits, which is reasonable since Thévenin circuits can easily be converted to Norton circuits via Equations 14.19 and 14.20.

There are a few constraints on these theorems. The elements in the network being reduced must be linear, and if there are multiple sources in the network they must be at the same frequency. As has been done in the past, the techniques for network reduction will be developed using phasor representation and hence will be limited to networks with sinusoidal sources. However, the approach is the same in the Laplace domain.

There are two approaches to finding the Thévenin or Norton equivalent of a general network. One is based on solving for the open-circuit voltage, v_{oc}, and the short-circuit current, i_{sc}. The other method evaluates only the open-circuit voltage, v_{oc}, and then determines R_T (or R_N) through network reduction. During network reduction, a source is replaced by its equivalent resistance, that is, short circuits ($R = 0$) substitute for voltage sources, and open circuits ($R = \infty$) substitute for current sources. Both network reduction methods are straightforward, but the open-circuit voltage/short-circuit current approach can be implemented on a computer. These approaches are demonstrated in the next two examples.

EXAMPLE 14.12

Find the Thévenin equivalent of the circuit in Figure 14.28 using both the v_{oc}–i_{sc} method and the v_{oc}–network reduction technique.

Solution: v_{oc}-Reduction method: First find the open-circuit voltage, v_{oc}, using standard network analysis. Convert all network elements to their phasor representations as shown in Figure 14.29.

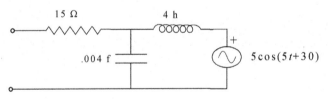

FIGURE 14.28 A network that will be reduced to a single impedance and source using two different strategies.

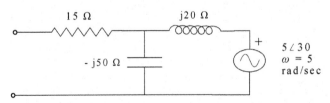

FIGURE 14.29 The network of Figure 14.28 after conversion to the phasor domain.

Since in the open-circuit case no current flows through the 15-Ω resistor, there is no voltage drop across this resistor, so the open-circuit voltage is the same as the voltage across the capacitor. The resistor is essentially not there with respect to open-circuit voltage. The resistor is not totally useless: it does play a role in determining the equivalent impedance and is involved in the calculation of short-circuit current. The open-circuit voltage, the voltage across the capacitor, can be determined by writing the mesh equation around the loop consisting of the capacitor, inductor, and source. Using the usual directional conventions, defining the mesh current as clockwise and going around the loop in a clockwise direction, note that the voltage source will be negative as there is a voltage drop going around in the clockwise direction.

$$-5\angle 30 - i(-j50 + j20) = 0$$

$$i = \frac{-5\angle 30}{-j50 + j20} = \frac{-5\angle 30}{-j30} = \frac{-5\angle 30}{30\angle -90} = -0.167\angle 120 \text{ A}$$

$$v_{oc} = iZ_C = -0.167\angle 120(-j50) = -0.167\angle 120(50\angle -90)$$

$$v_{oc} = -8.35\angle 30 \text{ V}$$

Note that the magnitude of the Thévenin equivalent voltage is actually larger than the source voltage. This is the result of a partial resonance between the inductor and capacitor.

Next, find the equivalent impedance by reduction. To reduce the network, essentially turn off the sources and apply network reduction techniques to what is left. Turning off a source does not mean you remove it from the circuit; rather, to turn off a source, you replace it by its equivalent resistance. For an ideal voltage source $R_T \to 0 \text{ }\Omega$, so the equivalent resistance of an ideal voltage source is $0 \text{ }\Omega$, i.e., a short circuit. To turn off a voltage source you replace it with a short circuit. After you replace the source with a short circuit, you are left with the network shown in Figure 14.30 on the left-hand side. Series–parallel reduction techniques will work for this network.

After replacing the voltage source with a short circuit, we are left with a parallel combination of inductor and capacitor. This combines to a single impedance, Z_P:

$$Z_P = \frac{Z_C Z_L}{Z_C + Z_L} = \frac{-j50(j20)}{-j50 + j20} = \frac{1000}{-j30} = \frac{1000}{30\angle -90}$$

$$Z_P = 33.33\angle 90 \text{ }\Omega$$

We are left with the series combination of Z_P and the 15-Ω resistor:

$$Z_T = 15 + 33.33\angle 90 = 15 + j33.33 = 36.55\angle 65.8 \text{ }\Omega$$

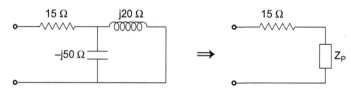

FIGURE 14.30 (Left) The network given in Figure 14.29 after effectively turning off the voltage source. The voltage source is replaced by its equivalent resistance, which is, since it is an ideal source, $0.0 \text{ }\Omega$ or a short circuit. (Right) The network after combining the two parallel impedances into a single impedance, Z_P.

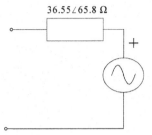

$36.55\angle 65.8\ \Omega$

FIGURE 14.31   The Thévenin equivalent of the network given in Figure 14.28 determined by the v_{oc}-reduction method.

Results: v_{oc}-Reduction method: The original circuit can be equivalently represented by the Thévenin circuit shown in Figure 14.31.

Solution $v_{oc}-i_{sc}$: In this method, we solve for the open-circuit voltage and short-circuit current. We have already found the open-circuit voltage so it is only necessary to find the short-circuit current. After shorting out the output and converting to phasor notation, the circuit is shown in Figure 14.32.

If we solve this using mesh analysis, it is a two-mesh circuit, but if we convert the inductor −source combination to a Norton equivalent, it becomes a single-node equation. To implement the conversion, use Equations 14.19 and 14.20:

$$I_N = \frac{V_T}{R_T} = \frac{5\angle 30}{j20} = \frac{5\angle 30}{20\angle 90} = 0.25\angle -60 \text{ V}; \quad R_N = R_T = j20\ \Omega$$

The new network is shown in Figure 14.33.

$$.25\angle -60 - v\left(\frac{1}{15} + \frac{1}{-j50} + \frac{1}{j20}\right) = 0$$

$$v = \frac{.25\angle -60}{\dfrac{1}{15} + \dfrac{1}{-j50} + \dfrac{1}{j20}} = \frac{.25\angle -60}{0.0667 + j0.02 - j0.05} = \frac{.25\angle -60}{0.0667 - j0.03}$$

$$v = \frac{.25\angle -60}{0.073\angle -24.2} = 3.42\angle -35.8 \text{ V}$$

$$i_{sc} = \frac{v}{15} = \frac{3.42\angle -35.8}{15} = 0.228\angle -35.8 \text{ A}$$

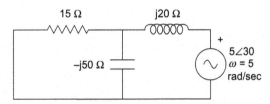

$15\ \Omega$ $j20\ \Omega$

$-j50\ \Omega$

$5\angle 30$
$\omega = 5$
rad/sec

FIGURE 14.32   The circuit shown in Figure 14.28 after shorting the output terminals and converting to phasor notation. Note that the capacitor and resistor are in parallel.

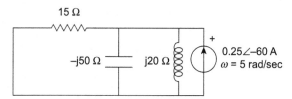

FIGURE 14.33   The circuit of Figure 14.32 after the voltage source and series impedance are converted to a Norton equivalent source. This is a one-node problem that can be solved from a single equation.

Now solve for R_T:

$$R_T = \frac{v_{oc}}{i_{sc}} = \frac{8.35 \angle 30}{0.228 \angle -35.8} = 36.6 \angle 65.8 \ \Omega$$

This is the same value for R_T found using the v_{oc}-reduction method and leads to the Thévenin equivalent shown in Figure 14.31. More complicated networks can be reduced using MATLAB as shown in the following example.

EXAMPLE 14.13

Find the Norton equivalent of the circuit of Figure 14.34 with the aid of MATLAB.

Solution: The open-circuit voltage and short-circuit current can be solved directly using mesh analysis in conjunction with MATLAB. In fact, the mesh equations in both cases (solving for v_{oc} and i_{sc}) are similar. The only difference is that when solving for the short-circuit current, the 20-Ω resistor will be short circuited and not appear in the equation.

First convert to phasor notation, and then encode the network directly into MATLAB. Since we are using MATLAB and the computational load is reduced, we keep ω as a variable in case we want to find the Norton equivalent for other frequencies.

After converting to phasor notation and assigning the mesh current, the circuit is shown in Figure 14.35. Note that the open-circuit voltage, v_{oc}, is the voltage across the 20-Ω resistor and the short-circuit current, i_{sc}, is just i_3 when the resistor is shorted.

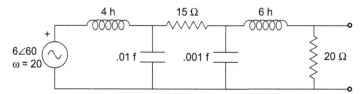

FIGURE 14.34   A complicated network that is reduced to a Norton equivalent in Example 14.13. To ease the mathematical burden, MATLAB is used to solve the equation.

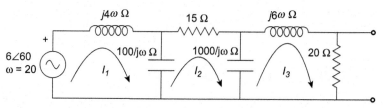

FIGURE 14.35   The circuit of Figure 14.34 after conversion to phasor notation. Since mesh analysis will be used, the mesh currents are shown.

Writing the KVL equations for the open-circuit condition:

$$
\begin{vmatrix} 6\angle 60 \\ \\ 0 \\ \\ 0 \end{vmatrix} = \begin{vmatrix} j4\omega + \dfrac{100}{j\omega} & -\dfrac{100}{j\omega} & 0 \\ -\dfrac{100}{j\omega} & 15 + \dfrac{100 + 1000}{j\omega} & -\dfrac{1000}{j\omega} \\ 0 & -\dfrac{1000}{j\omega} & 20 + j6\omega + \dfrac{1000}{j\omega} \end{vmatrix} \begin{vmatrix} i_1 \\ \\ i_2 \\ \\ i_3 \end{vmatrix}
$$

where $v_{oc} = 20\, i_3$.

The mesh equation for the short-circuit condition is quite similar:

$$
\begin{vmatrix} 6\angle 60 \\ \\ 0 \\ \\ 0 \end{vmatrix} = \begin{vmatrix} j4\omega + \dfrac{100}{j\omega} & -\dfrac{100}{j\omega} & 0 \\ -\dfrac{100}{j\omega} & 15 + \dfrac{100 + 1000}{j\omega} & -\dfrac{1000}{j\omega} \\ 0 & -\dfrac{1000}{j\omega} & j6\omega + \dfrac{1000}{j\omega} \end{vmatrix} \begin{vmatrix} i_1 \\ \\ i_2 \\ \\ i_3 \end{vmatrix}
$$

where $i_{sc} = i_3$.

The following is the program to solve these equations and find I_N and R_N:

```
% Example 14.13 Find the Norton equivalent of a three-mesh circuit
%
w = 20;                           % Define frequency
theta = 60*2*pi/360;
VS = 6*cos(theta) + 6*sin(theta)*j;   % Define Vs as rectangular
v = [VS 0 0]';                        % Define voltage vector
%
% Define open-circuit impedance matrix
Zoc = [4j*w+100/(j*w), -100/(j*w), 0; -100/(j*w), 15+1100/(j*w),...
       -1000/(j*w); 0, -1000/(j*w), 20+6j*w+1000/(j*w)];
ioc = Zoc\v;                      % Solve for currents
voc = 20*ioc(3);                 % and open-circuit voltage
%
% Define short-circuit impedance matrix and solve for short-circuit current
```

```
Zsc = [4j*w+100/(j*w), -100/(j*w), 0; -100/(j*w), 15+1100/(j*w),...
       -1000/(j*w); 0, -1000/(1j*w), 6j*w+1000/(j*w)];
i = Zsc\v                          % Solve for currents
isc = i(3);                        % Find isc
Zn = voc/isc;                      % Solve for Z_N (Equation 14.18)
%
% Output magnitude and phase of IN and RN
IN_mag = abs(isc)
IN_phase = angle(isc)*360/(2*pi)
ZN_mag = abs(Zn)
ZN_phase = angle(Zn)*360/(2*pi)
```

This program produces the following outputs:

$$I_N = 0.0031 \angle 20.6 \text{ A}$$
$$Z_N = 19.36 \angle 9.8 \ \Omega$$

It is easy to modify this program to find, and plot, the Norton element values over a range of frequencies. This exercise is given as one of the problems at the end of the chapter.

14.6 MEASUREMENT LOADING

We now have the tools to analyze the situation when two systems are connected together. For biomedical engineers not involved in electronic design, this situation most frequently occurs when making a measurement, so we analyze the problem in that context. However, the approach followed here applies to any situation when two systems are connected together.

14.6.1 Ideal and Real Measurement Devices

One of the important tasks of biomedical engineers is to make measurements, usually on a living system. Any measurement requires withdrawing some energy from the system of interest and that, in turn, alters the state of the system and the value of the measurement. This alteration is referred to here as "measurement loading." The word "load" is applied to any device that is attached to a system of interest, and "loading" is the effect the attached device has on the system. Measurement loading is a well-known phenomenon that extends down to the smallest systems and has a significant impact on fundamental principles in particle physics. The concepts developed earlier can be used to analyze the effect a measurement device has on the system being measured. In fact, an analysis of measurement loading is one of the major applications of Thévenin and Norton circuit models.

Just as there are ideal and real sources, there are ideal and real measurement devices or, equivalently, ideal and real loads. As mentioned previously, an ideal voltage source supplies a given voltage at any current and an ideal current source supplies a given current at any voltage. Since power is the product of voltage times current ($P = vi$), an ideal source can supply any amount of energy or power required: infinite if need be. Ideal measurement devices (or ideal loads) have the opposite characteristics: they can make a measurement without drawing any energy or power from the system being measured. Of course, we know from basic physics that such an idealization is impossible, but some measurement devices can provide nearly ideal measurements, at least for all practical purposes.

The goal in practical situations is to be able to make a measurement without significantly altering the system being measured. The ability to attain this goal depends on the characteristics of the source as well as the load; a given device might have little effect on one system, thus providing a reliable measurement, yet significantly alter another system, giving a measurement that does not reflect the underlying conditions. It is not just a matter of how much energy a measurement device requires (i.e., load impedance), but how much energy the system being measured can supply without a significant change in the quantity being measured (i.e., the source impedance).

Just as ideal voltage and current sources have quite different properties, ideal measurement devices for voltage and current differ significantly. A device that measures voltage is termed a "voltmeter." An ideal voltmeter would draw no power from the circuit being measured. Since $P = vi$ and v cannot be zero (that is what is being measured), an ideal voltmeter must draw no current while making the measurement. The current will be zero for any voltage only if the equivalent resistance of the voltmeter is infinite. An ideal voltmeter is effectively an open circuit, and the $v-i$ plot is a vertical line. Practical voltmeters do not have infinite resistances, but they do have very large impedances, of the order of 100s of megohms (1 megohm = 1 MΩ = 10^6 Ω), and this can be considered ideal for all but the most challenging conditions.

Current measuring devices are termed "ammeters." An ideal ammeter also needs no power from the circuit to make its measurement. Again, $P = vi$ and i cannot be zero in an ammeter, so now it is voltage that must be zero to draw no energy from the system being measured. This means that an ideal ammeter is effectively a short circuit having an equivalent resistance of 0.0 Ω. The $v-i$ plot of an ideal ammeter would be a horizontal line. Practical ammeters are generally not very ideal, having resistances approaching a tenth of an ohm or more. However, current measurements are rarely made in practice because that involves breaking a circuit connection to make the measurement unless special "clip-on" ammeters are used. The characteristics of ideal sources and loads are summarized in Table 14.1.

An illustration of the error caused by a less-than-ideal ammeter is given in the following example.

TABLE 14.1 Electrical Characteristics of Ideal Sources and Loads

Characteristics	Sources		Measurement Devices (Loads)	
	Voltage	Current	Voltage	Current
Impedance	0.0 Ω	∞ Ω	∞ Ω	0.0 Ω
Voltage	V_S	Up to ∞ A	$V_{measured}$	0.0 A
Current	Up to ∞ V	I_S	0.0 A	$I_{measured}$

EXAMPLE 14.14

A practical ammeter having an internal resistance of 2.0 Ω is used to measure the short-circuit current of the three-mesh network used in Example 14.13. How large is the error, that is, how much does the measurement differ from the true short-circuit current?

Solution: As with all issues of measurement loading, it is easiest to use the Thévenin or Norton representation of the system being loaded. The Norton equivalent of the three-mesh circuit is determined in Example 14.13 and is shown in Figure 14.36 loaded by the ammeter. From the Norton circuit, we know the true short-circuit current is: $i_{sc} = I_N = 0.0031\pi20.6$ A. The measured short-circuit current can be determined by applying nodal analysis to the circuit.

Applying KCL:

$$.0031 \angle 20.6 - v\left(\frac{1}{19.36\angle 9.8} + \frac{1}{2}\right) = 0$$

$$v = \frac{.0031\angle 20.6}{\dfrac{1}{19.36\angle 9.8} + \dfrac{1}{2}} = \frac{.0031\angle 20.6}{.052\angle - 9.8 + 0.5} = \frac{.0031\angle 20.6}{.051 - j.009 + 0.5} = \frac{.0031\angle 20.6}{0.55\angle - 0.9}$$

$$v = 0.0056\angle 21.5 \text{ V}$$

$$i_{sc\_measured} = \frac{v}{R} = \frac{.0056\angle 21.5}{2} = .0028\angle 21.5 \text{ A}$$

The measured short-circuit current is slightly less than the actual short-circuit current, which is equal to I_N. The error is:

$$\text{Error} = \frac{(.0031 - .0028)}{.0031}100 = 9.7\%$$

The difference in the measured current and the actual current is due to current flowing through the internal resistor. Since the external load is much less (an order of magnitude) than the internal resistor, most of the current still flows through the ammeter, so the error is small. The smaller the ammeter's impedance with respect to the system's internal impedance, the closer the measurement loading is to ideal. In this case, the load resistor is approximately one-tenth the value of the internal resistor, leading to an error of approximately 10%. For some measurements, this may be sufficiently accurate. Usually a ratio of 1:100 (i.e. two orders of magnitude or 1% error) is adequate for the type of accuracy required in biomedical engineering measurements.

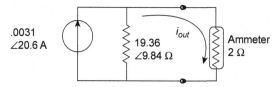

FIGURE 14.36　The output current of the Norton equivalent of the circuit shown in Figure 14.34. A less-than-ideal ammeter with an internal resistance of 2.0 Ω is being used to measure the short-circuit current. The effect of the nonzero resistance of the ammeter on the resultant current measurement is determined in Example 14.14.

The same rules regarding source and load resistance apply to voltage measurements, except now the load resistance should be much greater than the internal resistance (or impedance). In voltage measurements, if the load resistor is 100 or more times the internal resistance, the loading can usually be considered negligible and the measurement sufficiently accurate.

These general rules also apply whenever one network is attached to another. If voltages carry the signal (usually the case), the influence of the second network on the first can be ignored if the effective input impedance of the second network is much greater than the effective output impedance of the first network, Figure 14.37.

The relationship between source and load impedances has major consequences on system analysis and specifically the transfer function. Recall the two basic assumptions used in constructing a system's transfer function: (1) the input is an ideal source (i.e., $Z_1 \to 0$) and (2) the output is an ideal load ($Z_2 \to \infty$). Referring to Figure 14.37, if $Z_2 >> Z_1$, the transfer functions derived for each individual network independently will still work when the two are interconnected. If $Z_2 >> Z_1$ (say, 100 times), these assumptions are reasonably met and the original transfer functions can be taken as valid. The transfer function of the combined network is $TF_1(\omega)\, TF_2(\omega)$. Of course the input to the source network and the next load on loading network could still present problems that depend on the relative input and output impedances connected to the combined network. An analysis of voltage loading is found in the problems.

If the signal is carried as a current, then the opposite would be true. The output impedance of Network 1 should be much greater than the input impedance of Network 2, i.e., $Z_2 << Z_1$. In this situation, the current loading of Network 2 can be considered negligible with respect to Network 1. Signals are rarely viewed in terms of currents except for the output of certain transducers, particularly those that respond to light, and in these cases the signal is converted to a voltage by a special amplifier circuit (see Section 15.8.4).

What if these conditions are not met? For example, what if Z_2 is approximately equal to Z_1? In this case, you have two choices: calculate the transfer function of the two-network combination, or estimate the error that will occur owing to the interconnection as in the last example. Sometimes you actually want to increase the load on a system, purposely making Z_2 close to the value of Z_1. The motivation for such a strategy is explained in the next section.

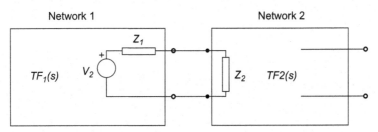

FIGURE 14.37 The input of Network 2 is connected to the output of Network 1. If the equivalent input impedance of Network 2, Z_L, is much greater than the output impedance of Network 1, Z_T, the output of Network 1 will not be altered by the connection to Network 2. The transfer functions derived for each of these networks separately are still valid.

14.6.2 Maximum Power Transfer

The goal in most measurement applications is to extract minimum energy from the system. This is also the usual goal when one system is connected to another because that way the original transfer function remains unchanged. Assuming voltage signals, the load impedance should be much greater than the internal impedance of the source, or vice versa if the signal is carried by current. This results in the minimum power and maximum voltage (or current) out of the source. But what if the goal is to extract maximum energy from the system? To determine the conditions for maximum power out of the system, we consider the Thévenin circuit with its load resistor in Figure 14.38.

To address this question, we assume that R_T is inside the source and cannot be adjusted, that is, it is an internal and inaccessible property of the source. If only R_L can be adjusted, the question becomes: what is the value of R_L that will extract maximum power from the system? The power out of the system is: $P = v_{out} i_{out}$ or $P = R_L i_{out}^2$. To find the value of the load resistor R_L that will deliver the maximum power to itself, we use the standard calculus trick for maximizing a function: solve for P in terms of R_L, and then take dP/dR_L and set it to zero:

$$P = R_L i^2; \quad i = \frac{V_T}{(R_L + R_T)}; \quad \text{hence,} \quad P = \frac{R_L V_T^2}{(R_L + R_T)^2}$$

Solving for $\dfrac{dP}{dR_L}$ by parts:

$$\frac{dP}{dR_L} = \frac{V_T^2 (R_L + R_T)^2 - 2V_T^2 R_L (R_L + R_T)}{(R_L + R_T)^4} = 0 \tag{14.23}$$

$$V_T^2 (R_L + R_T)^2 = 2V_T^2 R_L (R_L + R_T)$$

$$R_L + R_T = 2R_L;$$

$$R_L = R_T$$

So for maximum power out of the system, R_L should equal R_T (or, more generally, $Z_L = Z_T$), a condition known as "impedance matching." Since the power in a resistor is proportional to the resistance ($i^2 R$) and the two resistors are equal, the power transferred to R_L will be half the total power, and the other half is dissipated by R_T.

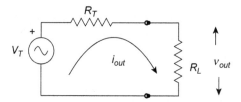

FIGURE 14.38 A Thévenin circuit is shown with a load resistor, R_L. For *minimum* power out of the system, R_L should be much greater than R_T. That way the output current and power approach 0. In this section we seek to determine the value of R_L that will extract maximum power from the Thévenin source.

Equation 14.23 is known as the "Maximum Power Theorem." Using this theorem, it is possible to find the value of load resistance that extracts maximum power from any network. Just convert the network to a Thévenin equivalent and set R_L equal to R_T. Recall that the maximum power theorem applies when R_T is fixed and R_L is varied. If R_T can be adjusted, then just by looking at Figure 14.38 we can see that maximum power will be extracted from the circuit when $R_T = 0$. When sinusoidal or other signals are involved, Equation 14.23 extends directly to impedances.

14.7 MECHANICAL SYSTEMS

All of the concepts described in this chapter are applicable to mechanical systems with only minor modifications. The concepts of equivalent impedances and impedance matching are often used in mechanical systems, particularly in acoustic applications. Of particular value are the concepts of real and ideal sources and real and ideal loads or measurement devices. As mentioned in Chapter 12, an ideal force generator produces a specific force irrespective of the velocity, just as an ideal voltage source produces the required voltage at any current. An ideal force generator will generate the same force at 0.0 velocity, or 10 mph, or 10,000 mph, or even at the speed of light (clearly impossible), if necessary. The force produced by a real force generator will decrease with velocity. This can be expressed in a force—velocity plot (analogous to the $v-i$ plot) as shown in Figure 14.39A.

An ideal velocity (or displacement) generator will produce a specific velocity against any force, be it 1 oz. or 1 ton, but for a real velocity generator, the velocity produced decreases as the force against it increases, Figure 14.39B. Again there is an ambiguity between real force and velocity generators. The device producing the force—velocity curve shown in Figure 14.39C could be interpreted as a nonideal force generator or a nonideal velocity generator: it is not possible to determine its true nature from the force—velocity plot alone.

Ideal measurement devices follow the same guiding principle in mechanical systems: they should extract no energy from the system being measured. For a force-measuring device, a "force transducer," velocity must be zero and position a constant. A constant position

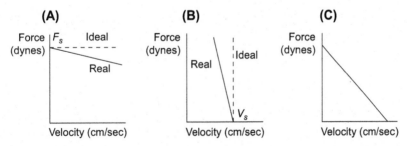

FIGURE 14.39   (A) The force—velocity plot of an ideal (*dashed line*) and a real (*solid line*) force generator. The force produced by a real generator decreases the faster it must move to generate that force. (B) The force—velocity plot of an ideal (*dashed line*) and a real (*solid line*) velocity generator. The velocity produced by a real generator decreases as the opposing force increases. (C) The force—velocity plot of a generator that may be interpreted as either a nonideal force or a nonideal velocity transducer.

condition is also known as an "isometric" condition. So an ideal force transducer requires no movement to make its measurement—it appears as a solid, immobile wall to the system being measured. Since mechanical impedance is defined as F/v in Equation 12.65, if v is zero for all F, then the mechanical impedance is infinite.

For an ideal velocity (or displacement) transducer, the force required to make a measurement would be zero. If F is zero for all v, then the mechanical impedance is zero. The characteristics of ideal mechanical sources and loads are given in Table 14.2 in a fashion analogous to the electrical characteristics in Table 14.1.

The concept of equivalent impedances and sources is useful in determining the alteration produced by a nonideal measurement device or load. The analog of Thévenin and Norton equivalent circuits can also be constructed for mechanical systems. The mechanical analog of a Thévenin circuit is a force generator with a parallel impedance (the configuration is reversed in a mechanical system), Figure 14.40A. Conversely, the mechanical equivalent analog of a Norton circuit is a velocity (or displacement) generator in series with the equivalent impedance, Figure 14.40B. To find the values for either of the two equivalent mechanical systems in Figure 14.40, we use the analog of the $v_{oc}-i_{sc}$ method, but now call it the $F_{iso}-v_{no\ load}$ method: find "isometric" force, the force when velocity is zero (position is constant), and the unloaded velocity, the velocity when no force is applied to the system.

Most practical lumped-parameter mechanical systems are not so complicated and do not usually require reduction to a Thévenin- or Norton-like equivalent circuit. Nonetheless, all of the concepts regarding source and load impedances developed previously apply to mechanical systems. If the output of the mechanical system is a force, the minimum load is one with a large equivalent mechanical impedance, a load that tends to produce a large opposing force

TABLE 14.2　Mechanical Characteristics of Ideal Sources and Loads

Characteristics	Sources		Measurement Devices (Loads)	
	Force	Velocity	Force	Velocity
Impedance	$0.0\ \Omega$	$\infty\ \Omega$	$\infty\ \Omega$	$0.0\ \Omega$
Force	F_S	Up to ∞ dyn	$F_{measured}$	0.0 dyn
Velocity	Up to ∞ cm/s	v_S	0.0 cm/s	$v_{measured}$

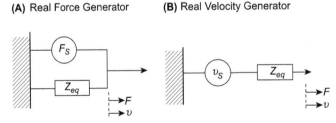

(A) Real Force Generator　　**(B)** Real Velocity Generator

FIGURE 14.40　(A) A mechanical analog of a Thévenin equivalent circuit consisting of an ideal force generator with a parallel mechanical impedance. (B) The mechanical analog of a Norton circuit consisting of an ideal velocity generator with series mechanical impedance. These configurations can be used to determine the effect of loading by a measurement device or by another mechanical system.

and allows very little movement. A solid wall is an example of high mechanical impedance. Conversely, if the output of the mechanical system is a velocity, the minimum load is produced by a system having very little opposition to movement, a small mechanical impedance. Air, or better yet a vacuum, is an example of a low-impedance mechanical system. Finally, if the goal is to transfer maximum power from the source to the load, the mechanical impedances of each should be equal. These principles are explored in the following examples.

EXAMPLE 14.15

The mechanical elements of a real force generator are shown on the left-hand side of Figure 14.41, whereas the mechanical elements of a real force transducer are shown on the right-hand side.

The left-hand side shows a real force generator consisting of an ideal force generator (F_S) in parallel with friction and mass. The right-hand side is a real force transducer consisting of a displacement transducer (marked X) with a parallel spring. This transducer actually measures the displacement of the spring, which is proportional to force ($F = k_e X$).

Find the force that would be measured by an ideal force transducer, that is, the force at the interface that would be produced by the generator if its velocity was zero. The force transducer actually measures displacement of the spring (the usual construction of a force transducer). What is the force that is actually measured by this nonideal transducer? How could this transducer be improved to make a measurement with less error? The system parameters are:

$$k_f = 20 \text{ dyn s/cm}; m = 5 \text{ g}; k_e = 2400 \text{ dyn/cm}; F_S = 10 \sin(4t).\text{dyn}$$

Solution: To find the force measured by an ideal force transducer, write the equations for the force generator and set velocity to zero as would be the case for an ideal force transducer. Note that F_S is negative based on its defined direction (the arrow pointing to the left).

$$F_{measured} = -F_S - v\,(j\omega m + kf) \quad \text{if } v = 0$$
$$F_{measured} = -F_S$$

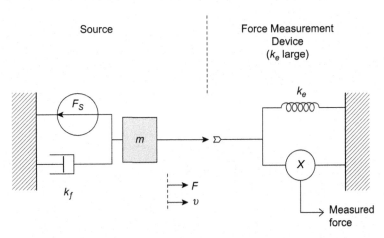

FIGURE 14.41 A realistic force generator on the left is connected to a realistic force transducer on the right. These combined systems are used to examine the effect of loading on a real force generator in Example 14.15.

As expected, the force measured by an ideal force transducer is just the force of the ideal component of the source. If the measurement device or load is ideal, then it does not matter what the internal impedance of the source is, the output of the ideal component is measured at the output. However, when the measurement transducer is attached to a less-than-ideal transducer, the velocity is no longer zero and some of the force is distributed to the internal impedance.

To determine the force out of the force generator under the nonideal load, write the equations for the combined system. Since there is the equivalent of only one node in the combined system, only a single equation will be necessary, but we need to pay attention to the signs.

$$-F_S - v\left(j\omega m + k_f + \frac{k_e}{j\omega}\right) = 0$$

$$-10 - v\left(j4(5) + 20 + \frac{2400}{j4}\right) = -10 - v(j20 + 20 - j600) = 0$$

$$v = \frac{-10}{20 - j580} = \frac{-10}{580.3\angle - 88} = -17.2 \times 10^{-3}\angle 88 \text{ cm/s}$$

$$F_{measured} = vZ_{k_e} = -17.2 \times 10^{-3}\angle 88(-j600) = -10.3\angle - 2 \text{ dyn}$$

Result: Hence the measured force is 3% larger than the actual force. The measured force is larger because of a very small resonance between the mass in the force generator and the elastic element in the transducer. Note that the elastic element is very stiff ($k_e = 2400$ dyn/cm) to make the transducer a good force transducer: the stiffer the elastic element, the closer the velocity will be to zero. To improve the measurement further, this element could be made even stiffer. For example, if the elasticity, k_e, were increased to 9600 dyn/cm, the force measured would be 10.08 $\angle -.5$ dyn, reducing the error to less than 1.0 percent. The determination of improvement created by different transducer loads is presented as a problem at the end of the chapter.

The next example involves the measurement of a velocity generator.

EXAMPLE 14.16

The mechanical elements of a real velocity generator are shown on the left side of Figure 14.42. The left side shows a real velocity generator consisting of an ideal velocity generator (V_S) in series with a friction and an elastic element. The right side is a real velocity transducer consisting of an ideal velocity transducer (marked X, its output is proportional to v_2) with a parallel spring. Assume V_S is a sinusoid with $\omega = 4$ rad/s and $k_f = 20$ dyn s/cm; $k_{e1} = 20$ dyn/cm; $k_{e2} = 2$ dyn/cm.

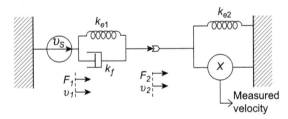

FIGURE 14.42 A realistic velocity generator is connected to a realistic velocity transducer. This configuration is used in Example 14.16 to examine the difference between the measured velocity and that produced by the ideal source.

Find the velocity that is measured by the velocity transducer on the right side of Figure 14.42. Note that this transducer is the same as the force transducer in the last example except the elasticity is very much smaller. Improvement in the accuracy of this transducer is given as one of the problems at the end of the chapter.

Solution: The solution proceeds in exactly the same manner as in the previous problem except that now there are two velocity points. However, since velocity v_1 is equal to V_S, it is not independent and only one equation needs to be solved.

Writing the sum of forces equation about the point indicated by v_2 point:

$$-\left(\frac{k_{e1}}{j\omega} + k_f\right)(v_2 - v_1) - \left(\frac{k_{e2}}{j\omega}\right)v_2 = 0 \quad \text{Substituting } v_1 = \mathscr{V}_s \text{ and other variables}$$

$$-\left(\frac{20}{j4} + 20\right)(v_2 - \mathscr{V}_s) - \left(\frac{2}{j4}\right)v_2 = (j5 - 20)(v_2 - \mathscr{V}_s) + j0.5(v_2) = 0$$

$$(-20 + j5.5)v_2 - (-20 + j5)\mathscr{V}_s = 0 \quad \text{Solving for } v_2$$

$$v_2 = \frac{(-20 + j5)\mathscr{V}_s}{-20 + j5.5} = \frac{(20.6\angle 165.9)\mathscr{V}_s}{20.7\angle 164.6} = (0.995\angle 1.3)\mathscr{V}_s \text{ cm/s}$$

The measured value of v_2 is very close to that of $\mathscr{V}_s$. That is the value of v_2 that would have been measured by an ideal velocity transducer, one that produced no resistance to movement. The measurement error is small because the impedance of the transducer, although not zero, is still much less than that of the source (i.e., $k_{e2} = .1\ k_{e1}$). The elasticity provides the only resistance to movement in the transducer, so if it is increased the error increases and if it is reduced the error is reduced. The influence of transducer impedance is demonstrated in several of the problems.

In some measurement situations, it is better to match the impedance of the transducer with that of the source. Matching mechanical impedances is particularly important in ultrasound imaging. Ultrasound imaging uses a high-frequency (1 MHz and up) pressure pulse wave that is introduced into the body and reflects off of various internal surfaces. The time of flight for the return signal is used to estimate the depth of a given surface. Using a scanning technique, many individual pulses are directed into the body at different directions and a two-dimensional image is constructed. Because the return signals can be very small, it is important that the maximum energy be sent into the body and maximum energy obtained from the return signals. The following example illustrates the advantage of matching acoustic (i.e., mechanical) impedances in ultrasound imaging.

EXAMPLE 14.17

An ultrasound transducer that uses a barium titinate piezoelectric device is applied to the skin. The transducer is round, with a diameter of 2.5 cm. Use an acoustic impedance of 24.58×10^6 kg-s/m^2 for barium titinate and an acoustic impedance of 1.63×10^6 kg-s/m^2 for the skin. What is the maximum percent power that can be transferred into the skin with and without impedance matching?

Solution: The interface between skin and transducer can be represented as two series mechanical impedances as shown in Figure 14.43. The idea is to transfer maximum power ($F\ v$) to Z_2. We know from the maximum power transfer theory that maximum power will be transferred to Z_2 when

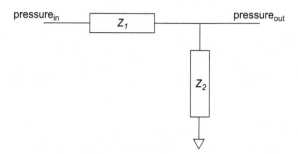

FIGURE 14.43 A model of the piezoelectric–skin interface in terms of acoustic (i.e., mechanical) impedance. Z_1 is the impedance of the transducer and Z_2 the impedance of the skin. In Example 14.16 the power transferred to Z_2 is determined when the two impedances are matched and when they are unmatched.

$Z_1 = Z_2$ (Equation 14.23). The problem also requires us to find how much power is transferred in both the matched and unmatched conditions.

The transducer is applied directly to the skin, so the area of the transducer and the area of the skin are the same and we can substitute pressures for forces since the areas cancel. In an electrical circuit, the two impedances act as a voltage divider, and in a mechanical circuit they act as a force (or pressure) divider:

$$\frac{F_{Z_2}}{F_{in}} = \frac{p_{Z_2}a}{p_{in}a} = \frac{p_{Z_2}}{p_{in}} = \frac{Z_2}{Z_1 + Z_2}$$

where F is forces and p is pressures. The power into the skin is given by $P_{Z2} = p_{Z2}\, v_{Z2}$ where p is the pressure across Z_2 and v is the velocity of Z_2.

$$\frac{P_{Z_2}}{P_{in}} = \frac{p_{Z_2}v}{p_{in}v} = \frac{p_{Z_2}}{p_{in}} = \frac{Z_2}{Z_1 + Z_2} \tag{14.24}$$

Note that the impedances are given in MKS units, not the cgs units used throughout this text. However, since they are used in a ratio, the units cancel so there is no need to convert them to cgs in this problem.

Result: When the impedances are matched, $Z_1 = Z_2$, then half the power is transferred to Z_2 as given by the maximum power transfer theorem, so the percent transferred is 50%. Under the unmatched conditions, the power ratio can be calculated as:

$$\frac{P_{Z_2}}{P_{in}} = \frac{Z_2}{Z_1 + Z_2} = \frac{1.63}{24.58 + 1.63} = 0.062$$

So the power ratio in the unmatched case is 6.2%, as opposed to 50% in the matched case. This approximately eightfold gain in power transferred into the body shows the importance of matching impedances to improve power transfer. Since we have no control over the impedance of the skin, we must adjust the impedance of the ultrasound transducer. In fact, special coatings are applied to the active side of the transducer to match tissue impedance. In addition, a gel having the same impedance is used to improve coupling and ensure impedance matching between the transducer and skin.

14.8 MULTIPLE SOURCES—REVISITED

In Chapter 6 we apply multiple sinusoids at different frequencies to a single input to find the frequency characteristics of a system. Superposition allows us to compute the transfer function for a range of frequencies. It assures us that if these multiple frequencies are applied to the system, the response is the summation of responses to individual frequencies. But what if the sources have both different frequencies and different locations?[4]

Even if the sources have different locations and different frequencies, superposition still can be used to analyze the network. We can solve the problem for each source separately knowing that the total solution will be the algebraic summation of all the partial solutions. We simply turn off all sources but one, solve the problem using standard techniques, and repeat until the problem has been solved for all sources. Then we add all the partial solutions for a final solution.

As stated previously, turning off a source does not mean removing it from the system: it is replaced by its equivalent impedance. Hence, voltage sources are replaced by short circuits and current sources by open circuits (so current sources are actually removed from the circuit). Similarly, force sources become straight connections and velocity sources essentially disappear. The following example uses superposition in conjunction with source equivalent impedance to solve a circuit problem with two sources having different frequencies.

EXAMPLE 14.18

Find the voltage across the 30-Ω resistor in the circuit of Figure 14.44.

Solution: First turn off the right-hand source by replacing it with a short circuit (its internal resistance), solve for the currents through the 30-Ω resistor, and then solve for the voltage across it. Then turn off the left-hand source and repeat the process. Note that the impedances of the inductor and capacitor will be different since the frequency is different. Add the two partial solutions to get the final voltage across the resistor.

Turning off the right-hand source and converting to the phasor domain gives the circuit in Figure 14.45.

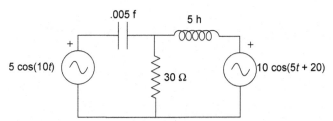

FIGURE 14.44 A two-mesh circuit that has both sources in different locations; each source has a different frequency. This type of problem can be solved using superposition.

---

[4]If the sources are at the same frequency but different locations, we do not have a problem, as the analysis techniques we have already developed can be used.

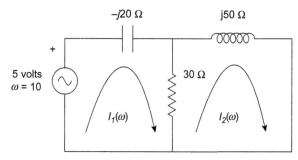

FIGURE 14.45 The circuit of Figure 14.44 with the right-hand source turned off. Since this source is a voltage source, turning it off means replacing it with a short circuit, the equivalent impedance of an ideal voltage source.

Applying KVL leads to a solution for the voltage across the center resistor:

$$\begin{vmatrix} 5 \\ 0 \end{vmatrix} = \begin{vmatrix} 30 - j20 & -30 \\ -30 & 30 + j50 \end{vmatrix} \begin{vmatrix} I_1(\omega) \\ I_2(\omega) \end{vmatrix}$$

Solving (MATLAB was used):

$$I_1(\omega = 10) = .217\angle 17; \quad I_2(\omega = 10) = .112\angle -42;$$
$$V_R(\omega = 10) = (I_1(\omega) - I_2(\omega))R = 5.57\angle 48 \text{ V}$$

Now, turning off the left-hand source and converting to the phasor domain leads to the circuit in Figure 14.46. Note that the impedances are different since the new source has a different frequency. Again apply the standard analysis.

$$\begin{vmatrix} 0 \\ -10\angle 20 \end{vmatrix} = \begin{vmatrix} 30 - j40 & -30 \\ -30 & 30 + j25 \end{vmatrix} \begin{vmatrix} I_1(\omega) \\ I_2(\omega) \end{vmatrix}$$

Solving:

$$I_1(\omega = 5) = 0.274\angle -156; \quad I_2(\omega = 5) = 0.456\angle 151;$$
$$V_R(\omega = 5) = (I_1(\omega) - I_2(\omega))R = 10.94\angle -65 \text{ V}$$

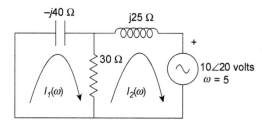

FIGURE 14.46 The circuit of Figure 14.44 with the left-hand source turned off.

The total solution is just the sum of the two partial solutions:

$$v_R(t) = 5.57 \cos(10t + 48) + 10.94 \cos(5t - 65) \text{ V}$$

This approach extends directly to any number of sources. It applies equally well to current sources as shown in Problem 20.

14.9 SUMMARY

Even very complicated circuits can be reduced using the rules of network reduction. These rules allow networks containing one or more sources (at the same frequency) and any number of passive elements to be reduced to a single source and a single impedance. This single source–impedance combination could be either a voltage source in series with the impedance, called a Thévenin source, or a current source in parallel with the impedance, a Norton source. Conversion between the two representations is straightforward.

One of the major applications of network reduction is to evaluate the performance of the system when circuits are combined together. The transfer function of each isolated network is determined based on the assumption that the circuit is driven by an ideal source and connected to an ideal load. This can be taken as true if the impedance of the source driving the network is much less than the network's input impedance and the impedance of the load is much greater than the network's output impedance.[5] Network reduction techniques provide a method for determining these input and output impedances.

The ratio of input to output impedance is particularly important when making physiological measurements. Often the goal is to make the input impedance of the measurement device as high as possible to minimally load the process being measured, that is, to draw minimum energy from the process. Sometimes it is desirable to transfer a maximum amount of energy between the process being measured and the measurement system. This is often true if the measurement device must also inject energy into the process to make its measurement. In such situations, an impedance matching strategy is used where the input impedance of the measuring device is adjusted to equal the output impedance of the process being measured.

All of the network reduction tools apply equally to mechanical systems. Indeed, one of the major applications of impedance matching in biomedical engineering is in ultrasound imaging, where the ultrasound transducer acoustic impedance must be matched with the acoustic impedance of the tissue.

This chapter concludes with the analysis of networks containing multiple sources at different frequencies. To solve these problems, the effect of each source on the network must be determined separately and the solutions for each source added together. Superposition is an underlying assumption of this summation-of-partial-solutions approach. When solving for the influence of each source on the network, the other sources are turned off by replacing them with their equivalent impedances: voltage sources are replaced by short circuits and current sources are replaced by open circuits.

---

[5]This assumes that the signals are carried as changes in voltage as is usually the case.

PROBLEMS

1. Find the combined values of the following elements.

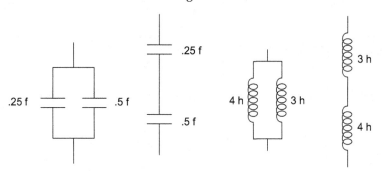

2. Find the value of R so the resistor combination equals 10 Ω. (Hint: use Equation 14.9.)

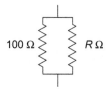

3. A two-terminal element has an impedance of 100 $\angle$ –30 Ω at $\omega = 2$ rad/s. The element consists of two components in series. What are they: two resistors, two capacitors, a resistor and capacitor, a resistor and inductor, or two inductors? What are their values?

4. An impedance, Z, has a value of 60 $\angle$ 25 Ω at $\omega = 10$ rad/s. What type of element should be added in series to make the combination look like a pure resistance at this frequency? What is the value of the added element at $\omega = 10$ rad/s? What is the value of the combined element at $\omega = 10$ rad/s?

5. Use network reduction to find the equivalent impedance of the following network between terminals A and B.

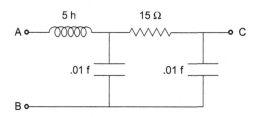

6. Use network reduction to find the equivalent impedance of the network in Problem 5 between the terminals A and C.

7. Find the equivalent impedance of the following network between terminals A and B.

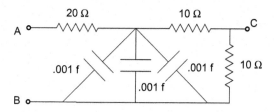

8. Plot the magnitude and phase of the impedance, Z_{eq}, in Example 14.3 over a range of frequencies from $\omega = 0.01$ to 1000 rad/s. Plot both magnitude and phase in log frequency and plot the phase in degrees. Use Bode plot techniques, not MATLAB.

9. The following $v-i$ characteristics were measured on a two-terminal device. Model the device by a Thévenin circuit and a Norton circuit.

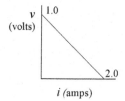

10. Plot the $v-i$ characteristics of this network at two different frequencies $\omega = 2$ and $\omega = 200$ rad/s. Note: the $v-i$ plot is a plot of the voltage magnitude against current magnitude. (Hint: find $|V_T|$ at the two frequencies.)

11. Two different resistors were placed across a two-terminal source (a battery) known to contain an ideal voltage source in series with a resistor. Applying two different load resistors to the terminals resulted in two different voltages: when $R = 1000\ \Omega$, $V = 8.5$ V, and when $R = 100\ \Omega$, $V = 8.2$ V. Find the load resistor, R_L, that extracts the maximum power from this source. What is the power dissipated in the load resistor?

12. The following magnitude v–i plot was found for a two-terminal device at the two frequencies shown. Model the device as a Thévenin equivalent. (Hint: to find R_T, use Equation 14.11 generalized for impedances.)

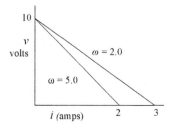

13. Find the Thévenin equivalent circuit of the following network. (Hint: modify the code in Example 14.13.)

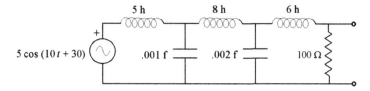

14. For the circuit in Problem 13, further modify the code in Example 14.13 to determine and plot the Norton current source and Norton impedance as a function of frequency. Plot both magnitude and phase of both variables for a range of frequency between 0.1 and 100 Hz.

15. The voltage of the following nonideal voltage source is measured with two voltmeters. One has an internal resistance of 10 MΩ, whereas the other, a cheapie, has an internal resistance of only 100 kΩ. What voltages will be read by the two voltmeters? How do they compare with the true Thévenin voltage? (Assume the voltmeters read peak-to-peak voltage, although voltmeters usually read the root mean square (RMS) voltage.) Note: The components in the Thévenin source have values commonly encountered in real circuits.

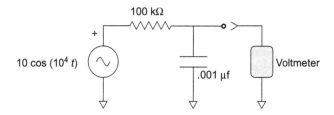

16. For the following network, what should the value of Z_L be to extract maximum power from the network? (Hint: use the same approach as in Example 14.13.)
    If the voltage source is increased to 15 cos (20t), what should the value of Z_L be to extract maximum power from the network?

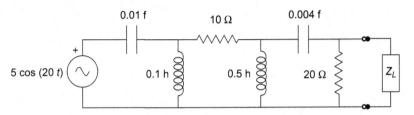

17. In the network shown in Problem 16, what is the impedance between nodes A and B assuming Z_L is not attached? Use the approach shown in Example 14.8. Apply a hypothetical 1-V source between these two points and use MATLAB-aided mesh analysis to find the current. Also you need to remove the influence of the 5-V source by turning it off, that is, replacing it by a short circuit. (Note that this can be reduced to a three-mesh problem, but even as a four-mesh problem it requires only a few lines of MATLAB code to solve.)

18. For the mechanical system shown in Example 14.15, find the difference between the measured force, $F_{measured}$, and F_S, the ideal source, if the friction in the source, k_f, is increased from 20 dyn-s/cm to 60 dyn cm/s. Find the difference between $F_{measured}$ and F_S with the increased friction in the source if the elasticity of the transducer is also increased by a factor of 3.

19. For the mechanical system shown in Example 14.16, find the difference between the measured velocity, v_2, and V_S if the elastic coefficient of the transducer, k_{e2}, is doubled, quadrupled, or halved.

20. For the mechanical system shown in Example 14.16, find the difference between the measured velocity, v_2, and V_S if the velocity transducer contains a mass of 4 g.

21. Find the voltage across the 0.01-f capacitor in the following circuit. Note that the two sources are at different frequencies.

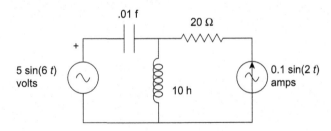

Basic Analog Electronics: Operational Amplifiers

15.1 GOALS OF THIS CHAPTER

Electronic circuits come in two basic varieties: analog and digital. Digital circuits feature electronic components that produce only two voltage levels, one high and the other low. This limits the signals they can handle to single bits that hold a 0 or 1. To transmit information, single bits are combined into groups; commonly an 8-bit binary number is called a "byte." Bytes can be combined to make larger binary numbers or can be used to encode alphanumeric characters, most often in a coding scheme known as ASCII code. Digital circuits form the basis of all modern computers and microprocessors. Although nearly all the medical instruments contain one or more small computers (i.e., microprocessors) along with related digital circuitry, bioengineers are not usually concerned with their design. They may be called upon to develop some, or all, of the software but not the actual electronics, except perhaps for interface circuits.

Analog circuit elements support a continuous range of voltages, and the information they carry is usually encoded as a time-varying continuous signal similar to those used throughout this text. All of the circuits described thus far have been analog circuits. Analog circuitry is a necessary part of most medical instrumentation because the biomedical sensors (biotransducers) usually produce analog electric signals. This includes devices that measure movement, pressure, bioelectric activity, sound and ultrasound, light, and other forms of electromagnetic energy. Bioelectric signals such as the EEG, ECG, and EMG are also analog signals.

Before analog signals are processed by a digital computer, some type of manipulation is usually required in the analog domain. This analog signal processing may not only consist of increasing the amplitude of the signal, but can also include filtering and other basic signal processing operations. Unlike digital circuitry, the design of analog circuits is often the responsibility of the biomedical engineer. After analog signal processing, the signal is usually converted to a digital signal using an analog-to-digital converter (ADC). The components of a typical biomedical instrument are summarized in Figure 15.1.

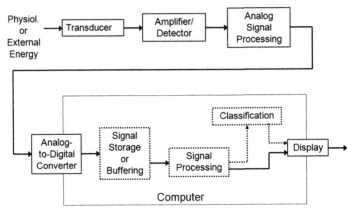

FIGURE 15.1   Basic elements of a typical biomedical instrument.

The goal of this chapter is to give you the basic tools to design commonly used analog circuits. In particular, you should understand when these circuits can be casually designed using an example circuit with standard off-the-shelf components or when special care must be taken to account for limitations in these real-world components.

Specific goals of this chapter include the following:

- Describing the concepts of an ideal amplifier
- Introducing the ideal op amp as a special case of an ideal amplifier
- Deriving the transfer function of inverting and noninverting amplifiers based on practical components known as operational amplifiers (op amps)
- Describing the limitations of practical (real-world) op amp components and determining when these become important in circuit design
- Describing the use of power supplies and potential noise sources
- Presenting and describing the most commonly used op amp circuits

15.2 THE AMPLIFIER

Increasing the amplitude or gain of an analog signal is termed "amplification" and is achieved using an electronic device known as an "amplifier." The properties of an amplifier are commonly introduced using a simplification called the "ideal amplifier." In this pedagogical scenario, the properties of a real amplifier are described as deviations from the idealization. In many practical situations, real amplifiers closely approximate the idealization; the limitations of real amplifiers and their deviations from the ideal become important only in more challenging applications. Nevertheless, the biomedical engineer involved in circuit design must know these limitations to understand when a typical amplifier circuit is being challenged.

An ideal amplifier is characterized by three properties:

1. It has a well-defined gain at all frequencies (or at least over a specific range of frequencies),

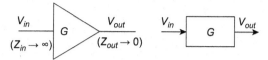

FIGURE 15.2 Schematic (left) and block diagram (right) of an ideal amplifier with a gain of G.

2. Its output is an ideal source ($Z_{out} = 0.0\ \Omega$), and
3. Its input is an ideal load ($Z_{in} \rightarrow \infty\ \Omega$).

An ideal amplifier is simply a pure gain term having ideal input (property 3) and output (property 2) characteristics. These properties make it the real-world embodiment of the systems gain element introduced in Section 6.5.1. The schematic and system representation of an ideal amplifier are shown in Figure 15.2.

The transfer function of this amplifier would be:

$$\frac{V_{out}(s)}{V_{in}(s)} = \frac{V_{out}(\omega)}{V_{in}(\omega)} = G \tag{15.1}$$

where G would usually be a constant, or a function of frequency.

Many amplifiers have a differential input configuration, that is, there are two separate inputs and the output is the gain constant times the difference between the two inputs. Stated mathematically:

$$V_{out} = G(V_{in2} - V_{in1}) \tag{15.2}$$

The schematic for such a "differential amplifier" is shown in Figure 15.3. Note that one of the inputs is labeled $+$ and the other $-$, to indicate how the difference is taken: the minus terminal subtracts its voltage from the plus terminal. (It is common to draw the negative input above the positive input.) The $+$ the terminal is also known as the "noninverting input," whereas the $-$ terminal is also referred to as the "inverting input."

Some transducers, and a few amplifiers, produce a differential signal—actually two signals that move in opposite directions with respect to the information they represent. For such differential signals, a differential amplifier is ideal, since it takes advantage of both input signals. Moreover, the subtraction tends to cancel any signal that is common to both inputs and undesirable noise is often common to both inputs. However, most of the time only a single signal is available. In these cases, a differential amplifier may still be used, but one of the

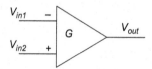

FIGURE 15.3 An amplifier with a "differential input" configuration. The output of this amplifier is G times $V_{in2} - V_{in1}$, i.e., $V_{out} = G(V_{in2} - V_{in1})$.

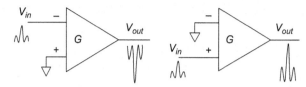

FIGURE 15.4   A differential amplifier configured as an inverting (left) and noninverting (right) amplifier.

inputs is set to zero by grounding. If the positive input is grounded and the signal is sent into the negative input, Figure 15.4 (left side), then the output will be the inverse of the input:

$$V_{out} = G(-V_{in}) \tag{15.3}$$

In this case the amplifier may be called an "inverting amplifier" for obvious reasons. If the opposite strategy is used and the signal is sent to the positive input while the negative input is grounded as in Figure 15.4 (right side), the output will have the same direction as the input. This amplifier is termed a "noninverting amplifier" (sort of a double negative).

15.3 THE OPERATIONAL AMPLIFIER

The "operational amplifier," or "op amp," is a basic building block for a wide variety of analog circuits. One of its first uses was to perform mathematical operations, such as addition and integration, in analog computers, hence the name operational amplifier. Although the functions provided by analog computers are now performed by digital computers, the op amp remains a valuable, perhaps the most valuable, tool in analog circuit design.

In its idealized form, the op amp has the same properties as the ideal amplifier described previously except for one curious departure: it has infinite gain. Thus an ideal op amp has infinite input impedance (ideal load), zero output impedance (an ideal source), and a gain, $A_v \rightarrow \infty$. (The symbols A_v and A_{VOL} are commonly used to represent the gain of an op amp.) Obviously an amplifier with a gain of infinity is of limited value, so an op amp is rarely used alone, but usually in conjunction with other elements that reduce its gain to a finite level.

Negative feedback can be used to limit the gain. Assume that the ideal amplifier is represented as A_V in the feedback system shown in Figure 15.5.

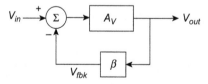

FIGURE 15.5   A basic feedback control system used to illustrate the use of feedback to set a finite gain in a system that has infinite feedforward gain, A_V.

The gain of the system can be found from the basic feedback equation introduced and derived in Example 6.1. When we insert A_V and β into the feedback equation, Equation 6.7, the overall system gain, G, becomes:

$$G = \frac{A_V}{1 + A_V \beta} \tag{15.4}$$

Now if we let the amplifier's gain, A_V, go to infinity:

$$G = \lim_{A_V \to \infty} \left| \frac{A_V}{1 + A_V \beta} \right| = \lim_{A_V \to \infty} \left| \frac{A_V}{A_V \beta} \right| = \frac{1}{\beta} \tag{15.5}$$

The overall gain expressed in decibel becomes:

$$G_{db} = 20 \log G = 20 \log \left(\frac{1}{\beta} \right) = -20 \log \beta \tag{15.6}$$

If $\beta < 1$, then the gain of the feedback system $G = V_{out}/V_{in}$ is >1 and the system increases the signal amplitude. If $\beta = 1$, then $G = 1$ and the amplitude of the output signal is the same as the input signal. This may seem useless, but there are times when such a system is used because we still get the benefits of the ideal input and output properties of the amplifier. We rarely make $\beta > 1$ because then $G < 1$, and the feedback system actually reduces the signal amplitude. If a reduction in gain is desired, it is easier to use a passive voltage divider, a series resistor pair with one resistor to ground. So in real op amp circuits, the feedback gain, β, is ≤ 1 and $G \geq 1$. This is fortunate, as all we need to produce a feedback gain <1 is a voltage divider network: one end of the two connected resistors goes to the output and the other end to ground, and the reduced feedback signal is taken from the intersection of the two resistors, Figure 15.6. A feedback gain of $\beta = 1$ is even easier: just feed the entire output back to the input.

The approach of beginning with an amplifier that has infinite gain then reducing that gain to a finite level with the addition of feedback seems needlessly convoluted. Why not design the amplifier to have a finite fixed gain to begin with? The answer is summarized in two

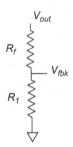

FIGURE 15.6 A voltage divider network that can be used to feed back a portion of the output signal as a negative feedback signal. To make the feedback signal, V_{fbk}, negative, it is fed to the inverting or negative input of an operational amplifier.

words: flexibility and stability. If feedback is used to set the gain of an op amp circuit, then only one basic amplifier needs to be produced: one with an infinite (or just very high) gain. Any desired gain can be achieved by modifying a simple two-resistor network. More importantly, the feedback network is almost always implemented using passive components: resistors and sometimes capacitors. Passive components are always more stable than transistor-based devices, that is, they are more immune to fluctuations owing to changes in temperature, humidity, age, and other environmental factors. Passive elements can also be more easily manufactured to tighter tolerances than active elements. For example, it is easy to buy resistors that have a 1% error in their values, whereas most common transistors vary in gain by a factor of two or more. Finally, back in the flexibility category, a wide variety of different feedback configurations can be used enabling one type of op amp to perform many different signal processing operations. Some of these different functions are explored in the section on op amp circuits at the end of this chapter.

15.4 THE NONINVERTING AMPLIFIER

In the noninverting amplifier, a two-resistor voltage divider feeds a reduced version of the output back to the inverting (i.e., negative) input of an op amp. Consider the voltage divider network in Figure 15.6. The feedback voltage, V_{fbk}, can be found by the simple application of Kirchhoff's voltage law (KVL). Assuming that V_{out} is an ideal source (which it is because it is the output of an ideal amplifier):

$$V_{out} - i(R_f + R_1) = 0; \quad i = \frac{V_{out}}{R_f + R_1}:$$

$$V_{fbk} = iR_1 = V_{out}\left(\frac{R_1}{R_1 + R_f}\right) \tag{15.7}$$

For the system diagram in Figure 15.5, we see that $\beta = V_{fbk}/V_{out}$. Using Equation 15.7 we can solve for β in terms of the voltage divider network:

$$\beta = \frac{V_{fbk}}{V_{out}} = \frac{R_1}{R_1 + R_f} \tag{15.8}$$

The transfer function of an op amp circuit that uses feedback to set the gain is just $1/\beta$ as given in Equation 15.5:

$$G \equiv \frac{V_{out}}{V_{in}} = \frac{1}{\beta} = \frac{1}{\dfrac{R_1}{R_1 + R_f}} = \frac{R_1 + R_f}{R_1} \tag{15.9}$$

An op amp circuit using this feedback network is shown in Figure 15.7. The gain of this amplifier is given by Equation 15.9. Since the input signal is fed to the positive side of the op amp, this circuit is termed a "noninverting amplifier."

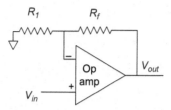

FIGURE 15.7   Noninverting operational amplifier (op amp) circuit. The gain of this op amp in terms of the feedback resistors is given by Equation 15.9.

The transfer function for the circuit in Figure 15.7 can also be found through circuit analysis, but a couple of helpful rules are needed.

1. Since the input resistance of the op amp approaches infinity, there is no current flowing into, or out of, either of the op amp's input terminals. Realistically, the input resistance of an op amp is so high that any current that does flow into the op amp's input is negligible.
2. Since the gain of the op amp approaches infinity, the only way the output can be finite is if the input is zero, that is, the difference between the plus input and the minus input must be zero. Stated yet another way, the voltage on the plus input terminal must be the same as the voltage on the minus input terminal of the op amp and vice versa. Realistically, the very high gain means that for reasonable output voltages, the voltage difference between the two input terminals is very small and can be ignored.

In practical op amps, the gain is large (up to 10^6) but not infinite, so the voltage difference in a practical op amp circuit might be a few millivolts. This small difference can generally be ignored. Similarly, the input resistance, although not infinite, is quite large: values of r_{in} (resistances internal to the op amp are denoted in lower case) are usually greater than 10^{12} Ω, so any input current will be very small and can be disregarded (especially since the input voltage must be zero or at least very small by Rule 2). We use these two rules to solve the transfer function of a noninverting amplifier in the following example.

EXAMPLE 15.1

Find the transfer function of the noninverting op amp circuit in Figure 15.7 using network analysis.

Solution: First, by Rule 2, the voltage between the two resistors, the voltage at the negative terminal, must equal V_{in} since V_{in} is applied to the lower terminal and the voltage difference between the upper and lower terminals is zero.

Define the three currents in and out of the node between the two resistors and apply Kirchhoff's current law (KCL) to that node as shown in Figure 15.8. Applying KCL:

By KCL: $-i_1 - i_{in} + i_f = 0$, but $i_{in} = 0$ according to Rule 1.

Hence if: $-i_1 - 0 + i_f = 0$, then $i_1 = i_f$

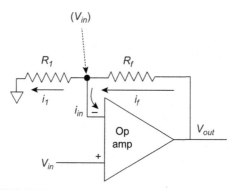

FIGURE 15.8 A noninverting operational amplifier circuit.

Applying Ohm's law to substitute the voltages for the currents and noting that the voltage at the node must equal V_{in} by Rule 2:

$$i_1 = \frac{V_{in}}{R_1} \text{ and } i_f = \frac{V_{out} - V_{in}}{R_f}$$

$$\text{Then since } i_1 = i_f\colon \frac{V_{in}}{R_1} = \frac{V_{out} - V_{in}}{R_f}$$

Solving for V_{out}:

$$V_{out} - V_{in} = V_{in}\left(\frac{R_f}{R_1}\right); \quad V_{out} = V_{in}\left(1 + \frac{R_f}{R_1}\right) = V_{in}\left(\frac{R_f + R_1}{R_1}\right)$$

and the transfer function becomes:

$$\frac{V_{out}}{V_{in}} = \frac{R_f + R_1}{R_1}$$

This is the same transfer function found using the feedback equation, Equation 15.9.

15.5 THE INVERTING AMPLIFIER

To construct an amplifier circuit that inverts the input signal, the ground and signal inputs of the noninverting amplifier are reversed as shown in Figure 15.9.

The transfer function of the inverting amplifier circuit is a little different from that of the noninverting amplifier, but can easily be found using the same approach (and tricks) used in Example 15.1.

EXAMPLE 15.2

Find the transfer function, or gain, of the inverting amplifier circuit shown in Figure 15.9.

Solution: Define the currents and apply KCL to the node between the two resistors, Figure 15.10. In this circuit the node between the two resistors must be at 0 volts by Rule 2, i.e., because the plus

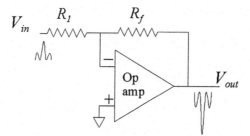

FIGURE 15.9 The operational amplifier circuit used to construct an inverting amplifier.

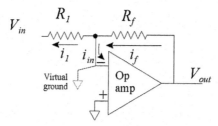

FIGURE 15.10 Inverting operational amplifier configuration showing currents assigned to the upper node. The upper node in this configuration must be at 0 volts by Rule 2 and is referred to as virtual ground.

side is grounded and the difference between the two terminals is 0, the minus side is effectively grounded. In fact, the inverting input terminal in this configuration is sometimes referred to as a "virtual ground," Figure 15.10.

As in Example 15.1, we apply KCL to the inverting terminal and find that $i_1 = i_f$ and by Ohm's law:

$$i_1 = \frac{0 - V_{in}}{R_1}; \quad i_f = \frac{V_{out} - 0}{R_f}; \quad \frac{-V_{in}}{R_1} = \frac{V_{out}}{R_f}$$

Again solving for V_{out}/V_{in}:

$$\frac{V_{out}}{V_{in}} = -\frac{R_f}{R_1} \tag{15.10}$$

The negative sign demonstrates that this is an inverting amplifier: the output is the negative, or inverse, of the input. The output is also larger than the input by a factor of R_f/R_1.

Equations 15.9 and 15.10 show that the gain of inverting and noninverting op amp circuits depend only on R_1 and R_f. The circuit designer can control the gain of these circuits simply by adjusting their values. If one of the resistors is variable (i.e., a potentiometer), then the amplifier would have a variable gain. The next example involves the design of a variable gain inverting amplifier.

EXAMPLE 15.3

Design an inverting amplifier circuit with a variable gain between 10 and 100. Assume you have a variable 1-MΩ potentiometer, that is, a resistor that can be varied between 0 and 1 MΩ. Also assume you have available a wide range of fixed resistors.

Solution: The amplifier circuit will have the general configuration of Figure 15.9. It is possible to put the variable resistance as part of either R_{in} or R_f, but let us assume that the potentiometer is part of R_f along with a fixed series resistance. This results in the circuit shown in Figure 15.11.

We vary feedback resistance, R_f, to get the desired gain variation. Assume that the variable resistor will be 0 Ω when the gain is 10 and 1 MΩ when the gain is 100. We can write two gain equations based on Equation 15.10 then solve for our two unknowns, R_1 and R_2.

For $G = 10$ the variable resistor is 0 Ω: $\frac{R_f}{R_1} = 10$; $\quad \frac{R_2 + 0}{R_1} = 10$; $\quad R_2 = 10R_1$

For $G = 100$ the variable resistor is 10^6 Ω:

$$\frac{R_2 + 10^6}{R_1} = 100; \quad \text{Substituting for } R_2; \quad 10R_1 + 10^6 = 100R_1$$

$$90R_1 = 10^6; \quad R_1 = 11.1 \text{ k}\Omega \quad R_2 = 111 \text{ k}\Omega \text{ and}$$

The final circuit is shown in Figure 15.12.

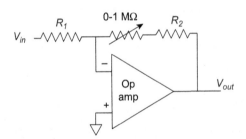

FIGURE 15.11 An inverting operational amplifier circuit with a variable resistor in the feedback path to allow for variable gain. In Example 15.3, the variable and fixed resistors are adjusted to provide an amplifier gain from 10 to 100.

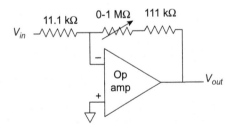

FIGURE 15.12 An inverting amplifier circuit with a gain between 10 and 100. In a real application the two fixed resistors would be rounded to 10 and 100 kΩ unless extreme precision was required.

The equation for the gain of noninverting and inverting op amp circuits can be extended to include feedback networks that contain capacitors and inductors. To modify these equations to include components other than resistors, simply substitute impedances for resistors. So the gain equation for a noninverting op amp circuit becomes:

$$\frac{V_{out}}{V_{in}} = \frac{Z_f + Z_1}{R_1} \qquad (15.11)$$

and the equation for an inverting op amp circuit becomes:

$$\frac{V_{out}}{V_{in}} = -\frac{Z_f}{Z_1} \qquad (15.12)$$

15.6 PRACTICAL OP AMPS

Practical op amps, the kind that you buy from electronics supply houses, differ in a number of ways from the idealizations used earlier. In many applications, perhaps most applications, the limitations inherent in real devices can be ignored. The problem is that any bioengineer who designs analog circuitry must know when the limitations are important and when they are not, and to do this he or she must understand the characteristics of real devices. Only the topics that involve the type of circuits the bioengineer is likely to encounter are covered here. Several excellent references can be found to extend these concepts to other circuits (Horowitz and Hill, 1989).

Deviations of real op amps from the ideal can be classified into three general areas: deviations in input characteristics, deviations in output characteristics, and deviations in transfer characteristics. Each of these areas is discussed in turn, beginning with the area likely to be of utmost concern to biomedical engineers: transfer characteristics.

15.6.1 Limitations in Transfer Characteristics of Real Operational Amplifiers

The most important limitations in the transfer characteristics of real op amps are bandwidth limitations and stability. In addition, real op amps have a large, but not infinite, gain. Bandwidth limitations occur because an op amp's magnitude gain is not only finite, but decreases with increasing frequency. The lack of stability results in unwanted oscillations and is due to the op amp's increased phase shifts with increasing frequency. Eventually these phase shifts can become so large that negative feedback turns into positive feedback and the circuit oscillates.

15.6.1.1 *Bandwidth*

The magnitude frequency characteristics of a popular op amp, the LF 356, are shown in Figure 15.13. Not surprisingly, even at low frequencies the gain of this op amp is less than infinity. This in itself would not be a cause for much concern as the gain is still quite high: approximately 106 dB or 199,530. The problem is that this gain is also a function of frequency

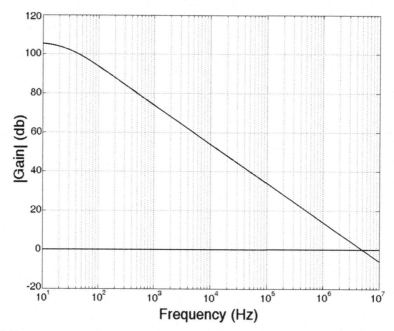

FIGURE 15.13   The open-loop magnitude gain characteristics of a popular operational amplifier, the LF 356.

so that at higher frequencies the gain becomes quite small. In fact, there is a frequency above which the gain is less than one. Thus at higher frequencies the transfer function equations no longer hold since they were based on the assumption that op amp gain is infinite. Since the bandwidth of a real op amp is limited, the bandwidth of an amplifier using such an op amp must also be limited. Essentially, the gain of an op amp circuit is limited by either the bandwidth limitations of the op amp or the feedback, whichever is lower.

An easy technique for determining the bandwidth of an op amp circuit is to plot the gain produced by the feedback circuit superimposed on the bandwidth curve of the real op amp. The former is referred to as the "closed-loop gain" since it includes the feedback, whereas the latter is termed the "open-loop gain" since it is the gain of the op amp without a feedback loop. The gain produced by the feedback network is, theoretically, $1/\beta$, Equation 15.5. The real transfer function gain is either this value or the op amp's open-loop gain, whichever is lower. (The gain in an op amp circuit can never be greater than what the op amp is capable of producing.) So to get the real gain, we plot $1/\beta$ superimposed on the open-loop curve. The real gain is simply the lower of the two curves. If the feedback network consists only of resistors, β will be constant for all frequencies, so $1/\beta$ will plot as a straight line on the frequency curve. (Although real resistors have some small inductance and capacitance, the effect of these "parasitic elements" can be ignored except at very high frequencies.)

For example, assume that $1/100$ of the signal is fed back to the inverting terminal of a real op amp. Then the feedback gain is:

$$\beta = 0.01 = -40 \text{ dB} \quad \text{and} \quad 1/\beta = 100 = 40 \text{ dB}$$

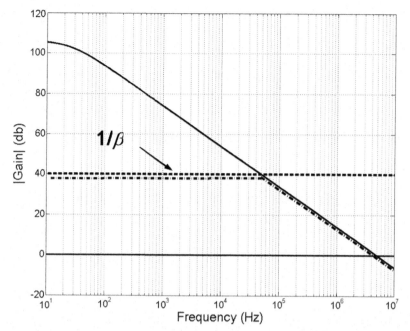

FIGURE 15.14 The open-loop magnitude spectrum of the LF 356 operational amplifier (*solid line*) with the frequency characteristics of $1/\beta$ (*dashed line* at 40 dB) superimposed. Since the feedback network contains only resistors, its spectrum is a constant at all frequencies. The gain of the amplifier circuit follows whichever curve is lower and is indicated by the heavy dashed line. The point occurs where the solid and dashed line intersect and indicates a bandwidth of approximately 50 kHz.

Figure 15.14 shows the open-loop gain characteristics of a typical op amp (LF 356) with the plot of $1/\beta$ superimposed (dashed line). The overall gain will follow the dashed line until it intersects the op amp's open-loop curve (solid line), where it will follow that curve (solid line) since this is less. The circuit's magnitude frequency characteristic will follow the heavy dash-dot lines seen in Figure 15.14. Given this particular op amp and this value of β, the bandwidth of the circuit, the intersection between the two curves, occurs at approximately 50 kHz. The intersection is taken as defining the bandwidth of the circuit, although the -3 dB point is at a slightly higher frequency. The feedback gain, β, is the same for both inverting and noninverting op amp circuits, so this approach for determining the amplifier bandwidth is the same in both configurations. The value $1/\beta$ is sometimes referred to as the "noise gain" because it is also the gain factor for input noise and errors, again irrespective of the specific configuration.

EXAMPLE 15.4

Find the bandwidth of the inverting amplifier circuit in Figure 15.15.

Solution: First determine the feedback gain, β, then plot the inverse ($1/\beta$) superimposed on the open-loop gain curve obtained from the op amp's specification sheets.

$$\frac{1}{\beta} = \frac{R_f + R_{in}}{R_{in}} = \frac{20 + 1}{1} \approx 20 = 26 \ \text{dB}$$

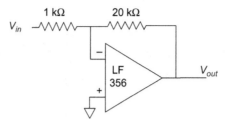

FIGURE 15.15 An inverting amplifier circuit. The bandwidth of this circuit is found in Example 15.4.

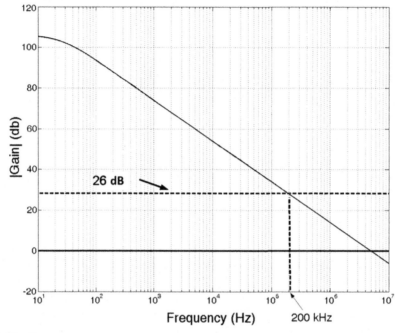

FIGURE 15.16 The open-loop gain curve of the LF356 with the $1/\beta$ line superimposed. This line intersects the operational amplifier's open-loop gain curve at around 200 kHz. This intersection defines the bandwidth of the amplifier circuit.

From the plot in Figure 15.16, showing the $1/\beta$ spectrum superimposed in the LF 356 open-loop spectrum, we see that the two curves intersect at approximately 200 kHz. Above 200 kHz the circuit gain drops off at 20 dB per decade. The intersection is used to define the circuit bandwidth, although the actual −3 dB point will be at a slightly higher frequency.

As shown in the open-loop gain curve of Figures 15.13, 15.14, and 15.16, a typical op amp not only has very high values of gain at low frequencies, but also a low cutoff frequency. Above this cutoff frequency, the gain curve rolls off at 20 dB/decade, the same as a first-order low-pass filter. Specifically, LF 356 has a low-frequency gain of around 106 dB but the open-loop bandwidth is only around 25 Hz. The addition of feedback dramatically lowers the overall gain, but just as dramatically increases the bandwidth. Indeed, the idea of using negative feedback in amplifier circuits was first introduced to improve bandwidth by trading reduced gain for increased bandwidth.

If the feedback gain is a multiple of 10, as is often the case, the bandwidth can be determined without actually plotting the two curves. The bandwidth can be determined directly from the amplifier gain and the frequency at which the open-loop curve intersects 0 dB. This frequency, where the op amp open-loop gain falls to 1.0 (i.e., 0 dB), is termed the "gain bandwidth product" (GBP). The GBP is given in the op amp specifications; for example, the GBP of the LF 356 is 5.0 MHz. To use the GBP to find the bandwidth of a feedback amplifier, start with the GBP frequency and knock off one decade in the bandwidth for every 20 dB in amplifier gain. Essentially you are sliding up the 20 dB/decade slope starting at 0 dB, the GBP. For example, if the amplifier has a gain of 1 (0 dB), the bandwidth equals the GBP, 5 MHz in the case of the LF356. For every 20 dB (or a factor of 10) above 0, the bandwidth is reduced by one decade. So an amplifier circuit with a gain of 10 (20 dB) would have a bandwidth of 500 kHz and an amplifier circuit with a gain of 100 (40 dB) would have a bandwidth of 50 kHz, assuming these circuits used the LF356. If the gain is 1000 (60 dB), the bandwidth would be reduced to 5 kHz. If the $1/\beta$ gain is not a power of 10, logarithmic interpolation would be required and it is easiest to use the plotting technique.

EXAMPLE 15.5

Using an LF 356, what is the maximum amplifier gain (i.e., closed-loop gain) that can be obtained with a bandwidth of 100 kHz.

Solution: From the open-loop curve given in Figure 15.13, the open-loop gain at 100 kHz is approximately 30 dB. This is the maximum close-loop gain that will reach the desired cutoff frequency. Designing the appropriate feedback network to attain this gain (and bandwidth) is straightforward using Equation 15.9 (noninverting) or Equation 15.10 (inverting).

15.6.1.2 Stability

Most op amp circuits use negative feedback: the feedback voltage, V_{fbk}, is fed to the inverting input of the op amp. Except in special situations, positive feedback is to be avoided. Positive feedback creates a vicious circle: the feedback signal enhances the feedforward signal, which enhances the feedback signal, which enhances the feedforward signal, and so on. A number of things can happen to a positive feedback network, most of them bad. The two most likely outcomes are that the output oscillates, sometimes between the maximum and minimum values the op amp can produce, or the output is driven into saturation and locked at the maximum or minimum level. When the word "stability" is used in context with an op amp circuit it means the absence of oscillation or other deleterious effects associated with positive feedback. The oscillation produced by positive feedback is a sustained repetitive waveform, which could be a sinusoid, but it may also be more complicated.

Positive feedback oscillation occurs in a feedback circuit where the overall gain or "loopgain" (i.e., the gain of the feedback and feedforward circuits) is greater than or equal to 1.0 and has a phase shift of 360 degrees:

$$\text{Loop gain(for oscillation)} \equiv \text{Feedforward Gain} \times \text{Feedback Gain} \; \exists \; 1.0\pi 360 \text{ degrees}$$

$$(15.13)$$

When this condition occurs, any small signal feeds back positively and grows to produce a sustained oscillation. Sometimes this amp oscillation will ride on top of the signal, sometimes it will overwhelm the signal, but in either case it is unacceptable.

Since the feedback signal is sent to the inverting input of the op amp, positive feedback should not occur.[1] But what if the op amp induces an additional phase shift of 180 degrees? Then the negative feedback becomes positive feedback since the total phase shift is 360 degrees. If the loop gain happens to be > 1 when this occurs, the conditions of Equation 15.13 are met and the circuit will oscillate. The base frequency of that oscillation will be equal to the frequency where the total phase shift becomes 360 degrees, that is, the frequency where the op amp contributes a phase shift of an additional 180 degrees. Hence, oscillation is a result of phase shifts induced by the op amp's phase characteristics at high frequencies.

A rigorous analysis of stability is beyond the scope of this text. However, since stability is so often a problem in op amp circuits, some discussion is warranted. Since the inverting input always contributes a180-degree phase shift (to make the feedback negative), to ensure stability we must make sure that everything else in the feedback loop contributes a phase shift that is less than 180 degrees, at least as long as the loop gain ≥ 1. If β is a constant, then any additional phase shift must come from the op amp, so all we need are op amps that never approach a phase shift of 180 degrees, at least for loop gains ≥ 1.0. Unfortunately, all op amps reach a phase shift of 180 degrees if we go high enough in frequency, so the only working strategy is to ensure that when the op amp reaches a phase shift of 180, the overall loop gain is < 1.0.

The overall loop gain is just $A_V(\omega) \beta(\omega)$. Putting the condition for stability in terms of the gain symbols we have used thus far, the condition for stability is:

$$\text{Loop gain}_{\text{stability}} = A_V(\omega)\beta(\omega) < 1.0 \angle 360 \text{ degrees.} \tag{15.14}$$

where A_V is the gain of the op amp at a specific frequency and β is the feedback gain. Alternatively, if we want to build an oscillator, the condition for oscillation would be as follows:

$$\text{Loop gain}_{\text{oscillation}} = A_V(\omega)\beta(\omega) \geq 1.0 \angle 360 \text{ degrees.} \tag{15.15}$$

With respect to β, Equation 15.14 shows that the worst case for stability occurs when β is large. In most op amp circuits β is < 1, but can sometimes be as large as 1. As mentioned, a feedback gain of $\beta = 1$ corresponds to the lowest op amp gain: a gain of 1 ($V_{out} = V_{in}$). Although it is somewhat counterintuitive, this means that stability is more likely to be a problem in low-gain amplifier circuits where β is large and most likely to be a problem when the gain is 1.0 since $\beta = 1$ in this case.

If the op amp gain, A_V, is less than 1.0 when its phase shift hits 180 degrees, and β is at most 1.0, then $Av\beta$ will be < 1, and the conditions for stability are met (Equation 15.14). Stated in terms of phase, the op amp's phase shift should be less than 180 degrees for all frequencies where its gain is > 1. In fact, most op amps have a maximum phase shift that is less than

---

[1] At any given frequency, a sinusoid shifted by 180 degrees is the inverse of an upshifted sinusoid: $\sin x = -\sin(x \pm 180)$. So the inverting input induces a phase shift of 180 degrees at all frequencies and the feedback signal is inverted, i.e., negative.

120 degrees to be on the safe side for gains ≥ 1.0. Such op amps are said to be "unity gain stable" because they will not oscillate even when $\beta = 1$ where the amplifier has unity gain. However, achieving this criterion requires some compromise on the part of the op amp manufacture, usually some form of phase compensation that reduces the GBP. In many op amp applications where the gain will be high and β correspondingly low, unity gain stability is overkill and the unity gain compensation results in a needless reduction of high frequency performance.

Op amp manufacturers have come up with two strategies to overcome the performance loss due to unity gain stability: (1) produce different versions of the same basic op amp, one that has a higher bandwidth but requires a minimum gain and another that is unity gain stable but with a lower bandwidth or (2) produce a single version but have the user supply the compensation (usually as an external capacitor) to suit the application requirements. The former has become more popular since it does not require additional circuit components. The LF 356 is an example of the former strategy. The LF 356 has a GBP of 5 MHz and is unity gain stable, whereas its "sister" chip, the LF 357, requires a $1/\beta$ gain of 5 or more, but has a GBP of 20 MHz.

Unfortunately, using an op amp with unity gain stability still does not guarantee a stable circuit. Stability problems can occur if the feedback network introduces a phase shift. A feedback network containing only resistors might be considered safe, but parasitic elements, small inductances, and capacitances can create an undesirable phase shift. Consider the feedback circuit in Figure 15.17 in which a capacitor is placed in parallel with one of the resistors. With a capacitor in the feedback network, β becomes a function of frequency and will introduce an additional phase shift in the loop. Will this additional capacitance present a problem with regard to stability?

To answer this question, we need to find the phase shift of the network at the frequency when the loop gain is 1.0. The loop gain will be 1.0 when:

$$A_V\beta = 1; \quad A_V = \frac{1}{\beta} \tag{15.16}$$

Hence the loop gain is 1 when $1/\beta$ equals the op amp gain A_V. On the plot of A_V and $1/\beta$ such as in Figure 15.16, this is when the two curves intersect. In the circuit presented in

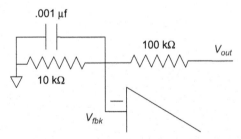

FIGURE 15.17 A feedback network that includes some capacitance either because of the intentional addition of a capacitor or because of parasitic elements. This feedback network will introduce an additional phase shift into the loop. This shift could be beneficial if it moves the overall loop phase shift away from 360 degrees or harmful if it moves it closer to 360 degrees.

Figure 15.17, the $1/\beta$ curve will not be a straight line, but can easily be found using the phasor circuit techniques.

For the voltage divider:

$$V_{fbk} = \frac{Z_1}{Z_1 + Z_f}V_{out}; \quad \text{Solving for } \beta: \quad \frac{V_{fbk}}{V_{out}} = \beta = \frac{Z_1}{Z_1 + Z_f}$$

$$\text{where}: \quad Z_1 = \frac{R_1\left(\dfrac{1}{j\omega C}\right)}{R_1 + \left(\dfrac{1}{j\omega C}\right)} = \frac{10^4\left(\dfrac{10^9}{j\omega}\right)}{10^4 + \left(\dfrac{10^9}{j\omega}\right)} = \frac{10^4}{1 + \dfrac{j\omega}{10^5}}$$

$$Z_f = 10^5 \Omega$$

Substituting in Z_1 and Z_f and solving for β:

$$\beta = \frac{\dfrac{10^4}{1 + \dfrac{j\omega}{10^5}}}{10^5 + \dfrac{10^4}{1 + \dfrac{j\omega}{10^5}}} = \frac{10^4}{10^5 + j\omega + 10^4} = \frac{.09}{1 + \dfrac{j\omega}{1.1 \times 10^5}} \tag{15.17}$$

This is just a low-pass filter with a cutoff frequency of 1.1×10^5 rad/s or 17.5 kHz. Figure 15.18 shows the magnitude frequency plot of β plotted from Equation 15.17 using standard Bode plot techniques.

Figure 15.19 shows the $1/\beta$ curve, the inverse of the curve in Figure 15.18, plotted superimposed on the open-loop gain curve of the LF 356. From Figure 15.19, the two curves intersect at around 85 kHz.

The phase shift contributed by the feedback network at 85 kHz can readily be determined from Equation 15.17. At 85 kHz, $\omega_1 = 2\pi (85 \times 10^3) = 5.34 \times 10^5$.

$$\beta = \frac{.09}{1 + \dfrac{j\omega}{1.1 \times 10^5}} = \frac{.09}{1 + \dfrac{j2\pi 85 \times 10^3}{1.1 \times 10^5}} = \frac{.09}{4.96 \angle 78} = 0.018 \angle -78$$

Thus the feedback network contributes a 78-degree phase shift to the overall loop gain. Although the phase shift of the op amp at 85 kHz is not known (detailed phase information is not often provided in op amp specifications), we can only be confident that the phase shift is no more than 120 degrees. Adding the worst case op amp phase shift to the feedback network phase shift $(120 + 78)$ results in a total phase shift that is >180 degrees, so coupled with the 180 degrees from the negative feedback this circuit is likely to oscillate. A good rule of thumb is that the circuit will be unstable if the $1/\beta$ curve breaks upward before intersecting the A_V line of the op amp. The reverse is also true: the circuit will be stable if the $1/\beta$ line intersects the A_V line at a point where it is flat or going downward.

If a feedback network can make the op amp circuit unstable, it stands to reason that it can also make it less unstable. This occurs when capacitance is added in parallel to the feedback

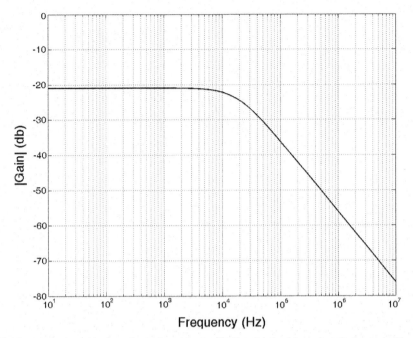

FIGURE 15.18 The magnitude frequency plot of the feedback gain, β, of the network shown in Figure 15.17. The equation for this curve (Equation 15.17) is found using standard phasor circuit analysis and plotted using Bode plot techniques.

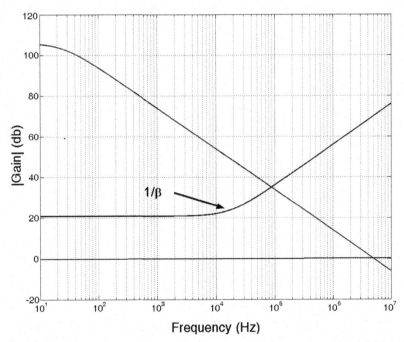

FIGURE 15.19 The inverse of the feedback gain given by Equation 15.17 (i.e., $1/\beta$) plotted superimposed against the A_V (i.e., open-loop) curve of the LF 356 operational amplifier.

resistor, the 100-kΩ resistor in Figure 15.17. In fact, the quick fix approach to oscillations in many op amp circuits is to add a capacitor to the feedback network in parallel with the feedback resistor. This usually works, although the capacitor might have to be fairly large. As shown in the section on filters, adding feedback capacitance reduces bandwidth: the larger the capacitance the greater the reduction in bandwidth. Sometimes a reduction in bandwidth is desired to reduce noise, but often it is disadvantageous. The influence of feedback capacitance on bandwidth is explored in the section on filters, whereas its influence on stability is demonstrated in a problem.

Stability problems are all too frequent and when they occur can be difficult and frustrating to correct. Possible problems include the feedback network, excessive capacitive load (which modifies the op amp's phase shift characteristics), use of an inappropriate op amp, and, most commonly, inadequate decoupling capacitors. The use of decoupling capacitors is discussed in Section 15.7.

15.6.2 Input Characteristics

The input characteristics of a real op amp can best be described as involved but not complicated. In addition to a large, but finite, input resistance, r_{in}, several voltage and current sources are found, Figure 15.20. The values of these elements are given for the LF 356 in parentheses.[2] The values assume the op amp is open loop; feedback may improve some values, so the values presented here can be taken as worst case. These sources have very small values and can often be ignored, but again it is important for the bioengineer to understand the importance of these elements to make intelligent decisions.

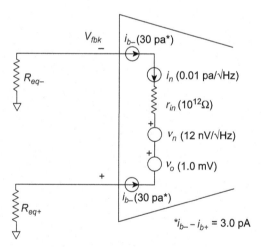

FIGURE 15.20 A schematic representation of the input elements of a practical operational amplifier. The input can be thought of as containing several voltage and current sources as well as a large input impedance. Each of these sources is discussed in the following sections.

---

[2]The curious units for the voltage and current noise sources, V/√Hz and pA/√Hz were introduced in the discussion of Johnson noise in Section 1.3.2.1.1 and are covered again later.

15.5.2.1 Input Voltage Sources

The voltage source, v_o, is a constant voltage source termed the "input offset voltage." It indicates that the op amp will have a nonzero output even if the input is zero, that is, even if $V_{in+} - V_{in-} = 0$. The output voltage produced by this small voltage source depends on the gain of the circuit. To find the output voltage under zero input conditions (the output offset voltage) simply multiply the input offset voltage by the $1/\beta$ gain. Again, this gain is the "noise gain" as explained later. Determination of the output offset voltage is demonstrated in the following example.

EXAMPLE 15.6

Find the offset voltage at the output of the amplifier circuit shown in Figure 15.21. In other words, determine the output, V_{out}, when $V_{in} = 0$.

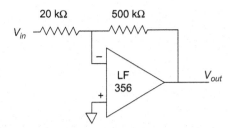

FIGURE 15.21 A typical inverting amplifier circuit. This amplifier will have a small output voltage even if $V_{in} = 0$. This offset voltage is determined in Example 15.6.

Solution: First find the noise gain, $1/\beta$, then multiply this by the input offset voltage, v_o, found in the LF 356 specifications sheet as 1.0 mV (see Appendix F). The value of $1/\beta$ is:

$$\frac{1}{\beta} = \frac{R_f + R_1}{R_1} = \frac{500 + 20}{20} = 26$$

Result: Given the typical value of v_o as 1.0 mV, V_{out} for zero input becomes:

$$V_{out} = v_o(26) = 1.0(26) = 26\,\text{mV}$$

The input offset voltage given in the specifications is a typical value, not necessarily the value of any individual op amp chip. This means that the input offset voltage will be around the value shown, but for any given chip it could be larger or smaller. It could also be either positive or negative, leading to a positive or negative output offset voltage. Sometimes, the specifications sheet also gives a maximum or "worst case" value.

The noise voltage source, v_n, specifies the noise normalized for bandwidth—more precisely, normalized to the square root of the bandwidth. This accounts for the strange units: $nV/\sqrt{Hz}$. Like Johnson noise and shot noise described in Chapter 1 (Section 1.3.2.1), noise in an op amp is distributed over the entire bandwidth of the op amp. To determine the actual noise in an amplifier it is necessary to multiply v_n by the square root of the circuit bandwidth as determined using the methods described previously. This value is then multiplied by the noise gain (i.e., $1/\beta$) to find the noise at the output. Finally we see why $1/\beta$ is also referred to as the noise gain: it multiples the noise at the input to produce the noise at the output, although it also multiplies offset voltage and other input errors. The next example shows how noise can be determined at the output of an amplifier circuit.

EXAMPLE 15.7

Find the noise at the output of the amplifier used in Example 15.6 that is due only to the op amp's noise voltage. (The resistors in the feedback network also contribute Johnson noise as does the input current noise source, i_n.)

Solution: There are multiple steps involved, but each is straightforward. Find the noise gain $1/\beta$; then determine the bandwidth from $1/\beta$ using the open-loop gain curve in Figure 15.13. Multiply the input noise voltage, v_n, by the square root of the bandwidth to find the noise at the input. Then multiply the result by the noise gain to find the value of the noise at the output.

From Example 15.6, the noise gain, $1/\beta$, was found to be 26, which corresponds to 28 dB. Referring to Figure 15.13, a $1/\beta$ line at 28 dB will intersect A_V at approximately 200 kHz. Taking the bandwidth as 200 kHz, the input noise voltage becomes:

$$v_{n\ input} = v_n \sqrt{BW} = 12 \times 10^{-9} \sqrt{200 \times 10^3} = 5.3\ \mu V$$

The noise at the output is this value multiplied by $1/\beta$:

$$V_{n\ output} = v_{n\ input} \left(\frac{1}{\beta}\right) = 5.3(26) = 137.8\ \mu V$$

The value of input voltage noise used in Example 15.7 is fairly typical for frequencies above 100 Hz. Op amp noise generally increases at lower frequencies and many op amp specifications include this information, including those of the LF 356. For the LF 356 the input voltage noise increases from $12\ nV/\sqrt{Hz}$ at 200 Hz and up to $60\ nV/\sqrt{Hz}$ at 10 Hz (see Appendix F). Presumably the voltage noise becomes even higher at lower frequencies, but values below 10 Hz are not given for this chip.

15.6.2.2 Input Current Sources

To evaluate the influence of input current sources, it is easiest to convert them to input voltages by multiplying by the equivalent resistance at the input terminals. Figure 15.20 notates the equivalent input resistances at the plus and minus terminals as R_{eq+} and R_{eq-}. These would have been determined by using the network reduction methods described in Chapter 14. The two current sources at the inputs, i_{b+} and i_{b-}, are known as the "bias currents." It may seem curious to show two current sources rather than one for both inputs, but this has to do with the fact that the offset currents at the two terminals are not exactly equal, although they do tend to be similar. Furthermore as with the bias voltage, the bias currents could be in either direction, in or out of their respective terminals.

To determine how these bias currents contribute to the output, we convert them to voltages by multiplying by the appropriate R_{eq}, and then multiplying that voltage by the noise gain. In most op amps, the two bias currents are approximately the same, so their influence on output offset voltage tends to cancel if the equivalent resistances at the two terminals are the same. Sometimes the op amp circuit designer tries to make the equivalent resistances at the two terminals the same just to achieve this cancellation. The amount by which the two bias currents are different, that is, the imbalance between the two currents, is called the "offset current" and is usually much less than the bias current. For example, in the LF 356, typical bias currents are 30 pA, whereas the offset current is only 3.0 pA, an order of magnitude less.

Figure 15.22 shows an inverting op amp circuit where a resistor has been added between the positive terminal and ground. The current flowing through this resistor is essentially zero (if you ignore the small bias currents) since the op amp's input impedance is quite large (recall Rule 1). So there is a negligible voltage drop across the resistor and the positive terminal is still at ground potential. This resistor performs no function in the circuit except to balance the voltage offset due to the bias currents. To achieve this balance, the resistor should be equal to the equivalent resistance at the op amp's negative terminal.

To determine the equivalent resistance at the negative terminal, we use the approaches described in Chapter 14 and make the usual assumption that the input to the op amp is an ideal source, Figure 15.23. We also assume that the output of the op amp is essentially an

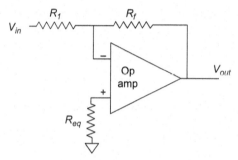

FIGURE 15.22   An inverting operational amplifier with a resistor added to the noninverting terminal to balance the bias currents. R_{eq} should be set to equal the equivalent resistance on the inverting terminal.

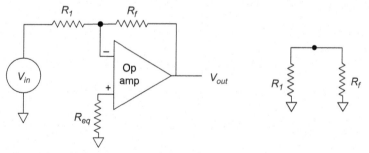

FIGURE 15.23   The left circuit is a typical inverting operational amplifier (op amp). If the op amp output, V_{out}, and the input, V_{in}, can be considered ideal sources, then they have effective impedances of 0 Ω. Since these two ideal sources are connected to ground, the ends of the two resistors are connected to ground, one through the ideal input and the other through the op amp's ideal. This makes them in parallel and the net resistance at the inverting input terminal is the parallel combination of R_1 and R_f.

ideal source. (Real op amp output characteristics are covered in the next section.) This means that both R_1 and R_f in Figure 15.23 are connected to ground at one end and the inverting input at the other. Since they are connected to the same points at both ends (ground and the inverting input) they are in parallel and the equivalent resistance hanging on the inverting terminal is the parallel combination of R_f and R_1, as shown on the right side:

$$R_{eq-} = \frac{R_f R_1}{R_f + R_1}; \quad \text{or more generally} : \quad Z_{eq-} = \frac{Z_f Z_1}{Z_f + Z_1} \qquad (15.18)$$

So to balance the resistances (or impedances) at the input terminals, the resistance at the positive terminal should be set to the parallel combination of the two feedback resistors (Equation 15.18). Sometimes, as an approximation, the resistance at the positive terminal is set to equal the resistor that has the smaller value of the two feedback network resistors (usually R_1). Another strategy is to make this resistor variable, and then adjust the resistance to cancel the output offset voltage from a given op amp. This has the advantage of removing any output offset voltage due to the input offset voltage, v_o, as well. The primary downside to this approach is that the resistor must be carefully adjusted after the circuit is built. If the feedback circuit contains capacitors, then Z_{eq} should be an impedance equivalent to the parallel combination of the feedback resistor and capacitor.

The current noise source is treated the same way as the bias currents: it is multiplied by the two equivalent resistances to find the input current noise, and then multiplied by the noise gain to find the output noise.

To find the total noise at the output, it is necessary to add in the voltage noise. Since the noise sources are independent, they add as the square root of the sum of the squares (Equation 1.9). In addition to the voltage and current noise of the op amp, the resistors will produce voltage noise as well. Repeating the equation for Johnson noise for a resistor from Chapter 1 (Equation 1.6):

$$V_J = \sqrt{4kT\,R\,BW} \quad V \qquad (15.19)$$

The three different noise sources associated with an op amp circuit are all dependent on bandwidth. The easiest way to deal with these three different sources is to combine them in one equation that includes the bandwidth:

$$V_{n\,in} = \left[\left(v_n^2 + \left(i_n (R_{eq+} + R_{eq-}) \right)^2 + 4kT(R_{eq+} + R_{eq-}) \right) BW \right]^{\frac{1}{2}} \qquad (15.20)$$

Op amp   Op amp   Johnson
voltage   current   noise
noise   noise

This equation gives the summed input noise. To find the output noise, multiply $V_{n\,in}$ by the noise gain.

$$V_{n\ out} = V_{n\ in}(\text{Noise Gain}) = V_{n\ in}\left(\frac{1}{\beta}\right) \tag{15.21}$$

The use of this approach to calculate the noise out of a typical op amp amplifier circuit is shown in Example 15.8.

EXAMPLE 15.8

Find the noise at the output of the op amp circuit shown in Figure 15.22 where $R_f = 500$, $R_1 = 10$, and $R_{eq} = 9.8$ kΩ.

Solution: First find the noise gain, $1/\beta$. From $1/\beta$ determine the bandwidth of the amplifier using the open-loop gain curves in Figure 15.13. Apply Equation 15.20 to find the total input voltage noise, including both current and voltage noise. Then multiply this voltage by the noise gain to find output voltage noise (Equation 15.21).

The noise gain is:

$$\frac{1}{\beta} = \frac{R_f + R_1}{R_1} = \frac{500 + 10}{10} = 51$$

From the specifications of the LF 356, $v_n = 12$ nV/$\sqrt{\text{Hz}}$ and $i_n = .01$ pA/$\sqrt{\text{Hz}}$. The equivalent resistance at the negative terminal is found from Equation 15.18 to be 9.8 kΩ (the parallel combination of 500 and 10 kΩ). For a $1/\beta$ of 51, the bandwidth is approximately 100 kHz (Figure 15.13). Using $T = 310$K, $4kT$ is 1.7×10^{-20} J, and Equation 15.20 becomes:

$$V_{n\ in} = \left[\left((12 \times 10^{-9})^2 + (0.01 \times 10^{-12}(19.6 \times 10^3))^2 + 1.7 \times 10^{-20}(19.6 \times 10^3)\right)10^5\right]^{\frac{1}{2}}$$

$$V_{n\ in} = \left[(1.44 \times 10^{-16} + 3.8 \times 10^{-20} + 3.33 \times 10^{-16})10^5\right]^{\frac{1}{2}}$$

$$V_{n\ in} = \left[(4.77 \times 10^{-16})10^5\right]^{\frac{1}{2}} = 6.9\ \mu\text{V}$$

Result: The noise at the output is found by multiplying by the noise gain:

$$V_{n\ out} = V_{n\ in}(\text{Noise Gain}) = 6.9 \times 10^{-6}(51) = 0.35\ \text{mV}$$

Analysis: One advantage to including all the sources in a single equation is that the relative contributions of each source can be compared. After converting to a voltage, the current noise source is approximately four orders of magnitude less than the other two voltage noise sources, so its contribution is negligible. The op amp's voltage noise does contribute to the overall noise, but most of the noise is coming from the resistors. Finding an op amp with a lower noise voltage would lower the noise, but of the 0.35 mV noise at the output, 0.29 mV is from the resistors. Of course, this is using the value of noise voltage for frequencies above 200 Hz. The noise voltage of the op amp at 10 Hz is 60 nV/$\sqrt{\text{Hz}}$, four times the value used in this example. If noise at the lower frequencies is a concern, another op amp should be considered. (For example, the Op-27 op amp features a noise voltage of only 5.5 nV/$\sqrt{\text{Hz}}$ at 10 Hz.)

15.6.2.3 *Input Impedance*

Although the input impedance of most op amps is quite large, the actual input impedance of the circuit depends on the configuration. The noninverting op amp has the highest input impedance, that of the op amp itself. In practice it may be difficult to attain the high impedance of many op amps because of leakage currents in the circuit board or wiring. Furthermore, the bias currents of an op amp will decrease its effective input impedance.

For an inverting amplifier, the input impedance is approximately equal to the input resistance, R_1 (Figure 15.9). This is because the input resistor is connected to "virtual ground" in the inverting configuration. Although this is much lower than the input impedance of the noninverting configuration, it is usually large enough for many applications. Where a very high input impedance is required, the noninverting configuration should be used. If even higher input impedances are required, op amps with particularly high input impedances are available, but the limitations on impedance are usually set by other components of the circuit such as the lead-in wires and circuit board.

15.6.3 Output Characteristics

Compared with the input characteristics, the output characteristics of an op amp are quite simple: a Thévenin source where the ideal source is $A_V (V_{in+} - V_{in-})$ and the resistance is r_{out}, Figure 15.24. For the LF 356, A_V is given as a function of frequency in Figure 15.13 and r_{out} varies between 0.05 and 50 Ω depending on the frequency and closed-loop gain.[3] The value of the output resistance is the lowest at lower frequencies and when the closed-loop gain is 1 (i.e., $\beta = 1$).

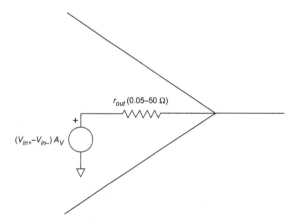

FIGURE 15.24   The output characteristics of an operational amplifier are those of a Thévenin source, including an ideal voltage source, $A_V (V_{in+} - V_{in-})$, and an output resistor, r_{out}.

---

[3]As with the input characteristics (specifically r_{in}), the value given in the specification sheet for r_{out} assumes the op amp is open loop. The presence of feedback reduces the effective output resistance.

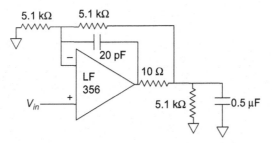

FIGURE 15.25 An operational amplifier (op amp) circuit that can be used to drive a large capacitive load. Note that this is a noninverting amplifier with a gain of only 2.0. This suggests that the main purpose of this circuit is to provide a drive to the capacitive load, not to amplify the signal. A number of additions to the classic noninverting amplifier (see Figure 15.6) are shown: a small (20-pf) feedback capacitor, a resistor in parallel with the capacitive load, and a small resistor in series between the op amp output and the load.

In some circumstances, other features of the output must be considered. Maximum voltage swing at the output is always a few volts less than the voltage that powers the op amp (see next section). The range of the output signal can become further limited at higher frequencies. In addition, many op amps have stability problems when driving a capacitive load. Figure 15.25 shows a circuit taken from the specifications sheet that can be used to drive a large capacitive load. In this circuit, the desired load is a 0.5-μf capacitor, which is considered fairly large in electronic circuits. Several strategies are used to reduce oscillation. A small feedback capacitor $(20 \, \text{pf} = 20 \times 10^{-12})$ is added to improve phase characteristics as mentioned previously. Another circuit addition is to place a resistor in parallel with the capacitor load so that the load is no longer purely capacitive. Yet another strategy is to place a small resistor at the output of the op amp before the feedback resistor providing a little isolation between the load and the op amp output. These strategies are often implemented on an ad hoc basis, but the design engineer should anticipate possible problems when capacitive (or inductive) loads are involved.

15.7 POWER SUPPLY

Op amps are active devices and require external power to operate. This external power is delivered as a constant voltage or voltages from a device known, logically, as a "power supply." Power supplies are commercially available in a wide range of voltages and current capabilities. Many op amps are "bipolar," that is, they handle both positive and negative voltages. (Unlike its use in psychology, in electronics the term bipolar has nothing to do with stability.) Bipolar applications require both positive and negative power supply voltages, and values of ±12 or ±15 V are common. The higher the power supply voltage, the larger the output voltages the op amp can produce, but all op amps have a maximum voltage limit. The maximum voltage for the LF 356 is ±18 V, but a special version, the LF 356B, can handle ± 22 V. High-voltage op amps are available as are low-voltage op amps for battery use. The latter also feature lower current consumption. (The LF 356 uses a nominal 5—10 mA and a few of them in a circuit will go through a 9-V battery fairly quickly.)

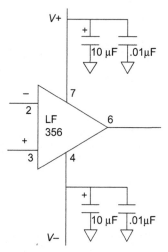

FIGURE 15.26 An LF356 operational amplifier connected to positive and negative power supply lines. The number next to each line indicates the pin number of a common version of the chip (the eight-pin DIP package). The capacitors are used to remove noise or fluctuations on the power supply lines that may be produced by other components in the system. These capacitors are called "decoupling capacitors."

The power supply connections are indicated on the op amp schematic by vertical lines coming from the side of the amplifier icon as shown in Figure 15.26. Sometimes the actual chip pins are indicated on the schematic as in this figure. Figure 15.26 also shows a curious collection of capacitors attached to the two supply voltages. Power supply lines often go to a number of different op amps or other analog circuitry.[4] These common power supply lines make great pathways for spreading signal artifacts, noise, positive feedback signals, and other undesirable fluctuations. One op amp circuit might induce fluctuations on the power supply line(s), and these fluctuations then pass to all the other circuits. Practical op amps do have some immunity to power supply fluctuations, but this immunity falls significantly with the frequency of the fluctuations. For example, the LF 356 will attenuate power supply variations at 100 Hz by 90 dB (a factor of 31,623), but this attenuation falls to 10 dB (a factor of 3) at 1 MHz.

A capacitor placed right at the power supply pin will tend to smooth out voltage fluctuations and reduce artifacts induced by the power supply. Since such a capacitor tends to isolate the op amp from power line noise, it is called a "decoupling capacitor." Figure 15.26 shows two capacitors on each supply line: a large 10-μf capacitor and a small 0.01-μf capacitor. Since the two capacitors are in parallel they are in theory equivalent to a single 10.01-μf capacitor. The small capacitor would appear to be contributing very little. In fact, the small capacitor is there because large capacitors have poor high-frequency performance: they look more like inductors than capacitors at higher frequencies. The small capacitor serves to reduce high-frequency fluctuations, whereas the large capacitor does the same at low frequencies. Although a given op amp circuit may not need both these decoupling capacitors,

[4]Digital circuits that may also be present in the system usually have their own power supply.

the 0.01 µF capacitor is routinely included by most design engineers on every power supply pin of every op amp. The larger capacitor may be added if strong low-frequency signals are present in the network.

15.8 OPERATIONAL AMPLIFIER CIRCUITS OR 101 THINGS TO DO WITH AN OPERATIONAL AMPLIFIER

Although there are more than 101 different signal processing operations that can be performed by op amp circuits, this is an introductory course so only a handful will be presented. However, they are the handful that you are most likely to need. For a look at the other 90+, see "*Art of Electronics*" by Horowitz and Hill (1989).

15.8.1 The Differential Amplifier

We have already shown how to construct inverting and noninverting amplifiers. Why not throw the two together to produce an amplifier that does both: a differential amplifier? As shown in Figure 15.27, a differential amplifier is a combination of inverting and noninverting amplifiers.

To derive the transfer function of the circuit in Figure 15.27, we once again employ the principle of superposition. Setting V_{in2} to zero effectively grounds the lower R_1 resistor, and the circuit becomes a standard inverting op amp with a resistance between the positive terminal and ground, Figure 15.28 left side. As stated previously, the only effect of this resistance is to balance the bias currents. For this partial circuit, the transfer function is:

$$V_{out} = -\frac{R_f}{R_1}V_{in1}$$

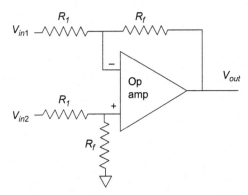

FIGURE 15.27   A differential amplifier circuit. This amplifier combines both inverting and noninverting amplifiers into a single circuit. As shown in the text, this circuit amplifies the difference between the two input voltages: $V_{out} = \frac{R_f}{R_{in}}(V_{in2} - V_{in1})$.

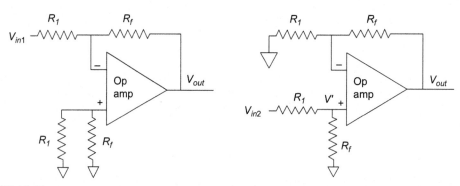

FIGURE 15.28 Superposition applied to the differential amplifier. Left circuit: The V_{in2} input is set to zero (i.e., grounded), leaving a standard inverting amplifier circuit. Right circuit: The V_{in1} input is grounded, leaving a non-inverting operational amplifier with a voltage divider circuit on the input. Each one of these circuits will be solved separately and the overall transfer function determined by superposition.

Setting V_{in1} to zero grounds the upper R_1 resistor, and the circuit becomes a noninverting amplifier with a voltage divider on the input. With respect to the voltage V' (Figure 15.28 right side), the circuit is a standard noninverting op amp:

$$V_{out} = \frac{R_f + R_1}{R_1} V'$$

The relationship between V_{in2} and V' is given by the voltage divider equation:

$$V' = \frac{R_f}{R_f + R_1} V_{in2}$$

Substituting and solving for V_{in2}:

$$V_{out} = \frac{R_f + R_1}{R_1} V' = \frac{R_f + R_1}{R_1} \frac{R_f}{R_f + R_1} V_{in2} = \frac{R_f}{R_1} V_{in2}$$

By superposition, the two partial solutions can be combined to find the transfer function when both voltages are present.

$$V_{out} = -\frac{R_f}{R_1} V_{in1} + \frac{R_f}{R_1} V_{in2} = \frac{R_f}{R_1} (V_{in2} - V_{in1}) \tag{15.22}$$

Thus the circuit shown in Figure 15.27 amplifies the difference between the two input voltages.

15.8.2 The Adder

If the sum of two or more voltages is desired, the circuit shown in Figure 15.29 can be used.

It is easy to show using an extension of the approach used in Example 15.2 that the transfer function of this circuit is:

$$V_{out} = \left(\frac{R_f}{R_1}\right)V_{in1} + \left(\frac{R_f}{R_2}\right)V_{in2} + \left(\frac{R_f}{R_3}\right)V_{in3} \qquad (15.23)$$

The derivation of this equation can be found as an exercise at the end of the chapter. This circuit can be extended to any number of inputs by adding more input resistors. If $R_1 = R_2 = R_3$, then the output is the straight sum of the three input signals amplified by R_f/R_1.

15.8.3 The Buffer Amplifier

At first glance the circuit in Figure 15.30 appears to be of little value. In this circuit all of the output is feedback to the inverting input terminal, so the feedback gain, β, equals 1. Since the gain of a noninverting amplifier is $1/\beta$, the gain of this amplifier is 1 and $V_{out} = V_{in}$. (This can also be shown using circuit analysis; this is an exercise in the problem section.) Although this

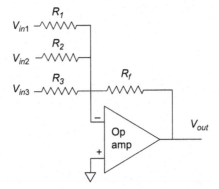

FIGURE 15.29 An operational amplifier circuit that takes a weighted sum of three input voltages, V_{in1}, V_{in2}, and V_{in3}.

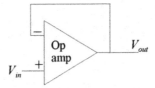

FIGURE 15.30 A "buffer amplifier" circuit. This amplifier provides no gain ($G = 1/\beta = 1$), but presents a very high impedance to the signal source, nearly an ideal load, and generates a low-impedance signal that looks like a nearly ideal source.

amplifier does nothing to enhance the amplitude of the signal, it does a great deal when it comes to impedance. Specifically, the incoming signal sees a very large impedance, the input impedance of the op amp ($>10^{12}\ \Omega$ for the LF 356), whereas the output impedance is very low (0.02 Ω at 10 kHz for the LF 356), approaching that of an ideal source. This circuit can take a signal from a high-impedance Thévenin source and provide a low-impedance, nearly ideal, source that can be used to drive several other devices. Although all noninverting op amp circuits have this impedance transformation function, the unity gain circuits are particularly effective and have the highest bandwidth since $1/\beta = 1$. Since this circuit provides a buffer between the high-impedance source and the other devices, it is sometimes referred to as a "buffer amplifier." The low output impedance also reduces noise pick up, and this circuit can be invaluable whenever a signal is sent over long wires or even off the circuit board. Many design engineers routinely use a buffer amplifier as the output to any signal that will be sent any distance, particularly if it is sent out of the instrument.

15.8.4 The Transconductance Amplifier[5]

Figure 15.31 shows another simple circuit that looks like an inverting op amp circuit except the input resistor is missing.

The input to this circuit is a current, not a voltage, and the circuit is used to convert this current into a voltage. Applying KCL to the negative input terminal, the transfer function for this circuit is easily determined.

$$i_n = i_f = \frac{V_{out}}{R_f};$$

Solving for V_{out}:

$$V_{out} = R_f i_n \qquad (15.24)$$

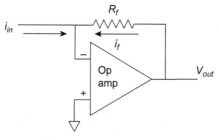

FIGURE 15.31   An operational amplifier circuit used to convert a current to a voltage. Because of this current-to-voltage transformation, this circuit is also referred to as a "transconductance amplifier."

---

[5]The transconductance amplifier described here takes in a current and puts out a voltage. There are also transconductance amplifiers that do the opposite: output a current proportional to an input voltage.

Some transducers produce current as their output and this circuit is used as the first stage to convert that current signal to a voltage. A common example in medical instruments is the photodetector transducer. These light-detecting transducers usually produce a current proportional to the light falling on them. These currents can be very small, in the nanoamps or picoamps. If R_f is chosen to be very large (10–100 MΩ), a reasonable output voltage is produced. Generally this output voltage requires additional amplification by a "second-stage" op amp amplifier.

Since the input to the op amp is current, noise depends only on current noise. This would include the noise current generated by both the op amp and the resistor. The net input current noise is then multiplied by R_f to find the output voltage noise. This approach is illustrated in the practical problem posed in Example 15.9.

EXAMPLE 15.9

A transconductance amplifier shown in Figure 15.32 is used to convert the output of a photodetector into a voltage. The traditional symbol for a photodetector is a diode with squiggly arrows as shown on the left side of Figure 15.32.

The photodetector has a sensitivity of R = 0.01 μA/μW. The R stands for "responsivity," which is the same as sensitivity. Sensitivity functions as the transfer function of a transducer describing the output for a given input. The photodiode has a noise current, which is called "dark current." The dark current of this photodiode is i_d = .05 pA/√Hz. What is the minimum light flux, φ, in microwatts that can be detected with an SNR of 20 dB and a bandwidth of 1 kHz?

Solution: This problem requires a number of steps, but the heart of the problem is how much current noise is generated at the input of the op amp. Once this is determined, the minimum signal current can be determined as 20 dB, or a factor of 10, times this current noise. The responsivity defines the output for a given input (R = φ/i), so once the minimum signal current is found, the minimum light flux, φ_{min}, can be calculated as i_{min}/R. To find the total noise current, modify Equation 15.20 for current noise rather than voltage noise. For resistor current noise, use Equation 1.7. Adding in the diode noise to the resistor and op amp current noise:

$$i_{n\ Total} = \left[\left(i_d^2 + i_n^2 + \frac{4kT}{R_f}\right)BW\right]^{\frac{1}{2}} = \left[\left(2q\,i_d + i_n^2 + \frac{4kT}{R_f}\right)BW\right]^{\frac{1}{2}} \tag{15.25}$$

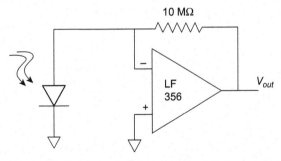

FIGURE 15.32 The output of a photodetector is fed to a transconductance amplifier. This circuit is used in Example 15.9 to determine the minimum light flux that can be detected with a signal to noise ratio (SNR) of 20 dB.

Note that the value of current noise decreases for increased values of R_f, so a good design would use as large a value of R_f as is practical and needed.

Solving into Equation 15.25 using the value of i_n from the LF 356 specifications sheet, the value of i_d from the photodetector, an R_f of 10 MΩ, then multiplying by the 1 kHz bandwidth:

$$i_{n\ Total} = \left[\left((2*1.6\times10^{-19})(1\times10^{-12}) + (.01\times10^{-12})^2 + \frac{1.7\times10^{-20}}{10^7}\right)10^3\right]^{\frac{1}{2}}$$

$$= \left[(1.6\times10^{-32} + 1\times10^{-28} + 1.7\times10^{-27})10^3\right]^{\frac{1}{2}} = 1.34\times10^{-12}\ \text{A} = 1.34\ \text{pA}$$

Thus the minimum signal required for an SNR of 20 dB is $10\times1.34 = 13.4$ pA. From the sensitivity of the photodetector, the minimum light flux that can be detected is:

$$\varphi_{min} = \frac{i_{min}}{R} = \frac{13.4\times10^{-12}}{0.01} = 1.34\times10^{-9}\ \text{W} = 1.34\times10^{-3}\ \mu\text{W}$$

Note that since R is in μA/μW and all the units used here are scaled versions of A/W, it is not necessary to scale this number. To determine the output voltage produced by this minimum signal, or any other signal for that matter, simply multiply this input current by R_f:

$$V_{out} = i_{in}R_f = 1.34\times10^{-9}(10^7) = 13.4\ \text{mV}$$

This is small signal, so it would be a good idea to increase the value of R_f. Since R_f is the largest contributor of noise, increasing its value by a factor of 10 would both increase the output signal by that amount and decrease the noise (since noise *current* is inversely related to the resistance). Modest additional improvement might then be obtained by using an op amp with a lower current noise since, as seen in the calculations, this is the second largest noise source.

15.8.5 Analog Filters

A simple single-pole low-pass filter can be constructed using an R-C circuit. An op amp can also be used to construct a low-pass filter with improved input and output characteristics and also provide increased signal amplitude. The easiest way to construct an op amp low-pass filter is to add a capacitor in parallel to the feedback resistor as shown in Figure 15.33.

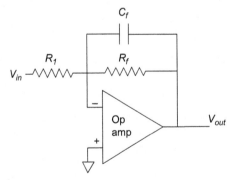

FIGURE 15.33 An inverting amplifier that also functions as a low-pass filter. As shown in the text, the low-frequency gain of this amplifier is R_f/R_1 and the cutoff frequency is $\omega = 1/R_f C_f$.

The equation for the transfer function of an inverting op amp with impedances in the feedback circuit is given in Equation 15.12 and repeated here:

$$\frac{V_{out}}{V_{in}} = -\frac{Z_f}{Z_1} \tag{15.26}$$

Applying Equation 15.26 to the circuit in Figure 15.33:

$$\frac{V_{out}}{V_{in}} = -\frac{Z_f}{Z_1} = -\frac{\frac{R_f \frac{1}{j\omega C_f}}{Rf + \frac{1}{j\omega C_f}}}{R_1} = \left(\frac{R_f}{R_1}\right) \frac{1}{(1 + j\omega R_f C_f)} \tag{15.27}$$

At frequencies well below the cutoff frequency, the second term goes to 1 and the gain is R_f/R_1. The second term is a low-pass filter with a cutoff frequency of $\omega = 1/R_f C_f$ rad/s or $f = 1/2\pi R_f C_f$ Hz. Design of an active low-pass filter is found in the problems.

It is also possible to construct a second-order filter using a single op amp. A popular second-order op amp circuit is shown in Figure 15.34.

Derivation of the transfer function requires applying KCL to two nodes and is provided in Appendix A. The transfer function is:

$$\frac{V_{out}}{V_{in}} = \frac{\frac{G}{(RC)^2}}{s^2 + \frac{3-G}{RC}s + \frac{1}{(RC)^2}} \quad \text{where}: \quad G = \frac{R_f + R_1}{R_1} \tag{15.28}$$

where G is the gain of the noninverting amplifier and equals $1/\beta$. Equating coefficients of Equation 15.28 with the standard Laplace transfer function of a second-order system (Equation 6.31):

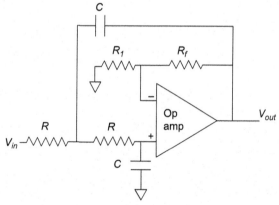

FIGURE 15.34 A two-pole active low-pass filter constructed using a single operational amplifier.

$$\omega_o = \frac{1}{RC}; \quad \delta = \frac{3 - G}{2} \tag{15.29}$$

EXAMPLE 15.10

Design a second-order filter with a cutoff frequency of 5 kHz and a damping of 1.0.

Solution: Since there are more unknowns than equations, several component values may be set arbitrarily with the rest determined by Equation 15.29. To find R and C, pick a value for C that is easy to obtain, then calculate the value for R (capacitor values are more limited than resistor values). Assume $C = 0.001$ µf, a common value. Then the value of R is:

$$\omega_o = 2\pi f = \frac{1}{RC} = 2\pi(5000) = 31,416 \text{ rad/s}$$

$$R = \frac{1}{31,416(0.001 \times 10^{-6})} = 31.8 \text{ k}\Omega$$

The value for G, the gain of the noninverting amplifier would be:

$$\delta = \frac{3 - G}{2} = 1.0; \quad G = 3 - 2\delta = 3 - 2 = 1$$

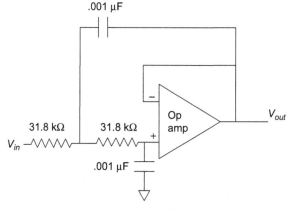

FIGURE 15.35 A two-pole low-pass filter with a cutoff frequency of 5 kHz and a damping factor of 1.0.

Hence for this particular damping, $G = 1/\beta = 1$ and $\beta = 1.0$. So all of the output is feedback to the noninverting input and a resistor divider network is not required. Other values of damping require the standard resistor divider network to achieve the desired gain. A second-order active filter having the desired cutoff frequency and damping is shown in Figure 15.35.

15.8.6 Instrumentation Amplifier

The differential amplifier shown in Figure 15.27 is useful in a number of biomedical engineering applications, specifically to amplify signals from biotransducers that produce a

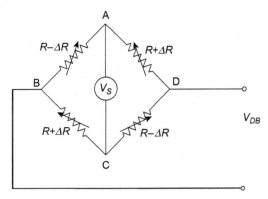

FIGURE 15.36   A bridge circuit that produces a differential output. The voltage at D moves in opposition to the voltage at B. The output voltage is best amplified by a differential amplifier.

differential output. Such transducers actually produce two voltages that move in opposite directions to a given input. An example of such a transducer is the strain gage bridge shown in Figure 15.36.

Here the strain gages are arranged in such a way that when a force is applied to the gages, two of them (A—B and C—D) undergo tension, whereas the other two (B—C and D—A) undergo compression. The two gages under tension decrease their resistance, whereas the two under compression increase their resistance. The net effect is that the voltage at B increases, whereas the voltage at D decreases an equal amount in response to the applied force. If the difference between these voltages is amplified using a differential amplifier such as that shown in Figure 15.27, the output voltage will be the difference between the two voltages and reflect the force applied. If the force reverses, the output voltage will change sign.

One of the significant advantages of this differential operation is that much of the noise, particularly noise picked up by the wires leading to the differential amplifier, will be common to both of the inputs and will tend to cancel. To optimize this kind of noise cancellation, the gain of each of the two inputs must be exactly equal in magnitude (but opposite in sign, of course). Not only must the two inputs be balanced, but the input impedance should also be balanced and often it is desirable that the input impedance be quite high. An "instrumentation amplifier" is a differential amplifier circuit that meets these criteria: balanced gain along with balanced and high input impedance. In addition, low noise is a common and desirable feature of instrumentation amplifiers.

A circuit that fulfills this role is shown in Figure 15.37. The output op amp performs the differential operation, and the two leading op amps configured as unity gain buffer amplifier provide similar high-impedance inputs. If the requirements for balanced gain are high, one of the resistors is adjusted until the two channels have equal but opposite gains. It is common to adjust the lower R_2 resistor. Since the two input op amps provide no gain, the transfer function of this circuit is just the transfer function of the second stage, which is shown in Equation 15.22 to be:

$$V_{out} = \frac{R_1 + R_2}{R_1}(V_{in2} - V_{in1}) \qquad (15.30)$$

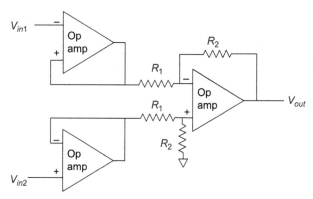

FIGURE 15.37 A differential amplifier circuit with high input impedance. Resistor R_1 can be adjusted to balance the differential gain so that the two channels have equal but opposite gains. In the interest of symmetry, it is common to reverse the position of the positive and negative operational amplifier (op amp) inputs in the upper input op amp.

There is one serious drawback to the circuit in Figure 15.37. To increase or decrease the gain it is necessary to change two resistors simultaneously: either both R_1's or both R_2's. Moreover, to maintain balance, they both have to be changed by exactly the same amount. This can present practical difficulties. A differential amplifier circuit that requires only one resistor change for gain adjustment is shown in Figure 15.38. The derivation for the input–output relationship of this circuit is more complicated than for the previous circuit, and is given in Appendix A:

$$V_{out} = \frac{R_4}{R_3}\left(\frac{R_1 + 2R_2}{R_1}\right)(V_{in2} - V_{in1}) \tag{15.31}$$

Since R_1 is now a single resistor, the gain can be adjusted by modifying this resistor. As this resistor is common to both channels, changing its value affects the gain of each channel

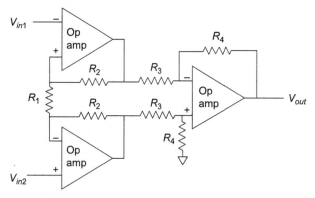

FIGURE 15.38 An instrumentation amplifier circuit. This circuit has all the advantages of the one in Figure 15.37 (i.e., balanced channel gains and high input impedance), but with the added advantage that the gain can be adjusted by modifying a single resistor, R_1.

equally and does not alter the balance between the gains of the two channels. It would be unusual to actually construct the circuit in Figure 15.38 since there are a number of integrated circuit instrumentation amplifiers that combine these components on a single chip. Such packages generally have very good balance between the two channels, very high input impedance, and low noise. For example, an instrumentation amplifier made by Analog Devices, Inc, the ADC624, has an input impedance of 10^9 Ω, a noise voltage of 4.0 nV/$\sqrt{\text{Hz}}$ at 1.0 kHz. Such chips also include a collection of highly accurate internal resistors that can be used to set specific amplifier gains by jumpers between selected pins with no need of external components.

The balance between the channels is measured in terms of V_{out} when the two inputs are at the same voltage. The voltage that is common (i.e., the same) to both input terminals is termed the "common mode voltage." In theory, the output should be zero no matter what the input voltage is so long as it is the same at both inputs. However, any imbalance between the gains of the two channels will produce some output voltage, and this voltage will be proportional to the common mode voltage. Since the idea is to have the most cancellation and the smallest output voltage to a common mode signal, the common mode voltage out of the amplifier is specified in terms of inverse gain. This inverse gain is called the "common mode rejection ratio" (CMRR), and is usually given in decibels.

$$V_{out} = \frac{V_{CM}}{\text{CMRR}} \tag{15.32}$$

The higher the CMRR the smaller the output voltage that results from the common mode voltage and the better the noise cancellation. The ADC624 has a CMRR of 120 dB. This means that the common mode gain is -120 dB. For example, if $+10$ V were applied to both input terminals (i.e., $V_{in1} = V_{in2} = 10$ V), V_{out} would be:

$$V_{out} = \frac{V_{CM}}{\text{CMRR}} = \frac{10}{120 \text{ dB}} = \frac{10}{10^{120/20}} = \frac{10}{10^6} = 10 \ \mu\text{V}$$

Although this value is not zero, it will be close to the noise level for most applications.

15.9 SUMMARY

There is more to analog electronics than just op amp circuits, but they do encompass the majority of analog applications. The ideal op amp is an extension of the concept of an ideal amplifier. An ideal amplifier has infinite input impedance, zero output impedance, and a fixed gain at all frequencies. An ideal op amp has infinite input impedance and zero output impedance, but has infinite gain. The actual gain of an op amp circuit is determined by the feedback network, which is generally constructed from passive devices. This provides great flexibility with a wide variety of design options and the inherent robustness and long-term stability of passive elements.

Real op amps come reasonably close to the idealization. They have very high input impedances and quite low output impedances. Deviations from the ideal fall into three categories:
deviations in transfer characteristics, deviations in input characteristics, and deviations in
output characteristics. The two most important transfer characteristics are bandwidth and
stability, where stability means the avoidance of oscillation. The bandwidth of an op amp circuit can be determined by combining the frequency characteristics of the feedback network
with the frequency characteristics of the op amp itself. The immunity of an op amp circuit
from oscillation can also be estimated from the frequency characteristics of the particular
op amp and those of the feedback network. Input errors include bias voltages and currents,
and noise voltages and currents. The bias and noise currents are usually converted to voltages
by multiplying them by the equivalent resistance at each of the input terminals. The effect of
these input errors on the output can be determined by multiplying all the input voltage errors
by the noise gain, $1/\beta$. Output deviations consist of small nonzero impedances and limitations on the voltage swing.

There is a wide variety of useful analog circuits based on the op amp. These include both
inverting and noninverting amplifiers, filters, buffers, adders, subtractors including differential amplifiers, transconductance amplifiers, and many more circuits not discussed here. The
design and construction of real circuits that use op amps is fairly straightforward, although
some care may be necessary to prevent noise and artifact from spreading through the power
supply lines. Decoupling capacitors, capacitors running from the power supply lines to
ground, are placed at the op amp's power supply feed to reduce the spread of noise through
the power lines.

PROBLEMS

1. Design a noninverting amplifier circuit with a gain of 500.
2. Design an inverting amplifier with a variable gain from 50 to 250.
3. What is the bandwidth of the following noninverting amplifier? If the same feedback
   network were used to design an inverting amplifier, what would be the bandwidth of
   this circuit?

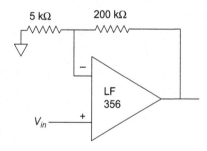

4. An amplifier has a GBP of 10 MHz. It is used in a noninverting amplifier where
   $\beta = 0.01$. What is the gain of the amplifier? What is the bandwidth?

5. An LF 356 is used to implement the variable gain amplifier in Example 15.3. What are the bandwidths of this circuit at the two extremes of the gain?

6. A .001µf capacitor is added to the feedback circuit of the following inverting op amp circuit. You can assume that before the capacitor was added the phase shift due to the amplifier when $Av\beta = 1$ was 120 degrees. (Recall the criterion for stability is that the phase shift induced by the op amp and the feedback network must be less than 180 degrees when $Av\beta = 1$.) After the capacitor is added, what is the phase shift of the op amp plus feedback network at the frequency where $Av\beta = 1$? Follow the example given in the section on stability.

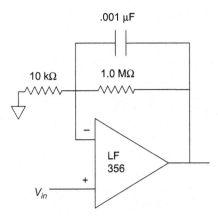

7. For the following circuit, assume $R_1 = 500$ kΩ, $R_f = 1.0$ MΩ, and $R_{eq} =$ the parallel combination of R_1 and R_f. What is the total offset voltage at the output? How much is this offset voltage increased if R_{eq} is replaced with a short circuit?

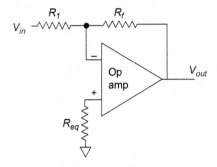

8. Derive the transfer function of the following adder circuit. Use KVL applied to node A.

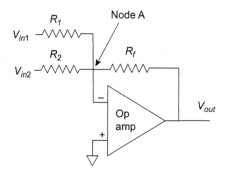

9. Find the total noise at the output of the following circuit. Identify the major source(s) of noise. What will the output noise be if the 50-kΩ ground resistor at the positive terminal is replaced with a short circuit? (Note this resistor adds to both the Johnson noise from the resistors and to the voltage noise generated by the op amp's noise current.)

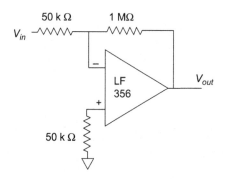

10. For the circuit in Problem 9, what is the minimum signal that can be detected with an SNR of 10 dB? What will be the voltage of such a signal at V_{out}?

11. An op amp has a noise current of 0.1 pA√Hz. This op amp is used as a transconductance amplifier (Figure 15.31). What should the minimum value of feedback resistance be so that the noise contribution from the resistor is less than the noise contribution from the op amp?

12. Design a one-pole low-pass filter with a bandwidth of 1 kHz. Assume you have capacitor values of 0.001, 0.01, 0.05, and 0.1 µF, and a wide range of resistors.

13. Design a two-pole low-pass filter with a cutoff frequency of 500 Hz and a damping factor of 0.8. Assume the same component availability as in Problem 15.

14. Design a two-pole high-pass filter with a cutoff frequency of 10 kHz and a damping factor of 0.707. (The circuit for a high-pass filter is the same as that for a low-pass filter except that the capacitors and resistors are reversed as shown in the following figure.)

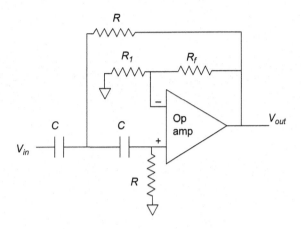

15. Design an instrumentation amplifier with a switchable gain of 10, 100, and 1000. (Hint: switch the necessary resistors in or out of the circuit as needed.)

APPENDICES

Appendix A
Derivations

A.1 DERIVATION OF EULER'S FORMULA

Assume a sinusoidal function:

$$x = \cos\theta + j\sin\theta \qquad (A.1)$$

(where $j = \sqrt{-1}$ as usual)

Differentiating with respect to θ produces:

$$\frac{dx}{d\theta} = j(\cos\theta + j\sin\theta) = jx \qquad (A.2)$$

Separating the variables gives:

$$\frac{dx}{x} = jd\theta \qquad (A.3)$$

and integrating both sides gives:

$$\ln x = j\theta + K$$

where K is the constant of integration. To solve for this constant, note that in Equation (A.1): $x = 1$ when $\theta = 0$. Applying this condition to Equation (A.3):

$$\ln 1 = 0 = 0 + K; \quad K = 0;$$

so Equation (A.3) becomes: $\ln x = j\theta$.

or

$$x = e^{j\theta} \qquad (A.4)$$

but since x is defined in Equation (A.1) as: $\cos\theta + j\sin\theta$

$$e^{j\theta} = \cos\theta + j\sin\theta \qquad (A.5)$$

Alternatively,

$$e^{-j\theta} = \cos\theta - j\sin\theta \qquad (A.6)$$

A.2 CONFIRMATION OF THE FOURIER SERIES

Fourier showed that a periodic function of period T can be represented by a series, possibly infinite, of sinusoids, or sine and cosines:

$$x(t) = \frac{a_0}{2} + \sum_{n=1}^{\infty}(a_n \cos n\omega_o t + b_n \sin n\omega_o t) \qquad (A.7)$$

where $\omega_o = 2\pi/T$ and a_n and b_n are the Fourier coefficients.

To derive the Fourier coefficients, let us begin with the a_0 or DC term. Integrating both sides of Equation (A.7) over a full period:

$$\int_0^T x(t)dt = \int_0^T \frac{a_0}{2}dt + \sum_{n=1}^{\infty} \int_0^T (a_n \cos n\omega_o t + b_n \sin n\omega_o t)dt \tag{A.8}$$

For $n = 0$, the second term on the right-hand side is zero since the sum begins at $n = 1$, and the equation becomes

$$\int_0^T x(t)dt = \int_0^T \frac{a_0}{2}dt; \quad \int_0^T x(t)dt = \frac{a_0 T}{2};$$

$$\tag{A.9}$$

$$a_0 = \frac{2}{T}\int_0^T x(t)dt$$

To find the other coefficients, multiply both sides of Equation (A.7) by $\cos(m\omega_o t)$, where m is an integer, and again integrate both sides.

$$\int_0^T x(t)\cos(m\omega_o t)dt = \int_0^T \frac{a_0}{2}\cos(m\omega_o t)dt + \sum_{n=1}^{\infty} \int_0^T (a_n \cos(m\omega_o t)\cos n\omega_o t$$

$$+ b_n \cos(m\omega_o t)\sin n\omega_o t)dt \tag{A.10}$$

Rearranging:

$$\int_0^T x(t)\cos(m\omega_o t)dt = \int_0^T \frac{a_0}{2}\cos(m\omega_o t)dt + \sum_{n=1}^{\infty} a_n \int_0^T \cos(m\omega_o t)\cos n\omega_o t\, dt$$

$$+ \sum_{n=1}^{\infty} b_n \int_0^T \cos(m\omega_o t)\sin n\omega_o t \tag{A.11}$$

Since m is an integer, the first and third terms on the right-hand side integrate to zero for all m. The second term also integrates to zero for all m except $m = n$. At $m = n$, the second term becomes:

$$an \int_0^T \cos^2(n\omega_o t)dt = \frac{\pi}{\omega_o}a_n = \frac{T}{2}a_n; \tag{A.12}$$

so that:

$$\frac{T}{2}a_n = \int_0^T x(t)dt$$

(A.13)

$$a_n = \int_0^T \cos(n\omega_o t)dt; \quad m = 1, 2, 3, \ldots$$

The bn coefficients are found in a similar fashion except Equation (A.7) is multiplied by $\sin(m\omega_o t)$, then integrated. In this case, all but the third term integrate to zero and the third term is nonzero only for $m = n$.

$$bn \int_0^T \sin^2(n\omega_o t)dt = \frac{\pi}{\omega_o}b_n = \frac{T}{2}b_n;$$

so that:

$$\frac{T}{2}b_n = \int_0^T x(t)dt$$

$$b_n = \int_0^T \sin(n\omega_o t)dt; \quad m = 1, 2, 3, \ldots$$

A.3 DERIVATION OF THE TRANSFER FUNCTION OF A SECOND-ORDER OP AMP FILTER

The op amp circuit for a second-order low-pass filter is shown in Figure A.1.

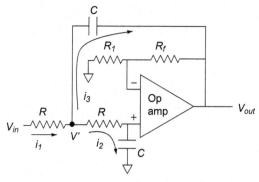

FIGURE A.1   A second-order op amp filter circuit.

This derivation applies not only to the following low-pass version, but also to the high-pass version where the positions of R and C are reversed.

Note that at node V' by KCL: $i_1 - i_2 - i_3 = 0$.

This allows us to write a nodal equation around that node:

$$\frac{V_{in} - V'}{R} - \frac{V' - V_{out}}{\frac{1}{Cs}} - i_2 = 0$$

where:

$$i_2 = \frac{V^+}{\frac{1}{Cs}} = V^+ Cs$$

Since the two terminals of an op amp must be at the same voltage, the voltage V^+ must be equal to V^-. Applying the voltage divider equation to the two feedback resistors, V^-, and hence V^+, can be found in terms of V_{out}:

$$V^+ = V^- = V_{out} G$$

where

$$G = \frac{R_1}{R_f + R_1}$$

So i_2 becomes: $V_{out}(Cs)/G$. Substituting i_2 into the nodal equation at V':

$$\frac{V_{in} - V'}{R} - (V_{out} - V')Cs - \frac{V_{out}(Cs)}{G} = 0$$

$$\frac{V_{in}}{R} - \frac{V'}{R} - V'Cs + V_{out}Cs - \frac{V_{out}Cs}{G} = 0$$

$$\frac{V_{in}}{R} - V'\left(\frac{1}{R} - V'Cs\right) + V_{out}Cs\left(1 - \frac{1}{G}\right) = 0$$

Note that V' can also be written in terms of just i_2:

$$V' = i_3\left(R + \frac{1}{Cs}\right) = \frac{V_{out}}{G}\left(R + \frac{1}{Cs}\right)$$

Substituting this for V' in the nodal equation:

$$\frac{V_{in}}{R} - \frac{V_{out}Cs}{G}\left(\frac{1}{R} + Cs\right)\left(R + \frac{1}{Cs}\right) + V_{out}Cs\left(1 - \frac{1}{G}\right) = 0$$

$$\frac{V_{in}}{R} = \frac{V_{out}Cs}{G}\left(3 + \frac{1}{RCs} + RCs - G\right)$$

Solving for V_{out}/V_{in}:

$$\frac{V_{out}}{V_{in}} = \frac{G}{RCs\left(1 + \dfrac{1}{RC} + RCs - g\right)} = \frac{G}{(RCs)^2 + (3-G)RCs + 1}$$

$$\frac{V_{out}}{V_{in}} = \frac{\dfrac{G}{(RC)^2}}{S2 + \dfrac{3-G}{RC}s + \dfrac{1}{(RC)^2}}$$

A.4 DERIVATION OF THE TRANSFER FUNCTION OF AN INSTRUMENTATION AMPLIFIER

The classic circuit for a three-op amp instrumentation amplifier is shown in Figure A.2.

To determine the transfer function, note that the voltage V_{in1} appears on both terminals of op amp 1, whereas V_{in2} appears on both terminals of op amp 2. The voltage out of op amp 1 will be equal to V_{in2} plus the voltage drop across the two resistors, R_2 and R_1:

$$V_{out1} = i_{12}(R_1 + R_2) + V_{in2};$$

but

$$i_{12} = \frac{V_{in1} - V_{in2}}{R_1}$$

$$V_{out1} = \frac{V_{in1} - V_{in2}}{R_1}(R_1 + R_2) + V_{in2}$$

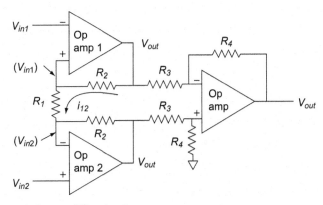

FIGURE A.2   Instrumentation amplifier circuit.

Applying the same logic to op amp 2, its output, V_{out2}, can be written as:

$$V_{out2} - V_{out1} = \frac{V_{in1} - V_{in2}}{R_1}(R_1 + R_2)2 - (V_{in1} - V_{in2})$$

$$= \left[\frac{2(R_1 + R_2)}{R_1} - 1\right](V_{in1} - V_{in2}) = \left[\frac{2R_1 + 2R_2 - R_1}{R_1}\right](V_{in1} - V_{in2})$$

$$V_{out2} - V_{out1} = \left(\frac{R_1 + 2R_2}{R_1}\right)(V_{in1} - V_{in2})$$

$$V_{out} = \left(\frac{R_4}{R_3}\right)(V_{out2} - V_{out1}) = \left(\frac{R_4}{R_3}\right)\left(\frac{R_1 + 2R_2}{R_1}\right)(V_{in1} - V_{in2})$$

The overall output, V_{out}, is equal to the difference of $V_{out2} - V_{out1}$ times the gain of the differential amplifier: R_4/R_3.

Appendix B
Laplace Transforms and Properties of the Fourier Transform

B.1 LAPLACE TRANSFORMS

Some useful Laplace transforms are given here. A more extensive list can be found in several tables available online.

1. 1 $\frac{1}{s}$

2. t^n $\frac{n!}{s^{n+1}}$

3. $e^{-\alpha t}$ $\frac{1}{s+\alpha}$

4. $(1 - e^{-\alpha t})$ $\frac{\alpha}{s(s+\alpha)}$

5. $\cos \beta t$ $\frac{s}{s^2+\beta^2}$

6. $\sin \beta t$ $\frac{\beta}{s^2+\beta^2}$

7. $\frac{1}{\beta^2}[1 - \cos(\beta t)]$ $\frac{1}{s\left(s^2+\beta^2\right)}$

8. $t - \frac{1}{\beta}\left(1 - e^{-\beta t}\right)$ $\frac{\beta}{s^2(s+\beta)}$

9. $e^{-\alpha t} - e^{-\gamma t}$ $\frac{\gamma-\alpha}{(s+\alpha)(s+\gamma)}$

10. $t - \frac{1}{\alpha}\left(1 - e^{-\alpha t}\right)$ $\frac{\alpha}{s^2(s+\alpha)}$

11. $\left(\frac{b\beta - b\alpha + c}{2\beta}\right)e^{-(\alpha-\beta)t} + \left(\frac{b\beta + b\alpha - c}{2\beta}\right)e^{-(\alpha+\beta)t}$ $\frac{bs+c}{s^2+2\alpha s+\alpha^2-\beta^2}$

12. $e^{-\alpha t}t[b + (c - b\alpha)t]$ $\frac{bs+c}{(s+\alpha)^2}$

13. $*e^{-\alpha t}\left(\frac{c-b\alpha}{\beta}\right)\sin \beta t + b \cos \beta t$ $\frac{bs+c}{s^2+2\alpha s+\alpha^2+\beta^2}$

14. $*1 - e^{-\alpha t}\left(\frac{\alpha-b}{\beta}\right)\sin \beta t + \cos \beta t$ $\frac{bs+\alpha^2+\beta^2}{s(s^2+2\alpha s+\alpha^2+\beta)}$

15. $*\frac{\omega_n}{\sqrt{1-\delta^2}}\left[e^{-\delta\omega_n t}\sin\left(\omega_n\sqrt{1-\delta^2}\,t\right)\right]$ $\frac{\omega_n^2}{s^2+2\delta\omega_n+\omega_n^2}$

16. $*1 - \frac{e^{-\delta\omega_n t}}{\sqrt{1-\delta^2}}\sin\left(\omega_n\sqrt{(1-\delta^2)}\,t + \theta\right)$ $\frac{\omega_n^2}{s\left(s^2+2\delta\omega_n+\omega_n^2\right)}$

     where $\theta = \tan^{-1}\left(\frac{\sqrt{1-\delta^2}}{\delta}\right)$

*Roots are complex.

B.2   PROPERTIES OF THE FOURIER TRANSFORM

The Fourier transform has a number of useful properties. A few of the properties discussed in the book are summarized here.

Linearity:

$$z(t) = ax(t) + by(t) \Rightarrow Z(\omega) = aX(\omega) + bY(\omega)$$

Differentiation:

$$\frac{dx(t)}{dt} \Rightarrow j\omega X(\omega)$$

Integration:

$$\int_{-\infty}^{t} x(\tau)d\tau \Rightarrow \frac{X(\omega)}{j\omega}$$

Time shift:

$$x(t - \tau) \Rightarrow X(\omega)e^{-j\omega\tau}$$

Time scaling:

$$x(at) \Rightarrow \frac{1}{a}X\left(\frac{\omega}{a}\right)$$

Convolution:

$$\int x(\tau)y(t - \tau)d\tau \Rightarrow X(\omega)Y(\omega)$$

Multiplication:

$$x(t)y(t) \Rightarrow \frac{1}{2\pi}\int X(v)Y(\omega - v)dv$$

where ω and v are frequencies.

Appendix C
Trigonometric and Other Formulae

$\sin(-x) = -\sin x$

$\cos(-x) = \cos x$

$\tan x = \dfrac{\sin x}{\cos x}$

$\sin(\omega t) = \cos(\omega t - 90)$

$\cos(\omega t) = \sin(\omega t + 90)$

$\sin(x \pm y) = \sin x \cos x \pm \cos x \sin y$

$\cos(x \pm y) = \cos x \cos y \mp \sin x \sin y$

$\sin 2x = 2 \sin x \cos x$

$\cos 2x = \cos^2 x - \sin^2 x = 2 \cos^2 x - 1 = 1 - 2 \sin^2 x$

$\sin^2 x = \dfrac{1 - \cos 2x}{2}$

$\cos^2 x = \dfrac{1 + \cos 2x}{2}$

$\sin^2 x + \cos^2 x = 1$

$\sin x \pm \sin y = 2 \sin \dfrac{1}{2}(x \pm y) \cos \dfrac{1}{2}(x \mp y)$

$\cos x + \cos y = 2 \cos \dfrac{1}{2}(x + y) \cos \dfrac{1}{2}(x - y)$

$\cos x - \cos y = -2 \sin \dfrac{1}{2}(x + y) \sin \dfrac{1}{2}(x - y)$

$\sin x \cos y = \dfrac{1}{2}[\sin(x + y) + \sin(x - y)]$

$\cos x \cos y = \dfrac{1}{2}[\cos(x + y) + \cos(x - y)]$

$$\sin x \sin y = \frac{1}{2}[\cos(x+y) - \cos(x-y)]$$

$$e^{\pm j\theta} = \cos\theta \pm j\sin\theta$$

$$\cos\theta = \frac{e^{j\theta} + e^{-j\theta}}{2}$$

$$\sin\theta = \frac{e^{j\theta} - e^{-j\theta}}{2j}$$

Appendix D
Conversion Factors: Units

TABLE 1 Metric Conversions

1 cc (10^3 mm^3)	1×10^{-6} m^3
1 kg wt	9.80665 N
1 kg wt	9807×10^5 dyn
1 kg wt	980.665 dyn
1 kg m	9.80665 J
1 kg m	8.6001 g cal
1 g	0.001 kg
1 J	1 W s
1 J	10^7 erg
1 J	2.778×10^{-7} kw h
1 Pa	1 dyn/cm^2
1 rad	56.296 degrees
1 rad/s	57.296 degrees/s
1 N	10^5 dyn
1 dyn	0.0010197 g wt
1 erg (1 dyn cm)	1×10^{-7} J
1 km	10^5 cm
1 m	10^{-10} Å
1 m	10^{-12} microns
1 L	1,000,027 (0.10) cm^3 (cc)
1 W	1 J/s

(Continued)

1 W	10^7 erg/s
1 W	10^7 dyn cm/s
1 amp	1 C/s
1 C (amp/s)	6.281×10^{18} electronic charges
1 C	3×10^9 electrostatic units
1 Ω	1 V/amp
1 V	1 J/C
1 H	1 V s/amp
1 F	1 C/V

TABLE 2   English-to-Metric Conversions

1 in. (U.S.)	2.540 cm
1 ft	0.3048 m
1 yd	0.9144 m
1 fathom (6 ft)	1.829 m
1 mi	1.609 km
1 in.^{3}	16.387 cm^3
1 in.^{2}	6.4516 cm^2
1 pint (0.5 qt)	0.473 L
1 gal (0.013368 ft^3)	3.7854 L
1 gal	3785.4 cm^3
1 gal wt (8.337 lb)	3.785 kg
1 ct	6.2 g
1 oz (troy)	31.134 g
1 oz (avdps)	28.35 g
1 lb (troy)	373.24 g
1 lb (troy)	0.3782 kg
1 lb (avdps)	453.59 g
1 lb (avdps)	0.45359 kg

1 lb wt	4.448×10^5 dyn
1 lb wt	4.448 N
1 BTU	1054.8 J
1 hp	0.7457 kw
1 rpm	6 degrees/s
1 rpm	0.10472 rad/s

PRESSURE CONVERSIONS

1 cm Hg (0°C)	1.333×10^4 dyn/cm^2
1 cm Hg (0°C)	135.95 kg/m^2
1 cm Hg (4°C)	980.368 dyn/cm^2
1 Pa	1 dyn/cm^2

TABLE 3   Metric-to-English Conversions

1 cm	0.3937 in.
1 m	3.281 ft
1 km	0.6214 mi
1 km	3280.8 ft
1 km	1.0567×10^{-13} light year
1 g	0.0322 oz (troy)
1 kg	2.2046 lb (avdps)
1 degrees/s	0.1667 rpm
1 cm/s	0.02237 mi/h
1 cm/s	3.728×10^{-4} mi/min

TABLE 4   Constants

Gravitational constant: g	32.174 ft/s^2
Gravitational constant: g	980.665 cm/s^2
Speed of light: vc	2.9986×10^{10} cm/s
Dielectric constant (vacuum): ε_0	8.85×10^{-12} col^2/mm^2

TABLE 5 American Standard Wire Gage
or Brown and Sharpe Gage

No.	Diameter (in.)
1	0.2893
2	0.2576
3	0.2294
4	0.2043
5	0.1819
6	0.1620
7	0.1443
8	0.1285
9	0.1144
10	0.1019
11	0.0907
12	0.0808
13	0.0720
14	0.0641
15	0.0571
16	0.0508
17	0.0453
18	0.0403
19	0.0359
20	0.0320
21	0.0285
22	0.0253
23	0.0226
24	0.0201
25	0.0179
26	0.0159
27	0.0142
28	0.0126
29	0.0113
30	0.100
31	0.00893
32	0.00795

TABLE 6 Scaling Prefixes

Scale	Prefix	Examples
10^{-15}	femto	femtoseconds
10^{-12}	pico	picoseconds, picoamps
10^{-9}	nano	nanoseconds, nanoamps, nanotechnology
10^{-6}	micro	microsecond, microvolts, microfarads
10^{-3}	milli	milliseconds, millivolts, milliamps, millimeters
10^{3}	kilo	kilohertz, kilohms, kilometers
10^{6}	mega	megahertz, megaohms, megabucks
10^{9}	giga	gigahertz, gigawatt
10^{12}	tera	terahertz

Appendix E
Complex Arithmetic

Complex numbers and complex variables consist of two numbers rolled into one. However, they are not completely independent, as most operations affect both numbers in some manner. The two components of a complex number are the real part and the imaginary part, the latter called so because it is multiplied by $\sqrt{-1}$. The imaginary part is denoted by the symbol i in mathematical circles, but the symbol j is used in engineering since i is reserved for current. A typical complex number would be written as: $a + jb$ where a is the real part and jb is the imaginary part.

Complex numbers are visualized as lying on a plane consisting of a real horizontal axis and an imaginary vertical axis, as shown in Figure E.1.

A complex number is one point on the real–imaginary plane and can be represented either in rectangular notation as a real and imaginary coordinate (Figure E.1, dashed lines), or in polar coordinates as a magnitude C and angle θ (Figure E.1, solid line). In this text the polar form is written using a shorthand notation: $C \angle \theta$.

To convert between the two representations, refer to the geometry of Figure E.1. To go from polar to rectangular, apply standard trigonometry:

$$a = C \cos \theta \quad b = C \sin \theta$$

To go in the reverse direction:

$$C^2 = a^2 + b^2$$

$$\tan \theta = \frac{b}{a}$$

$$\theta = \tan^{-1}\left(\frac{b}{a}\right)$$

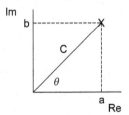

FIGURE E.1   A complex number can be visualized as one point on a plane.

These operations are very useful in complex arithmetic because addition and subtraction are done using the rectangular representation, whereas multiplication and division are easier using the polar form. Care must be taken with regard to signs and the quadrant of the angle, θ.

E.1   ADDITION AND SUBTRACTION

To add two or more complex numbers, add real numbers to real, and imaginary numbers to imaginary:

$$(a + jb) + (c + jd) = (a + c) + j(b + d)$$

Subtraction follows the same strategy:

$$(a + jb) - (c + jd) = (a - c) + j(b - d)$$

If the numbers are in polar form (or mixed) covert them to rectangular form first.

EXAMPLE E.1

Add the following complex numbers:

$$8\angle - 30 + 6\angle 60$$

Solution: Convert both numbers to rectangular form following the above-mentioned rules. Note that the first term is in the fourth quadrant so it will have the general form: $a - jb$.

For the first term:
$a = 8 \cos(-30); b = 8 \sin(-30);$
$a = 6.9; b = -4$

For the second term:
$c = 6 \cos(60) = 3; d = 6 \sin(60) = 5.2$
Sum $= 6.9 - j4 + 3 + j5.2 = 9.9 + j1.2$

This could then be converted back to polar form if desired.

E.2   MULTIPLICATION AND DIVISION

These arithmetic operations are best done in polar form, although they can be carried out in rectangular notation. For multiplication, multiply magnitudes and add angles:

$$C\angle\theta(D\angle\phi) = CD\angle(\theta + \phi)$$

For division, divide the magnitudes and subtract the angles:

$$\frac{C \angle \theta}{D \angle \phi} = \frac{C}{D} \angle (\theta - \phi)$$

EXAMPLE E.2

Perform the indicated multiplications or divisions:

A. $8 \angle -30(6 \angle 120)$;
B. $(10 - j6)(1 + j10)$;
C. $\frac{7 \angle 305}{6 \angle -80}$;
D. $\frac{6 \angle 50}{8 - j6}$

Solution: For the numbers already in polar coordinates, simply follow the rules given earlier. Otherwise convert to polar form where necessary.

A. $8 \angle -30(6 \angle 120) = 48 \angle 90$
B. $(10 - j6)(5 + j10) = 11.66 \angle -31(11.2 \angle +63) = 130.5 \angle 32$
C. $\frac{7 \angle 305}{6 \angle -80} = 1.166 \angle 385 = 1.166 \angle 25$
D. $\frac{6 \angle 50}{8 - j6} = \frac{6 \angle 50}{10 \angle -36.9} = 0.6 \angle 86.9$

More involved arithmetic operations can call for combinations of these conversions.

EXAMPLE E.3

Add the two complex fractions:

$$\frac{5 + j6}{3 - j7} + \frac{-8 + j6}{-3 - j8}$$

Solution: Convert all the rectangular representations to polar form, carry out the division, then convert back to rectangular form for the addition. When converting from rectangular to polar form, note that each number is in a different quadrant, so care must be taken with the angles.
Evaluate each fraction in turn:

$$\frac{5 + j6}{3 - j7} = \frac{7.8 \angle 50}{7.6 \angle -67} = 1.03 \angle 117 = -0.47 + j0.92$$

$$\frac{-8 + j6}{-3 - j8} = \frac{10 \angle 143}{8.5 \angle -110} = 1.18 \angle 253 = -0.34 - j1.13$$

So the sum becomes:

$$-0.47 + j0.92 - 0.34 - j1.13 = -0.81 - j0.21 = 0.83 \angle -165$$

It is a good idea to visualize where each number falls in the real–imaginary plane, or at least what quadrant of the plane, to help keep angles and signs straight.

Multiplication or division by the number j has the effect of rotation of the complex point by $\pm$ 90 degrees. This is apparent if the number j, which is in rectangular form, is converted to polar form: $j = 1\angle 90$. So multiplying or dividing by j adds or subtracts 90 degrees from a number:

$$j(C\angle\theta) = 1\angle 90(C\angle\theta) = C\angle(\theta + 90)$$

$$\frac{C\angle\theta}{j} = \frac{C\angle\theta}{1\angle 90} = C\angle(\theta - 90)$$

Appendix F
LF356 Specifications

Description	Specification
Bias current	30 pA
Offset current	3.0 pA
Offset voltage	3.0 mV
Gain bandwidth product	5.0 MHz (LF356)
Gain bandwidth product	20 MHz (LF357; minimum $1/\beta = 5$)
Input impedance	$10^{12}\,\Omega$
Open loop gain (DC)	106 dB
Common mode rejection ratio	100 dB
Maximum voltage	± 18 V (LF356); ± 22 V (LF356 B)
Noise current	0.01 pA/$\sqrt{\text{Hz}}$
Noise voltage (1000 Hz)	12 nV/$\sqrt{\text{Hz}}$
Noise voltage (100 Hz)	15 nV/$\sqrt{\text{Hz}}$
Noise voltage (10 Hz)	60 nV/$\sqrt{\text{Hz}}$

Only those specifications useful for analyzing circuits and problems in, Chapter 15 are given here. For a detailed set of specifications, see the various manufacturer specification sheets. These specifications are for the LF356 except as noted, and typical values are given.

Appendix G
Determinants and Cramer's Rule

The solution of simultaneous equations can be greatly facilitated by matrix algebra. When the solutions must be done by hand, the use of determinants is helpful, at least when only two or three equations are involved. A determinant is a specific single value defined for a square array of numbers. Given a 2×2 array, the determinant would be found by the application of the so-called diagonal rule where the product of the main diagonal (solid arrow) is subtracted by the product of the off-diagonal (dotted arrow):

$$\det = \begin{vmatrix} a_{11} & a_{12} \\ a_{21} & a_{22} \end{vmatrix}$$

This gives rise to the equation:

$$\det = \begin{vmatrix} a_{11} & a_{12} \\ a_{21} & a_{22} \end{vmatrix} = a_{11}a_{22} - a_{12}a_{21} \tag{G.1}$$

For a 3×3 array, the determinant is found by an extension of the diagonal rule. One way to visualize this extension is to repeat the first two columns at the right side of the array. Then the diagonals can be drawn directly:

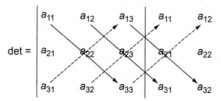

This procedure produces the equation:

$$\det = \begin{vmatrix} a_{11} & a_{12} & a_{13} \\ a_{21} & a_{22} & a_{23} \\ a_{31} & a_{32} & a_{33} \end{vmatrix} = (a_{11}a_{22}a_{33} + a_{12}a_{23}a_{31} + a_{13}a_{32}a_{21}) - (a_{13}a_{22}a_{31} + a_{23}a_{32}a_{11} + a_{33}a_{21}a_{12})$$

$$\tag{G.2}$$

It is a lot easier using MATLAB where the determinate is obtained by the command: det(A), where A is the matrix.

Cramer's rule is used to solve simultaneous equations using determinants. The equations are first put in matrix format (shown here using electrical variables):

$$\begin{vmatrix} v_1 \\ v_2 \end{vmatrix} = \begin{vmatrix} Z_{11} & Z_{12} \\ Z_{21} & Z_{22} \end{vmatrix} \begin{vmatrix} i_1 \\ i_2 \end{vmatrix} \tag{G.3}$$

The current i_1 is found using:

$$i_1 = \frac{\det Z_1}{\det Z} = \frac{\det \begin{vmatrix} v_1 & Z_{12} \\ v_2 & Z_{22} \end{vmatrix}}{\det \begin{vmatrix} Z_{11} & Z_{12} \\ Z_{21} & Z_{22} \end{vmatrix}} = \frac{v_1 Z_{22} - v_2 Z_{12}}{Z_{11} Z_{22} - Z_{21} Z_{12}} \tag{G.4}$$

And in a similar fashion, the current i_2 is found by:

$$i_2 = \frac{\det Z_1}{\det Z} = \frac{\det \begin{vmatrix} Z_{11} & v_1 \\ Z_{21} & v_2 \end{vmatrix}}{\det \begin{vmatrix} Z_{11} & Z_{12} \\ Z_{21} & Z_{22} \end{vmatrix}} = \frac{Z_{11} v_2 - Z_{21} v_1}{Z_{11} Z_{22} - Z_{21} Z_{12}} \tag{G.5}$$

Extending Cramer's rule to 3×3 matrix equation:

$$\begin{vmatrix} v_1 \\ v_2 \\ v_3 \end{vmatrix} = \begin{vmatrix} Z_{11} & Z_{12} & Z_{13} \\ Z_{21} & Z_{22} & Z_{23} \\ Z_{31} & Z_{32} & Z_{33} \end{vmatrix} \begin{vmatrix} i_1 \\ i_2 \\ i_3 \end{vmatrix}$$

The three currents are obtained as:

$$i_1 = \frac{\det \begin{vmatrix} v_1 & Z_{12} & Z_{13} \\ v_2 & Z_{22} & Z_{23} \\ v_3 & Z_{32} & Z_{33} \end{vmatrix}}{\det \begin{vmatrix} Z_{11} & Z_{12} & Z_{13} \\ Z_{21} & Z_{22} & Z_{23} \\ Z_{31} & Z_{32} & Z_{33} \end{vmatrix}} \quad i_2 = \frac{\det \begin{vmatrix} Z_{11} & v_1 & Z_{13} \\ Z_{21} & v_2 & Z_{23} \\ Z_{31} & v_3 & Z_{33} \end{vmatrix}}{\det \begin{vmatrix} Z_{11} & Z_{12} & Z_{13} \\ Z_{21} & Z_{22} & Z_{23} \\ Z_{31} & Z_{32} & Z_{33} \end{vmatrix}} \quad r_3 = \frac{\det \begin{vmatrix} Z_{11} & Z_{12} & v_1 \\ Z_{21} & Z_{22} & v_2 \\ Z_{31} & Z_{32} & v_3 \end{vmatrix}}{\det \begin{vmatrix} Z_{11} & Z_{12} & Z_{13} \\ Z_{21} & Z_{22} & Z_{23} \\ Z_{31} & Z_{32} & Z_{33} \end{vmatrix}}$$

where each determinant would be evaluated using Equation G.2 or using MATLAB.

Bibliography

Bruce, E.N., 2001. Biomedical Signal Processing and Signal Modeling. John Wiley and sons, Inc., NY.

Davasahayam, S.R., 2000. Signals and Systems in Biomedical Engineering: Signal Processing and Physiological Systems Modeling. Kluver Academic/Plenum Pub. NY, NY.

Goldberger, A.L., Amaral, L.A.N., Glass, L., Hausdorff, J.M., Ivanov, P.C., Mark, R.G., Mietus, J.E., Moody, G.B., Peng, C.-K., Stanley, H.E., 2000. PhysioBank, PhysioToolkit, and PhysioNet: components of a new research resource for complex physiologic signals. Circulation 101 (23), e215–e220. Circulation Electronic Pages. http://circ.ahajournals.org/cgi/content/full/101/23/e215.

Haynes, W., 2011. Handbook of Physics and Chemistry, ninty second ed. CRC Press, Boca Raton, FL.

Horowitz, P., Hill, W., 1989. The Art of Electronics, second ed. Cambirdge Univeristy Press, NY, NY.

Johnson, D., 1998. Applied Multivariate Methods for Data Analysis. Brooks/Cole Pub., Inc., Pacific Grove, CA.

Koo, M.C.K., 2000. Physiological Control Systems: Analysis, Simulation, and Estimation. IEEE Press, Piscataway, NJ.

Lathi, P.B., 2005. Linear Systems and Signals, second ed. Oxford University Press, New York/Oxford.

Northrop, R.B., 2003. Signals and Systems Analysis in Biomedical Engineering. CRC Press, Boca Raton, FL.

Northrop, R.B., 2004. Analysis and Appliction of Analog Electronic Circuits to Biomedical Instrumentation. CRC Press, Boca Raton, FL.

Ridout, V.C., 1991. Mathematical and Computer Modeling of Physiological Systems. Prentice Hall, Englewood Cliffs, NJ.

Semmlow, J.L., Griffel, B., 2014. Biosignal and Biomedical Image Processing, third ed. CRC Press, Inc., Boca Raton, FL.

Smith, S.W., 1997. The Scientist and Engineer's Guide to Digital Signal Processing. California Technical Publishing, San Diego, CA.

Index